D0933583

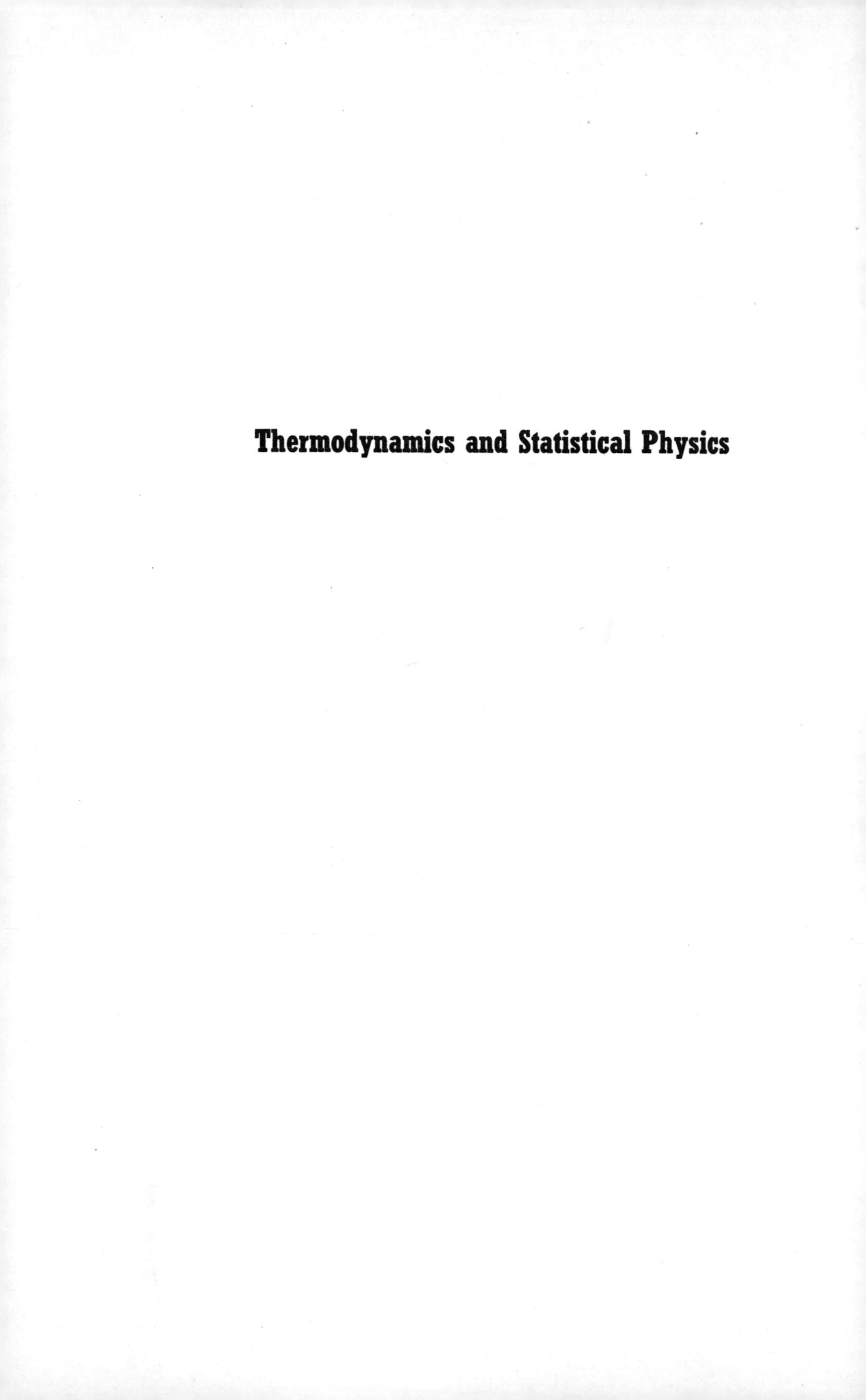

Thermodynamics and Statistical Physics

Thermodynamics and Statistical Physics

AN ELEMENTARY TREATMENT
WITH CONTEMPORARY APPLICATIONS

DON C. KELLY

Miami University
Oxford, Ohio

ACADEMIC PRESS

New York San Francisco London

A Subsidiary of Harcourt Brace Jovanovich, Publishers

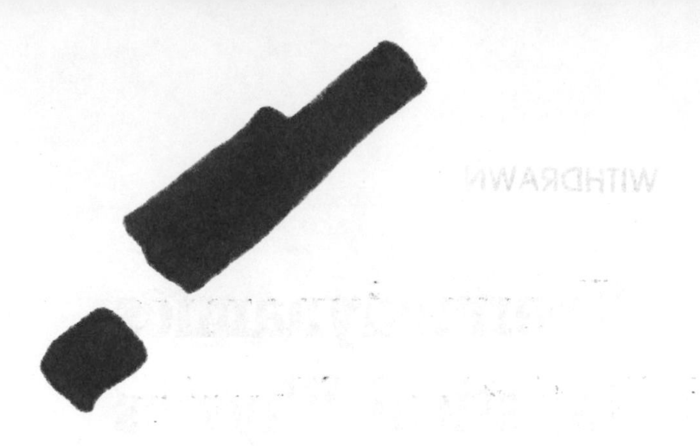

ACADEMIC PRESS, INC.
111 Fifth Avenue, New York, New York 10003

United Kingdom Edition published by
ACADEMIC PRESS, INC. (LONDON) LTD.
24/28 Oval Road, London NW1

LIBRARY OF CONGRESS CATALOG CARD NUMBER: 72-88354

PRINTED IN THE UNITED STATES OF AMERICA

To Jane

Contents

Chapter 4 **Thermodynamic Processes and Thermodynamic Work**

Chapter 5 **The Second Law of Thermodynamics**

Chapter 6 **Entropy**

Chapter 7 **The Third Law of Thermodynamics**

Chapter 8 **Thermodynamic Potentials**

Chapter 14 **Mathematics for Classical Statistics**

Chapter 15 **Classical Statistical Mechanics**

Chapter 16 **Applications of Statistical Mechanics**

Chapter 17 **Quantum Physics and Quantum Statistics**

Chapter 18 **Applications of Quantum Statistical Mechanics**

Preface

This book evolved from the author's experience in teaching a course in thermodynamics and statistical physics at Miami University to undergraduate students majoring in physics or engineering. Their previous exposure to the subject was a very brief survey in the freshman course. All portions of the text have been class tested in mimeographed note form—many of them several times.

The primary object of the book is a modest one. It is to help the student arrive at the point where his conceptions of the fundamental ideas are not only correct and precise, but where he feels comfortable with them as well. A lasting concept must seem "handmade."

The scope of the material is limited. The kinetic theory treatment stops short of the Boltzmann equation and its ramifications. The chapters on statistical mechanics make no reference to ensembles. The Boltzmann equation and ensemble theory are not elementary topics to the typical student.

The mathematical level grows in sophistication as the book progresses. Chapters 2, 10, and 14 serve to introduce and/or review various mathematical techniques.

Problems are interspersed within the text so that the student can immediately see how strong a grasp of the material he has. Numerous examples, serving various ends, are included.

References at the end of each chapter range from popular expositions suited to student use to journal articles which the instructor can digest for use as supplementary material.

My thanks are extended to the many students who made helpful suggestions. The comments of Mr. Kent Eschenberg were particularly helpful. The author is especially grateful to Professor George B. Arfken for reading the manuscript and for offering important suggestions. Above all I wish to thank my wife Jane—who was an inspiration in the writing of the book.

Overview

Thermodynamics and statistical physics deal with the properties of bulk matter under circumstances where the notions of temperature and heat cannot be ignored. In their present form the laws of thermodynamics and related principles comprise the essence of 150 years of experimentation and theoretical interpretation. Largely because of its empirical nature thermodynamics remains a formidable subject, especially at an elementary level. In part, the difficulty stems from the almost universal applicability of these laws. The great generality of the laws of thermodynamics requires that they be independent of the detailed workings of any particular physical system. Instead, they are based on universal principles such as the conservation of energy (first law of thermodynamics). The definitions of thermodynamic variables, though operational† and straightforward, may not convey much of a feel for the quantity so defined. This stems from the fact that such variables are nonmechanical. Historically, thermodynamics did not evolve on the basis of any atomic or mechanical model of matter which lets you see how it works.

There are four laws of thermodynamics, labeled zeroth, first, second, and third. Chronologically, only the third law is correctly numbered. The second law was formulated in 1824. The first law was stated about 20 years later. The third and zeroth laws were put forth in the 20th century.

---

† An operational definition defines a quantity by describing how to measure it.

The zeroth law relates the ideas of temperature equality and thermodynamic equilibrium, without suggesting what temperature measures. The zeroth law formalizes an important experimental fact, namely, that the mutual thermodynamic equilibrium of two systems demands equality of but one property—the property we call "temperature."

The first law recognizes heat as a form of energy, and is interpreted as a statement of energy conservation.

The second law places limits on the extent to which heat may be converted into mechanical energy (work) and other useful forms. The second law is linked with the irreversibility of processes which occur spontaneously in nature, such as the flow of heat.

The third law concerns the absolute zero of temperature. One consequence of the third law is that all schemes by which absolute zero is approached suffer a severe case of diminishing returns, rendering this temperature unattainable. The third law has not discouraged low-temperature research, as temperatures within a few millionths of a degree of absolute zero have been achieved.

By contrast, statistical physics is more transparent. One starts with a microscopic model of the physical system. For example, a rarified gas is envisioned as a collection of microscopic particles (atoms or molecules) subject to the laws of mechanics. Physically observable quantities are identified as average values of certain microscopic quantities. For example, we will see that the temperature of a gas is a measure of the average kinetic energy of its molecules. One can follow the transition from the microscopic realm of atoms and molecules to the macroscopic level of bulk matter and thus develop an intuitive feeling for the quantities introduced in the process.

In general, all but the simplest models present enormously complicated mathematical problems. These complexities arise from the many-body nature of the problems. In solids, liquids, and dense gases, any given atom or molecule is most likely to be found being pushed and pulled by many neighboring atoms. Future successes in many areas of statistical physics appear to be tied to the development of improved methods for handling the many-body problem.

If the successes of statistical physics were restricted to a few idealized models, one might seriously question its methods or even its foundations. However, one overpowering result of great generality ensures our confidence in the formalism: It is possible to deduce the laws of thermodynamics by applying the principles of statistical physics. Thus statistical physics can guarantee that a Dewar of liquid helium will obey the laws of thermodynamics. In addition it provides a formalism which, in principle (but not yet fully in practice), allows us to relate the many measurable properties of bulk helium to the microscopic structure of a single helium atom.

Chapter 1

A Framework for Thermodynamics

1.1 PRELIMINARY NOTIONS

"Pick a system" is advice offered freely to physics students. The *system* is the focus of attention—that chunk of the universe whose behavior we strive to understand. The *surroundings* of a system are the remaining portions of the universe which directly affect the behavior of the system.

Favorite mechanical systems include apples which fall vertically under the influence of gravity and blocks which slide down frictionless planes. Mechanics emphasizes a system's response to external forces. A favorite system in thermodynamics and statistical physics is a rarified gas housed in a cylinder fitted with a piston. In thermodynamics we are generally concerned with the effects of energy exchanges between the system and its surroundings.

The primary virtue of systems like the falling apple and sliding block is their remarkable simplicity. By applying the principles of mechanics—Newton's laws of motion—one can predict the behavior of a falling apple. The predicted behavior is in accord with both experiment and common sense—our everyday experience. Success in dealing with such transparent problems strengthens our confidence in the principles of Newtonian mechanics.

The gas-piston system of Figure 1.1c serves the same purpose in thermodynamics; it illustrates the principles at an intuitive level. Be forewarned, however, that your deductions as beginners in thermodynamics often will not seem to be a matter of common sense. In large part the difficulty in acquiring

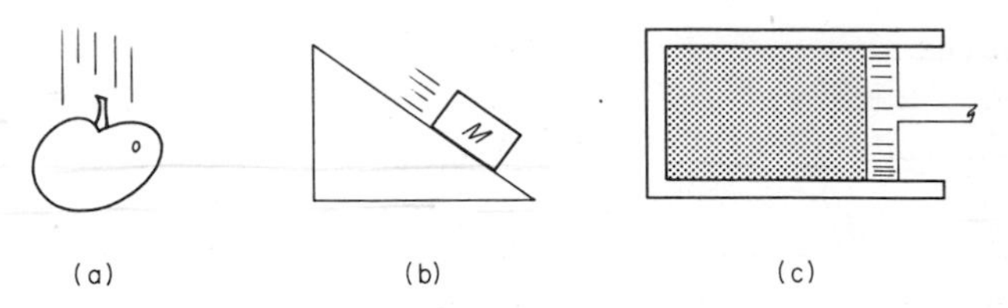

Figure 1.1. Favorite systems in mechanics: (a) apple falling vertically due to gravity, (b) sliding block on a frictionless plane, and in statistical physics, (c) gas–piston system.

a feel for thermodynamics stems from a lack of firsthand experience with the thermal aspects of even the simplest systems, such as gas-filled cylinders.

As a first step toward understanding the behavior of a physical system we must choose quantities which describe the system and the effects of the surroundings. If these physical quantities can be *measured*, either directly or indirectly, they are termed *macroscopic*. For example, if the system is a column of mercury in a thermometer, the density and column length, both measurable, qualify as macroscopic properties. For a confined gas, like that of Figure 1.1c, temperature, density, pressure, and volume are macroscopic variables. The macroscopic quantities used to describe the condition of a thermodynamic system are generally referred to as *thermodynamic variables*.

One may also describe a physical system at the microscopic level. This requires that we first introduce a conceptual model of the system. The model is constructed at the atomic level and employs *microscopic* variables—quantities which are not directly measurable. For example, a gas is envisioned as a collection of an enormous number of atoms which careen about wildly. The mass and kinetic energy of an atom are microscopic variables. The total mass of the gas is measurable, but not the mass of a particular atom.

Thermodynamics is formulated in terms of macroscopic variables. Kinetic theory and statistical mechanics employ microscopic variables in schemes which correlate various features of models with the macroscopic properties of physical systems.

••••••

1.1.1 Which of the quantities listed below are macroscopic and which are microscopic?

(a) the pressure which drives a piston in an automobile engine,

(b) the force exerted by an atom in a gas when it collides with the container wall,

(c) the wavelength of a water wave,

(d) the speed of a supersonic aircraft,

(e) the speed of a neutron in an atomic nucleus.

••••••

Thermodynamic variables are further classified according to their dependence on the size of the system (as measured by its mass or volume). A variable is termed *extensive* if its value depends on the size of the system. A variable is referred to as *intensive* if its value is independent of the size of the system. For example, the mass of a system is an extensive property. Its mass density (mass per unit volume) is an intensive quantity.

•••••• ••••••

1.1.2 Indicate which of the following are extensive quantities:

(a) the electrical resistance of a length of copper wire,
(b) the electrical resistivity of a length of copper wire,
(c) the kinetic energy of an automobile,
(d) the index of refraction of glass,
(e) the moment of inertia of a flywheel.

•••••• ••••••

Thermodynamic systems are classified further with respect to the exchange of matter and energy with their surroundings. In this connection it is necessary to regard heat as a form of energy which is transferred from one region to another by virtue of a temperature difference between them. A more precise and operational definition of heat is presented in Chapter 3.

An *isolated* system can exchange neither matter nor energy with its surroundings. The only candidate for a truly isolated system is the universe. However, it is not certain that all of the laws of thermodynamics apply on a cosmic scale. Despite the fact that an isolated system is an abstraction which cannot be realized in practice, a system may be regarded as isolated whenever matter and energy exchanges do not significantly disturb the processes occurring within the system. For example, a fluid in a container with rigid impermeable walls which are covered with insulation approximates an isolated system. The rigidity of the walls prevents any compressive work being done on the fluid and the insulation minimizes heat transfer. Thus no matter and a minimal amount of energy can be exchanged between the fluid and its surroundings.

A *closed* system is capable of exchanging energy, but not matter, with its surroundings. The walls of a building closely approximate a closed system. They engage in the transfer of thermal energy with the environment without an appreciable exchange of matter.

An *open* system is capable of exchanging both matter and energy with its surroundings. A living animal is a good example of an open system. Matter is exchanged with the surroundings by the acts of eating, breathing, and excreting. Energy exchanges include the absorption and rejection of heat and the performance of mechanical work by body muscles.

In connection with heat flow, we note that extrapolation from experience suggests the idea of a perfect insulator. The technical name for this wondrous device is *adiabatic wall*. A system enclosed by adiabatic walls is unaffected by thermal changes (as opposed to mechancial changes) in its surroundings, and is said to be "thermally isolated." A system is said to be "in thermal contact with its surroundings" whenever its boundaries do not function as adiabatic walls. Two systems in thermal contact are capable of exchanging heat and are often said to be "separated by diathermic or conducting walls."

Students are often disturbed when the concept of an adiabatic wall is thrust upon them, hailed as fundamental, and then used frequently. They are made uneasy by a feeling that thermodynamics is based on a fantasy. How can thermodynamics deal meaningfully with the real world by introducing an abstraction such as the adiabatic wall; that is, something which cannot be "built"? The notion of an adiabatic wall is valid and useful in thermodynamics in precisely the same way that the idea of a frictionless surface is helpful in mechanics: Both can be approached, that is, almost achieved. Mechanical friction cannot be eliminated completely. Nevertheless it is often possible to reduce frictional forces to the point where they are ignorable by comparison with other forces. The lack of complete thermal isolation is likewise often ignorable.

•••••• ••••••

1.1.3 Construct a brief list of physical systems, classifying them as open, closed, or isolated. Be prepared to explain your choices of classification, including the extent to which your entries under closed and isolated are approximations.

•••••• ••••••

1.2 STATES—MECHANICAL AND THERMODYNAMIC

Consider a system composed of N particles. The particles might be the atoms comprising your brain or the molecules of the air in some classroom. The mechanical state of the system is fixed by a specification of the position and velocity of each particle. Six variables are required to specify the position and velocity of each particle—three components of the position vector and three components of the velocity vector. Thus, a total of $6N$ microscopic variables is needed to completely specify the mechanical state of an N-particle system. If the values of these $6N$ variables are known at some instant, Newton's laws of motion provide a scheme which, in principle only, allows the mechanical state of the system to be determined at any later time.

Statistical physics is concerned with systems for which N is so large that we are unable, as a practical matter, to specify the mechanical state of the system; that is, there are too many microscopic variables of which to keep track.

Typically, N is of the order of Avogadro's number[†]

$$N_0 = 6.023 \times 10^{23}$$

The mere contemplation of specifying the mechanical state of such an enormous number of particles ends there—as a very brief contemplation.

Fortunately, complete knowledge of the mechanical state is unnecessary when we are concerned with understanding the macroscopic behavior of the system. The measurable properties reflect the combined actions of many atoms or molecules. The behavior of any particular atom is inconsequential. The gas–piston system of Figure 1.1c is typical of those of interest in statistical physics in that it consists of many particles for which a mechanical description is not feasible and, happily, unnecessary. We resort to either a thermodynamic or a statistical description.

In contrast to the huge number of microscopic variables required to specify a mechanical state, a thermodynamic state is defined by specifying values for only a few macroscopic variables. In general, experiment dictates which are the relevant thermodynamic variables for any specific system. In principle, every measurable property of a system must stand ready to serve as a thermodynamic variable. In practice, the behavior of most systems is adequately described by a small number of variables. Experience also reveals that the thermodynamic variables are not all independent. An equation relating two or more thermodynamic variables is called an *equation of state*. As an example, suppose that the system is a simple gas[‡] confined within some definite volume. The thermodynamic states are determined by specifying the density, pressure, volume, and temperature of the gas.[§] Only three of these four thermodynamic variables are independent, being related by one equation of state. Equations of state may be established empirically, that is, by fitting an equation to the experimental data. They also may be arrived at theoretically by applying the techniques of kinetic theory or statistical mechanics.

1.2.1 The English physicist William Thomson (who later became Lord Kelvin) had a favorite way of impressing upon people the enormity of Avogadro's number. His recipe was this: Imagine a glass of water in which each molecule is "tagged" in such a way as to make it distinguishable. Next, dump this glass of tagged water molecules into the oceans of Earth and stir.

[†] One gram-molecular weight constitutes 1 mole of a substance. According to Avogadro's hypothesis 1 mole of any substance contains the same number of molecules. This number, whose value was unknown to Avogadro, is called *Avogadro's number*.

[‡] A simple system is one composed of a single chemical species, as opposed to a mixture of chemically distinct components.

[§] More precisely, the equilibrium states are determined by these thermodynamic variables. The notion of thermodynamic equilibrium is discussed in Section 1.3.

Once the tagged molecules have been thoroughly mixed with the untagged water, reach in and dip out a glassful of water. If the mixing has distributed the tagged molecules uniformly, the glass will contain approximately *ten thousand* of the tagged molecules.

Thomson's example is indeed impressive. Devise your own example illustrating the enormity of N_0. The secret of constructing such examples lies in finding some way of diluting N_0 to obtain an easily comprehended number.

We note that by using radioactive isotopes, unknown in Thomson's day, it is possible to tag molecules. The use of these tracer isotopes is now widespread in medicine, biophysics, and ecological studies.

1.2.2 (a) Estimate the number of molecules you inhale in one breath. Your estimate may be rough but indicate the basis for it.

(b) The density of air is approximately 1.3×10^{-3} gm/cm^3. The total mass of our atmosphere is roughly 5×10^{21} gm. Assume that all of the molecules exhaled by George Washington in his dying breath are now distributed uniformly throughout our atmosphere. Make a rough estimate of the number of these molecules which you inhale with each breath.

•••••• ••••••

1.3 THERMODYNAMIC EQUILIBRIUM AND TEMPERATURE

Experience has shown that isolated systems tend toward states of *thermodynamic equilibrium.* For a system to be in thermodynamic equilibrium three conditions must be satisfied:

(a) The system must be in mechanical equilibrium, that is, free of any net external force and torque.

(b) The system must be in chemical equilibrium. All chemical reactions must have run their course.

(c) The measurable properties of the system must be spatially uniform throughout and must not change with time.

The uniformity conditions must be amended in situations where the system is intrinsically heterogeneous. For example, a closed insulated container filled with a liquid and its vapor is in thermodynamic equilibrium provided both the vapor and the liquid satisfy the conditions for equilibrium.

•••••• ••••••

1.3.1 Give three examples of everyday systems which are in thermodynamic equilibrium or very nearly so. Indicate any features which prevent attainment of complete thermodynamic equilibrium.

•••••• ••••••

"Temperature" is a word which means something to every student before he studies thermodynamics. However, the everyday meaning of the word is largely qualitative and is restricted by the range of human sense perceptions. It is necessary to give an operational definition of temperature to render it a useful thermodynamic variable. In the case of temperature it is logically necessary to first give operational status to the idea of temperature *equality*.

Imagine two isolated systems, each in thermodynamic equilibrium. Let thermal contact be established between the two in such a way that the overall combination is isolated. In general, both component systems would undergo changes. When the changes subside the two systems are said to be in *mutual thermodynamic equilibrium*. For simplicity we say the systems are in thermal equilibrium. We also say that their temperatures are equal. Evidently, thermal equilibrium is synonymous with temperature equality.

It should be noted that this definition of temperature equality does not require a scale of temperature. In mechanics Newton's first law of motion presents a similar situation. The first law stands as an operational definition of zero net force. We do not need an operational definition of force (supplied by Newton's second law) to recognize a force-free condition; zero acceleration defines zero force. In a similar fashion, mutual thermodynamic equilibrium defines temperature equality, regardless of what temperature scale is adopted.

1.4 THE ZEROTH LAW OF THERMODYNAMICS

In mathematics the transitive property is introduced as a postulate or as an elementary consequence of other postulates. The transitive law takes the form:

$$\text{If } A = C \text{ and } B = C, \text{ then } A = B.$$

An analogous law holds in thermodynamics, although it is neither a postulate nor a logical truth, but rather the statement of an experimental fact, namely, thermal equilibrium obeys a transitive law. This fact, often referred to as the *zeroth law* of thermodynamics, may be stated as follows: Two systems separately in thermal equilibrium with a third are in thermal equilibrium with each other. Figure 1.2 depicts a schematic demonstration of the zeroth law. Suppose that when thermal contact is established between A and C there are no changes observed in their thermodynamic coordinates†; similarly when B and C are placed in thermal contact. If A and B are then placed in thermal contact, it is observed that no changes occur in their thermodynamic states. They are in mutual equilibrium when contact is first established. As Figure 1.2 suggests,

† The expressions "thermodynamic coordinate" and "thermodynamic variable" are synonymous.

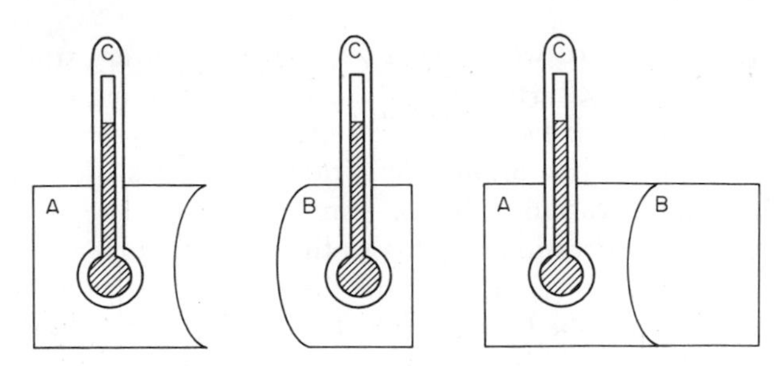

Figure 1.2. Schematic test of the zeroth law of thermodynamics.

C might be a mercury-in-glass thermometer, the height of the mercury column being the thermometric variable. The fact that the height of the column is the same whether the thermometer is in thermal contact with A or B leads one to expect that the thermodynamic state of A and B will remain unchanged when thermal contact is established between them. Experiment bears out the correctness of this expectation thereby establishing the zeroth law of thermodynamics.

Since equality of temperature is synonymous with mutual thermodynamic equilibrium we can phrase the zeroth law in equation form. Even though we have not set about the task of defining a temperature scale we can write

$$T_A = T_C$$

as a symbolic way of indicating that A and C are in thermal equilibrium. This allows the zeroth law to be expressed as the statement:

$$\text{If } T_A = T_C \text{ and } T_B = T_C\text{, then } T_A = T_B.$$

1.5 PRESSURE—AN IMPORTANT THERMODYNAMIC VARIABLE

Pressure is among the more prominent thermodynamic variables, and as such is encountered throughout thermodynamics and statistical physics. Pressure measures force per unit area. For example, consider an aquarium containing 200 lb of water. If the base of the aquarium has an area of 4 sq ft, the pressure on the base due to the water is

$$P = \frac{200 \text{ lb}}{4 \text{ sq ft}} = 50 \frac{\text{lb}}{\text{sq ft}}$$

The weight of our atmosphere is supported by the surface of the earth. The resulting pressure is approximately 10^6 dyn/cm^2, or roughly 15 lb/sq in. This means that a vertical column of our atmosphere having a cross-sectional area of 1 sq in., weighs about 15 lb. The familiar mercury barometer is generally used to measure the atmospheric pressure. The so-called barometer formula provides the basis for its operation. Consider a column of fluid of height h

and cross-sectional area A. The volume of the fluid is $V = hA$. If ρ denotes the mass density of the fluid, the mass within the column is

$$m = \rho V = \rho hA$$

The weight of the column is

$$W = mg = \rho ghA$$

where g is the acceleration of gravity. The pressure at the base of the column due to the weight of the fluid is

$$\frac{W}{A} = \rho gh \tag{1.1}$$

Notice that A, the area of the column, drops out. Pressure is an intensive variable associated with the extensive variable, force. Equation (1.1) gives the change in pressure produced by the weight of the liquid. If P_0 is the pressure at the surface of the fluid, the total pressure at depth h is

$$P = P_0 + \rho gh \tag{1.2}$$

This is the barometer equation.

••••••

1.5.1 Estimate the mass and weight of the Earth's atmosphere.

••••••

The mercury barometer consists of a narrow tube, closed at one end. The tube is filled with mercury and inverted in a reservoir (see Figure 1.3). The

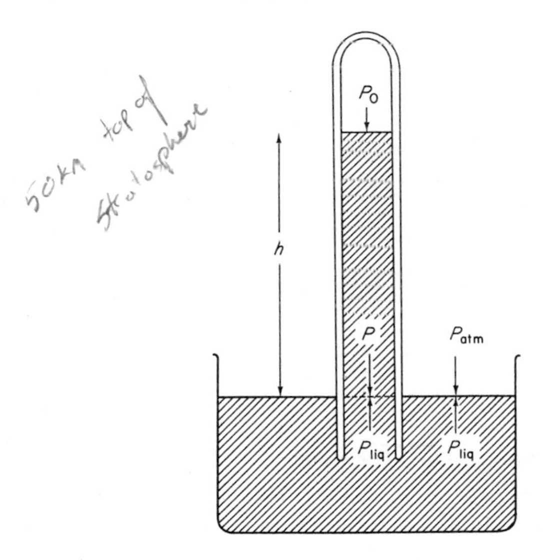

Figure 1.3. Elementary barometer.

...re at the surface of the mercury reservoir is atmospheric (P_{atm}). Inside ...e capillary the pressure at the level of the reservoir is $P_0 + \rho g h$. The pressure P_0 at the top of the mercury column is not quite zero because a small amount of mercury vapor fills the otherwise evacuated space. The weight of this vapor produces the pressure P_0, which is ignorable except in precise scientific work. The height of the column is such that the pressures at the level of the reservoir surface are the same inside and outside. This is necessary for mechanical equilibrium. Thus[†]

$$P_{\text{atm}} = P_0 + \rho g h \simeq \rho g h$$

There are several units of pressure currently in use. In addition to the pound per square inch (lb/sq in.) and the dyne per square centimeter (dyn/cm^2), there is the atmosphere (atm)

$$1 \text{ atm} \equiv 1.0132 \times 10^6 \, \frac{\text{dyn}}{\text{cm}^2}$$

••••••

1.5.2 At what depth is the total pressure acting on a scuba diver equal to 2 atm?

••••••

Because pressure can be determined by measuring the height of a barometer column it is often indicated in terms of millimeters of mercury. What is meant by 1 mm of mercury of course is the pressure corresponding to a column of mercury one millimeter in height. The millimeter of mercury gradually is being replaced by the *Torr*. The Torr is an absolute unit of pressure defined as $(1.01325 \times 10^6/760)$ dyn/cm^2, and is very nearly equal to 1 mm of mercury. The bar, defined as 10^6 dyn/cm^2, is another unit of pressure. The millibar (1 mb $= 10^{-3}$ bar) is frequently used in meteorological work. Newspaper weather maps often show isobars (lines of equal pressure) with the pressures indicated in millibars. The pressures indicated are invariably near 1000 mb, that is, near 1 atm.

[†] In addition to the equality sign we shall have occasion to employ the following symbols: $\equiv$ which is "identically equal to"; (also means "defined as"). For example,

$$x^2 - y^2 \equiv (x-y)(x+y)$$

$$\cosh Z \equiv \frac{e^Z + e^{-Z}}{2}$$

$\simeq$ which is "approximately equal to." For example,

$$\pi \simeq 3.1$$

$\sim$ which is "roughly equal to" or "of the same order of magnitude." For example,

$$\tfrac{4}{3}\pi r^3 \sim r^3$$

••••••

1.5.3 A student decides to make a barometer by simply placing the capillary tube in the reservoir, without first filling the tube with mercury. He has heard his instructor explain that the barometer in Figure 1.3 operates because the external (atmospheric) pressure supports the weight of the mercury column. The student argues that if the atmosphere is willing to support the column it should not mind forcing the mercury upward into the column as well. Explain why this will not happen.

1.5.4 The barometer was invented by Torricelli, one of Galileo's students. Torricelli used many different types of liquids, even wine. If the pressure is 1 atm and $g = 980 \text{ cm/sec}^2$, what is the height of the column if the density of the wine is 0.9 gm/cm^3? This exercise illustrates why one does not see many wine barometers!

••••••

Example 1.1 Archimedes' Principle

Archimedes' principle states that an object which displaces a fluid experiences a buoyant force equal to the weight of the fluid it displaces. It is this buoyant force which supports a floating object. To establish the principle, consider a fluid in hydrostatic equilibrium and focus attention on some arbitrary volume element. The net force on the fluid element vanishes when it is in hydrostatic equilibrium. Two types of forces act on the element; a gravitational force equal to the weight of the element and the forces exerted on its surface by the surrounding fluid. The net surface force must just balance the weight of the element in order for it to be in equilibrium. This surface force is a buoyant force—it prevents the fluid element from sinking. If a solid body displaces the fluid element, the same surface forces act on it as acted on the now displaced fluid. Hence, the object experiences a buoyant force equal to the weight of the fluid it displaced.

From Archimedes' principle it follows that an object will float in any liquid whose density is greater than the average density of the object. Ice floats in water because the density of ice (at 0°C) is 0.92 gm/cm^3. Let ρ_0 denote the average density of an object of volume V which floats in a fluid of density ρ_f. The fraction of the volume V which is submerged is such as to make the buoyant force equal to the weight of the object. We can write

$$mg = \rho_0 V g = \text{weight of object}$$

If f is the fraction of the volume V submerged, the buoyant force is

$$B = \rho_f f V g$$

Setting $B = mg$ gives

$$f = \frac{\rho_o}{\rho_f}$$

As an example, the density of ice is 0.92 gm/cm^3 and the density of liquid water is 1.00 gm/cm^3. Thus $f = 0.92$, which means that only about 8% of an ice cube or an iceberg is above the surface. ◇

•••••• ••••••

1.5.5 The figure of 8% ignores the buoyant force exerted by the air displaced by the exposed part of the ice cube. Explain why this does not introduce any significant error.

1.5.6 (a) Make a rough estimate of your mass density without reference to your volume.

(b) Would you float in a pool of mercury?

(c) Do you think you could walk across a pool of mercury? Consider the question of stability.

1.5.7 The molecular weight of helium is 4 gm/mole. The molecular weight of air is 28.9 gm/mole. A helium-filled balloon rises because it displaces a weight of air in excess of its own weight. Recall that the molar volume is approximately the same for all gases, 22,400 cm^3 at 32 °F and 1 atm (STP).

(a) Estimate the mass and volume of helium required to lift a mass of 100 kg.

(b) In view of your answer for part (a) would it be feasible to market miniblimps? In particular, would a one-man blimp fit in a one-car garage?

1.5.8 A hollow, airtight, brass sphere is suspended from one end of an equal-arm balance. A solid brass cylinder is suspended from the other end, and it is observed that they balance, that is, the balance arm is horizontal. The apparatus is then placed inside a large bell jar and the air is pumped out. Describe and explain what happens to the balance.

•••••• ••••••

A pair of entertaining toys rely on Archimedes' principle. Known as the Cartesian diver, the first consists of a hollow plastic figure (the diver) placed in a water-filled bottle. The bottle is fitted with a stopper. Small holes in the side of the diver allow it to be partially filled with water, to the point where it has a density very slightly less than the density of the water and so just floats. This leaves the diver with a small amount of air trapped inside. The bottle is filled so as to leave little or no air space. If the "solid" bottle is squeezed—the demonstration being more spectacular if only one hand is used—the diver plummets to the bottom. Squeezing the bottle increases the pressure in the water which forces water into the diver, increasing the density of the diver. The diver sinks when his density rises above that of the water. The same results may be achieved by warming the water.

The lava lamp is another device which relies on Archimedes' principle. Large globules of paraffin are submerged in a cylinder filled with machine oil. The density of the paraffin at room temperature is just slightly greater than that of the oil at the same temperature. An electric light bulb illuminates and heats the base of the cylinder. The temperature of the oil is greatest at the bottom of the cylinder. The heat causes the paraffin to expand more than the oil, to the point where it becomes less dense than the oil, whereupon it rises toward the surface. As it rises it is cooled by the surrounding oil. It contracts until it becomes more dense than the surrounding oil, whereupon it sinks. The cohesive nature of the paraffin results in a strking slow motion rise and fall.

1.6 TEMPERATURE AND THERMOMETRY

We wish to move the notion of temperature from its status as a sense perception to one of a reliable and useful scientific concept. This requires an operational definition of temperature, that is, we must specify how temperature is measured.

Thermometers are characterized by a thermometric substance and a thermometric variable. For example, in the familiar household thermometer, the thermometric substance is a liquid, usually colored alcohol or mercury. The length of the liquid column is the thermometric variable.

The most fundamental scale of temperature is based on the laws of thermodynamics. A discussion of this scale, known as the thermodynamic scale, is presented in Chapter 5. The upshot of the development presented there is this: On the thermodynamic scale, temperature is defined in terms of heat. A temperature so defined is independent of the properties peculiar to any thermometric substance. This is clearly a most desirable characteristic. The recorded temperature depends on the state of the system but not on the material used as a thermometric substance.

An empirical scale is arrived at by arbitrarily defining a temperature in terms of some thermometric variable. A mathematically elementary relationship is generally adopted. For the constant volume gas thermometer (Figure 1.4) and several others, the relation is linear. The temperature is defined in terms of the gas pressure P by

$$T(P) = bP$$

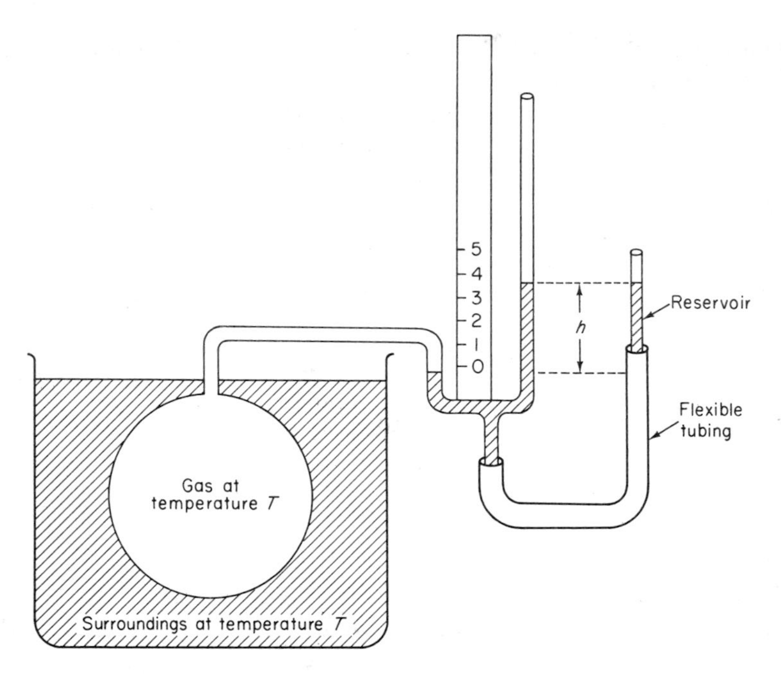

Figure 1.4. Schematic of a constant volume gas thermometer. The gas is maintained at a constant volume by means of the mercury reservoir. If the temperature of the bath T rises, the level of the reservoir h must be increased to maintain the same volume. In practice many refinements and corrections are necessary. Among these are accounting for the change in volume of the gas container and the fact that the gas in the tube leading to the manometer is not at the same temperature as the gas in the main vessel.

More generally we can write

$$T(X) = bX \tag{1.3}$$

as a linear relation defining a temperature $T(X)$ in terms of the value of some thermometric variable X. The value of the proportionality constant b is fixed by assigning a temperature to some standard fixed point.

By virtue of international agreement, the temperature of water at its triple point[†] has been assigned the value of 273.16 K, where K is the abbreviation for the unit called the kelvin. Accessibility and ease of reproduction are the primary virtues of using the triple point of water as the fundamental fixed point.

[†] The solid, liquid, and gaseous phases of water can coexist in thermodynamic equilibrium at a unique pressure (4.58 Torr) and temperature. This is referred to as the triple point of water.

If X_3 denotes the value of the thermometric variable at the triple point of water, we have

$$273.16\,\text{K} = bX_3$$

or

$$b = \frac{273.16\,\text{K}}{X_3}$$

The empirical temperature scale is now defined as

$$T(X) = 273.16\,\text{K}\,\frac{X}{X_3} \tag{1.4}$$

In particular, for the constant volume gas thermometer,

$$T(P) = 273.16\,\text{K}\,\frac{P}{P_3} \tag{1.4a}$$

••••••

1.6.1 Using a constant volume gas thermometer of modest precision, a student finds a value of 1.37 for the ratio of the steam-point pressure and the triple-point pressure. What is the corresponding steam-point temperature?

••••••

The constant volume gas thermometer is of special interest because it can be used to establish the fundamental thermodynamic scale over an important range of temperatures. Constant volume gas thermometers, employing different gases, generally will not be in agreement except at the triple point of water. Further, at any temperature other than that of the triple point of water, the thermometer reading depends slightly on the gas pressure at the triple point P_3. For example, suppose that oxygen is used as the gas, and that the thermometer is filled so as to give a triple-point pressure of 200 Torr ($P_3 =$ 200 Torr). If the thermometer is then used to determine the steam point† via (1.4a),

$$T_{\text{stm}} = 273.16\,\text{K}\,\frac{P_{\text{stm}}}{P_3}$$

a value near, but in excess of, 373.15 K would be recorded. Suppose next that some of the oxygen is removed until the triple-point pressure is 100 Torr. The

† The boiling point is the temperature at which a liquid and its vapor are in thermodynamic equilibrium. The normal boiling point is the equilibrium temperature when the pressure of the vapor is exactly 1 atm. The normal boiling point of pure water is called the steam point.

temperature of the steam point, computed according to the above ralation, would be different than the value found with $P_3 = 200$ Torr. It would, in fact, be nearer 373.15 K. If the triple-point pressure is reduced again and the steam point determined, the temperature recorded is again different, and still nearer 373.15 K. A sequence of such measurements, using progressively lower triple-point pressures enables one to extrapolate the results to the point where $P_3 = 0$. It is, of course, impossible to operate a constant volume gas thermometer at zero pressure. The extrapolated value of T_{stm}, using oxygen as the thermometric substance, is 373.15 K. If a different gas, say helium or nitrogen, is used instead of oxygen, a similar variation in T_{stm} is noted as the triple-point pressure is reduced. However, and this is the vitally significant point, when the results for helium, nitrogen, and so on, are extrapolated to $P_3 = 0$, they give a common temperature for the steam point, namely 373.15 K. Extrapolating the data to the point where $P_3 = 0$ gives a value for T_{stm} which is the same regardless of the gas used. This is precisely the requirement for a thermodynamic scale—a thermometer whose readings are independent of the thermometric substance. Figure 1.5 suggests the relationship between the values of P_3 and T_{stm}, as given by

$$T_{stm} = 273.16\,\mathrm{K}\,\frac{P_{stm}}{P_3}$$

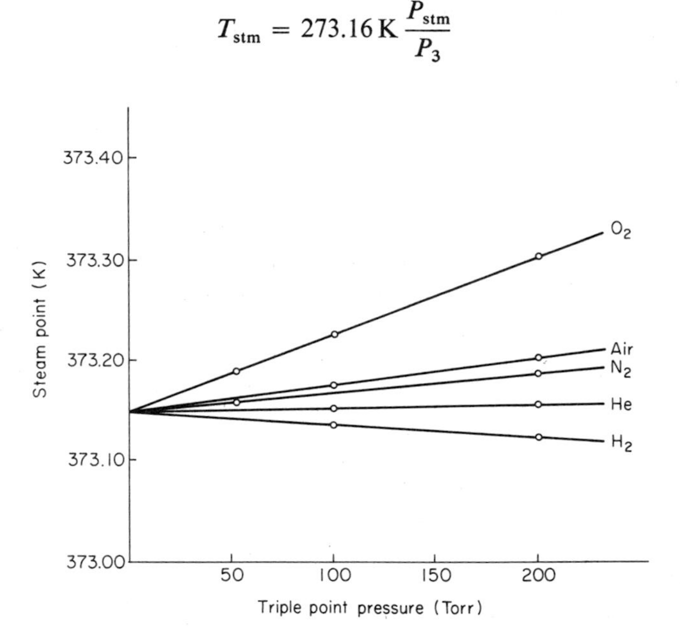

Figure 1.5. The steam-point temperature recorded by a constant volume gas thermometer. For any particular gas, the steam-point temperature depends on the triple-point pressure. However, when the data are extrapolated to zero pressure a unique steam point 373.15 K is indicated. [Adapted from D. Halliday and R. Resnick, "Physics." Wiley, New York (1960).]

A common temperature for the steam point is reached only when the data are extrapolated to $P_3 = 0$.

The temperature resulting from such extrapolation is often referred to as the ideal gas temperature

$$T_{\text{I.G.}} = \lim_{P_3 \to 0} 273.16\,\text{K}\,\frac{P}{P_3} \tag{1.5}$$

where I.G. stands for ideal gas. The concept of an ideal gas stems from the fact that all real gases exhibit similar behaviors at sufficiently low pressures. The ideal gas equation of state has the form

$$PV = nRT \tag{1.6}$$

where P, V, and T are the gas pressure, volume, and temperature, respectively and n denotes the number of moles of gas and R is the universal gas constant. In the cgs system of units R has the value

$$R = 8.32 \times 10^7\ \text{erg/K} \tag{1.7}$$

An alternate form of the ideal gas equation of state employs the number of molecules N rather than the number of moles. In (1.6) we replace n by the ratio of the total number of molecules and Avogadro's number

$$n = \frac{N}{N_0} \tag{1.8}$$

If, further, we define a constant k by

$$k = \frac{R}{N_0} \tag{1.9}$$

the ideal gas equation of state becomes

$$PV = NkT \tag{1.10}$$

The quantity k is called the Boltzmann constant and has the value

$$k = 1.38 \times 10^{-16}\ \text{erg/K}$$

•••••• ••••••

1.6.2 Let M and ρ denote the molecular weight and mass density of a gas, respectively. Show that the ideal gas equation of state can be written as

$$P = \frac{R\rho T}{M}$$

•••••• ••••••

In addition to the Kelvin scale, three others are in widespread use today. These are the Celsius, Fahrenheit, and Rankine scales. On the Celsius scale, temperature t_C is related to the Kelvin temperature T by

$$t_C = (T - 273.15)\,^\circ\mathrm{C} \tag{1.11}$$

with °C being the abbreviation for degrees Celsius. Thus, for example, the Celsius temperature of the triple point of water is 0.01 °C, since $T_3 = 273.16$ K. Notice that the kelvin and the Celsius degree have the same size. The scales differ only in the location of their zeros.

Throughout most English-speaking countries, the Fahrenheit scale of temperature is widely used. The symbol °F on the household thermometer stands for degrees Fahrenheit.

Temperatures on the Fahrenheit scale are related to those of the Celsius scale by the equation

$$t_F = \left(32 + \frac{9}{5} t_C\right)^\circ\mathrm{F} \tag{1.12}$$

This places the ice and steam points at 32 °F and 212 °F. The Fahrenheit and Celsius scales evidently differ in the size of their units and in the location of their zeros.

••••• •••••

1.6.3 Derive the inverse of (1.12), that is, the equation expressing t_C in terms of t_F.

••••• •••••

The Rankine temperature scale is widely used in engineering. The Rankine and Kelvin scales have a common zero (absolute zero), but the size of their units differ. The Rankine degree is equal to the Fahrenheit degree, and thus is only 5/9 as large as the kelvin and Celsius degree,

$$1\,^\circ\mathrm{R} = \frac{5}{9}\,\mathrm{K} \tag{1.13}$$

As a consequence, a temperature on the Rankine scale is 9/5 as great as the corresponding Kelvin temperature

$$t_R = \left(\frac{9}{5} T\right)^\circ\mathrm{R} \tag{1.14}$$

For example, the Rankine temperature of the ice point of water is

$$\frac{9}{5}(273.15)\,^\circ\mathrm{R} = 491.57\,^\circ\mathrm{R}$$

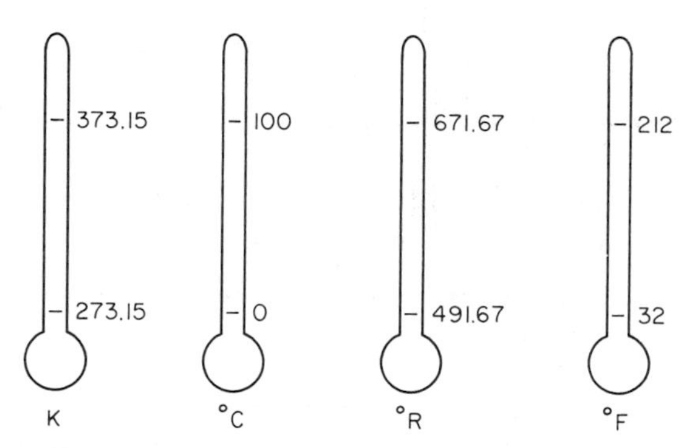

Figure 1.6. Four identically constructed thermometers calibrated in kelvins (K), degrees Celsius (°C), Rankine (°R), and Fahrenheit (°F). The figures indicate the temperature values for the ice and steam points.

The Rankine scale bears the same relationship to the Fahrenheit scale as does the Kelvin scale to the Celsius scale, that is, their temperature units are of equal size but the locations of the origins differ. Figure 1.6 summarizes the relationships among the four scales.

•••••• ••••••

1.6.4 Determine the Fahrenheit temperature of absolute zero (0 °R).

1.6.5 Derive formulas expressing

(a) Rankine temperature in terms of Celsius temperature, and
(b) Fahrenheit temperature in terms of Kelvin temperature.

↳ most common

•••••• ••••••

REFERENCES

Three splendid texts which have been favorites for many years are:

E. Fermi, "Thermodynamics." Dover, New York (1956).

M. W. Zemansky, "Heat and Thermodynamics," 5th ed. McGraw-Hill, New York (1968).

This edition includes material on kinetic theory and statistical mechanics not found in earlier ones.

G. N. Lewis and M. Randall, "Thermodynamics," 2nd ed. McGraw-Hill, New York (1961).

Strongly recommended supplements, aimed at the novice, are:

H. C. Van Ness, "Understanding Thermodynamics." McGraw-Hill, New York (1969).

and

M. W. Zemansky, "Temperatures Very Low and Very High." Van Nostrand-Reinhold, Princeton, New Jersey (1964).

Brief glimpses of the underlying atomic structure of matter and its relation to the macroscopic world of our senses are presented in:

M. Born, "The Restless Universe," Chapter 1. Dover, New York (1951).

and

R. P. Feynman, R. B. Leighton, and M. Sands, "The Feynman Lectures on Physics," Volume 1, Chapters 1 and 2. Addison-Wesley, Reading, Massachusetts (1963).

A problem-solver volume is:

L. Pincherle, "Worked Problems in Heat, Thermodynamics, and Kinetic Theory for Physics Students." Pergamon, Oxford (1966).

An authoritative and exhaustive trio covering virtually every aspect of thermometry is:

"Temperature, Its Measurement and Control in Science and Industry," Volume III. Van Nostrand-Reinhold, Princeton, New Jersey (1962).

F. G. Brickwedde (ed.), Part 1. Provides detailed expositions of thermometry scales and methods.

A. I. Dahl (ed.), Part 2. Is devoted to thermometry principles and instrumentation.

J. D. Hardy (ed.), Part 3. Deals with applications in the fields of biology and medicine.

Many practical aspects of thermometry, including the most recent international standards of temperature, are reviewed in:

W. T. Gray and D. I. Finch, How Accurately Can Temperature Be Measured? *Phys. Today* **24**, 32 (Sept. 1971).

Chapter 2

Mathematical Preliminaries

This chapter is to serve as an introduction or as a review of the mathematics needed for the development of thermodynamics. Sections 2.1 and 2.2 present the concepts of the differential and the partial derivative, and stress their geometrical significance. Two commonly-used partial derivative relations are derived and illustrated in Section 2.3. In Section 2.4 the coefficient of expansion and the compressibility are introduced to illustrate one way in which partial derivatives are used in thermodynamics.

2.1 DIFFERENTIALS

On a macroscopic scale the observable properties of systems are generally continuous. For example, the pressure produced by the atmosphere of the Earth at a given locality varies smoothly with altitude. The temperature of the atmosphere also exhibits spatial continuity. The continuity of physical properties allows a very concise mathematical formulation of thermodynamics. Two mathematical constructs are especially useful: the differential and the partial derivative. Our aim here is to introduce these two quantities and establish their geometric significance. An appreciation of their geometrical content greatly strengthens their physical interpretations.

The differential is useful in relating changes of two or more interdependent quantities. Let x and $y(x)$ denote corresponding values of the independent and dependent variables, respectively. We write

$$\Delta y = y(x+dx) - y(x) \tag{2.1}$$

for the change in y resulting from a change of dx in x. We call dx the differential of x. It denotes a change in the value of the variable x. The size of dx is arbitrary—it need not be small in any sense of the word. The differential of the dependent variable is defined by

$$dy \equiv \left(\frac{dy}{dx}\right) dx \tag{2.2}$$

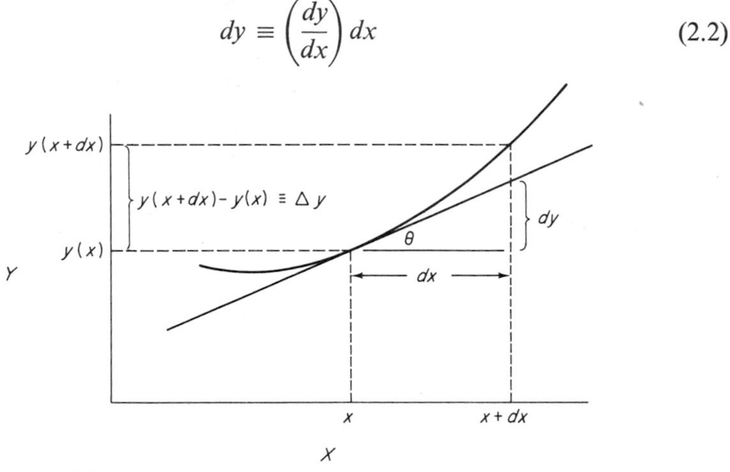

Figure 2.1. The differential dy is a linear approximation to the change Δy, with $dy = \tan\theta\, dx = (dy/dx)\, dx$.

Parentheses have been used to indicate that (dy/dx) is the derivative of y with respect to x. Reference to Figure 2.1 reveals the geometrical significance of dy. The differential dy gives a linear approximation to the change Δy. For sufficiently small dx, dy is a suitable replacement for Δy

$$\Delta y \simeq \left(\frac{dy}{dx}\right) dx \tag{2.3}$$

What is the criterion for "sufficiently small"? An example will help to establish a feeling for this.

Example 2.1 The Fractional Change in g

Treat the Earth as a sphere of radius R, having a uniform distribution of mass. The gravitational acceleration a distance r from the center of the Earth, and above its surface, is

$$g(r) = g_0 \frac{R^2}{r^2}$$

where g_0 is the acceleration of gravity at the surface. We have

$$\left(\frac{dg}{dr}\right) = -2g_0\frac{R^2}{r^3} = \frac{-2}{r}g(r)$$

so that the differential of g is

$$dg = -2g(r)\frac{dr}{r}$$

Observe that $dg/g = -2\,dr/r$. If the fractional change in r, dr/r, is much less than unity, so will dg/g be. For small dr/r then, we expect that $\Delta g/g$, the fractional change in g, is adequately approximated by dg/g. A direct calculation of $\Delta g/g$ bears out this expectation.

$$\Delta g = g(r+dr) - g(r) = g_0\,R^2\left\{\frac{1}{(r+dr)^2} - \frac{1}{r^2}\right\}$$

Minor algebra produces

$$\frac{\Delta g}{g} = -2\frac{dr}{r}\frac{\left(1+\dfrac{1}{2}\dfrac{dr}{r}\right)}{\left(1+\dfrac{dr}{r}\right)^2} = \frac{dg}{g}\frac{\left(1+\dfrac{1}{2}\dfrac{dr}{r}\right)}{\left(1+\dfrac{dr}{r}\right)^2}$$

As expected $\Delta g/g$ is closely approximated by dg/g when dr/r is much less than unity. Because of the continuous nature of $g(r)$ we can conceive of arbitrarily small values of dr, whereupon dg represents the change in g. ◇

The example illustrates a general relationship between differentials and differences, namely that the change in y for sufficiently small dx may be represented by the differential

$$dy = \left(\frac{dy}{dx}\right)dx \simeq \Delta y \tag{2.4}$$

Thermodynamics is concerned with situations where x and $y(x)$ represent measurable quantities, that is, physical properties of matter. The continuous nature of such properties allows us to envision changes sufficiently small to permit the use of

$$dy \equiv \left(\frac{dy}{dx}\right)dx \tag{2.2}$$

to relate changes in x and y.

......

2.1.1 Determine the relation between $\Delta y/y$ and dy/y for the parabola $y = x^2$. If dx/x is 0.01, by what amount do $\Delta y/y$ and dy/y differ?

......

2.2 PARTIAL DERIVATIVES

The need for partial derivatives arises when more than one independent variable enters. Consider a function z of two independent variables, x and y,

$$z = z(x, y)$$

Geometrically $z(x, y)$ may be thought of as the height of a surface above the x–y plane. A three-dimensional surface can be represented by an equation of the form $f(x, y, z) = 0$. For example, the equation of a sphere of radius b centered at the origin may be written as

$$f(x, y, z) \equiv x^2 + y^2 + z^2 - b^2 = 0$$

Here we regard z as the dependent variable. However, we are free to choose any pair as independent and the third as dependent.

We can form derivatives of $z(x, y)$ with respect to both x and y. The derivatives are called *partial* derivatives and are defined by

$$\left(\frac{\partial z}{\partial x}\right) \equiv \lim_{\Delta x \to 0} \frac{z(x+\Delta x, y) - z(x, y)}{\Delta x} \tag{2.5}$$

$$\left(\frac{\partial z}{\partial y}\right) \equiv \lim_{\Delta y \to 0} \frac{z(x, y+\Delta y) - z(x, y)}{\Delta y} \tag{2.6}$$

In forming $(\partial z/\partial x)$, the definition (2.5) shows that the independent variable y is treated as a constant. As an example, consider the function

$$z(x, y) = 2x^2 y$$

where x and y are independent variables. The partial derivatives of $z(x, y)$ are

$$\frac{\partial z}{\partial x} = 4xy, \qquad \frac{\partial z}{\partial y} = 2x^2$$

......

2.2.1 Verify for the function

$$z(x, y) = 2x^2 y$$

that

$$\frac{\partial}{\partial x}\left(\frac{\partial z}{\partial y}\right) = \frac{\partial}{\partial y}\left(\frac{\partial z}{\partial x}\right)$$

2.2.2 For the functions in (i) and (ii) below

(a) evaluate $\partial f/\partial x$ and $\partial f/\partial y$,
(b) verify that

$$\frac{\partial}{\partial y}\left(\frac{\partial f}{\partial x}\right) = \frac{\partial}{\partial x}\left(\frac{\partial f}{\partial y}\right)$$

that is, show that the order of partial differentiation does not matter.

(i) $f(x,y) = x^2y - xy^2$,
(ii) $f(x,y) = ye^x$.

••••• •••••

To avoid confusion the partial derivatives are often subscripted to indicate which variables are independent of the differentiation variable. In the previous example we might have written

$$\left(\frac{\partial z}{\partial x}\right)_y = 4xy, \qquad \left(\frac{\partial z}{\partial y}\right)_x = 2x^2$$

The subscripts tell us which variables are being treated as constant during the differentiation. Figure 2.2 suggests the geometrical significance of the partial derivative.

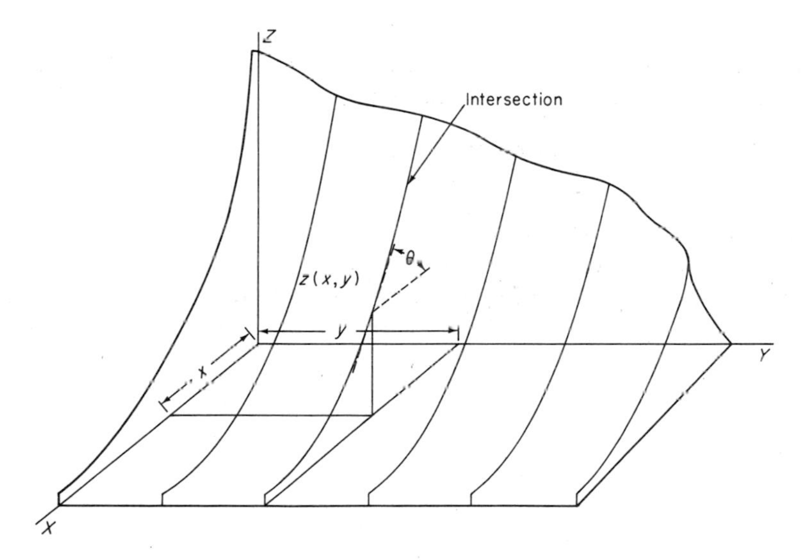

Figure 2.2. The partial derivative $(\partial z/\partial x)_y = \tan\theta$ gives the slope of a line formed by the intersection of the plane $z = z(x,y)$ and the plane $y = \text{constant}$.

•••••• ••••••

2.2.3 With reference to Figure 2.2, explain the geometrical significance of $(\partial z/\partial y)_x$.

•••••• ••••••

We define the differential of z by

$$dz = \left(\frac{\partial z}{\partial x}\right)_y dx + \left(\frac{\partial z}{\partial y}\right)_x dy \tag{2.7}$$

It follows from the geometrical significance of the partial derivatives that, for sufficiently small dx and dy, dz gives the change in z between $z(x, y)$ and $z(x+dx, y+dy)$. Thus

$$\Delta z \equiv z(x+dx, y+dy) - z(x, y) \simeq dz \tag{2.8}$$

To demonstrate this more convincingly we subtract and add $z(x, y+dy)$ on the right side, forming two differences,

$$\Delta z = \{z(x+dx, y+dy) - z(x, y+dy)\} + \{z(x, y+dy) - z(x, y)\}$$

For sufficiently small dx and dy the differences may be replaced by the corresponding differentials. Thus

$$\Delta z \simeq \left(\frac{\partial z}{\partial x}\right)_{y+dy} dx + \left(\frac{\partial z}{\partial y}\right)_x dy$$

The first derivative is evaluated at $y+dy$. For small dy this differs from the derivative at y by an amount proportional to dy. Without sacrificing the accuracy of the approximation we can set

$$\left(\frac{\partial z}{\partial x}\right)_{y+dy} dx \simeq \left(\frac{\partial z}{\partial x}\right)_y dx$$

which establishes the desired result

$$\Delta z \simeq \left(\frac{\partial z}{\partial x}\right)_y dx + \left(\frac{\partial z}{\partial y}\right)_x dy = dz \tag{2.8a}$$

As an illustration, consider

$$z(x, y) = yx^3$$

We have

$$\left(\frac{\partial z}{\partial x}\right)_y = 3x^2y, \qquad \left(\frac{\partial z}{\partial y}\right)_x = x^3$$

whereupon (2.7) gives

$$dz = 3x^2y\,dx + x^3\,dy$$

•••••• ••••••

2.2.4 The entropy S of an ideal gas is given in terms of the pressure and temperature by†

$$S(T, P) = C_p \ln T - R \ln P + S_0$$

Show that

$$dS = \frac{C_p}{T} dT - \frac{R}{P} dP$$

•••••• ••••••

2.3 TWO IMPORTANT PARTIAL DERIVATIVE RELATIONS

Let $z(x, y)$ denote a function of the independent variables x and y. We can write

$$dz = \left(\frac{\partial z}{\partial x}\right)_y dx + \left(\frac{\partial z}{\partial y}\right)_x dy \tag{2.9}$$

We are equally free to regard x as a function dependent on the independent variables z and y. The differential of $x(z, y)$ is then expressible as

$$dx = \left(\frac{\partial x}{\partial z}\right)_y dz + \left(\frac{\partial x}{\partial y}\right)_z dy \tag{2.10}$$

Both (2.9) and (2.10) hold simultaneously, that is, both are simply alternate ways of relating changes in the interdependent trio x, y, and z. If (2.10) is inserted into (2.9), there results

$$dz = \left\{\left(\frac{\partial z}{\partial x}\right)_y \left(\frac{\partial x}{\partial z}\right)_y\right\} dz + \left\{\left(\frac{\partial z}{\partial x}\right)_y \left(\frac{\partial x}{\partial y}\right)_z + \left(\frac{\partial z}{\partial y}\right)_x\right\} dy \tag{2.11}$$

This relation is valid for all conceivable changes in y and z. If y is held constant while z varies, so that $dy = 0$, $dz \neq 0$, (2.11) yields the first of the two sought after relations

$$\left(\frac{\partial z}{\partial x}\right)_y \left(\frac{\partial x}{\partial z}\right)_y = 1 \tag{2.12}$$

An alternate form

$$\left(\frac{\partial z}{\partial x}\right)_y = \frac{1}{\left(\frac{\partial x}{\partial z}\right)_y} \tag{2.13}$$

† We denote by $\ln x$ the logarithm of x to the *natural base e*. Any other base will be specifically noted, for example, $\log_{10} x$, $\log_2 x$, and so on.

reveals it as an inversion theorem. For a situation where z remains constant, but for which y changes, $dz = 0$ and $dy \neq 0$, and (2.11) becomes

$$0 = \left\{ \left(\frac{\partial z}{\partial x}\right)_y \left(\frac{\partial x}{\partial y}\right)_z + \left(\frac{\partial z}{\partial y}\right)_x \right\} dy$$

Since dy is not zero, the expression in the braces must vanish. This gives the second relation, a pseudo chain rule,[†]

$$\left(\frac{\partial z}{\partial x}\right)_y \left(\frac{\partial x}{\partial y}\right)_z = -\left(\frac{\partial z}{\partial y}\right)_x \tag{2.14}$$

An alternate form, obtained by applying (2.12), is

$$\left(\frac{\partial x}{\partial y}\right)_z \left(\frac{\partial y}{\partial z}\right)_x \left(\frac{\partial z}{\partial x}\right)_y = -1 \tag{2.15}$$

Note that each of the derivatives in (2.15) involves the three variables in "cyclic order"

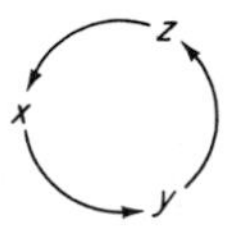

······ ······

2.3.1 Convert (2.14) into (2.15) by making use of (2.12).

2.3.2 Verify the relation

$$\left(\frac{\partial x}{\partial y}\right)_z \left(\frac{\partial y}{\partial z}\right)_x \left(\frac{\partial z}{\partial x}\right)_y = -1$$

for the variables x, y, and z related by

$$x^2 + y^2 + z^2 - a^2 = 0$$

Hint: Show that $(\partial x/\partial y)_z = -y/x$ and then evaluate the other two partial derivatives either by inspection or by a cyclic permutation argument.

······ ······

[†] Recall that if $g = g(x)$ and $x = x(y)$, then the usual chain rule of differential calculus gives

$$\left(\frac{dg}{dy}\right) = \left(\frac{dg}{dx}\right)\left(\frac{dx}{dy}\right)$$

2.4 USING PARTIAL DERIVATIVES IN THERMODYNAMICS

Evaluating partial derivatives of a specified mathematical function is a routine operation. Manipulating partial derivatives in calculations is a skill polished by practice. In applications of thermodynamics it often is necessary to use quantities which are represented by partial derivatives. The coefficient of volume expansion is one such quantity. Most substances undergo a change in volume when they are heated at constant pressure. The volume of a simple fluid or solid may be regarded as a variable dependent on the two independent variables pressure and temperature

$$V = V(P, T)$$

If the temperature is changed slightly, from T to $T+\Delta T$, while the pressure remains constant, the volume of the sample changes by an amount proportional to ΔT. The change in volume is also proportional to the volume itself. We can therefore express the change in volume as

$$\Delta V = \beta V \Delta T \tag{2.16}$$

The proportionality factor β is termed the coefficient of volume expansion. From the experimental viewpoint, β is given by

$$\beta = \frac{1}{V}\frac{\Delta V}{\Delta T} \tag{2.17}$$

The measured ratio $\Delta V/\Delta T$ corresponds to the partial derivative $(\partial V/\partial T)_P$. More precisely,

$$\left(\frac{\partial V}{\partial T}\right)_P = \lim_{\Delta T \to 0} \frac{\Delta V}{\Delta T}$$

For small changes in T and V then, the partial derivative $(\partial V/\partial T)_P$ is adequately represented by the measured ratio $\Delta V/\Delta T$, and vice versa. Consequently, in theoretical developments the coefficient of volume expansion is identified as the quantity

$$\beta = \frac{1}{V}\left(\frac{\partial V}{\partial T}\right)_P \tag{2.18}$$

A quantity closely related to β is the coefficient of linear expansion. If $L(T, F)$ is the length of an object subject to a force F, the coefficient of linear expansion is defined as

$$\alpha = \frac{1}{L}\left(\frac{\partial L}{\partial T}\right)_F \tag{2.19}$$

Comparing this with (2.18) we see that α is the linear analog of β; length appears in place of the volume, force in place of pressure. Experimentally, α is determined by measuring the temperature variations in length. If the length of the sample changes from L to $L+\Delta L$ as the temperature changes from T to $T+\Delta T$,

$$\alpha = \frac{1}{L}\frac{\Delta L}{\Delta T} \tag{2.20}$$

In practice, (2.20) is used to relate the changes in length and temperature, that is,

$$\Delta L = \alpha L \, \Delta T \tag{2.21}$$

•••••• ••••••

2.4.1 In 1956, a high-school athlete hurled a discus 185 ft $2\frac{3}{8}$ in. at the Ohio State Championship meet. The distance was measured with a tape made of invar, a high-grade steel, with the (very small) coefficient of expansion $\alpha = 0.7 \times 10^{-6}/°\mathrm{C}$. At that time the National High School record was 185 ft $2\frac{1}{2}$ in. The tape was calibrated at a temperature of 25 °C. The temperature on the day of the track meet was 85 °F. Because of thermal expansion the actual distance was greater than that measured with the tape. Determine if the actual distance exceeded the national record.

2.4.2 A brass sphere ($\alpha = 1.9 \times 10^{-5}/°\mathrm{C}$) has a diameter of 5.00 cm at 25 °C. A steel ring ($\alpha = 1.1 \times 10^{-5}/°\mathrm{C}$) has a diameter of 4.995 cm at 25 °C. At what temperature will the sphere just be able to pass through the ring

(a) if the ring is heated, but not the sphere?
(b) if both the ring and the sphere have the same temperature?

2.4.3 Derive the result $\beta = 3\alpha$ by considering the change in volume of a cube of side L.
Hint: A quick derivation makes use of

$$V = L^3, \qquad dV = 3L^2\, dL$$

2.4.4 Determine the change in density of a solid copper sphere when its temperature rises from 20 °C to 60 °C. The density of copper at 20 °C is 8.96 gm/cm³. You may wish to derive first the result

$$\frac{\Delta\rho}{\rho} = -\beta\,\Delta T$$

where $\Delta\rho$ and ΔT are the changes in density and temperature.

•••••• ••••••

TABLE 2.1

Coefficients of Linear Expansion for Selected Liquids and Solids[a]

Liquids		Solids	
Substance	$10^3\alpha$ (°C^{-1})	Substance	$10^5\alpha$ (°C^{-1})
Acetic acid	1.07	Silver (Ag)	1.89
Acetone	1.49	Aluminum (Al)	2.80
Benzene	1.24	Gold (Au)	1.42
Carbon		Calcium (Ca)	2.50
tetrachloride	1.24	Iron (Fe)	1.17
Ether	1.66	Sodium (Na)	7.1
Mercury	0.18	Lead (Pb)	2.91
Pentane	1.61	Rubidium (Rb)	9.0
Water	0.21	Zinc (Zn)	3.3
		Carbon steel	1.15
		Glass	0.90

[a] Values refer to measurements at a pressure of 1 atm and for temperatures near 20 °C.

$$\alpha \equiv \frac{1}{L}\left(\frac{\partial L}{\partial T}\right)_F$$

Values for liquids adapted from J. H. Perry, "Chemical Engineer's Handbook." Copyright by McGraw-Hill, New York (1950). Values for solids taken from "International Critical Tables of Numerical Data: Physics, Chemistry, and Technology." Nat. Res. Council, Washington, D.C., 1923–1933.

For solids, measurements of α are much simpler than determinations of β. Because of this, and in view of the simple proportionality of β and α,[†] values of β for solids are seldom tabulated. Typical values of α are listed in Table 2.1. It will be noted that the listed values of α are positive and that they refer to a common temperature of 20 °C. For most solids α varies smoothly with temperature and is nearly independent of pressure changes. The tabulated values of α cluster near 10^{-5}/°C for solids and 10^{-3}/°C for liquids. This means that the volume of a solid changes by about one part in 10^5 for each degree change in temperature. A liquid expands by about one part in 10^3 for the same change in temperature. A few substances exhibit negative values of α and β over certain ranges of temperature. Most notable of these is ice for which β is negative at temperatures below 63 K and (for water) over the narrow range of

[†] For isotropic solids $\beta = 3\alpha$ (see Problem 2.4.3).

temperature from 0 °C to 3.98 °C. Thus, if water at 1 °C is warmed to 2 °C, its volume decreases and its density increases. The maximum density of water occurs at a temperature of 3.98 °C (at a pressure of 1 atm). Type metal, an alloy of lead and antimony, also exhibits a negative value of β. After the type is formed the mold is inverted and heated slightly. The type shrinks and falls free of the mold. At one time a similar alloy was used for plugs in ceiling fire sprinkler systems.

•••••• ••••••

2.4.5 Explain why ice forms first at the surface of a lake rather than at its bottom.

2.4.6 Suppose that ice did not float and reflect on some possible consequences.

•••••• ••••••

For gases, β is strongly temperature-dependent, being very nearly equal to the reciprocal of the Kelvin temperature

$$\beta_{\text{gas}} \simeq \frac{1}{T}$$

For the ideal gas, β is exactly $1/T$ (compare Problem 2.4.8).

Another quantity related to a partial derivative is the isothermal compressibility κ. The analytical definition of κ is

$$\kappa = -\frac{1}{V}\left(\frac{\partial V}{\partial P}\right)_T \tag{2.22}$$

Because $(\partial V/\partial P)_T$ is negative for most substances (they shrink when squeezed) the minus sign is inserted to make κ a positive quantity. Thus κ measures the extent to which the volume of a system is diminished by an increase in pressure, hence the name compressibility. Experimentally, κ is determined as the ratio

$$\kappa = -\frac{1}{V}\frac{\Delta V}{\Delta P} \tag{2.23}$$

In practice, (2.23) is used to relate the changes in volume and pressure

$$\Delta V = -\kappa V \Delta P \tag{2.23a}$$

•••••• ••••••

2.4.7 A student of mass 100 kg balances himself on a cube of lead (Pb). The volume of the cube is 1 cm^3 before the students stands on it. What is the fractional change in its volume while he balances atop it?

•••••• ••••••

TABLE 2.2

Compressibilities of Selected Liquids and Solids[a]

Liquids		Solids	
Substance	$10^{12}\kappa$ (cm²/dyn)	Substance	$10^{6}\kappa$ (atm⁻¹)
Acetic acid	0.91	Aluminum (Al)	1.38
Acetone	1.28	Aluminum oxide (Al_2O_3)	0.40
Benzene	0.94	Gold (Au)	0.60
Carbon disulfide	0.93	Calcium (Ca)	5.89
Carbon tetrachloride	1.05	Calcium carbonate ($CaCO_3$)	1.36
Ether	1.87	Flourite (CaF_2)	1.23
Ethyl bromide	1.29	Iron (Fe)	0.61
Mercury	0.41	Pyrite (FeS_2)	1.70
Methanol	1.21	Lithium (Li)	8.98
Toluene	0.90	Lithium fluoride (LiF)	1.53
Water	0.46	Lead (Pb)	2.38
		Galena (PbS)	1.91
		Rubidium (Rb)	40.5
		Zinc (Zn)	1.76

[a] Values refer to pressures near 1 atm. The reference temperature is 20 °C for liquids and 30 °C for solids. Since 1 atm $= 1.013 \times 10^6$ dyn/cm², 10^{-6} atm^{-1} is very nearly equal 10^{-12} cm²/dyn and the numerical values for liquids and solids are directly comparable

$$\kappa \equiv -\frac{1}{V}\left(\frac{\partial V}{\partial P}\right)_T$$

Values for liquids taken from "Handbook of Chemistry and Physics," 51st ed. Chem. Rubber Publ. Co., Cleveland, Ohio (1970). Values for solids taken from "International Critical Tables of Numerical Data: Physics, Chemistry, and Technology." Nat. Res. Council, Washington, D.C., 1923–1933.

Typical values of κ are displayed in Table 2.2. It is evident from the table that 10^{-12} cm²/dyn $\simeq 10^{-6}$/atm is a typical value of κ. A pressure increase of 1 atm therefore produces a fractional change in volume of approximately one part in a million. For gases the compressibility is very nearly the reciprocal of the pressure

$$\kappa_{\text{gas}} \simeq \frac{1}{P}$$

In fact, for the ideal gas κ is exactly $1/P$ (compare Problem 2.4.8). For a pressure of 1 atm we have

$$\kappa_{\text{gas}} \simeq 1 \text{ atm}^{-1}$$

about one million times greater than the compressibility of a typical solid. The relatively large compressibilities of gases is in accord with everyday experience. A balloon filled with air or helium is readily compressed; a steel ball bearing seems virtually incompressible. The fact that κ_{gas} is inversely proportional to the pressure is also consistent with experience. A half-flat tire is deformed much more by the weight of an auto than one which is properly inflated.

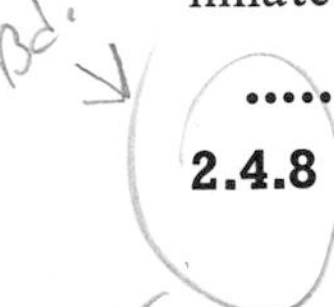

•••••

2.4.8 Use the ideal gas equation of state $PV = nRT$ to show that

$$\beta_{\text{I.G.}} = \frac{1}{T}, \qquad \kappa_{\text{I.G.}} = \frac{1}{P}$$

•••••

Another derivative of physical significance is $(\partial P/\partial T)_V$, which measures the change in pressure produced by a change in temperature at constant volume. It would be extremely difficult to measure $(\partial P/\partial T)_V$ directly. Fortunately, it is also unnecessary. By applying the pseudo chain rule (2.14) and the inversion theorem (2.13) one finds

$$\left(\frac{\partial P}{\partial T}\right)_V = \frac{\beta}{\kappa} \tag{2.24}$$

This is typical of the way the pseudo chain rule and the inversion theorem are used in thermodynamics. A quantity difficult to measure is expressed in terms of more readily observable parameters.

REFERENCES

A concise introduction to much of the required mathematics is provided in:

M. H. Hull, Jr., "The Calculus of Physics." Benjamin, New York (1969).

The derivative concept is covered in Chapter 1. Partial differentiation receives attention in Chapter 5.

An abbreviated "self-propelled" volume which offers a concise introduction or rapid review of calculus fundamentals is:

D. Kleppner and N. Ramsey, "Quick Calculus." Wiley, New York (1965).

Chapter 3

The First Law of Thermodynamics

3.1 THE MANY FACES OF ENERGY

Energy is a concept which shows us many faces. Mechanics has acquainted us with the kinetic energy of a moving object, the gravitational potential energy of a raised weight, and the elastic potential energy of a stretched spring. Electromagnetism reveals still other forms of energy. Among these are the energy transported by electromagnetic radiation. The electric field between the plates of a charged capacitor harbors energy as does a magnetic field. A charged battery stands ready to convert chemical energy into the energy of motion of electric charges.

Our notion of energy is unlikely to change very much from the elementary idea that energy measures the capacity to perform work. Our appreciation of energy grows as we find it thrust forward as a unifying concept in all areas of physics, indeed, throughout all science. Two primary attributes of energy quickly become apparent: Energy can be converted from one form to another, and energy is conserved. The first law of thermodynamics quantifies these attributes of energy. All forms of energy, mechanical, electrical, chemical, nuclear, and so on, are recognized together with the possibility of conversions from one type to another. In particular, the recognition of heat as a form of energy is necessary in order to preserve the conservation of energy. The most general statement of the first law of thermodynamics is this:

The total energy of the universe is constant.

Considerable analysis is needed to convert this sweeping statement into a working law of nature.

A mountain of *failures* stand as the experimental foundation of the first law. No one has succeeded in building a perpetual motion machine of the first kind—a device which produces more energy than it consumes. Many have tried, including crackpots and swindlers, who would hoax the gullible.

•••••• ••••••

3.1.1 It is one thing to glibly pronounce that a perpetual motion machine will not work because it violates the first law of thermodynamics. It is quite another matter to understand precisely where a scheme goes wrong. Two proposed perpetual motion machines of the first kind are shown in Figure 3.1. Give a qualitative explanation of why they will not operate as suggested below. In other words, find the "fly in the ointment."

The machine in part (a) consists of a wheel whose axle is to act as the drive-shaft of a vehicle. The spokes of the wheel are permanent magnets. The magnets are mounted so that their south poles all lie on the periphery of the wheel. Mounted above the center of the wheel is another permanent magnet. Symmetry considerations show that the net torque on the wheel is zero. To change this balance of torques, a magnetic shield is erected, so as to eliminate

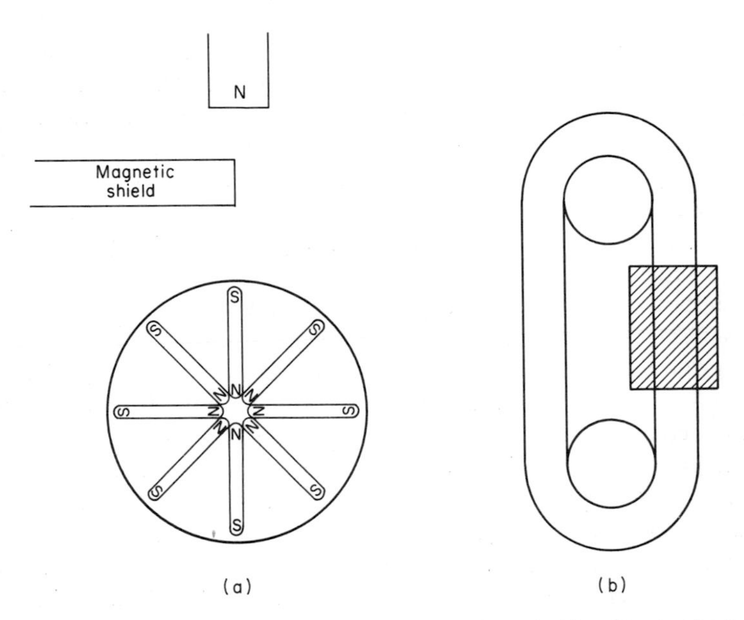

Figure 3.1. Two perpetual motion machines of the first kind: (a) magnetic wheel and (b) airtight tube.

the magnetic forces (and the torques) on the left side of the wheel. (A very crude magnetic shield might be simply a piece of soft iron. More elaborate magnetic shielding is used regularly, for example, to cancel the magnetic field of the earth in certain precision experiments.) With the shielding in position the net torque on the unshielded right side of the wheel should produce a counterclockwise rotation of the wheel.

The second device in part (b) consists of a flexible airtight tube designed to rotate around two pulley hubs. The axle of the lower hub serves as the driveshaft of a vehicle. Standing alone the gravitational force on the left side of the tube balances the force on the right, that is, there is no net torque and no rotation. By surrounding the right side of the tube with a "silo" filled with water a buoyant force is brought into play, that is, the portion of the tube inside the silo displaces water and is buoyed up by a force equal to the weight of the water displaced. With the buoyant force acting upward, the right side of the tube moves upward. Because of its continuous loop construction, a fixed volume of tubing is always submerged and the hubs rotate continuously, delivering power to the driveshaft.

•••••• ••••••

Example 3.1 The Sun, Energy Conversion, Nuclear Fusion, and the Bomb

The Sun serves as the source of energy by which life is maintained. The relentless stream of sunlight drives the winds, rains, and ocean currents and supplies energy for the complex processes of photosynthesis. The energy carried by sunlight is absorbed by green plants and used to convert carbon dioxide and water into oxygen and carbohydrates. Some of the absorbed energy sustains the plant, the remainder is converted into chemical energy—the energy associated with interatomic bonds. Green plants supply both the food we eat and the oxygen we breathe. The oxygen aids the release of chemical energy during metabolic processes.

Strangely enough the photosynthetic and metabolic processes which spend the energy delivered by sunlight are not understood as completely as the processes which release energy deep within the Sun. The source of stellar energy is nuclear *fusion.* At ordinary temperatures two hydrogen atoms can unite to form a hydrogen molecule. In the process they release energy to the surroundings. A similar sort of union, not of atoms but of atomic nuclei, releases energy in stellar interiors. The temperature in the core of the Sun is near 10 million kelvins. The primary nuclear fuel in the core is hydrogen, but it is *ionized* hydrogen. The high temperature there acts to strip away the atomic electron of the hydrogen atom. In most instances the hydrogen nucleus is a proton. Just as the union of two hydrogen atoms releases energy, the fusion of two protons is energetically favorable. Nuclear fusion results in the release of energy. The scale of the energy developed in fusion reactions is quite substantial, about a million times that for the reaction between two hydrogen atoms.

According to Einstein's theory of relativity, mass is just another aspect of energy. Mass and energy are equivalent. The relation between mass m and energy E is given by that famous equation

$$E = mc^2$$

where c is the speed of light. In a reaction which releases energy ΔE, the masses of the reactants and products differ by Δm, where

$$\Delta E = \Delta m c^2$$

The form of the released energy is not determined by this equation, only its total amount. In fusion reactions, the released energy appears as kinetic energy of the products. Much of this in turn is converted into electromagnetic energy—sunlight—which then escapes from the solar surface.

The reactions between nuclei in stellar interiors are referred to as *thermonuclear* reactions. High temperature and density are very necessary conditions for achieving fusion reactions. In order for two protons to fuse they must approach one another very closely. Only when they approach within distances on the order of 10^{-13} cm can the attractive nuclear force effectively act to promote fusion. The attractive nuclear force is opposed by the electrical force of repulsion between the positively charged protons. In order to overcome this electrical repulsion the two protons must have a high relative kinetic energy as they approach. It is the high temperature which provides the kinetic energy, that is, the higher the temperature the faster the protons move. High density aids by making collisions between protons more frequent.

To date technological efforts to produce a controlled thermonuclear reaction have been unsuccessful. The high temperatures involved require that the reactants be confined by a magnetic field which keeps them away from the solid walls of the reactor vessel. Unfortunately, the magnetically confined configurations tend to be unstable. They quickly become turbulent and disperse. To date the only successful man-made thermonuclear reactions have been uncontrolled—they are better known as *hydrogen bombs*.

The explosion of TNT converts chemical energy into kinetic energy, heat, and light (radiant energy). A hydrogen bomb converts mass into other forms of energy, such as the kinetic energy of the reaction products and radiant energy. The hydrogen bomb is such an awesome energy converter that the energy release is reckoned in the equivalent number of millions of tons of TNT. The energy released by the explosion of one million tons of TNT is 4×10^{22} ergs. A unit of energy, the megaton, is defined as

$$1 \text{ megaton} \equiv 4 \times 10^{22} \text{ ergs}$$

The weight of a hydrogen bomb which releases 1 megaton of energy is presently a "secret."

One of the most valuable techniques a scientist can develop is the art of making order-of-magnitude estimates. These aid in deciding the feasibility of an idea. We illustrate the technique by making an order of magnitude estimate of the size of a hydrogen bomb.

The customary unit for expressing energies in nuclear physics is the million electron volt (MeV)

$$1 \text{ MeV} = 10^6 \text{ eV} = 1.6 \times 10^{-6} \text{ erg}$$

The mass of a proton is 1.67×10^{-24} gm. The mass energy of a proton $m_p c^2$ is 1.5×10^{-3} erg, or slightly less than 1000 MeV. The energy released by the fusion of two protons is approximately 1.4 MeV. In round numbers then nuclear fusion releases about 1/1000 of the mass energy of the reacting particles. It follows that the mass energy of a 1-megaton hydrogen bomb must be at least 1000 megaton, or 4×10^{25} ergs. Setting

$$mc^2 \simeq 4 \times 10^{25} \text{ ergs}$$

gives, for an estimate of the mass of the bomb,

$$m \simeq 4 \times 10^4 \text{ gm}$$

Scaling up, we can say that the mass of a 25-megaton hydrogen bomb is at least one million gm. A mass of 10^6 gm weighs very nearly 1 ton. We conclude that 25-megaton hydrogen bombs cannot be carried about in briefcases or vest pockets.

A hydrogen bomb is an uncontrolled thermonuclear fusion reaction. For over 25 years scientists have been striving to achieve a controlled fusion reaction. A fusion power reactor would have three distinct advantages over the now operative fission reactors. First, the fusion reaction leaves no radioactive ashes. In contrast, the by-products of fission reactions remain radioactive for many years and must be buried deep in the earth. Second, a good share of the energy released in the fusion reaction is in the form of kinetic energy of charged particles. By judiciously guiding these charges in a magnetic field it would be possible to induce currents in an external circuit. The fusion reactor would thereby permit the direct production of electrical power. In contrast, a fission reactor liberates energetic neutrons and leads to a less direct and less efficient production of electric power. The neutron kinetic energy is converted to heat, which is used to produce steam, which drives a turbine, which powers an electrical generator. The third point on which fusion reactors promise to outrank fission reactors concerns the nuclear fuel. The fissile isotopes of uranium and plutonium are scarce. Even with sophisticated "breeder" reactors which generate additional fissile material the projected energy demands of the world will bankrupt our fission fuels in a few centuries. The basic fusion reactor fuel is deuterium.† Our supply of deuterium is virtually unlimited. We literally have oceans of it! About one out of every 6000 water molecules contains a deuterium atom, that is, it is DHO rather than H_2O. The recovery of deuterium from sea water is a relatively simple operation. Thus a supply of fusion fuel will be available as long as the oceans survive. ◇

•••••• ••••••

3.1.2 The oxidation of 1 mole of sucrose ($C_{12}H_{22}O_{11}$) liberates 5.67×10^{10} ergs, in the form of heat.

(a) What is the decrease in mass of the system?

(b) What is the fractional mass to energy conversion ($\Delta m/m$ = mass converted/initial mass)?

3.1.3 In the year 1908, in central Siberia, a gigantic fireball crashed to Earth. The fireball was visible for many hundreds of kilometers. Strangely, no meteoric fragments were found. Within recent years it has been speculated that the fireball was composed of antimatter; a fragment of some antiplanet or anticomet, perhaps. A meeting of matter and antimatter is like a "kiss-of-death." They annihilate—their mass energy (mc^2) is converted into other forms of energy. For example, an antielectron (positron) interacts with an electron. The two particles disappear, annihilate, and in their place is an equivalent amount of electromagnetic energy in the form of γ rays. Antiprotons annihilate with protons, and so on. Most of the mass energy is quickly converted into heat and radiant energy.

In order to reach the ground, a chunk of antimatter would have to tunnel through our atmosphere. It would annihilate matter in its path and generate a monumental shock wave. (The Siberia shock wave made two transits around the earth.)

(a) Assume that a piece of antimatter travels vertically downward. Estimate the minimum mass which it must have in order to reach the surface of the earth.

(b) (Optional and a bit tricky) Treat the fireball as a sphere of uniform density. As it descends, annihilations reduce its radius. Show that the minimum

---

† There are three isotopes of hydrogen, differing in their nuclear composition. The nucleus of ordinary hydrogen consists of a single proton. Deuterium and tritium are the heavy isotopes. The nucleus of a deuterium atom consists of one proton and one neutron. The deuterium nucleus is called a *deuteron*. The heaviest isotope of hydrogen is tritium. The nucleus of a tritium atom is called a *triton* and contains one proton and two neutrons. The nuclear physicist is likely to designate these nuclei as p, d, and t or by the symbols ^1_1H, ^2_1H, and ^3_1H. The atoms are often designated as H, D, and T.

initial radius which the fireball must have in order to reach the surface of the earth is

$$R = \frac{P}{4\rho_f g}$$

where P is the atmospheric pressure at the surface of the earth, g is the acceleration of gravity, and ρ_f is the mass density of the fireball.

(c) Compute the mass of the fireball of part (b), assuming it is composed of anti-iron ($\rho_f = 7.9$ gm/cm^3). Remembering that an equal mass of the atmosphere annihilates, compute the energy release, in megatons, resulting from the descent of such a fireball.

•••••• ••••••

3.2 CALORIMETRIC DEFINITION OF HEAT

The concept of heat as a measurable quantity evolved during the 17th and 18th centuries. It was not until the middle of the 19th century that heat was recognized as a form of energy. The earliest quantitative theory viewed heat as a fluid, called caloric. This fluid was invisible and weightless. If the temperature of a body increased, it was because caloric had been absorbed. Cooling was accompanied by the loss of caloric. In order to explain certain calorimetry experiments it was necessary to postulate that caloric was conserved, that is, it could neither be created nor destroyed. The caloric theory was immensely successful. It could explain a great many thermal phenomena, such as the expansion of heated objects. The caloric theory was eventually discarded because it could not offer satisfactory explanations of certain phenomena. For instance, if two ice cubes are rubbed together in vacuum, they gradually melt. The caloric theory cannot adequately explain this phenomenon.

The language of thermodynamics contains many reminders of the caloric theory. Among these is the word "calorie," a unit of heat. The calorimetric definition of the unit of heat called the *calorie* is this:

> One calorie (1 cal) is the amount of heat required to raise the temperature of one gram of water from 14.5 °C to 15.5 °C.

A kilocalorie (1 kcal) is 1000 calories. In dietetics, the kilocalorie is generally referred to as the Calorie

$$1 \text{ kcal} \equiv 1 \text{ Cal} = 1000 \text{ cal}$$

Although heat was not originally interpreted as a form of energy, it was recognized as an extensive quantity—1 cal raises the temperature of 1 gm by 1 °C, but 5 cal are required to raise the temperature of 5 gm by 1 °C. For a given mass of material it is also found that the amount of heat absorbed is directly proportional to the temperature increase. This direct proportionality holds true only for small changes of temperature and only so long as the heated substance does not undergo a change of phase, such as melting.

Let Q stand for the heat absorbed or rejected by a body of mass m and let ΔT denote its change in temperature. The observation that Q is proportional to $m\,\Delta T$ lets us insert a proportionality factor C and obtain the equation

$$Q = Cm\,\Delta T \tag{3.1}$$

The factor C is called the *specific heat capacity* and is a property of the physical and chemical composition of the material.[†] Equation (3.1) stands as a calorimetric definition of heat; but what is heat?! Equation (3.1) cannot answer. It tells us how to measure heat, not what heat measures.

•••••• ••••••

3.2.1 If we consider 1 gm of a substance ($m = 1$ gm) the calorimetric definition of heat, Eq. (3.1), is

$$Q = C\,\Delta T$$

(a) Present an interpretation which justifies the designation of heat capacity for C.

(b) Present a second interpretation which reveals C as a measure of the thermal inertia of the material, its ability to resist a change of temperature. (Recall from mechanics that the mechanical inertia of a body, as measured by its mass, measures the object's resistance to acceleration.) You may wish to answer parts (a) and (b) by comparing two substances with different values of C.

3.2.2 One British thermal unit (Btu) is defined as the amount of heat required to raise the temperature of 1 lb of water by 1 °F. Show that 1 Btu equals approximately 252 cal.

3.2.3 Radio astronomers have provided us with a detailed picture of our own galaxy, the Milky Way, as well as glimpses of other distant galaxies. It has been estimated that the total energy received to date by all radio telescopes is less than 1 erg. To appreciate how minute an amount of energy this is, determine

(a) the temperature increase of 1 gm of water when it receives 1 erg of heat energy;

[†] Where no confusion can result we often shorten "specific heat capacity" to "specific heat."

(b) the height to which you could raise an ant having a mass of 40 mg by expending work of 1 erg.

•••••• ••••••

At ordinary temperatures, the specific heats of most solids and liquids vary only slightly with pressure and volume. For gases it is much more important to distinguish the conditions under which heat is absorbed or rejected. For example, the specific heat of a gas maintained at a constant pressure is greater than the specific heat measured when the gas is held at a constant volume. It was this curious fact concerning the specific heats of gases which led Robert Mayer to first hypothesize that heat was a form of energy and that it could be converted into other forms of energy. For this singular stroke of genius, Mayer shares (with James Prescott Joule) the credit for formulating the first law of thermodynamics. As a prelude to stating the first law in equation form the important concept of work is reviewed in the next section. We conclude this section with a brief survey of calorimetry, the science of heat measurement.

The basic question in calorimetry is this: If two objects, initially at different temperatures, are placed in thermal contact and permitted to reach equilibrium, what is their final temperature? As a specific example, consider a situation in which a 500-gm block of aluminum at a temperature of 400 K is placed in a tank containing 1000 gm of water at a temperature of 300 K. The specific heat capacities of aluminum and water are 0.215 cal/gm-°C and 1.000 cal/gm-°C, respectively. Let T_f denote the final temperature. From (3.1) we find the heat rejected by the aluminum to be

$$Q_A = m_A C_A \Delta T_A = 107.5(T_f - 400\text{ K}) \quad \text{cal}$$

The heat absorbed by the water is

$$Q_W = m_W C_W \Delta T_W = 1000(T_f - 300\text{ K}) \quad \text{cal}$$

The magnitudes of Q_A and Q_W are equal, that is, the heat rejected by the aluminum is absorbed by the water. However, Q_A is negative (heat rejected) and Q_W is positive (heat absorbed). We can therefore write $Q_A + Q_W = 0$ and solve for T_f. The result is $T_f \simeq 310\text{ K}$.

The method is readily generalized. Thus, if we have N substances, with masses $m_1, m_2, \ldots, m_N$, initially at temperatures $T_1, T_2, \ldots, T_N$, we can write

$$\sum_{=1}^{N} Q_j = 0, \qquad Q_j = m_j C_j (T_f - T_j)$$

thereby obtaining

$$T_f = \frac{\sum_{j=1}^{N} m_j C_j T_j}{\sum_{j=1}^{N} m_j C_j}$$

Problem 3.2.4 indicates how the method of mixtures can be used to measure heat capacities.

•••••• ••••••

3.2.4 The method of mixtures is a basic calorimetric technique for determining specific heat capacities. It relies on the ideas presented above. However, the unknown quantity is one of the specific heat capacities. The final temperature is an experimental datum.

(a) Show that the specific heat capacity of the component labeled 1 is given by

$$C_1 = \frac{\sum_{j=2}^{N} m_j C_j (T_f - T_j)}{m_1 (T_1 - T_f)}$$

(b) Place 600 gm of lead (Pb) shot in an aluminum calorimeter, which holds 100 gm of water. The water and calorimeter are initially at a temperature of 80 °C. The initial temperature of the lead is 25 °C. If the final temperature of the mixture is 72 °C, determine the specific heat of lead. The mass of the calorimeter is 20 gm and its specific heat capacity is 0.215 cal/gm-°C.

3.2.5 The air temperature above coastal areas is profoundly influenced by the large specific heat of water (1 cal/gm-°C). One reason is that the heat released when 1 cm^3 of water cools by 1 °C will heat an enormously greater volume of air by this same amount. Estimate the volume of air which can be heated by 1 °C by the 1 cal released by the cooling of 1 cm^3 of water by 1 °C. The specific heat of air is approximately $\frac{1}{4}$ cal/gm-°C. The density of air under standard conditions is 0.00129 gm/cm^3.

•••••• ••••••

3.3 WORK IN THERMODYNAMICS

The concepts of work and energy are related closely. It will be recalled from mechanics that a force **F**, acting through a displacement **ds**, performs an amount of work dW given by the scalar product of **F** and **ds**,

$$dW = \mathbf{F} \cdot \mathbf{ds} \tag{3.2}$$

To determine the total work performed over a finite displacement, the differential elements of work must be summed, that is, integrated. We can write, for the total work,

$$W = \int_A^B dW = \int_A^B \mathbf{F} \cdot \mathbf{ds} \tag{3.2a}$$

The symbolism $\int_A^B dW$ means "add up all elements of work performed between

A and B." In order to evaluate the integral it is generally necessary to specify the path from A to B. We say that W is "path-dependent."

••••• •••••

3.3.1 An object moves in the x–y plane under the influence of a force which has only an x component

$$F_x = xy$$

The work done by this force acting along a path leading from the origin to the point $x = 1$, $y = 1$ depends on the nature of the path. Demonstrate this fact by showing that the work done along the straight-line path $y = x$ is not the same as the work done along the parabolic path $y = x^2$.

••••• •••••

The mechanical work of a force is the most familiar form of work. The work performed on a thermodynamic system may take other forms. In thermodynamics, *work* is broadened to include *all forms of energy transfer excepting heat.*† Suppose, for example, that the system is an electric cell—a battery. While the cell is being charged, electrical work is being done by an external electromotive force (emf). This work is stored within the cell as internal energy. In our study of thermodynamics we shall be particularly concerned with two forms of work. One is the work performed by a fluid as it expands or contracts. The other is the work done in altering the magnetic structure of certain types of solids.

Work of expansion and magnetic work are studied in depth in the next chapter. For now it suffices to recognize that thermodynamic systems can perform work of various forms. There are situations where it is more natural to envision the surroundings as performing work on the system. This can be described by saying that the system performs a "negative amount of work." We shall use W (or dW) to designate work performed *by* the system against its surroundings. This amounts to nothing more than choosing a name for the work done by the system and W is the name! In some instances W may be negative. For example, an expanding gas performs a positive amount of work against its surroundings. If a gas is compressed, it does negative work, that is, the surroundings do a positive amount of work. It should therefore come as no surprise when a calculation produces a negative value of W.

••••• •••••

3.3.2 If W is the name for "work done by the system against its surroundings," what is $-W$ the name for?

••••• •••••

† This statement refers to *closed* systems. For open systems the transfer of matter also results in the transfer of energy.

3.4 INTERNAL ENERGY

Consider a system which is thermally isolated from its surroundings. The jargon of thermodynamics describes this situation by saying that the system is "surrounded by adiabatic walls." If the system is insulated, it can perform work only at the expense of its energy, but what, precisely, is meant by "the energy of the system"? If the surroundings perform a positive amount of work on the system, we instinctively feel that the energy of the system has been increased. This intuitive feeling can be parlayed into an operational definition of internal energy. Let W denote the work done by the system during a process which carries it from some initial equilibrium state to some final equilibrium state. Let U designate the internal energy of the system—the quantity we wish to define operationally. An appeal to the conservation of energy lets us write

Internal energy of the system in its initial state	=	Internal energy of the system in its final state	+	Work done by the system on its surroundings

In symbolic form this becomes

$$U_i = U_f + W$$

where the subscripts refer to the initial and final states. We already have an operational definition of work. This relation therefore serves to operationally define the *change* in the internal energy of the system. By rearranging terms it may be written

$$U_f - U_i = -W \tag{3.3}$$

where $U_f - U_i$ is the change in the internal energy of the system and $-W$ is the work done *on* the system by its surroundings. Equation (3.3) is often referred to as the adiabatic form of the first law of thermodynamics.

Two remarks on (3.3) are worthy of mention. First, only the difference $U_f - U_i$ is operationally defined. It is permissible to select some particular state and arbitrarily assign it the value of zero internal energy. The *standard state* of an element or compound is defined as its most stable physical form at a pressure of one atmosphere and a temperature of 25 °C ($\simeq$298 K). For reasons to be given in Chapter 8 the value zero is assigned not to the internal energy of the standard state, but rather to the quantity (P is pressure, V is volume)

$$H \equiv U + PV$$

known as the *enthalpy*. By definition then, the enthalpies of 1 liter of gaseous helium, 1 kg of liquid mercury, and 1 pinch of table salt are all zero at a pressure of 1 atm and a temperature of 25 °C.

The second comment bears on the nature of internal energy. A definite value of U is associated with each equilibrium state of a system. For this reason U is referred to as a *state function.* The value of U for a given state is independent of the previous history of the system. Stated otherwise, the thermodynamic "path" leading to a given state does not affect the value of U for that state. In this respect, internal energy is analogous to wealth. A man who is a millionaire today may have been born in poverty or with a silver spoon in his mouth. He may have amassed and lost several fortunes. He may have inherited the money only the day before. Which, if any, of these financial paths he followed is irrelevant in so far as his present wealth is concerned.

Our definition of internal energy, though operational, may not leave us with much of a feel for the quantity. Our previous experience suggests that U be regarded as the sum of the kinetic and potential energies of the parts—the atoms and molecules which constitute the microscopic structure of the system. While this is a healthy viewpoint it is not thrust upon us by thermodynamics. Thermodynamics operates at the macroscopic level. If we want to see inside a system and understand what makes it tick, we must erect a microscopic model and a theory to go with it. We must believe in atoms and puzzle over the forces which bind them together to form molecules and solid objects. The disappointment of learning that thermodynamics cannot answer certain questions inspired men like Rudolph Clausius, James Clerk Maxwell, Ludwig Boltzmann, and Josiah Willard Gibbs to develop the kinetic theory of gases and statistical mechanics. These two disciplines, which permit events to be followed at the microscopic level in a statistical or average sort of fashion, are developed in the second half of this text.

•••••• ••••••

3.4.1 Gases are notoriously poor heat conductors. For this reason the rapid expansion or compression of a gas may be regarded as an adiabatic process, that is, it is over before the gas has a chance to absorb or reject a significant amount of heat.

A man exerts a constant force of 1000 N in compressing the gas in a pump cylinder. The force acts over a distance of 50 cm.

(a) How much work is done on the gas?

(b) What is the change in the internal energy of the gas? What assumptions have you made in arriving at your answer?

•••••• ••••••

3.5 THE FIRST LAW OF THERMODYNAMICS

Credit for formulating the first law of thermodynamics is shared by Robert Mayer and James Prescott Joule. Working independently and in response to quite different motivations, both arrived at the same conclusions, to wit:

(a) Heat is one of many different forms of energy.

(b) Energy is conserved—it may be transformed, but not created or destroyed.

In 1842 Mayer presented the first determination of the mechanical equivalent of heat, the amount of mechanical energy equivalent to 1 cal. Mayer's value was within 20% of the presently accepted value of

$$1 \text{ cal} = 4.186 \text{ joule} = 4.186 \times 10^7 \text{ erg} = 0.0040 \text{ Btu}$$

James Prescott Joule, an English brewer, based his belief in the conservation of energy on religious grounds,† "... the grand agents of nature are, by the Creator's fiat, indestructible; ... (1843, p. 435)." In a series of experiments spanning 40 years Joule measured the mechanical equivalent of heat. He demonstrated that the dissipation of a given amount of mechanical or electrical energy always resulted in the evolution of a proportional amount of heat.

The general form of the first law of thermodynamics recognizes that a change in the internal energy of a system can be accomplished by the performance of work, by the exchange of heat, or by a combination of both. Let Q denote the quantity of heat added to a system during a process which carries it from some initial equilibrium state to some final equilibrium state. In general such a process changes the internal energy of the system and results in the performance of work. The adiabatic form of the first law, Equation (3.3), may be generalized to take account of heat exchange by stating that

$$\begin{Bmatrix}\text{heat added}\\ \text{to a system}\end{Bmatrix} \text{ appears as } \begin{Bmatrix}\text{a change in the}\\ \text{internal energy}\end{Bmatrix} \text{ and/or } \begin{Bmatrix}\text{is used to}\\ \text{do work}\end{Bmatrix}.$$

The equation expressing this statement, the general form of the first law of thermodynamics, is

$$Q = U_{\text{f}} - U_{\text{i}} + W \tag{3.4}$$

We remark that Q is the *name* for the net amount of heat added to the system. If Q should turn out to be negative, it simply means that the system rejected a net amount of heat. In an adiabatic process $Q = 0$ and (3.4) reverts to the adiabatic form (3.3).

Example 3.2

A system absorbs 1.2×10^6 cal of heat and performs 1.7×10^{14} ergs of work against its surroundings. Determine the change in the internal energy of the system.

---

† Joule's accounts of his first determination of the mechanical equivalent of heat were published in the *Philosophical Magazine, Series 3*, **23**, 263, 347, 435 (1843).

We rewrite (3.4) as

$$U_f - U_i = Q - W$$

Since the heat is absorbed by the system,

$$Q = +1.2 \times 10^6 \text{ cal}$$

Positive work is done by the system against the surroundings

$$W = +1.7 \times 10^{14} \text{ ergs}$$

If we elect to express $U_f - U_i$ in ergs, Q must be converted to ergs. The student should verify that this conversion gives

$$Q = 0.50 \times 10^{14} \text{ ergs}$$

Inserting these values of Q and W into the first law gives

$$U_f - U_i = -1.2 \times 10^{14} \text{ ergs}$$

The minus sign means that the final internal energy is less than the initial by 1.2×10^{14} ergs. Internal energy is sacrificed because the work performed exceeded the heat absorbed. ◇

•••••• ••••••

3.5.1 The surroundings perform 1.7×10^{14} ergs of work on the system of Example 3.2. If the system absorbs 1.2×10^6 cal of heat, compute the change in its internal energy.

3.5.2 The Joule constant, $J = 4.186$ joule/cal, is often referred to as the "mechanical equivalent of heat." At the end of his 1842 essay Mayer expresses his results for the mechanical equivalent of heat by stating that the warming of a given weight of water from 0 °C to 1 °C is equivalent to the mechanical energy acquired by the same weight of water in falling from a height of 365 m. Show that the value of J for these data is approximately

$$J = 3.58 \times 10^7 \text{ ergs/cal}$$

Assume that the specific heat of water is 1 cal/gm-°C and take the acceleration of gravity to be 981 cm/sec².

•••••• ••••••

In Section 3.3 it was observed that work is generally a path-dependent quantity. In Section 3.4 it was emphasized that internal energy is a property of the system. The value of U for a particular thermodynamic state is independent of the prior history of the system. Therefore, the change $U_f - U_i$ depends solely on the initial and final states. It is independent of the thermodynamic path joining the two states. It now follows from the first law that Q is also a path-dependent quantity, that is, Q is the sum of a path-independent quantity $U_f - U_i$ and a path-dependent quantity W. The path dependence of

Q follows still more directly from the first law if we consider a *cyclic process.* In a cyclic process the initial and final states are the same so that $U_f = U_i$, and the first law reduces to

$$Q = W \tag{3.4a}$$

that is, it states that the total work done by the system over the cycle equals the net heat absorbed by the system. Since W depends on the nature of the cyclic path the heat Q is likewise path-dependent.

•••••• ••••••

3.5.3 For many years the Patent Office was deluged with applications for patents on various types of perpetual motion machines. This flood subsided abruptly and completely when the rules were amended to require that a working model accompany each application. Suppose that you were employed in the Patent Office prior to this ruling and that an application for a patent on a heat engine was presented to you, the performance figures claimed for the engine being as follows:

Net heat consumption per cycle: 10 kcal
Mechanical work output per cycle: 10^{12} ergs
Electrical work output per cycle: 10^5 watt-sec

Would it be necessary to give detailed consideration to the application? Explain.

•••••• ••••••

In situations where a thermodynamic process proceeds smoothly we can safely envision it as a continuous sequence of small changes. Mathematically we can represent the quantities appearing in the first law by infinitesimals. The miniaturized form of the first law is

$$dQ = dU + dW \tag{3.5}$$

Of course, the physical content of the law is unaltered by rewriting it in this form. Furthermore, this form presents a pitfall of which one must be aware. This concerns the path dependence of heat and work. The fact that we represent heat and work as dQ and dW does not imply the existence of state functions, Q and W, which measure the heat and work content of a system. Thus dQ and dW are mathematically infinitesimal, but they are not true differentials†; they denote small amounts of heat and work.

In order to fully exploit the rich content of the first law we must determine under what conditions it is permissible to express dQ, dU, and dW in terms of the thermodynamic coordinates (P, V, T, and so on), their changes (dP, dV,

† Some authors use a special notation such as $đW$ or δQ as a reminder of this pitfall.

dT, and so on), and other measurable properties, such as the specific heat capacities. Such a description is possible for processes which qualify as reversible. The concept of reversibility is discussed in the next chapter. Also in the next chapter, we develop expressions for dU and dW in terms of the thermodynamic coordinates and their changes. The final step in the program, expressing dQ in a similar fashion, carries us to the second law of thermodynamics.

REFERENCES

A beneficial use of hydrogen bombs, for the propulsion of interstellar vehicles, is discussed in:

F. J. Dyson, Interstellar Transport, *Phys. Today* **21**, 41 (October 1968).

The arguments against building such a spacecraft are related in:

F. J. Dyson, Death of a Project, *Sci.* **149**, 141 (1965).

The multifaceted nature of energy is surveyed in spectacular fashion in:

M. Wilson, "Energy." Time Inc., New York (1963).

The importance of energy to all technologies is brought home in:

F. J. Dyson, The Search for Extraterrestrial Technology, *in* R. Marshak (ed.), "Perspectives in Modern Physics." Wiley (Interscience), New York (1966).

An entertaining account of attempts to violate the first and second laws of thermodynamics is given in:

S. W. Angrist, Perpetual Motion Machines, *Sci. Am.* **218**, 115 (Jan. 1968).

See also the "dynamaforce generator" described by:

M. Gardner, Mathematical Games. *Sci. Am.* **226**, 100 (Feb. 1972).

A convenient source of biographical data on Joule, Mayer, and other pioneers is:

W. F. Magie, "A Source Book in Physics." McGraw-Hill, New York (1935).

The caloric theory of heat died a lingering death. See:

S. C. Brown, The Caloric Theory of Heat, *Am. J. Phys.* **18**, 367 (1950).

Professor Brown is the foremost authority on the life and works of Count Rumford, a remarkable individual who sparked the rise of the kinetic view of heat. See:

S. J. Brown, "Count Rumford, Physicist Extraordinary." Doubleday, Garden City, New York (1962).

Discussions of vector algebra and path integrals are given by:

M. H. Hull, Jr., "The Calculus of Physics," Chapters 3 and 4. Benjamin, New York (1969).

Chapter 4

Thermodynamic Processes and Thermodynamic Work

In this chapter the important concepts of quasi-static and reversible processes are developed. The notion of thermodynamic work is introduced and illustrated in terms of the work performed by an expanding fluid and the work performed when a material is subjected to a magnetic field. The chapter concludes with an exploration of the relationship between specific heat capacities and the internal energy.

4.1 THERMODYNAMIC PROCESSES

A thermodynamic process results in changes in the thermodynamic coordinates and/or changes in the surroundings.† As examples of thermodynamic processes we may cite:

(a) a chemical reaction in which a precipitate forms;
(b) the explosive combination of hydrogen and oxygen to form water;
(c) the diffusion of perfume vapor—an effect observed quite often by girl watchers;

† A cyclic process leaves the state of the system unchanged; all thermodynamic coordinates return to their original values.

(d) the compression of a fluid by a piston;
(e) the stretching of a piece of wire;
(f) the magnetization of a chromium potassium alum pill.

The first three processes are irreversible. Irreversible processes occur spontaneously within a system which is not in thermodynamic equilibrium. Heat flow is one type of irreversible process. Heat flows spontaneously, that is, without any outside help from regions of higher temperature to regions of lower temperature. Experience reveals that such processes tend to produce a state of mutual equilibrium between the system and its surroundings. Once attained, an equilibrium state persists as long as the surroundings remain unchanged. The designation irreversible is most apt, for such processes act in an undirectional fashion—to drive a system toward an equilibrium state. The study of irreversible processes properly belongs to the field of non-equilibrium thermodynamics. Equilibrium thermodynamics is concerned with quasi-static reversible processes.

4.2 QUASI-STATIC PROCESSES

A quasi-static process is one in which the surroundings and thermodynamic coordinates of the system change slowly—so slowly, that the process may be viewed as one in which the system passes through a succession of equilibrium states. Geometrically, a quasi-static process may be represented by a path which joins the initial and final equilibrium states by a succession of intermediate equilibrium states. If the equation of state can be represented by a real surface (for example, the P–V–T surface of an ideal gas), the path of the quasi-static process can actually be traced on this surface.

Among the many quasi-static processes which a thermodynamic system can undergo, members of the "iso" family are favorites, for example,

$$\text{Isothermal:} \quad T = \text{constant}$$

$$\text{Isobaric:} \quad P = \text{constant}$$

$$\text{Isovolumic:} \quad V = \text{constant}$$

Regardless of the nature of a quasi-static process it is important to distinguish between the equation which specifies the path of the process and the equation of state. Thus, for example, an ideal gas obeys the equation of state $PV = nRT$ at every stage of a quasi-static process whether the process is isobaric or isothermal, that is, regardless of the path. The equation of state defines the nature of the thermodynamic *surface*. The particular quasi-static process

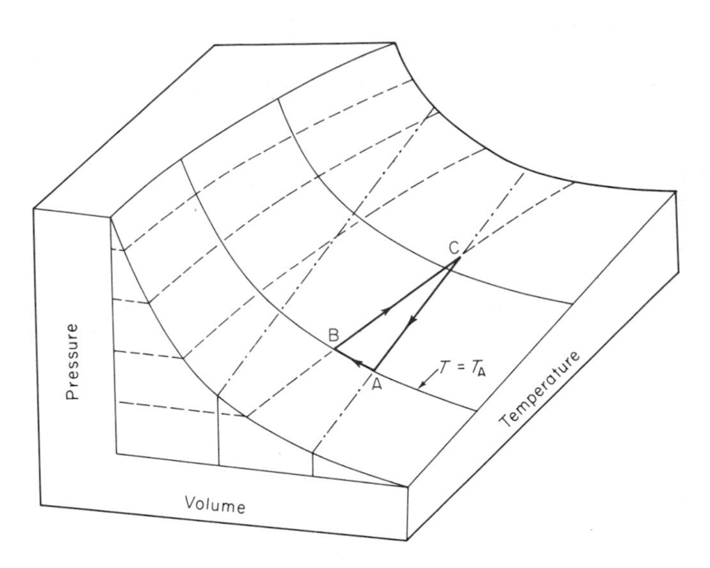

Figure 4.1. The *P–V–T* surface for an ideal gas. The path from A to B describes an isothermal (constant temperature) compression of the gas. The path from B to C describes an isobaric (constant pressure) expansion of the gas. The path from C to A describes an isovolumic (constant volume) decompression.

defines a *path on that surface*. Figure 4.1 shows the thermodynamic surface for the ideal gas. The three paths AB, BC, and CA describe isothermal, isobaric, and isovolumic processes, respectively.

It is frequently desirable to represent a *P–V–T* surface in two dimensions. This is usually done by projecting the *P–V–T* surface onto the *P–V* plane. Figure 4.2 shows the *P–V* diagram for the ideal gas surface of Figure 4.1.

••••••

4.2.1 Figure 4.2 shows the projection of the *P–V–T* surface of Figure 4.1. The projection of Figure 4.2 is onto the *P–V* plane. Make sketches of the projections onto the *P–T* and *V–T* planes. Include lines showing the path of the cyclic process A → B → C → A.

••••••

The importance of quasi-static processes is understandable. We wish to describe processes in terms of the thermodynamic coordinates of the system, and these are well defined only for equilibrium states. A semiquantitative criterion for quasi-static behavior may be stated as follows: Let τ_p denote a time characteristic of the process; for example, the time required for the

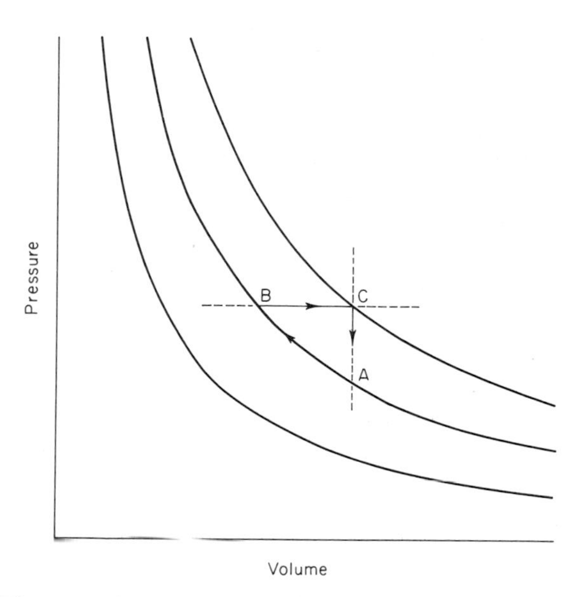

Figure 4.2. The *P–V* plane projection of the ideal gas surface. The cyclic process, A → B → C → A, is the same as that indicated in Figure 4.1.

thermodynamic coordinates to undergo a measurable change. The rate at which the system approaches equilibrium is also characterized by a time, which we denote by τ_E. For a process to qualify as quasi-static we demand

$$\tau_E \ll \tau_p \tag{4.1}$$

that is, the time needed to reach equilibrium must be small compared to the time required for the system to change its thermodynamic state. The condition $\tau_E \ll \tau_p$ means that the irreversible processes work fast, always keeping the system in near-equilibrium states. For example, consider a process in which a piece of metal is slowly heated. Heat conduction is the irreversible process which strives to achieve temperature equality throughout the metal. If the heat is added very slowly, by maintaining the heating element at a temperature just slightly above that of the metal, the temperature will rise uniformly throughout the block. In other words, thermometers positioned at different locations in the metal would give identical readings if heat conduction works fast in comparison to the rate of heating. For heat conduction, τ_E is the time for a temperature change to spread throughout the block. The time required to add enough heat to produce a detectable change in temperature is τ_p. Rapid heat conduction and the slow addition of heat cooperate to make $\tau_E \ll \tau_p$ and make the process quasi-static.

4.3 REVERSIBLE PROCESSES AND THERMODYNAMIC WORK

The example of a gas being compressed quasi-statically by a piston serves admirably to introduce the notions of a *reversible process* and *thermodynamic work*. Figure 4.3 suggests how a quasi-static compression might be accomplished. Consider first the behavior of the system when there is no friction between the piston and the cylinder walls. Adding or removing a grain of sand is to be regarded as producing an infinitesimal change in the surroundings. At any stage we can halt the compression and initiate expansion by removing a grain of sand, provided there is no friction. If the sand is removed grain by grain, so as to reverse the order in which it was added, the gas will retrace the thermodynamic path it followed during the compression. A significant point is that the reversal can be initiated by an infinitesimal change in the surroundings.

The quasi-static frictionless compression just described is an example of a *reversible process*. A general definition is this:

> A reversible process is a quasi-static process which can be reversed by an infinitesimal change in the surroundings.

Please note that a reversible process is not necessarily one which goes one way then back again. Reversible implies it *can* be reversed.

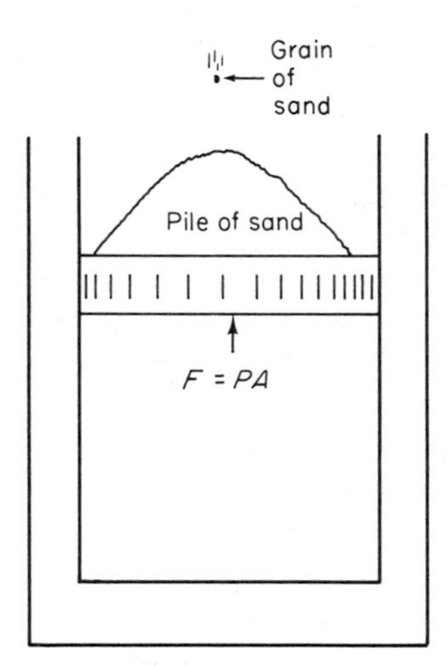

Figure 4.3. An example of the quasi-static compression of a gas. This can be achieved by slowly piling sand on the piston. If the piston moves without friction the removal of one grain of sand (an infinitesimal change) causes the piston to reverse its motion. If there is friction present, the removal of one grain need not produce a reversal of the piston.

The presence of friction is fatal to reversibility. If there is friction, we can still compress the gas quasi-statically by adding sand slowly but—and here is where the reversibility is lost—the compression cannot be reversed by an infinitesimal change in the surroundings. Removing one grain of sand—an infinitesimal change—will not reverse the process. It requires a finite change in the surroundings to revert from compression to expansion.

It is, of course, not possible to eliminate friction. Nevertheless, it often is possible to render friction negligible by comparison with the force exerted by the gas. Very frequently one deals with a solid or a liquid whose surface expands against the surrounding atmosphere. Under such circumstances the motion of the piston, that is, the surface, is as nearly frictionless as one can imagine.

An important consequence of reversibility is this: The work done by the system can be expressed in terms of its thermodynamic variables and their changes. We illustrate this point using the gas–piston system. If P is the gas pressure and A the cross-sectional area of the piston, the force exerted on the piston by the gas is

$$F = PA \tag{4.2}$$

Let the piston move outward (expansion) a distance ds. The work done by the gas is

$$dW = F\,ds = PA\,ds \tag{4.3}$$

However, $A\,ds = dV$ is the change in the volume of the gas. Thus, the work done by the gas during a quasi-static change of volume is

$$dW = P\,dV \tag{4.4}$$

The expression (4.4) is so simple that its most remarkable characteristic may escape our notice. We have succeeded in expressing the work solely in terms of the thermodynamic coordinates of the system. The nature of the external force and other characteristics of the surroundings do not appear in the expression for dW. This remarkable result is possible only to the extent that the compression is quasi-static and friction-free. Friction would introduce an additional term in (4.4) representing the work done by frictional forces. The work $dW = P\,dV$ is often referred to as *thermodynamic work*, as it is expressed entirely in terms of the thermodynamic coordinates of the system. By contrast, the work done by a frictional force would not be classified as thermodynamic work because such a force is not a thermodynamic coordinate of a system. As only thermodynamic work will be of much concern we generally do not retain the over worked adjective "thermodynamic" in referring to such work.

In Chapter 3, we called attention to the path dependence of work and heat. Having an explicit expression for the work of expansion

$$dW = P\,dV$$

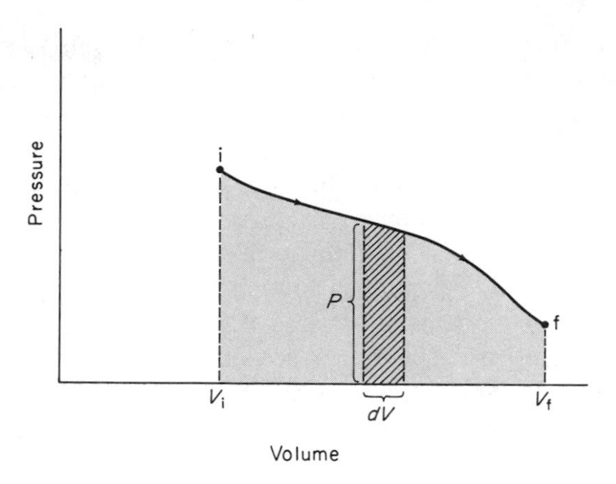

Figure 4.4. Geometrically, the work $dW = P\,dV$ corresponds to the hatched area. The total work done by the gas in the expansion from V_i to V_f corresponds to the total area beneath the P–V path.

affords us the opportunity to illustrate the path dependence of thermodynamic work. The work done by the system during a finite change of volume is found by integrating $dW = P\,dV$. We write

$$W = \int_i^f P\,dV \tag{4.5}$$

with i and f denoting the initial and final states, respectively. A geometric representation of the integral makes it clear that W depends on the thermodynamic path between states. Geometrically W corresponds to the shaded area of Figure 4.4. This area clearly depends on the nature of the thermodynamic path connecting the initial and final states.

Example 4.1 The Path Dependence of Work

To illustrate the path dependence of W we calculate the thermodynamic work done during the expansion of an ideal gas. Two types of expansion (that is, two different paths) are considered. Figure 4.5 shows the two paths. The first path consists of an isobaric expansion at a pressure P_1 followed by an isovolumic decompression at volume V_2. The other path considered is the isothermal expansion along the isotherm $T = T_1$. For the isobaric expansion

$$W_{\text{isob}} = \int_{V_1}^{V_2} P_1\,dV = P_1(V_2 - V_1) \tag{4.6}$$

where isob is isobaric. There is no work done during the isovolumic decompression, so that the total work for the path 1→A→2 is

$$W_{1A2} = P_1(V_2 - V_1)$$

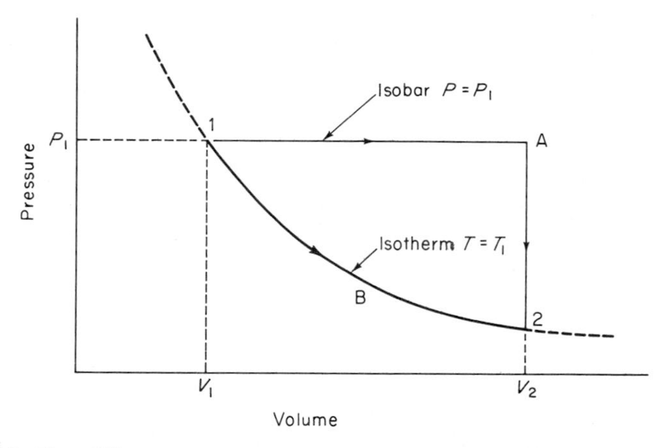

Figure 4.5. Two different paths connecting the initial and final states. The work for the path $1 \to A \to 2$ exceeds that for the isothermal path $1 \to B \to 2$.

A check of Figure 4.5 verifies that this work is represented geometrically by the area of the rectangle of height P_1, and length $V_2 - V_1$. Consider next the work performed by the gas during an isothermal expansion. With $T = T_1$ throughout the expansion the pressure is related to the volume by

$$P = \frac{RT_1}{V}$$

The isothermal work is†

$$W_{1B2} = RT_1 \int_{V_1}^{V_2} \frac{dV}{V} = RT_1 \ln\left(\frac{V_2}{V_1}\right) \tag{4.7}$$

Geometrically the work W_{1B2} corresponds to the area beneath the isothermal path $1 \to B \to 2$ in Figure 4.5.

It is obvious from the inspection of Figure 4.5 that $W_{1A2} > W_{1B2}$ even though both paths connect the same initial and same final states. The general conclusion to be drawn from this specific example is that the *thermodynamic work depends on the path of the process* connecting the initial and final states. ◇

One other aspect of the geometrical interpretation of thermodynamic work should be noted. A cyclic process may result in the performance of net thermodynamic work by the system. Consider, for example, a process which takes 1 mole of an ideal gas through the cycle (compare Figure 4.5) $1 \to A \to 2 \to B \to 1$.

---

† Recall that $\int_a^b dx/x = \ln(b/a)$, where $\ln x$ refers to the natural logarithmic base.

The total work done by the system over the cycle, abbreviated cyc, is

$$W_{\text{cyc}} = W_{1A2} + W_{2B1} = P_1(V_2 - V_1) - RT_1 \ln\left(\frac{V_2}{V_1}\right)$$

Geometrically this work is represented by the net area enclosed by the path of the cycle on the P–V diagram. This result is generally valid: the quasi-static work done by the pressure forces over a full cycle can be written

$$W = \oint P\, dV \tag{4.8}$$

The symbolism $\oint$ signifies that the integration is over a full cycle. Geometrically, $\oint P\, dV$ represents the area enclosed by the path describing the cycle on a P–V diagram. Figure 4.6a–d illustrates this point. During expansion the volume increases so that dV and thus $dW = P\, dV$ are positive. The total work

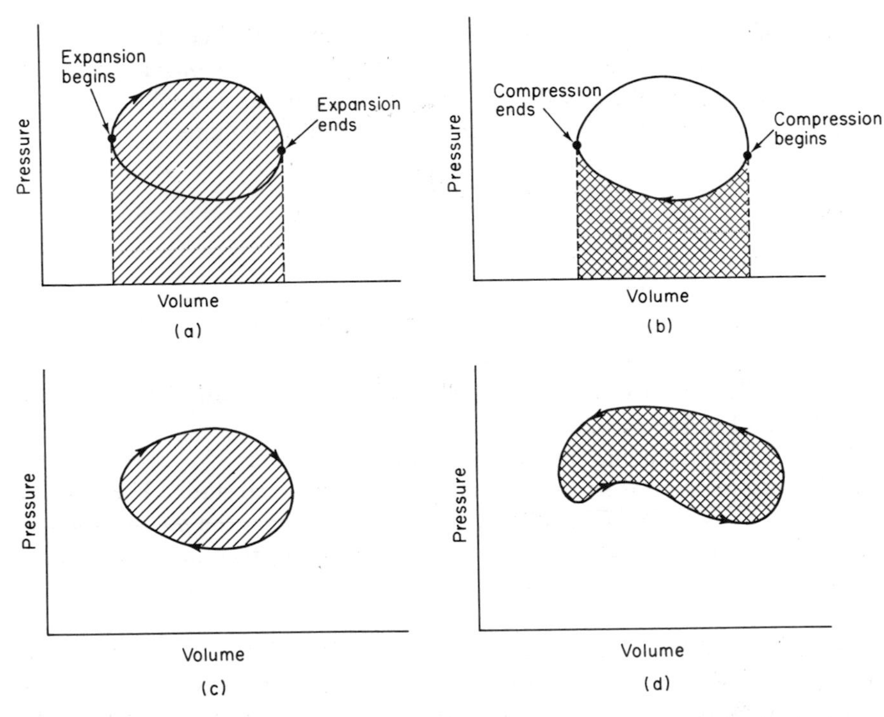

Figure 4.6. (a) The work of expansion is positive; it corresponds to the hatched area. (b) The work of compression is negative; its magnitude corresponds to the cross-hatched area. (c) The net work is positive. It corresponds to the area enclosed by the clockwise path. (d) A cycle for which the net work done is negative; its magnitude corresponds to the enclosed area. Note that the cycle generates a counterclockwise path.

of expansion is thus positive and its magnitude corresponds to the hatched area of Figure 4.6a. The compression from V_2 to V_1 reduces the volume so that dV and dW are negative. The total work of compression is negative but its magnitude is given by the cross-hatched area of Figure 4.6b. The net work done by the pressure force over a full cycle is evidently the sum of the works of expansion and compression. Geometrically the net work corresponds to the difference between the corresponding areas. Figure 4.6c shows that this is just the area enclosed by the path of the cycle. In this case the net work done by the system is positive, the pressure during expansion being greater than that during compression. Note that the cyclic process generates a clockwise path on the P–V diagram. A clockwise path on a P–V diagram always corresponds to positive work.

Figure 4.6d shows a cyclic process for which the net work is negative. The work of compression is negative and its magnitude exceeds the positive work of expansion. Note that the cycle traces out a counterclockwise path, which on a P–V diagram always corresponds to negative work. The magnitude of the net work still corresponds to the enclosed area.

In the next section we study magnetic work and show how it may be expressed in a form analogous to (4.4).

•••••• ••••••

4.3.1 The paths for two cyclic processes are indicated in the accompanying figure. Indicate for each part whether the net work done is positive, negative, or zero. Explain the basis for your choice.

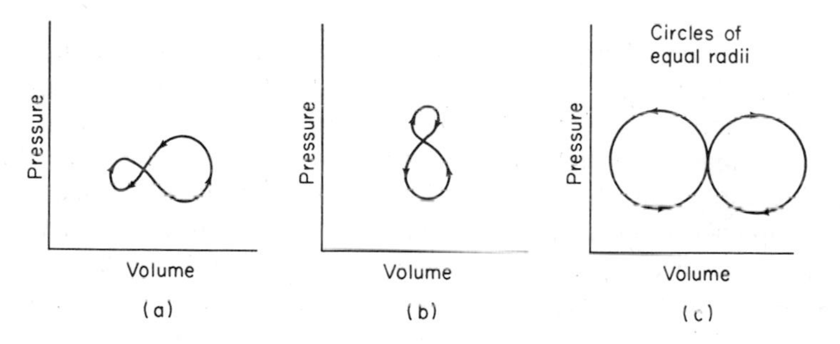

4.3.2 Certain classes of stars show periodic variations of their luminosity. This behavior has been traced to periodic pulsations of the star. A detailed study of the pulsations of such stars has been made by Robert F. Christy.† In order to follow the temporal evolution of the pulsations, Christy treats the star as a set of concentric shells of mass. He solves a set of nonlinear equations

† R. F. Christy, Lectures on Variable Stars, delivered at the *3rd Summer Inst. Astronomy and Astrophysics, Stony Brook, New York* (*1969*). See also *Rev. Mod. Phys.* **36**, 555 (1964).

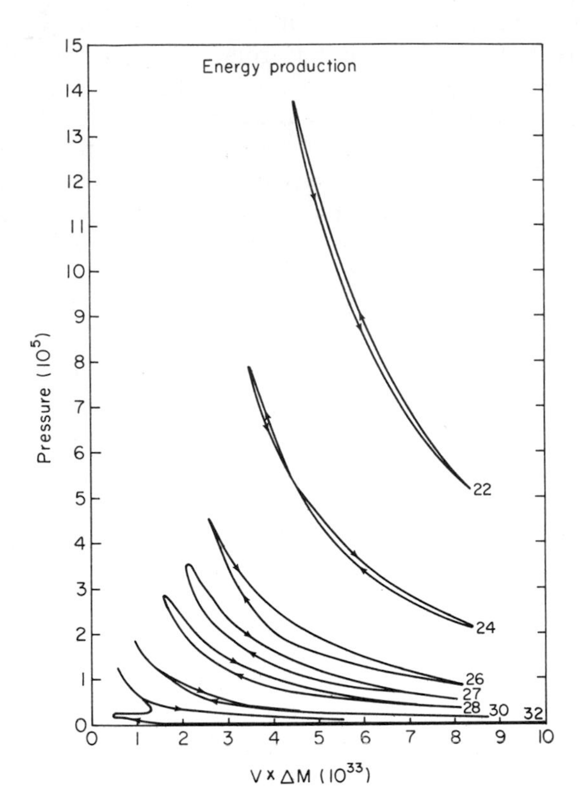

giving the pressure, density, and so on, in each shell. In order to understand what drives the pulsation, as well as what inhibits it, Christy computes the amount of mechanical work done per cycle (that is, per complete pulsation) within each mass shell. If $\oint P\,dV$ is positive for a shell, it means that the shell is feeding energy into the pulsation. If $\oint P\,dV$ is negative, the mass shell is soaking up energy—it is performing negative work as it acts to dissipate the pulsation. The accompanying figure (after R. F. Christy) shows the P–V cycles for several mass shells. Determine the shell(s) which lead to dissipation of the pulsation and those which lead to its excitation. Numbers on the P–V cycles refer to different mass shells of star, the central shell being number 1, the surface shell, number 40.

4.3.3 An inequality important in statistical physics is

$$x \geqslant \ln(1+x)$$

Prove this relation by evaluating the areas beneath the graphs of

$$P_1(V) = \frac{1}{V} \qquad \text{and} \qquad P_2(V) = 1$$

over the interval from $V = 1$ to $V = 1 + x$. Sketch the functions and the areas. (You must sketch them in order to see the inequality!)

•••••• ••••••

4.4 MAGNETIC WORK

When a gas expands it performs $P\,dV$ work against its surroundings. This work of expansion is a familiar example of thermodynamic work. Magnetic work is a far less familiar example, but one of growing importance. Whereas a gas can be cooled by being allowed to expand and perform work at the expense of its internal energy, certain types of solids can be cooled by allowing them to perform magnetic work. The technique of magnetic cooling is especially valuable at temperatures below 1 K. A detailed study of magnetic cooling is presented in Chapter 9. In this section we derive the magnetic analog of $P\,dV$, an expression showing how magnetic work can be expressed in terms of thermodynamic variables.

A long thin solenoid can be used to set up a strong and reasonably uniform magnetic field. The relationship between the field strength H and the solenoid current I follows from the integral form of Ampere's law

$$\oint \mathbf{H} \cdot \mathbf{dl} = \frac{4\pi}{c} I_{\text{enc}} \tag{4.9}$$

where I_{enc} is the total current enclosed by the loop indicated in Figure 4.7. The

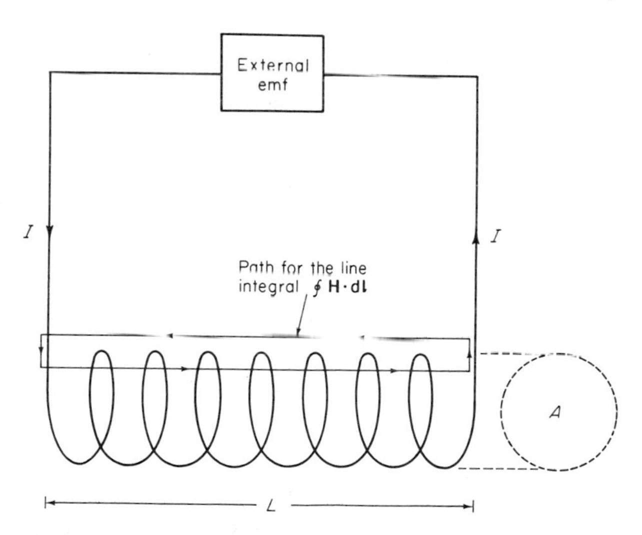

Figure 4.7. A long thin solenoid, having many turns per unit length. The leakage of the magnetic field from within the solenoid is negligible. The cross-sectional area of the solenoid is A.

total current enclosed is NI, where N is the total number of turns of the solenoid. The loop integral has the value,

$$\oint \mathbf{H} \cdot \mathbf{dl} = HL$$

where L is the length of the solenoid. This assumes that the magnetic field is negligible outside the solenoid and uniform within. The magnetic field is thus

$$H = \frac{4\pi NI}{cL} \tag{4.10}$$

If the volume V of the solenoid is filled with a magnetic material, the magnetic induction B therein is

$$B = H + \frac{4\pi M}{V} \tag{4.11}$$

where M is the total magnetic moment of the material.†

If the external emf is changed so that the current increases from I to $I+dI$, three things happen:

(a) the magnetic field changes, as per (4.10),

(b) the change in H causes M to change. If we denote these changes by dH and dM, it follows from (4.11) that the change in B is

$$dB = dH + \frac{4\pi\, dM}{V} \tag{4.11a}$$

(c) the magnetic flux threading the solenoid changes by an amount

$$d\Phi = NA\, dB \tag{4.12}$$

The changing flux induces a "back" emf (Lenz on guard!). It is this back emf which performs work. The emf is given by Faraday's law

$$\mathscr{E}_{\text{back}} = -\frac{1}{c}\frac{d\Phi}{dt}$$

The rate at which the system does work is‡

$$\frac{dW}{dt} = \mathscr{E}_{\text{back}} I = -\frac{I}{c}\frac{d\Phi}{dt}$$

† Here M arises through the collective response (to H) of the magnetic moments of the electrons, ions, and so on. Thermodynamics does not require that we introduce any such microscopic picture—it merely accepts M as a relevant and measurable quantity. The magnetic moment per unit volume M/V is often referred to as the magnetization.

‡ Recall that electric power is the product of voltage and current.

Using (4.12) gives

$$dW = -\frac{NIA}{c}dB \tag{4.13}$$

as the work done by the system as the induction changes from B to $B+dB$. From (4.10) we have

$$\frac{NIA}{c} = \frac{HLA}{4\pi} = \frac{HV}{4\pi}$$

where LA is the solenoid volume V, whence,

$$dW = -\frac{VH\,dB}{4\pi} \tag{4.13a}$$

This is one form for the work done *by* the system as a consequence of a change in the magnetic induction. An alternate form, which follows by using (4.11a) to replace dB by $dH+4\pi\,dM/V$, is

$$dW = -V\frac{H\,dH}{4\pi} - H\,dM \tag{4.14}$$

The first term on the right side represents the additional energy stored in the magnetic field by virtue of the increase in the solenoid current. The second term is the work done against the external source in changing the magnetic moment of the material. There is a question as to what should appear for the magnetic work done by the system in the differential form of the first law. This question is readily answered once we decide what constitutes the system. If we take the system to be the magnetic material plus the magnetic field, then (4.13a) gives the magnetic work. Often, however, the material alone is regarded as the thermodynamic system and the term $VH\,dH/4\pi$, representing energy stored in the magnetic field, is dropped. We adopt the latter viewpoint—the material alone constitutes the system. The magnetic work done *by* the system is thus

$$dW = -H\,dM \tag{4.15}$$

The magnetic analog of $P\,dV$ is therefore $-H\,dM$. The correspondences are

$$H \to P \qquad \text{(both intensive variables)}$$

$$-M \to V \qquad \text{(both extensive variables)}$$

The minus sign appearing in $-H\,dM$ arises because a positive amount of work is performed by the system when M decreases. For a magnetic system the first law of thermodynamics reads

$$dQ = dU - H\,dM + P\,dV \tag{4.16}$$

In practice the $P\,dV$ work is often ignorable compared with $-H\,dM$.

••••• •••••

4.4.1 (a) Argue that the total work performed by a magnetic material during a reversible process is given by

$$W = -\int_{\mathrm{i}}^{\mathrm{f}} H\, dM$$

Present a geometrical interpretation of the integral.

(b) Explain why the total heat absorbed (reversibly) by a magnetic material during a cyclic process is given by

$$Q = -\oint H\, dM$$

4.4.2 The relation between M and H is generally written

$$\frac{M}{V} = \chi H$$

where χ is the magnetic susceptibility. In general χ depends on both H and T. However, for weak fields and temperatures well above the liquid helium range many paramagnetic† materials obey Curie's law,

$$\chi = \frac{C}{T}$$

where C is the Curie constant, not the heat capacity.

Consider the following reversible processes, both of which result in the doubling of M:

(i) an isothermal process in which the magnetic field strength is doubled,
(ii) a process carried out with H held constant, during which the temperature is reduced to half its original value.

Assume that the volume remains fixed during both processes.

(a) Indicate the paths of the two processes on an H–M diagram (the magnetic analog of a P–V diagram).

(b) Determine which of the two processes results in the larger performance of work by the surroundings. You may do this analytically or by an appeal to the geometrical representation of magnetic work.

••••• •••••

† Paramagnetic materials have small positive values of χ and are attracted by the poles of a strong magnet. Diamagnetic materials have small negative values of χ and are repelled by a magnetic field.

4.5 HEAT CAPACITIES AND INTERNAL ENERGY

Under most circumstances the temperature of a system changes when it absorbs heat. Exceptions arise during a change of phase, as when a solid melts and becomes a liquid. The first law shows that the heat is used to change the internal energy and/or to perform work. The equations $dW = P\,dV$ and $dW = -H\,dM$ reveal how the thermodynamic work can be expressed in terms of the thermodynamic variables. In this section we investigate the relationship between internal energy changes and the other thermodynamic variables. We shall discover that the internal energy is closely related to the heat capacity.

In Chapter 3 we introduced the specific heat capacity C which relates the heat absorbed and the temperature change of a sample of mass m, namely,

$$Q = mC\,\Delta T \tag{3.1'}$$

C is an intensive quantity—the heat capacity per gram. Two quantities, closely related to C, are frequently encountered. The total heat capacity is simply the mass of the system times the heat capacity per gram,

$$\text{total heat capacity of } m \text{ grams} = mC$$

When the mass m equals the molecular weight the total heat capacity is referred to as the molar heat capacity. In all future developments we shall write C for the total heat capacity instead of mC. In certain instances the molar heat capacity will also be designated simply as C. Where confusion might arise the meaning of the symbol C will be noted explicitly. With this convention in force (3.1′) would appear as $Q = C\,\Delta T$, or in differential form as $dQ = C\,dT$. The total heat absorbed or rejected in a finite change of temperature is

$$Q = \int dQ = \int_{T_i}^{T_f} C\,dT \tag{4.17}$$

In general, the heat capacity depends on T. If the heat capacity does not change significantly over the temperature range T_i to T_f, C may be treated as a constant in the integration which then gives

$$Q = C(T_f - T_i) = C\,\Delta T \tag{4.18}$$

Happily, many substances have nearly constant heat capacities over sizeable temperature ranges, permitting use of the elementary relation $Q = C\,\Delta T$ in lieu of the more involved integral relation (4.17).

••••• •••••

4.5.1 (a) Determine the amount of heat required to raise the temperature of 50 gm of aluminum from 25 °C to 35 °C. Assume the heat capacity is

TABLE 4.1

Specific Heat Capacity at Constant Pressure[a]

Substance	C_p (cal/gm-°C)	Substance	C_p (cal/gm-°C)
Silver (Ag)	0.056	Argon (Ar)	0.125
Aluminum (Al)	0.215	Benzene (C_6H_6)	0.388
Gold (Au)	0.031	Bromine (Br)	0.107
Bismuth (Bi)	1.029	Carbon dioxide (CO_2)	0.229
Copper (Cu)	0.092	Chlorine (Cl_2)	0.115
Iron (Fe)	0.107	Ethyl alcohol (C_2H_5OH)	0.580
Mercury (Hg)	0.033	Helium (He)	1.252
Potassium (K)	0.184	Hydrogen (H_2)	3.393
Magnesium (Mg)	0.248	Methyl alcohol (CH_3OH)	0.603
Sodium (Na)	0.295	Neon (Ne)	0.246
Nickel (Ni)	0.103	Nitrogen (N_2)	0.249
Lead (Pb)	0.031	Oxygen (O_2)	0.219
Zinc (Zn)	0.084	Silicon (Si)	0.173

[a] Values refer to a pressure of 1 atm and a temperature of 25 °C for solids and liquids; 15 °C for gases. Adapted with permission from "International Critical Tables of Numerical Data: Physics, Chemistry, and Technology." Nat. Res. Council, Washington, D.C., 1923–1933, and E. H. Kennard, "Kinetic Theory of Gases." McGraw-Hill, New York (1938).

constant. Consult Table 4.1 for data. Bear in mind that Table 4.1 lists specific heats.

(b) If the amount of heat computed in part (a) were added to 50 gm of lead what would be its temperature change?

4.5.2 At sufficiently low temperatures the heat capacities of many crystalline solids are proportional to T^3. This empirical fact finds theoretical justification in the Debye theory of solids which predicts a molar heat capacity given by

$$C = 464\left(\frac{T}{\theta}\right)^3 \quad \frac{\text{cal}}{\text{mole-K}}$$

where θ is known as the *Debye temperature* and is a characteristic of the material. For temperatures low enough to permit use of the Debye form of C show that the heat absorbed in raising the temperature of one mole from T_i to T_f is

$$Q = 116\,\theta\left[\left(\frac{T_f}{\theta}\right)^4 - \left(\frac{T_i}{\theta}\right)^4\right]$$

•••••• ••••••

In most laboratory situations heat exchange takes place at constant pressure. On the other hand, kinetic theory and statistical mechanics generate expressions for the heat capacity for a system of constant volume. Still other variants of the heat capacity are encountered in dealing with magnetic systems. We will use subscripts to designate which thermodynamic variable remains constant during the heat exchange. Thus

$$(dQ)_V = C_v dT \tag{4.19}$$

indicates a constant volume exchange, C_v being the heat capacity at constant volume. Similarly,

$$(dQ)_P = C_p dT \tag{4.20}$$

signifies a heat transfer at constant pressure. The specific heat capacities at constant pressure for a number of substances are displayed in Table 4.1.

Consider a simple system of fixed mass for which P, V, and T serve as the thermodynamic variables. These three variables are related by an equation of state so that only two are independent. Let V and T be taken as the independent pair. The pressure P is thereby relegated to the status of a dependent variable. The internal energy is a state function and therefore can be expressed in terms of V and T. The change in U, resulting from changes dT and dV, is given by

$$dU = \left(\frac{\partial U}{\partial T}\right)_V dT + \left(\frac{\partial U}{\partial V}\right)_T dV \tag{4.21}$$

Inserting this form of dU into the first law, $dQ = dU + P\,dV$, gives

$$dQ = \left(\frac{\partial U}{\partial T}\right)_V dT + \left[P + \left(\frac{\partial U}{\partial V}\right)_T\right] dV \tag{4.22}$$

First, consider a reversible process in which heat is added while holding the volume constant. On the left side we can set

$$dQ = (dQ)_V = C_v dT$$

On the right side we have $dV = 0$, that is, there is no change of volume. This leaves

$$C_v dT = \left(\frac{\partial U}{\partial T}\right)_V dT$$

whereupon we conclude that the heat capacity at constant volume is related to the internal energy by

$$C_v = \left(\frac{\partial U}{\partial T}\right)_V \tag{4.23}$$

The manipulations used to obtain this relation will be employed again. It is recommended that the student become familiar with the procedures.

Next, consider the addition of heat at constant pressure. The first law may be written

$$dQ_P = C_p\,dT = C_v\,dT + \left[P + \left(\frac{\partial U}{\partial V}\right)_T\right]dV \tag{4.24}$$

Since we must determine what happens at constant pressure we change our minds and regard T and P as the independent variables. The volume then becomes a variable dependent on P and T and we can write

$$dV = \left(\frac{\partial V}{\partial T}\right)_P dT + \left(\frac{\partial V}{\partial P}\right)_T dP$$

At constant pressure, $dP = 0$ and

$$(dV)_P = \left(\frac{\partial V}{\partial T}\right)_P dT$$

Inserting this into (4.24) produces the result

$$C_p = C_v + \left[P + \left(\frac{\partial U}{\partial V}\right)_T\right]\left(\frac{\partial V}{\partial T}\right)_P \tag{4.25}$$

Remarkably enough, the tangle of terms appearing beyond C_v in (4.25) can be expressed rather simply in terms of directly measurable properties of the system. The relationship between C_p and C_v is especially simple for an ideal gas. The ideal gas is "defined" by two equations:

(a) the equation of state

$$PV = nRT$$

(b) the condition that the internal energy is independent of the volume†

$$\left(\frac{\partial U}{\partial V}\right)_T = 0 \tag{4.26}$$

The second condition reduces (4.25) to

$$C_p = C_v + P\left(\frac{\partial V}{\partial T}\right)_P$$

† The experimental basis for this assumption is provided by the classic free expansion experiments of Gay-Lussac and Joule and Kelvin. These are discussed in Chapter 6.

From the equation of state we find, by differentiating with P held constant,

$$P\left(\frac{\partial V}{\partial T}\right)_P = nR$$

If we take $n = 1$ mole, there results

$$C_p = C_v + R \tag{4.27}$$

in which C_p and C_v now denote molar heat capacities.

For an ideal gas, (4.23) and (4.26) combine with (4.21) to give

$$dU = C_v \, dT \tag{4.28}$$

Experimentally it is found that gases which exhibit the "ideal" characteristics defined by (a) and (b) also have nearly constant values of C_v. With C_v constant, (4.28) can be integrated to give the internal energy

$$U = C_v T \tag{4.29}$$

In all future encounters with the ideal gas we take $U = C_v T$ for the internal energy.

The fact that C_p exceeds C_v is readily understood. Note that C_v and C_p are numerically equal to the amount of heat required to raise the temperature of 1 mole of gas by 1 °C at constant volume and constant pressure, respectively. When the gas is heated at constant volume all of the heat is used to raise the temperature of the gas. The gas does no work when maintained at constant volume $dW = P\,dV = 0$. If the same change of temperature is to be accomplished at constant pressure, more heat is required because the gas expands and does work. It was the fact that C_p exceeds C_v which led Robert Mayer to the realization that heat, like work, is but another of the many forms assumed by energy. Mayer saw that heat was being converted into work. Such a transformation implies that heat and work are merely different aspects of the same entity—energy.

Example 4.2 Adiabatic Transformations of an Ideal Gas

An adiabatic process is one in which there is no heat exchange between the system and its surroundings. We now show that the path followed by an ideal gas during an adiabatic expansion or compression is determined by the equation

$$TV^{\gamma-1} = \text{constant}$$

If we set $dQ = 0$, the first law reads

$$dU + dW = 0 \tag{4.30}$$

For an ideal gas

$$dU = C_v \, dT \tag{4.28}$$

For dW we write $P\,dV$ and use the equation of state to replace P by RT/V. Inserting these results into (4.30) and dividing by $C_v T$ leaves

$$\frac{dT}{T} + \frac{R}{C_v}\frac{dV}{V} = 0 \tag{4.31}$$

We recall the results

$$d(\ln x^\alpha) = \alpha \frac{dx}{x}$$

and

$$\ln \mathrm{AB} = \ln \mathrm{A} + \ln \mathrm{B}$$

to recognize (4.31) as

$$d(\ln[TV^{\gamma-1}]) = 0 \tag{4.32}$$

with

$$\gamma \equiv 1 + \frac{R}{C_v} \tag{4.33}$$

It follows that $\ln(TV^{\gamma-1})$ and thus $TV^{\gamma-1}$ itself remain constant during the adiabatic expansion or compression. This relationship between T and V describes the path of the process. We can write

$$TV^{\gamma-1} = T_0 V_0^{\gamma-1} \tag{4.34}$$

where T_0 and V_0 are the temperature and volume of the gas for some particular state. For example, if T_0 and V_0 denote the initial temperature and volume, (4.34) relates the temperature and volume for all other states reached by an adiabatic expansion or compression. By using the equation of state, the path can be described in terms of any two of the three variables P, V, and T (see Problem 4.5.3). ◇

······ ······

4.5.3 Using the ideal gas equation of state, show that an adiabatic path can also be described by

$$PV^\gamma = P_0 V_0^\gamma$$

and by

$$P^{1-\gamma} T^\gamma = P_0^{1-\gamma} T_0^\gamma$$

4.5.4 Wreckage of a submarine was photographed at a great depth. The wreckage showed distinct signs of charring and other indications of fire. If plates on the vessel failed, the inrushing water would adiabatically compress the trapped air. If the initial temperature was 300 K, determine the temperature when the volume of the air decreased to $\frac{1}{8}$ its initial value. Compare the temperature with the melting point of lead (600 K). Determine the pressure in atmospheres, assuming an initial pressure of 1 atm. Take $\gamma = \frac{7}{5}$, a value appropriate for an ideal diatomic gas.

4.5.5 An ideal gas expands, doubling its volume. Among the many ways this can be accomplished are: (i) isobaric; (ii) isothermal; (iii) adiabatic. Determine which of these types of expansion:

(a) results in the largest and smallest changes in the gas temperature;
(b) results in the gas performing the largest and smallest amounts of work against the surroundings;
(c) requires the largest and smallest heat absorption by the gas.

Where possible, use qualitative arguments based on the P–V diagram, knowledge of $U(T)$, and so on, rather than quantitative calculations.

•••••• ••••••

Example 4.3 Dalton's Law of Partial Pressures

In a mixture of gases, each molecular species contributes to the total pressure. Dalton's law states that if the gases are sufficiently dilute (so that each behaves as an ideal gas), the total pressure is the sum of the *partial pressures*. The partial pressure of each component species is defined as the pressure which it would exert if it alone filled the container. If P_i denotes the partial pressure of the ith component, we can state Dalton's law of partial pressures as (K is the number of component gases)

$$P = \sum_{i=1}^{K} P_i \tag{4.35}$$

Dalton's law is valid only to the extent that each component of the mixture behaves as an ideal gas. If there are n_i moles of the ith component, its partial pressure is given by

$$P_i V = n_i RT$$

It then follows from the statement of Dalton's law (4.35) that the mixture obeys the ideal gas equation of state, that is,

$$PV = \left(\sum_i P_i\right) V = \left(\sum_i n_i\right) RT = nRT$$

where n is the total number of moles of gas in the mixture

$$n = \sum_i n_i$$

Thus, so long as each component behaves as an ideal gas the mixture itself exhibits an ideal gas character. ◇

•••••• ••••••

4.5.6 Show that the partial pressure of the ith component of an ideal gas mixture may be expressed as

$$P_i = x_i P$$

where x_i is the mole fraction of the ith component, $x_i \equiv n_i/n$.

•••••• ••••••

REFERENCES

The concept of reversibility is explored in:

H. C. Van Ness, "Understanding Thermodynamics," Chapter 3. McGraw-Hill, New York (1969).

A careful and exhaustive development of the concept of thermodynamic work is given by:

M. W. Zemansky, "Heat and Thermodynamics," 5th ed., Chapter 3. McGraw-Hill, New York (1968).

The thermodynamics of magnetic systems is notorious for its ambiguous treatments. The following note, which makes special reference to superconductors, provides a convenient comparison and summary of thermodynamic relations in a magnetic field:

J. L. Mundy and V. L. Newhouse, Thermodynamic Systems Involving Magnetic Fields, *Am. J. Phys.* **34**, 1195 (1966).

A broad survey of the theoretical and experimental aspects of specific heats is given in:

E. S. R. Gopal, "Specific Heats and Low Temperatures." Plenum, New York (1966).

Despite the title, the treatise is not limited to the low-temperature regime. A treatment of magnetic specific heats is included.

Chapter 5

The Second Law of Thermodynamics

5.1 INTRODUCTION

The first law of thermodynamics places a quantitative restriction on energy-conversion processes. Such transformations can alter the form but not the amount of energy. The second law of thermodynamics imposes limits on processes which convert heat into work. The conversion of heat into work is a matter of vital importance to all phases of technology. The upshot of the second law is that the complete conversion of heat into work is not possible.

The discussions which follow make frequent reference to heat engines. A *heat engine* is any device which, operating in a cycle, absorbs heat, converts part of it into work, and rejects the remainder. Phrased in terms of the limitations of heat engines, the second law states that there can be no heat engine which can absorb a given amount of heat and convert it completely into work. This is the *practical* thrust of the second law. A more precise statement of the second law is given in the following section where we present the formulations of Lord Kelvin and Rudolph Clausius.

The concept of a heat reservoir must be assimilated as a preliminary step. A *heat reservoir* is a body of uniform temperature throughout, whose mass is sufficiently large that its temperature is unchanged by the absorption or rejection of heat. If an ice cube is tossed into the Atlantic Ocean, it produces no observable change in the temperature of the ocean. The Atlantic Ocean thereby qualifies as a heat reservoir. If the same ice cube were dropped into a cup of hot tea, a noticeable temperature change would occur. If a snowflake were to fall into the tea, no temperature change would be evident. The cup of tea therefore qualifies as a heat reservoir in the latter case, but not in the former.

5.2 THE KELVIN–CLAUSIUS FORMULATIONS OF THE SECOND LAW OF THERMODYNAMICS

Precise statements of the second law were formulated by William Thomson (later Lord Kelvin) and Rudolph Clausius in the early 1850's. Although worded differently, their statements are equivalent.

The Kelvin statement of the second law is this:

> It is impossible to devise a process whose *only* result is to convert heat, extracted from a *single* reservoir, entirely into work.

Evidently the word "only" in Kelvin's statement is an important qualifier. Many processes can be performed whereby a system converts heats completely into work. However, all such processes result in a final state which differs from the initial state. There is some *other* result beyond the conversion of heat into work. "Single" is also a key word. Heat engines convert heat into work, that is, into other forms of energy. A heat engine extracts heat from high temperature reservoirs, converts *part* of it into work, and rejects the remainder to lower-temperature reservoirs. Converting the net heat absorbed into work does not violate the second law, and is in fact demanded by the first law. A violation of Kelvin's statement of the second law would result if a heat engine could extract heat from a single reservoir and convert it *entirely* into work—without the rejection of a net amount of heat to other reservoirs.

The Clausius statement of the second law is this:

> It is impossible to devise a process whose *only* result is to extract heat from a reservoir and transfer it to a reservoir at a higher temperature.

The equivalence of the two statements may be stated somewhat cryptically as

> If Kelvin false, then Clausius false,
> and
> If Clausius false, then Kelvin false.

Of course, demonstrating their logical equivalence does not prove either version of the second law. The proof is very definitely "in the eating of the pudding." Along with other laws of nature, the second law of thermodynamics survives only because it is in accord with experiment. No amount of experimental evidence could ever establish it as true. The evidence supporting the second law is the failure of all attempts to construct perpetual motion machines of the second kind.[†] A few examples best illustrate why fame, fortune, and scientific immortality would descend instantly upon anyone who managed to violate the second law.

The crust of our Earth, its atmosphere, and its oceans are giant heat reservoirs. Imagine an engine which could extract heat from the atmosphere and convert it into mechanical energy. The mechanical energy could be used to operate machinery and to generate electric currents. Air taken in and cooled by the extraction of heat could be used as a refrigerant. Air-conditioning systems would be a by-product of such an engine! The supply of energy in the atmosphere is virtually limitless, its content being maintained by the Sun. It would be unnecessary to mine coal or drill for gas and oil—*if* there were such engines. However, there are none! The oceans of the Earth likewise contain enormous quantities of energy. Imagine a ship with a power plant which could extract heat from sea water. The heat could be converted into work and used to propel the ship. Such wondrous heat engines are destined to remain figments of the imagination.[‡]

•••••• ••••••

5.2.1 One mole of an ideal gas expands isothermally.

(a) If the gas doubles its volume, show that the work of expansion is

$$W = RT \ln 2.$$

(b) Since the internal energy of an ideal gas depends solely on its temperature, there is no change in U during the expansion. It follows from the first law that the heat absorbed by the gas during the expansion is converted completely into work. Why does this not violate the second law?

•••••• ••••••

If the complete conversion of heat into mechanical energy is not possible, the obvious question is, "What is the optimum design for a heat engine?" The answer to this question was supplied by Sadi Carnot, a French engineer. Carnot's brilliant analysis of the problem gained for him credit for discovery

---

[†] Recall that a perpetual motion machine of the first kind is one which would manufacture energy—it would violate the first law of thermodynamics.

[‡] The interior of the Earth is hotter than its surface. The interior is a bona fide high-temperature reservoir and is now being tapped at several geothermal power plants. See J. Barnea, Geothermal Power, *Sci. Am.* **226**, 70 (January 1972).

of the second law of thermodynamics. Carnot's efforts are especially noteworthy because his views were set forth in 1824, nearly 20 years prior to the Mayer–Joule formulation of the first law. Carnot died 8 years later in a cholera epidemic, at the age of 36. Because Carnot employed the caloric theory of heat his applications of the second law were not always sound. Carnot's notebooks, unpublished until long after his death, revealed his suspicion of the caloric theory and his resolve to perform experiments similar to those carried out by Joule.

There are two remarkable aspects of Carnot's researches. First, he found that the efficiency with which heat can be converted into other forms of energy depends on the nature of the cyclic process employed but not at all on the working substance—the material which undergoes the cyclic transformations. Second, Carnot discovered the most efficient cyclic process for converting heat into other forms of energy. This cycle, now known as the Carnot cycle, is discussed in Section 5.3. Carnot's achievements are truly monumental. Without benefit of a correct theory of heat he set forth the ultimate standard of heat engine efficiency.

There are of course many types of heat engines in use today. Basically, a heat engine operates by (a) extracting heat from a high-temperature reservoir, (b) converting part of the heat into work, and (c) rejecting the remainder to a low-temperature reservoir. It was Carnot's investigation of the efficiency of heat engines which led him to an understanding of the second law. A brief look at the Carnot cycle paves the way for a demonstration of the equivalence of the Kelvin and Clausius formulations of the second law.

•••••• ••••••

5.2.2 It was emphasized in Section 4.5 that specific heat capacities are never negative. Adding heat to a system never causes its temperature to decrease. This is an experimental fact. Construct a scheme which would violate the second law of thermodynamics by assuming you have a material with a negative heat capacity.

5.2.3 Capillary action causes liquids to rise against the force of gravity. Suppose you have a tube in which water will rise to a height of 2 ft. Next, suppose the tubing is bent so as to form an inverted "j" 1-ft-high. Capillary action therefore has an easy time filling the entire tube. Next, one places a small paddle wheel beneath the end of the tube. As water drips onto the paddle wheel it turns and generates mechanical energy. Water which strikes the paddle wheel falls back into the reservoir with less kinetic energy than it would have had it not struck the wheel. Some of its kinetic energy has been converted into the mechanical energy of the rotating paddle wheel. This would result in a cooling of the water were it not for the fact that the reservoir is maintained at a constant temperature by virtue of thermal contact with the Earth. So—the only result of the device is to extract heat from a reservoir (the Earth) and convert it into work. This violates the second law and so there is no point in trying to build one. Explain why the device does not operate as just advertised. [*Hint:* In case you have not already discovered the flaw it will be helpful to know that the capillary tube will not even function as a perpetual motion machine of the third kind. (A perpetual motion machine of the third kind is a device which "just runs." A frictionless wheel which spins forever once started or a pendulum which swings without damping typify perpetual motion machines of the third kind.)]

•••••• ••••••

5.3 THE CARNOT CYCLE

The Carnot cycle is a series of reversible processes. It consists of two adiabatic legs connected by a pair of isothermal legs. Figure 5.1 illustrates a Carnot cycle in which a gas serves as the working substance. The arrows indicate the sequence of steps. During the isothermal expansion from 1 to 2,

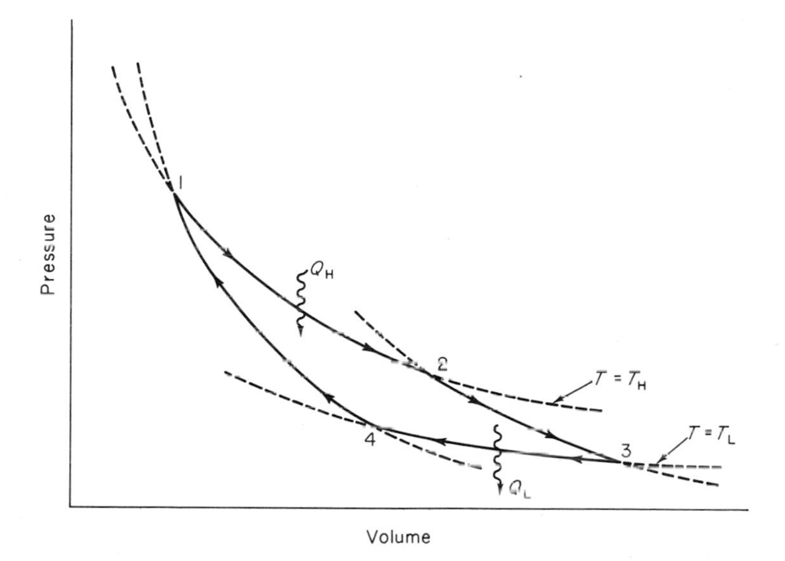

Figure 5.1. The Carnot cycle for a gas as it appears on a *P–V* diagram, where T_H and T_L refer to the high and low temperature reservoirs. Heat Q_H is absorbed from a reservoir as the gas expands isothermally from 1 to 2. The expansion from 2 to 3 is performed adiabatically. Heat Q_L is rejected during the isothermal compression from 3 to 4. The cycle is completed by an adiabatic compression from 4 to 1.

heat Q_H is absorbed from a high-temperature reservoir. The gas expands performing work against the surroundings. The expansion of the gas continues from 2 to 3 under adiabatic conditions, that is, the gas is insulated from its surroundings during the expansion. An isothermal compression from 3 to 4 results in the performance of positive work by the surroundings and the rejection of heat Q_L by the gas to the low-temperature reservoir. The cycle is completed from 4 to 1 by adiabatic compression. There is no heat transfer and the positive work performed by the surroundings increases the internal energy of the gas. The increase shows up as a temperature rise from T_L to T_H.

The overall result of the cycle is the conversion into work of some of the heat extracted from the high-temperature reservoir. To avoid possible misunderstandings we point out that our sign convention, for heat absorbed by the system and by the surroundings, has been temporarily abandoned, with Q_H and Q_L denoting positive quantities of heat. The rejection of heat by the system will be indicated explicitly by a minus sign. For the Carnot cycle just described the system absorbs heat Q_H and rejects an amount Q_L. The net amount of heat received from the surroundings is $Q_H - Q_L$. By the same token, the net work W done by the system over the cycle is composed of two positive contributions and two negative contributions. During the expansions, 1 to 2 and 2 to 3, the system performs positive work. During the compressions, 3 to 4 and 4 to 1, the system does negative work, that is, the surroundings perform positive work. Since the system returns to its initial thermodynamic state there is no net change in its internal energy; $\Delta U = 0$ for a cyclic process. The first law

$$Q = \Delta U + W$$

when applied to the Carnot cycle becomes

$$Q_H - Q_L = W$$

This merely states that the net work done by the system over the cycle equals the net heat intake.

Observe that only a portion of the heat absorbed at the higher temperature is converted into work. The second law of thermodynamics would crumble if it were possible to make Q_L zero.

•••••• ••••••

5.3.1 In Figure 5.1, the adiabat legs are steeper than the isotherm segments of the Carnot cycle. This may be verified provided the equation of state and internal energy of the gas are known functions of P, V, and T. For an ideal gas (1 mole)

$$PV = RT; \qquad U = C_v T$$

where C_v is the molar heat capacity at constant volume and is a constant.

The slope dP/dV of the ideal gas adiabat is larger by a factor of $\gamma = 1+R/C_v$ than the slope of an isotherm. Verify this by showing that

$$\left(\frac{dP}{dV}\right)_{\text{isot}} = -\frac{P}{V}; \quad \left(\frac{dP}{dV}\right)_{\text{adia}} = -\gamma\frac{P}{V}$$

where isot is isotherm and adia is adiabat. Recall that an adiabat is defined by the condition that $dQ = 0$ and that in general $dQ = dU + P\,dV$ for a gas.

5.3.2 Suppose that an ideal gas is used as the working substance in the Carnot cycle depicted in Figure 5.1. The internal energy of an ideal gas depends on the temperature alone $U = U(T)$.

(a) If the gas absorbs $Q_H = 3$ kcal during the isothermal leg from 1 to 2, how much work is performed by the gas?

(b) If the gas performs 1 kcal of work during the adiabatic expansion from 2 to 3, what is the change in its internal energy?

(c) How much work must be performed on the gas from 3 to 4 for it to reject $Q_L = 2$ kcal of heat?

(d) How much work must be performed on the gas to take it from 4 to 1?

(e) What is the net amount of work performed by the gas during the cycle?

•••••• ••••••

5.4 THERMODYNAMIC EFFICIENCY

We would say that a process is efficient if it converts a sizable fraction of the heat absorbed into useful work, rejecting only a small fraction. The *thermodynamic efficiency* η is defined as the fraction of the total heat absorbed which is converted into work (over a cycle)

$$\eta = \frac{W}{Q_{\text{abs}}} \tag{5.1}$$

An alternate expression for η follows from the fact that the work performed during a cyclic process equals the net heat exchange: Thus $W = Q_{\text{abs}} - Q_{\text{rej}}$, so that

$$\eta = \frac{Q_{\text{abs}} - Q_{\text{rej}}}{Q_{\text{abs}}} = 1 - \frac{Q_{\text{rej}}}{Q_{\text{abs}}} \tag{5.2}$$

For the Carnot cycle, Q_H is the total heat absorbed and Q_L is the heat rejected. The Carnot efficiency is therefore

$$\eta_c = 1 - \frac{Q_L}{Q_H} \tag{5.2a}$$

In the next section we show that no heat engine operating between two reservoirs can be more efficient than a Carnot engine operating between the same two reservoirs.

Example 5.1 Calculation of Thermodynamic Efficiency

As an elementary illustration showing how thermodynamic efficiencies are determined we consider the three-legged cycle shown in the accompanying figure. The working substance is taken to be an ideal gas. The isobaric expansion of the gas from 1 to 2 leads to an increase in temperature and the absorption of heat

$$Q_{\text{abs}} = C_p \Delta T = C_p(T_2 - T_1)$$

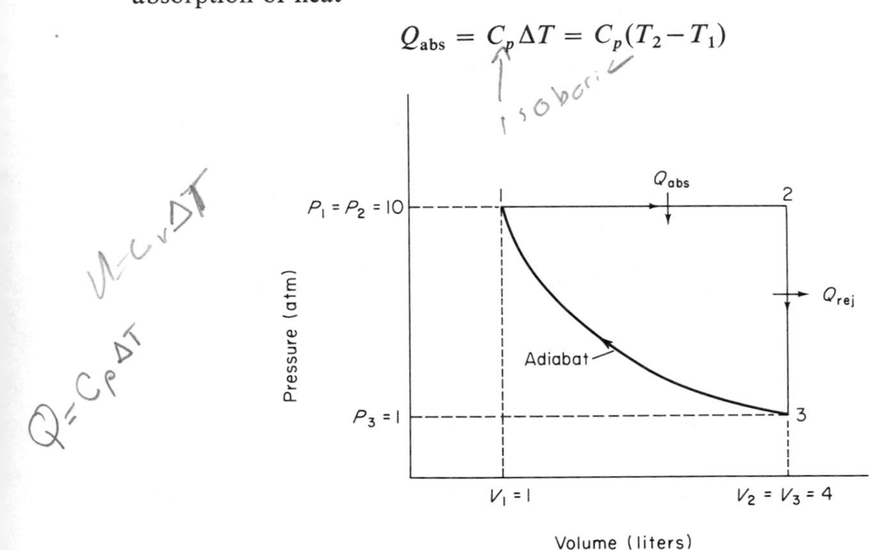

The isovolumic decompression from 2 to 3 results in a temperature drop and the rejection of heat

$$Q_{\text{rej}} = -C_v \Delta T = C_v(T_2 - T_3)$$

The cycle is completed along the adiabat from 3 to 1, there being no heat exchange. From (5.2)

$$\eta = 1 - \frac{Q_{\text{rej}}}{Q_{\text{abs}}} = 1 - \frac{C_v(T_2 - T_3)}{C_p(T_2 - T_1)}$$

Recalling that $C_p/C_v = \gamma$ we have

$$\eta = 1 - \frac{1}{\gamma} \frac{\left(1 - \dfrac{T_3}{T_2}\right)}{\left(1 - \dfrac{T_1}{T_2}\right)} \tag{5.3}$$

The temperature ratios may be expressed in terms of the volume and pressure ratios by using the equation of state. As η is an intensive property of a system we consider 1 mole of gas. It follows from $PV = RT$ that an alternate expression for the efficiency is

$$\eta = 1 - \frac{1}{\gamma} \frac{\left(1 - \frac{P_3}{P_2}\right)}{\left(1 - \frac{V_1}{V_2}\right)} \quad \diamondsuit \tag{5.4}$$

•••••• ••••••

5.4.1 (a) Use the ideal gas equation of state to show that (5.4) follows from (5.3).

(b) Determine the efficiency of the cycle of Example 5.1, and the P–V data therein, for the cases of ideal monatomic and diatomic gases ($\gamma_{\text{mon}} = 5/3$; $\gamma_{\text{diat}} = 7/5$).

5.4.2 Show that the efficiency for the cycle shown in the accompanying figure is

$$\eta = \frac{\gamma - 1}{\frac{\gamma P_2}{P_2 - P_1} + \frac{V_1}{V_2 - V_1}}$$

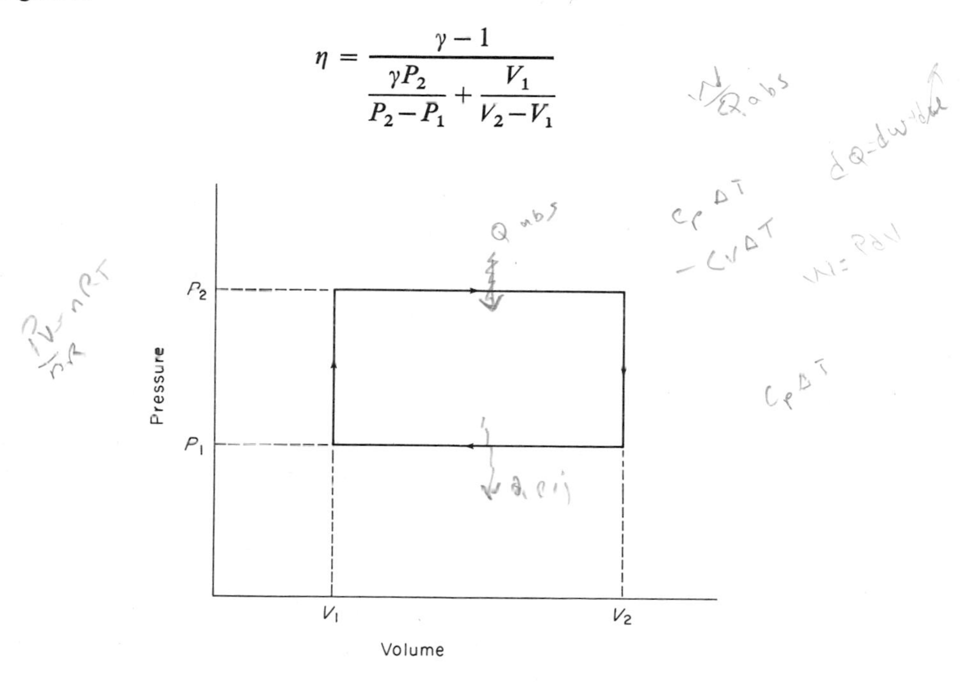

•••••• ••••••

It was emphasized that the Carnot cycle consists of a sequence of reversible processes. As described earlier, the Carnot cycle functions as a heat engine; it converts heat into work. If the sequence of steps is reversed, the Carnot cycle functions as a refrigerator. The working substance of a refrigerator extracts heat from a low-temperature reservoir, has (positive) work performed on it by the surroundings, and rejects heat at a higher temperature. It therefore pumps heat uphill. The efficiency of a refrigerator is defined as the fraction of the heat rejected which was received as work

$$\eta = \frac{W}{Q_{\mathrm{abs}}} = 1 - \frac{Q_{\mathrm{rej}}}{Q_{\mathrm{abs}}}$$

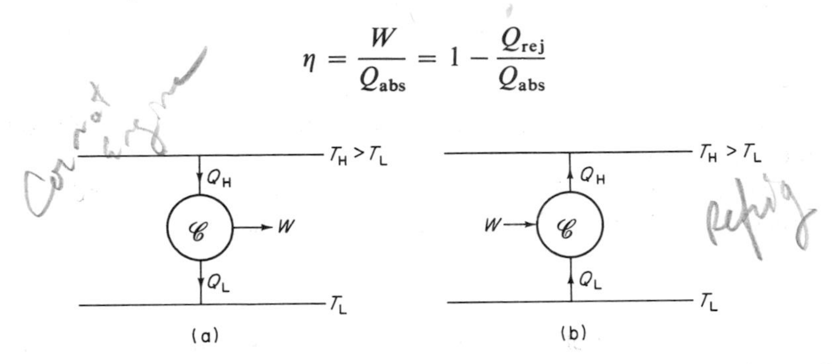

Figure 5.2. The Carnot engine (a) and refrigerator (b). An arrow directed toward the engine indicates heat absorption *by* or work performed *on* the engine.

This definition guarantees that a reversible cycle has the same efficiency whether it operates as a heat engine or as a refrigerator. Figures 5.2a and b tell the Carnot cycle story forward and backward. The definition of the thermodynamic efficiency of a refrigerator is inconvenient in practice. A more useful measure of the capabilities of a refrigerator is the coefficient of performance defined as

$$\zeta \equiv \frac{\text{heat absorbed}}{\text{work performed}} = \frac{Q_{\mathrm{abs}}}{W}$$

Optimally, a refrigerator extracts a sizable amount of heat from a low-temperature reservoir with the expenditure of a minimal amount of work, that is, it has a large coefficient of performance. The coefficient of performance is inversely proportional to the basic economic variable of refrigeration—the cost per calorie extracted.

We discuss the Carnot cycle further in the next section. We conclude this section by demonstrating the equivalence of the Kelvin and Clausius formulations of the second law. We must show that if Kelvin's statement is false, so then is Clausius', and vice versa.

Suppose first that Kelvin's statement is false, that is, suppose that a process is possible whose only result is to convert heat extracted from a reservoir completely into work. The work so obtained could then be completely converted back into heat at any desired temperature, via friction, for example. In particular, it could be deposited in a heat reservoir whose temperature exceeds that of the reservoir from which the heat was extracted. The only result of the scheme is to transfer a quantity of heat from a lower to a higher temperature, which is a violation of Clausius' postulate. Thus, if Kelvin is false, so is Clausius.

To complete the proof of their logical equivalence we must also show that if Clausius' version of the second law is false, so also is Kelvin's statement. Let us therefore assume that Clausius' statement is false and that there exists a scheme whose only result is to extract an amount of heat Q and transfer it to a reservoir at a higher temperature. We set up a Carnot engine which operates between the two reservoirs. Let the engine extract heat Q from the higher-temperature reservoir, convert part of it into work, and reject the remainder to the low-temperature reservoir. The size of the engine is designed such that it executes an integral number of cycles during the process. The engine therefore undergoes no change. The high-temperature reservoir likewise has undergone no change. It absorbed and subsequently rejected heat Q. If η is the efficiency of the engine, the work output is ηQ, and the heat returned to the low-temperature reservoir is $(1-\eta)Q$. Overall then the only result is that the low-temperature reservoir has given up heat ηQ, and this has been converted into work. The only result of the process is to convert into work heat extracted from a single reservoir—a violation of Kelvin's statement of the second law. This completes the proof of the equivalence of Clausius' and Kelvin's statement of the second law.

We now turn to Carnot's Theorem which, among other things, provides the basis for a thermodynamic temperature scale—a scale independent of the thermometric substance.

5.5 CARNOT'S THEOREM AND THE THERMODYNAMIC TEMPERATURE SCALE

The Carnot cycle is of special interest for two reasons. The first concerns Carnot's Theorem, which deals with the efficiency of heat engines. Carnot's Theorem proves that a heat engine which extracts heat from one reservoir, performs work, and rejects heat to a second reservoir cannot have an efficiency which exceeds that of a Carnot engine operating between the same two reservoirs. The efficiency of a Carnot engine therefore presents a theoretical limit which cannot be exceeded by clever design.

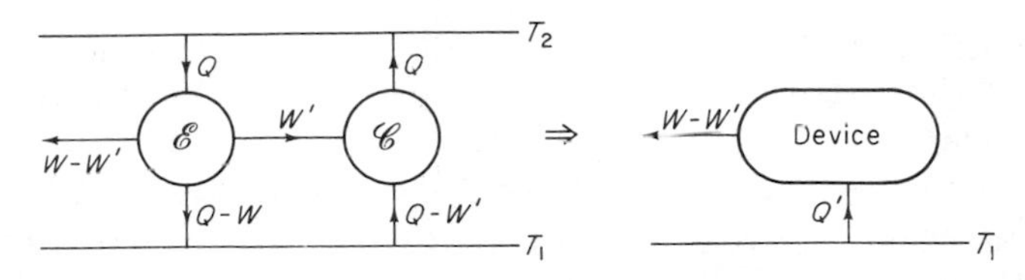

Figure 5.3. If the efficiency of $\mathscr{E}$ exceeds the Carnot efficiency, it is possible to use a portion of the work output of $\mathscr{E}$ to drive a Carnot refrigerator so as to leave the high-temperature reservoir unchanged. The only result of $\mathscr{E}$ and $\mathscr{C}$ in tandem would be to convert heat completely into work—a violation of the second law.

Carnot's Theorem is this:

> No heat engine operating between two heat reservoirs can be more efficient than a Carnot engine operating between the same two reservoirs.

The proof of the theorem follows these lines: One shows that if the theorem is false, it is possible to violate the second law. Since it is not possible to circumvent the second law, the theorem is not false. Figure 5.3 shows how the proof is formalized. We assume that the engine $\mathscr{E}$ on the left is more efficient than the Carnot engine $\mathscr{C}$ on the right. Let the engine execute an integral number of cycles in the process of extracting heat Q from the high-temperature reservoir, converting part of it into work W, and rejecting heat $Q-W$ to the low-temperature reservoir. We divide the work output into two parts, W' and $W-W'$. The work W' is used to drive the Carnot engine, operating as a refrigerator, through an integral number of cycles and such that it rejects heat Q to the high-temperature reservoir. We can always arrange to have the Carnot refrigerator go through an integral number of cycles (for a given value of Q) by adjusting its size. The efficiencies are given by

$$\eta_{\mathscr{E}} = \frac{W}{Q}, \qquad \eta_{\mathscr{C}} = \frac{W'}{Q}$$

Since, by assumption, $\eta_{\mathscr{E}} > \eta_{\mathscr{C}}$, we must have $W > W'$, and so $W-W'$ is positive, that is, the $\mathscr{E}-\mathscr{C}$ tandem does a positive amount of work against the surroundings. This series of operations leaves the high-temperature reservoir unchanged—it rejected and reabsorbed heat Q. Both $\mathscr{E}$ and $\mathscr{C}$ execute an integral number of cycles so that they are left unchanged. The low-temperature reservoir received heat $Q-W$ and rejected heat $Q-W'$ so that overall it rejected an amount of heat equal to $W-W'$. Therefore, the low-temperature reservoir rejected an amount of heat equal to the net work output of the $\mathscr{E}-\mathscr{C}$ tandem. The only result of the process is the conversion into work of heat extracted from a reservoir. However, this violates Kelvin's form of the second law. Since this is not possible we conclude that the original assumption is false—it is not possible for $\eta_{\mathscr{E}}$ to exceed $\eta_{\mathscr{C}}$. Thus, Carnot's Theorem is proved.

A corollary to Carnot's Theorem states that the efficiency of all Carnot engines (working between the same two reservoirs) is the same regardless of their working substances. To prove the corollary, merely replace the engine by a Carnot engine $\mathscr{C}_2$, whose working substance is different from that of the Carnot engine which operates as a refrigerator $\mathscr{C}_1$. Carnot's Theorem shows that

$$\eta_{\mathscr{C}_2} \text{ is not greater than } \eta_{\mathscr{C}_1},$$

that is, the "newly arrived" Carnot engine cannot have an efficiency which exceeds that of the "original" Carnot engine. Since the Carnot cycle is reversible we can turn everything around and let $\mathscr{C}_1$ operate as a heat engine which drives $\mathscr{C}_2$ now operating as a refrigerator. Carnot's Theorem now states that

$$\eta_{\mathscr{C}_1} \text{ is not greater than } \eta_{\mathscr{C}_2},$$

We conclude that

$$\eta_{\mathscr{C}_1} = \eta_{\mathscr{C}_2}$$

which proves the corollary.

Carnot's Theorem and Corollary are very important. They impose a fundamental limit on the conversion of heat into work. No amount of ingenuity can devise a heat engine more efficient than the Carnot engine. In practice it is not generally feasible to use a Carnot cycle. Among other reasons, it may be prohibitively expensive to maintain adiabatic conditions because of the elaborate insulation required. The Carnot efficiency therefore serves as a standard one aims to approach using a more economical cycle.

•••••• ••••••

5.5.1 The Stirling engine was invented by Robert Stirling in 1816. The working substance was air. The accompanying figure shows an idealized

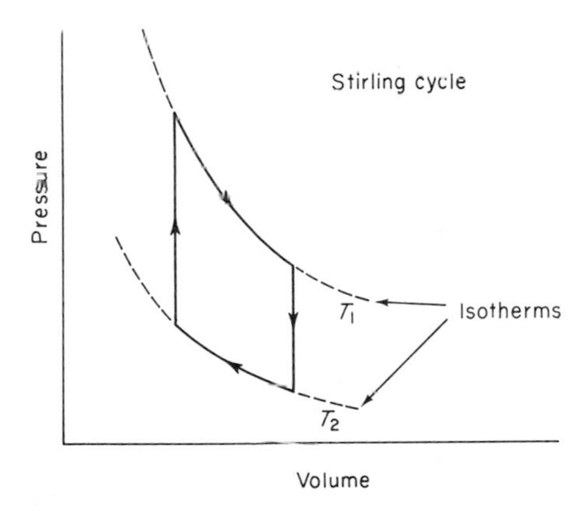

version of the Stirling cycle. There is renewed interest in the Stirling engine today because it holds promise as an ecologically clean automotive power plant. Note that in the Stirling cycle two isovolumic legs replace the pair of adiabatic legs of the Carnot cycle. The design of the engine is such that no heat is exchanged with the surroundings during the isovolumic steps. Instead, there is an internal exchange of heat. There are two gas-driven pistons, connected by a linkage which maintains a temperature difference while allowing the exchange of gas between piston cylinders. During the isovolumic phases of the cycle one piston moves downward, the other upward. There is an exchange of gas, and heat, between cylinders. There is no change in total volume and no heat exchange with the surroundings. With these facts in mind, verify that the efficiency of the Stirling cycle equals that of the Carnot cycle. Assume that the working substance is an ideal gas.

•••••• ••••••

The fact that the Carnot efficiency is a universal property, independent of the working substance used in the engine, brings us to the second reason for taking another look at the Carnot cycle. In Chapter 3 we noted that the thermodynamic scale of temperature is based on the laws of thermodynamics—that it is independent of the thermometric substance used to construct a thermometer. The Carnot efficiency is independent of the working substance, but does depend on the temperatures of the heat reservoirs. This fact makes it possible to use the Carnot efficiency to define a thermodynamic temperature scale.

The efficiency of a Carnot engine which absorbs heat Q_1 along one isothermal leg and rejects Q_2 along another is

$$\eta_c = 1 - \frac{Q_2}{Q_1}$$

A thermodynamic temperature scale is defined by taking $Q_2/Q_1 = T_2/T_1$ or

$$\frac{T_2}{Q_2} = \frac{T_1}{Q_1} \tag{5.5}$$

This definition tells us that the thermodynamic temperature T is proportional to Q. Heat serves as the thermometric variable for the thermodynamic scale. This definition leaves open the size of the temperature unit. As noted in Chapter 1, the temperature of the triple point of water has been assigned the value 273.16 K. Thus if we operate a Carnot engine with one reservoir at the triple point, the temperature of the other reservoir is

$$T = 273.16 \frac{Q}{Q_3} \text{K} \tag{5.5a}$$

The thermodynamic scale is completely satisfactory from an operational viewpoint because an operational definition of heat has been provided by the first law of thermodynamics. As a consequence of (5.2) and (5.5a), the efficiency of a Carnot engine working between reservoirs at T_H and T_L is given by

$$\eta_c = 1 - \frac{T_L}{T_H} \tag{5.6}$$

In the next section we use (5.5), applied to a special Carnot cycle, to derive the Clausius–Clapeyron equation, which is extremely useful in studies of phase equilibrium.

5.6 THE CLAUSIUS–CLAPEYRON EQUATION

Our atmosphere contains water in the gaseous phase. The contribution of the water molecules to the total atmospheric pressure is called the "water vapor[†] pressure." For any given temperature there is a maximum water vapor pressure known as the "saturated water vapor pressure." The concept of a vapor pressure is not restricted to water nor to our atmosphere. The region above any liquid will contain its vapor along with, perhaps, other gases. The vapor pressure is that portion of the total pressure contributed by the vapor. There is a maximum vapor pressure, the saturated vapor pressure, whose value depends on the temperature and the physical nature of the molecules, but not on the presence or absence of other gases. The saturated vapor pressure depends solely on the temperature in a way dictated by the Clausius–Clapeyron equation. A special miniaturized Carnot cycle will be used to derive the Clausius–Clapeyron equation. We preface the derivation with a brief description of the liquid–vapor transition.

The vaporization of a liquid is readily understood from the microscopic viewpoint. By virtue of their kinetic energy some molecules are able to overcome the attractive intermolecular forces, escape the liquid, and become part of the vapor atmosphere above the liquid surface. Of course, molecules continually fall back into the liquid. Equilibrium exists—we say the vapor is *saturated*—when the rates of escape and return are equal.

The vapor exerts a pressure on the liquid. Macroscopically, this pressure arises because the liquid surface must support the weight of the vapor "piled" on it. Microscopically, the pressure results because the surface experiences an impulsive force each time a molecule falls back into the liquid. A similar force arises when a molecule "leaps" upward into the vapor atmosphere.

[†] The words "vapor" and "gas" are completely synonymous, although "vapor" is frequently used in referring to the gaseous phase in equilibrium with its liquid or solid phase.

The vapor pressure increases with temperature. The temperature at which the vapor pressure equals 1 atm (760 Torr) is termed the *normal boiling point*, and the temperature at which the vapor pressure equals the ambient pressure above the liquid is called the *boiling point*. By reducing the pressure above the liquid (climb a mountain, reduce pressure via pump, and so on), the boiling point may be lowered. For example, the normal boiling point of liquid helium is 4.2 K. If the pressure of the liquid is reduced by pumping away the vapor rapidly, the temperature of the liquid is reduced, the thermal energy being carried away with the vapor as *latent heat* of vaporization.† Temperatures down to about 0.8 K may be produced in this fashion. Below 0.8 K, an unusual film creep phenomenon peculiar to ^{4}He prevents further cooling. The rare isotope ^{3}He may be cooled by forced evaporation to temperatures near 0.3 K.

••••• •••••

5.6.1 How does the presence of water vapor affect the air pressure? For equal air temperatures, would the air pressure be greater over an arid desert-like surface, or over an ocean?

••••• •••••

To understand the vapor–liquid phase transition, consider a gas at a temperature T confined within a cylinder by a piston. The gas is to be compressed isothermally. This is accomplished by maintaining thermal contact between the contents of the cylinder and a heat reservoir at temperature T. The isothermal compression leads to an increase in pressure. The relationship between P and V along an isotherm of the gaseous phase is qualitatively similar to that for an ideal gas

$$PV = nRT$$

In particular, an isothermal compression increases the pressure. Provided the temperature is not too high, a real gas will reach a pressure at which condensation begins. Referring to Figure 5.4 we can follow the compression of the vapor from 1 to 2 along the isotherm labeled T. At 2 we have a saturated vapor.

† The latent heat of vaporization is the amount of heat required to convert a substance (at its boiling point) from a liquid to a vapor. The adjective latent was introduced by Joseph Black early in the 19th century. Ordinarily the addition of heat produces a temperature rise. This heat, Black referred to as a sensible heat. Black also discovered that the addition of heat could produce a change of phase, but that no temperature change occurred during the change of phase. To describe this, Black coined the terminology "latent heat." Latent heats are associated with vaporization, melting, and sublimation. The latent heat of fusion is the amount of heat required to convert a substançe (at its melting point) from a solid to a liquid. The latent heat of sublimation is the amount of heat required to convert a substance from the solid phase directly to the gaseous phase.

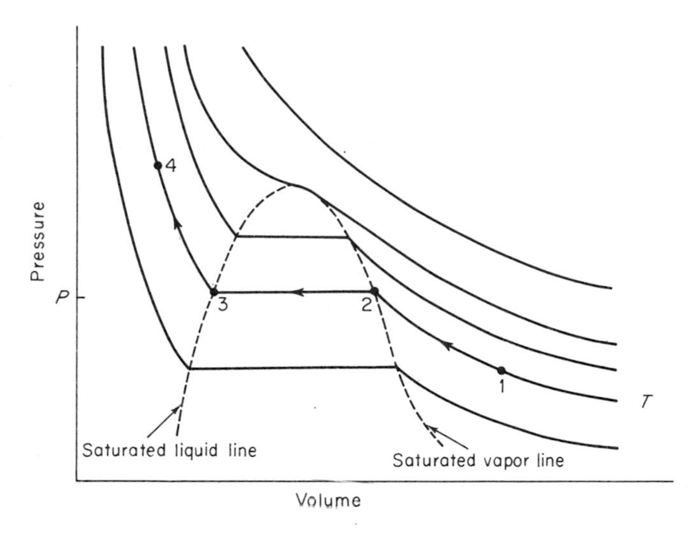

Figure 5.4. Typical isotherms in the condensation region. (See text for details.)

As the piston reduces the volume of the system, vapor continues to condense, while the pressure and temperature remain constant. The constancy of pressure and temperature is a general characteristic of condensation. The horizontal segment of the isotherm represents states in which both liquid and vapor are present. The pressure for the horizontal segment (2 to 3) of the isotherm is called the "saturated vapor pressure at a temperature T." Continued compression eventually reduces the volume to the point 3 where there is no vapor, that is, the vapor has condensed completely and the volume of the system equals the volume of the liquid. At 3 we have a saturated liquid. The isotherm rises sharply for the liquid phase, 3 to 4, because liquids are nearly incompressible.

The process of condensation is reversible. By expanding the volume of the system, the liquid may be evaporated. Bear in mind that the system must be in thermal contact with a heat reservoir which supplies or absorbs whatever heat is required to maintain a constant temperature—it is not just a matter of moving the piston to expand or compress the system.

The locus of saturated vapor points forms the saturated vapor line of Figure 5.4. Likewise, the saturated liquid line is the locus of saturated liquid states. The *critical point* marks their common terminus. At the critical point there is no distinction between liquid and vapor—their properties being identical. At temperatures above the critical temperature there is no condensation—compression leads to a continuous transition from the gaseous to liquid state. Further discussion of phase transitions is presented in Chapter 9.

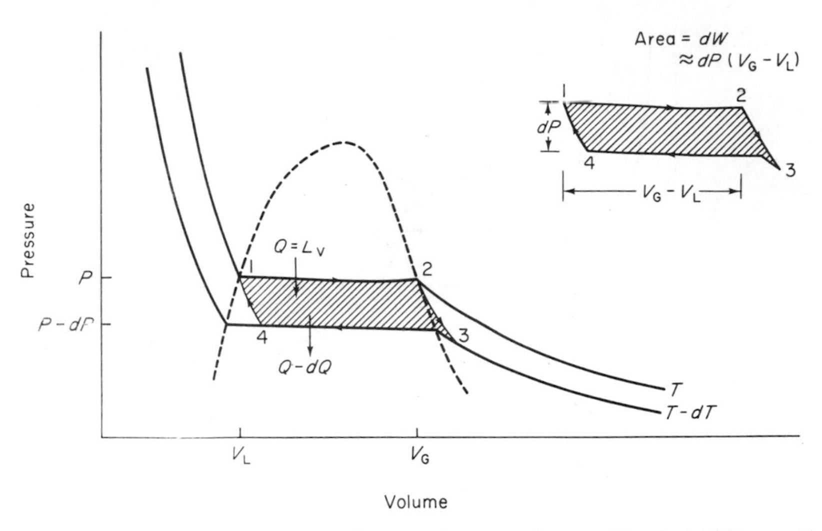

Figure 5.5. Special Carnot cycle and approximate work area. The dashed line outlines the dome portion of the *P–V–T* surface where liquid and vapor can coexist in equilibrium.

The Clausius–Clapeyron equation relates the vapor pressure P and temperature T, by giving the slope dP/dT of the *P–V–T* surface over the liquid–vapor coexistence region. We now derive the Clausius–Clapeyron equation by considering a very special Carnot cycle. Let 1 mole of the fluid be carried through the Carnot cycle indicated in Figure 5.5. The cycle is started at 1, with 1 mole of liquid (volume V_L) and no vapor present. The isothermal expansion from 1 to 2 requires the addition of heat Q. At 2, the liquid has completely evaporated so that we have 1 mole of vapor (volume V_G) and no liquid. The heat added Q is therefore equal to L_v the (molar) latent heat of vaporization. The adiabatic expansion $(2 \rightarrow 3)$ lowers the temperature and pressure infinitesimally (to $P-dP$ and $T-dT$). The gas is then compressed isothermally $(3 \rightarrow 4)$ and the cycle completed by an adiabatic compression to the initial state. The area on the *P–V* diagram enclosed by the cycle equals the work done by the system during the cycle. Approximating† this area by a rectangle (compare Figure 5.5) gives

$$dW = dP(V_G - V_L) \tag{5.7}$$

where V_G and V_L are the volumes occupied by 1 mole of gas and liquid, respectively. Because the system undergoes a Carnot cycle between

† The approximation would become exact if a more rigorous but cumbersome limiting process were employed.

temperatures T and $T-dT$ we have by (5.5),

$$\frac{Q-dQ}{Q} = \frac{T-dT}{T}$$

which gives

$$dQ = Q\frac{dT}{T}$$

However, in a cyclic process, $dQ = dW$, that is, the net heat absorbed equals the work done. Thus

$$Q\frac{dT}{T} = dP(V_G - V_L) \tag{5.8}$$

Remembering that $Q = L_v$, the latent heat of vaporization, and solving for dP/dT gives the Clausius–Clapeyron equation

$$\frac{dP}{dT} = \frac{L_v}{T(V_G - V_L)} \tag{5.9}$$

The same sort of analysis can be applied to the solid–liquid transition. The temperature variation of the melting pressure is governed by

$$\frac{dP}{dT} = \frac{L_f}{T(V_L - V_S)} \tag{5.9a}$$

where L_f is the latent heat of fusion, T is the melting point, and $V_L - V_S$ is the solid to liquid change of volume.

•••••• ••••••

5.6.2 Compute the change (including sign) in the temperature at which water freezes when the pressure is raised from 1 atm to 100 atm.

•••••• ••••••

It is apparent from Figure 5.5 that V_G and V_L depend on temperature. Not so apparent is the fact that the latent heat L_v also varies with temperature. By invoking some reasonable approximation, or using experimental data, it is possible to integrate the Clausius–Clapeyron equation and obtain a relation between P and T. A crude but often adequate approximation is to ignore V_L in comparison with V_G and then assume that the vapor obeys the ideal gas law $PV_G = RT$. With these approximations inserted into (5.9) we obtain

$$\frac{dP}{P} = \left(\frac{L_v}{R}\right)\frac{dT}{T^2} \tag{5.10}$$

If L_v is assumed constant, integration gives

$$\ln P = -\frac{L_v}{RT} + \text{constant} \tag{5.11}$$

A more refined version of this equation serves as a useful low-temperature thermometer. The vapor pressure above liquid helium is readily measured and, since it depends on temperature alone, serves to indicate the liquid temperature.

Example 5.2 Latent Heat of Vaporization of Mercury

According to (5.11), a plot of $\ln P$ versus $1/T$ should give a straight line of slope $-L_v/R$. Thus by measuring both the vapor pressure and the temperature it is possible to determine the latent heat of vaporization. Let (P_1, T_1) and (P_2, T_2) denote corresponding pairs of data. From (5.11)

$$L_v = \frac{R \ln\left(\dfrac{P_2}{P_1}\right)}{\dfrac{1}{T_1} - \dfrac{1}{T_2}}$$

We use P–T data for mercury taken from the Chemical Rubber Co. Handbook,†

$$P_2 = 14.8, \qquad T_2 = 470\ \text{K}$$

$$P_1 = 0.1, \qquad T_1 = 354\ \text{K}$$

These data give $L_v \simeq 14.4$ kcal/mole, tolerably close to the precision experimental value of 13.99 kcal/mole. ◇

The next section presents Maxwell's demon and a distinct change of pace. Our concern still lies with the second law, but our viewpoint shifts to the microscopic, molecular level. On to the demon!

•••••• ••••••

5.6.3 Consult tables giving vapor pressure data for water. Compute the latent heat of vaporization per mole and per gram. The experimental value at 100 °C is 539 cal/gm.

5.6.4 The "drinking duck" appears to be a perpetual motion machine. The accompanying figure shows the internal structure of the duck. The beak of the

† "Handbook of Chemistry and Physics," 51st ed. Chem. Rubber Pub. Co., Cleveland, Ohio (1970).

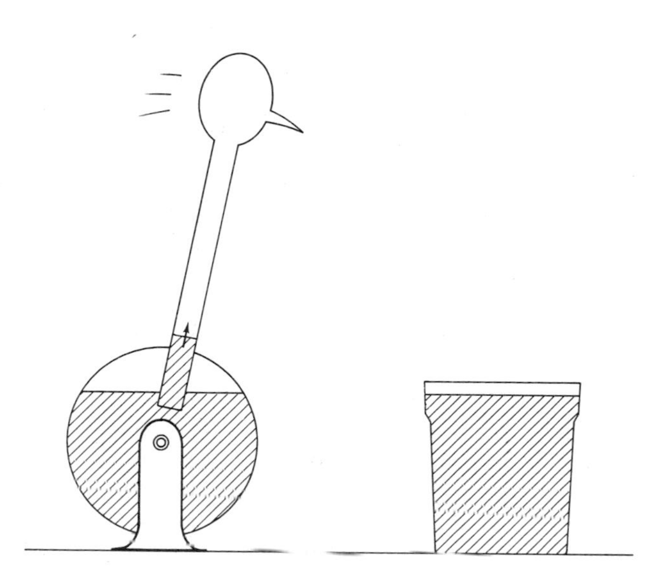

duck is covered with a thin layer of very absorptive fabric, much like an ink blotter. If the beak is moistened and atmospheric conditions are such as to permit the liquid to evaporate, the duck begins to sway on its perch, and finally tips forward to "drink." It then rights itself. So long as the beak is moist the drinking action continues. The fluid filling the duck is highly volatile (usually ethyl ether). Explain the principle of the duck's operation.

5.6.5 The air in a person's lungs is saturated with water vapor. What is the water vapor pressure in the lungs? (This problem will require the student to consult tables not contained in this text.)

•••••• ••••••

5.7 MAXWELL'S DEMON

The labor saving perpetual motion machines conjured up by wishful thinking in Section 5.2 revealed the monumental consequences of any scheme which could violate the second law. A great many clever people have sought to circumvent the second law. Many of these people were crackpots with at most a very superficial understanding of physics. A most notable exception was the scheme proposed by James Clerk Maxwell, the brilliant physicist who synthesized the theory of electromagnetism. Maxwell invented a "demon." Rightfully, it is known as "Maxwell's demon." The demon would violate the second law by making heat flow from a cold body to a hot body. Maxwell's proposal was not idle speculation. He was not trying to set up a straw man (demon!) for others to joust with. Maxwell's demon relied on the kinetic theory of gases. At the time, kinetic theory was in its infancy, but was developing fruitfully in the hands of Clausius, Maxwell, and Boltzmann. Kinetic theory operates at the microscopic level. The second law seemed firmly established at the macroscopic level, but what about the underlying microcosm? Perhaps a new wealth of science lurked deep within the breadbox-sized objects with which we deal from day to day. Perhaps it would be possible to violate the second law by judicious engineering at the microscopic level. Toward the end of his famous textbook Maxwell therefore posed the question as to whether or not it might be possible to sort the molecules of a gas in such a way as to violate the second law. Maxwell says, "But if we conceive a being whose faculties are so sharpened that he can follow every molecule in its course, ...,"† and so the demon was born. We now describe Maxwell's demon and explain how its *modus operandi* supposedly violates the second law.

Two results of the kinetic theory of gases must be inserted here without proof. Their justification comes later, in Chapters 11 and 12. First, there is a *distribution* of molecular speeds in a gas, that is, not all molecules move at the same speed. Some travel at very low speeds, some at very high speeds; the majority travel at speeds comparable with the average molecular speed. Second, we need to know that the temperature of a gas is increased by any mechanism which increases the average speed of the molecules—the higher the average speed, the hotter the gas. Let us accept these two kinetic theory results and put the demon to work.

When the demon reports for work he finds a situation such as that shown in Figure 5.6. Two heat reservoirs are indirectly connected by a closed gas-filled system. Thermal contact maintains each end of the gas system at the temperature of the adjacent reservoir. Thus, the gas at one end is hot, that at the other end is cold. The demon operates a partition which may be opened and closed. Because his faculties are "so sharpened" the demon can observe each molecule as it approaches the partition and determine its speed.

The demon sets out to pump heat from the cold reservoir to the hot reservoir. He does this as follows: The demon opens the partition whenever letting the molecule pass will raise the average molecular speed of the hot gas and lower the average molecular speed of the cold gas. There are two ways of doing this: (a) Let fast molecules excape from the cold gas, thereby cooling it and heating the hot gas. (b) Let slow molecules escape from the hot side, thereby increasing the average speed of those remaining, and thus also, raising the temperature. The demon in effect sorts and trades molecules. Equal numbers of molecules are passed in both directions so that the masses of the hot and

† J. C. Maxwell, "Theory of Heat," p. 338. Longmans, Green, New York (1894).

Figure 5.6. The demon controls the flow of molecules between the hot and cold reservoirs.

cold gases are not changed. Further, the volume containing the gases does not change so that no work is done by the gas. Now for the point! The action of the demon would lower the temperature of the cold gas were it not connected to a cold reservoir. The reservoir supplies whatever heat is required to maintain the temperature of the cold gas. Similarly, the hot gas rejects heat thereby maintaining its temperature at a constant value. The heat rejected by the hot gas might be used to power a heat engine.

The net effect of the demon's actions are to transfer heat from a cold reservoir to a hot reservoir. The gases on either side of the barrier are unchanged. Thus, we have a device—cyclic in nature—whose *only* effect is to transfer heat from a low temperature to a higher temperature. Such a demon could fry eggs with ice cubes, heat a home by extracting heat energy from the atmosphere, and run all the machinery, power plants, and so forth, on Earth by extracting thermal energy from the oceans or the Earth itself. Unhappily, there is one minor problem—it will not work! Why not? Because we forgot the demon. We did not include the demon as part of the system. The demon cannot be regarded as part of the surroundings since he must communicate with the molecules in order to measure their speeds. The measurements made by the demon (or any instrumentalized version of the demon) require that he exchange energy with the molecules. Some of the energy being pumped uphill must be fed to the demon—he does not work free. Once the system is enlarged to include the demon one discovers that the heat pump has sprung a leak—a large leak. The demon is so hungry that he eats himself out of a job. His appetite completely wipes out the potential gain of his molecule-sorting action. To demonstrate this quantitatively is not easy. Basic to any detailed analysis of Maxwell's demon is the concept of entropy.† Entropy is a thermodynamic state function, a property of a system. A quantitative development of thermodynamics—not just the analysis of Maxwell's demon—necessitates the introduction of entropy. In the next chapter we study this essential thermodynamic state function from a macroscopic viewpoint. A microscopic view of entropy is forthcoming in Chapter 15.

REFERENCES

Carnot's paper is translated and interpreted in:

E. Mendoza (ed.), "Reflections on the Motive Power of Fire and Other Papers on the Second Law of Thermodynamics." Dover, New York (1960).

In particular, Mendoza notes in the Introduction that French physicists interpreted the caloric theory as meaning (a) heat is conserved and (b) heat and work are equivalent.

Attempts to circumvent the second law of thermodynamics by means of fluctuations which drive unilateral devices have been advanced from time to time. One of these devices—the ratchet and pawl—is dissected and demolished by:

R. P. Feynman, R. B. Leighton, and M. Sands, "The Feynman Lectures on Physics," Volume 1, Lecture 46. Addison-Wesley, Reading, Massachusetts (1963).

For details of the Stirling engine see:

M. W. Zemansky, "Heat and Thermodynamics," 5th ed., Chapter 7. McGraw-Hill, New York (1968).

---

† Analyses of Maxwell's demon from an entropy viewpoint are listed in the References at the end of this chapter.

A versatile demonstration model of the Stirling engine (operable as engine or refrigerator) is available from LaPine Scientific Co., Chicago, Illinois.

A vivid picture of what a skillful demon could accomplish is presented in:

G. Gamow, "Mr. Tompkins in Paperback," Chapter 9. Cambridge Univ. Press, London and New York (1971).

Maxwell's demon is discussed at some length by:

W. Herenberg, Maxwell's Demon, *Sci. Am.* **217**, 103 (Nov. 1967).

A historical survey of Maxwell's demon is presented in:

L. Brillouin, "Science and Information Theory," 2nd ed., Chapter 13. Academic Press, New York (1962).

Chapter 6

Entropy

In this chapter we introduce *entropy*, a thermodynamic state function. The introduction of entropy is vital to the quantitative development of thermodynamics. In particular, we can state the second law of thermodynamics in a most succinct fashion in terms of entropy. In Sections 6.1 and 6.2 we concern ourselves with "discovering" entropy and with seeing how it aids the mathematical development of thermodynamics. The entropy formulation of the second law is presented in Section 6.3. The chapter concludes with a discussion of the role of the second law in arguments which predict a "heat death" for the universe.

6.1 ENTROPY AND THE CARNOT CYCLE

In the last chapter, we discussed the Carnot cycle, a series of reversible processes which involve heat transfer. Recall that a reversible process is one which may be reversed by an infinitesimal change in the surroundings. A system and its surroundings may exchange heat reversibly if their temperatures differ infinitesimally. To establish this fact we first observe that heat flows into or out of a system depending upon whether its temperature is less than or greater than the temperature of the surroundings. The heat exchange is reversible only so long as an infinitesimal change in the temperature of the

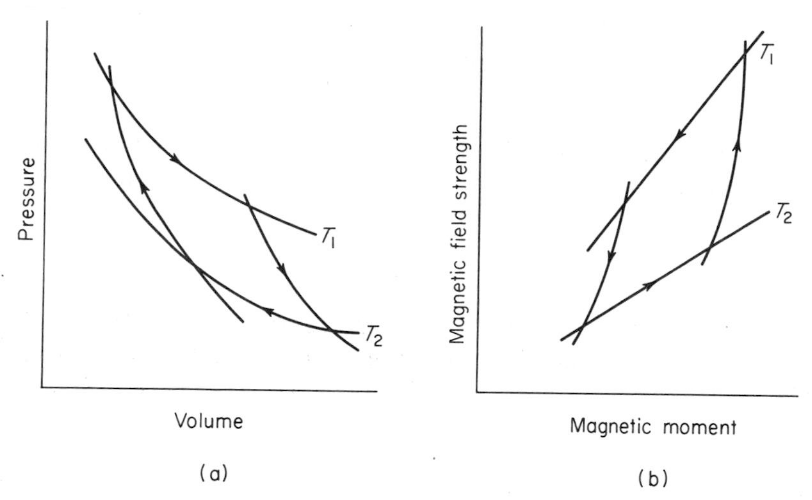

Figure 6.1. Carnot cycles for (a) an ideal gas and (b) for a paramagnetic solid. For the sequence of steps indicated by the arrows, both cycles result in the absorption of heat Q_1 by the system at a temperature T_1, and the rejection of heat Q_2 at a lower temperature T_2.

surroundings can reverse the flow of heat. However, this is possible only if the temperatures of the system and its surroundings differ infinitesimally. Figures 6.1a and b show the steps in Carnot cycles as executed by an ideal gas and a paramagnetic substance.

••••• •••••

6.1.1 Refer to Figure 6.1. Both Carnot cycles absorb heat at the higher temperature T_1 and reject heat at the lower temperature T_2. Both convert heat into work. Explain why the P–V and H–M cycles are traced in opposite senses.

•••••• ••••••

The quantities of heat absorbed and rejected along the isotherms of the Carnot cycle are related to the corresponding temperatures, namely,

$$\frac{Q_1}{T_1} = \frac{Q_2}{T_2} \tag{6.1}$$

In Chapter 5 we temporarily repealed our sign convention on heat; both Q_1 and Q_2 are positive quantities in (6.1). We now reinstate our earlier convention that heat absorbed by the system is positive and that heat rejected by the system is negative. In (6.1), Q_2 denotes heat rejected so that we should now write

$$\frac{Q_1}{T_1} = \frac{-Q_2}{T_2} \qquad (Q_2 < 0)$$

or

$$\frac{Q_1}{T_1} + \frac{Q_2}{T_2} = 0 \tag{6.2}$$

There are four legs to the Carnot cycle: two isotherms and two adiabats. Because there is no heat exchange along the adiabats we can rewrite (6.2) as

$$\sum_{\text{all legs}} \frac{Q_i}{T_i} = 0 \tag{6.2a}$$

Furthermore, since the cycle must be traversed slowly to guarantee reversibility we can envisage the finite heat exchanges as being built up as sequences of infinitesimal exchanges. The summation over the cycle is then replaceable by an integral, namely,

$$\oint \frac{dQ}{T} = 0 \tag{6.3}$$

Equation (6.3), which has been deduced for the Carnot cycle, is in fact valid for any reversible process. A general proof is unnecessarily involved. For our purposes it suffices to consider Figure 6.2, which suggests how the result of (6.3) is established for a fluid system for which P, V, and T serve as the thermodynamic variables. The arbitrary cyclic process is mentally dissected into a succession of Carnot cycles. Each segment of the cycle path forms one of the isothermal legs of a Carnot cycle. Traversing all of the Carnot cycles once is equivalent to traversing the actual cycle path once. This is because each interior adiabat forms a leg of two adjacent Carnot paths. All such legs are

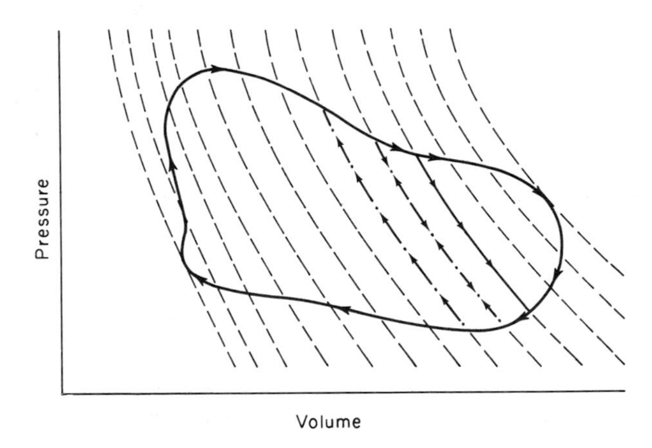

Figure 6.2. An arbitrary cyclic path may be chopped into a series of equivalent Carnot cycles to show that $\oint dQ/T = 0$ for a reversible process. The dashed lines represent the adiabats.

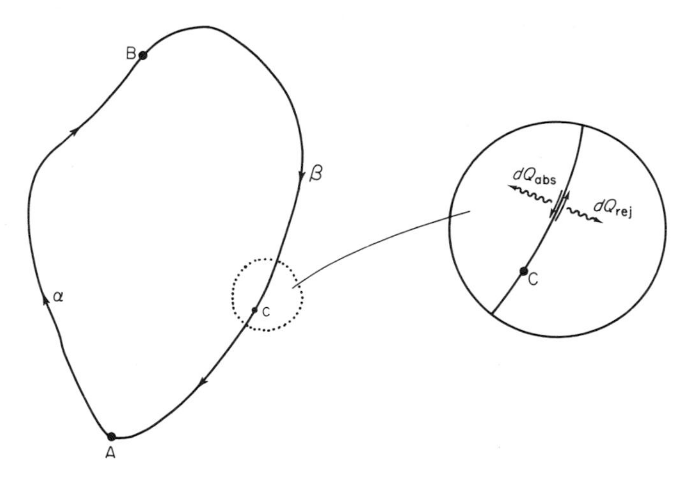

Figure 6.3. Symbolic representation of a reversible cyclic process, leading from A to B along path α and from B to A along path β. Heat absorbed (rejected) during each step would be rejected (absorbed) if the process were reversed.

traversed twice, in opposite directions, and in effect cancel out of the analysis. Since (6.3) applies to each Carnot cycle individually it is also valid for them collectively, and this in turn suggests that it holds for all reversible processes.

As a consequence of (6.3) we can infer the existence of a state function, called the entropy. To see how such an inference is made, consider a cyclic reversible process which carries a system through a pair of thermodynamic states labeled A and B. We can think of A and B as two points joined by a pair of paths (compare Figure 6.3). We can write the integral of (6.3) as a sum of two integrals. Thus

$$\oint \frac{dQ}{T} = \left(\int_A^B \frac{dQ}{T}\right)_\alpha + \left(\int_B^A \frac{dQ}{T}\right)_\beta$$

which may be reexpressed as

$$\left(\int_A^B \frac{dQ}{T}\right)_\alpha - \left(\int_B^A \frac{-dQ}{T}\right)_\beta \tag{6.4}$$

Now, because the path is reversible, we can write for the integral on the right side

$$\left(\int_B^A \frac{-dQ}{T}\right)_\beta = \left(\int_A^B \frac{dQ}{T}\right)_\beta \tag{6.5}$$

Physically, (6.5) follows because if heat dQ is absorbed along some segment as the system proceeds from A to B along β, an equal amount will be rejected along the same segment as the system moves from B to A. Mathematically

(6.5) may be thought of as a special case of (6.4) where α and β denote the same path. Comparing (6.4) and (6.5) shows that the integral of dQ/T between A and B is the same for both paths,

$$\left(\int_A^B \frac{dQ}{T}\right)_\alpha = \left(\int_A^B \frac{dQ}{T}\right)_\beta \tag{6.6}$$

The only restriction on α and β is that they be reversible paths connecting A and B. Equation (6.6) says that the value of the integral is independent of the reversible path from A to B. If one "walks" from A to B and adds up dQ/T along each step, then, no matter which reversible path one takes, the total is the same. We can therefore drop the subscripts which designate different paths. The value of the integral *does* depend on the states A and B and we can write

$$\int_A^B \frac{dQ}{T} = \int_A^B dS = S(B) - S(A) \tag{6.7}$$

where

$$dS = \frac{dQ}{T} \tag{6.8}$$

is the *entropy change* resulting from the reversible exchange of heat dQ with a source of temperature T. The entropy S is a thermodynamic state function—a property of the system—implicitly defined for all equilibrium states. From the relation $dS = dQ/T$ we see that the absorption of heat ($dQ > 0$) produces an entropy increase ($dS > 0$). It is important to remember that (6.8) holds only for a reversible process.

••••••

6.1.2 The heat capacity $C(T)$ for a system is given by

$$C(T) = AT^3$$

where A is a constant. If the entropy of the system is taken to be zero at $T = 0$, integrate (6.8) to show that the entropy at temperature T is

$$S(T) = \tfrac{1}{3}AT^3 = \tfrac{1}{3}C(T)$$

••••••

The units of entropy are those of energy divided by temperature; for example, calories per kelvin. The abbreviation eu stands for entropy unit, where one entropy unit is defined as one calorie per kelvin,

$$1 \text{ eu} \equiv 1 \text{ cal/K}$$

For reversible processes we can use (6.8) to express dQ as

$$dQ = T\,dS \tag{6.8a}$$

This is a noteworthy result. It expresses the heat absorbed solely in terms of the thermodynamic coordinates of the system. No reference is made to the surroundings. A similar situation arises in connection with the work dW performed by a system. Only for reversible processes is it possible to express dW in terms of

$$dW = P\,dV \tag{4.4}$$

or

$$dW = -H\,dM \tag{4.15}$$

Using (6.8a) and (4.4), the first law becomes

$$T\,dS = dU + P\,dV \tag{6.9}$$

a form suited only to a description of reversible processes.

Entropy is an extremely important physical property. It is probably also unfamiliar. Our immediate experience does not seem to contain the entropy concept. In fact, however, entropy is known to us all, but under other names. Entropy, it turns out, is the scientific alias for the familiar idea of *disorder*. The next section is devoted to exposing entropy as disorder under another name, and to presenting examples of entropy computations.

6.2 ENTROPY—EXAMPLES AND INSIGHTS

The concept of entropy will remain vague until we have an answer to the question, "What does entropy measure?" We shall emphasize the concept of entropy which has evolved from the microscopic domains of kinetic theory and statistical mechanics. Ludwig Boltzmann was the first to emphasize that the entropy of a system is a measure of its randomness or disorder. Disordering at the molecular level registers macroscopically as an entropy increase.

In this section we present examples which demonstrate how changes in entropy are computed. The examples selected illustrate the idea that entropy measures disorder.

Example 6.1 Entropy, Phase Changes, and Disorder

Consider a 1-gm ice cube at its melting point. If we add the latent heat of fusion, about 80 cal,[†] the ice cube melts. This change of phase takes place at

---

[†] The various latent heats vary with pressure. The numerical values quoted refer to a pressure of 1 atm.

constant temperature. The process is reversible so that we may integrate

$$dS = \frac{dQ}{T} \tag{6.8}$$

to obtain

$$\Delta S_{sl} = \frac{L_f}{T} \tag{6.10}$$

where

$$\Delta S_{sl} \equiv S_{\text{liquid}} - S_{\text{solid}} \tag{6.11}$$

is the entropy change, and L_f is the latent heat of fusion, the heat required to melt 1 gm of the solid. In the case at hand, the latent heat of fusion is approximately 80 cal and the melting point is roughly 273 K, whence,

$$\Delta S_{sl} \simeq \frac{80}{273} \; \frac{\text{cal}}{\text{K}} = 0.29 \text{ eu}$$

First we observe that ΔS_{sl} is positive—melting results in an entropy increase. Secondly we point out that the molecular structure of the final state (liquid) is highly randomized by comparison with the initial state (solid). In a crystalline solid the molecules form an ordered periodic structure. In a liquid only a short-range order is evident. By short-range order, we mean that any given molecule is only able to influence the behavior of nearby molecules. Melting results in an entropy increase and molecular disordering. Entropy measures disorder.

The computed value, $\Delta S_{sl} = 0.29$ eu, standing alone, is rather barren. To provide a basis for comparison we next compute entropy changes for two other processes involving the gram of water; (a) raising its temperature isobarically to the boiling point and (b) vaporizing it.

If heat is added to the liquid at constant pressure, its temperature rises until it reaches the boiling point. The heat added in changing the temperature by an amount dT is

$$dQ = C_p \, dT \tag{6.12}$$

where C_p is the specific heat at constant pressure. The corresponding entropy change is

$$dS = \frac{C_p \, dT}{T} \tag{6.13}$$

For many liquids, and water in particular, C_p is nearly constant between the melting and boiling points. If C_p is treated as constant, the total entropy change between the melting and boiling points is

$$\Delta S_{\text{mb}} = C_p \int_{T_{\text{m}}}^{T_{\text{b}}} \frac{dT}{T} = C_p \ln\left(\frac{T_{\text{b}}}{T_{\text{m}}}\right) \tag{6.14}$$

For 1 gm of water, $C_p \simeq 1$ cal/K and $\ln(T_{\text{b}}/T_{\text{m}}) \simeq 0.31$. These data give

$$\Delta S_{\text{mb}} \simeq 0.31 \text{ eu}$$

This is only slightly larger than the entropy change of melting. Raising the temperature of the water increases the molecular disorder. This additional randomization is due largely to the fact that the average molecular speed increases with temperature. As the liquid is heated, molecular agitation grows. The chance that a given molecule influences the behavior of a nearby molecule decreases as their relative speed increases—they pass too swiftly to communicate effectively. Thus, as the temperature increases, molecular correlations diminish and disorder rises.

Having raised the water to its boiling point, it can be vaporized by adding still more heat. The entropy change in going from the liquid to the vapor phase is readily found by the same arguments leading to (6.11). The result is

$$S_{\text{vapor}} - S_{\text{liquid}} \equiv \Delta S_{l\text{v}} = \frac{L_{\text{v}}}{T} \tag{6.15}$$

where L_{v} is the latent heat of vaporization and T is the boiling point. For 1 gm of water at 1 atm, $L_{\text{v}} = 539$ cal, $T = 373$ K, and $\Delta S_{l\text{v}} \simeq 1.44$ eu. The entropy change for the liquid–vapor transition is roughly five times that of the two previous processes. What disruption at the molecular level can be identified as the cause of this entropy increase? What sort of disordering has occurred? A partial answer is to be found by noting that at the normal boiling point, 1 gm of liquid water occupies a volume of about 1 cm^3. At the same pressure and temperature the vapor occupies a volume of approximately 1700 cm^3. The same collection of molecules which started as a liquid is enormously spread out as a vapor. The molecules are greatly disorganized by virtue of having a larger volume in which to roam. It is somewhat like the disorder which would result if the door to a chicken coop were left open, permitting the chickens freedom to move about in a barnyard area 1700 times as large as that of the coop.

The entropy change in the liquid–vapor transition is not wholly accounted for by the increase in volume. There is also a contribution stemming from the breaking of molecular bonds. A molecule which escapes from the liquid to take its place in the vapor leaves behind a less ordered structure. There is a "hole" in the liquid. Such defects represent increased disorder and contribute to the total entropy change. ◇

......

6.2.1 At a pressure of 1 atm, liquid helium boils at 4.2 K. The latent heat of vaporization is 4.9 cal/gm. Determine the entropy change (per gram) resulting from vaporization. Express your result in entropy units per gram.

6.2.2 The specific heat of copper (at constant pressure) is 0.092 cal/gm-°C.

(a) If the temperature of 20 gm of copper is raised from 0 °C to 20 °C, determine the entropy change of copper. Express your results in entropy units.

(b) How much heat is absorbed by the copper in the process?

(c) If all of the heat computed in part (b) were absorbed by the copper at the average temperature (10 °C), what would be the entropy change?

6.2.3 The latent heat of fusion of lead (Pb) is 1.12 kcal/mole. The normal melting point is 327.5 °C. Determine the entropy change of 100 gm of lead when it melts at a pressure of 1 atm.

......

We have stated that S is a thermodynamic state function. The definition $dS = dQ/T$ does not tell us how S depends on the thermodynamic variables T, P, V, and so on. The functional form of S can be determined using the first law once we know (a) the equation of state and (b) the functional form of the internal energy. Example 6.2 demonstrates how the first law may be used to establish the form of S. Problem 6.2.4 gives the student an opportunity to try his hand at determining the entropy of a photon gas.

Example 6.2 Determination of the Entropy Using the First Law

The internal energy of a hypothetical substance is given by

$$U = aVT^2, \qquad a = \text{constant}$$

and its equation of state is

$$PV = U$$

or

$$P = aT^2$$

We wish to determine the entropy of the system as a function of T and V. Starting with the relation

$$dS = \frac{dQ}{T}$$

we use the given information and the first law to express dQ in terms of dT and dV. This expression for dQ leads to a form for dQ/T which then must be recognized as the differential of some explicit function of T and V. From the equation of state we find

$$P\,dV = aT^2\,dV$$

Forming the differential of the expression for U gives

$$dU = aT^2\,dV + 2aVT\,dT$$

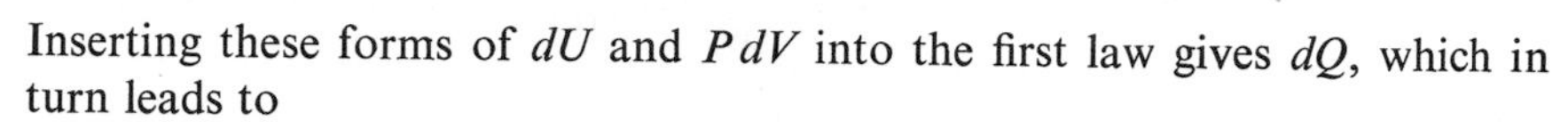

Inserting these forms of dU and $P\,dV$ into the first law gives dQ, which in turn leads to

$$dS = \frac{dQ}{T} = 2aT\,dV + 2aV\,dT$$

The right-hand side must be recognized as the differential of $2aTV$

$$dS = d(2aTV)$$

The functional form of S is therefore

$$S(T, V) = 2aTV + S_0$$

where S_0 is a constant of integration.

In the next section we use this technique to determine the entropy of an ideal gas. ◇

•••••• ••••••

6.2.4 The internal energy of a photon gas is given by

$$U = bVT^4$$

where b is a constant. The equation of state is

$$PV = \tfrac{1}{3}U$$

Derive an expression for the entropy of the photon gas. Choose the constant of integration to be such that the entropy is zero at $T = 0$.

•••••• ••••••

6.3 THE IDEAL GAS, FREE EXPANSION, AND GIBBS' PARADOX

An expression for the entropy of an ideal gas is readily derived and proves useful in many applications. Consider 1 mole of an ideal gas. Using the first law we can write

$$dS = \frac{dQ}{T} = \frac{dU + P\,dV}{T}$$

For an ideal gas,

$$dU = C_v\,dT$$

and

$$PV = RT$$

The heat capacity C_v is a constant. The expression for dS converts to

$$dS = C_v \frac{dT}{T} + R \frac{dV}{V}$$

This relation may be integrated to give

$$S = C_v \ln T + R \ln V + S_0 \tag{6.16}$$

where S_0 is an integration constant. The value of S_0 is of no interest so long as we concern ourselves only with entropy changes. For example, in a process which carries 1 mole of an ideal gas from (T_1, V_1) to (T_2, V_2), the entropy change is

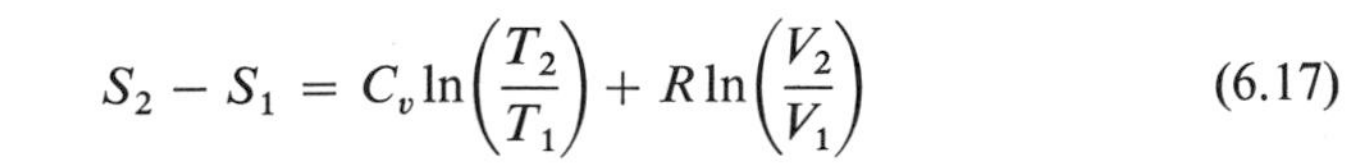

$$S_2 - S_1 = C_v \ln\left(\frac{T_2}{T_1}\right) + R \ln\left(\frac{V_2}{V_1}\right) \tag{6.17}$$

The unknown constant S_0 has dropped out.

••••••

6.3.1 Equation (6.16) gives the entropy for 1 mole of an ideal gas. Carry through the corresponding analysis to show that the entropy of n moles is simply n times that of 1 mole, that is, verify that entropy is an extensive quantity.

••••••

There are two rather instructive applications of (6.17). These are the free expansion of a gas and Gibbs' paradox. Consider the situation schematicized in Figure 6.4. A vessel with rigid, insulated walls is divided by a partition.

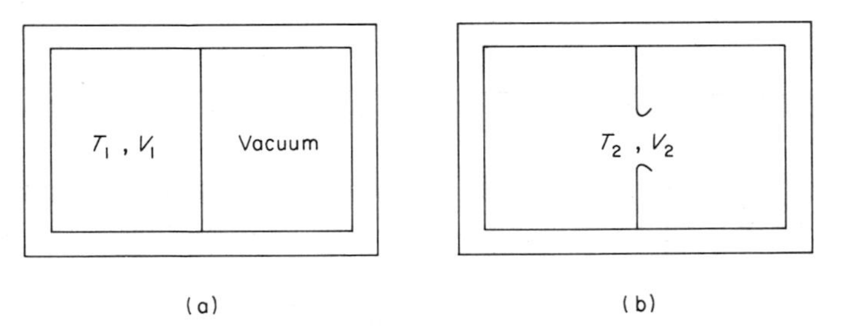

Figure 6.4. Initially, (a) one portion of the rigid, insulated container is a vacuum while an ideal gas occupies the other section. After the partition ruptures, (b) the gas expands to fill the container. The insulation and rigid walls guarantee that the system is isolated.

Initially, one side contains an ideal gas while the other side is a vacuum. Consider what happens when the partition ruptures, allowing the gas to expand and fill the entire container. The insulation prevents heat exchange. Thus, $dQ = 0$ for each stage of the expansion. The rigid walls guarantee that no work is done on the system, that is, there can be no change in the volume of the system and thus no $P\,dV$ work by the surroundings. The process is called a free expansion because of the absence of any external work as the gas rushes (or oozes) into the evacuated region. We therefore have the results that $dQ = 0$ and $dW = 0$ at each stage of the expansion. It follows from the first law that $dU = 0$; the internal energy of the gas does not change. However, for an ideal gas $U = C_v T$, the internal energy depends on T alone. If U remains constant, so must the temperature; thus $T_2 = T_1$. The final volume of the gas V_2 is clearly greater than the initial volume V_1, so that $\ln(V_2/V_1)$ is a positive number. From (6.17) we then have the result that

$$S_2 - S_1 = R \ln\left(\frac{V_2}{V_1}\right) > 0$$

that is, the final entropy exceeds the initial entropy. The gas has expanded but its temperature has not changed; the gas spreads out over a larger volume, resulting in further disorder. The predicted entropy increase concurs with the notion that entropy measures disorder. Very straightforward—or is it? Let us retreat momentarily to the point where we prescribed that the container was insulated. We noted that the insulation guaranteed that $dQ = 0$ at each stage of the expansion. From

$$dS = \frac{dQ}{T} \tag{6.8}$$

it might seem that we should conclude that $dS = 0$ at each stage, and therefore that the entropy of the gas remains constant. Not so! The entropy does change. The expression $dS = dQ/T$ does not apply to processes such as the free expansion. The free expansion is an irreversible process. It cannot be reversed by any infinitesimal change in the surroundings. As we warned earlier the relation $dS = dQ/T$ holds only for reversible transformations.

Students are apt to encounter two types of paradoxes. On the one hand, there are the pseudo-paradoxes. These are pedagogically designed and motivated. By resolving them the student gains insight and understanding—he's educated! The various perpetual motion machines suggested earlier are examples of such paradoxes. They are hoaxes; understanding precisely why they do not work is instructive.

The second type of paradox can lay some claim to legitimacy. At some point in history it was a genuine paradox—a facet of nature which could not be explained. Historically, the resolution of such paradoxes have brought about modifications of existing physical laws. Gibbs' paradox is a case in point. Gibbs' paradox is no longer paradoxical. It can be understood—explained—via quantum statistical mechanics. Consider again the insulated container of Figure 6.4. In the initial equilibrium state, one section contains ideal gas A, the other section contains ideal gas B. For simplicity we suppose that the gases have equal temperatures and that they occupy equal volumes. The initial entropy of the system is the sum of the entropies of A and B.

$$S_\mathrm{i} = C_{v\mathrm{A}} \ln T + C_{v\mathrm{B}} \ln T + 2R \ln V + S_{0\mathrm{A}} + S_{0\mathrm{B}}$$

The partition separating the two gases is now removed, permitting the two gases to mix. In the final equilibrium state the now-mixed gases are at temperature T (the temperature does not change) and occupy a volume $2V$, the total container volume. The final entropy differs from the initial entropy only in that both A and B occupy a volume $2V$ rather than V. Thus

$$S_\mathrm{f} = C_{v\mathrm{A}} \ln T + C_{v\mathrm{B}} \ln T + 2R \ln 2V + S_{0\mathrm{A}} + S_{0\mathrm{B}}$$

The entropy change

$$S_\mathrm{f} - S_\mathrm{i} = 2R \ln 2V - 2R \ln V = 2R \ln 2 \tag{6.18}$$

is positive. The entropy increases because the gases have mixed; mixing results in greater disorder. In fact, the entropy increase computed above is referred to as the entropy of mixing.

So far, there is no paradox! The result is in accord with the idea that entropy measures disorder. However, Josiah Willard Gibbs made the following observation, from which springs the paradox: Suppose that gases A and B are *identical*. The initial and final states of the system are then identical. There is no operational way of detecting any mixing, for none has occurred. If this point is not clear, mentally replace the atoms with marbles. Suppose first that the marbles are distinguishable by their color. If the A marbles are white and the B marbles are black, any mixing is readily detected. However, if the marbles are all black, no mixing can be detected even though marbles change their positions. Regardless of their physical nature, so long as the A and B objects are identical no experimental test can distinguish between the initial and final states.

Now for the paradox! When A and B are identical there is no mixing; the initial and final states are identical. Hence there is no entropy change. The paradox, first pointed out by Gibbs, is that the entropy change predicted by (6.18) is the same ($2R \ln 2$) even when A and B are identical gases. The very straightforward calculation performed earlier predicts an increase in entropy even when the gases are identical—when in fact there is no entropy change. Something is wrong! There were no mathematical errors. We have a genuine paradox. A correct result is obtained so long as A and B are different. It is the physics which has gone wrong. The expression for the entropy of an ideal gas must be modified in some way to account for the discrepancy. The modification should be such that the entropy of mixing becomes zero when the gases are identical.

The correct expression for the entropy was deduced by Gibbs. However, his was an *ad hoc* recipe. He saw how to obtain a satisfactory answer but could not justify his procedure. Little wonder! The fundamental basis for the corrected entropy formula lies within modern quantum theory, a discipline that was to develop 30 years after Gibbs' insight led him to the correct expression. Problem 6.3.4 presents the corrected entropy formula and an invitation to verify its superiority.

•••••   •••••

6.3.2 Explain why no temperature change occurs during the expansion and mixing described in connection with Gibbs' paradox.

6.3.3 Two vessels, both containing 1 mole of an ideal gas, are brought into thermal contact. The initial temperatures of the gases are 300 and 450 K. The gases are identical, and occupy equal volumes which remain constant.

(a) What is their final common temperature?
(b) Compute the entropy change.
(c) Optional: Consider the more general case where the initial temperatures are T_1 and T_2 and prove that the entropy always increases provided $T_1 \neq T_2$. [*Hint:* The arithmetic mean of two unequal numbers is greater than the geometric mean $\frac{1}{2}(x+y) > \sqrt{xy}$.]

6.3.4 Gibbs observed that his paradox was resolved if the entropy for n moles of an ideal gas is written as

$$S_{\text{cor}} = nC_v \ln T + nR \ln\left(\frac{V}{n}\right) + nS_0$$

rather than as

$$S = nC_v \ln T + nR \ln V + nS_0$$

The two expressions differ in that S_{cor} (corrected) has V/n, the volume per mole, in place of V. When A and B are identical the volume per mole of gas remains unchanged in the mixing experiment described earlier. Verify that (6.18) again results when A and B are different (using S_{cor}), but that there is no entropy change when A and B are identical. Take $n_A = n_B = 1$ mole for convenience.

•••••   •••••

6.4 *T–S* DIAGRAMS

We have seen how a *P–V* diagram may be used to interpret the work done by a fluid system. A temperature–entropy (*T–S*) diagram serves a similar purpose for the heat absorbed or rejected in a process. Suppose a system undergoes a reversible transformation during which the temperature and entropy vary as shown in Figure 6.5. From the figure we see that the heat

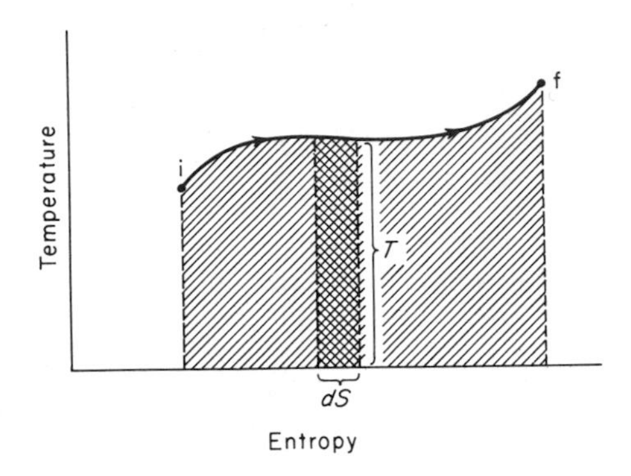

Figure 6.5. A reversible transformation takes the system from i to f. The solid line indicates the relation between T and S during the process. The cross-hatched area is $T\,dS$. The total hatched area beneath the solid line corresponds to the heat absorbed by the system during the process.

absorbed by the system can be given a geometrical interpretation. For an entropy change dS, the heat absorbed is

$$dQ = T\,dS \tag{6.8a}$$

As the figure shows, $T\,dS$ corresponds to the cross-hatched area of the strip of height T and width dS. The hatched area in Figure 6.5 corresponds to the total heat absorbed during the process

$$Q = \int_{\mathrm{i}}^{\mathrm{f}} T\,dS$$

Figure 6.6a shows the *T–S* diagram for a Carnot cycle. The two adiabatic legs of a Carnot cycle are also isentropic because the process is reversible.† Figure 6.6b shows the corresponding steps on a *P–V* diagram.‡ During the

† Recall that $dS = dQ/T$ applies to reversible processes. If $dQ = 0$ (adiabatic), then $dS = 0$ as well (isentropic).

‡ The *P–V* diagram shows a Carnot cycle for an ideal gas. The *T–S* plot is universal in that its shape is the same regardless of the working substance undergoing the Carnot cycle.

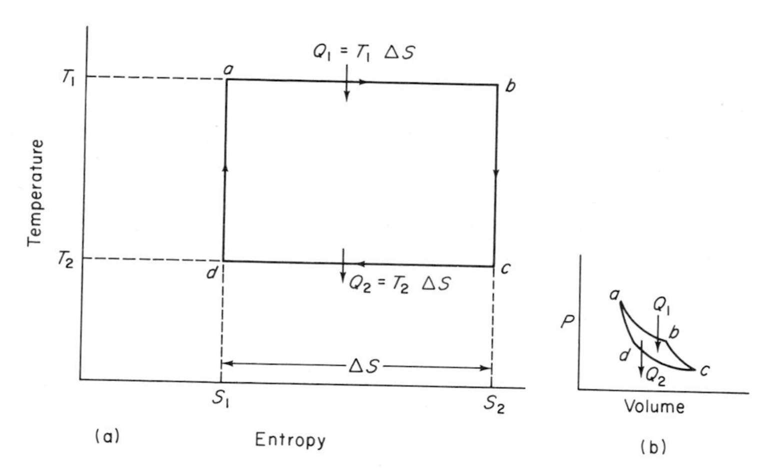

Figure 6.6. (a) The T–S diagram for the Carnot cycle. (b) The corresponding P–V diagram for an ideal gas.

isothermal transition from a to b an amount of heat

$$Q_1 = T_1 \Delta S$$

is absorbed by the system. Geometrically, $T_1 \Delta S$ is the area of the rectangle of height T_1 and width ΔS. Along the isotherm from c to d the system rejects heat of magnitude

$$Q_2 = T_2 \Delta S$$

which also corresponds to a rectangular area. The net amount of heat absorbed by the system is $Q_1 - Q_2$. From Figure 6.6 we see that this corresponds to the rectangular area generated by the cyclic path on the T–S diagram. Observe also that the cycle path is traced clockwise and that the system absorbs a net amount of heat. If the Carnot cycle were reversed, the enclosed area would then correspond to the net heat rejected to the surroundings.

It follows from the first law that $W = Q_1 - Q_2$ is the work done by the system during the cycle, that is, the enclosed area represents the work done by the system. We have used the Carnot cycle because it is a familiar example. The foregoing statements remain valid for all reversible processes.

In the case of the Carnot cycle, the T–S diagram leads to an instant derivation of its thermodynamic efficiency. From its definition,

$$\eta = \frac{\text{work performed}}{\text{heat absorbed}} = \frac{W}{Q_1}$$

From the T–S diagram,

$$W = Q_1 - Q_2 = T_1 \Delta S - T_2 \Delta S$$

Inserting this gives the desired result

$$\eta = \frac{(T_1 - T_2)\,\Delta S}{T_1\,\Delta S} = 1 - \frac{T_2}{T_1}$$

••••• •••••

6.4.1 Show that the thermodynamic efficiency of the circular cycle shown in the accompanying figure is

$$\eta = \frac{100\pi}{400 + 50\pi}$$

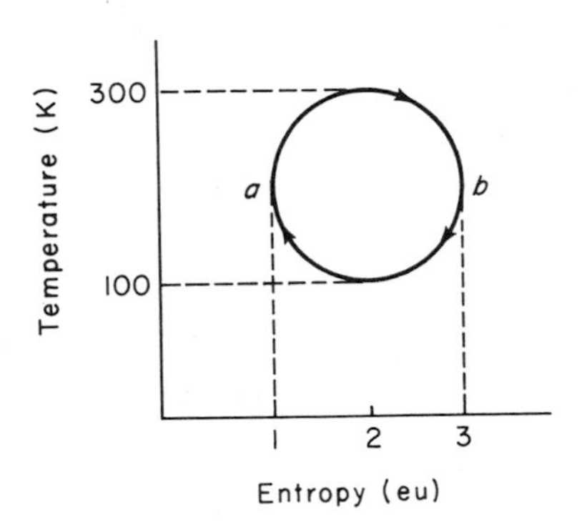

••••• •••••

Before moving on to an entropy formulation of the second law of thermodynamics, two remarks are appropriate. First, our insistence that entropy measures disorder has, at best, been made plausible. Little more could be expected, since we have not attempted to define disorder. A quantitative approach to the disorder–entropy relation is possible only at the microscopic level. In so far as the formal manipulations and applications of thermodynamics are concerned, we may regard entropy as just another useful physical property. Secondly, we have quietly, and purposely, ignored the question of how entropy is determined experimentally. This is related to another point swept under the rug—the entropy constant which appeared in (6.16). In principle the measurement of entropy change is quite straightforward. By adding or extracting heat dQ reversibly, the entropy change follows from

$$dS = \frac{dQ}{T}$$

Until a value of entropy is assigned to some standard state we cannot tabulate anything beyond entropy changes. Why not just pick some convenient state and assign a value of (say) zero to the entropy of a compound in that state?

The answer is that nature will not let us pick just any state. Experiment shows that the entropy of all systems approach a common value as the temperature tends toward absolute zero. This is the gist of the third law of thermodynamics. (We shall state it more precisely later.) The practical problem of tabulating entropies therefore involves the third law, and the behavior of substances at temperatures approaching absolute zero. We will meet these problems head-on in the next chapter.

6.5 ENTROPY FORMULATION OF THE SECOND LAW OF THERMODYNAMICS

In Section 6.3 we studied two processes involving isolated systems; the free expansion of a gas and the mixing of two gases. Both resulted in an entropy increase, and both are irreversible processes. These two processes exemplify an alternate statement of the second law of thermodynamics, advanced by Rudolph Clausius:

The entropy of an isolated system never decreases.

This statement is generally referred to as the *entropy principle*. Stated mathematically this version of the second law reads,

$$\Delta S_{\text{I.S.}} \geqslant 0 \tag{6.19}$$

where ΔS denotes the change in entropy brought about by any thermodynamic processes which connect equilibrium states and I.S. is isolated system. If an isolated system undergoes a reversible process, $dS = dQ/T = 0/T = 0$; there is no change in its entropy. This statement is contained in the equality portion of (6.19). If an isolated system is not in equilibrium, irreversible processes operate to drive it toward equilibrium. Irreversible processes generate entropy. They mix, disorder, randomize, and so on. The inequality in (6.19) refers to nonequilibrium situations in which irreversible processes generate entropy. From these remarks it follows that equilibrium corresponds to a state of maximum entropy—the entropy increases until the system attains equilibrium, after which the entropy remains constant.

Care must be exercised in applying the entropy principle. An increase in entropy is claimed only for an isolated system. A healthy 10-year-old boy converts mashed potatoes into muscle; disorder into order, but the 10-year-old is not an isolated system. If any living system is isolated, its entropy increases. Disorder mounts; the system withers and dies.

On the grounds that thermodynamics can be fun if it is not made laborious we shall not present a mathematical proof of the entropy principle. The interested student should consult the references. Instead we turn our attention to a question which has intrigued scientists for over a century. Does the second law of thermodynamics apply to the largest isolated system—the universe?

6.6 ENTROPY AND UNIVERSAL ACCIDENTS

The entropy principle, as formulated by Rudolph Clausius, states that the entropy of an isolated system never decreases. Entropy increases until it reaches a maximum value. The irreversible processes responsible for the increase of entropy are the same mechanisms which drive the system toward equilibrium. Thus, thermodynamic equilibrium corresponds to a state of maximum entropy.

The entropy principle is sometimes advanced in arguments for a "heat death" of the universe. The irreversible processes of nature result in an increase of entropy in the universe. The ultimate state of maximum entropy is one of thermodynamic equilibrium—a universe of uniform temperature.

However, the myriad of irreversible processes which perform so ably in establishing equilibrium in subcosmic chunks of the universe seem to lose their punch when applied to that part of the universe visible to us. The astronomers, astrophysicists, cosmologists, and anyone with one good eye can give evidence that the portion of the universe in our immediate neighborhood is not in equilibrium, nor are there any signs that it is tending toward equilibrium. This situation is a sharp contrast to those we normally observe, in which an isolated system rather quickly reaches equilibrium. The universe is certainly isolated—why does it not tend toward equilibrium?

One suggested resolution of this paradox is known as the fluctuation hypothesis. The fluctuation hypothesis was suggested by Ludwig Boltzmann, one of the fathers of statistical mechanics. All systems in thermodynamic equilibrium exhibit fluctuations. A reasonably strong analogy exists between fluctuations and auto accidents. Imagine a busy highway, with the traffic moving in a regular fashion. Occasionally two cars collide. Accidents involving three or more cars are less likely. Accidents involving many cars are very unusual—that is why newspapers report such incidents.

A fluctuation in a thermodynamic system is the result of an accident on the atomic level involving many atoms. In the atmosphere a density fluctuation occurs when many more (or many less) than the average number of molecules happen to populate some particular volume. A temperature fluctuation arises when a group of very fast (or very slow) molecules occupy some particular volume. Such fluctuations occur by chance—they are accidents of nature. For this reason, fluctuations within small volumes, which involve relatively few particles, are more likely to occur than fluctuations over larger volumes, which necessarily involve many more particles. Big fluctuations are like big accidents—both are very unlikely.

Boltzmann's fluctuation hypothesis suggests that the universe is in equilibrium but that the portion which we observe is part of a gigantic fluctuation—the granddaddy of all accidents. At first sight, the argument for such a hypothesis seems to be a strong one. Some sort of fluctuation is required to ensure the existence of observers (you and me!), that is, biological development requires special conditions—conditions of a distinctly nonequilibrium nature. Thus the very fact that such biological development has occurred—that I write and you read—seems to be strong evidence for the fluctuation hypothesis. However, the argument is unsound. It is enormously more likely that such a fluctuation would occur over a small volume, say the size of our solar system, and leave the rest of the immediate universe in equilibrium. To pursue the traffic analogy, being involved in an accident is not unusual, but we would generally be able to see beyond the wreckage and discern the equilibrium flow of traffic. The chance that the fluctuation hypothesis is true is less than the likelihood of an accident involving every car on the road today. Such states of true chaos seem most unlikely.

Boltzmann's fluctuation hypothesis is untenable. How then can we explain the fact that we live in a restless universe? One possible answer is suggested by the theory of gravitation. This theory, better known as general relativity, deals with systems in variable gravitational fields. On a cosmic scale we can no longer assume a spatially uniform gravitational field. Further, we must allow for the time variation of gravitational effects. Our day to day observations of systems which tend toward equilibrium are restricted to situations in which the surroundings do not vary with time, or vary at a rate which is slow by comparison with the speed at which the system approaches equilibrium. The general theory of relativity teaches us that the geometry of space is determined by the masses within the space. As the distribution of mass in our universe varies so do the geometrical properties of space. The situation is roughly analogous to a gas in a container whose surface wiggles, twists, and dilates. Such a system would not approach equilibrium because the external conditions are changing. The variable nature of the gravitational forces within our universe makes it behave like a container whose volume wiggles, twists, and dilates. In a sense, our universe is not an isolated system.

You may ask, "What sort of thermodynamic history is predicted for our universe by the general theory of relativity?" The answer depends on what sort of *model* you pick for our universe. It also depends on the answer to a question astronomers are now trying to answer, namely, what is the average mass density of the universe? The present (estimated) value for the average mass density is 7×10^{-31} gm/cm^3. The observed expansion of the universe will continue indefinitely unless the density is at least 2×10^{-29} gm/cm^3. Life in a forever expanding universe seems less attractive than in one which is "closed." For this and other (less psychologically motivated) reasons, astronomers are still looking for additional matter in the vast expanses of the universe.

REFERENCES

An interesting and enlightening extension of Gibbs' paradox is presented in:

M. J. Klein, Note on a Problem Concerning the Gibbs Paradox, *Am. J. Phys.* **26**, 80 (1958).

A careful and extensive development of the entropy concept, from both the classical and quantum viewpoints, is contained in:

J. S. Dugdale, "Entropy and Low Temperature Physics." Hutchinson, London (1966).

Chapter 13 includes comments on Maxwell's demon.

A rigorous proof of the entropy principle is given by:

E. Fermi, "Thermodynamics," Chapter 4. Dover, New York (1956).

Cosmological consequences of the "missing mass" are touched on in:

W. H. McCrea, Cosmology Today, *Am. Sci.* **58**, 521 (1970).

and

J. A. Wheeler, Our Universe: The Known and Unknown, *Am. Sci.* **56**, 1 (1968).

The origin of the irreversibility in nature is still a matter of debate. For a critical review, see:

B. Gal-Or, The Crisis about the Origin of Irreversibility and Time Aniosotropy, *Science* **176**, 11 (1972).

Chapter 7

The Third Law of Thermodynamics

7.1 ENTROPY, ABSOLUTE ZERO, AND THE THIRD LAW

While the remarks of this chapter aim at the third law, the reader should be reminded of the first two. The first law may be phrased as a warning[†]: "You cannot win. You can only break even." The first law is a statement of energy conservation. It recognizes the impossibility of winning with perpetual motion machines of the first kind—cyclic devices which would export more energy than they import. The second law is even more pessimistic about one's chances of winning. The second law recognizes the impossibility of constructing a cyclic device which converts heat completely into work. In terms of the thermal efficiency

$$\eta = \frac{\text{work output}}{\text{heat input}}$$

the second law may be stated

$$\eta < 1$$

For example, the efficiency of the Carnot cycle is

$$\eta_c = 1 - \frac{T_{\text{cold}}}{T_{\text{hot}}}$$

[†] Compare, *Am. Sci.* **52**, 40A (1964).

It appears that if $T_{\text{cold}} = 0\,\text{K}$, then $\eta_c = 1$, and we would have a device which converts heat completely into work. This has led to the following one line statement of the second law: "You can break even only at absolute zero." However, is it possible to reach absolute zero? Experiment decides whether or not one can reach the state $T = 0\,\text{K}$. Experiment reveals that all schemes for lowering temperature suffer an attack of diminishing returns upon approaching absolute zero. They become less and less effective as the temperature decreases. The third law, which comes as a disappointment to those who aspire to break even may be stated as "You cannot reach absolute zero." The immediate conclusion to be drawn from the first, second, and third laws is that you can neither win nor break even. This makes thermodynamics a very expensive pastime.

There is an alternate way of stating the third law. It is phrased in terms of the entropy. The entropy interpretation of the third law reinforces the viewpoint stressed in Chapter 6—that entropy measures disorder.

The entropy change dS resulting from the reversible transfer of heat dQ at temperature T is

$$dS = \frac{dQ}{T}$$

This is an operational definition, not of entropy, but of the change in entropy. This did not strike us as unusual when it was first advanced because we have encountered quite similar situations before. For example, in mechanics only the change of potential energy is operationally defined. We are free to choose the position of zero potential energy at our convenience. In thermodynamics there is likewise a choice as to the state of zero internal energy. The first law defines only the change of internal energy.

Despite the fact that only dS is operationally defined we do not have complete freedom in the choice of the state of zero entropy. Nature rules otherwise. The third law of thermodynamics expresses this further restriction on the entropy. It was Walther Nernst who first saw the necessity of inventing a new law of thermodynamics. In 1907 Nernst advanced his heat theorem. Now enlarged, revised, and bolstered by additional experimental evidence, Nernst's Heat Theorem stands as the third law of thermodynamics. In its most unqualified form the third law is this:

> The entropy of a system approaches zero as its temperature tends toward absolute zero.

This statement must be qualified to make it precise and universally valid—to put it in the category of physical law. However, after all the qualifying "weasel" words are inserted, the essential content of the third law remains

$$S \to 0 \quad \text{as} \quad T \to 0$$

In the next section we cite evidence for the third law by showing how experiment confirms its predictions. In Section 7.3 we note certain exceptions, interpret them, and modify the original statement accordingly. In Section 7.4 we illustrate how the third law renders absolute zero unattainable. To fully appreciate the third law we must continue to cultivate the notion that entropy measures disorder. The statement that entropy tends to zero with temperature implies that absolute zero is the ultimate if not fully attainable state of order.

7.2 THE THIRD LAW IN ACTION

Let us accept the unrestricted statement of the third law, deduce some of its consequences, and then check to see if these are borne out by experiment. If we utilize one of the Maxwell relations,†

$$\left(\frac{\partial V}{\partial T}\right)_P = -\left(\frac{\partial S}{\partial P}\right)_T \tag{8.18}$$

the coefficient of volume expansion, $\beta = (1/V)(\partial V/\partial T)_P$, can be expressed as

$$\beta = -\frac{1}{V}\left(\frac{\partial S}{\partial P}\right)_T$$

Using the third law we can show that $(\partial S/\partial P)_T$ falls toward zero with the temperature. Therefore, the third law predicts that β tends toward zero as the temperature approaches absolute zero. To show that $(\partial S/\partial P)_T$ tends to zero we note that it corresponds to the experimentally observed quantity

$$\frac{\Delta S}{\Delta P} = \frac{S(T, P+\Delta P) - S(T, P)}{\Delta P}$$

where $\Delta S/\Delta P$ is a function of T and P. If the entropy falls to zero as T approaches zero, it follows that the entropy change ΔS also must tend to zero, regardless of the value of ΔP. Thus the ratio $\Delta S/\Delta P$ goes to zero

$$\frac{\Delta S}{\Delta P} = \left(\frac{\partial S}{\partial P}\right)_T \to 0 \qquad \text{as} \qquad T \to 0$$

and it follows that β tends to zero with the temperature. Figure 7.1 shows a plot of α ($\alpha = \frac{1}{3}\beta$) versus temperature for gold. The figure reveals that α approaches zero as the temperature tends toward absolute zero, in accord with the third law.

---

† Maxwell's relations, including (8.18), are derived in Chapter 8.

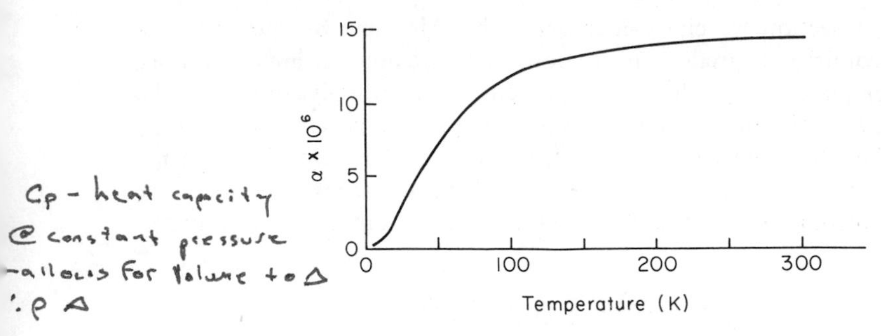

Figure 7.1. Coefficient of linear expansion α for gold. (After H. M. Rosenberg, "Low Temperature Solid State Physics." Oxford Univ. Press, London and New York (1963).

A second consequence of the third law concerns the difference of the heat capacities $C_p - C_v$. We noted earlier that C_p and C_v differ because materials tend to expand or contract when heated at constant pressure. However, as β goes to zero the volume ceases to change with the temperature. We therefore expect C_p and C_v to become equal as T approaches zero. In fact, examination of (4.26) shows that the difference of C_p and C_v is proportional to

$$\left(\frac{\partial V}{\partial T}\right)_P = \beta V$$

a quantity which vanishes as $T \to 0$. Figure 7.2 shows the low-temperature confluence of C_p and C_v for solid neon.

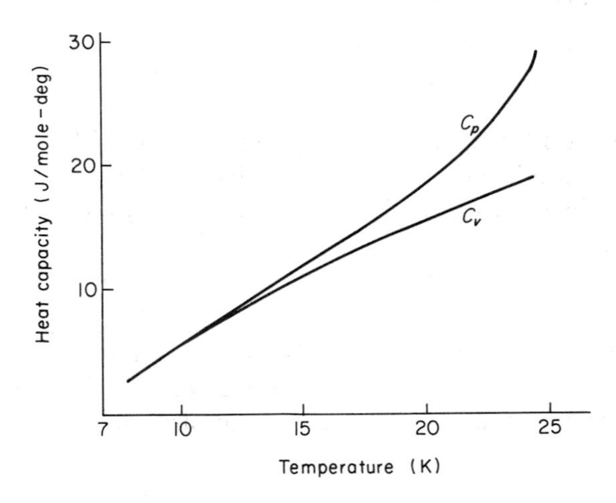

Figure 7.2. Heat capacities of solid neon. [After C. H. Fagerstroem and A. C. Hollis Hallet, "Low Temperature Physics," Part B. Plenum, New York (1965).]

A third experimental confirmation of the third law is provided by low-temperature heat capacity measurements. The third law not only predicts that C_p and C_v become equal, it also predicts that each vanishes as T falls toward absolute zero. To show this we write

$$dQ = C\,dT$$

where C is the appropriate heat capacity. The corresponding entropy change is

$$dS = \frac{C\,dT}{T} \tag{7.1}$$

If we integrate this equation from absolute zero to some temperature T, we obtain

$$S(T) - S(0) = \int_0^T \frac{C\,dT}{T} \tag{7.2}$$

At this point we seize an opportunity to qualify our original statement of the third law. A more precise statement is that the entropies of all systems approach the *same* value as the temperature tends toward zero. The choice of this common value is left open, and the scientific community has settled on zero. With the choice $S(0) = 0$, the integral in (7.2) fixes the entropy at temperature T as

$$S(T) = \int_0^T \frac{C\,dT}{T} \tag{7.3}$$

The behavior of C as T approaches zero is severely restricted by the condition that

$$S \to 0 \quad \text{as} \quad T \to 0 \tag{7.4}$$

To see the necessity of this restriction, suppose that C did not vanish with T, but instead reached a constant value as the temperature approached zero. We could then remove C from the integral in (7.3) and obtain

$$S(T) = C \int_0^T \frac{dT}{T}$$

However, the integral is infinite![†] To put it another way, the entropy $S(T)$ does not exist if C tends toward a nonzero value as T approaches zero. In order for $S(T)$ to have a definite value the heat capacity must vanish with the temperature. Experiment confirms this restriction on C. For electrically insulating solids it is found that C is proportional to T^3 at low temperatures

$$C_{\text{ins}} = bT^3 \tag{7.5}$$

[†] Thus, $\int_{T_0}^{T} dT/T = \ln(T/T_0) \to \infty$ as $T_0 \to 0$.

For electrically conducting solids, which have electrons free to roam about, the low-temperature behavior of C has the form

$$C_{\text{cond}} = aT + bT^3 \tag{7.6}$$

The aT term is contributed by the conduction electrons. As the student can readily verify, both (7.5) and (7.6) lead to entropies which vanish as T tends to zero.

••••••

7.2.1 (a) Using the heat capacities of (7.5) and (7.6) evaluate the low-temperature entropies of insulating and conducting solids. Verify that $S(T) \to 0$ as $T \to 0$.

(b) Show that the entropy vanishes at $T = 0$ provided

$$C \propto T^q$$

where q is any number greater than zero.

••••••

Yet another confirmation of the third law is furnished by measurements of the melting pressure of ^{4}He. Under atmospheric pressure ^{4}He remains a liquid at the lowest temperature. However, if sufficient pressure is applied, the liquid solidifies. The melting curve marks the liquid–solid boundary on a pressure versus temperature plot. Figure 7.3 shows the behavior of the melting curve for ^{4}He. The slope of the melting curve is given by the Clausius–

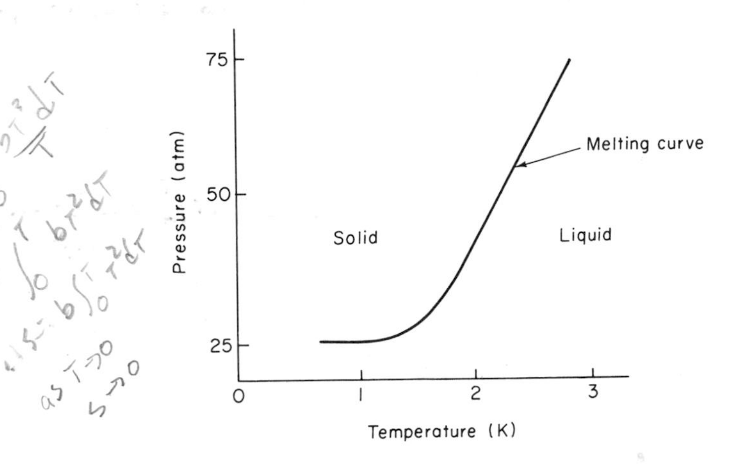

Figure 7.3. Melting pressure of ^{4}He. The slope of the curve dP/dT is given by the Clausius–Clapeyron equation (7.7). [After K. Mendelssohn, "Cryophysics." Wiley (Interscience), New York (1960).]

Clapeyron equation

$$\frac{dP}{dT} = \frac{L_f}{T(V_l - V_s)} \tag{5.9a}$$

where L_f is the latent heat of fusion at temperature T and $V_l - V_s$ is the volume change in the solid to liquid transition. Because melting is an isothermal process, L_f/T is just the entropy change of melting

$$S_l - S_s = \frac{L_f}{T} \tag{6.10}$$

The slope of the melting curve is therefore

$$\frac{dP}{dT} = \frac{S_l - S_s}{V_l - V_s} \tag{7.7}$$

Both S_l and S_s tend to zero as T is lowered toward 0 K. It follows that

$$S_l - S_s \to 0 \qquad \text{as} \qquad T \to 0$$

However, $V_l - V_s$, the volume change which accompanies melting, remains nonzero to the lowest temperature. It then follows from (7.7) that the slope of the melting curve must become zero near absolute zero. Figure 7.3 shows a striking confirmation of this prediction, the melting pressure being nearly constant below 1 K.

7.3 EXCEPTIONS AND RESTATEMENT

Despite an impressive string of successes, the unqualified statement of the third law is not universally valid. There are systems for which the entropy does not tend toward zero at low temperatures and which force us to restate the third law. The exceptions may be traced to irreversible processes which, instead of driving the system to equilibrium, leave it stranded in a nonequilibrium state.

The resulting difficulties are not heaped solely on the third law. Strictly speaking, the first and second laws no longer apply either. The reason for this is that thermodynamic properties, entropy, internal energy, and so on, are operationally defined only for equilibrium states. Thermodynamics therefore can make no positive statements about transformations which begin or end in nonequilibrium states.

To understand why these exceptions are not disastrous for the third law we must appreciate yet another aspect of entropy. The entropy and energy of a system are alike in one important respect. Both are sums of contributions from various sources. For example, consider a gas of diatomic molecules. In addition to the familiar translational kinetic energy molecules can possess rotational and vibrational energies, all of which contribute to the internal energy of the gas. Now, the parallel between energy and entropy is this. Each and every mechanism that donates to the energy of a system also contributes to its entropy. The internal energy of a diatomic gas is a sum of the energies of rotation, vibration, translation, etc. Similarly, the entropy is the sum of a translational entropy, a vibrational entropy, and so on. This one-to-one correspondence of entropy and energy should not seem mysterious or miraculous. In a reversible process in which no work is performed the first law reduces to

$$dQ = dU$$

The entropy change is

$$dS = \frac{dQ}{T} = \frac{dU}{T}$$

It is immediately clear that if U is expressed as a sum of terms, then S may be expressed by a similar sum. The multiplicity of entropy and energy contributions is described by saying that a system possesses different *degrees of freedom*. In a well-behaved system, lowering the temperature reduces the energy and entropy associated with each degree of freedom. In particular, each contribution to the entropy tends toward zero with the temperature.

In the exceptional cases, where the system is in a nonequilibrium state, not every degree of freedom is involved. Generally, only the entropy contributions of a few degrees of freedom fail to drop toward zero. We say their entropy is "frozen in," an appropriate terminology in view of the temperature of the surroundings. The other degrees of freedom behave normally—their entropy contributions vanish at absolute zero. It is helpful to think of a system as a collection of subsystems. Each degree of freedom is a subsystem—a part of the whole. The exceptions to the unqualified form of the third law should be regarded as situations wherein the various subsystems are not all in mutual equilibrium. Those subsystems which are in mutual equilibrium obey the unqualified form of the third law. Their entropy contributions vanish at absolute zero. The rebellious subsystems are not in equilibrium. They defy a thermodynamic description. The subsystem concept lets us restate the third law in a universally valid form. Our earlier unqualified statement (The entropy of a system approaches zero as its temperature tends toward absolute zero.) is replaced by the following restatement, essentially as stated by Simon[†]:

[†] F. E. Simon, "Year Book of the Physical Society," p. 1. The Physical Society, London (1956). Adapted with permission of The Institute of Physics and The Physical Society.

> The contribution to the entropy of a system from each subsystem which is in internal thermodynamic equilibrium disappears at the absolute zero.

The Simon formulation of the third law has much to recommend it. It advises that we may apply the unqualified form of the third law to those subsystems (degrees of freedom) which are in internal (mutual) equilibrium. The sense of our original unqualified form is thereby maintained, and lets us salvage many results based upon it.

Exceptions to the unqualified form of the third law are to be found in the behavior of certain supercooled liquids. With care and patience it is possible to cool certain liquids below their customary melting point. Glycerol is an outstanding example for it can exist as a supercooled liquid down to the lowest temperatures, even though its normal melting point is about 19 °C. Even though the supercooled liquid is in a nonequilibrium state it is possible to measure its heat capacity. The specific heats of crystalline and supercooled liquid glycerol differ significantly from the melting point down to about 180 K. Below 180 K the heat capacities are nearly equal and both tend to zero at absolute zero. The difference in the two heat capacities may be traced to the translational degrees of freedom. In the liquid, the molecules are relatively free to roam about. In the crystalline solid the molecules form a lattice, thereby curtailing translational motion. The extra degrees of freedom give the supercooled liquid greater energy and entropy than the solid.

It is instructive to contrast the behavior of glycerol with that of a normal substance—one which obeys the unqualified form of the third law. To draw close parallels we must choose a normal substance which can exist in more than one form down to the lowest temperatures. Tin, sulfur, phosphine (PH_3), and cyclohexanol ($C_6H_{12}O$) are four such substances. For example, there are two forms of tin which, though chemically identical, have different crystalline structures. Grey tin is the stable form at low temperatures and white tin is the stable form at high temperatures. The transition temperature is 19 °C (=292 K). Grey tin at 19 °C gradually (in a few hours) changes into white tin, releasing a heat of transition of 535 cal/mole.

In practice one tests whether or not the unqualified version of the third law is followed by comparing two determinations of the entropy difference. One determination follows from the heat of transition. A second evaluation is made using the measured heat capacities. We now outline this procedure and then illustrate with the data for tin and glycerol.

If form 1 undergoes a reversible transformation to form 2 at a temperature T, releasing a heat of transition $L_{1\to2}$, the entropy change is

$$S_2(T) - S_1(T) = \frac{L_{1\to2}}{T} \tag{7.8}$$

Next, we integrate $dS_2 = C_2\, dT/T$ from 0 to T, obtaining

$$S_2(T) - S_2(0) = \int_0^T C_2 \frac{dT}{T}$$

Combining this with the corresponding equation for form 1 gives a second expression for $S_2(T) - S_1(T)$,

$$S_2(T) - S_1(T) = S_2(0) - S_1(0) + \int_0^T (C_2 - C_1) \frac{dT}{T} \tag{7.9}$$

Comparing (7.8) and (7.9) shows that the entropy difference at absolute zero is

$$S_2(0) - S_1(0) = \frac{L_{1\to 2}}{T} - \int_0^T \frac{(C_2 - C_1)}{T}\, dT \tag{7.10}$$

The quantities on the right side of (7.10) are subject to measurement. If the right side is zero (to within the experimental uncertainty of about 0.1 eu/mole), we conclude that the substance obeys the unqualified form of the third law. Otherwise, we conclude that there is a frozen-in entropy.

For the grey-to-white tin transition

$$\frac{L_{g\to w}}{T} = \frac{535}{292}\ \frac{\text{eu}}{\text{mole}} = 1.83\ \frac{\text{eu}}{\text{mole}}$$

Measurements of the heat capacities give

$$\int_0^{292} \frac{(C_w - C_g)}{T}\, dT = 1.77\ \frac{\text{eu}}{\text{mole}}$$

These results establish that $S_w(0)$ and $S_g(0)$ are equal to within the experimental uncertainty and that tin obeys the unqualified form of the third law.

By contrast the data for crystalline and supercooled glycerol give†

$$\frac{L_{s\to l}}{T} \simeq 15\ \frac{\text{eu}}{\text{mole}}, \qquad \int_0^{292} \frac{(C_l - C_s)\, dT}{T} \simeq 10\ \frac{\text{eu}}{\text{mole}}$$

Thus

$$S_l(0) - S_s(0) \simeq 5\ \frac{\text{eu}}{\text{mole}}$$

The entropy difference here is well beyond the experimental uncertainties—the entropies of the supercooled liquid and the crystalline solid differ at absolute zero. The entropies differ because the liquid is more disordered than the crystalline solid phase. The supercooled liquid preserves the excess disorder—it appears experimentally as a frozen-in entropy difference.

† It is coincidental that the transition temperature is 292 K for both tin and glycerol.

Those exceptional substances whose entropy does not approach zero at absolute zero do not permit the ultimate low temperature to be attained. Their entropy is frozen-in. It is associated with degrees of freedom which are beyond external control. Only those degrees of freedom which remain in mutual equilibrium can be manipulated at the macroscopic level, and their contributions to the entropy vanish at absolute zero.

•••••• ••••••

7.3.1 The molar heat capacities of two forms of a (hypothetical) substance are given by

$$C_\alpha = 0.09T + 0.003T^3 \quad \text{eu}$$

$$C_\beta = 0.01T \quad \text{eu}$$

The β-form undergoes a phase transition to the α-form at a temperature of 20 K. The measured heat of transition is $L_{\beta\to\alpha} = 190$ cal/mole. Determine whether or not the substance obeys the unqualified version of the third law, that is, whether both the α and β forms have the same entropy at absolute zero. Assume that the data for $L_{\beta\to\alpha}$ and the heat capacities are reliable to within $\pm 2\%$.

•••••• ••••••

7.4 THE UNATTAINABILITY OF ABSOLUTE ZERO

Our earliest statement of the third law was simply that "You cannot reach absolute zero." The unattainability of absolute zero is a consequence of the fact that the entropy tends to zero with the temperature. In layman's language, it is a fatal attack of diminishing returns which prevents any scheme from lowering the temperature to absolute zero. While a general proof of unattainability can be formulated we shall be content to illustrate the idea with a particular example.

The illustration has been constructed so as to be formally identical to the following riddle: A man sets out to drive from Pittsburgh to New York City. He drives half way during the first day. On each succeeding day he drives half of the distance which remained at the end of the previous day. How many days are required to complete the journey? The answer is that the trip is never completed. So long as the distance traveled each day is a fraction of the remaining distance the trip can never be completed. An analogous fate befalls methods designed to lower temperatures to absolute zero. These schemes invariably lead to temperature changes proportional to the temperature itself—they lower the temperature by some fraction of the remaining "distance" to absolute zero.

It is not difficult to appreciate that any optimized method of lowering temperatures includes adiabatic and reversible transformations. Thus, if the system performs a positive amount of work adiabatically it does so at the expense of its internal energy—and a decrease in U lowers the temperature. It is also desirable to employ reversible processes. Irreversible processes generate entropy. But we know that entropy and temperature tend toward zero together. Entropy production via irreversible processes therefore opposes temperature reduction.

For our illustration we suppose that the entropy is directly proportional to the product of temperature and volume

$$S = aTV \qquad a = \text{constant}$$

With the system initially in a state described by $T = T_0$, $V = V_0$, we perform a series of two-step processes, as indicated in Figure 7.4. The two straight lines on the S–T diagram are plots of $S = aTV$ for $V = V_0$ and $V = \frac{1}{2}V_0$. First, the system is isothermally compressed from V_0 to $\frac{1}{2}V_0$. This decreases the entropy, as indicated in the figure. The compression is followed by an adiabatic (and *isentropic*) expansion which returns the system to its initial volume. Since

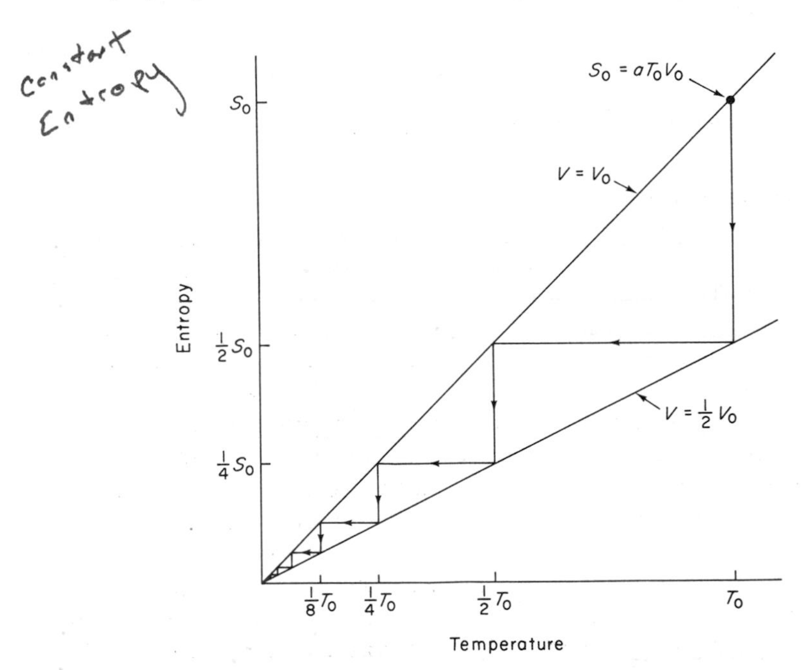

Figure 7.4. Isothermal compression reduces the volume and entropy. Isentropic expansion lowers the temperature. Because the temperature reduction is a fraction of the starting temperature no finite number of repetitions can reach absolute zero.

$S = aTV$ remains constant during the isentropic process, the volume change from $\frac{1}{2}V_0$ to V_0 must be accompanied by a temperature decrease from T_0 to $\frac{1}{2}T_0$. Overall, the two-step process returns the system to its initial volume and cuts its temperature in half. By repeating the two-step sequence the temperature can be halved again. And so it might go—*ad infinitum.* Each repetition halves the existing temperature. However, just as the man who forever goes halfway never reaches New York City, so the temperature never reaches absolute zero.

The third law places no restrictions on how closely one may approach absolute zero. A recently developed "dilution" refrigerator, which utilizes a mixture of ^{3}He and ^{4}He permits experiments to be performed in the milli-degree (10^{-3} K) range. Details of dilution refrigerators are presented in Chapter 18.

REFERENCES

The "inside story" of the third law is told by:

F. Simon, The Third Law of Thermodynamics, "Year Book of the Physical Society," pp. 1–21. The Physical Society, London (1956).

Among the historical tidbits included is the revelation that Nernst never thought in terms of entropy!

A thorough discussion of all aspects of the third law is given in:

J. Wilks, "The Third Law of Thermodynamics." Oxford Univ. Press, London and New York (1961).

Some portions make use of elementary quantum mechanics. Nevertheless, the beginner will find helpful insights throughout the book.

A general proof of the unattainability of absolute zero can be found in:

M. W. Zemansky, "Heat and Thermodynamics," 5th ed., p. 497. McGraw-Hill, New York (1968).

Chapter 8

Thermodynamic Potentials

In the preceding chapter we completed our program of converting the first law into the mathematically viable form

$$dU = TdS - PdV \tag{8.1}$$

The internal energy U has proved to be a most useful thermodynamic property. Conceptually, we picture U as the sum of the kinetic and potential energies of the component parts of the system. Despite its conceptual clarity, the internal energy is ill-suited for the analysis of many thermodynamic processes. In this chapter we introduce three additional quantities, related directly to U, which find wide application in thermodynamics. These are the enthalpy H, the Gibbs free energy G, and the Helmholtz free energy F. Along with U, these functions are referred to as thermodynamic potentials. Each has the dimensions of energy. The potentials give thermodynamic theory greater flexibility and coherence. They also provide more direct links with experiment than could be managed using only the internal energy.

A major shortcoming of our formulation of thermodynamics is removed by introducing a fifth potential. In the interest of clarity we have confined our attention almost exclusively to closed systems consisting of a single chemical species. In practice it often becomes necessary to deal with open systems, which exchange matter with their surroundings, and with mixtures whose components undergo chemical or nuclear reactions. Both situations force us to deal with systems having a variable number of particles. The thermodynamic parameter introduced to shoulder this task is called the chemical potential. The importance of the chemical potential is further enlarged by the fact that it is used extensively in statistical mechanics.

In the first section we present a brief introduction to the thermodynamic potentials. Later sections offer more leisurely and interpretive discussions of the potentials.

8.1 INTRODUCTION

The differential form of the first law (8.1) suggests that the entropy S and volume V be regarded as independent variables and that U be treated as a function of S and V

$$U = U(S, V)$$

If this is done, temperature and pressure become dependent variables whose relation to U, S, and V is readily determined. As a purely mathematical consequence of the fact that U is a function of the independent variables S and V we can write

$$dU = \left(\frac{\partial U}{\partial S}\right)_V dS + \left(\frac{\partial U}{\partial V}\right)_S dV \tag{8.2}$$

Comparing this with the first law (8.1) we see that[†]

$$T = \left(\frac{\partial U}{\partial S}\right)_V, \qquad P = -\left(\frac{\partial U}{\partial V}\right)_S \tag{8.3}$$

Entropy and volume are not necessarily the most suitable variables, a fact abundantly clear to anyone who has rummaged about a laboratory in search of an entropy meter. Let us therefore define a new thermodynamic potential in which entropy is relegated to the role of a dependent variable. This can be accomplished by adding and subtracting $S\,dT$ on the right side of (8.1). We then obtain

$$dU = T\,dS + S\,dT - S\,dT - P\,dV$$

The first pair of terms on the right side form a complete differential

$$T\,dS + S\,dT = d(TS)$$

[†] The equating of coefficients in (8.1) and (8.2) is mandatory because S and V are independent variables.

Transferring $d(TS)$ to the left side leaves

$$d(U-TS) = -S\,dT - P\,dV \tag{8.4}$$

The combination $U-TS$ is known as the Helmholtz free energy

$$F = U - TS \tag{8.5}$$

Equation (8.4) shows that F is a function whose differential involves dT and dV, but not dS

$$dF = -S\,dT - P\,dV \tag{8.5a}$$

Therefore F is to be regarded as a function of the independent variables T and V. The entropy and pressure become dependent variables. To see how S and P are determined we note that as a purely mathematical consequence of its dependence on T and V we can write

$$dF = \left(\frac{\partial F}{\partial T}\right)_V dT + \left(\frac{\partial F}{\partial V}\right)_T dV \tag{8.6}$$

Comparing this with (8.5a) shows that S and P are determined from

$$S = -\left(\frac{\partial F}{\partial T}\right)_V, \qquad P = -\left(\frac{\partial F}{\partial V}\right)_T \tag{8.7}$$

The Helmholtz free energy is of great importance in statistical mechanics. Once $F(T,V)$ has been obtained, the relation

$$P = -\left(\frac{\partial F}{\partial V}\right)_T$$

yields the equation of state of the system.

••••••

8.1.1 The free energy of a gas is given by

$$F = -RT\ln(V-b) - \frac{a}{V} + CT$$

where R, b, a, and C are constants. Show that the gas obeys the van der Waals equation of state

$$\left(P+\frac{a}{V^2}\right)(V-b) = RT$$

••••••

If we choose to eliminate V rather than S, we are led to the enthalpy function. Starting with the first law, $dU = T\,dS - P\,dV$, we play "complete the differential," this time by subtracting and adding $V\,dP$.[†] This gives

$$dU = T\,dS + V\,dP - P\,dV - V\,dP$$

The last two terms form a complete differential so that

$$d(U+PV) = T\,dS + V\,dP$$

The enthalpy H is defined as

$$H = U + PV \tag{8.8}$$

whence,

$$dH = T\,dS + V\,dP \tag{8.9}$$

and H emerges as a function of S and P. The fact that H may be regarded as a function of S and P leads to the results

$$T = \left(\frac{\partial H}{\partial S}\right)_P, \qquad V = \left(\frac{\partial H}{\partial P}\right)_S \tag{8.10}$$

••••••

8.1.2 Using the fact that the enthalpy may be regarded as a function of the two variables S and P, obtain (8.10).

••••••

The final thermodynamic potential, the Gibbs free energy, may be generated from H or from F. Thus adding and subtracting $S\,dT$ on the right side of (8.9) leads to

$$d(H-TS) = -S\,dT + V\,dP$$

Alternately, adding and subtracting $V\,dP$ on the right side of (8.5a) results in

$$d(F+PV) = -S\,dT + V\,dP$$

Equivalent definitions of the Gibbs free energy are therefore

$$G = H - TS \tag{8.11}$$

and

$$G = F + PV$$

In terms of the internal energy,

$$G(T,P) = U + PV - TS \tag{8.12}$$

[†] Observe that an independent variable is converted to dependent status by completing the differential in which it stands.

The differential expressions above show that T and P are the natural choices for independent variables

$$dG = -S\,dT + V\,dP \tag{8.13}$$

Treating G as a function of T and P leads to the relations

$$S = -\left(\frac{\partial G}{\partial T}\right)_P, \qquad V = \left(\frac{\partial G}{\partial P}\right)_T \tag{8.14}$$

•••••   •••••

8.1.3 Using (8.13) and the fact that the Gibbs free energy is best regarded as a function of the independent variables T and P, derive (8.14).

8.1.4 The technical name for the game we call "complete the differential" is Legendre transformation. Suppose that the differential of a function has the form

$$d\Psi = u\,dx + v\,dy$$

In this form, x and y appear as the independent variables. It is desired to replace x by u as an independent variable.

(a) Show that this may be accomplished by completing the $u\,dx$ differential.
(b) Write down the differential of the function $\Omega(u, y) \equiv \Psi - ux$.
(c) Construct a function whose differential involves du and dv rather than dx and dy.

8.1.5 In advanced mechanics the Lagrangian L is regarded as a function of the variables x and v. Its differential therefore has the form

$$dL = f\,dx + p\,dv, \qquad f = \left(\frac{\partial L}{\partial x}\right)_v, \quad p = \left(\frac{\partial L}{\partial v}\right)_x$$

It is often desirable to deal with the Hamiltonian H, a function of p and x.

(a) By completing a differential, show that the expression $pv - L$ emerges as a function of p and x. (You will probably discover that $L - pv$ emerges somewhat more naturally, but $pv - L$ is the Hamiltonian!)

(b) The Lagrangian for a particle of mass m in a uniform gravitational field (gravitational acceleration g) is given by

$$L = \tfrac{1}{2}mv^2 - mgx$$

(v is the speed of the particle and x is its distance above some reference level of zero gravitational potential energy). Express the Hamiltonian as a function of p and x. Note that p is defined as $p = (\partial L/\partial v)_x$. What is the physical significance of p?

•••••   •••••

8.2 THE MAXWELL RELATIONS

The differentials of the four thermodynamic potentials are given by

$$dU = T\,dS - P\,dV \tag{8.1}$$

$$dF = -S\,dT - P\,dV \tag{8.5a}$$

$$dH = T\,dS + V\,dP \tag{8.9}$$

$$dG = -S\,dT + V\,dP \tag{8.13}$$

Equation (8.1) is the first law. The next three follow in straightforward fashion from our "complete the differential" program. Any number of mnemonics have been constructed to aid in remembering these equations. One such scheme uses the diagram of Figure 8.1. The four potentials label sides of a

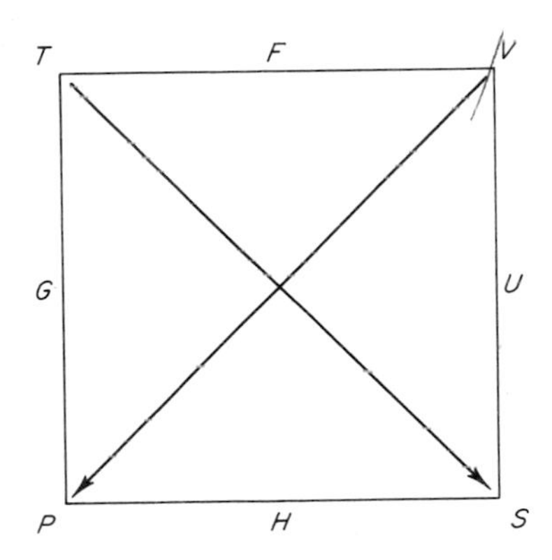

Figure 8.1. "The Friendly Vagabond Usually Shares His Precious Gifts."

square. The four variables S, T, P, and V label corners. A mnemonic phrase is used to remember the order of the symbols. As an example, there is the expression, "*The Friendly Vagabond Usually Shares His Precious Gifts.*" The student is invited to make up his own mnemonic. Note that each thermodynamic potential is flanked in the diagram by the two variables whose differentials appear in the expression for the differential of that potential. For example, U is flanked by S and V, and (8.1) shows that $dU = T\,dS - P\,dV$. The two arrows aid in the assignment of plus and minus signs. Part of a picture is worth at least 500 words.... As in the accompanying figure, going "with" an arrow rates a plus sign, opposing an arrow merits a minus sign. Test yourself!

$$dG = ?$$

dG = -SdT + VdP

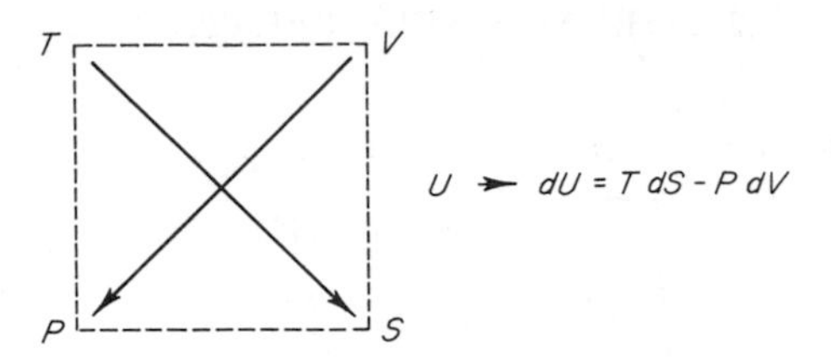

By comparing two expressions for dU, (8.1) and (8.2), we arrived at (8.3),

$$T = \left(\frac{\partial U}{\partial S}\right)_V, \qquad P = -\left(\frac{\partial U}{\partial V}\right)_S \tag{8.3}$$

By cross differentiation we can obtain the first of the Maxwell relations. Thus, if we differentiate with respect to V on both sides of the first equation in (8.3), we obtain

$$\left(\frac{\partial T}{\partial V}\right)_S = \frac{\partial^2 U}{\partial V \partial S}$$

Similarly, applying $(-\partial/\partial S)_V$ to both sides of the second equation in (8.3) produces

$$-\left(\frac{\partial P}{\partial S}\right)_V = \frac{\partial^2 U}{\partial S \partial V}$$

The order of differentiation makes no difference

$$\frac{\partial^2 U}{\partial V \partial S} = \frac{\partial^2 U}{\partial S \partial V}$$

so that we obtain the result,

$$\left(\frac{\partial T}{\partial V}\right)_S = -\left(\frac{\partial P}{\partial S}\right)_V \tag{8.15}$$

This is one of the Maxwell relations. These relations are of considerable utility in thermodynamics. They permit the replacement of quantities which are difficult if not impossible to measure by quantities open to direct measurement. For example, $(\partial P/\partial S)_V$ is not directly measureable, whereas $(\partial T/\partial V)_S$ can be measured. Thus, if an insulated system undergoes a reversible process, its entropy remains constant ($dS = dQ/T = 0/T = 0$). If the change in volume ΔV and temperature ΔT are measured, we have

$$\frac{\Delta T}{\Delta V} = \left(\frac{\partial T}{\partial V}\right)_S$$

The remaining Maxwell relations follow by cross differentiation from (8.7), (8.10), and (8.14). They are

$$\left(\frac{\partial S}{\partial V}\right)_T = \left(\frac{\partial P}{\partial T}\right)_V \tag{8.16}$$

$$\left(\frac{\partial T}{\partial P}\right)_S = \left(\frac{\partial V}{\partial S}\right)_P \tag{8.17}$$

$$-\left(\frac{\partial S}{\partial P}\right)_T = \left(\frac{\partial V}{\partial T}\right)_P \tag{8.18}$$

······ ······

8.2.1 By performing cross differentiations, derive the three Maxwell relations (8.16)–(8.18).

······ ······

The mnemonic diagram can be used to recall the Maxwell relations. Again, part of a picture is worth... Consider the accompanying figures. Test yourself! [Check your answers against (8.16) and (8.18).]

To illustrate the utility of the Maxwell relations, we show how (8.18) is used to derive an expression for $T\,dS$ needed later. If S is regarded as a function

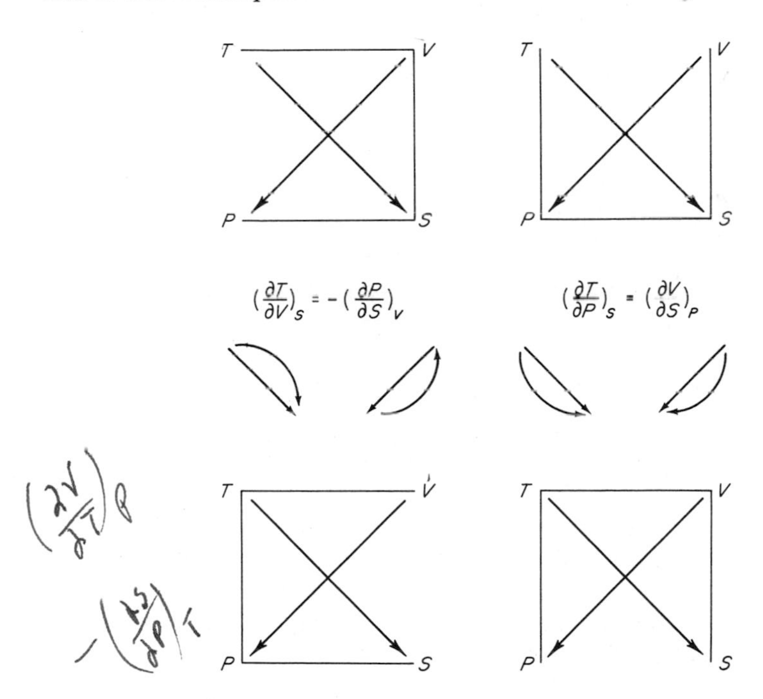

of T and P,

$$dS = \left(\frac{\partial S}{\partial T}\right)_P dT + \left(\frac{\partial S}{\partial P}\right)_T dP \tag{8.19}$$

In a reversible process, TdS equals the heat absorbed. If this heat exchange takes place at constant pressure, the heat capacity C_p enters as

$$(TdS)_P = dQ_P = C_p dT \tag{8.20}$$

From (8.19), with $dP = 0$,

$$(TdS)_P = T\left(\frac{\partial S}{\partial T}\right)_P dT \tag{8.21}$$

Comparing these two expressions for $(TdS)_P$ shows that

$$C_p = T\left(\frac{\partial S}{\partial T}\right)_P \tag{8.22}$$

Inserting this result back into (8.19) yields

$$TdS = C_p dT + T\left(\frac{\partial S}{\partial P}\right)_T dP$$

If we now use the fourth Maxwell relation (8.18), we obtain the desired result

$$TdS = C_p dT - T\left(\frac{\partial V}{\partial T}\right)_P dP \tag{8.23}$$

By an argument similar to that leading to (8.23), and using (8.16), one can obtain another very useful expression for TdS. This task is left as an exercise for the student. The result is

$$TdS = C_v dT + T\left(\frac{\partial P}{\partial T}\right)_V dV \tag{8.24}$$

•••••• ••••••

8.2.2 Use (8.19) and (8.23) to show that

$$\mu_{JT} \equiv \left(\frac{dT}{dP}\right)_H = \frac{1}{C_p}\left[T\left(\frac{\partial V}{\partial T}\right)_P - V\right]$$

where $(dT/dP)_H$ denotes the ratio of dT and dP for an isenthalpic process (H is constant) and μ_{JT} is called the Joule–Thomson coefficient. It measures the temperature change produced by a change in pressure under isenthalpic conditions. We encounter μ_{JT} again in the next chapter when we study the thermodynamics of gas liquefaction.

8.2.3 If the entropy is regarded as a function of T and V, its differential is

$$dS = \left(\frac{\partial S}{\partial T}\right)_V dT + \left(\frac{\partial S}{\partial V}\right)_T dV$$

Show that this form of dS leads to (8.24). You will need the Maxwell relation (8.16).

8.2.4 Show that (8.23) may be expressed in the form

$$T\,dS = C_p\,dT - \beta VT\,dP$$

where β is the coefficient of volume expansion.

....

8.3 HELMHOLTZ FREE ENERGY

In a purely mechanical system, where the concepts of heat and temperature may be disregarded, the adiabatic form of the first law ($Q = 0$) applies

$$W = -(U_f - U_i) = -\Delta U \tag{3.3'}$$

We interpret this by saying that the performance of work by the system is accomplished through a sacrifice of internal energy. Pressure, the thermodynamic variable which corresponds directly to the mechanical variable force, is related to the internal energy (8.3)

$$P = -\left(\frac{\partial U}{\partial V}\right)_S \tag{8.3'}$$

In a reversible adiabatic process the entropy remains constant, and we note that the derivative in (8.3) is evaluated holding S constant. Thus for strictly mechanical phenomena, the internal energy is the natural energy variable. The pressure, which embodies the dynamics of the system, follows from U via (8.3). However, thermodynamics flourishes because many processes are not adiabatic. Many important processes take place isothermally. In an isothermal process, the Helmholtz free energy $F = U - TS$ becomes the logical energy variable. To justify this statement we must show how the change in F is related to the work done in an isothermal process.

If T is held constant, we have from $F = U - TS$

$$\Delta F = \Delta U - T\Delta S$$

If the process is reversible $T\Delta S = Q$, the heat evolved. From the first law

$$-W = \Delta U - Q$$

we obtain

$$-\Delta F = W \quad \text{(isothermal reversible process)} \tag{8.25}$$

We interpret this by saying that, in a reversible isothermal process, the work done by a system equals the decrease in its Helmholtz free energy. If the isothermal process is irreversible, we have $T\Delta S > Q$ and the same steps leading to (8.25) give instead $-\Delta F > W$. Thus, in general

$$-\Delta F \geqslant W \quad \text{(isothermal process)} \tag{8.26}$$

The inequality in (8.26) stems from the fact that irreversible processes generate entropy over and above that needed to satisfy the first law. Contrasting (8.26) with the adiabatic form of the first law (3.3) reveals that the Helmholtz free energy is the natural energy variable in an isothermal process. This view is strengthened by (8.7)

$$P = -\left(\frac{\partial F}{\partial V}\right)_T \tag{8.7'}$$

This relation, which replaces (8.3), shows that the dynamics contained in P now follow most naturally from F rather than from U. This shift of importance from the internal energy to the Helmholtz free energy followed because our attention switched from adiabatic processes to isothermal processes.

Whereas (3.3) is an equality, the equality in (8.26) holds only for reversible processes. We interpret (8.26) by saying that work performed isothermally and reversibly is "paid for" by a decrease of free energy. For an irreversible process, only part of the decrease in free energy is transformed into work ($W < -\Delta F$). The remainder of the free energy decrease is degraded by the irreversible processes responsible for entropy production.

•••••• ••••••

8.3.1 (a) Show that the Helmholtz free energy for 1 mole of an ideal gas can be expressed in the form

$$F = C_v T[1 - \ln T + (1-\gamma)\ln V - S_0]$$

where S_0 is a constant and $\gamma = 1 + R/C_v$.

(b) Verify the relation

$$P = -\left(\frac{\partial F}{\partial V}\right)_T$$

that is, show that it leads to the ideal gas equation of state.

(c) One mole of an ideal gas at temperature T undergoes a free expansion which doubles its volume. Show that the free energy change is

$$\Delta F = -RT \ln 2$$

(d) Is the condition $W \leqslant -\Delta F$ satisfied? Can the process be identified as reversible or irreversible?

....

8.4 ENTHALPY AND THERMOCHEMISTRY

The four thermodynamic potentials were invented to make thermodynamics "easy." Each potential is the natural energy variable for certain classes of physical processes. The enthalpy finds perhaps its most important application in the field of thermochemistry.

Thermochemistry concerns the thermodynamic aspects of processes which convert heat into chemical energy, or vice versa. By chemical energy we mean the energy associated with the electrical forces which bind atoms together to form molecules. The basic idea is quite simple. One wants to learn how and how much energy is released in chemical reactions. Conveniently, much of the energy is released as heat. We can gain valuable insights into reaction energetics simply by measuring the heat evolved.

The most elementary aspects of thermochemistry involve only the first law. A central concept in thermochemistry is the heat of reaction. A chemical reaction may either liberate or absorb energy. If the reaction releases energy, the temperature of the materials in the reaction vessel increases, and heat is subsequently rejected to the surroundings. In this case the reaction is said to be *exothermic*. The converse, where heat is absorbed by the system, is referred to as an *endothermic* reaction. Very loosely, the heat of reaction is the heat transferred between the reacting system and its surroundings. In order to have an unambiguous definition of the heat of reaction it is necessary to refer the reactants and products to the same pressure and temperature. Thus the heat of reaction is defined as the heat absorbed or rejected in going from some reactant state of temperature T and pressure P to some final product state at the same temperature and pressure. Careful measurements of heats of reaction have aided in establishing tables of enthalpies, internal energies, and other thermodynamic potentials.

The majority of chemical reactions are conducted at a constant pressure of approximately 1 atm. If the reaction proceeds at constant pressure, the heat of reaction equals the enthalpy change. This follows from the first law of thermodynamics and the definition of enthalpy. Thus from the definition $H = U + PV$ we have, at constant pressure,

$$(dH)_P = dU + P\,dV \tag{8.27}$$

However, from the first law

$$dQ = dU + P\,dV$$

from which we conclude $(dH)_P = dQ$; the heat absorbed or rejected in a reversible isobaric process equals the enthalpy change. In a finite process,

$$Q = \Delta H \equiv H_f - H_i \tag{8.28}$$

This elementary observation is the foundation of thermochemistry. By measuring the heat of reaction we can determine the enthalpy change.

Experiment lets us measure enthalpy *changes*. There remains some freedom which is exploited by assigning a value of zero enthalpy to the elements when they are in the so-called standard state. The standard state of a substance is defined to be its most stable form at a pressure of 1 atm and a temperature of 25 °C. The enthalpy of each element in its standard state is zero—by definition. In principle the enthalpies of the elements in other states and of chemical compounds can be determined by measuring appropriate heats of reaction. (Enthalpy tables are frequently labeled enthalpy of formation or heat of formation.) Thus, we may conceive of forming 1 mole of sodium chloride by starting with 1 mole of metallic sodium and 1 mole of gaseous chlorine—each in its standard state—and permitting them to unite chemically. The initial enthalpy of the system is zero. If we took care to see that the sodium chloride ended up at a temperature of 25 °C and a pressure of 1 atm, the heat evolved would be approximately 98.23 kcal. This is the standard heat of formation of sodium chloride. The value of 98.23 kcal is not determined by carrying out the aforementioned reaction, but rather by indirect measurements. The standard heat of formation is generally indicated as ΔH_f°. For example, the reaction equation

$$C + 2H_2 \rightarrow CH_4, \qquad \Delta H_f^\circ = -17.88 \text{ kcal}$$

signifies that the standard heat of formation (denoted here by the subscript f) for methane is 17.88 kcal/mole. The negative sign reveals that heat is rejected in the reaction. In reactions other than those leading to the formation of a compound from its elements the heat of reaction is generally indicated by ΔH. For example, the oxidation of methane would be described by

$$CH_4 + 2O_2 \rightarrow CO_2 + 2H_2O, \qquad \Delta H = -192 \text{ kcal}$$

In all cases ΔH (or ΔH_f°) is defined as the enthalpy of the products (pr) less that of the reactants (re)

$$\Delta H = H_{pr} - H_{re} \tag{8.29}$$

A little experience quickly convinces one that the chemical symbols appearing in the reaction equation can be regarded as enthalpies and treated algebraically.

Thus the reaction

$$AB + CD \rightarrow AC + BD$$

implies (assuming standard conditions for reactants and products)

$$\underbrace{(\Delta H_f^\circ)_{AB} + (\Delta H_f^\circ)_{CD}}_{H_{re}} = \underbrace{(\Delta H_f^\circ)_{AC} + (\Delta H_f^\circ)_{BD}}_{H_{pr}} - \Delta H$$

This is merely a definition of ΔH. Its utility will become evident shortly when we encounter it disguised as Hess's law.

The development of useful enthalpy tables relies on two facts. First, enthalpy is a state function, that is, a property of the thermodynamic state of a system which is independent of how the state is reached. One mole of sodium chloride in its standard state has an enthalpy of -98.23 kcal regardless of how it reached that state. This fact makes it possible to determine the enthalpies of certain compounds by indirect means and is the basis for Hess's law to be discussed shortly. The second fact is an empirical one, namely, that the enthalpy of a system depends strongly on temperature but only weakly on pressure. If we hold the temperature of a system constant and increase the pressure by 10 atm, the enthalpy change generally would be less than that produced by a 1 °C change in temperature. We can therefore often ignore pressure changes in enthalpy computations. A notable exception is encountered in gas liquefaction schemes where the slight pressure variation of enthalpy is crucial. We shall generally ignore the pressure dependence of enthalpy. This permits us to treat a chemical reaction as an isobaric process.

In a constant pressure reaction the heat rejected equals the enthalpy change

$$Q_p = H_{pr} - H_{re} \equiv \Delta H \tag{8.28'}$$

If the reactants are in their standard states, $H_{re} = 0$ and the heat rejected to the calorimeter equals the enthalpy of the products. For example, in the reaction

$$\tfrac{1}{2}H_2 + \tfrac{1}{2}Br_2 \rightarrow HBr$$

the heat rejected to the calorimeter is 8.66 kcal/mole of HBr formed. Thus $(\Delta H_f^\circ)_{HBr} = -8.66$ kcal/mole. Similarly, the complete oxidation of 1 mole of carbon releases 94.05 kcal, according to the reaction

$$C + O_2 \rightarrow CO_2$$

from which we conclude $(\Delta H_f^\circ)_{CO_2} = -94.05$ kcal/mole.

In many instances the enthalpy of formation must be determined indirectly. It is here that (8.28') finds extensive application. By systematically combining observed heats of reaction it is possible to determine an enthalpy of formation not open to direct measurement. Various difficulties may rule out such a direct measurement. For example, several reactions may proceed simultaneously. The measured heat of reaction is then a composite for the set of reactions. In other instances the reaction may proceed too slowly or too violently to permit a direct measurement of the heat of reaction. In still other situations the compound of interest may form only as an unstable intermediate which is subsequently transformed into a more stable compound. In such instances the enthalpy of formation must be determined by applying Hess's law, which states that the enthalpy change in a reaction is independent of the path by which the reaction proceeds. The validity of Hess's law is guaranteed by the fact that enthalpy is a state function. The enthalpy change depends on the initial and final states, but not on their connecting path. Algebraically, Hess's law is applied by adding or subtracting heats of reaction in such a way as to yield the unknown enthalpy. To illustrate the application of Hess's law, consider the problem of determining the enthalpy of formation of CO. A direct measurement of the heat of reaction for

$$C + \tfrac{1}{2}O_2 \rightarrow CO$$

is not possible because the carbon monoxide will be further oxidized to carbon dioxide. However, we can determine the enthalpy of formation of CO by regarding it as an intermediate step in the formation of CO_2, that is, we consider the reaction

$$C + O_2 \rightarrow CO_2, \qquad \Delta H_1 = -94.05 \text{ kcal}$$

as taking place according to the two-step sequence,

$$C + \tfrac{1}{2}O_2 \rightarrow CO, \qquad \Delta H_2 = \,?$$

$$CO + \tfrac{1}{2}O_2 \rightarrow CO_2, \qquad \Delta H_3 = -67.64 \text{ kcal}$$

To apply Hess's law one merely replaces the chemical symbols by the corresponding enthalpies and recalls the definition $\Delta H = H_{pr} - H_{re}$ given by (8.29). For example,

$$CO + \tfrac{1}{2}O_2 \rightarrow CO_2$$

translates into

$$(\Delta H_f^\circ)_{CO} + \tfrac{1}{2}(\Delta H_f^\circ)_{O_2} = (\Delta H_f^\circ)_{CO_2} - \Delta H_3$$

The enthalpy of O_2 is zero so

$$(\Delta H_f^\circ)_{CO_2} - (\Delta H_f^\circ)_{CO} = \Delta H_3 = -67.64 \text{ kcal}$$

In similar fashion one finds from the enthalpy equations for the other two reactions,

$$(\Delta H_f^\circ)_{CO_2} = \Delta H_1 = -94.05 \text{ kcal}$$

$$(\Delta H_f^\circ)_{CO} = \Delta H_2$$

From this set of equations and the measured values of ΔH_1 and ΔH_3 we find

$$(\Delta H_f^\circ)_{CO} = \Delta H_2 = \Delta H_1 - \Delta H_3 = -26.41 \text{ kcal}$$

These relationships may be expressed diagrammatically. The accompanying figure shows how Hess's law follows from the path independence of enthalpy change.

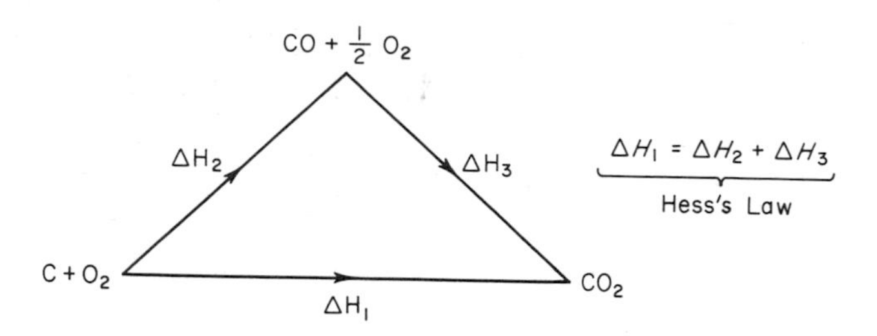

8.4.1 Determine ΔH for the reaction

$$C + H_2O \rightarrow CO + H_2$$

from the following information:

$$C + \tfrac{1}{2}O_2 \rightarrow CO, \qquad \Delta H = -26.4 \text{ kcal}$$

$$H_2 + \tfrac{1}{2}O_2 \rightarrow H_2O, \qquad \Delta H = -57.8 \text{ kcal}$$

8.4.2 Apply Hess's law to determine ΔH for the reaction

$$Mg + 2HCl \rightarrow MgCl_2 + H_2$$

from the measured heats of reaction for

$$Mg + Cl_2 \rightarrow MgCl_2, \qquad \Delta H = -153.2 \text{ kcal}$$

$$\tfrac{1}{2}H_2 + \tfrac{1}{2}Cl_2 \rightarrow HCl, \qquad \Delta H = -22.1 \text{ kcal}$$

8.5 GIBBS FREE ENERGY

The differential expression

$$dG = -S\,dT + V\,dP \tag{8.13}$$

establishes the Gibbs free energy as a function of the two independent variables P and T. The importance of G in chemistry and biology stems from the fact that many chemical reactions proceed at constant pressure and temperature.

If such reactions are reversible, (8.13) applies and shows that the Gibbs free energy is conserved. The Gibbs free energy is the natural energy variable in isothermal, isobaric processes. The quantity of signal importance in such a reaction is the change in G

$$\Delta G = G_{\text{pr}} - G_{\text{re}}$$

As we demonstrate below, a reaction will "go" spontaneously at constant temperature and pressure provided ΔG is negative, that is, if the system releases free energy to its surroundings. As a preliminary step then the chemist or biologist must ascertain whether or not a proposed reaction would result in a decrease in G. If ΔG should prove to be positive the reaction can still be made to go, but only through the addition of energy, perhaps in the form of heat. Such a change in G is not spontaneous and the pressure and/or temperature of the system are likely to be altered.

The decrease in G for a spontaneous isothermal and isobaric process is established as follows: From the definitions of F and G we find

$$G = F + PV$$

The change in G during an isobaric process is

$$\Delta G = \Delta F + P\Delta V$$

However, in an isobaric process

$$P\Delta V = P(V_{\text{f}} - V_{\text{i}})$$

equals the work done by the system, whence,

$$\Delta G = \Delta F + W \tag{8.30}$$

As we established in Section 8.3, the decrease in the Helmholtz free energy during an isothermal process is always greater than the work performed

$$-\Delta F \geqslant W \tag{8.26'}$$

or

$$\Delta F + W \leqslant 0$$

Comparing this with (8.30) establishes the desired result

$$\Delta G \leqslant 0 \tag{8.31}$$

the Gibbs free energy cannot increase during an isobaric and isothermal process. The equality in (8.31) corresponds to a reversible process, the inequality to an irreversible process. The spontaneous irreversible processes drive a system toward equilibrium, where its Gibbs free energy is a minimum.

An equilibrium state for a system at the same temperature and pressure as its surroundings is stable. Any spontaneous processes, which might upset the equilibrium, can occur only if they would decrease G. For a system already in the state of minimum G no such spontaneous processes can occur—the system is in equilibrium and the equilibrium is stable. Of course, equilibrium can be upset by external means, for example, by a change in the temperature of the surroundings. This causes the system to move toward a new equilibrium state. The free energy of the new state is again a minimum, but it is a minimum referring to the new temperature. An equilibrium value of G is a minimum compared to its values for other thermodynamic states of the same pressure and temperature.

Example 8.1

Two reactions of biological importance are the oxidation of glucose and photosynthesis. The overall result of the first of these may be represented by the scheme

$$C_6H_{12}O_6 + 6O_2 \rightarrow 6CO_2 + 6H_2O$$

The values of ΔG and ΔH, referred to the standard temperature of 298.16 K, are (per mole of glucose $C_6H_{12}O_6$)

$$\Delta G = -688 \text{ kcal/mole}, \qquad \Delta H = -673 \text{ kcal/mole}$$

Photosynthesis is equivalent to the reaction[†]

$$CO_2 + H_2O \rightarrow \{CH_2O\} + O_2$$

and has

$$\Delta G = 115 \text{ kcal/mole}, \qquad \Delta H = 112 \text{ kcal/mole}$$

For the oxidation of glucose, ΔG is negative and so the reaction proceeds spontaneously, liberating free energy. For photosynthesis, ΔG is positive and the reaction does not proceed spontaneously. The conversion of the radiant energy of sunlight into free energy drives the photosynthesis reactions. ◇

•••••• ••••••

8.5.1 The two reactions of Example 8.1 result in entropy changes. Compute the entropy changes and explain why there is no violation of the second law. [*Hint:* Use (8.11) to relate ΔH and ΔG, noting that the reactions proceed isothermally.]

•••••• ••••••

[†] The notation $\{CH_2O\}$ denotes the basic composition of a carbohydrate.

Another important class of processes which occur at constant pressure and temperature are changes of phase. Melting and vaporization are two familiar phase changes. Changes of phase are reversible processes, so that the Gibbs free energy is conserved, that is, G does not change. The subject of phase changes is important in many fields such as chemistry, physics, metallurgy, meteorology, and so on. Chapter 9 provides an introduction to the thermodynamics of phase changes.

In chemical and nuclear reactions, the number of particles of each reacting species can change. The thermodynamic potentials of the system generally change if the chemical composition is altered. In the next section we introduce the chemical potentials, quantities which describe the dependence of the thermodynamic potentials on the chemical composition.

8.6 CHEMICAL POTENTIAL

We have expressed the Gibbs free energy change by

$$dG = -S\,dT + V\,dP \tag{8.13}$$

This expression is incomplete under certain circumstances. For example, suppose we take a small pan of water as our system. We expose the water to the surrounding atmosphere on a warm day when the humidity is very low—the sort of day which makes you wish you did not have to study thermodynamics. If the water is at the same temperature and pressure as the surroundings, there will be no change in T or P. Under these conditions, (8.13) predicts that $dG = 0$, that is, that G is a constant. However, G is not constant! If we watch the system, the water in the pan, it disappears before our eyes—by evaporation. Thus G is an extensive quantity; it is directly proportional to the amount of water present. As water evaporates, G decreases. When the water has evaporated completely, $G = 0$. The deficiency of (8.13) is that it holds only for closed systems whose chemical composition does not change. This shortcoming is easily rectified. Consider an open system comprised of one phase of a single chemical species.† The pan of water is an example of such a system. Let N denote the number of molecules in the system, and note that N can change because the system is open. The Gibbs free energy depends on T, P, and N. If T, P, and N change by differential amounts, the change in G is

$$dG = \left(\frac{\partial G}{\partial T}\right)_{P,N} dT + \left(\frac{\partial G}{\partial P}\right)_{T,N} dP + \left(\frac{\partial G}{\partial N}\right)_{P,T} dN \tag{8.32}$$

† By a phase we mean matter which is homogeneous and which has definite boundaries.

The derivatives $(\partial G/\partial T)_{P,N}$ and $(\partial G/\partial P)_{T,N}$ are evaluated with N held constant. The constancy of N was implicit in the developments of Section 8.1 where we arrived at the results

$$\left(\frac{\partial G}{\partial T}\right)_P = -S, \qquad \left(\frac{\partial G}{\partial P}\right)_T = V$$

We now make this explicit by writing

$$\left(\frac{\partial G}{\partial T}\right)_{P,N} = -S, \qquad \left(\frac{\partial G}{\partial P}\right)_{T,N} = V \tag{8.33}$$

This converts (8.32) into

$$dG = -S\,dT + V\,dP + \left(\frac{\partial G}{\partial N}\right)_{T,P} dN \tag{8.34}$$

For a closed system, N is constant so that $dN = 0$ and (8.34) reduces to (8.13). The quantity

$$\left(\frac{\partial G}{\partial N}\right)_{T,P} \equiv \mu \tag{8.35}$$

is called the chemical potential; μ is a measure of how G is altered by a change in the number of particles present. With this definition, (8.34) becomes

$$dG = -S\,dT + V\,dP + \mu\,dN \tag{8.36}$$

It is important to note that the chemical potential is an intensive quantity. Thus μ is independent of N although it can vary with T and P. To see this, imagine that we double the size of the system—we mentally slap two of them together thereby making every extensive quantity twice as large without changing any of the intensive variables. The extensive quantities in (8.36) are S, V, dG, and dN. Each of these is doubled. In order for (8.36) to remain valid dT, dP, and μ must remain unchanged, that is, T, P, and μ must be intensive quantities. The fact that μ is an intensive variable means that it cannot depend on the amount of material in the system. Explicitly, μ cannot depend on N, although it will in general vary with T and P;

$$\mu = \mu(T, P), \qquad \left(\frac{\partial \mu}{\partial N}\right)_{T,P} = 0 \tag{8.37}$$

To account for the presence of several species of particles or more than one phase is routine from the bookkeeping point of view. For a system of C species we can write

$$G = G(T, P, N_1, N_2, \ldots, N_C)$$

where N_i denotes the number of molecules of the ith species. In place of (8.36) there then appears

$$dG = -S\,dT + V\,dP + \sum_{i}^{C} \mu_i\,dN_i \tag{8.38}$$

where

$$\mu_i = \left(\frac{\partial G}{\partial N_i}\right)_{T,P} \tag{8.39}$$

is the chemical potential for the ith species. Each chemical potential is an intensive quantity, that is, μ_i can depend on T and P and the other μ's but not on the N_i, which are extensive variables. Suppose we consider a process which proceeds at constant pressure and temperature. Then $dT = 0$, $dP = 0$ in (8.38), and

$$dG = \sum_{i} \mu_i\,dN_i \tag{8.40}$$

We can "integrate" this equation to obtain

$$G = \sum \mu_i N_i \tag{8.41}$$

This integration is purely formal, and the result (8.41) follows directly from (8.40) upon remembering that both G and the N_i are extensive variables, whereas the μ_i are intensive.

We have seen how the introduction of the chemical potentials modified (8.13) by the addition of the term $\sum_i \mu_i\,dN_i$, giving (8.38). As G is related directly to the other thermodynamic potentials, they are modified in a similar manner. The expressions (8.1), (8.5a), and (8.9) are replaced by

$$dU = T\,dS - P\,dV + \sum_{i=1}^{C} \mu_i\,dN_i \tag{8.42}$$

$$dF = -S\,dT - P\,dV + \sum_{i=1}^{C} \mu_i\,dN_i \tag{8.43}$$

$$dH = T\,dS + V\,dP + \sum_{i=1}^{C} \mu_i\,dN_i \tag{8.44}$$

From these expressions it is evident that μ_i can be defined not only by (8.39), but also as a derivative of U, F, or G,

$$\mu_i = \left(\frac{\partial U}{\partial N_i}\right)_{S,V} = \left(\frac{\partial F}{\partial N_i}\right)_{T,V} = \left(\frac{\partial H}{\partial N_i}\right)_{S,P} \tag{8.45}$$

With four equations featuring the chemical potentials there is now a choice of interpretations. The viewpoint adopted is likely to depend on the "ax one has to grind."

The biologist is likely to interpret μ_i from the standpoint of the Gibbs free energy. He is concerned with chemical reactions in which the temperature and pressure remain constant, and for which the Gibbs free energy is the natural energy variable.

The theoretical physicist or chemist concerned with the microscopic structure of matter is apt to interpret μ_i in terms of the Helmholtz free energy F. Indeed, we will adopt this viewpoint in our study of statistical mechanics.

The thermochemist and the low-temperature physicist who find it most convenient to deal with the enthalpy are prone to interpret μ_i in terms of their favorite thermodynamic potential.

There is even an interpretation of μ_i tailored to the needs of the student! Having struggled diligently to assimilate the concepts of heat and internal energy he now finds that the first law (with $T\,dS = dQ$) appears as

$$dU = dQ - P\,dV + \sum_{i=1}^{C} \mu_i\,dN_i \tag{8.42a}$$

This form permits an obvious interpretation of the chemical potentials. As its name suggests, the internal energy accounts for *all* forms of energy stored within the system. The right side of (8.42a) accounts for the different ways by which U can change. Thus the addition of heat dQ or the performance of mechanical work $-P\,dV$ can alter U, but changes in U also can arise through reactions which alter the composition of the system. It is precisely these changes in U which are reflected in the term $\sum_{i=1}^{C} \mu_i\,dN_i$ in (8.42a); the chemical potential simply reveals another of the many faces of energy.

......

8.6.1 An ideal monatomic gas has a molar heat capacity of $C_v = \frac{3}{2}R$. Recall that $k = R/N_0$ is the Boltzmann constant, and use the result of Problem 8.3.1 in conjunction with (8.45) to show that the chemical potential for an ideal monatomic gas may be expressed as

$$\mu = \frac{F}{N} = \tfrac{3}{2}kT[1 - \ln T - \tfrac{2}{3}\ln V - S_0]$$

......

Example 8.2 Phase Equilibrium

We wish to demonstrate that equilibrium between two (or more) phases of a substance requires equality of their chemical potentials. Thus, if a liquid and its vapor are to be in equilibrium it is necessary that $\mu_l = \mu_v$. Since phase transformations occur at constant temperature and pressure, (8.40) applies. We write (for two phases in equilibrium)

$$dG = \mu_1\,dN_1 + \mu_2\,dN_2 \tag{8.40a}$$

where 1 and 2 refer to the two phases. In Section 8.5 we established that G is a minimum for an equilibrium state. We envision a transformation in which a small amount of one phase converts to the other, while maintaining the equilibrium; for example, a drop of water vaporizes. On the one hand we must have $dG = 0$ to maintain G at a minimum, while on the other we must have

$$dN_1 = -dN_2 \tag{8.46}$$

to conserve mass. It then follows from (8.40a) that

$$\mu_1 = \mu_2 \tag{8.47}$$

Phase equilibrium requires equality of the chemical potentials. For multi-component systems a generalization of (8.47) holds. The chemical potential of each species is the same for phases in joint equilibrium. For example, a pure substance at its triple point would have

$$\mu(\text{solid}) = \mu(\text{liquid}) = \mu(\text{gas}) \tag{8.47a}$$

A solution of bromine and water in equilibrium with their vapors would satisfy

$$\mu_{\text{Br}}(\text{liquid}) = \mu_{\text{Br}}(\text{gas}), \qquad \mu_{\text{H}_2\text{O}}(\text{liquid}) = \mu_{\text{H}_2\text{O}}(\text{gas}) \tag{8.47b}$$

Example 8.3 The Condition for Chemical Equilibrium

Chemical equilibrium is one of the conditions for thermodynamic equilibrium. In Section 8.5 we showed that the Gibbs free energy is a minimum for a state of thermodynamic equilibrium. We can use this fact to establish the general condition for chemical equilibrium.

For the purposes of illustration, consider the reaction whereby hydrogen and chlorine combine to form hydrogen chloride. This reaction may be represented in equation form by

$$H_2 + Cl_2 - 2HCl = 0$$

In general chemical reactions can be expressed in the form

$$\sum \nu_i A_i = 0 \tag{8.48}$$

in which the A_i denote chemical symbols and the ν_i are either positive or negative integers. For the hydrogen chloride reaction, $\nu_{H_2} = \nu_{Cl_2} = 1$ and $\nu_{HCl} = -2$.

If thermodynamic equilibrium prevails, the Gibbs free energy is a minimum. Suppose we alter the amounts of the various components under conditions of constant temperature and pressure. For processes in which only the N_i are altered, (8.40) gives the change in G as

$$dG = \sum_{i=1}^{C} \mu_i \, dN_i \tag{8.40}$$

We want to apply this relation to a special situation, namely, where the N_i change in such a way as to maintain the existing chemical equilibrium. If G is a minimum at equilibrium, the changes in the N_i must be such that this minimum value is preserved, that is, we must have $dG = 0$ if the equilibrium is to persist

$$\sum_{i=1}^{C} \mu_i \, dN_i = 0 \tag{8.49}$$

The dN_i in (8.49) are related in a very special way to the ν_i in (8.48). To see just how these quantities are related, consider the hydrogen chloride reaction. Suppose we start by mixing 1 mole of hydrogen and 3 moles of chlorine. There will be many molecular reactions in which hydrogen chloride is produced. Eventually the reactions cease and we have chemical equilibrium. Now there is one very obvious way in which we can change the N_i without upsetting the existing equilibrium—and it is the only way. We can add, say, 1 mole of hydrogen and 1 mole of chlorine and remove 2 moles of hydrogen chloride. This does not unbalance the equilibrium because the added H_2 and Cl_2 combine to form the two moles of HCl which are removed. In general chemical equilibrium is maintained so long as

$$dN_i = K\nu_i \tag{8.50}$$

where K is an arbitrary constant, but must be the same for all reaction participants. In the example just cited, $K = N_0$. Inserting (8.50) into (8.49) leads to the desired relation—the condition necessary for chemical equilibrium

$$\sum_i \mu_i \nu_i = 0 \tag{8.51}$$

This equation illustrates quite clearly an earlier remark that the chemical potentials are not all independent variables.

REFERENCES

The relationships between the second law and organic evolution are explored in:

H. F. Blum, "Time's Arrow and Evolution," 2nd ed. Harper, New York (1962).

Blum reveals the complementary aspects of the second law and Darwin's theory of natural selection. Note that Blum uses F to denote the Gibbs free energy G and E for the internal energy U.

Whether a thermodynamic system is open or closed may have unexpected consequences. Evidence supporting the "myth" that hot water cools faster than cold water is presented in:

G. S. Kell, Cooling of Hot and Cold Water, *Am. J. Phys.* **37**, 565 (1969).

An extensive treatment of thermochemistry will be found in:

G. W. Castellan, "Physical Chemistry." Addison-Wesley, Reading, Massachusetts (1964).

A more elementary approach is offered in:

B. H. Mahan, "Elementary Chemical Thermodynamics," Chapter 2. Benjamin, New York (1964).

Chapter 9

Phase Changes and Low-Temperature Physics

9.1 PHASE CHANGES AND PHASE DIAGRAMS

In the preceding chapter we defined a phase as matter which is homogeneous and which has definite boundaries. For example, the air in a balloon is in the gaseous phase and is separated from its surroundings by a definite boundary. Each chemically or physically distinct species present in a system qualifies as a component. The air is a single-phase multicomponent system, being a homogeneous mixture of nitrogen, oxygen, water vapor, and other gases. A container filled with water and water vapor constitutes a two-phase single-component system, there being a definite boundary between the liquid and gaseous phases as well as between the combined system and its surroundings.

A solid may be heated until its melts and becomes a liquid. The liquid in turn may be heated until it reaches its boiling point and vaporizes. Such phase changes are isobaric and isothermal processes. Heat is absorbed but the temperature does not rise until the phase change is complete. The bulk of the heat absorbed is used to alter the microscopic structure of the system. In melting for example, the atomic structure changes from the highly ordered array of the solid phase to the randomized liquid phase in which atoms behave like people in a panic-stricken crowd, jostling their neighbors as they move about erratically. In the melting process, atomic bonds must be broken. The heat of fusion, L_f, is the amount of heat absorbed in converting 1 gm of a substance from solid to liquid. A large heat of fusion is indicative of strong atomic binding. Phase changes are reversible processes. When a liquid solidifies an amount of heat L_f is released.

•••••• ••••••

9.1.1 Explain:

(a) why you may experience a chill upon emerging from a shower and
(b) how the air temperature may rise at the onset of a snowfall.

•••••• ••••••

A heating curve illustrates the relationship between temperature and heat absorbed during a phase change. Figure 9.1 shows the heating curve for ice and water near the melting point. It plots temperature versus the amount of heat added Q. If the heat is added at a constant rate, Q is directly proportional to the heating time t. The heating curve shows that the sample starts as ice at $-5\,°C$. The addition of heat at a constant rate results in a uniform temperature rise until the ice reaches the melting point of $0\,°C$. No further increase of temperature results until all of the ice has melted. For a sample of m gm, the total heat absorbed during the phase change is mL_f, as indicated in the figure. Once all of the ice has melted, the addition of more heat causes the temperature rise to resume.

Example 9.1 A Latent Heat Measurement

If the heating rate is known, a graph of temperature versus heating time (that is, a heating curve) permits a determination of the latent heat. For example, suppose that heat is added at a constant rate of $\dot{Q} = 100$ W to

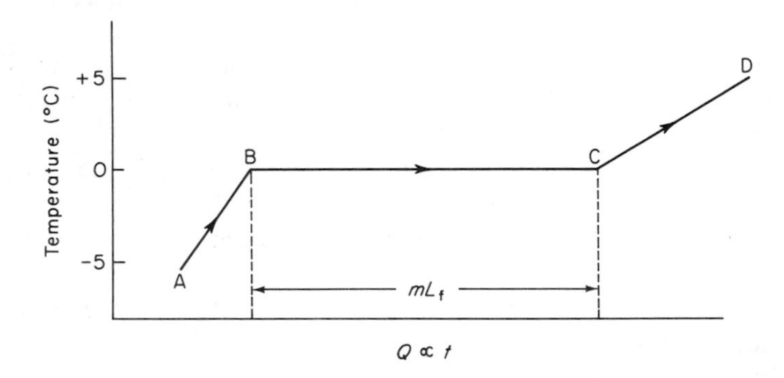

Figure 9.1. The heating curve for a sample of m gm of ice. The initial temperature is $-5\,°C$. Heat is added at a constant rate making Q proportional to t, the heating time. At B the ice has reached the melting point. During the change of phase (melting) the temperature remains constant ($0\,°C$). At C the ice has melted completely.

$m = 20$ gm of ice, and that the time during which the temperature remained constant (melting time) was 66 sec. The total heat absorbed during melting may be expressed as

$$mL_f = \dot{Q}t_{melt}$$

which gives

$$L_f = \frac{(100 \text{ joule/sec})(66 \text{ sec})}{(20 \text{ gm})} \frac{1 \text{ cal}}{4.18 \text{ joule}} = 79.1 \frac{\text{cal}}{\text{gm}} \quad \diamond$$

•••••• ••••••

9.1.2 Explain how the slope of the heating curve is related to the specific heat of the sample. Is your explanation consistent with Figure 9.1, where the slope of the segment AB is approximately twice that of the segment CD? How do the specific heats of ice and liquid water compare?

•••••• ••••••

A liquid at its boiling point may be vaporized by adding its heat of vaporization L_v. Part of the energy so supplied is used to free molecules from the surface of the liquid. A large portion of L_v is converted into work of expansion —the $P\,dV$ work done against the surrounding atmosphere.

Sublimation—the solid-to-vapor change of phase—is quite similar to vaporization, it being a process of boiling atoms from the surface of a solid rather than a liquid. The sublimation of dry ice (solid carbon dioxide) is a familiar example.

Example 9.2 The Method of Mixtures Again

A 50-gm ice cube at −2 °C is placed in 840 gm of water at 12 °C. The mixture is contained in an aluminum calorimeter with a mass of 200 gm and a specific heat of 0.215 cal/gm-°C. The initial temperature of the calorimeter is 12 °C. The specific heat of ice is 0.47 cal/gm-°C. The final temperature of the mixture is 7 °C. Using these data we wish to determine the heat of fusion of ice.

If we ignore all work of expansion, we can equate the heat rejected by the calorimeter and water with the heat absorbed by the ice cube. The 50 gm of ice is warmed to 0 °C. It then absorbs the heat of fusion and melts and, as water, is warmed to the final temperature of 7 °C. The heat absorbed by the 50 gm is

$$Q_{abs} = 50 \text{ gm}\left[\left(0.47 \frac{\text{cal}}{\text{gm-°C}}\right)(2\,°\text{C}) + L_f + \left(1.00 \frac{\text{cal}}{\text{gm-°C}}\right)(7\,°\text{C})\right]$$

The heat rejected by the water and calorimeter in cooling 5 °C is

$$Q_{rej} = 840 \text{ gm}\left(1.00 \frac{\text{cal}}{\text{gm-°C}}\right)(5\,°\text{C}) + 200 \text{ gm}\left(0.215 \frac{\text{cal}}{\text{gm-°C}}\right)(5\,°\text{C})$$

Equating Q_{abs} and Q_{rej} gives the heat of fusion as

$$L_f = 80.4 \text{ cal/gm}$$

The accepted value is 79.7 cal/gm suggesting that some form of heat or energy exchange has been ignored. In this type of experiment the heat absorbed or rejected by the thermometers, stirrers, and so on, must be included and the effects of radiation minimized or accounted for. ✧

•••••• ••••••

9.1.3 An immersion heater is placed in a mixture of 100 gm of ice and 150 gm of water (at 0 °C). The heater converts electrical energy into heat at a rate of 3000 W. Assume that all of the heat is transferred to the water-ice mixture. (This ignores the heat absorbed by the heater coil and container.) Determine the time required:

(a) to melt the ice,
(b) to bring the water to its normal boiling point,
(c) to completely evaporate the water.

•••••• ••••••

Melting and vaporization are termed first-order phase changes. The hallmark of a first-order phase transition is an entropy change. As a consequence of the entropy change there is an accompanying latent heat. For example, when ice melts the entropy of the resulting liquid exceeds that of the ice. The latent heat of fusion L_f is related to the entropy change ΔS and melting point T by

$$L_f = T\Delta S = T(S_l - S_s)$$

In the so-called higher-order phase changes the entropy is the same for both phases but other thermodynamic properties change. There is no latent heat.

•••••• ••••••

9.1.4 Starting from the thermodynamic relation

$$dS = \frac{dQ}{T}$$

derive the result

$$L_f = T\Delta S$$

Would the result be valid if a change of phase were not a reversible process?

•••••• ••••••

In Chapter 5 we indicated how the Clausius–Clapeyron equation could be used as a thermometer, or as the basis of a scheme for determining latent heats.

For one-component systems the Clausius–Clapeyron equation also may be used to map out important boundaries on what are known as *phase diagrams*. One-component systems such as the simple fluid or the paramagnetic solid are elementary in that their equations of state relate only three thermodynamic variables. For a fluid the equation of state may be stated as a relation giving the pressure in terms of the volume and temperature. Examples include our favorite—the ideal gas—

$$P = \frac{RT}{V}$$

and the van der Waals equation of state,

$$P = \frac{RT}{V-b} - \frac{a}{V^2}$$

A paramagnetic solid which obeys Curie's Law has an equation of state of the form (the volume V is regarded as a constant)

$$M(H,T) = CV\frac{H}{T}$$

Regardless of the functional form of the equation of state it can be represented geometrically as a surface so long as it relates just three variables. Figure 4.1 shows a segment of the P–V–T surface for the ideal gas. By relabeling the axes appropriately the same surface portrays the Curie paramagnet equation of state.

•••••• ••••••

9.1.5 (a) Relabel the P–V–T axes of Figure 4.1 so that the surface represents the Curie paramagnet equation of state

$$M(H,T) = CV\frac{H}{T}$$

Note that this relabeling can be accomplished in more than one way. Of the three quantities P, V, and T, only one is an extensive variable. Likewise, of M, H, and T, just one is an extensive variable. Choose the relabeling scheme which places the extensive variables in correspondence. (You may wish to review Section 4.4.)

(b) Describe the three magnetic processes which correspond to the paths A→B→C→A in Figure 4.1.

•••••• ••••••

The ideal gas, the van der Waals gas, and the Curie paramagnet share one common property. Unlike real substances they do not exhibit any changes

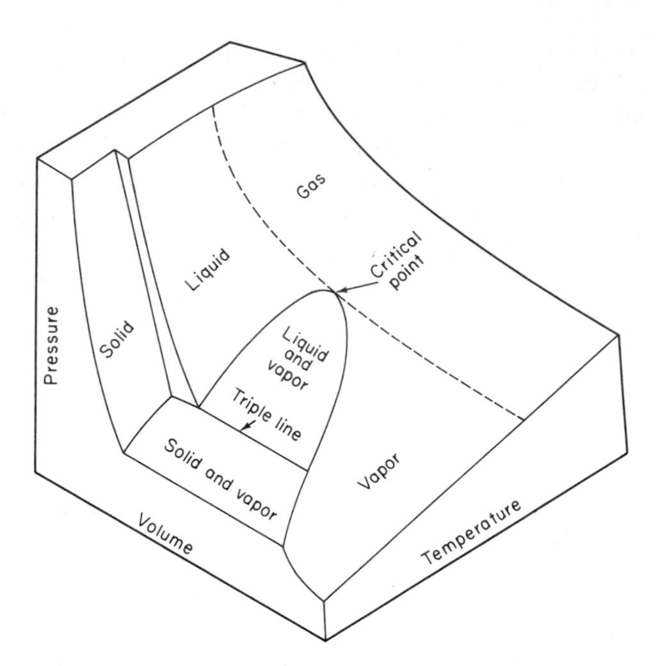

Figure 9.2. The *P–V–T* surface for carbon dioxide. The dashed line marks the critical isotherm. The appearance of the surface, when viewed along the volume axis, is shown in Figure 9.3.

of phase. Figure 9.2 shows the *P–V–T* surface of carbon dioxide (CO_2). The high-temperature low-pressure segments of the surface are very similar to that of the ideal gas. Various regions of the surface are labeled solid, liquid, and gas. These correspond to ranges of P, V, and T for which only a single phase can exist. Other portions of the surface locate states where two (or three) phases may coexist in a stable equilibrium. The line labeled "triple line" identifies a unique pressure and temperature and a range of volume (or density) over which all three phases can simultaneously exist. The point labeled "critical point" is of special significance. The critical point for a pure substance is located at a unique pressure, temperature, and density (P_c, T_c, ρ_c) which marks the upper terminus of the liquid–vapor coexistence region. The isotherm passing through the critical point is called the "critical isotherm." If a gas is compressed isothermally at a temperature in excess of the critical temperature (that is, along an isotherm above the critical point) it never undergoes condensation—it remains a gas. If it is compressed isothermally at a temperature less than T_c, a separation into liquid and vapor phases is observed. It is possible to start with a substance in the gaseous phase at a temperature $T_i < T_c$, raise its temperature above T_c, compress it isothermally,

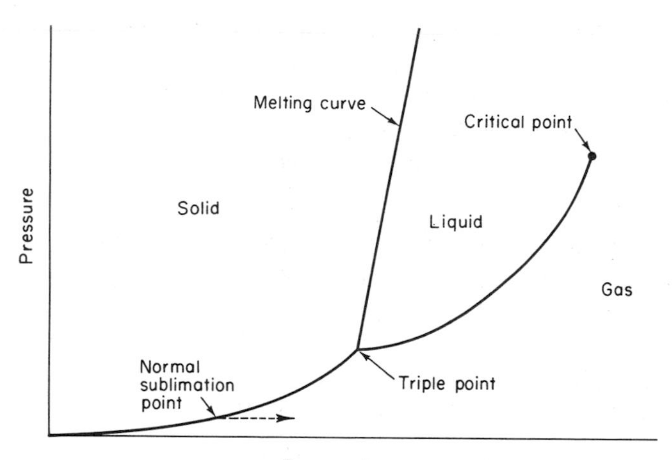

Figure 9.3. *P–T* diagram for carbon dioxide (not to scale). The lines define conditions for stable equilibrium between two phases. Their intersection is the triple point (5.1 atm, 216.6 K). The slopes of the three curves are determined by the Clausius–Clapeyron equation. Carbon dioxide is typical of most pure substances in that it expands upon melting. As a consequence the slope of its melting curve is positive ($dP/dT > 0$). The horizontal arrow at the normal sublimation point indicates that CO_2 sublimes at 1 atm (194.6 K). If the pressure is raised above 5.1 atm, heating the solid may result in melting. The values at the critical point are 72.9 atm and 309.1 K.

and then cool it to T_i, ending with a liquid. At no time would any condensation be evident. There simply would be a continuous transition from the gaseous to liquid phase.

If the P–V–T surface is projected onto the P–T plane we obtain a P–T phase diagram.† Figure 9.3 shows the P–T diagram for carbon dioxide. The triple line projects onto the P–T diagram as the triple point. The portion of the P–V–T surface labeled "liquid and vapor" appears as a line on the P–T diagram—the vaporization curve. The other coexistence regions likewise appear as lines on the P–T plot. The solid–liquid region of the surface projects onto the P–T plane as the line designated melting curve (also termed the "fusion line"). The solid–gas segment translates into the sublimation curve. The Clausius–Clapeyron equation determines the slopes of these three curves. The Clausius–Clapeyron equation was derived in Chapter 5

$$\frac{dP}{dT} = \frac{L_v}{T(V_g - V_l)} \tag{5.9a'}$$

† We shall use the explicit designation P–T diagram since there are many diagrammatic ways of representing the phases of a system.

where L_v is the latent heat of vaporization, the heat required to convert one mole of liquid to vapor and V_g and V_l denote the molar volumes of gas and liquid, respectively. A generalized form of (5.9a′), applicable to sublimation and melting as well as vaporization, is

$$\frac{dP}{dT} = \frac{L_{21}/T}{(V_2 - V_1)} \tag{9.1}$$

where L_{21} is the molar latent heat for the change from phase 1 to phase 2. The numerator L_{21}/T is the molar entropy change. In all three phase changes, melting, vaporization, and sublimation, the entropy increases. Stated differently, the associated latent heat is positive in all three cases. Vaporization and sublimation result in volume increases so that $V_2 - V_1$ is positive. It follows from (9.1) that dP/dT is positive for these two transitions

$$\left(\frac{dP}{dT}\right)_v > 0, \qquad \left(\frac{dP}{dT}\right)_{sub} > 0$$

Most substances, including carbon dioxide, expand upon melting, in which case

$$\left(\frac{dP}{dT}\right)_{melt} > 0, \qquad V_l > V_s$$

Although the form of the CO_2 diagram is typical its scale is not. The triple-point vapor pressure is unusually high—over 5 atm. As a consequence, the normal melting point lies well below the triple point and the solid sublimes—converts from solid to vapor—at a pressure of 1 atm. For this reason solid CO_2 has become better known as dry ice.

There are a few substances which contract upon melting. For these materials the Clausius–Clapeyron equation reveals that the melting curve has a negative slope

$$\left(\frac{dP}{dT}\right)_{melt} < 0, \qquad V_l < V_s$$

Water is among the substances which contract upon melting. Figure 9.4 shows the P–T diagram for water and reveals the negative slope of the melting curve. The dashed line in Figure 9.4 reveals that an increase in pressure can cause melting. Pressure-induced melting permits glaciers to flow around obstacles. The refreezing which follows the removal of pressure is called regelation. This phenomenon may be demonstrated using a block of ice, a large weight, and a fine wire. The wire is looped over the block of ice with the weight attached beneath. The pressure beneath the wire causes the ice to melt and the wire slowly falls through the block. The water refreezes behind the wire. Thus, even though the wire passes through it, the block ends up unchanged.

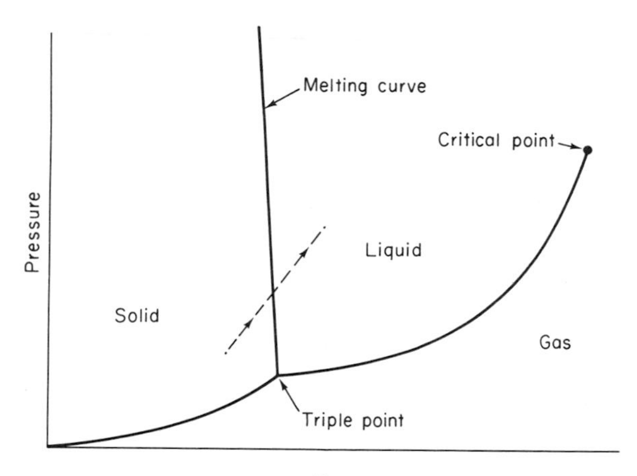

Figure 9.4. The P–T diagram for water (not to scale). Only a small portion of the P–T diagram is shown. Not indicated are the many high-pressure phases of ice. Because water expands upon freezing the slope of the melting curve is negative ($dP/dT < 0$). An increase in pressure (as suggested by the dashed line) can thereby induce melting. The values at the critical point are $P_c = 218.2$ atm and $T_c = 647.3\,\text{K}$. The values at the triple point are $P_3 = 4.58$ Torr and $T_3 = 273.16\,\text{K}$.

Example 9.3 The Heat of Sublimation of CO_2—An Estimate

In Chapter 5 we showed how the heat of vaporization could be determined by applying the Clausius–Clapeyron equation to vapor pressure data. Here we indicate how it may be used to estimate the heat of sublimation of solid carbon dioxide. If we have two data points on the sublimation curve, reasonably close together, we can determine L_{sub} by setting

$$\frac{dP}{dT} \simeq \frac{\Delta P}{\Delta T} = \frac{P_2 - P_1}{T_2 - T_1}$$

where (P_1, T_1) and (P_2, T_2) locate the two pressure–temperature data points. This approximation treats the sublimation curve as a straight line between the points (P_1, T_1) and (P_2, T_2). The latent heat of sublimation then follows from (9.1) as

$$L_{sub} = T(V_g - V_s)\frac{\Delta P}{\Delta T}$$

Since

$$V_g \gg V_s$$

we may safely ignore V_s by comparison to V_g. Furthermore, if we treat the gas as ideal and use $PV_g = RT$, we obtain

$$L_{sub} \simeq \frac{RT^2}{P}\frac{\Delta P}{\Delta T}$$

This expression gives the molar heat of sublimation since we inserted the volume of 1 mole of gas.

Figure 9.3 can be used to make an estimate of L_{sub} for CO_2. The two points used are the normal sublimation point

$$P_1 = 1 \text{ atm}, \qquad T_1 = 194.6 \text{ K}$$

and the triple point

$$P_2 = 5.1 \text{ atm}, \qquad T_2 = 216.6 \text{ K}$$

For T and P we use the average values between the two points

$$P \simeq 3 \text{ atm}, \qquad T = 206 \text{ K}$$

The resulting value of L_{sub} is

$$L_{sub} \simeq 5.3 \text{ kcal/mole}$$

If we divide L_{sub} by the molecular weight of CO_2 (44 gm/mole), we obtain the specific heat of sublimation as 121 cal/gm. This is in fair agreement with the experimental value of 136.9 cal/gm. ◇

•••••• ••••••

9.1.6 (a) Determine the entropy change when 1 mole of CO_2 sublimes.

(b) Is the majority of the heat of sublimation used to change the internal energy or to perform work of expansion? Use the first law to make a rough (order-of-magnitude) calculation to support your answer.

•••••• ••••••

9.2 LIQUEFACTION AND CRYOGENICS

Cryogenics is the science of low temperatures. It is a mushrooming field which continually finds new and wider applications in research and technology. The history of cryogenics is delightfully recounted in Mendelssohn's popular work, "The Quest for Absolute Zero."† The early history of cryogenics is largely the story of pioneering efforts to liquefy gases. The liquefaction of oxygen by Cailletet and Pictet in 1877 was followed by a parade of successes. In succession, nitrogen, hydrogen, and finally helium were liquefied. The

† K. Mendelssohn, "The Quest for Absolute Zero." McGraw-Hill, New York (1966).

TABLE 9.1

Boiling Points and Heats of Vaporization of Cryogenic Fluids

Substance	Normal boiling point (K)	Heat of vaporization (cal/gm)
^4He	4.2	4.9
H_2	20.3	108.0
Ne	27.1	20.5
N_2	77.3	47.6
F_2	85.0	41.2
Ar	87.2	39.0
O_2	90.1	50.9
CH_4	111.6	121.9

normal boiling points of several cryogenic liquids are given in Table 9.1, along with their latent heats of vaporization. We will discuss only the basic physics underlying liquefaction schemes. Detailed descriptions of liquefiers and their operation will be found in the references.

There are two recognized techniques for liquefying gases. These are the external work and the internal work methods. Both methods act to reduce the kinetic energy of the molecules to a point where the attractive intermolecular forces cause the molecules to coalesce as liquid droplets. In the external work scheme a gas is expanded so as to perform work against its surroundings. If the expansion is adiabatic or nearly so, the work is done at the expense of the internal energy of the gas, and the temperature decreases. There are two primary drawbacks to the external work method. An expansion engine necessarily involves moving parts, the lubrication of which is difficult at low temperatures. Equally important is the fact that the temperature drop produced by an adiabatic expansion decreases in direct proportion to the initial temperature.

••••••　　　　••••••

9.2.1 The equation describing the adiabatic expansion of an ideal gas is (see Example 4.2 and Problem 4.5.3)

$$\left(\frac{T}{T_0}\right)^{\gamma} = \left(\frac{P}{P_0}\right)^{\gamma-1}$$

Take $\gamma = 7/5$, a value appropriate for diatomic gases, and compute the final temperature for an adiabatic expansion which starts at $P = 100$ atm, $T = 300$ K, and ends at $P = 1$ atm. While real gases do not exhibit ideal behavior at 100 atm the resulting temperature is not unrealistic. How does your result compare with the normal boiling points (NBP) of oxygen and nitrogen?

••••••　　　　••••••

The internal work method employs a throttling process, in which gas is forced through a narrow valve. To understand the meaning of the phrase internal work, we must anticipate two results of the kinetic theory of gases (Chapter 11). One result is that the temperature of a gas is a measure of the kinetic energy of the molecular motions. The second result is that, on the average, the intermolecular forces are attractive. During a throttle the average distance between molecules increases. Work is done by the molecules against the attractive intermolecular forces. The work increases the potential energies of the molecules at the expense of their kinetic energies. The decrease in kinetic energies registers as a temperature drop. The work is internal because the forces act between molecules in the gas, in contrast to the external work method where work is performed by forces pushing against the surroundings.

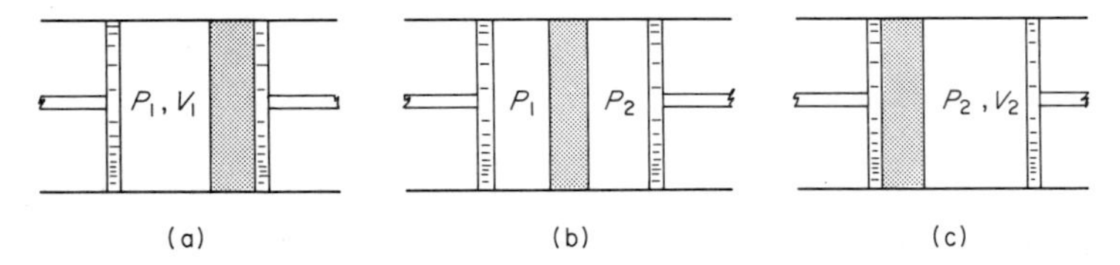

Figure 9.5. Three stages of the Joule–Thomson porous-plug experiment. Initially (a), the gas is to the left of the porous plug (dotted area) at a high pressure. The pressure forces the gas to seep through the pores of the plug. At an intermediate stage (b) a portion of the gas has wiggled through the plug. The piston on the right side expands, maintaining a constant pressure P_2. The left piston maintains a constant pressure P_1. The final state (c) finds all of the gas on the right side of the plug at the lower pressure P_2.

Thermodynamically, the throttling process is equivalent to the Joule–Thomson porous-plug experiment. In this experiment, gas is forced through a porous ceramic plug (see Figure 9.5). The gas starts on the left side of the plug at P_1, V_1 [part (a)]. As the piston forces gas through the plug, a second piston travels to the right. The pressure on the left side is maintained constant at P_1 throughout the process. The pressure on the right side is maintained constant at P_2. The final configuration [part (c)] has the gas on the right side of the plug at P_2, V_2. The entire system is insulated so that no heat is absorbed or rejected during the process. The total work done by the gas is (recall that work done isobarically is $P\Delta V$)

$$W = P_1(0-V_1) + P_2(V_2-0) = P_2V_2 - P_1V_1$$

With $Q = 0$, the first law gives

$$0 = \Delta U + W$$

Letting U_1 and U_2 denote the initial and final internal energies we find

$$U_1 + P_1V_1 = U_2 + P_2V_2 \tag{9.2}$$

However, $U+PV$ is the enthalpy H. We may therefore conclude that the enthalpy of the gas is left unchanged,

$$H_1 = H_2 \qquad \text{(porous-plug experiment or throttling process)} \quad (9.2a)$$

We should not jump to the conclusion that the throttling process is isenthalpic. It is not—because the process is irreversible. The intermediate states of the gas during the throttle are not equilibrium states. No infinitesimal change can reverse the gas flow through the porous plug (or throttle valve). The gas always flows from the high-pressure side to the low-pressure side. Nevertheless it is essential to realize that the enthalpy is left unchanged by the throttling process.

In a gas liquefaction apparatus the porous plug is replaced by a narrow valve. The gas pressure is perhaps 200 atm on one side, and only slightly above 1 atm on the other. If gases were ideal, there would be no temperature change upon throttling. This follows from the fact that the enthalpy of an ideal gas depends on the temperature alone. If there is no enthalpy change in an ideal gas, there can be no temperature change. In real gases, H depends on the pressure as well as on the temperature. The throttling of a real gas may result in cooling or heating, depending upon the value of the Joule–Thomson coefficient

$$\mu_{JT} = \left(\frac{\partial T}{\partial P}\right)_H$$

Experimentally μ_{JT} is determined as follows: An initial temperature and (high) pressure (T_1, P_1) are selected. The throttling process is then performed, carrying the gas to some chosen post-throttle pressure. The post-throttle temperature is measured. These data give a pair of T–P points which lie on the same isenthalp, that is, the two data points correspond to the same enthalpy. The process is then repeated, starting with the same initial T–P values throttling down to a different final pressure. In this manner a series of T–P points are obtained, all lying on the same isenthalp. The slope of this isenthalpic curve equals μ_{JT} for that value of the enthalpy. If the series of measurements is repeated for a different set of pre-throttle T–P values, a different isenthalp can be constructed, its slope giving the corresponding values of μ_{JT}. Figure 9.6 shows the typical form of such isenthalps. As the figure reveals, the Joule–Thomson coefficient is positive over a definite range of T–P values. This region is bounded by the dashed line known as the inversion curve on which μ_{JT} is zero. Cooling always results if the initial T–P point for the throttle lies within the region of nonnegative values of μ_{JT}. This conclusion follows most clearly from Figure 9.6 itself. It must be borne in mind that throttling always carries the gas to a lower pressure, that is, to the left on the T–P plot.

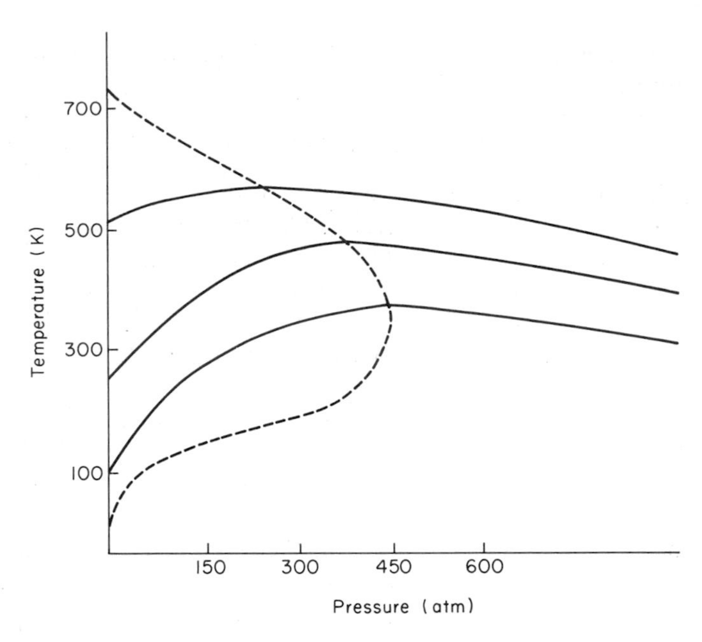

Figure 9.6. Typical form of isenthalps (solid lines) and inversion curve (dashed line, $\mu_{JT} = 0$). The throttle always carries the gas to the left, that is, to a lower pressure. If the throttle starts on the inversion curve, cooling results.

The essential elements of a Joule–Thomson liquefaction system are shown in Figure 9.7. First, the gas is compressed. This raises its temperature. The high-pressure gas is cooled and then throttled. When the system is operating in a steady state the throttling process causes a fraction of the gas to condense into the liquid phase. The remainder flows back to the compressor through a heat exchanger. The heat exchanger effects an internal exchange of heat. In the heat exchange operation, the pre-throttle gas is cooled by guiding it toward the J–T valve through coils bathed in a stream of cooler post-throttle gas which is thereby warmed as it returns to the compressor.

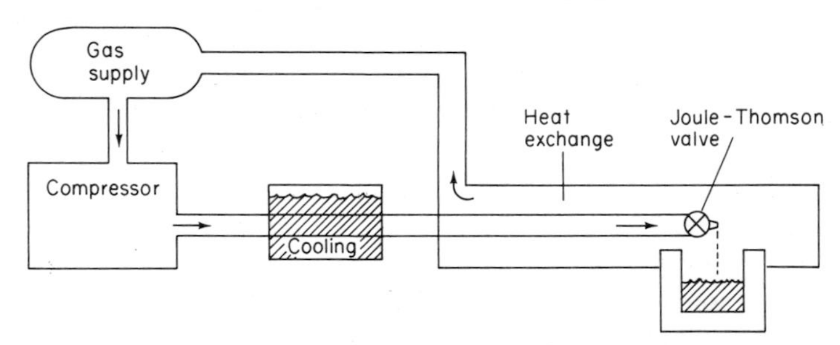

Figure 9.7. Schematic of Joule–Thomson liquefier. (See text for details.)

Low-temperature physics was an academic subject until the development of a commercial helium liqufier. The pioneering design was that of Kapitza in 1934. Kapitza used a combination of Joule–Thomson throttles and an expansion engine to liquefy helium. In 1946 Collins perfected a modification of Kapitza's liquefier. The Collins Machine turns out liquid helium on a commercial scale, providing the refrigerant for low-temperature research and technology.

Ultra-low temperatures have been achieved by using a magnetic version of the external work method. Magnetic cooling is discussed in the next section. A recent development which shows promise is the dilution refrigerator, which employs a mixture of liquid ^{3}He and ^{4}He. The ^{3}He–^{4}He refrigerator, whose operation is based on quantum mechanical effects, has made it possible to conduct experiments at temperatures near 10 millidegrees (0.01 K). Dilution refrigeration is discussed in Chapter 18.

9.3 MAGNETIC COOLING

The adiabatic expansion of a gas lowers its temperature. From a microscopic viewpoint the cooling stems from a conversion of molecular kinetic energy into external mechanical work. The adiabatic demagnetization of a paramagnetic substance also results in cooling. The energy converted into work is the potential energy of the magnetized material. Whereas an expanding gas sacrifices energy by pushing outward against atmospheric pressure, a paramagnetic substance lowers its internal energy by doing work against the currents and emf's which maintain its magnetic "atmosphere."

The energy decrease stemming from adiabatic demagnetization follows at once from the magnetic version of the first law of thermodynamics (4.16). In magnetic cooling experiments, the magnetic work $-H\,dM$ greatly exceeds the mechanical work $P\,dV$. Accordingly, we ignore the mechanical work term in (4.16) which then reads

$$dQ = dU - H\,dM \qquad (4.16')$$

where H is the applied magnetic field and M is the magnetic moment of the material. For an adiabatic process, $dQ = 0$, whence

$$dU = H\,dM \qquad (9.3)$$

A paramagnetic material has a positive susceptibility χ

$$\frac{M}{V} = \chi H, \qquad \chi > 0 \qquad (9.4)$$

It follows that a demagnetization—a decrease in H—diminishes M and thus U.

The reduction of the internal energy is revealed as a drop in temperature. The possibility of cooling by adiabatic demagnetization was suggested in 1926 by Debye and, independently, by Giauque, who also performed the first successful experiments.

The paramagnetic substance must have a comparatively large magnetic susceptibility. A number of hydrated salts containing paramagnetic ions have proved satisfactory in this respect. For example, in chromium potassium alum $K_2SO_4 \cdot Cr_2(SO_4)_3 \cdot 24H_2O$ the Cr^{3+} ions are the paramagnets which give rise to the large susceptibility. In another widely used salt, cerium magnesium nitrate, cerium ions are the magnetically active particles. Whereas an ordinary paramagnetic substance has a magnetic susceptibility of $\chi \sim 10^{-6} - 10^{-5}$ cgs units, the salts employed in magnetic coolers have $\chi \sim 10^{-2} - 1$ cgs units, over their operating temperature range of roughly 0.01–1 K.

The process of magnetic cooling starts with the salt pill suspended inside an evacuated container. The container in turn is cooled to near 1 K by bathing it in liquid helium. A strong magnetic field is applied and a small amount of gaseous helium is admitted to the sample container. The application of the magnetic field would increase the internal energy of the salt if the gaseous helium were not present. The gaseous helium conducts the added energy, as heat, to the surrounding liquid helium bath.† The magnetization is very nearly an isothermal process. This series of steps leaves the salt pill at the bath temperature (~ 1 K) in a strong magnetic field (10–50 kG). The gaseous helium is then pumped out of the sample container, leaving the paramagnetic sample thermally isolated from its surroundings. The magnetic field is then removed suddenly. The adiabatic demagnetization cools the salt. It is important to understand that the cooling is a result of an adiabatic process—the salt does not reject any heat to its surroundings. It is cooled because it performs magnetic work against its surroundings at the expense of its internal energy.

The thermodynamics of the cooling is best interpreted in terms of an entropy–temperature (S–T) diagram. Figure 9.8 indicates the behavior of the entropy as a function of the temperature for several different values of H, the applied magnetic field strength. At a fixed temperature the entropy decreases as H increases. This is in accord with the "entropy measures disorder" idea. The field acts to align the magnetic moments of the paramagnetic ions, an order-producing action. The stronger the magnetic field, the more complete is the alignment. In Figure 9.8, the directed arrows indicate the key steps in magnetic cooling. The isothermal magnetization $1 \rightarrow 2$ is followed by the adiabatic demagnetization $2 \rightarrow 3$. The adiabatic demagnetization is isentropic

† The gaseous helium serves only to conduct heat from the sample to the bath. It remains gaseous because its pressure is less than the saturated vapor pressure of helium at the bath temperature.

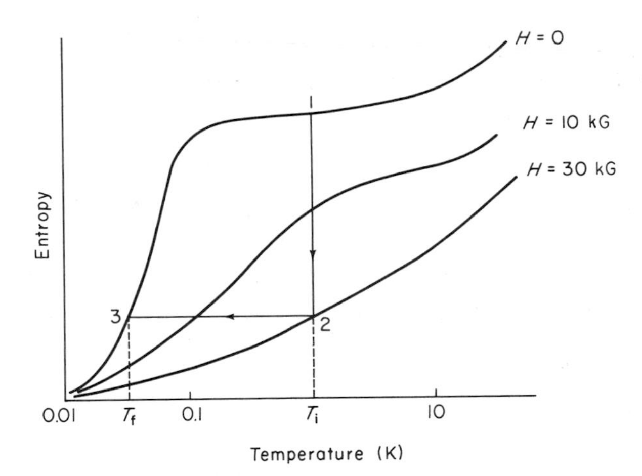

Figure 9.8. Magnetic cooling proceeds via the two-step process indicated. An isothermal magnetization at a temperature of 1 K carries the system to a state of lower entropy. The adiabatic demagnetization lowers the temperature to less than 0.1 K.

because it is (very nearly) a reversible process, that is, $dQ = T\,dS$ applies to a reversible process, and if $dQ = 0$ (adiabatic), then $dS = 0$ (isentropic). It is essential to start the demagnetization from as low a temperature as possible because the final temperature is directly proportional to the initial temperature.

••••••

9.3.1 (a) Assume that the internal energy U depends on T alone and show that the isothermal magnetization results in an entropy decrease given by

$$\Delta S \equiv S(H,T) - S(0,T) = -\frac{1}{T}\int_0^H H\,dM$$

(b) Show that the entropy decrease for a material obeying Curie's law ($\chi = C/T$) is

$$\Delta S = -\frac{CVH^2}{2T^2}$$

••••••

The thermodynamics of adiabatic demagnetization can be developed swiftly by exploiting analogies between magnetic and mechanical variables. In Chapter 4 we derived an expression for the thermodynamic work performed by a magnetic system

$$dW_{\mathrm{mag}} = -H\,dM \tag{4.15'}$$

Adding to this the mechanical work $P\,dV$ gives the total thermodynamic work

$$dW = P\,dV - H\,dM \tag{9.5}$$

In Chapter 8 the magnetic work term was ignored. In the present situation it is quite legitimate to ignore the $P\,dV$ term in favor of the magnetic work $-H\,dM$. Observe that P and H are both intensive variables; V and M are both extensive variables. From (9.5) it follows that the analogs are

$$P \leftrightarrow H, \qquad V \leftrightarrow -M \tag{9.6}$$

A number of useful relations among the thermodynamic variables were derived in Chapter 8. Among these were the Maxwell relations. In particular, the derivatives of the entropy and volume were found to be related by

$$-\left(\frac{\partial S}{\partial P}\right)_T = \left(\frac{\partial V}{\partial T}\right)_P \tag{8.18}$$

Replacing P and V by their magnetic analogs results in

$$\left(\frac{\partial S}{\partial H}\right)_T = \left(\frac{\partial M}{\partial T}\right)_H \tag{9.7}$$

By regarding the entropy as a function of the independent variables T and H we are led to the magnetic version of (8.23). Thus,

$$dQ = T\,dS = T\left(\frac{\partial S}{\partial T}\right)_H dT + T\left(\frac{\partial S}{\partial H}\right)_T dH \tag{9.8}$$

The heat capacity at constant H is defined by

$$(dQ)_H \equiv C_H\,dT \tag{9.9}$$

Evidently

$$C_H = T\left(\frac{\partial S}{\partial T}\right)_H \tag{9.10}$$

which also follows from (8.22) if P is replaced by its magnetic analog H. Returning to (9.8) we have, upon using (9.7) and (9.10),

$$dQ = C_H\,dT + T\left(\frac{\partial M}{\partial T}\right)_H dH \tag{9.11}$$

which is the magnetic counterpart of (8.23). For an adiabatic process this yields the relation between changes in H and T

$$dT = -\frac{T}{C_H}\left(\frac{\partial M}{\partial T}\right)_H dH \tag{9.12}$$

For paramagnetic materials $(\partial M/\partial T)_H$ is negative. To see this, note that if H is held constant, a temperature increase results in a decrease of M. That is, a temperature increase means greater agitation at the atomic level. Thermal motions tend to destroy the alignment of atomic moments which produce a net magnetic moment M. Thus, higher T means lower M so that $(\partial M/\partial T)_H$ is negative. It follows from (9.12) that dT and dH have the same sign; an adiabatic demagnetization decreases the temperature.

The unusually large magnetic susceptibilities of certain paramagnetic salts make possible the production of ultra-low temperatures. They also provide a means of measuring them. The susceptibility χ is a temperature-dependent quantity and may therefore serve as a thermometer. The paramagnetic salt functions as the core of an inductance coil. The temperature decrease resulting from demagnetization causes an increase in the permeability of the core. This in turn gives rise to an induced emf which is directly proportional to the change in χ.

At sufficiently high temperatures (above 1 K) the susceptibility follows Curie's law

$$\chi = \frac{C}{T} \tag{9.13}$$

where C is the Curie constant for the material, not its specific heat. This simple T dependence of χ does not persist at the lowest operating temperatures. Nevertheless it has become general practice to assume that χ has the Curie form, but with the absolute temperature T replaced by a magnetic temperature T^*. The relation between T^* and T must then be determined experimentally for each paramagnetic material. A technique for converting T^* measurements into T values was invented by Simon and Kurti. Their method, which has always impressed the author as a diabolically clever application of thermodynamics, is discussed later in this section.

•••••• ••••••

9.3.2 Show that

$$\left(\frac{\partial M}{\partial T}\right)_H = -\frac{M}{T}$$

for a substance which obeys Curie's law.

•••••• ••••••

The low temperature produced by demagnetization is in fact the temperature of the gas of ion magnetic moments, not that of the host lattice. In order for the lattice to be cooled to nearly the same temperature as the ions, two conditions must be satisfied. First, the heat capacity of the lattice must be small compared to that of the ion moments. This enables the lattice and ion moments to reach equilibrium at a temperature close to the temperature of the moments immediately after the demagnetization. Second, it is necessary that the ion–lattice equilibrium be established quickly, before external heat leaks (such as radiation and vibration) restore equilibrium between the sample and the helium bath. Happily, both of these conditions are satisfied by a number of paramagnetic salts. The lattice specific heat C_{lat} is given by the Debye formula

$$C_{\text{lat}} = 464\left(\frac{T}{\theta}\right)^3 \frac{\text{cal}}{\text{mole-K}} \tag{4.19}$$

where θ is the Debye temperature, a characteristic of the lattice. If we take $\theta \sim 300$ K, we find $C_{\text{lat}} \sim 10^{-5}$ cal/mole-K at $T \sim 1$ K. The specific heat of the ion moments displays an anomalously large value at some low temperature. The temperature at which the peak occurs is a characteristic of the material, but typical values fall in the 0.01–0.1 K range. The heat capacity of the ion moments may reach 1 cal/mole-K, some 100 million times that of the lattice at the same temperature. This means that even a dilute concentration of ions can cool the lattice without undergoing a serious temperature increase. The high heat capacity of the ion moments also serves as a thermal anchor which maintains the low temperature against heat leaks and, more importantly, can serve as a refrigerant to cool other materials.

The practice of introducing a magnetic temperature T^* merits further discussion. We define T^* in terms of the susceptibility and the Curie constant

$$\chi = \frac{C}{T^*} \tag{9.14}$$

Evidently $T^* = T$ so long as the material obeys Curie's law. The absolute thermodynamic temperature is defined by

$$T = \left(\frac{\partial U}{\partial S}\right) \tag{8.3$'$}$$

Experimentally, the partial derivative appears as $\Delta U/\Delta S$, the ratio of corresponding changes in U and S. Thus

$$T = \frac{\Delta U}{\Delta S} \tag{9.15}$$

Suppose that the demagnetization measurements indicate a magnetic temperature change of ΔT^*. Let us note that (9.15) may be rewritten as

$$T = \frac{\Delta U/\Delta T^*}{\Delta S/\Delta T^*} \tag{9.16}$$

The temperature T refers to the final ($H = 0$) state. Thus if a means can be devised for evaluating

$$\left(\frac{\Delta U}{\Delta T^*}\right)_{H=0} \quad \text{and} \quad \left(\frac{\Delta S}{\Delta T^*}\right)_{H=0}$$

the absolute temperature can be determined. The numerator of (9.16) is simply the zero-field heat capacity

$$C^* \equiv \left(\frac{\Delta U}{\Delta T^*}\right)_{H=0} \tag{9.17}$$

where C^* is measured by irradiating the sample with a γ-ray source—a neat way of adding energy to a substance surrounded by liquid helium. The change in internal energy equals the energy deposited by the absorbed γ rays

$$\Delta U = Q_\gamma \tag{9.18}$$

The quantity ΔT^* is measured by the change in susceptibility. The rate at which energy is deposited in the sample by the γ rays depends only on the strength of the source, the geometry of the sample, and so on. Hence a determination of the rate at which γ rays deposit energy can be conducted at a higher temperature where the heat capacity is known from calorimetric measurements.

To see how the ratio $\Delta S/\Delta T^*$ is determined, consider Figure 9.9. Plotted there on a T^*–H diagram are the paths of two sequential processes. The

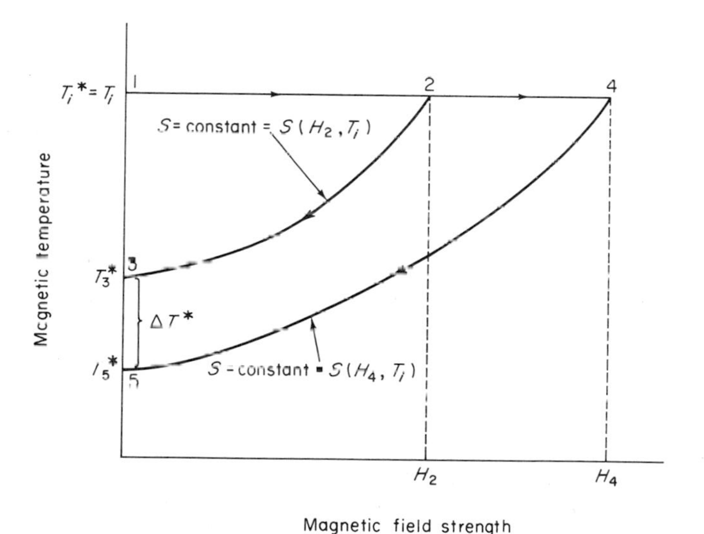

Figure 9.9. T^*–H diagram indicating how the ratio $\Delta S/\Delta T^*$ can be inferred from adiabatic demagnetization data.

processes are identical with those involved in the magnetic cooling sequence depicted by Figure 9.8. Each of the two paths consists of an isothermal magnetization followed by an adiabatic (and isentropic) demagnetization. The temperature at which the isothermal magnetizations proceed is sufficiently high that $T_i^* = T_i$ can be measured by conventional low-temperature thermometry (for example, vapor pressure). The final T^* temperatures, T_5^* and T_3^*, are measured via the susceptibility, so that

$$\Delta T^* = T_3^* - T_5^*$$

is thereby determined. The entropy difference follows from the result (see Problem 9.3.1)

$$S(H, T_i) - S(0, T_i) = -\frac{1}{T_i}\int_0^H H\,dM \tag{9.19}$$

Since T_i is high enough that $T_i^* = T_i$, the susceptibility is given by $\chi = C/T_i$, from which it follows that $M = CVH/T_i$. This leads to the result

$$S(H_4, T_i) - S(0, T_i) = -\frac{CVH_4^2}{2T_i^2} \tag{9.20}$$

The entropy change required to determine T is

$$\Delta S = S(H_2, T_i) - S(H_4, T_i) = \frac{CV(H_4^2 - H_2^2)}{2T_i^2} \tag{9.21}$$

To recap, the absolute temperature is defined by (8.3) which, experimentally, is equivalent to (9.16). Using (9.17) gives

$$T = \frac{\Delta U/\Delta T^*}{\Delta S/\Delta T^*} = \frac{C^*\,\Delta T^*}{\Delta S} \tag{9.22}$$

The magnetic temperature follows from measurements of the susceptibility. The heat capacity C^* is determined by using a calibrated γ-ray source and ΔS is inferred from (9.21).

REFERENCES

The liquid–solid transition in nitrogen provides a spectacular demonstration. See:

R. L. Wild and D. C. McCollum, Dramatic Demonstration of Change of Phase, *Am. J. Phys.* **35**, 540 (1967).

A modification of the demonstration is suggested by:

H. C. Jensen, Freezing Nitrogen: A Modification, *Am. J. Phys.* **36**, 919 (1968).

Virtually every aspect of low-temperature physics is discussed at length in:

G. K. White, "Experimental Techniques in Low-Temperature Physics." Oxford Univ. Press, London and New York (1959).

See also:

J. S. Dugdale, "Entropy and Low Temperature Physics," Chapter 15. Hutchinson, London (1966).

A very readable account of cryogenic applications is given in:

R. J. Allen, "Cyrogenics." Lippincott, Philadelphia, Pennsylvania (1964).

Included are discussions of liquefiers, superfluidity, and superconductivity.
An entertaining account of cryogenic principles and applications is given in:

D. K. C. MacDonald, "Near Zero." Doubleday, Garden City, New York (1961).

Historical commentary on magnetic cooling can be found in:

C. G. B. Garrett, "Magnetic Cooling." Harvard Univ. Press, Cambridge, Massachusetts (1954).

Chapter 6 provides much detail on the structures of the paramagnetic salts. See also

C. T. Lane, "Superfluid Physics," Chapter 1. McGraw-Hill, New York (1962).

G. K. White, "Experimental Techniques in Low-Temperature Physics," Chapter 8. Oxford Univ. Press, London and New York (1959).

H. B. G. Casimir, "Magnetism and Very Low Temperatures." Dover, New York (1961).

A detailed development which describes magnetic cooling with superconductors and antiferromagnetic salts as well as with the usual paramagnetic salts is given in:

H. Weinstock, Thermodynamics and Statistical Aspects of Magnetic Cooling, *Am. J. Phys.* **36**, 36 (1968).

A note of caution concerning the reversibility of adiabatic demagnetization is voiced in:

R. Baierlein and B. Bertman, Reversibility in Magnetic Cooling, *Am. J. Phys.* **37**, 101 (1969).

Chapter 10

Elementary Statistical Concepts

Both kinetic theory and statistical mechanics require an appreciation and working knowledge of certain aspects of probability theory. In order not to break the continuity of later analyses we present here some rudimentary notions needed in the development of kinetic theory and statistical mechanics. Additional mathematical machinery required expressly for statistical mechanics will be introduced in Chapter 14. Our immediate needs are minimal compared with what are generally considered to be the elements of probability theory.

10.1 MEAN VALUES AND PROBABILITIES

The physical variables dealt with in practice fall into two classes, discrete and continuous. A discrete variable Z may assume only certain definite values, such as $Z_1, Z_2, Z_3, \ldots$. For example, if Z denotes the number on the face of a die, the possible values are

$$Z_1 = 1, \quad Z_2 = 2, \quad Z_3 = 3, \quad Z_4 = 4, \quad Z_5 = 5, \quad \text{and} \quad Z_6 = 6$$

A continuous variable may take on any value within some range. For example, the speeds of nitrogen molecules in our atmosphere range from zero to thousands of kilometers per second. In this and the immediately following sections the basic ideas are developed in terms of discrete variables. The corresponding treatment for continuous variables is detailed in Section 10.3.

A quantity of prime importance, both physically and mathematically, is the mean value of a statistical variable. Suppose a series of N trials, that is, measurements of Z, results in N_1 occurrences of Z_1, N_2 occurrences of Z_2, and so on. The mean value of Z for the series of trials is denoted as $\langle Z \rangle$ and is defined by

$$\langle Z \rangle \equiv \frac{Z_1 N_1 + Z_2 N_2 + Z_3 N_3 + \cdots}{N_1 + N_2 + N_3 + \cdots} \tag{10.1}$$

Using summation notation

$$\langle Z \rangle = \frac{\sum_i Z_i N_i}{\sum_i N_i} \tag{10.2}$$

The denominator is just the total number of trials N

$$\sum_i N_i = N$$

Thus,

$$\langle Z \rangle = \frac{\sum_i Z_i N_i}{N} \tag{10.3}$$

••••• •••••

10.1.1 A series of 28 rolls of a die yields the results $N_1 = 5$, $N_2 = 5$, $N_3 = 5$, $N_4 = 6$, $N_5 = 3$, and $N_6 = 4$, where N_k is the number of times the integer k came up. Show that the mean for the series is approximately 3.32.

••••• •••••

The Nielsen television ratings and the Gallup political polls sample preferences of the American population. A series of measurements may be thought of as a sample of an indefinitely large assembly the population of all measurements. In (10.3), N_i/N is the fraction of the sample measurements which give Z_i. The ratio N_i/N is called the relative frequency of Z_i. Experience suggests that the relative frequency approaches a definite limit as the sample size increases. The student may very well have tested this convergence in an error-analysis experiment. The limit of N_i/N as N tends to infinity is called the probability of (the occurrence of) Z_i and is denoted as P_i

$$P_i \equiv \lim_{N \to \infty} \frac{N_i}{N} \tag{10.4}$$

Note that the probabilities are subject to a normalization condition,

$$\sum P_i = 1 \tag{10.5}$$

We instinctively feel that the limit in (10.4) exists, that is, that each N_i increases in such a fashion that N_i/N approaches a definite value. In statistical physics we are more likely to be concerned with predicting the outcome of a contemplated experiment than in describing the results of a completed one. For example, suppose we are asked to predict the mean value of a series of N measurements of a discrete variable Z. We must first assign (predict) the values of N_i/N and then compute $\langle Z \rangle$ using

$$\langle Z \rangle = \frac{\sum_i Z_i N_i}{N} \tag{10.3}$$

Our confidence in the existence of the limit of N_i/N defined by (10.4) influences our prediction. In the absence of any information to the contrary the most plausible guess we can make is that the ratio N_i/N will equal its limiting value P_i. By most plausible we mean that no more plausible estimate of N_i/N could be assigned on the basis of the information provided. The expected (that is, predicted) value of $\langle Z \rangle$ is then

$$\langle Z \rangle = \sum_i Z_i P_i \tag{10.6}$$

It is in this sense that statistical physics utilizes probability theory. The task in statistical physics is to first *predict* the P_i and then to *compute* related mean values using (10.6).

In practice, measurements of a statistical variable Z are used to determine the mean value of some function $Q(Z)$ of the variable. For example, we may wish to determine the mean area of circular disks from measurements of their diameters. If Z denotes the disk diameter,

$$\text{area} \equiv Q(Z) = \frac{\pi Z^2}{4}$$

is the desired physical quantity whose mean value $\langle Q \rangle$ is sought. If a series of N measurements result in N_1 occurrences of Z_1, N_2 occurrences of Z_2, and so on, there are corresponding numbers of occurrences of

$$Q_1 = \frac{\pi {Z_1}^2}{4}, \qquad Q_2 = \frac{\pi {Z_2}^2}{4}, \qquad \cdots$$

The mean value of $Q(Z)$ is thus

$$\langle Q \rangle = \frac{Q_1 N_1 + Q_2 N_2 + \cdots}{N_1 + N_2 + \cdots}$$

or

$$\langle Q \rangle = \sum_i Q_i \frac{N_i}{N} \tag{10.7}$$

If we are asked to predict $\langle Q \rangle$, we will (somehow!) assign probability estimates for the N_i/N, obtaining

$$\langle Q \rangle = \sum_i Q_i P_i \tag{10.8}$$

Evidently, it is immaterial whether we talk of the probability of the occurrence of Z_i or of the probability of the occurrence of $Q_i = Q(Z_i)$. It also should be clear that (10.7) and (10.8) are valid for any physical quantity $Q(Z)$.

Example 10.1 RMS Values

The rms value is defined as the square *root* of the *mean square* value,

$$Z_{\text{rms}} \equiv \sqrt{\langle Z^2 \rangle}$$

To illustrate, we determine the rms value for the die roll data of Problem 10.1.1. The mean square value of Z for the 28 tosses is given by

$$\langle Z^2 \rangle = \frac{\sum_i Z_i^2 N_i}{N}$$

The result is $\langle Z^2 \rangle = 385/28$, which yields $Z_{\text{rms}} = 3.71$. Note that $Z_{\text{rms}} > \langle Z \rangle$, since the larger values of Z carry greater weight in determining $\langle Z^2 \rangle$ in consequence of the squaring operation.

It may happen that the mean value $\langle Z \rangle$ is zero. While this is a nontrivial attribute of the quantity it gives no hint as to the magnitudes of the values assumed by the quantity. The rms value then becomes a useful measure of the magnitudes of the quantity. It would be equally legitimate to deal with the mean magnitude $\langle |Z| \rangle = \sum_i |Z_i| (N_i/N)$. This is seldom done because certain mathematical properties make Z_{rms} much more manageable than $\langle |Z| \rangle$. ✧

•••••• ••••••

10.1.2 Prove in general that $Z_{\text{rms}} \geqslant \langle Z \rangle$. [*Hint:* The mean square value of a variable is never negative. Your proof should make use of the fact that $\langle \zeta^2 \rangle \geqslant 0$, where $\zeta \equiv Z - \langle Z \rangle$.]

10.1.3 The standard deviation σ is defined by

$$\sigma \equiv \sqrt{\sum_i (Z_i - \langle Z \rangle)^2 P_i}$$

Show that σ may be expressed as

$$\sigma = \sqrt{\langle Z^2 \rangle - \langle Z \rangle^2}$$

•••••• ••••••

Statistical physics enlarges and extends the layman's conception of probability. Everyday usages of the word "probability" are typified by the following situations: With reference to the outcome of one flip of a coin we have the statement, "The probability of heads is $\frac{1}{2}$." With reference to drawing one card from a standard deck one hears statements such as, "The probability of drawing the ace of spades is $\frac{1}{52}$." In connection with the roll of a die we have statements such as, "The probability of throwing a four is $\frac{1}{6}$." Used in these contexts, the probabilities are referred to as *a priori* probabilities. The terminology *a priori* probability refers to the probability *in the absence of prior knowledge.* Thus, for example, suppose you were given a die whose six faces bear the numbers 1–6. The die might be loaded so that the face bearing the "3" turns up in a suspiciously large percentage of the rolls. However, in the absence of any data pertaining to the die's roll characteristics, one would be led to postulate equal probabilities of $\frac{1}{6}$ for the likelihood that a 1, 2, 3, 4, 5, or 6 turn up. Whenever we lack evidence which favors certain possible outcomes over others, it is plausible to assume equal probabilities for all possible outcomes. The acquisition of information may change the probabilities. For example, suppose a ruled line divides a sheet of paper into two parts. You are then told that a single pencil mark dots the paper and asked to assign probabilities for its appearance in either portion. In the absence of any further information the most plausible assignment of probabilities is $\frac{1}{2}$ for either side. If you receive the additional information that the area on one side of the ruled line is twice that on the other side, the most plausible assignment of probabilities becomes $\frac{1}{3}$ and $\frac{2}{3}$.

••••• •••••

10.1.4 An urn contains six white balls and six black balls.

(a) What is the probability that a ball drawn from the urn will be white?
(b) If a white ball is drawn from the urn (and not returned), what is the probability that a second draw will yield another white ball?

10.1.5 An underwater cable 300 km long breaks.

(a) What is the probability that the break occurred within 2 km of the middle?
(b) What is the probability that the break occurred within 2 km of a specified end?
(c) What is the probability that the break occurred within 2 km of a point located 1 km from a specified end?

10.1.6 The serial number of a one dollar bill has eight digits. A complete run of bills consists of serial numbers 00000001–99999999.

(a) What is the probability that all eight digits are 9's?

(b) What is the probability that the last two digits are 9's?

(c) What is the probability that all eight digits are the same (11111111, 22222222, ...)? Note that there is no bill with eight zeros.

•••••• ••••••

The assignment of equal *a priori* probabilities in situations like the die tossing example illustrates Laplace's Principle of Insufficient Reason:

In the absence of evidence to the contrary, all possible outcomes are equally probable.

It was the great French mathematician Laplace who stated that "probability theory is nothing but common sense reduced to calculation." When we assign *a priori* probabilities we are using common sense. Thus if you are asked for the probability of drawing an ace from a standard deck of cards, your common sense answers 4/52. On the other hand, if you have evidence which favors one outcome over another, you certainly would not assign them equal probabilities. For example, if you know that one deuce and one king have been removed from the deck, common sense and very little calculation leads to a probability of 4/50 for drawing an ace, but only 3/50 for the chance of drawing a deuce.

Example 10.2 Star–Star Collisions

To illustrate the power of common sense probability arguments, we bring them to bear on a problem related to the collision of two galaxies. Galaxies come in two shapes. Young ones tend to have a pinwheel (or spiral arm) structure. Old ones are usually ellipsoidal. For our purposes we invent a third type, cubical in shape. Such liberties as this may introduce uncertainties of a factor of two or three into the final result. This is within the spirit of the calculation; we are after only a rough estimate.

We suppose that two cubical galaxies meet head-on and pass through each other. The question we ask is this: How many collisions between stars can be expected? To obtain a definite answer we need some astronomical data. Astronomy reveals that most galaxies are about the same size as our own. The characteristic dimension of a galaxy is 10^4 pc, where 1 pc (parsec) is close to 3×10^{18} cm.† Furthermore there appear to be roughly the same number of stars, about 10^{11}, in most galaxies. For our estimate we shall take each galaxy to be a cube of side 10^4 pc, containing 10^{11} stars.

† One parsec is approximately equal to 3 light years.

The average radius of a star is also needed. Our own star, the Sun, is nearly typical, its radius being 7×10^{10} cm. To simplify the arithmetic we assume that each star has a radius of 3×10^{11} cm $= 10^{-7}$ pc. We further assume that the stars are spread randomly throughout their cubical volumes.

Because of the powerful attraction of gravity two stars can disrupt each other without actually touching. We demand a first class smash—we count only collisions which would occur in the absence of gravitational attraction. This requires that the distance between the centers of the two stars be less than the diameter. Equivalently we can double the radius of all the stars in one galaxy and regard all those in the other as points or, perhaps more picturesquely, as darts. One galaxy becomes a set of 10^{11} identical targets. The other becomes a swarm of darts. In addition to the collective motions which carry the galaxies together there are stellar motions within the galaxies. These motions render each star in one galaxy a potential target for each of the dart stars.

We first ascertain the chance that one of the 10^{11} darts strikes a star during its trip through the target galaxy. This is the same as asking for the probability that a dart strike one of 10^{11} targets spread over an area of 10^8 pc^2, that is, we can place all of the target stars on one face of the cube since the incident star has the opportunity to travel across the entire galaxy. This (common sense!) probability is simply the fraction of the area of the face covered by target stars. The target area of each star is $\pi(2 \times 10^{-7})^2$ pc^2, or about 10^{-13} pc^2. With 10^{11} target stars the total target area is 10^{-2} pc^2, so that

$$\begin{array}{c}\text{probability that any one incident} \\ \text{star will strike a target star}\end{array} = \frac{10^{-2}\ \text{pc}^2}{10^{8}\ \text{pc}^2} = 10^{-10}$$

However, there are 10^{11} incident stars, each having the same chance of colliding with a target star. The expected number of star–star collisions therefore works out to 10. Thus even with 10^{11} stars shooting at another 10^{11}, only about 10 collisions are likely—there is a lot of empty galaxy out there. ◇

The predictive aspects of statistical physics are typified by the following problem: A crystal contains N atoms. There are many different possible energies open to each atom. We label the possible energies $\varepsilon_1, \varepsilon_2, \varepsilon_3, \ldots, \varepsilon_M$. The number of possible energies M is enormous. Let N_i denote the number of atoms with energy ε_i. Determine the most probable values of the N_i subject to the constraints.

$$\sum_{i=1}^{M} N_i = N \qquad \text{(total number of atoms is } N\text{)}$$

and

$$\sum_{i=1}^{M} \varepsilon_i N_i = E \qquad \text{(total energy of atoms is } E\text{)}$$

The first constraint means only that the number of atoms must be the same for all proposed schemes of distribution. The second condition might refer to the physical situation in which the crystal is isolated. In such a case the total energy of the atoms E is constant. Not every conceivable way of assigning energies to the N atoms is consistent with $\sum \varepsilon_i N_i = E$. From an algebraic viewpoint there are two equations and M unknowns, and $M \gg 2$. Clearly the N_i are not uniquely determined by the two equations of constraint. Since the number of equations is far fewer than the number of unknowns we require some principle —a guiding light—which lets us assign values to the N_i. Our later study of statistical mechanics will reveal the guiding light to be none other than the second law of thermodynamics. In short, the scheme by which statistical mechanics would have the atoms distributed is such that the entropy of the crystal is maximized.

It is important to understand the significance of the N_i which statistical mechanics "cranks out." These N_i are not the measured numbers of atoms having energy ε_i. Such a direct measurement is not possible. The value of N_i assigned by statistical mechanics is in effect a prediction of the number of atoms which possess energy ε_i. This prediction is made in compliance with all constraints ($\sum N_i = N$; $\sum \varepsilon_i N_i = E$; entropy maximized). There is no claim that a crystal satisfying the constraints will in fact have the prescribed distribution of atoms. What statistical mechanics says is that if

$$\sum N_i = N \qquad \text{and} \qquad \sum \varepsilon_i N_i = E$$

and if the entropy is a maximum, then the most likely distribution of the atoms is such and such.†

This situation is comparable to saying that the *a priori* probability of rolling a four is $\frac{1}{6}$. There is no guarantee that a four will turn up in one-sixth of the rolls, but it is most likely that this will occur—no other outcome is more likely.

Suppose that we are given the maximum entropy prediction of the N_i. Then $P_i = N_i/N$ is the probability that a crystal atom have an energy ε_i. With the P_i in hand we can proceed to compute mean values according to

$$\langle Q \rangle = \sum_{i=1}^{M} Q_i P_i, \qquad Q_i = Q(\varepsilon_i) \tag{10.8'}$$

† The actual form of N_i is not important here. Later in the game we will discover that N_i is proportional to $\exp(-\varepsilon_i/kT)$, where T is the absolute temperature and k is the Boltzmann constant.

This $\langle Q \rangle$ is the expected (predicted) mean value of the quantity Q. In practice one computes the mean values of observable quantities so that the predicted value of $\langle Q \rangle$ can be compared with experiment.

10.2 CONDITIONAL PROBABILITIES

In many instances it is necessary to determine the probability for compound events or propositions. For example, the *a priori* probability for heads on the flip of a coin is $\frac{1}{2}$. Suppose we wish to determine the probability that heads will turn up on two consecutive flips. The set of two flips comprise a compound event. To cite another example, suppose it is known that two aces have been removed from a standard deck of cards. (Which two is unknown.) The problem: What is the probability of drawing the ace of spades? The proposition before us is compounded of two elementary propositions: (a) the ace of spades is in the deck and (b) the ace of spades will be drawn, given that it is in the deck. In the discussion which follows we use the following notation: Capital letters are used to designate propositions (events). Thus, H might denote the proposition, "The flip of a coin results in heads." The corresponding probability is designated as $P(\mathrm{H})$:

$$P(\mathrm{H}) = \text{probability of heads}$$

More generally,

$$P(\mathrm{A}) = \text{probability that event A occurs}$$

The product AB denotes the compound proposition that both A and B occur and $P(\mathrm{AB})$ is the probability that both A and B occur. A basic problem is this: How is $P(\mathrm{AB})$ related to $P(\mathrm{A})$ and $P(\mathrm{B})$? To establish such a relation requires the introduction of conditional probabilities. As the name implies, the conditional probability for some event refers to a probability based on prior knowledge—a condition which affects its likelihood. The notation $P_{\mathrm{B}}(\mathrm{A})$ denotes "the probability of A, given that B has occurred."

As an example of a conditional probability, consider the following situation. Let A denote the proposition "an ace will be drawn" and let B denote the proposition "two aces are replaced by two deuces," both statements in reference to a standard deck of 52 cards. The *a priori* probability that an ace will be drawn is 4/52,

$$P(\mathrm{A}) = \frac{4}{52}$$

However, if it is known that two deuces have replaced a pair of aces, the probability of selecting an ace is reduced to 2/52. Thus the conditional probability $P_B(A)$ is

$$P_B(A) = \frac{2}{52}$$

••••• •••••

10.2.1 If two aces are replaced by two deuces (B) and if C denotes the proposition "a deuce will be drawn," what is $P_B(C)$?

••••• •••••

We can now state the relation between $P(AB)$, $P_B(A)$, and $P(A)$, namely

$$P(AB) = P(B) \cdot P_B(A) \tag{10.9}$$

This relation is in fact a definition of the conditional probability $P_B(A)$. To illustrate the use of (10.9), consider a deck of cards from which two aces have been removed, leaving 50 cards. Which two aces remain is not known. Let A denote the proposition, "the ace of spades will be drawn" and let B denote the proposition "the ace of spades will be in a deck from which two aces have been removed." Since we do not know which two aces remain we have $P(B) = 2/4$ for the probability that the ace of spades is in the partial deck of 50 cards. If we know that the ace of spades is present, the probability of drawing it is 1/50

$$P_B(A) = \frac{1}{50}$$

Thus using (10.9)

$$P(AB) = P(B)\,P_B(A) = \frac{1}{100}$$

•••••• •••••

10.2.2 Determine the conditional probability that a six will turn up on the roll of a die, given that an even number turns up.

10.2.3 An urn contains six white balls and six black balls. What is the probability that two draws will yield two white balls when:

(a) the first ball drawn is returned to the urn before the second draw;
(b) the first ball drawn is not returned to the urn before the second draw.

••••• •••••

A very special and important use of (10.9) arises where the occurrence of A does not depend at all on the occurrence of B. If the probability for A is not affected by B, we have

$$P_B(A) = P(A)$$

whereupon (10.9) becomes

$$P(AB) = P(A) \cdot P(B) \tag{10.10}$$

If the probability for A is not affected by B, we say that A and B are independent. Thus for independent events, A and B, the joint probability $P(AB)$ is simply the product of $P(A)$ and $P(B)$ as expressed by (10.10). As a very simple example consider a pair of flips of a coin. Let

$$A = \text{heads on first flip}$$

and

$$B = \text{heads on second flip}$$

Then

$$AB = \text{heads on both flips}$$

Since $P(A) = P(B) = \frac{1}{2}$, (10.10) gives the common sense result $P(AB) = \frac{1}{4}$.

We will have occasion to apply (10.10) in Section 10.4. Conditional probabilities are required in many areas of statistical physics.

10.3 CONTINUOUS VARIABLES AND DISTRIBUTION FUNCTIONS

In the case of continuous statistical variables it is unwise (from an operational viewpoint) to speak of the probability of a value of Z. Any series of trials will record a finite number of values of Z, yet there exist a nondenumerable (uncountable) infinity of possible values. Hence the probability (number of occurrences divided by number of trials) is zero for all but a few values of Z—those that turned up.

For example, consider the task of determining the probability that a 10-year-old boy in Denmark has a specified height. We can imagine a horde of doctors and nurses invading schools and homes throughout Denmark, measuring and recording the heights of all 10-year-old boys. For each observed height there could be assigned a probability, namely,

$$P_i = \frac{\text{number of 10-year-old Danish boys of height } h_i}{\text{total number of 10-year-old Danish boys}}$$

This would assign nonzero values of P_i only to those heights observed. This approach would make P_i zero for all heights not observed. Certainly, the fact that the heights 1.384 m, 1.386 m, and 1.388 m were observed, but not the heights 1.385 m, 1.387 m, and 1.389 m, does not suggest the impossibility of the latter heights. Suppose that instead of P_i we were to specify the fraction of 10-year-old Danish boys whose heights fall in a specified range of heights. The result would be a description much more in line with our intuitive notions about the likelihood of observing any particular height.

To this end we find it both meaningful and useful to introduce a probability *distribution function* for the values of a statistical variable. The distribution function measures the likelihood that a value of a statistical variable falls within a specified range of values. Let $dP(Z)$ denote the probability that the value lies in the range from Z to $Z+dZ$. We define the probability distribution function $F(Z)$ by

$$dP = F(Z)\,dZ \tag{10.11}$$

In terms of the distribution function we can give $dP(Z)$ a very helpful geometrical interpretation. Figure 10.1 displays the form typical of several distribution functions.

From Figure 10.1 we see that $dP(Z) = F(Z)\,dZ$ equals the area of the strip of width dZ between the abscissa and the graph of $F(Z)$. This interpretation is extremely useful. For example, we can immediately conclude that the probability that Z lie in the range from Z_A to Z_B is given by the total area under the $F(Z)$ plot from Z_A to Z_B. Formally,

$$\text{probability that } Z \text{ lies in the range from } Z_A \text{ to } Z_B = \int_{Z_A}^{Z_B} F(Z)\,dZ \tag{10.12}$$

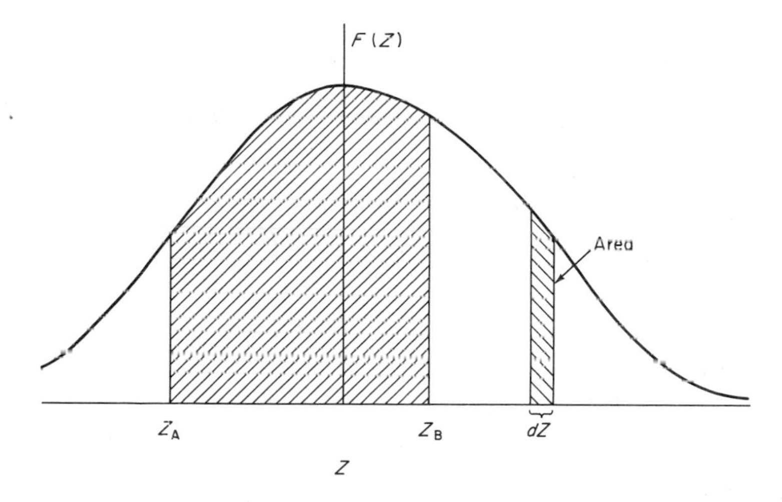

Figure 10.1. Probability is measured by the area beneath a plot of the distribution function (area $= F(Z)\,dZ = dP$). The area between Z_A and Z_B gives the probability that a measurement of Z will fall in that range.

In the discrete variable case, the sum of all probabilities totaled unity

$$\sum_i P_i = 1 \tag{10.5}$$

For continuous variables the summation is replaced by an integration over the full range of possible values of Z. Thus the analog of (10.5) for continuous variables is the normalization condition

$$\int_{Z_L}^{Z_U} dP(Z) = \int_{Z_L}^{Z_U} F(Z)\, dZ = 1 \tag{10.13}$$

where Z_L and Z_U designate the lower and upper limits which Z can attain. Geometrically, (10.13) is simply a statement of the fact that the total area beneath the plot of $F(Z)$ equals unity.

In general one can make the transition from discrete to continuous probability formulas by replacing a summation over P_i by an integration over dP. In particular, the mean value equation

$$\langle Q \rangle = \sum_i Q(Z_i)\, P_i \tag{10.8'}$$

converts to

$$\langle Q \rangle = \int_{Z_L}^{Z_U} Q(Z)\, dP \tag{10.14}$$

••••• •••••

10.3.1 The distribution function for a continuous variable x is

$$F(x) = A(1-x^2), \qquad -1 \leqslant x \leqslant 1$$

$$F(x) = 0, \qquad |x| > 1$$

(a) Show that $A = \frac{3}{4}$ is required for a properly normalized distribution function $\int_{-\infty}^{+\infty} F(x)\, dx = 1$.

(b) Evaluate $\langle x \rangle$ and $\langle x^2 \rangle$.

(c) What can be said about $\langle x^N \rangle$, where N is an odd integer? Mental calculation is in order rather than the pencil-and-paper variety.

••••• •••••

An especially important distribution function is encountered in the kinetic theory of gases. There is a distribution of molecular speeds in a gas. The corresponding probability distribution function is denoted $g(v)$. We write

$$dP(v) = g(v)\, dv \tag{10.15}$$

where $dP(v)$ is the probability that the speed of a molecule lies in the range from v to $v + dv$. We could equivalently state that $g(v)\, dv$ is the fraction of the molecules with speeds in the range v to $v + dv$.

The Maxwellian[†] form of the molecular distribution function is

$$g_M(v) = 4\pi\left(\frac{m}{2\pi kT}\right)^{3/2} v^2 \exp\left(\frac{-mv^2}{2kT}\right) \tag{10.16}$$

The Maxwellian form of $g(v)$ describes the distribution of speeds in a gas in thermal equilibrium at a temperature T. The quantity k is the Boltzmann constant and m is the mass of a molecule. The numerical factor in $g_M(v)$ guarantees that the probability $dP(v)$ is normalized to unity, that is, that

$$\int_{\text{all speeds}} dP(v) = \int_{v=0}^{\infty} g_M(v)\, dv = 1 \tag{10.17}$$

•••••• ••••••

10.3.2 Verify that $\int_{v=0}^{\infty} g_M(v)\, dv = 1$. Refer to Appendix A for assistance in evaluating the definite integral.

10.3.3 (a) Show that the mean molecular speed for the Maxwellian distribution is

$$\langle v \rangle = \sqrt{\frac{8kT}{\pi m}}$$

(b) Show that the rms speed for the Maxwellian distribution is

$$v_{\text{rms}} = \sqrt{\frac{3kT}{m}}$$

The relevant integrals are furnished in Appendix A.

(c) By writing $k = R/N_0$ and noting that the molar mass is $M = N_0 m$ show that these results may be expressed as

$$\langle v \rangle = \sqrt{\frac{8RT}{\pi M}}, \qquad v_{\text{rms}} = \sqrt{\frac{3RT}{M}}$$

10.3.4 A tabletop is ruled with parallel lines spaced 1 in. apart. Toothpicks, each 1 in. in length, are dropped onto the tabletop in random fashion.

(a) What is the probability $T(\theta)$ that a toothpick crosses (or touches) a line, given that it makes an angle θ with the lines?

(b) The angle θ is a random variable over the range $\theta = 0$ to $\theta = \pi/2$, that is, the method of dropping does not favor any particular angle. What is

---

[†] Named after James Clerk Maxwell, who first deduced its form through theoretical arguments.

the probability $dP(\theta)$ that a toothpick makes an angle with the lines which lies in the range from θ to $\theta+d\theta$? (It is also possible to consider the variable θ as ranging from 0 to π, or even from 0 to 2π, but be careful of signs if you make such a choice. Remember, probabilities cannot be negative.)

(c) Show that the mean probability $\langle T\rangle$ that a toothpick cross or touch a line equals $2/\pi$. This is the fraction of toothpicks which cross or touch a line. [*Note:* Assuming that a large number of trials gives the assumed random distribution, this gives a statistical method for determining π. Thus if N drops result in M crossings or touchings, we have $M/N = 2/\pi$ or $\pi = 2N/M$. Try it yourself with, say, 157 toothpicks. A favorable outcome would yield 100 crossings or touchings.]

•••••• ••••••

10.4 FREE PATH LENGTHS AND THEIR DISTRIBUTION

The atoms of a gas are incessantly in motion. Collisions between atoms cause deflections and result in irregular zig-zag paths. The distance between collisions is called the *free path length.* Collisions occur randomly so that the length of individual free paths cannot be predicted. A useful measure of free path length is the average or mean distance between collisions, the so-called *mean free path* λ. We can estimate λ as follows: Let the effective radius of an atom be denoted by R_0. By effective we mean that a collision occurs if the centers of two atoms approach to within $2R_0$ or less. It is customary to denote the effective cross-sectional area of an atom as σ. If we take $R_0 \simeq 2\times10^{-8}$cm, the typical cross section is

$$\sigma \equiv \pi R_0{}^2 \simeq 10^{-15}\ \text{cm}^2$$

As the atoms speed about between collisions they sweep out cylindrical volumes. Over a free path length r, an atom sweeps out a volume σr. The mean volume swept out is $\sigma\lambda$, where λ is the mean free path. Each atom is eagerly vying for its share of the container volume. The mean volume $\sigma\lambda$ is therefore one atom's share of the container volume. With N atoms in a volume V we have

$$\sigma\lambda = \text{volume per atom} = \frac{V}{N} \tag{10.18}$$

Under standard conditions, 1 mole of a gas occupies a volume of 22.4 liters and contains Avogadro's number of atoms. The volume per atom is therefore

$$\frac{V}{N} = \frac{22{,}400\ \text{cm}^3}{6.023\times10^{23}} \simeq 4\times10^{-20}\ \text{cm}^3$$

It follows that $\lambda \simeq 4 \times 10^{-5}$ cm. Note that λ is about 2000 times as great as the effective atomic diameter. Thus under ordinary conditions the mean free path is gigantic by comparison with atomic dimensions. A gas atom spends the majority of its time in free flight.

The distance between successive collisions cannot be predicted. However, a probability for a given free path length can be assigned. We write

$$W(r) = \begin{array}{c}\text{probability that particle will travel a distance } r\\ \text{without suffering a collision}\end{array} \tag{10.19}$$

The distance of travel is measured from the last collision. The collisions occur at random. At random means that there is no rule or law which can specify how far any given atom will travel before suffering a collision. Translated into mathematical terms, at random means that the probability that a collision occur between r and $r+dr$ is proportional to dr, but independent of r.

$$\begin{array}{c}\text{probability that a collision will occur}\\ \text{between } r \text{ and } r+dr\end{array} = \alpha\, dr \tag{10.20}$$

The proportionality constant α generally depends on factors such as gas pressure, temperature, and composition. The important point is that α does not depend on the distance r.

In traveling a distance dr the particle either does or does not suffer a collision. Thus

$$\begin{array}{c}\text{probability that a collision does not}\\ \text{occur between } r \text{ and } r+dr\end{array} = 1 - \alpha\, dr \tag{10.21}$$

It is now possible to write two equations for $W(r+dr)$, the probability that the particle travel a distance $r+dr$ without undergoing a collision. For sufficiently small dr, $W(r+dr)$ can differ only slightly from $W(r)$. Denoting this change by dW we write,

$$W(r+dr) = W(r) + dW \tag{10.22}$$

A second equation for $W(r+dr)$ follows by noting that a collision-free path of length $r+dr$ consists of two independent steps; one of length r followed by one of length dr. As independent events, the overall probability $W(r+dr)$ is just the product of the separate probabilities.

$$W(r+dr) = \begin{bmatrix}\text{probability that a particle}\\ \text{will travel a distance } r \text{ without}\\ \text{suffering a collision}\end{bmatrix} \cdot \begin{bmatrix}\text{probability that no}\\ \text{collision will occur}\\ \text{between } r \text{ and } r+dr\end{bmatrix} \tag{10.23}$$

The first factor is $W(r)$, while the second is given by (10.21). Thus

$$W(r+dr) = W(r)[1-\alpha\, dr] \tag{10.24}$$

Equating (10.22) and (10.24) leads to

$$\frac{dW}{W} = -\alpha\, dr \tag{10.25}$$

which may be integrated to give

$$W(r) = Ae^{-\alpha r} \tag{10.26}$$

••••• •••••

10.4.1 The constant A equals one, being fixed by the condition $W(r=0) = 1$. Explain the physical significance of this condition.

••••• •••••

Setting $A = 1$ gives the final result,

$$W(r) = e^{-\alpha r} \tag{10.27}$$

To check (10.27) and pave the way for determining the physical significance of α we introduce

$$dP(r) \equiv \text{probability of a free path length in the range } r \text{ to } r+dr \tag{10.28}$$

A free path length between r and $r+dr$ also consists of two independent events, a collision-free trip of length r followed by a collision between r and $r+dr$. Thus

$$dP = \begin{bmatrix} \text{probability that a particle} \\ \text{will travel } r \text{ without} \\ \text{suffering a collision} \end{bmatrix} \cdot \begin{bmatrix} \text{probability that a} \\ \text{collision will occur between} \\ r \text{ and } r+dr \end{bmatrix} = W\alpha\, dr$$

Using (10.27),

$$dP(r) = e^{-\alpha r}\alpha\, dr \tag{10.29}$$

Knowledge of $dP(r)$ enables one to determine the physical significance of α. The mean free path, designated earlier by λ, follows from (10.14) with $Q = r$

$$\lambda \equiv \langle r \rangle = \int_{r=0}^{\infty} r\, dP(r) = \int_{r=0}^{\infty} re^{-\alpha r}\alpha\, dr \tag{10.30}$$

An integration by parts shows that the integral equals $1/\alpha$. Thus

$$\text{mean free path} = \lambda = \frac{1}{\alpha}$$

The relevant probabilities, $W(r)$ and $dP(r)$, may now be rewritten in terms of λ.

$$W(r) = e^{-r/\lambda}$$

$$dP(r) = \frac{e^{-r/\lambda}\, dr}{\lambda} \tag{10.31}$$

The exponential factor $e^{-r/\lambda}$ shows that there is but a slim chance of a free path length many times λ, the mean value.

•••••• ••••••

10.4.2 (a) An atom in a gas suffers collisions at random. Show that the probability $W(t)$ that the atom travel for a time t (measured from time of last collision) without undergoing a collision is given by

$$W(t) = e^{-\nu t}$$

where ν is some constant.

(b) Show that $\nu = 1/\tau$, where τ is the mean free time, the mean time between collisions. The definition of τ is

$$\tau \equiv \langle t \rangle = \int_0^\infty t\, dP(t)$$

where $dP(t)$ is the probability that the atom travel without a collision for a time t and then suffer a collision between t and $t+dt$.

•••••• ••••••

REFERENCES

There are many fine textbooks in probability theory. The following contain introductory chapters especially suitable for the beginner:

B. V. Gnedenko and A. Ya. Khinchin, "An Elementary Introduction to the Theory of Probability." Dover, New York (1961).

J. V. Upensky, "Introduction to Mathematical Probability." McGraw-Hill, New York (1937).

V. E. Gmurman, "Fundamentals of Probability Theory and Mathematical Statistics." American Elsevier, New York (1968).

A recent review of Maxwell's kinetic theory research is contained in:

S. G. Brush, James Clerk Maxwell and the Kinetic Theory of Gases: A Review Based on Recent Historical Studies. *Am. J. Phys.* **39**, 631 (1971).

Chapter 11

Gases : Models and Equipartition

In this chapter we introduce a model of a gas. The model pictures a gas as a vast collection of particles—atoms and molecules—which move about incessantly. This model leads to the ideal gas equation of state and provides a mechanical interpretation of temperature.

The degree of freedom concept is developed. Crudely stated, the number of degrees of freedom equals the number of independent ways in which the system may store energy.

The equipartition of energy theorem is presented; the theorem states that the energy of a system is distributed equally among the degrees of freedom, that is, there is an equal partitioning of energy.

We apply the equipartition of energy theorem to the ideal gas model. This leads to a prediction of the heat capacities, C_p and C_v, which prove to be in good agreement with the experimental values.

Finally, we consider deviations from the ideal behavior. The concept that pressure measures energy per unit volume is introduced and used to deduce the van der Waals and virial equations of state.

11.1 THE IDEAL GAS

We have previously noted that all gases may be described by the same equation of state, provided that the pressure is not too high and the temperature not too low. This is the by now familiar, ideal gas equation of state. For n moles of gas, the pressure, temperature, and volume are related by

$$PV = nRT \tag{11.1}$$

where R is the universal gas constant whose value depends on the system of units adopted

$$\begin{aligned} R &= 8.317 \times 10^7 \text{ erg/K} \\ &= 8.317 \text{ joule/K} \\ &= 1.987 \text{ cal/K} \\ &= 0.08208 \text{ liter-atm/K} \end{aligned}$$

An alternate form of the equation of state is also useful, namely,

$$PV = NkT \tag{11.2}$$

in which N is the total number of molecules present in the volume V, and $k = R/N_0$ is the Boltzmann constant.

Example 11.1 Atomic Size versus Container Size

We cannot see the molecules which comprise the air we breathe. Atoms and molecules presumably are small. How small? What is the size of a typical atom? To answer this question does not require an electron microscope or an understanding of quantum mechanics. A reliable estimate of atomic size can be based on the observation that solids and liquids are virtually incompressible. Extremely high pressures are required to alter the volume of a liquid by a few percent. Many solids can withstand very large compressive stresses without significant distortion. By contrast, a gas is readily compressed. These facts suggest that the atoms in a gas are relatively far apart while those in liquid or solid form are packed rather tightly. As we now demonstrate, the dimension characteristic of atoms is 10^{-8} cm. This minute length is called an angstrom and abbreviated as 1 Å.

To estimate atomic sizes we assume that the volume of a solid or liquid equals the total volume of the constituent atoms. That is, we choose to ignore the space between atoms. This seems a reasonable assumption, for if there were much space between atoms it should be possible to squeeze them closer together by applying pressure. However, as already noted, squeezing solids and liquids scarcely affects their volumes.

Consider 1 mole of a substance. Let V denote its volume and let M signify the gram-molecular weight. The number of molecules in 1 mole is Avogadro's number $N_0 \simeq 6 \times 10^{23}$. Next, we envision the volume V as composed of N_0 subvolumes of equal size—one for each molecule. More precisely, we picture each molecule as residing in a cubical room. If the length of the edge of such a cube is d, its volume is d^3. With such a picture d becomes a measure of atomic size. Taken collectively the N_0 cubes comprise the total volume V. Thus

$$V = N_0 d^3$$

Next we observe that the mass density ρ may be expressed by

$$\rho = \frac{\text{mass of 1 mole}}{\text{volume of 1 mole}} = \frac{M}{V} = \frac{M}{N_0 d^3}$$

Solving for d gives

$$d = \left(\frac{M}{\rho N_0}\right)^{1/3} \tag{11.3}$$

Provided M is in grams and ρ is in grams per cubic centimeter we find

$$d \simeq \left(\frac{5M}{3\rho}\right)^{1/3} \text{Å}, \qquad 1\ \text{Å} = 10^{-8}\ \text{cm} \tag{11.3a}$$

As Table 11.1 shows, the factor $(5M/3\rho)^{1/3}$ is essentially the same for many elemental solids and liquids. Other materials furnish similar results. In view of these results we are led to conclude that the angstrom characterizes atomic

TABLE 11.1

Characteristic Atomic Dimensions

	Element	M (gm)	ρ (gm/cm^3)	$\left(\frac{5M}{3\rho}\right)^{1/3}$ (Å)
Solids	Li	6.9	0.53	2.8
	Ne	20.2	1.00	3.2
	Na	23.0	0.97	3.4
	Al	27.0	2.70	2.6
	K	29.1	0.86	4.2
	Ar	39.9	1.65	3.4
	Zn	65.4	7.1	2.5
	Rb	85.5	1.53	4.5
	Ag	107.9	10.5	2.6
	Cs	132.9	1.87	4.9
	Pt	195.2	21.4	2.5
	Au	197.0	19.3	2.6
	Pb	207.2	11.3	3.1
Liquids	He	4.0	0.12	3.8
	Ne	20.2	1.21	2.9
	Ar	39.9	1.39	3.6
	Kr	83.8	2.61	3.7
	Xe	131.3	3.06	4.1
	Hg	200.6	13.6	2.9

sizes. The corresponding atomic volume is roughly 10^{-24} cm^3. Let us see how this atomic volume compares with the container volume.

One mole of atoms in the gaseous state under standard conditions occupies a volume of 22.4 liters, about 2×10^4 cm^3. The number of atoms in the molar volume is $N_0 \simeq 6 \times 10^{23}$. If each atom has a volume of 10^{-24} cm^3, the total atomic volume of 1 mole of atoms is close to 1 cm^3. Comparing this with the container volume of over 2×10^4 cm^3 we conclude that atoms in the gaseous state are dispersed throughout a volume more than 10,000 times as great as their atomic volume. Stated another way, if all of the atoms were piled in one corner of their container, they would fill only about 1/10,000 of its volume. ◇

The next step is to put forth a model of a gas. Consideration of the equation of state for the model leads to an important interpretation: The temperature of a gas is a measure of the average kinetic energy of its atoms and molecules.

11.2 IDEAL GAS MODEL AND CRITIQUE

The following set of four assumptions define a particular model of a gas. We shall show that this model leads to the ideal gas equation of state. A critique of the model follows, indicating the experimental basis for each assumption and the extent to which each is realistic. Our intention of deducing the equation of state implicitly assumes that the gas is in thermal equilibrium with the container walls.

The Model:

(a) The gas consists of a large number of particles (atoms†) which obey Newton's laws of motion.

(b) The motion of the atoms is random.

(c) The volume of the atoms themselves is negligible by comparison with the volume of the container. The atoms of a monatomic gas are treated as point masses.

(d) The atoms collide elastically with the container walls, but do not collide with one another. For a monatomic gas, this means that the atoms possess only translational kinetic energy, except during the instant of impact in a wall collision.

Critique of the Model:

(a) Experimental evidence for assuming that the gas contains many atoms is furnished by Perrin's determinations of Avogadro's number. In a variety

† For the present we consider a monatomic gas.

of experiments, Perrin established that $N_0 \simeq 6 \times 10^{23}$. At a moderate pressure, even a fraction of a cubic centimeter of gas contains an enormous number of atoms.

Choosing to operate at the level of classical mechanics is motivated by an important historical fact. Kinetic theory successfully evolved within a classical framework. Newton's laws of motion are adequate for the development of most aspects of the theory. Of course, atoms "really" obey the laws of quantum mechanics. Indeed, certain paradoxes arise in kinetic theory if one insists on applying classical mechanics at every turn. In particular, the heat capacities of many gases exhibit certain anomalies which may be explained only through a quantum mechanical treatment. The heat capacity anomalies are historically significant for they sounded early warnings that something was wrong with the foundations of 19th century physics.

For the moment we concentrate on the successes of kinetic theory as they stemmed from the framework of classical physics. Later we will find out what went wrong. Still later we will see how quantum physics can correct shortcomings of the classical theory.

(b) The assumption of random motion is quite general and would be incorporated in more sophisticated models, whether classical or quantum mechanical. The experimental evidence for random motion is based on observations of Brownian motion. The earliest observations of Brownian motion involved colloidal suspensions and demonstrated the random motion of atoms in the liquid state. The random motion of atoms in a gas was demonstrated by Fletcher and Millikan in 1911. They observed the Brownian motion of electrically charged oil droplets, using many of the same techniques developed by Millikan to determine the charge of the electron.

(c) Only in a few gross respects does an atom behave like a point mass. However, it is these same few characteristics which are most crucial for a determination of the equation of state. The translational kinetic energy of the atom can be written as $\frac{1}{2}mv^2$, where m is the atomic mass and v is the speed of the center of mass. The linear momentum is given by mv. With respect to translational kinetic energy and linear momentum, any structure of the atom is irrelevant—it behaves like a point mass. However, as pointed out earlier, atoms do take up space, an effect which becomes more important as the gas is compressed. For rarified gases, the atomic volume effect can be accounted for by a little "handwaving." One simply replaces the container volume V by $V - V_{ex}$, where V_{ex} is the so-called excluded volume. The idea is that the entire container volume is not available to any given atom; the presence of the other atoms excludes a certain volume. This point is developed further in Section 11.6.

Liquid-to-vapor transition data justify the neglect of the atomic volume in a rarified gas. For example, 1 gm of water at the normal boiling point occupies a volume of approximately 1 cm^3. After vaporization it occupies a volume of nearly 1700 cm^3, suggesting that each molecule in the vapor has about 1700 times as much room to move about in as it had in the liquid state. In this respect, the datum for water is typical.

(d) The assumption of elastic collisions is artificial. A realistic picture is much more complicated. Some atoms collide with the wall and adhere to it (adsorption). However, the assumption of elastic collisions does not lead to any error so long as the gas is in thermal equilibrium with the container walls. Thermal equilibrium requires that no net transfer of energy occur between the gas and the walls. The elastic collision assumption rules out *any* energy transfer between the gas and the walls, automatically ensuring that no net energy is transferred.

Perhaps the most significant assumption is that the atoms do not collide with one another. Even in a rarified gas, atoms experience mutual interatomic forces. These interatomic forces depend largely on the structure of the atom. Thus, for example, the force between a pair of hydrogen atoms is distinctly different from the force between a pair of helium atoms. As later analysis will show, it is these interatomic forces which give rise to the nonideal characteristics of real gases. We deal with this shortcoming of our model in Section 11.6.

11.3 EQUATION OF STATE

Having chosen a model, the next step is to establish the equation of state. This requires that we compute the gas pressure, the average normal force per unit area arising from the collisions of the atoms with the container walls. The gas is enclosed in a rectangular container having sides of length L_x, L_y, and L_z.

Let v_x denote the x component of velocity of an atom which subsequently collides with the wall (see Figure 11.1). The atom rebounds with an x

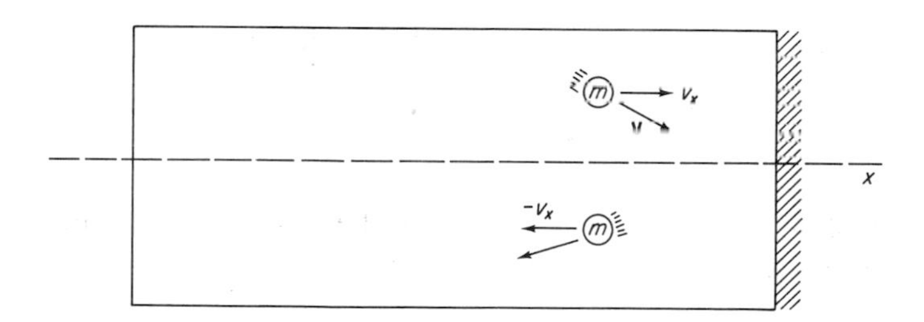

Figure 11.1. An atom undergoes an elastic collision with the wall. The normal component of its velocity changes sign. Its speed is unchanged.

component of velocity equal to $-v_x$. The change in its x component of momentum is

$$(p_x)_{\text{aft}} - (p_x)_{\text{bef}} = -2mv_x$$

with aft standing for after and bef for before. A change of equal magnitude and opposite direction is experienced by the container wall. The x momentum transferred to the wall is

$$\begin{array}{c}\text{momentum transfer to wall as a}\\ \text{result of one collision}\end{array} \equiv \Delta p_x = 2mv_x \tag{11.4}$$

This momentum is parceled out among the many atoms comprising the container walls. The momentum transferred by a single atom is not observable—measuring instruments respond to the combined effect of many such collisions.

An atom with velocity component v_x travels the length L_x of the container in a time L_x/v_x. The time Δt between collisions with a particular wall is twice as great

$$\Delta t = \frac{2L_x}{v_x} \tag{11.5}$$

The average rate at which such an atom transfers momentum to the wall is

$$\frac{\Delta p_x}{\Delta t} = \frac{2mv_x}{2L_x/v_x} = \frac{m{v_x}^2}{L_x} \tag{11.6}$$

By Newton's second law, the rate of change of momentum equals force

$$\frac{\Delta p_x}{\Delta t} = f_x \tag{11.7}$$

Thus the average force exerted by the impacts of an atom having an x-velocity component v_x is

$$f_x = \frac{m{v_x}^2}{L_x} \tag{11.8}$$

The total force on the wall F_x is obtained by summing over all particles.

$$F_x = \sum_{\text{a.a.}} f_x = \sum_{\text{a.a.}} \frac{m{v_x}^2}{L_x} \tag{11.9}$$

where a.a. stands for all atoms. The area of the container wall normal to the x axis is $A = L_y L_z$. The pressure P on this area is

$$P = \frac{F_x}{A} = \sum_{\text{a.a.}} \frac{m{v_x}^2}{L_x L_y L_z} \tag{11.10}$$

Since $L_x L_y L_z$ is the volume V of the container we can write

$$PV = \sum_{\text{a.a.}} m v_x^2 \tag{11.11}$$

Now, the pressure of the gas is the same on every wall; there is no preferred direction of motion. Thus we also have the results

$$PV = \sum_{\text{a.a.}} m v_y^2 = \sum_{\text{a.a.}} m v_z^2 \tag{11.12}$$

Further, since

$$v_x^2 + v_y^2 + v_z^2 = v^2 \tag{11.13}$$

we can write

$$PV = \tfrac{1}{3} \sum_{\text{a.a.}} m(v_x^2 + v_y^2 + v_z^2) = \tfrac{1}{3} \sum_{\text{a.a.}} m v^2 \tag{11.14}$$

The total kinctic energy K_{tot} of the gas is

$$K_{\text{tot}} = \sum_{\text{a.a.}} \tfrac{1}{2} m v^2 \tag{11.15}$$

whereupon we have

$$PV = \tfrac{2}{3} K_{\text{tot}} \tag{11.16}$$

To maintain the particle point of view we write K_{tot} as N times the mean kinetic energy per atom $\langle K \rangle$

$$K_{\text{tot}} = N\langle K \rangle$$

and thence,

$$PV = \tfrac{2}{3} N\langle K \rangle \tag{11.17}$$

As it stands, (11.17) is the equation of state which the model has produced.

At this juncture it is worthwhile to note that the analysis applies equally well to diatomic and polyatomic gases. Only considerations involving the translational kinetic energy and linear momentum of the center of mass enter the pressure calculation. Since we follow only the center of mass motion, any structure of the atoms or molecules is irrelevant in so far as the equation of state is concerned. Thus all ideal gases obey the same equation of state regardless of any microscopic structure attributed to their atoms and molecules. This conclusion is in accord with the experimental data for real gases. Experimentally, the equation of state obeyed (approximately) by all gases is

$$PV = NkT \tag{11.2}$$

Comparing (11.17) and (11.2) we are led to conclude that the temperature of an ideal gas is proportional to the mean kinetic energy of the atoms.

Specifically, we must have

$$\langle K \rangle = \langle \tfrac{1}{2}mv^2 \rangle = \tfrac{3}{2}kT \tag{11.18}$$

if the model is to reproduce the empirical form of the equation of state. Note that (11.18) is *not a definition of temperature*, since we cannot measure the mean kinetic energy of an atom, that is, such a definition would not be operational. Our model of the ideal gas does present us with a microscopic interpretation of temperature. It takes us beyond the semiquantitative statement "equality of temperature is a condition for thermal equilibrium," and gives us a mechanical feel for temperature. The temperature of a gas measures the translational kinetic energy of the atomic motions. Clearly this revelation would never be forthcoming from any macroscopic thermodynamic analysis. In passing we note that the random atomic motions are referred to the container. Thus, for example, the temperature of the gas in a balloon is not altered by transporting it in an auto at 60 mph, in a jet at 600 mph, or in a spacecraft at 6000 mph.

••••• •••••

11.3.1 (a) Compute the mean kinetic energy of a monatomic gas atom for a temperature of 300 K. Express the result in electron volts. ($1\,\text{eV} = 1.6 \times 10^{-12}$ erg.) Note that the mean energy depends on the temperature of the gas, but not on the mass of the atoms.

(b) Compute the rms speed of a helium atom for a temperature of 300 K. It will be recalled that the rms value of a quantity is the square root of the mean value of the square of the quantity. Thus

$$v_{\text{rms}} = \sqrt{\langle v^2 \rangle}$$

The mass of a helium atom is 6.66×10^{-24} gm.

11.3.2 One of the attractive features of "nearby" outer space is the essentially complete vacuum existing there. A total vacuum is not achieved because of the solar wind. Vast quantities of ionized hydrogen—electrons and protons—are spewed off the Sun. This material streams outward, filling the solar system with a stream of particles. The number density and rms speed of the solar wind protons have been measured by Mariner 2 and other space probes. The values are

$$n_{\text{p}} \simeq 3/\text{cm}^3, \qquad v_{\text{rms}} \simeq 500 \text{ km/sec}$$

These data refer to regions "near" Earth, but at altitudes well above its permanent atmosphere.

(a) Compute the corresponding kinetic pressure of the solar wind, assuming that the protons behave as an ideal gas [see (11.17)].

(b) How does the solar wind pressure compare with the best man-made vacuum achieved on Earth (approximately 10^{-10} Torr)?

•••••• ••••••

At this point, kinetic theory lets us reap another benefit from the model. In our thermodynamic treatment, the ideal gas was defined by two relations. The first was the equation of state (11.1). The second relation specified the internal energy to be (for n moles),

$$U = nC_v T \tag{11.19}$$

with the molar heat capacity C_v being a constant. In Section 11.5 we show that our model leads to this same relation. Furthermore, the model predicts the value of C_v. Here then, kinetic theory will make its first modest prediction, thereby allowing a comparison with experiment. It is first necessary to introduce the notion of degrees of freedom and to formulate the equipartition of energy theorem.

11.4 THE EQUIPARTITION OF ENERGY

Several examples can best serve to illustrate the idea of a *degree of freedom.* The system depicted in Figure 11.2 consists of a bead (regarded as a point mass)

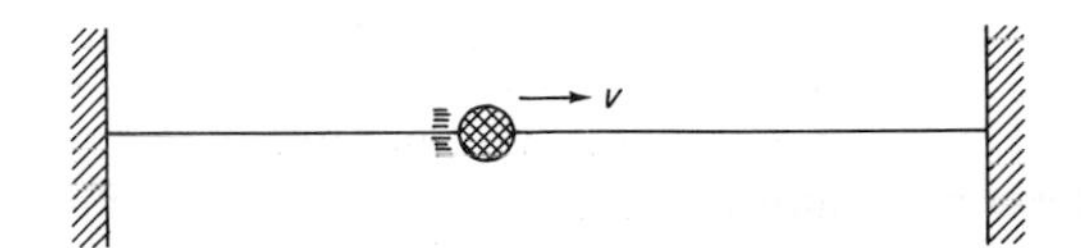

Figure 11.2. Bead on a wire—a system with one degree of freedom.

of mass m which is free to move along a horizontal wire. The constraints imposed by the wire permit only one-dimensional motion. The total energy of the bead is kinetic,† $E_{\text{bead}} = \frac{1}{2}mv^2$. Only one variable, the speed of the bead, is needed to specify the energy of the bead. Note that E is proportional to v^2; we say that E is a quadratic function of v. We also say that the bead possesses one degree of freedom—its total energy depends on but one variable. In general the

† We take the gravitational potential energy of the bead to be zero. Other choices could be made. The essential point is that the gravitational potential is a constant, independent of the position or speed of the bead.

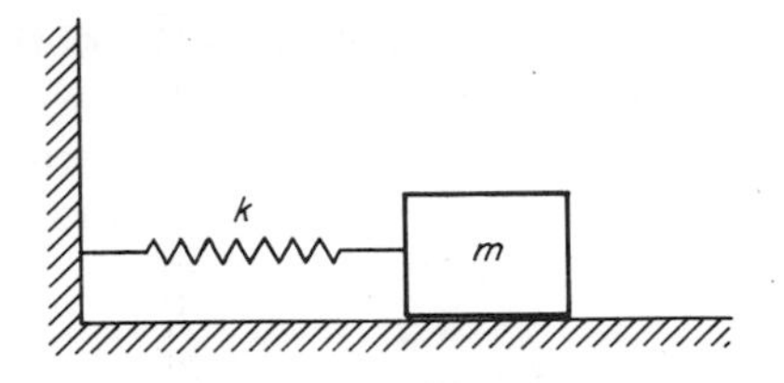

Figure 11.3. The simple harmonic oscillator has two degrees of freedom.

number of degrees of freedom possessed by a system† equals the number of independent variables needed to specify its energy.

An often-used system in physics is the simple harmonic oscillator (SHO). Figure 11.3 shows one realization of the SHO. The spring–mass system is capable of undergoing one-dimensional motion—oscillations along the x axis. The total energy of the one-dimensional SHO is

$$E = \tfrac{1}{2}mv^2 + \tfrac{1}{2}kx^2$$

As such the SHO possesses two degrees of freedom. Two independent variables, v and x, are needed to specify the kinetic and potential energies of the oscillator. We shall use f to denote the number of degrees of freedom of a system, so that $f = 2$ for the SHO. The SHO is an extremely important system throughout physics. At the elementary level it is useful in illustrating Newtonian dynamics and the essentials of oscillatory motion. In quantum mechanics it turns out that certain systems (for example, electromagnetic waves and sound waves) behave essentially like collections of SHO's.

Figure 11.4 shows a system with six degrees of freedom. It is a three-dimensional SHO. The mass is free to move in any direction from its equilibrium position. Let x, y, and z denote the components of its displacement and v_x, v_y, and v_z the corresponding velocity components. The total energy of the system is

$$E = \tfrac{1}{2}mv_x{}^2 + \tfrac{1}{2}mv_y{}^2 + \tfrac{1}{2}mv_z{}^2 + \tfrac{1}{2}k_x x^2 + \tfrac{1}{2}k_y y^2 + \tfrac{1}{2}k_z z^2$$

so that $f = 6$ for the three-dimensional SHO. Note that the total energy is a quadratic function of each of the six independent variables x, y, z, v_x, v_y, and v_z.

The atoms of a crystalline solid can be visualized as the masses of a three-dimensional SHO. The springs represent interatomic forces which bind the atoms together and give the solid its elastic quality. A macroscopic chunk of such a solid thus appears as a gigantic bedspring-like arrangement with masses (atoms) at the intersections of the spring network.

Another important system is the free point mass—a point mass subject to no

---

† Here the word "system" refers to the component parts (for example, atoms, molecules, and so on) which comprise a thermodynamic system.

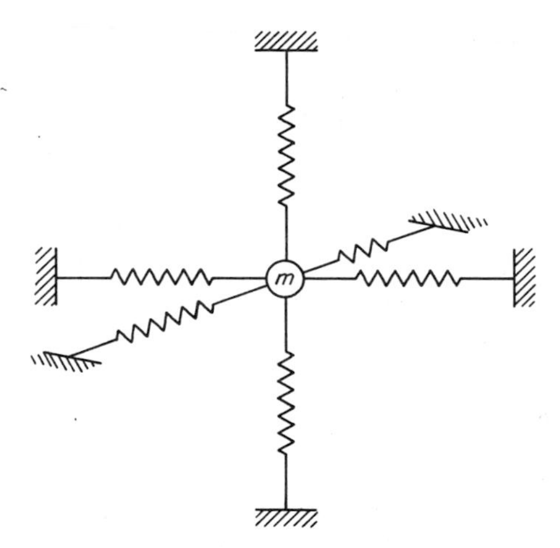

Figure 11.4. A three-dimensional oscillator possesses six degrees of freedom.

external forces. The free point mass has already served us as a model for the monatomic gas atom. The total energy of the point mass is

$$E = \tfrac{1}{2}mv_x^2 + \tfrac{1}{2}mv_y^2 + \tfrac{1}{2}mv_z^2$$

Three independent variables, v_x, v_y, and v_z, are required to specify the total energy so that $f = 3$ for the free point mass.

If the point mass experiences a uniform gravitational acceleration g in the negative z direction, it acquires an additional degree of freedom. The total energy is

$$E = \tfrac{1}{2}mv_x^2 + \tfrac{1}{2}mv_y^2 + \tfrac{1}{2}mv_z^2 + mgz$$

with mgz being the gravitational potential energy. Unlike the previous examples E is not a quadratic function of all four independent variables.

Another system of interest is the so-called dumbbell molecule. This model pictures a diatomic molecule as a pair of point masses bound together by a rigid rod. As Figure 11.5 suggests, the dumbbell molecule appears to be a

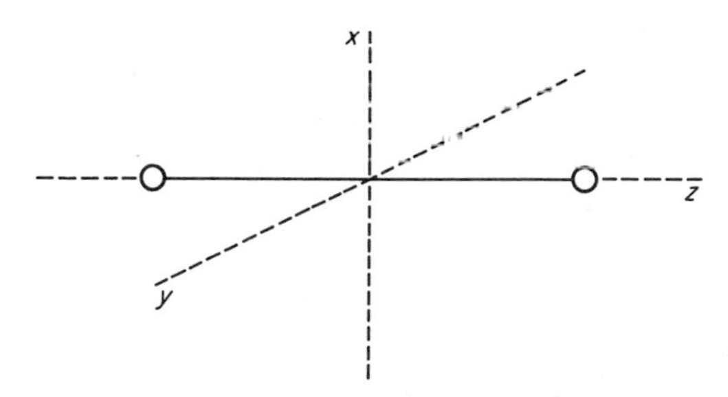

Figure 11.5. The dumbbell molecule has five degrees of freedom.

miniaturization of the circus strongman's equipment. For the dumbbell molecule we have $f = 5$. Three degrees of freedom are accounted for by the translational kinetic energy of the molecule. If m is the total mass of the molecule and v_x, v_y, and v_z are the center of mass velocity components, the translational kinetic energy is

$$E_{\text{trans}} = \tfrac{1}{2}mv_x^2 + \tfrac{1}{2}mv_y^2 + \tfrac{1}{2}mv_z^2$$

The dumbbell molecule is free to rotate about its center of mass and thus can possess rotational energy as well. Taking the z axis to lie along the rod we can write

$$E_{\text{rot}} = \tfrac{1}{2}I_x\omega_x^2 + \tfrac{1}{2}I_y\omega_y^2$$

where I_x and I_y are the moments of inertia about the x and y axes while ω_x and ω_y are the corresponding angular velocities about these axes. The assumption of point masses means that $I_z = 0$. The total energy is the sum of the translational and rotational contributions

$$E_{\text{dumb}} = \tfrac{1}{2}mv_x^2 + \tfrac{1}{2}mv_y^2 + \tfrac{1}{2}mv_z^2 + \tfrac{1}{2}I_x\omega_x^2 + \tfrac{1}{2}I_y\omega_y^2$$

The variables v_x, v_y, v_z, ω_x, and ω_y are independent so that $f = 5$ for the dumbbell molecule.

As a final example we consider a somewhat more realistic model of a diatomic molecule. There is a vast amount of spectroscopic evidence to support the reality of the rotational degrees of freedom ascribed to the dumbbell molecule. Additionally, there is a wealth of spectroscopic data suggesting that a diatomic molecule also possesses vibrational degrees of freedom. The two atoms in the molecule are not bound rigidly, but can move relative to one another in a vibratory fashion. These vibrational capabilities can be incorporated by using a model in which the rigid rod of the dumbbell is replaced by a stiff spring (see Figure 11.6). The resulting system—a vibrating dumbbell molecule—has seven degrees of freedom. In so far as the vibrational motions are concerned, the molecule behaves like a one-dimensional SHO, which has

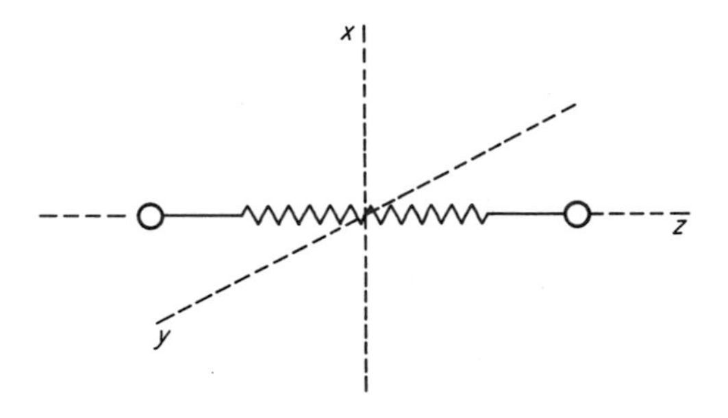

Figure 11.6. The vibrating dumbbell molecule has two vibrational degrees of freedom.

two degrees of freedom. Therefore, the vibrational capability adds two degrees of freedom to the three translational and two rotational degrees of freedom of the rigid dumbbell model. Furthermore, the vibrational energy E_{vib} can be expressed in the form

$$E_{\text{vib}} = \tfrac{1}{2}m'v_\zeta^{\,2} + \tfrac{1}{2}k\zeta^2$$

where m' is a mass (a particular combination of the masses of the two atoms) and k is an effective spring constant describing the electronic force which binds the atoms. The independent variables ζ and v_ζ can be related to the displacements and velocities of the two atoms. We need not indicate these relations since the only fact germane to our later analysis is that E_{vib} is a quadratic function of both ζ and v_ζ.

These examples serve to illustrate the notion of a degree of freedom. A system possesses one degree of freedom for each independent variable needed to determine the total energy. Except for the gravitational potential energy encountered in one example, the total energy proved to be a quadratic function of these independent variables. The equipartition theorem is concerned with the relation between the temperature and the mean energy associated with each degree of freedom.

In Section 11.3 we noted the fact that there was no preferred direction of motion for the atoms so that

$$\langle \tfrac{1}{2}mv_x^{\,2} \rangle = \langle \tfrac{1}{2}mv_y^{\,2} \rangle = \langle \tfrac{1}{2}mv_z^{\,2} \rangle \tag{11.20}$$

Since

$$\langle \tfrac{1}{2}mv_x^{\,2} \rangle + \langle \tfrac{1}{2}mv_y^{\,2} \rangle + \langle \tfrac{1}{2}mv_z^{\,2} \rangle = \langle \tfrac{1}{2}mv^2 \rangle = \tfrac{3}{2}kT \tag{11.18$'$}$$

it follows that equal mean energies are ascribed to each of the three degrees of freedom. Each degree of freedom contributes $\frac{1}{2}kT$ to the total mean energy

$$\langle \tfrac{1}{2}mv_x^{\,2} \rangle = \langle \tfrac{1}{2}mv_y^{\,2} \rangle = \langle \tfrac{1}{2}mv_z^{\,2} \rangle = \tfrac{1}{2}kT \tag{11.21}$$

This applies on a per particle basis. The energy per mole attributable to each degree of freedom is

$$N_0(\tfrac{1}{2}kT) = \tfrac{1}{2}RT$$

The equipartition of energy theorem states that each degree of freedom contributes $\frac{1}{2}kT$ to the mean energy. Thus if the particles (atoms or molecules) of a gas possess f degrees of freedom, the mean energy per particle is $\frac{1}{2}fkT$. The thermal energy of the system is partitioned equally among the degrees of freedom—thus the name of the theorem.

The theorem is subject to a number of restrictions. It is true only for those degrees of freedom which contribute quadratic terms to the total energy. Thus, for example, the point mass in a uniform gravitational field has four degrees of freedom, but only three contribute quadratic terms to the energy,

$$E = \tfrac{1}{2}mv_x^2 + \tfrac{1}{2}mv_y^2 + \tfrac{1}{2}mv_z^2 + mgz$$

The mean energy is

$$\langle E \rangle = \tfrac{3}{2}kT + \langle mgz \rangle$$

The equipartition theorem decrees that each of the three translational degrees of freedom contributes equally to $\langle E \rangle$, but is silent on the relation between $\langle mgz \rangle$ and the temperature. Statistical mechanics does enable one to relate $\langle mgz \rangle$ and T, but has little to say, in a general sense, about those degrees of freedom which do not contribute quadratic terms to the energy.

For future reference we note that for both monatomic and diatomic ideal gases, each of the degrees of freedom contributes a quadratic term to the total energy so that the mean energy of an ideal gas atom or molecule may be written

$$\langle E \rangle = \tfrac{1}{2}fkT \tag{11.22}$$

•••••• ••••••

11.4.1 The equipartition energy is $\frac{1}{2}kT$ per particle.

(a) Prove that the corresponding energy per mole is $\frac{1}{2}RT$, where R is the gas constant.

(b) Assuming that an equipartition of energy prevails, what is the total energy of 1 mole of an ideal gas (temperature T) whose molecules behave like rigid dumbbells.

•••••• ••••••

11.5 INTERNAL ENERGY AND HEAT CAPACITIES

The internal energy is taken to be the total energy of the particles (atoms or molecules) comprising the gas. Two sources contribute to the internal energy of a real gas. First, each particle possesses its own private energy. For monatomic gases this energy is wholly kinetic. Polyatomic molecules may also possess rotational and vibrational energy. The second source of internal energy in a real gas is the mutual potential energy of the particles—generally referred to as an interaction energy to stress the fact that two or more particles are involved. The interaction energy is electrical in nature. It arises because the charges of neighboring atoms and molecules exert mutual Coulomb forces. In a rarified gas, these forces and the associated potential energies are ordinarily negligible because of the overall electrical neutrality of the particles. However, when two atoms collide, that is, pass within a few angstroms of each other, their electron atmospheres clash and the Coulomb forces and potential energies become significant. The mean distance which an atom travels between collisions in a rarified gas may be several thousand angstroms. Comparing this distance with the collision range of a few angstroms shows that the atoms spend most of their time in free flight. Only rarely are they interacting with other atoms. It was this fact which justified ignoring collisions in the ideal gas model. To ignore collisions is completely equivalent to neglecting the interaction energy. When this is done the internal energy of an ideal gas is just the sum of the private energies carried by each particle. We can now formulate these ideas quantitatively and use the equipartition theorem to deduce the internal energy of an ideal gas.

Let $\langle E \rangle$ denote the mean energy of a particle and let E_{INT} denote the interaction energy of the entire gas. The internal energy U is

$$U = N\langle E \rangle + E_{\text{INT}} \tag{11.23}$$

where $N = nN_0$ is the number of particles in the gas. For an ideal gas, E_{INT} is ignored so that we have simply

$$U = N\langle E \rangle, \qquad \text{ideal gas} \tag{11.24}$$

The mean energy is related to the temperature by the equipartition theorem. For particles with f degrees of freedom,

$$\langle E \rangle = \tfrac{1}{2}fkT \tag{11.22}$$

whence,

$$U = \tfrac{1}{2}fNkT \tag{11.25}$$

An alternate form for U is obtained by noting that $Nk = nR$, whence

$$U = n\left(\frac{fR}{2}\right)T \tag{11.26}$$

In our thermodynamic dealings with the ideal gas the form assumed for internal energy was

$$U = nC_v T \tag{11.19}$$

Comparison of (11.19) with (11.26) shows that the *molar* heat capacity predicted by our model is

$$C_v = \frac{fR}{2} \tag{11.27}$$

•••••

11.5.1 Show that (11.27) follows from the thermodynamic relation (4.24)

$$C_v = \left(\frac{\partial U}{\partial T}\right)$$

•••••

Remembering that $R = 1.987$ cal/K we arrive at the handy form

$$C_v \simeq f \text{ cal/K} \tag{11.27a}$$

that is, each degree of freedom contributes approximately 1 cal/K to C_v. As we showed earlier C_p and C_v are related (for an ideal gas) by

$$C_p = C_v + R$$

so that the model gives

$$C_p = \left(\frac{f+2}{2}\right) R \tag{11.28}$$

The ratio of the heat capacities

$$\frac{C_p}{C_v} = \gamma \tag{11.29}$$

is also subject to a check. The quantity γ may be determined from measurements of the speed of sound in the gas.† From (11.27) and (11.28) it follows that for an ideal gas γ has the value

$$\gamma = \frac{f+2}{f} \tag{11.30}$$

Note that these results are deductions—predictions—coaxed from the kinetic theory model. For a monatomic gas, $f = 3$ so that the model predicts

$$C_p = \frac{5}{2} R = 4.98 \ \frac{\text{cal}}{\text{K}} \tag{11.31}$$

$$\gamma = \frac{5}{3} = 1.67$$

For the diatomic dumbbell molecule ($f = 5$) we obtain

$$C_p = \frac{7}{2} R = 6.97 \ \frac{\text{cal}}{\text{K}} \tag{11.32}$$

$$\gamma = \frac{7}{5} = 1.40$$

† The relation between γ and the speed of sound is

$$(\text{sound speed}) = \sqrt{\frac{\gamma P}{\rho}}$$

where P is the pressure and ρ the mass density of the gas.

TABLE 11.2

Molar Heat Capacity C_p and γ for Monatomic and Diatomic Gases[a]

Gas	C_p (cal/K)	γ
Helium (He)	5.00	1.66
Argon (A)	5.00	1.67
Neon (Ne)	—	1.64
Krypton (K)	—	1.68
Xenon (Xe)	—	1.66
Hydrogen (H_2)	6.87	1.41
Carbon monoxide (CO)	6.94	1.40
Nitrogen (N_2)	6.94	1.40
Oxygen (O_2)	7.03	1.40
Hydrogen chloride (HCl)	7.07	1.40
Chlorine (Cl_2)	8.15	1.36

[a] The values refer to a pressure of 1 atm and a temperature near 15 °C. Adapted with permission from E. H. Kennard, "Kinetic Theory of Gases." McGraw-Hill, New York (1938).

The vibrating dumbbell molecule ($f = 7$) yields the results

$$C_p = \frac{9}{2}R = 9.96 \ \frac{\text{cal}}{\text{K}} \tag{11.33}$$

$$\gamma = \frac{9}{7} = 1.28$$

Table 11.2 summarizes the results of measurements for several gases. The values quoted are for measurements made at a pressure of 1 atm and temperatures near 300 K. Agreement between experiment and theory is excellent for the monatomic gases. The monatomic gas values of C_p and γ exhibit only very slight variation with temperature. The results for diatomic gases tell a sadder tale. The experimental values are close to those predicted for either the dumbbell model or the vibrating dumbbell model. However, the table does not reveal a most disturbing fact; namely that C_p exhibits a definite temperature variation. This is completely at variance with the ideal gas models—all of which predict temperature-independent values of C_p and γ. Figure 11.7 shows the temperature variation of C_p for molecular hydrogen H_2. Hydrogen exhibits a heat capacity characteristic of a monatomic gas at low temperatures. At higher temperatures it shows a value of C_p near $(7/2)\,R$, the value predicted for an ideal gas of dumbbell molecules. At still higher temperatures it appears that

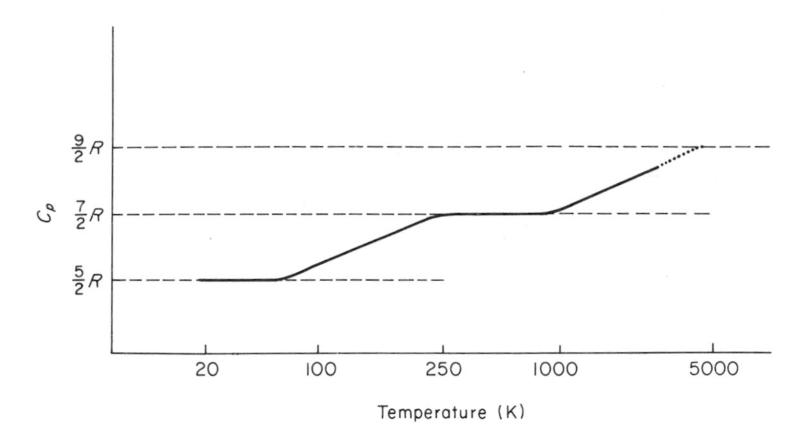

Figure 11.7. The molar heat capacity of molecular hydrogen showing the apparent excitation of additional degrees of freedom as the temperature rises. The temperature scale is logarithmic. The high-temperature data suggest—but do not confirm—that C_p is tending toward $(9/2)R$. A significant fraction of the molecules dissociate before the temperature reaches 5000 K.

C_p is approaching the value of $(9/2)R$, appropriate for the vibrating dumbbell model. The hydrogen molecules appear to acquire additional degrees of freedom—even pieces of them—as the temperature increases.

Try as they might, physicists could not reconcile the temperature variation of the heat capacity with any sort of mechanical model. By the beginning of the 20th century there was a gradual awakening. Physicists began to explore the possibility that it was the *physics, not the model*, which was in need of modification. Their efforts culminated in the development of quantum mechanics and quantum statistical mechanics. Thus the rules of the game—the laws of Newtonian mechanics—were found to have their limitations. At the atomic level, new rules were called for—the laws of quantum mechanics. The models have come through unscathed. Indeed, the dumbbell and vibrating dumbbell models are successfully used today in conjunction with quantum statistical mechanics. The quantum statistical formalism explains the observed temperature variation of C_p for H_2 and for other diatomic molecules.

•••••• ••••••

11.5.2 Figure 11.8 shows a plot of the molar heat capacities of several crystalline solids versus absolute temperature. Observe that the plots of C_v tend toward a value of roughly 6 cal/mole-K with increasing temperature.

(a) How many degrees of freedom must be ascribed to each atom in a solid if it is to have a molar heat capacity of approximately 6 cal/mole-K?

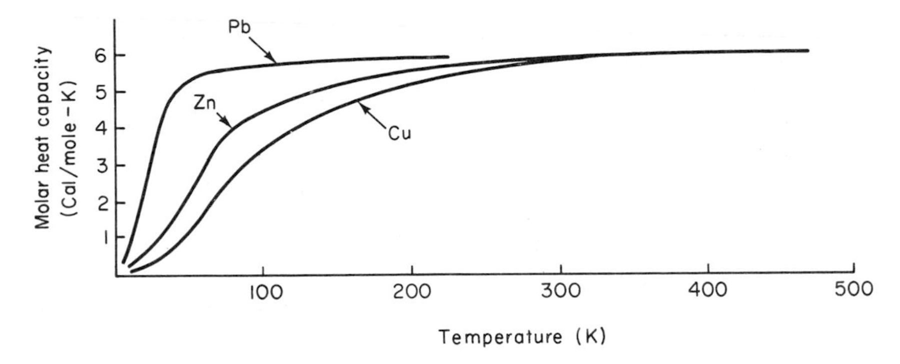

Figure 11.8. Molar heat capacities of three solids.

(b) What model(s) of a solid would accomplish this? Note the rapid decline in C_v at low temperatures. This behavior cannot be understood in terms of the equipartition of energy. However, the quantum theory again comes to the rescue and provides a theory of specific heats which is in excellent agreement with the low-temperature experimental results.

11.5.3 Rigid adiabatic walls surround a system which consists of 1 mole of a gas and 1 mole of a solid. The gas is ideal and monatomic. The atoms of the solid behave like three-dimensional oscillators, that is, they possess six degrees of freedom. Initially a rigid adiabatic wall separates the gas and solid. The initial temperatures are $T_s = 150\,\text{K}$, $T_g = 300\,\text{K}$. The adiabatic wall separating the solid and gas is then removed. Determine the final equilibrium temperature of the system.

11.5.4 We have constructed the equation of state of an ideal gas by consideration of atoms bouncing about inside a container. The resulting ideal gas equation of state was

$$PV = NkT = \tfrac{2}{3}U, \qquad U = \tfrac{3}{2}NkT$$

Electromagnetic radiation (light waves) exhibits particle characteristics. These particles are called *photons*. Photons display the following properties:

(a) A photon travels in a vacuum with the speed of light c.

(b) The photon has an energy ε given by $\varepsilon = h\nu$, where ν is the frequency of the source of radiation and h is Planck's constant.

(c) The photon has a momentum p whose magnitude is $p = \varepsilon/c$. The photon momentum is parallel to its velocity, just as for ordinary particles. Using these properties show that the equation of state for a photon gas is given by

$$PV = \tfrac{1}{3}U$$

where P is the photon (light) pressure, V is the volume of the container, and U is the internal energy of the photon gas. Your calculation should mimic the procedure followed in deriving the ideal gas equation of state. In particular you are to assume that the photons suffer elastic collisions with the walls of the container. [*Hint:* Recall that $A_x B_x + A_y B_y + A_z B_z = \mathbf{A} \cdot \mathbf{B}$ and note that the momentum and velocity of the photon are parallel and satisfy

$$\mathbf{p} \cdot \mathbf{v} = pc = \varepsilon$$

11.5.5 Show that the equation of state of an ideal gas can be expressed in the form

$$P = (\gamma - 1)\frac{U}{V}$$

Note that this form reveals pressure as a measure of internal energy density U/V.

•••••• ••••••

11.6 PRESSURE, ENERGY DENSITY, AND DEPARTURES FROM IDEAL GAS BEHAVIOR

The ideal gas model has one serious deficiency—it takes no account of intermolecular forces. As noted earlier, neglecting intermolecular forces is equivalent to ignoring the mutual potential energy of the molecules—the so-called interaction energy. Taking account of the interaction energy leads to a more realistic equation of state. In the process we uncover a new and useful aspect of pressure.

If we compare (11.2) and (11.25), we see that[†]

$$PV = \tfrac{2}{3}U \tag{11.34}$$

In terms of the internal energy density (internal energy per unit volume)

$$u \equiv \frac{U}{V} \tag{11.35}$$

we have

$$P = \tfrac{2}{3}u \tag{11.36}$$

The fact that pressure and energy density are proportional is most significant.

[†] Throughout this discussion we confine ourselves to monatomic gases.

Indeed, if we were to insist on interpreting pressure only as a force per unit area, we might be led to such *erroneous* conclusions as "there is a pressure on the container walls but nowhere else within the gas," or—*equally incorrect*—"a gas not confined by material walls has no pressure." Such erroneous ideas can be squelched by adopting the viewpoint that

pressure measures energy density.

This point of view is extremely fruitful in considerations involving systems which possess energy in different forms. The energy density associated with each form of energy makes a contribution to the total pressure. For example, a large cloud of interstellar hydrogen gas may contract under the action of its own gravitational forces. The system possesses energy of three distinct types. The hydrogen atoms possess kinetic energy by virtue of their motion. The gravitational forces responsible for the contraction also give rise to a gravitational potential energy. Such a protostar also radiates as it contracts, that is, the hydrogen atoms emit radiant energy. The radiant energy constitutes a photon gas and makes a contribution to the total pressure of the system.

To amplify our plea that pressure be regarded as a measure of energy density we show that the pressure acting on a charged conducting surface just equals the energy density of the electrostatic field. The energy density associated with an electric field E is

$$u_E = \frac{E^2}{8\pi} \tag{11.37}$$

Consider a conducting surface carrying a net charge. The charges distribute themselves over the surface in a fashion dependent upon its shape, but in general they try to get away from one another.[†] In moving away from each other the charges literally stretch the conducting surface. They make it a little bit larger, thereby decreasing the surface charge density and the energy stored in the electric field. We wish to show that the electrostatic force per unit area—the pressure resulting from the surface charge—is precisely $E^2/8\pi$, that is, that pressure measures energy density. To calculate this pressure, or electrical stress as it is sometimes referred to, we pick a small element of the surface. It follows from Gauss' law that the electric field strength just above the surface is

$$E_{\text{abo}} = \frac{4\pi q}{A} \tag{11.38}$$

[†] More precisely, they spread out so as to minimize the total energy stored in the electric field which they create.

where q is the charge on the element and A is its area. Because we deal with a static field and a conducting surface, the field just below the surface, that is, inside the conductor, is zero.

$$E_{\text{bel}} = 0$$

The field just outside the area A is a superposition of two fields, the field E_q produced by the charge q and the field E' set up by the other charges, not on A. The same is true for the field just below the surface. It follows from the fact that $E_{\text{bel}} = 0$ that

$$E' = E_q = \tfrac{1}{2} E_{\text{abo}}$$

Below the surface E_q has the same magnitude, but opposite direction, and the net field is zero. The field E' exerts a force on the charge q which occupies the area A. This force is

$$F = qE' = \tfrac{1}{2} q E_{\text{abo}}$$

It is this force which stretches the surface. The electrical pressure is

$$P_E = \frac{F}{A} = \frac{1}{2} \frac{q E_{\text{abo}}}{A}$$

However, from (11.38), $q/A = (1/4\pi) E_{\text{abo}}$, whence

$$P_E = \frac{1}{8\pi} E_{\text{abo}}^2 = u_E \tag{11.39}$$

that is, the electrostatic pressure just equals the energy density in the electrostatic field.

A similar result holds for magnetic fields, the energy density in a field B being

$$u_B = \frac{B^2}{8\pi} \tag{11.40}$$

There is a magnetic pressure, equal to u_B, which occasionally manifests itself in spectacular fashion by tearing apart the windings of an electromagnet.

The pressure of a light wave, the so-called radiation pressure, may be related to the combined electric and magnetic field energy densities. It may also be interpreted in terms of photon impacts (see Problem 11.5.4).

As an application of the "pressure measures energy density" concept we turn to the problem of accounting for the nonideal behavior of gases. For purposes of comparison we need the ideal gas equation of state

$$PV = nRT \tag{11.1}$$

which may be rewritten in the form

$$\frac{Pv}{kT} = 1 \tag{11.41}$$

in which $v = V/N$ is the specific volume, the volume per atom.

Two important equations of state, frequently used to represent the behavior of real gases, are the van der Waals equation

$$\left(P + \frac{a}{v^2}\right)(v - b) = kT \tag{11.42}$$

and the virial equation of state

$$\frac{Pv}{kT} = 1 + \frac{B}{v} + \frac{C}{v^2} + \cdots \tag{11.43}$$

The coefficients B, C, ... are termed the second, third, ... virial coefficients. The virial coefficients can be determined experimentally and it is found that they vary with temperature. In a rarified gas the volume per atom v is large enough that successive terms in the virial equation become progressively smaller. The terms B/v, C/v^2, and so on, measure the extent to which the gas deviates from ideal behavior. The van der Waals equation (11.42) also approximates the ideal gas law for sufficiently large v. The van der Waals and virial equations of state may be understood—even derived—if we interpret pressure as a measure of the total energy density of the gas.

The ideal gas equation evolved from a model in which collisions were ignored. As a consequence, the atoms possessed kinetic energy only. Suppose now that we turn on the interatomic forces. These forces do work and convert some of the kinetic energy of colliding atoms into electrical potential energy. The potential energy shared by the atoms is stored only briefly—for the duration of the collision. As the atoms recede from one another their potential energy is converted back into kinetic energy. As a consequence of these collisions a certain amount of energy is stored in the form of electrical potential energy. Because the energy of the gas atoms is not wholly kinetic we expect that any improvement on the ideal gas model must somehow take account of the potential energy. For the ideal gas we found that the pressure was directly proportional to the internal energy density, namely,

$$P = \frac{2}{3}u = \frac{N}{V}kT \tag{11.36$'$}$$

the internal energy consisting entirely of the kinetic energy of the atoms. The internal energy of a real gas is the sum of the potential and kinetic energies of the atoms. We therefore modify (11.36) by writing

$$P = \frac{N}{V}kT + \Phi \tag{11.44}$$

the term Φ being proportional to the potential energy density of the gas. Postulating the form (11.44) for P permits us to deduce both the van der Waals equation and the virial equation. To establish the van der Waals equation (11.42), we argue that

(a) Φ is negative, and
(b) Φ is proportional to $(N/V)^2$.

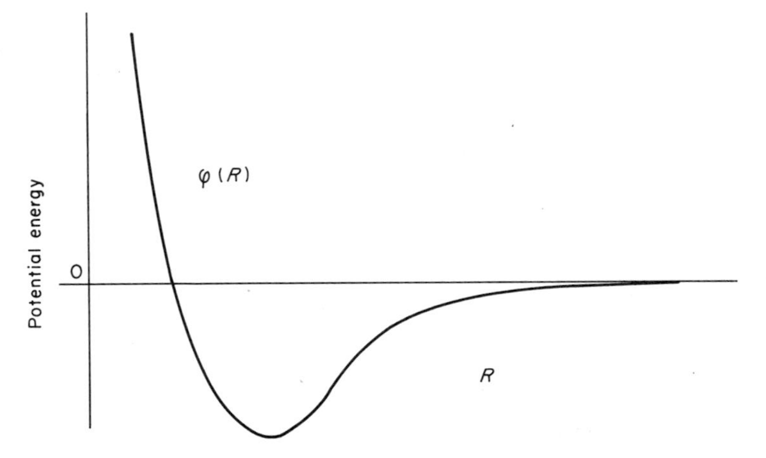

Figure 11.9. Typical form of the interatomic potential. The quantity R denotes the distance between the centers of the two interacting atoms.

To establish (a) we must know how the mutual potential energy of a pair of atoms varies with the distance between them. Figure 11.9 displays the typical form of this potential energy. For large separations (several angstroms) the potential energy is negative and very small in magnitude. The associated force is attractive. Smaller separations result in a more negative potential energy and a stronger attractive force. For sufficiently small separations, the atomic charge clouds overlap to such an extent that the two positively charged nuclei feel each other's presence. When this occurs the force becomes strongly repulsive and the potential energy rises sharply.

The vast majority of encounters between two atoms result in attractive forces for which the potential energy is negative. Only rarely do two atoms collide with such gusto that they attain a positive potential energy—here identified with a repulsive force. Since an overwhelming majority of the collisions involve a negative potential energy we conclude that the mean potential energy is also negative (a) $\Phi < 0$. To justify (b) we assume that Φ, the interatomic potential energy contribution to the pressure, is proportional to the mean potential energy density of the gas. Let ψ denote the mean potential energy per atom. With N atoms in a volume V we can write the potential energy per unit volume as

$$\Phi = A\frac{N}{V}\psi \tag{11.45}$$

where A is some dimensionless proportionality factor. The question remaining is, "How does ψ depend on N/V?" To answer this we mentally follow a typical atom about and compute its mean potential energy—then ask how the result would be changed if N/V were twice as large. If doubling N/V leaves ψ unchanged, then ψ is independent of N/V. If doubling N/V doubles ψ, then ψ is directly proportional to N/V, and so on. In following the test atom we would discover that most of its collisions are binary—the test atom interacts with only one atom at a time. Occasionally a three-body collision will occur in which the test atom interacts simultaneously with two other atoms. Still rarer are collisions which involve four atoms, and so on. Let us restrict ourselves to consideration of binary collisions for the moment since they are the most frequent type and therefore most important in determining the mean potential energy ψ. By restricting ourselves to binary collisions it is not difficult to show that ψ is directly proportional to N/V. This may be seen as follows: Let N/V be doubled by doubling the number of atoms. With N doubled, the test atom would on the average suffer twice as many binary collisions in any given time interval because the number of targets is twice as great. Doubling the number of collisions doubles the mean potential energy. Reflection shows that doubling N/V by any means—such as halving V with N fixed—has the same effect; the binary collision rate and thus the mean potential energy ψ are doubled. Thus ψ is directly proportional to N/V

$$\psi \propto \frac{N}{V} \tag{11.46}$$

whereupon we conclude that Φ is proportional $(N/V)^2$

$$\Phi = A\frac{N}{V}\psi \propto \left(\frac{N}{V}\right)^2 \tag{11.47}$$

Having established earlier that Φ is negative we can write

$$\Phi = -a\left(\frac{N}{V}\right)^2, \qquad a > 0 \tag{11.48}$$

where a is a positive proportionality factor.

The total pressure $P = (NkT/V) + \Phi$ thus has the form

$$P = \frac{NkT}{V} - \frac{aN^2}{V^2} \tag{11.49}$$

With $v = V/N$ this converts to

$$\left(P + \frac{a}{v^2}\right)v = kT \tag{11.50}$$

which is almost van der Waals' equation of state. We can convert (11.50) into the van der Waals equation (11.42) by remedying one further defect of the ideal gas model. We assumed that the atoms of an ideal gas were mass points. In fact, however, the atoms take up space so that the available volume is less than that of the empty container. Let b denote the average volume excluded by one atom. Then N atoms claim a volume Nb and the effective volume of the container is $V - Nb$. For a rarified gas $V \gg Nb$, that is, the atoms take up only a small fraction of the total volume of the container. Replacing V by $V - Nb$ is equivalent to replacing v by $v - b$, so that (11.50) becomes

$$\left[P + \frac{a}{(v-b)^2}\right](v-b) = kT \tag{11.51}$$

The term $a/(v-b)^2$ already represents a small correction and since $v \gg b$ we can legitimately write

$$\frac{a}{(v-b)^2} \simeq \frac{a}{v^2} \tag{11.52}$$

consistent with the accuracy carried elsewhere. Using this approximation in (11.51) gives the van der Waals equation of state

$$\left(P + \frac{a}{v^2}\right)(v-b) = kT \tag{11.42}$$

Our microscopic approach has yielded the following interpretation of van der Waals' equation of state: The quantity b measures the volume taken up by an individual atom. The term a/v^2 measures the potential energy density and arises because of the predominantly attractive interatomic forces. For given values of v and T, (11.42) shows that the total pressure P is less than it would be in the absence of interatomic forces. This result can be understood from two viewpoints. On the one hand we can think in terms of energy density and postulate that the pressure is proportional to the total energy density. Our arguments then show that the potential energy density contribution to P is negative $(-a/v^2)$. We can also understand this reduction in P from a more conventional point of view. Suppose there were no forces between the gas atoms. There would be a certain pressure on the container walls resulting from atomic impacts. If we now imagine "turning on" the predominantly attractive interatomic forces, it follows that the atoms would not strike the container walls quite so vigorously. The pressure would be somewhat smaller because the attractive forces tend to restrain the motions of the atoms.

The arguments advanced here to obtain the van der Waals equation may be extended to deduce the virial equation. The extension requires inclusion of the effects of interactions between three, four, etc., particles. In general the mean potential energy per atom ψ has the form

$$\psi = \psi_2 + \psi_3 + \psi_4 + \cdots$$

where ψ_s is the contribution arising from the simultaneous interaction of s atoms. Whereas ψ_2 is proportional to N/V, ψ_3 varies as $(N/V)^2$, and ψ_4 as $(N/V)^3$, and so on. Recognition of this point then leads directly to the virial equation of state (11.43).

REFERENCES

A thumbnail history of the kinetic theory of gases will be found in:

E. H. Kennard, "Kinetic Theory of Gases," Chapter 1. McGraw-Hill, New York (1938).

The concept of a degree of freedom and the equipartition of energy theorem are discussed in detail in:

W. Kauzmann, "Kinetic Theory of Gases," Chapter 3. Benjamin, New York (1966).

A general proof of the equipartition theorem for classical systems is given in:

F. C. Andrews, "Equilibrium Statistical Mechanics," Chapter 16. Wiley, New York (1963).

Nonideal gas equations of state are discussed from the microscopic viewpoint in:

W. Kauzmann, "Kinetic Theory of Gases," Chapter 2. Benjamin, New York (1966).

R. D. Present, "Kinetic Theory of Gases," Chapter 6. McGraw-Hill, New York (1958).

F. W. Sears, "An Introduction to Thermodynamics, The Kinetic Theory of Gases, and Statistical Mechanics," 2nd ed., Chapter 11. Addison-Wesley, Reading, Massachusetts (1953).

Chapter 12

Velocity Space Distribution Functions

In this chapter we introduce the concepts of velocity space and velocity space distribution functions. Some of the experimental techniques used to determine velocity distributions are presented in Section 12.6.

12.1 INTRODUCTORY IDEAS

Consider a gas composed of N atoms which occupy a volume V. For simplicity it is assumed that the atoms are identical, for example, all helium atoms or all neon atoms. Further, it is assumed that the gas is in equilibrium with the container walls.

A measuring instrument, being macroscopic by nature, cannot register the effects of a single atom. Any observed meter reading measures a response to the combined action of many atoms. For this reason kinetic theory does not require a detailed microscopic description of the gas. Something less sharp than a specification of the mechanical state will be adequate. Nevertheless, we begin by envisioning a *geometrical* representation of the mechanical state of the gas. The resulting picture suggests a type of description which will meet the needs of kinetic theory. To specify the mechanical state requires that the position

$$\mathbf{r} = (x, y, z)$$

and velocity

$$\mathbf{v} = (v_x, v_y, v_z)$$

be given for each of the N atoms. Geometrically this information may be represented by N points in *position space* and N points in *velocity space*. For a gas of $N \sim 10^{23}$ atoms we can appreciate that there is a high density of points in both position space and velocity space. The high density of such points is important. Many atoms have positions which differ only slightly, and many atoms have velocities which are nearly the same. Also at hand is the experimental fact that instruments respond only to the combined action of many atoms. This suggests that kinetic theory be developed in a fashion which focuses attention on groups of atoms which occupy nearly the same position and which have nearly equal velocities. The key object in such a theory is the particle distribution function. The distribution function concept was introduced in Chapter 10. In general, kinetic theory is concerned with distribution functions which refer to both the positions and the velocities of particles. By dealing with systems which are spatially uniform we abandon half of this conceptual burden—we must "worry" only about the distribution of velocities. From a geometric viewpoint a distribution function measures the density of points, each representing an atom, in a velocity space. Before introducing such distribution functions we digress to develop a familiarity with volume elements in position space and in velocity space.

Consider a volume element (see Figure 12.1)

$$d^3r = dx\,dy\,dz \tag{12.1}$$

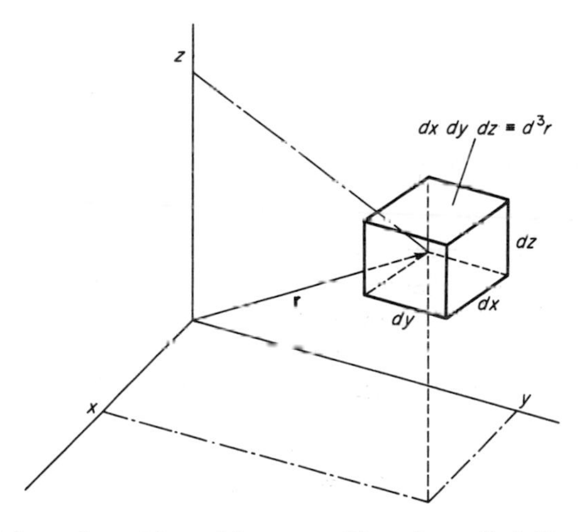

Figure 12.1. Volume element in position space. The volume d^3r is big on a microscopic scale but small by macroscopic standards.

located at

$$\mathbf{r} = (x, y, z)$$

An atom is said to be "in d^3r" if its coordinates lie in the ranges x to $x+dx$, y to $y+dy$, and z to $z+dz$. The volume element d^3r is big on a microscopic scale, but small on a macroscopic scale. Big microscopically means that it contains many atoms—at least most of the time†—thereby allowing a meaningful statistical analysis. Small macroscopically means that measurable properties of the atoms (for example, temperature, pressure) do not vary throughout d^3r. Thus, for example, a meter designed to measure gas pressure at a point records the average effect of many particles over a macroscopic region. Such macroscopic sampling volumes are assumed large by comparison with d^3r.

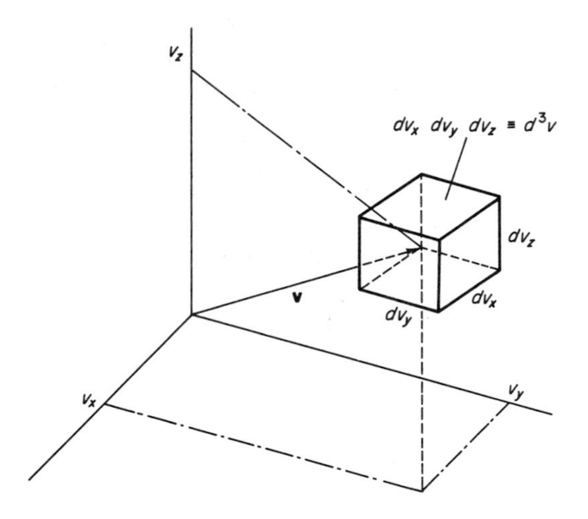

Figure 12.2. Volume element in velocity space. The volume d^3v is big on a microscopic scale but small by macroscopic standards.

At this point we must make a short conceptual leap. Envision first the particles in some volume element d^3r. In general, the particle velocities will be spread over a wide range. We can mentally sort the particles of d^3r into groups having similar velocities. To facilitate this bookkeeping we slice up velocity space in the same fashion we dissected position space. By analogy with d^3r we define a volume element in velocity space by

$$d^3v = dv_x\, dv_y\, dv_z \tag{12.2}$$

† A fluctuation might depopulate the volume element briefly.

In velocity space, d^3v is located at

$$\mathbf{v} = (v_x, v_y, v_z)$$

An atom is said to be "in d^3v" if its velocity vector has components which lie in the range v_x to $v_x + dv_x$, v_y to $v_y + dv_y$, and v_z to $v_z + dv_z$. Figure 12.2, strongly reminiscent of Figure 12.1, shows a typical d^3v. As with d^3r, the volume d^3v is big on a microscopic scale, but small by macroscopic standards. This means that d^3v encompasses a range of velocities large enough to include many atoms. Dealing with volume elements which contain many particles allows us to ignore fluctuations.

12.2 A DISTRIBUTION FUNCTION FOR VELOCITY SPACE

Consider a gas of N atoms in thermodynamic equilibrium. For a system in thermodynamic equilibrium there is no spatial variation in composition—the number of atoms per cubic centimeter is the same throughout the gas. The situation is quite different when we inquire after the distribution of atoms with respect to velocity. Elementary considerations rule out distributions which have every atom moving with the same velocity, or even the same speed. If all atoms had the same speed at some instant, interparticle collisions would quickly result in a spread of atomic speeds. Such distributions cannot represent equilibrium because they are altered by collisions. The equilibrium distribution of velocities is one which is produced and maintained by collisions.

Because the number density is uniform we can ease our conceptual burden by dealing with the full volume V of the gas, rather than with some miniscule chunk of it. If and when it should become necessary to focus our attention on some particular spatial volume element d^3r we may presume that the fraction d^3r/V of all atoms reside therein.

To describe the distribution of molecular velocities we divide velocity space into conveniently shaped and sized elements, named d^3v. Let dN denote the number of molecules in d^3v; we write

$$dN = f(\mathbf{v})\, d^3v \tag{12.3}$$

thereby defining a function $f(\mathbf{v})$ which measures the density of molecules in velocity space. We shall refer to $f(\mathbf{v})$ as the "velocity space distribution function" or, when no confusion can result, simply as the "distribution function." Equation (12.3) is of the utmost importance. You must develop a "seat of the pants" feeling for what it says to be able to master kinetic theory. In arriving at (12.3) we mentally sliced up velocity space. By reversing this procedure—by reassembling the pieces of velocity space and adding the numbers of atoms in each—we arrive at the normalization condition on $f(\mathbf{v})$. This synthesis is formally accomplished by integrating dN over all segments of velocity space. Thus summing the populations of all velocity space volume elements gives the total number of particles N

$$N = \int_{\mathbf{v}} dN = \int_{\mathbf{v}} f(\mathbf{v})\, d^3v \tag{12.4}$$

The notation $\int_{\mathbf{v}}$ is a symbolic instruction to integrate over all volume elements of velocity space.

••••• •••••

12.2.1 Explain the nature of $f(\mathbf{v})$ by analogy with the mass density $\rho(\mathbf{r})$.

••••• •••••

One of the immediate questions is this: "What is the functional form of $f(\mathbf{v})$ for a gas in thermodynamic equilibrium?" The answer was first provided over 100 years ago by James Clerk Maxwell, the man with the equations and the demon! The Maxwellian distribution function for N atoms in thermodynamic equilibrium at a temperature T is

$$f(v) = N\left(\frac{m}{2\pi kT}\right)^{3/2} \exp\left(\frac{-\frac{1}{2}mv^2}{kT}\right) \tag{12.5}$$

The Maxwellian velocity distribution function is unique—it is the only possible distribution of velocities for a gas in thermodynamic equilibrium.† Two points regarding the Maxwellian distribution should be noted.

(a) The Maxwellian distribution is spherically symmetric in velocity space, that is, $f(v)$ depends on the speed of an atom but not on its direction of motion. This fact will permit us to employ a spherical velocity space in many applications.

(b) The second and more significant feature of the Maxwellian distribution is that the only intrinsic property of the atoms appearing in $f(v)$ is their mass m. Evidently the precise manner in which the atoms interact is immaterial. However, the fact that the atoms do collide is crucially important. Interatomic collisions produce and maintain the Maxwellian distribution. In a very real sense, collisions between atoms drive a gas toward thermodynamic equilibrium and subsequently maintain the equilibrium state.

We have not presented any sort of derivation of the Maxwellian distribution. Such derivations as might be presented now are either cumbersome, or shaky,

† This remark is applicable only within the framework of nonrelativistic, classical kinetic theory. Relativistic and quantum effects modify the Maxwellian form of the equilibrium distribution function.

or both! Statistical mechanics furnishes a rigorous and straightforward derivation of the Maxwellian distribution (compare Chapter 15). Some of the experiments designed to confirm the Maxwellian distribution of velocities are discussed in Section 12.6.

Finally it must be remarked that the uniqueness of the Maxwellian velocity distribution function is limited to the thermodynamic equilibrium domain. Nonequilibrium kinetic theory furnishes numerous examples of non-Maxwellian distribution functions.

12.3 MEAN VALUES

Let $Q(\mathbf{v})$ denote some physical property depending on the velocity of an atom. Prime examples are linear momentum $m\mathbf{v}$ and the kinetic energy $\frac{1}{2}mv^2$. A fundamental hypothesis of kinetic theory is that the measurable quantity corresponding to any such property is Q_{tot}, the sum of the Q's for the N atoms. The $dN(\mathbf{v})$ atoms in the velocity space volume element d^3v each contribute $Q(\mathbf{v})$ to Q_{tot}. The contribution to Q_{tot} from the atoms in d^3v is therefore,

$$Q(\mathbf{v})\,dN(\mathbf{v}) = Q(\mathbf{v})f(\mathbf{v})\,d^3v$$

Integrating over all elements of velocity space gives Q_{tot},

$$Q_{\text{tot}} = \int_{\mathbf{v}} Q(\mathbf{v})f(\mathbf{v})\,d^3v \tag{12.6}$$

Frequently it is desirable to present results on a per particle basis. The mean value per particle $\langle Q\rangle$ is defined by

$$\langle Q\rangle = \frac{Q_{\text{tot}}}{N} \tag{12.7}$$

In terms of $f(\mathbf{v})$ we have

$$Q = \frac{1}{\text{N}}\int_{\mathbf{v}} Q(\mathbf{v})f(\mathbf{v})\,d^3v \tag{12.8}$$

Example 12.1 Total Momentum of a Selected Group of Molecules

Let us determine the total momentum of the molecules in a gas whose velocities are in the positive x direction, assuming they have a Maxwellian distribution. In this case

$$Q = mv_x$$

and Q_{tot} is to be computed for those molecules with positive values of v_x. The velocity space integration is over all values of v_y and v_z. Thus using (12.6)

$$Q_{\text{tot}} = \int_{v_x=0}^{+\infty} \int_{v_y=-\infty}^{+\infty} \int_{v_z=-\infty}^{+\infty} m v_x f(v)\, dv_x\, dv_y\, dv_z$$

Inserting the Maxwellian form of $f(v)$ given by (12.5) and noting that

$$v^2 = v_x^{\,2} + v_y^{\,2} + v_z^{\,2}$$

gives (the v_y and v_z integrations are identical),

$$Q_{\text{tot}} = N\left(\frac{m}{2\pi kT}\right)^{3/2} \int_0^{\infty} m v_x \exp\left(-\frac{1}{2}\frac{m v_x^{\,2}}{kT}\right) dv_x \left[\int_{-\infty}^{+\infty} \exp\left(-\frac{1}{2}\frac{m v_y^{\,2}}{kT}\right) dv_y\right]^2$$

Appendix A furnishes expressions for the two integrals, yielding

$$Q_{\text{tot}} = \frac{1}{2} N \sqrt{\frac{2mkT}{\pi}}$$

Half of the N molecules have positive values of v_x, so that the quantity $\sqrt{2mkT/\pi}$ is the mean momentum per particle in the positive x direction. ✧

•••••• ••••••

12.3.1 (a) Determine (by symmetry considerations, preferably) the mean velocity $\langle \mathbf{v} \rangle$ for a gas whose atoms have a Maxwellian distribution of velocities.

(b) The flow of heat in a gas is directly proportional to the mean value of the quantity $\frac{1}{2}mv^2\mathbf{v}$. Determine $\langle \frac{1}{2}mv^2\mathbf{v} \rangle$ (again by symmetry considerations, preferably) for a gas whose atoms have a Maxwellian distribution of velocities. Are your conclusions in accord with the fact that the Maxwellian distribution describes a gas in thermodynamic equilibrium?

•••••• ••••••

We have previously noted that the Maxwellian distribution function is spherically symmetric in velocity space. That is, the Maxwellian form of $f(\mathbf{v})$ depends on the magnitude of $\mathbf{v}$ but not on its direction. It will be to our everlasting advantage to momentarily turn aside from kinetic theory proper and study spherical coordinate and velocity spaces.

12.4 SPHERICAL COORDINATE SPACE AND SPHERICAL VELOCITY SPACE

Frequently the symmetries inherent in a physical situation dictate the use of coordinates other than the familiar rectangular system. Of particular importance is the system of spherical coordinates and the associated elements of area

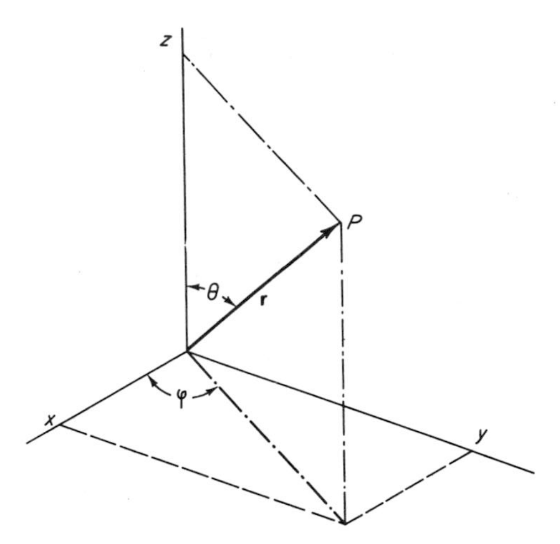

Figure 12.3. Spherical coordinates. The relationships between the spherical coordinates (r, θ, φ) and the Cartesian coordinates (x, y, z) are given by (12.9).

and volume. Figure 12.3 shows the relationships between the Cartesian and spherical coordinates of a point P. The position vector $\mathbf{r}$ directed from the origin to P may be specified by giving the values of its Cartesian components x, y, and z. An equivalent specification is rendered in spherical coordinates by giving the values of r, θ, and φ. From Figure 12.3 we see that the two sets of coordinates are related by

$$\begin{aligned} x &= r \sin\theta \cos\varphi \\ y &= r \sin\theta \sin\varphi \\ z &= r \cos\theta \end{aligned} \tag{12.9}$$

and that

$$r^2 = x^2 + y^2 + z^2$$

A differential element of area in spherical coordinates is generated by sweeping the radius vector along two mutually perpendicular directions: First through an angle $d\theta$, from θ to $\theta + d\theta$ with φ held constant, and second through an angle $d\varphi$, from φ to $\varphi + d\varphi$ with θ held constant. As Figure 12.4 illustrates, these steps generate two sides of the differential element of area

$$dA = r^2 \sin\theta \, d\theta \, d\varphi$$

Note that an integration over the surface of a sphere is effected by integrating from $\varphi = 0$ to $\varphi = 2\pi$ and from $\theta = 0$ to $\theta = \pi$. For example, the total surface area of a sphere is given by

$$A_{\text{sph}} = r^2 \int_{\theta=0}^{\pi} \sin\theta \, d\theta \int_{\varphi=0}^{2\pi} d\varphi = 4\pi r^2$$

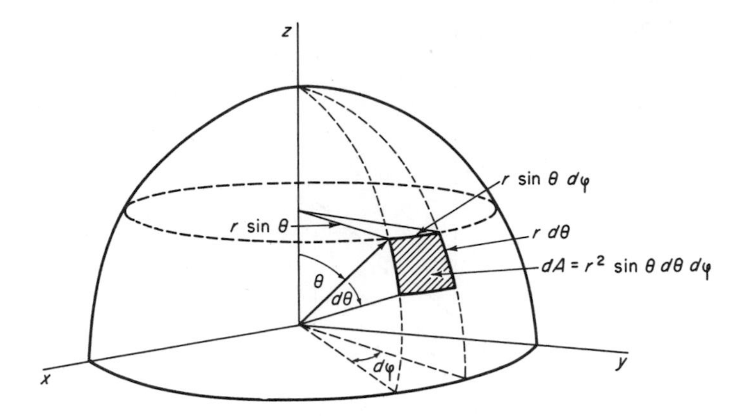

Figure 12.4. Differential element of area in spherical coordinates.

The volume element in spherical coordinates is obtained by multiplying the surface area element by dr. Thus†

$$d^3r = dA\,dr = r^2\,dr \sin\theta\,d\theta\,d\varphi \tag{12.10}$$

Figure 12.5 shows d^3r. The volume of a spherical shell is

$$A_{\text{sph}}\,dr = 4\pi r^2\,dr \tag{12.11}$$

obtained by integrating d^3r over the surface of the sphere, that is, over θ and φ.

The use of spherical coordinates is readily extended to velocity space. In place of the rectangular components of $\mathbf{v}$

$$\mathbf{v} = (v_x, v_y, v_z)$$

we can use the spherical components

$$\mathbf{v} = (v, \theta, \varphi)$$

The velocity space volume element becomes

$$d^3v = v^2\,dv \sin\theta\,d\theta\,d\varphi \tag{12.12}$$

The pictures are the same as those of Figures 12.3–12.5 with x replaced by v_x, r replaced by the speed v, and so on. The volume of a spherical shell of radius v and thickness dv is

$$v^2\,dv \int_{\theta=0}^{\pi} \sin\theta\,d\theta \int_{\varphi=0}^{2\pi} d\varphi = 4\pi v^2\,dv \tag{12.13}$$

---

† The symbolism d^3r is a generic one used to designate three-dimensional volume elements. It is not restricted to indicate the rectangular volume element $dx\,dy\,dz$.

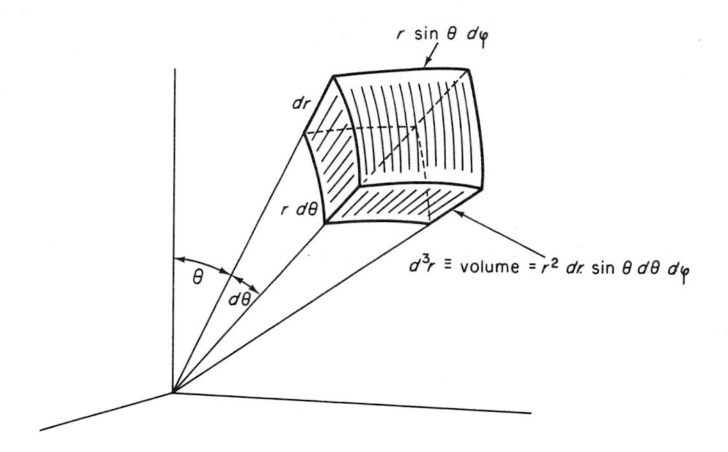

Figure 12.5. Volume element for spherical coordinates.

When the symmetry of velocity space dictates its use, we may take $d^3v = 4\pi v^2\,dv$ and integrate over the speed alone.

We shall make liberal use of spherical velocity space throughout the remainder of our work in kinetic theory.

Example 12.2 Mean Kinetic Energy for the Maxwellian Distribution

As an illustration of the use of spherical velocity space we compute the mean kinetic energy in a gas by using (12.8) with $Q = \frac{1}{2}mv^2$. We do this using the Maxwellian velocity space distribution function, given by (12.5). Because both $Q(v) = \frac{1}{2}mv^2$ and the distribution function are spherically symmetric we can integrate over spherical shells of volume $d^3v = 4\pi v^2\,dv$. Thus

$$\left\langle \frac{1}{2}mv^2 \right\rangle = \frac{1}{\mathrm{N}} \int_{v=0}^{\infty} \frac{1}{2}mv^2 f(v)\, 4\pi v^2\, dv$$

Inserting the Maxwellian form of $f(v)$ and introducing the dimensionless variable $u^2 = \frac{1}{2}mv^2/kT$ converts this to

$$\left\langle \frac{1}{2}mv^2 \right\rangle = \frac{4kT}{\sqrt{\pi}} \int_0^{\infty} u^4 \exp(-u^2)\, du$$

The Gaussian integral has the value $\frac{3}{8}\sqrt{\pi}$ (see Appendix A), which leaves the result

$$\langle \tfrac{1}{2}mv^2 \rangle = \tfrac{3}{2}kT$$

The total kinetic energy is $K = N\langle \frac{1}{2}mv^2 \rangle = \frac{3}{2}NkT$. It should be noted [compare Equation (11.23)] that this is the result we were obliged to assume in our derivation of the ideal gas law. ◇

••••••　　　　　　　　　　　　　　　　　　　　　　••••••

12.4.1 Assuming a Maxwellian distribution of velocities, evaluate the standard deviation of the kinetic energy,

$$\sigma(K) = \sqrt{\langle K^2 \rangle - \langle K \rangle^2}$$

••••••　　　　　　　　　　　　　　　　　　　　　　••••••

12.5 THE DISTRIBUTION OF MOLECULAR SPEEDS

For distribution functions which are spherically symmetric in velocity space, such as the Maxwellian, the quantity

$$4\pi v^2 f(v)\, dv = dN \tag{12.14}$$

gives the number of atoms with speeds in the range from v to $v+dv$. Thus the quantity $4\pi v^2 f(v)$ gives the distribution of atomic speeds. Figure 12.6 is a plot of $4\pi v^2 f(v)$ versus v for the Maxwellian distribution. Using the Maxwellian form of $f(v)$ gives

$$4\pi v^2 f(v) = 4\pi N \left(\frac{m}{2\pi kT}\right)^{3/2} v^2 \exp\left(\frac{-mv^2}{2kT}\right) \tag{12.15}$$

The most immediate advantage of dealing with the distribution of speeds is evidenced by Figure 12.6 itself. The distribution of speeds is readily visualized and plotted. This promotes other advantages. For example, if we ask for the most probable atomic speed, a glance at Figure 12.6 reveals that this is the speed for which $4\pi v^2 f(v)$ is a maximum. The equation defining the most

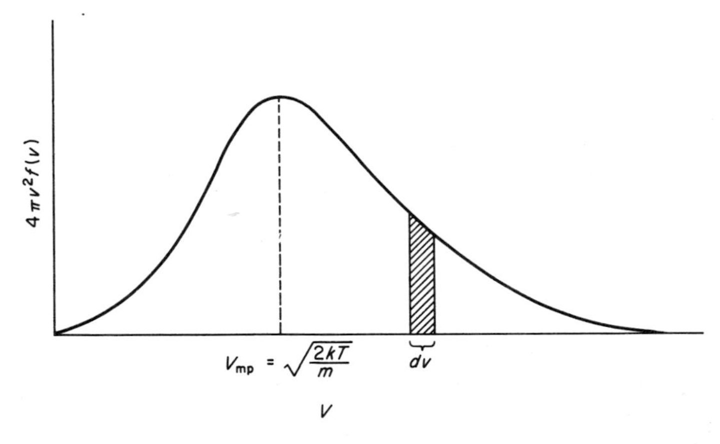

Figure 12.6. The Maxwellian distribution of atomic speeds. Shaded area $= 4\pi v^2 f(v)\, dv$.

probable speed is thus

$$\frac{d}{dv}(4\pi v^2 f(v)) = 0$$

For the Maxwellian distribution we obtain (A = constant)

$$\frac{d}{dv} Av^2 \exp\left(\frac{-\frac{1}{2}mv^2}{kT}\right) = 2Av\left(1 - \frac{mv^2}{2kT}\right)\exp\left(\frac{-\frac{1}{2}mv^2}{kT}\right) = 0$$

The equation determining the most probable speed v_{mp} is thus

$$2Av\left(1 - \frac{mv^2}{2kT}\right)\exp\left(\frac{-\frac{1}{2}mv^2}{kT}\right) = 0$$

This yields

$$v_{\text{mp}} = \sqrt{\frac{2kT}{m}} \tag{12.16}$$

Example 12.3 Mean Kinetic Energy for an Electron Gas

Metals such as silver and copper are excellent conductors of electricity. Their high electrical conductivity stems from the fact that each atom in the atomic lattice is readily "coaxed" into volunteering an electron for the task of carrying current. The conducting material can be regarded as a gas of electrons confined within a container-like lattice of atoms and positively charged ions. If the electrons are assumed to have a Maxwellian distribution of speeds, grave troubles arise (see Chapter 17). Instead one finds that $f(v)$ is essentially constant from $v = 0$ up to a characteristic speed $v = v_F$, the so-called Fermi speed. Beyond v_F, $f(v)$ is zero. Thus

$$f(v) = \begin{cases} A, & 0 \leqslant v \leqslant v_F \\ 0, & v > v_F \end{cases} \tag{12.17}$$

The value of A is fixed by the normalization condition

$$N = \int_{\text{all speeds}} dN = \int_0^\infty 4\pi v^2 f(v)\, dv$$

This establishes the relationship of N, v_F, and A

$$N = \frac{4\pi}{3} {v_F}^3 A \tag{12.18}$$

Let us compute the mean kinetic energy of an electron, and express it in terms of the Fermi energy

$$\varepsilon_F \equiv \tfrac{1}{2} m {v_F}^2$$

which is the maximum energy which any electron can have according to the distribution (12.17). We have

$$\langle K\rangle = \frac{1}{N}\int \tfrac{1}{2}mv^2\,dN = \frac{1}{N}\int_0^{v_F} \tfrac{1}{2}mv^2\,4\pi v^2 A\,dv$$

Using (12.18) to express A/N in terms of v_F gives

$$\langle K\rangle = \tfrac{3}{5}(\tfrac{1}{2}mv_F^{\,2}) = \tfrac{3}{5}\varepsilon_F \tag{12.19}$$

Thus the mean kinetic energy is a very healthy fraction of the maximum energy ε_F. ◇

....

12.5.1 Sketch $4\pi v^2 f(v)$ for the electron gas distribution function given by Equation (12.17).

(a) What is the mean electron speed?
(b) What is the most probable electron speed?

The Fermi speed v_F is related to the number density of electrons n_e by the equation

$$v_F = \frac{h}{m}\left(\frac{3n_e}{8\pi}\right)^{1/3}$$

where $h = 6.624 \times 10^{-27}$ erg-sec is Planck's constant and $m = 0.91 \times 10^{-27}$ gm is the electron mass. Assume that each atom in metallic silver donates one electron, so that n_e is equal to the number of silver atoms per unit volume. The density of silver at room temperature is 10.5 gm/cm^3 and the mass of a single silver atom is approximately 1.8×10^{-22} gm.

(c) Compute v_F and ε_F. How does v_F compare with the speed of sound? How does ε_F compare with the characteristic thermal energy $\frac{3}{2}kT$, at $T = 300$ K? Would you describe the electron gas as hot or cold by comparison with its surroundings?

12.5.2 Compute the average speed of molecules at $T = 300$ K in

(a) He
(b) N_2

assuming a Maxwellian distribution of velocities.

(c) The velocity of sound is given by

$$v_s = \sqrt{\frac{\gamma kT}{m}} = \sqrt{\frac{\gamma RT}{M}}$$

where γ is the ratio of heat capacities ($\gamma = C_p/C_v$). Using the ideal gas values of γ for monatomic and diatomic gases determine the ratio of the speeds of sound in helium and nitrogen (both gases at same temperature). The relation

$$f\lambda = v_s$$

relates the frequency f, wavelength λ, and speed of sound. The human "audio system" may be thought of as an open-ended resonant chamber in which the wavelengths of sound waves which can propagate are fixed by the geometry of the throat and mouth. There is a maximum wavelength which can propagate. To this maximum wavelength there corresponds a minimum frequency called the fundamental frequency. Lower frequencies (longer wavelengths) are severely attenuated. They are cut off and do not propagate. The fundamental frequency of the adult male is about 100 Hz. (1 Hz = 1 cycle/sec), and determines the pitch of the voice. The waves propagating in the throat and mouth are a superposition of waves of the fundamental frequency and integral multiples thereof (overtones, harmonics, partials).

(d) If the lungs are filled with helium the resulting speech has a high-pitched character (demonstration recommended). Explain this and estimate the minimum frequency emanating from the mouth.

•••••• ••••••

12.6 EXPERIMENTAL VERIFICATION OF THE MAXWELLIAN DISTRIBUTION

A number of investigators have verified the Maxwellian distribution of speeds. We discuss two of these efforts; the novel gravitational deflection study by Estermann, Simpson, and Stern (1947) and the more recent and precise experiment by Miller and Kusch (1955). A review and comprehensive bibliography of other investigations is given by Ramsey.

Figure 12.7 shows a schematic of the apparatus used by Estermann, Simpson, and Stern. A small sample of cesium in an oven heated to about 450 K provided a source of atoms. The oven slit O and a collimating slit S produced a narrow beam of atoms. The detector D was a hot tungsten wire. A

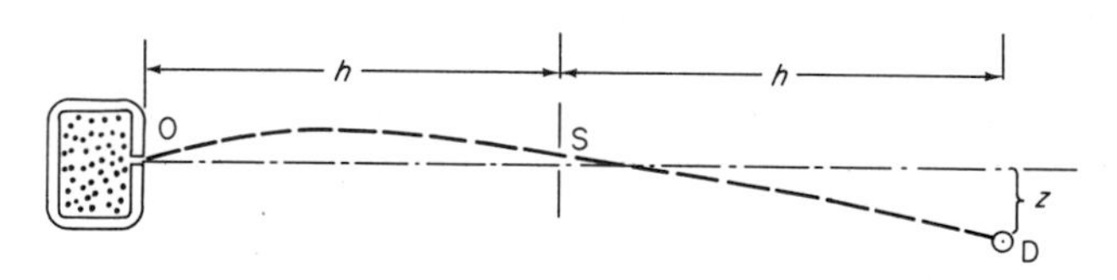

Figure 12.7. Schematic for velocity distribution measurements.

cesium atom striking the wire would be ionized and the resulting positive cesium ions collected by a negatively charged cylinder surrounding the wire. The resulting ion current was thus directly proportional to the number of cesium atoms which reached the detector. The vertical position of the detector could be varied. The slit S was positioned midway between the oven slit and the detector. An atom leaving the oven with the speed v would fall a distance $z = gh^2/v^2$ below the level of the slits, with g denoting the acceleration of gravity and $2h$ the oven-to-detector distance. The distribution of atomic velocities in the beam could thus be inferred by measuring the relative numbers (relative ion currents) for different values of z. The experimental results confirmed the Maxwellian distribution law although there were significant discrepancies for the low speed atoms resulting from scattering of the beam at the slits.†

•••••• ••••••

12.6.1 With reference to Figure 12.7:

(a) Show that the distance of fall is $z = gh^2/v^2$. Note that an atom which reaches the detector must leave the oven along a slightly elevated trajectory such that it falls back to the same level in travelling the distance h. Thus the atoms are travelling downward as they pass through the slit S.

(b) Compute the distance z for a cesium atom which travels with the most probable atomic speed ($\sqrt{2kT/m}$) for a temperature of 450 K. The distance h was 100 cm in the experiments of Estermann, Simpson, and Stern.

(c) The slit and the detector wire had widths of 0.02 mm. How would these finite widths affect the measurements?

•••••• ••••••

Miller and Kusch used a long rotating cylinder (see Figure 12.8) with a large number of helical grooves cut along its surface. The angular pitch of the grooves was φ_0. The cylinder was positioned so as to act as a shutter in the path of an atomic beam. Only atoms with particular speeds could pass through the grooves to the detector beyond. When the cylinder rotates with an angular velocity ω, atoms with speed

$$v_0 = \frac{\omega L}{\varphi_0} \tag{12.20}$$

† The distribution of speeds in the oven is presumably Maxwellian. The distribution of atomic speeds in the beam differs from that in the oven. The faster moving atoms are more likely to escape the oven with the result that the distribution of speeds in the beam is proportional to the product of the distribution function and the speed v. This point is discussed further in Chapter 13, after the concept of particle flux has been developed (see Problem 13.3.1).

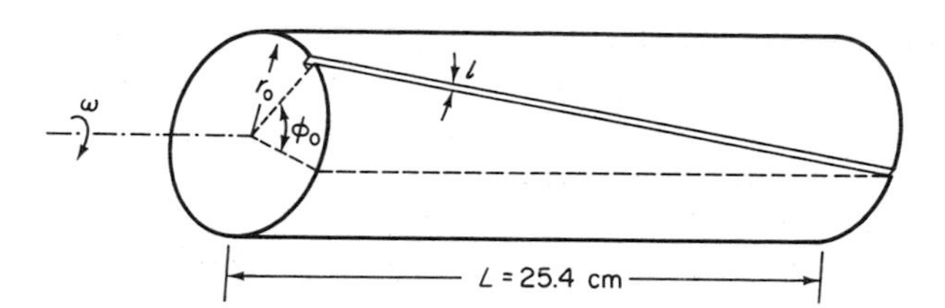

Figure 12.8. Rotating cylinder with grooves used as a velocity selector. L = 25.4 cm; l = width of groove.

can pass without striking the walls of the groove. By varying the angular velocity ω the relative distribution of atomic speeds in the beam could be ascertained. The experiment gave excellent agreement with the expected Maxwellian distribution law for beams of thallium and potassium atoms emanating from an oven operated at temperatures ranging from 466 K to 944 K.

•••••• ••••••

12.6.2 Derive Equation (12.20). Note that if the atom is to pass through the groove without striking the walls, it must travel the length of the cylinder L in the same time required for the cylinder to rotate through the angle φ_0.

12.6.3 The finite width l of the grooves in the Miller–Kusch apparatus allowed atoms with a narrow range of speeds to pass through the velocity selector for a fixed angular speed. The limiting values of the speeds correspond to atoms which enter a groove at one side and leave at the opposite side. If r_0 is the radius of the cylinder, show that the maximum and minimum speeds which pass are

$$v_{\max} = \frac{v_0}{1-\gamma}, \qquad v_{\min} = \frac{v_0}{1+\gamma}$$

where $\gamma = l/r_0\,\varphi_0$. Compute the velocity spread $\Delta v = v_{\max} - v_{\min}$ for $v_0 = 6.3 \times 10^4$ cm/sec, $l = 0.0424$ cm, $r_0 = 10.00$ cm, $\varphi_0 = 2\pi/74.7$ rad.

•••••• ••••••

REFERENCES

The concept of velocity space is basic to the development of kinetic theory. See, for example:

E. H. Kennard, "Kinetic Theory of Gases," Chapter 2. McGraw-Hill, New York (1938).

The distribution function concept and its relation to mean values is developed in the reference above as well as in:

A. F. Brown, "Statistical Physics," Chapter 3. Edinburgh Univ. Press, Edinburgh (1968).

Two experiments which verify the Maxwellian distribution are reported in:

I. Esterman, O. C. Simpson, and O. Stern, The Free Fall of Atoms and the Measurement of the Velocity Distribution in a Molecular Beam of Cesium Atoms, *Phys. Rev.* **71**, 238 (1947).

and

R. C. Miller and P. Kusch, Velocity Distribution in Potassium and Thallium Atomic Beams, *Phys. Rev.* **99**, 1314 (1955).

Detailed accounts of velocity distribution measurements are given in:

N. F. Ramsey, "Molecular Beams," Chapter 2. Oxford Univ. Press, London (1956).

Chapter 13

Transport Processes

13.1 INTRODUCTION

Emphasis was laid previously on the fact that real processes are irreversible. A distinguishing characteristic of irreversible processes is the transport of molecular properties. Molecular properties of prime importance include mass, momentum, and energy. The molecules of a gas, in moving about, carry or transport these molecular properties. If the gas is in equilibrium, there are no net flows of mass, momentum, or energy. That is, if an observer could watch the molecules enter and leave some small volume element, he would find that equal quantities of mass enter and leave, so that no net change occurs. The inward and outward flows of energy and momentum are similarly balanced. A nonequilibrium situation results in a net flow of one or more molecular properties. For example, diffusion results in a net flow of particles. Momentum transport is responsible for the viscosity (fluid friction) of gases and liquids, and heat conduction results in the transport of internal energy.

The basic flow concept is that of a flux. The fluxes of mass, momentum, and energy are illustrated in Section 13.2. In Section 13.3 we use the velocity space distribution function to determine the flux of molecules through a leak, that is, the rate at which a confined gas escapes through a small hole in the container wall. In Section 13.4 the flux of internal energy is related to the process of heat conduction.

13.2 FLUXES

The concept of a particle flux is basic to the discussion of transport phenomena. Consider a group of particles, all of which move with the same velocity $\mathbf{v}_0$. Let n denote the number density of the particles, the number per unit volume,†

$$n = \frac{N}{V}$$

The particle flux is a vector quantity defined by

$$\mathbf{S} = n\mathbf{v}_0 \tag{13.1}$$

To establish the physical significance of the flux, imagine that the particles are moving along a cylindrical pipe of cross-sectional area A. The number of particles passing any cross section in time t is

$$nAv_0 t = SAt$$

This result may be established as follows: Particles which pass a given cross section during a time t lie somewhere within a cylinder of length $v_0 t$ and thus of volume $Av_0 t$ (see Figure 13.1). As the number density is n, the number of

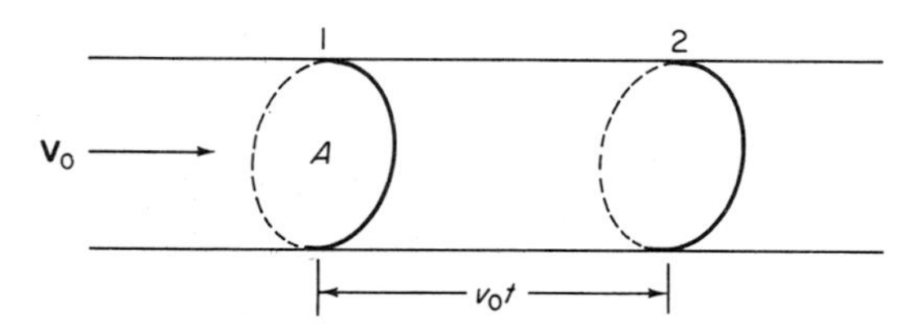

Figure 13.1. Particles which pass point 1 during a time interval t lie somewhere within the cylindrical volume $Av_0 t$.

particles in the volume $Av_0 t$ is $nAv_0 t = SAt$. The particles in this volume are those which entered in time t; particles there originally have moved out. Thus the number of particles which cross area A in time t is given by SAt, with S being the particle flux. This interpretation requires that the area A be oriented perpendicular to the flow of particles. Had we chosen a different surface (see Figure 13.2) of area A' we would have concluded that the number crossing in time t is

$$SA' \cos\theta t = SAt, \qquad A' \cos\theta = A$$

The area $A = A' \cos\theta$ is the projection of A' normal to the particle flow.

---

† Note that this use of n differs from an earlier usage where the same symbol denoted the number of moles.

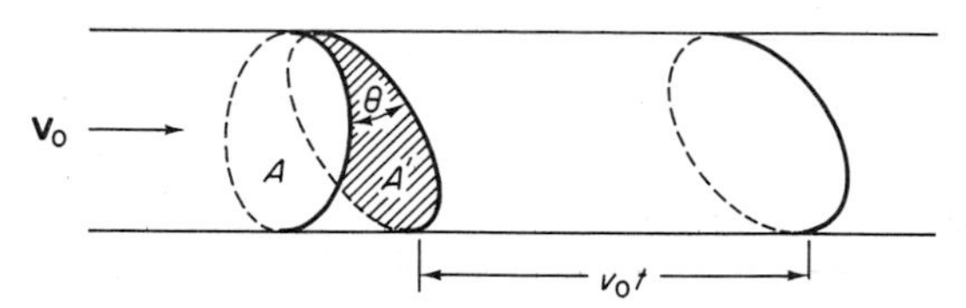

Figure 13.2. It is the projection of A' normal to the flow of particles ($A' \cos\theta$) which determines the number of particles crossing A' in time t. The volume of the slanted cylinder is $A' \cos\theta \, v_0 t = A v_0 t$.

If SAt gives the number crossing A in time t, we can interpret the particle flux as the number of particles that cross per unit area per unit time. The flux S is numerically (but not dimensionally) equal to the number of particles which cross unit area in unit time. The flux is an intensive quantity characterizing the flow of molecules, whereas the total number of particles passing through the area A in time t is an extensive quantity, being directly proportional to both A and t. It also follows that the rate at which particles cross an area A normal to the flow is SA.

$$SA = nv_0 A = \frac{\text{number of particles crossing the normal area } A}{\text{sec}}$$

Example 13.1 Molecular Flux

A pipe transports 60 gal of water/min. The cross-sectional area of the pipe is 40 cm^2. What is the flux of water molecules? We must first determine the number of molecules in 1 gal of water. Recalling that noble Anglo-Saxon refrain, "A pint's a pound the world around," reminds us that 1 pt of water weighs 1 lb so that 1 gal weighs 8 lb. A weight of 1 lb has a mass of approximately 450 gm. Therefore,

$$(\text{mass of 1 gal of water}) \simeq 8 \text{ lb} \frac{450 \text{ gm}}{\text{lb}} = 3600 \text{ gm}$$

The mass of 1 mole of water is 18 gm. Thus

$$(\text{number of molecules in 1 gal of water})$$

$$\simeq 3600 \frac{\text{gm}}{\text{gal}} \frac{1 \text{ mole}}{18 \text{ gm}} 6 \times 10^{23} \frac{\text{molecules}}{\text{mole}}$$

$$= 1.20 \times 10^{26} \frac{\text{molecules}}{\text{gal}}$$

Because 60 gal/min is 1 gal/sec, it follows that 1.2×10^{26} molecules flow across 40 cm^2 each sec. The flux is therefore 3×10^{24} molecules/sec. ◇

••••• •••••

13.2.1 The density of the water in Example 13.1 is 1 gm/cm^3. Show that the speed of the flow is approximately 90 cm/sec.

13.2.2 A sports arena has a capacity of 13,200. It takes 15 min to empty a capacity crowd using 24 doors. Each door has a usable emptying area of 15 sq. ft. What is the people flux when a capacity crowd exits as specified?

••••• •••••

We must extend the notion of a particle flux to the microscopic domain. If $f(\mathbf{v})$ is the distribution function, the number of atoms with velocities in d^3v is $dN = f(\mathbf{v})\,d^3v$ and the corresponding particle flux is $\mathbf{v}\,dN/V$

$$\mathbf{dS} = \mathbf{v}f(\mathbf{v})\frac{d^3v}{V} \tag{13.2}$$

The net particle flux is found by integrating over the full range of velocities

$$\mathbf{S} = \int_{\substack{\text{all}\\ \text{veloc.}}} \mathbf{dS} = \int_{\mathbf{v}} \mathbf{v}f(\mathbf{v})\frac{d^3v}{V} \tag{13.3}$$

It is frequently desirable to introduce other fluxes. For example, the mass flux is obtained by multiplying the particle flux by the mass per particle. This gives the directed rate of mass flow per unit area. The portion of the mass flux contributed by atoms in d^3v is

$$\mathbf{dM} \equiv m\,\mathbf{dS} = m\mathbf{v}f(\mathbf{v})\frac{d^3v}{V}$$

where m is the mass of a single atom. The total mass flux is

$$\mathbf{M} = \int_{\substack{\text{all}\\ \text{veloc.}}} \mathbf{dM} = \frac{1}{V}\int_{\mathbf{v}} \mathrm{m}\mathbf{v}f(\mathbf{v})\,d^3v = m\mathbf{S}$$

The mass flux is evidently the rate at which mass flows per unit area. In general, if $\Phi(\mathbf{v})$ represents some molecular property, the corresponding flux of the property resulting from the particles in d^3v is

$$\mathbf{d}\mathscr{F} \equiv \Phi(\mathbf{v})\,\mathbf{dS} = \Phi(\mathbf{v})\,\mathbf{v}f(\mathbf{v})\frac{d^3v}{V}$$

and the total flux of the quantity is

$$\mathscr{F} = \frac{1}{V}\int_{\substack{\text{all}\\ \text{veloc.}}} \Phi(\mathbf{v})\,\mathbf{v}f(\mathbf{v})\,d^3v \tag{13.4}$$

The fluxes of mass, momentum, and energy are of primary importance in the kinetic theory of gases. In fact, the hydrodynamic equations for a gas may be derived from consideration of just these three fluxes. In a plasma (an assembly of ionized particles) the flux of electric charge also must be considered. By combining the equations of kinetic theory with Maxwell's equations one can develop magnetohydrodynamics, the hydrodynamics of electrically conducting fluids.

Example 13.2 Kinetic Energy Flux

A hailstorm produces a vertical flux of hailstones. Each hailstone has the same mass (12 gm) and reaches the ground with the same speed (800 cm/sec). The number density of hailstones is $0.001/\text{cm}^3$. Determine the particle flux **S** and the kinetic energy flux of the hailstones.

The vector **S** is directed vertically downward. Its magnitude is

$$S = nv_0 = \left(\frac{0.001}{\text{cm}^3}\right)\left(800\ \frac{\text{cm}}{\text{sec}}\right) = \frac{0.8}{\text{cm}^2\text{-sec}}$$

The flux of kinetic energy is also downward. Since each hailstone has the same mass and speed the flux of kinetic energy $\mathscr{F}_k$ is the kinetic energy of one stone times the particle flux

$$\mathscr{F}_k = \tfrac{1}{2}mv_0^{\,2}S = 3.072 \times 10^6\ \text{ergs/cm}^2\text{-sec} \quad \diamond$$

••••••

13.2.3 Determine the fluxes of

(a) mass,
(b) linear momentum,

for the hailstones of Example 13.2.

••••••

13.3 MOLECULAR EFFUSION

The escape of gas through a small hole in a container is called *effusion*. In general, gas atoms proceed in both directions through such a hole. The net rate of effusion depends on the difference of the particle fluxes into and out of the container. For simplicity we assume that a vacuum surrounds the container so that only the escape of molecules need be considered. We also assume that the hole is large in comparison to atomic sizes. An atom, upon arriving at the opening, is able to pass through freely. On the other hand the hole must not be so large that the outward flow materially alters the distribution of velocities

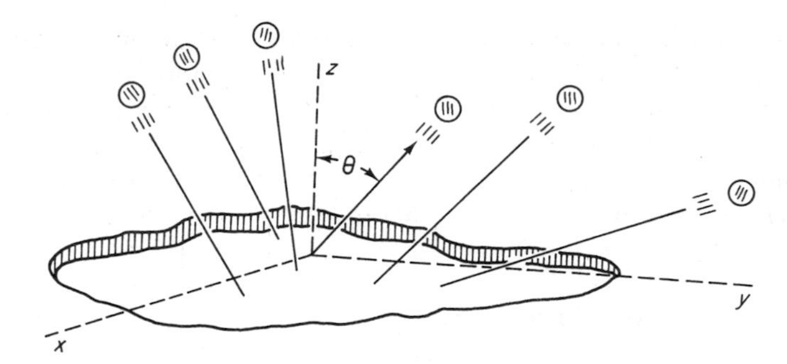

Figure 13.3. Geometry for computing the rate of molecular effusion. The polar angle θ is measured from the direction of the positive z axis. Only molecules with θ ranging from 0 to $\pi/2$ are moving outward.

of the remaining atoms. Collisions between atoms will be sufficiently numerous to maintain the form of the velocity distribution provided the dimensions of the hole do not exceed the mean free path length. For the moment we assume only that the distribution function is spatially uniform and spherically symmetric in velocity space. Later the Maxwellian distribution is called forth to provide a numerical estimate of the rate of effusion.

Figure 13.3 displays the geometry to be used. The positive z axis is directed out of the containing vessel. The opening lies in the x–y plane. The spherical symmetry of the velocity distribution guarantees that the x and y components of the flux are zero. Furthermore, only atoms with positive z components of velocity can escape. From (13.3) it follows that the flux is

$$S_z = \frac{1}{V}\int_{\substack{\text{all veloc.}\\ \text{with } v_z > 0}} v_z f(v)\, d^3v \tag{13.3a}$$

Spherical velocity space variables are called for

$$v_z = v\cos\theta$$
$$d^3v = v^2\, dv \sin\theta\, d\theta\, d\varphi$$

We are concerned only with atoms moving in the positive z direction. Consequently the integration over the polar angle θ ranges from 0 to $\pi/2$. The azimuth angle φ runs from 0 to 2π as usual. With these provisions

$$S_z = \frac{1}{V}\int_0^\infty vf(v)\, v^2\, dv \int_0^{\pi/2} \cos\theta \sin\theta\, d\theta \int_0^{2\pi} d\varphi$$

The angular integrations result in the factor π

$$S_z = \pi\int_0^\infty vf(v)\, v^2 \frac{dv}{V} = \frac{N}{4V}\frac{\int_0^\infty vf(v)\, 4\pi v^2\, dv}{N}$$

Noting that $N/V = n$, the number density, and that $(1/N)\int_0^\infty vf(v)\,4\pi v^2\,dv = \langle v\rangle$ is the mean atomic speed leaves us with

$$S_z = \tfrac{1}{4}n\langle v\rangle \tag{13.5}$$

Experiments by Knudsen in 1909 indirectly verified this expression for the effusive flux.

Recalling that $S = nv_0$ for a group of atoms with the same velocity, one more or less expects the $n\langle v\rangle$ dependence of S_z. Many aspects of such processes may be understood qualitatively by assuming that each atom moves with the mean speed or that each possesses the average energy. In this vein, it is helpful to introduce the idea of an average atom—one which moves with the mean speed or which possesses the average energy. In the next section we use the notion of the average atom to discuss thermal conduction. A rigorous treatment of transport phenomena in gases requires the use of the Boltzmann equation, a somewhat fearsome integro-differential equation.

•••••• ••••••

13.3.1 The average speed in a beam of atoms effusing from an oven at temperature T is greater than the average speed of atoms within the oven, because the faster atoms are more likely to escape. Show that the average speed of an atom in the beam is $3\pi/8$ greater than the average speed of an atom in the oven, assuming a Maxwellian distribution of speeds in the oven. [*Hint:* The probability that an atom of speed v will escape from the oven is directly proportional to its speed. It is also proportional to the probability that there is an atom of speed v in the oven. In other words, the probability of finding an atom in the beam with its speed in the range v to $v+dv$ is a conditional probability.]

•••••• ••••••

13.4 THERMAL CONDUCTION

We may understand the phenomenon of thermal conduction in a gas in terms of a flux of internal energy. For simplicity we assume that the temperature of the gas varies only along the z direction. Experimentally it is observed that the heat flux (heat flow per unit area per sec) is proportional to the temperature gradient $\partial T/\partial z$. With q_z denoting the heat flux we may write

$$q_z = -\kappa \frac{\partial T}{\partial z} \tag{13.6}$$

The constant κ is called the *thermal conductivity*. The minus sign expresses the fact that heat flows from a higher to a lower temperature. Equation (13.6) is Fourier's law of heat conduction.

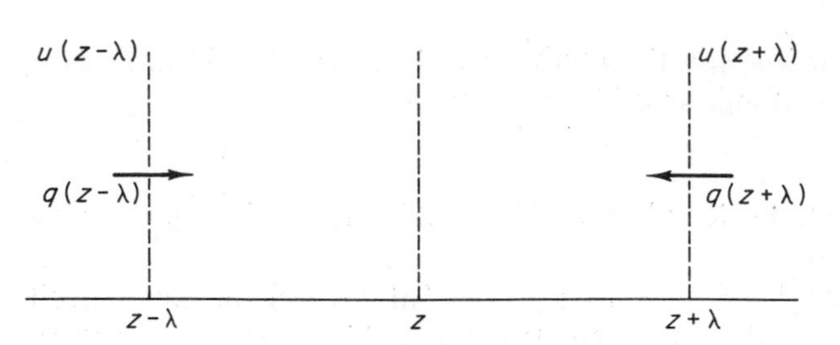

Figure 13.4. The average atom crossing the plane $z = \text{constant}$ originates from a plane at $z+\lambda$ or at $z-\lambda$.

13.4.1 Can a system in thermodynamic equilibrium conduct heat? Explain.

We can deduce (13.6) by considering the net flux of energy across an imaginary plane perpendicular to the z axis (see Figure 13.4). On the average, an atom crossing this plane will travel one mean free path λ before suffering a collision. It follows that atoms reaching this plane have traveled a distance λ, on the average, since undergoing their last collision. The general effect of collisions is to randomize the velocities of the colliding particles. On the average then, the momentum and energy of an atom are determined by its most recent collision. We therefore picture the average atoms which reach z as emerging from planes at $z+\lambda$ and at $z-\lambda$ and having energies and momenta appropriate to these planes.

Let u denote the mean internal energy per atom. Now u depends on T and thus indirectly on position z. In order to focus all attention on the thermal conduction we assume that there is no net particle flux, that is, no diffusion. We take the particle flux (directed toward z) to be $\frac{1}{4}n\langle v\rangle$ at both $z+\lambda$ and $z-\lambda$. This gives zero net particle flux at z. The internal energy flux of the average atoms is u times the particle flux

$$q = \frac{n\langle v\rangle}{4} u \tag{13.7}$$

The net internal energy flux at z is the sum of the oppositely directed fluxes emanating at $z \pm \lambda$. We have

$$q(z-\lambda) = \frac{n\langle v\rangle}{4} u(z-\lambda)$$

$$q(z+\lambda) = -\frac{n\langle v\rangle}{4} u(z+\lambda)$$

It is the net flux $q(z+\lambda) + q(z-\lambda)$, which we identify as the heat flux q_z

$$q_z = \frac{n\langle v\rangle}{4}[u(z-\lambda) - u(z+\lambda)] \tag{13.8}$$

Note that there is no net flux of atoms; equal numbers pass z in either direction. Nevertheless a net flow of thermal energy results because of a difference in the internal energy at $z+\lambda$ and at $z-\lambda$. Atoms coming from the region of higher temperature carry the greater energy. A Taylor series expansion for the internal energy gives

$$u(z \pm \lambda) = u(z) \pm \lambda\left(\frac{\partial u}{\partial z}\right) + \cdots \tag{13.9}$$

In practice, u can vary only slightly over distances comparable to λ so that the first two terms in the series provide an adequate approximation. Inserting the series into (13.8) gives

$$q_z = -\frac{1}{2}n\langle v\rangle\lambda\frac{\partial u}{\partial z} \tag{13.10}$$

Using the chain rule we can write

$$\frac{\partial u}{\partial z} = \left(\frac{\partial u}{\partial T}\right)\left(\frac{\partial T}{\partial z}\right)$$

Inserting this into (13.10) gives

$$q_z = -\frac{1}{2}n\langle v\rangle\lambda\left(\frac{\partial u}{\partial T}\right)\left(\frac{\partial T}{\partial z}\right) \tag{13.10a}$$

Comparing this with (13.6) shows that the thermal conductivity is given by

$$\kappa = \frac{1}{2}n\langle v\rangle\lambda\frac{\partial u}{\partial T} \tag{13.11}$$

This may be expressed in terms of more familiar quantities. Let U denote the internal energy per mole. The internal energy per atom is then $u = U/N_0$. Thus

$$\frac{\partial u}{\partial T} = \frac{1}{N_0}\frac{\partial U}{\partial T}$$

The derivative $\partial U/\partial T$ is identified as C_v, the molar heat capacity,

$$C_v = \left(\frac{\partial U}{\partial T}\right)_v$$

and thus

$$\frac{\partial u}{\partial T} = \frac{C_v}{N_0}$$

Using this result in (13.11) gives

$$\kappa = \frac{1}{2} n \langle v \rangle \lambda \frac{C_v}{N_0} \tag{13.12}$$

The factor of $\frac{1}{2}$ is not to be relied upon because of our crude average atom approach. More refined treatments lead to slightly different numerical factors. However, the underlying idea is sound; what one observes macroscopically as a heat flux is a consequence of an internal energy flux on the microscopic level.

Example 13.3 Thermal Conductivity of Air

Using the results obtained in previous chapters we can estimate the thermal conductivity of air. Under standard conditions we have

$$n \simeq 2.5 \times 10^{19} \frac{\text{molecules}}{\text{cm}^3}, \qquad \langle v \rangle \simeq 3 \times 10^4 \frac{\text{cm}}{\text{sec}}, \qquad \lambda \simeq 4 \times 10^{-5} \text{ cm}$$

We take $C_v = (5/2)\,R$, the value appropriate for an ideal diatomic gas. Noting that $R/N_0 = k$, the Boltzmann constant, we have

$$\frac{C_v}{N_0} = \frac{5}{2} k \simeq 3 \times 10^{-16} \frac{\text{erg}}{\text{K}}$$

Inserting these values into (13.12) gives

$$\kappa \simeq 5 \times 10^3 \frac{\text{erg}}{\text{K-cm-sec}} \simeq 10^{-4} \frac{\text{cal}}{\text{K-cm-sec}}$$

The experimental value of κ for air at 0 °C is 0.57×10^{-4} cal/K-cm-sec, within a factor of two of our crude estimate. ✧

······ ······

13.4.2 Compare the values of κ for air and glass. A layer of air 1-cm-thick supports a temperature difference of 20 °C. Estimate the thickness of glass needed to maintain the same temperature difference (for equal heat fluxes).

13.4.3 Assume that the atoms or molecules of a gas possess cross-sectional areas of size σ. Further, treat the gas as ideal, with f degrees of freedom. Assume a Maxwellian distribution of velocities for the purpose of determining $\langle v \rangle$. Show that

$$\kappa = \frac{fk}{4\sigma} \sqrt{\frac{8RT}{\pi M}}, \qquad M = \text{molecular weight}$$

······ ······

REFERENCES

A kinetic theory text, selected at random, is virtually certain to contain a semiquantitative development of transport phenomena along the lines of this chapter. In particular the following texts provide such discussion:

W. Kauzmann, "The Kinetic Theory of Gases," Chapter 5. Benjamin, New York (1966).

F. L. Friedman and L. Sartori, "The Classical Atom," Chapter 2. Addison-Wesley, Reading, Massachusetts (1965).

E. H. Kennard, "Kinetic Theory of Gases," Chapter 4. McGraw-Hill, New York (1938).

Chapter 14

Mathematics for Classical Statistics

In Chapter 10 we introduced several mathematical ideas and techniques of special utility in statistical physics. The material developed there satisfied the demands of kinetic theory. Statistical mechanics makes further demands and we must now add to our mathematical repertoire.

Statistical mechanics concerns itself with the likes of atoms, molecules, and subatomic particles. Fortunately there is a common thread of statistical ideas. This is especially true in classical statistical mechanics where atoms and billiard balls are subject to the same statistical laws. The developments of this chapter are designed to guide us through classical statistical mechanics. Quantum mechanics introduces its own very special constraints and modifies the classical statistics.†

14.1 PROBABILITY VERSUS NUMBER OF WAYS

Figure 14.1 shows the geometrical representation of the probabilities for the possible outcomes in the game of craps. With a pair of dice there are 36 possible outcomes, each equiprobable, so that each square has an area of 1/36. From the figure it is evident how one determines the probability of rolling a four, a

† Classical statistics is an abbreviated appellation for classical statistical mechanics.

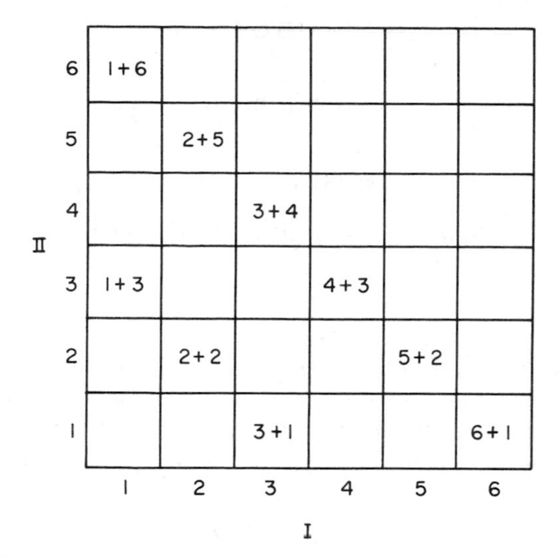

Figure 14.1. Area measures probability. The area of each square is 1/36. From the figure it is evident that three squares correspond to rolling a four, while six squares correspond to rolling a seven. The probability of rolling a four is thus $3(1/36) = 1/12$; the probability of rolling a seven is 1/6.

seven, a twelve, and so on. Precisely the same information is conveyed by letting the 36 squares be of unit area. Each square then corresponds to one possible outcome of a roll of the dice. From this viewpoint there are six ways of rolling a seven, three ways of rolling a four, and so on. In many instances an analysis in terms of the number of ways of achieving each outcome is preferred to one phrased in terms of the probability for each outcome.

•••••• ••••••

14.1.1 The possible outcomes for the flip of a quarter are a head H or a tail T. (If the coin stands on edge, it is flipped again!) What outcomes can result from the flip of three quarters? What is the probability that flips of the three quarters result in two heads and one tail? What is the number of ways for such a result to occur?

14.1.2 An urn contains three black balls and two white balls. What is the probability that two draws will result in the withdrawal of two black balls? (One ball is withdrawn on each draw.) The first ball drawn is not returned to the urn. Enumerate all possible outcomes of the two draws. What fraction of the possible outcomes result in the withdrawal of two black balls?

•••••• ••••••

14.2 FACTORIALS

The product of the N integers from 1 through N is called "N factorial" and written $N!$

$$N! = (1)\cdot(2)\cdot(3)\cdot(4)\cdots(N-2)\cdot(N-1)\cdot(N) \tag{14.1}$$

A more general definition of $N!$ is furnished by the relation

$$N! = \int_0^\infty x^N e^{-x}\,dx \tag{14.2}$$

The equivalence of (14.1) and (14.2) when N is a positive integer may be demonstrated by repeated integration by parts of (14.2). The garden variety factorials we ordinarily deal with are special cases of what are termed gamma functions. The gamma function $\Gamma(N)$ is defined by an integral

$$\Gamma(N) = \int_0^\infty x^{N-1} e^{-x}\,dx, \qquad N \neq 0, -1, -2, -3, \ldots \tag{14.3}$$

Comparing (14.2) and (14.3) we see that

$$N! = \Gamma(N+1) \tag{14.4}$$

For the present we need only the ordinary factorials.

•••••• ••••••

14.2.1 Use (14.2) to show that $0! = 1$.

14.2.2 Prove that

$$N\,\Gamma(N) = \Gamma(N+1)$$

[*Hint:* Integrate by parts. Assume N to be such that all integrals converge.]

•••••• ••••••

14.3 DISTINGUISHABLE OBJECTS AND DISTRIBUTIONS

The objects of a statistical analysis may be distinguishable in some respects but not others. A dozen white spheres may all have different radii. They are thus distinguishable on the basis of size but not color. The most direct method for sorting distinguishable objects is to number them. In the simplest situations it is possible to forget the objects and manipulate the numbers which they carry. Thus our objects are generally called $1, 2, 3, \ldots$, and so on.

A pair of distinguishable objects, 1 and 2, allows two distinct arrangements or orderings, namely 1 2 and 2 1. For three objects 1, 2, and 3, there are six distinct arrangements, namely

$$\begin{matrix} 123 & 213 & 312 \\ 132 & 231 & 321 \end{matrix}$$

In general there are $N!$ distinct arrangements of N distinguishable objects. To establish this fact note that the first object in line may be chosen in N ways—any one of the N objects may stand first. For each of these N choices any one of the remaining $N-1$ objects may stand second in line, so that there are a total of $N(N-1)$ distinct arrangements of the first two objects. Each of the $N(N-1)$ arrangements of the first two objects may be combined with any one of the $N-2$ possible choices for the third object for a total of $N(N-1)(N-2)$ distinct arrangements of the first three objects. The enumeration continues in this same fashion, the total number of distinct arrangements of N objects being

$$N\cdot(N-1)\cdot(N-2)\cdot(N-3)\cdots(3)\cdot(2)\cdot(1) = N!$$

Having ascertained that there are $N!$ distinct arrangements of N distinguishable objects we turn to a related problem. Suppose the N distinguishable objects are arranged in M groups. The groups themselves are also distinguishable, that is, they are numbered. For example, the N objects might be the guests in a hotel with M floors. A *distribution* of the N objects is defined by specifying how many reside in each of the M groups.

Distribution: A distribution of N objects in M groups is specified by the numbers

$$N_1, N_2, N_3, \ldots, N_M$$

where N_i is the number of objects in the ith group.

If we regard the N objects as the guests in a hotel having M floors, a distribution specifies the numbers $N_1, N_2, \ldots, N_M$ of guests on each floor. We use the symbolisms $\{N_i\}$ and $\{N_1, N_2, \ldots, N_M\}$ to designate distributions. Note that a distribution involves a certain degree of ignorance. It specifies *how many* but *not which* objects are in each group.

The *total* number of ways of achieving a distribution of N objects is $N!$. Not all of these are *distinct* ways because many of the $N!$ ways are merely rearrangements of the order *within* groups. For the distribution $\{N_1, N_2, \ldots, N_M\}$ there are a total of

$$(N_1!)(N_2!)(N_3!)\ldots,(N_M!)$$

ways of ordering objects within groups. It follows that the number of distinct ways of achieving the distribution $\{N_i\}$ is

$$D\{N_i\} = \frac{\text{total number of ways}}{\text{number of reorderings within groups}} = \frac{N!}{(N_1!)(N_2!)\cdots(N_M!)} \tag{14.5}$$

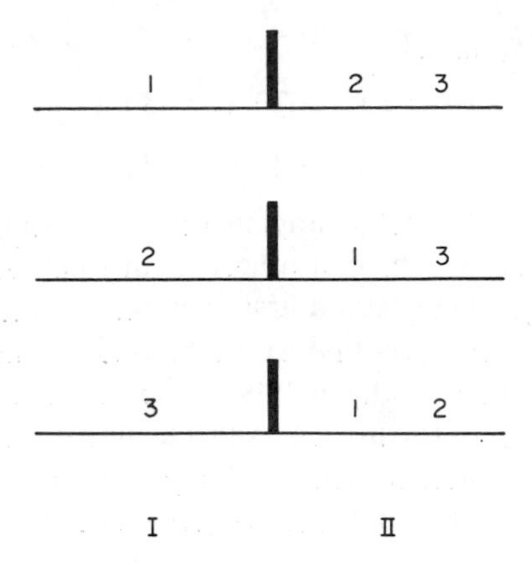

Figure 14.2. The three distinct floor assignments for the distribution; $\{N_1 = 1; N_2 = 2\}$.

To illustrate this result, consider the case $N = 3$, $M = 2$, and the distribution $\{N_1 = 1;\ N_2 = 2\}$. From (14.5), $D\{N_i\} = 3!/(1!)(2!) = 3$. The three distinct ways of achieving the distribution are depicted in Figure 14.2. In this instance, half of the total of $N! = 6$ arrangements are distinct—the other half correspond to rearrangements of the order within the second group. In terms of our animation, the order in which guests 2 and 3 sign the register has no bearing on the distribution if both are assigned to the same floor. Consequently 1 ▌2 3 and 1 ▌3 2 represent the same way of achieving the distribution $\{N_1 = 1;\ N_2 = 2\}$.

•••••• ••••••

14.3.1 Explain why it is the product and not the sum of the number of rearrangements within groups which gives the total number of rearrangements.

14.3.2 Consider the case $N = 4$, $M = 2$, and the distribution

$$\{N_1 = 2;\ N_2 = 2\}$$

Use a diagram to indicate the six distinct ways of achieving the distribution. For one of the six ways indicate the four rearrangements which give the same way of achieiving the distribution.

•••••• ••••••

Products like the one in the denominator of (14.5) occur so frequently in statistical considerations that a special notation is introduced,

$$N_1!\,N_2!\,N_3!\cdots N_M! \equiv \prod_{i=1}^{M} N_i! \tag{14.6}$$

the symbol $\prod_{i=1}^{M}$ being an instruction to form the product of M factors. In terms of the product notation,

$$D\{N_i\} = \frac{N!}{\prod_{i=1}^{M} N_i!} \tag{14.5a}$$

••••••

14.3.3 Evaluate the following products:

(a) $\prod_{i=1}^{4} i^2$, (c) $\prod_{i=0}^{3} 2$,

(b) $\prod_{s=1}^{3} (2s+1)$, (d) $\prod_{s=1}^{4} s^2$.

••••••

Next, we must generalize our development to allow for structure within the groups which comprise a distribution. For example, suppose that we refine the hotel problem so as to account for room assignments as well as floor assignments. We number the rooms on each floor and distinguish an arrangement which has, say, guest 1 in room 1 on floor 1 from one which has guest 1 in room 2 on floor 1, and so on. In a more general context we ask, "What is the total number of ways of achieving a distribution if the groups are partitioned into cells and we count as distinct arrangements which give different cell populations?" Let Q_i denote the number of cells in the ith group. If we think in terms of the hotel problem, Q_i is the number of rooms on the ith floor. The first of the N_i guests assigned to the ith floor can occupy any one of the Q_i rooms. For each—repeat—for each of these Q_i possible dispositions of the first guest, an equal number are open to the second guest, so that Q_i^2 assignments are possible for the pair.† Each time another guest is considered the existing number of assignments possible is increased by a factor of Q_i. Thus there are $Q_i, Q_i^2, Q_i^3, \ldots$, and finally, $Q_i^{N_i}$ possible room assignments for the N_i guests in the Q_i rooms on the ith floor. For the entire hotel—M floors in all—there are

$$Q_1^{N_1} Q_2^{N_2} Q_3^{N_3} \cdots Q_M^{N_M} = \prod_{i=1}^{M} Q_i^{N_i} \equiv C\{N_i\} \tag{14.7}$$

different room assignments for a given floor assignment.

••••••

14.3.4 Consider the case $N = 2$, $M = 2$, $Q_1 = Q_2 = 2$, and the distribution $N_1 = N_2 = 1$.

---

† Thus Q_i of these Q_i^2 assignments would place the two guests in the same room.

(a) Calculate and identify $D\{N_i\}$.

(b) For one of the distinct floor assignments identify the different room assignments.

•••••• ••••••

This same product applies in general—$C\{N_i\}$ equals the number of different cell assignments for a given distribution.

Let $W\{N_i\}$ denote the total number of ways of achieving the distribution including the added distinctions brought in by counting different cell assignments. Evidently $W\{N_i\}$ is the product of $D\{N_i\}$ and $C\{N_i\}$. That is,

$$W\{N_i\} = \begin{pmatrix}\text{number of distinct ways of realizing a given distribution}\\ = D\{N_i\}\end{pmatrix}$$

$$\cdot\begin{pmatrix}\text{number of distinct ways of distributing objects over cells}\\ \text{for the given distribution} = C\{N_i\}\end{pmatrix}$$

Combining (14.5a) and (14.7) we obtain

$$W\{N_i\} = N! \prod_{i=1}^{M} \frac{Q_i^{N_i}}{N_i!} \tag{14.8}$$

It is fair to ask just how the foregoing considerations find application in statistical physics. In Chapter 15 we consider distributions of atoms. The atoms are analogous to the hotel guests. The atoms are distributed with respect to energy. Many different energies are open to the individual atoms. The different energies correspond to the different hotel floors. It is possible for atoms to exist in different states (cells) which have equal energies. That is, the states are distinguishable on the basis of properties other than energy. The different states of equal energy correspond to different rooms on the same floor.

Statistical mechanics is concerned almost exclusively with the *most probable distribution*, which is defined as the distribution which can be formed in the largest number of ways. Thus we state as a definition:

The most probable distribution is the one for which $W\{N_i\}$ is largest.

In statistical mechanics we shall be concerned with determining the most probable distribution of particles. For example, our first venture will be to deduce the most probable spatial distribution of gas atoms. Later we will seek the most probable distribution with respect to particle energy. In all cases the problem is one of determining the most probable distribution subject to the constraints imposed by the surroundings. Generally these constraints define the thermodynamic nature of the system. For example, to represent an isolated thermodynamic system we require that the total energy and the total number of particles have fixed values. Before pushing ahead with the physics we concern ourselves with one final mathematical topic, the saddle-point integration technique.

14.4 SADDLE-POINT INTEGRATION

In developing the formalism of statistical mechanics one frequently encounters expressions which involve the logarithms of very large factorials. Certain aspects of the theoretical analysis are greatly simplified by using Stirling's formula for such logarithms. Stirling's formula—an approximation—is

$$\ln(N!) \simeq N \ln N - N + \ln\sqrt{2\pi N}$$

Stirling's formula may be derived by a so-called saddle-point integration. The saddle-point method is quite versatile, and a worthy addition to any scientist's bag of tricks. Accordingly its development is not relegated to the appendixes.

The saddle-point method is applicable to an important class of definite integrals,

$$A = \int_a^b f(x)\, dx \tag{14.9}$$

in which the integrand $f(x)$ exhibits a sharply peaked behavior, as suggested by Figure 14.3, in which the hatched area represents the integral A. The saddle-point method approximates this area by the area under a Gaussian curve.

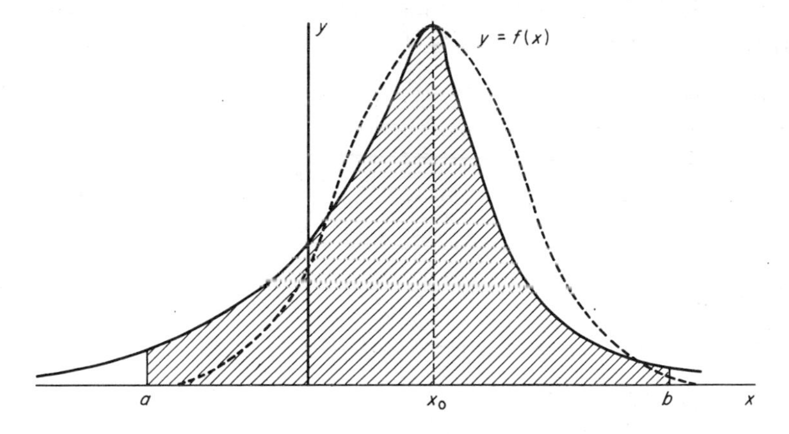

Figure 14.3. Typical form of integrand for which the saddle-point method is useful. The dashed line shows the Gaussian curve which is fitted to $f(x)$.

We begin by expressing $f(x)$ in exponential form. Let

$$g(x) = -\ln f(x) \tag{14.10}$$

so that

$$f(x) = e^{-g(x)} \tag{14.11}$$

Next, $g(x)$ is expanded in a Taylor series about the point $x = x_0$

$$g(x) = g_0 + g_0'(x-x_0) + \tfrac{1}{2}g_0''(x-x_0)^2 + \cdots \tag{14.12}$$

where

$$g_0 = g(x_0), \qquad g_0' = \left(\frac{dg}{dx}\right)_{x=x_0}, \qquad g_0'' = \left(\frac{d^2g}{dx^2}\right)_{x=x_0} \tag{14.13}$$

The first part of the approximation consists of retaining only the first three terms of the Taylor series. The success of the method depends on the precision with which the three-term series represents $g(x)$ in the range near $x = x_0$. If $f(x)$ is sharply peaked and x_0 is taken to be the peak point, the resulting approximation of $g(x)$ is generally adequate. The point x_0 is defined by the condition that $g(x)$ have a minimum there; in other words, x_0 is found by solving

$$\left(\frac{dg}{dx}\right)_{x=x_0} = 0 \tag{14.14}$$

The point x_0 is called the *saddle point* for reasons not at all in evidence here. (The method we use here is a simplification of the full saddle-point technique used in a complex variable theory.) From the relation $f(x) = e^{-g(x)}$ it is clear that $f(x)$ is a maximum where $g(x)$ is a minimum, and that the peak of $f(x)$ falls at $x = x_0$. With $g_0' = 0$, (14.12) becomes

$$g(x) = g_0 + \tfrac{1}{2}g_0''(x-x_0)^2 \tag{14.15}$$

so that

$$A = \int_a^b \exp[-g(x)]\,dx = \exp(-g_0)\int_a^b \exp[-\tfrac{1}{2}g_0''(x-x_0)^2]\,dx \tag{14.16}$$

Noting that

$$\exp[-g(x_0)] = f(x_0)$$

we can write

$$A = f(x_0)\int_a^b \exp[-\tfrac{1}{2}g_0''(x-x_0)^2]\,dx \tag{14.17}$$

The second and final approximation consists of replacing the upper and lower limits by $+\infty$ and $-\infty$. This introduces little error provided most of the area—the major contribution to the integral—lies under the central peak of the Gaussian curve. Making this change of limits and introducing the change of variable

$$z = x - x_0, \qquad dz = dx$$

transforms (14.17) into

$$A = f(x_0) \int_{-\infty}^{+\infty} \exp(-\tfrac{1}{2} g_0'' z^2)\, dz \tag{14.18}$$

The integral is of the standard Gaussian form (see Appendix A)

$$\int_{-\infty}^{+\infty} \exp\left(-\frac{1}{2} g_0'' z^2\right) dz = 2G_0\left(\frac{1}{2} g_0''\right) = \sqrt{\frac{2\pi}{g_0''}} \tag{14.19}$$

Thus the saddle-point approximation assumes the concise form

$$A = \sqrt{\frac{2\pi}{g_0''}}\, f(x_0) \tag{14.20}$$

where

$$g(x) = -\ln f(x) \tag{14.10}$$

and the saddle point x_0 is defined by the condition

$$\left(\frac{dg}{dx}\right)_{x=x_0} = 0 \tag{14.14}$$

To illustrate the method we use it to approximate two integrals which may be evaluated exactly. These examples serve to encourage a cautious optimism in the use of the saddle-point technique. As a third example we use the technique to derive Stirling's approximate formula for $\ln(N!)$.

Example 14.1

Consider the integral

$$A = \int_{-\infty}^{+\infty} \frac{dx}{x^2 + a^2}$$

The exact value of A is π/a. Here,

$$f(x) = \frac{1}{x^2 + a^2}$$

For $g(x)$ we have

$$g(x) = -\ln f(x) = +\ln(x^2+a^2)$$

and

$$g' = \frac{2x}{x^2+a^2}, \qquad g'' = \frac{2}{(x^2+a^2)} - \frac{4x^2}{(x^2+a^2)^2}$$

The condition fixing x_0 is

$$g_0' = 0 = \frac{2x_0}{x_0^2+a^2}$$

which has the solutions

$$x_0 = 0 \qquad \text{and} \qquad x_0 = \pm\infty$$

It is clear that $x_0 = 0$ is the saddle point; the solutions at $\pm\infty$ give the minima of $f(x)$. (It is always advisable to sketch the integrand roughly to aid in locating the saddle point.) Thus

$$x_0 = 0 \qquad \text{and} \qquad f(x_0) = \frac{1}{a^2}$$

Further, $g_0'' = 2/a^2$, giving for the saddle point approximation

$$A = \sqrt{\frac{2\pi}{g_0''}}\, f(x_0) = \frac{\sqrt{\pi}}{a}$$

The saddle-point approximation is too small by a factor of $\sqrt{\pi}$, the exact result being π/a. The poorest example was presented first—we hope to encourage *cautious* optimism. The saddle-point method works best when there is an honest exponential present in the integrand—an exponential which gives the integrand a sharp peak. ◇

Example 14.2

Consider next the integral

$$A = \int_0^{\infty} x^2 e^{-\beta x^2}\, dx$$

The exact value of this integral is $(\sqrt{\pi}/4)\beta^{-3/2}$ (see Appendix A). The saddle-point method proceeds from

$$f(x) = x^2 e^{-\beta x}$$

$$g(x) = -\ln f(x) = \beta x^2 - 2\ln x$$

$$g' = 2\beta x - \frac{2}{x}$$

The saddle point is determined by the equation

$$g_0' = 0 = 2\beta x_0 - \frac{2}{x_0}$$

which gives

$$x_0 = \beta^{-1/2} \quad \text{and} \quad f(x_0) = \beta^{-1}e^{-1}$$

Further, one finds $g_0'' = 4\beta$ and then

$$A = \frac{e^{-1}}{\sqrt{2}}\sqrt{\pi}\,\beta^{-3/2}$$

The numerical factor $e^{-1}/\sqrt{2}$ has the value 0.259. Thus the saddle-point method gives

$$A = 0.259\sqrt{\pi}\,\beta^{-3/2}$$

which compares quite favorably with the exact value

$$A = 0.250\sqrt{\pi}\,\beta^{-3/2}$$

This example was chosen to encourage cautious *optimism*. We now turn to a third example, a derivation of Stirling's approximation for $\ln(N!)$. ◇

Example 14.3

The Stirling approximation for $\ln(N!)$ is

$$\ln(N!) = N\ln N - N + \ln\sqrt{2\pi N} \tag{14.21}$$

To bring to bear our saddle-point machinery requires the integral representation of $N!$

$$N! = \int_0^\infty x^N e^{-x}\,dx \tag{14.2}$$

We now evaluate the integral in (14.2) via the saddle-point method. This yields the Stirling approximation. We have

$$f(x) = x^N e^{-x}$$

$$g(x) = -\ln f(x) = x - N\ln x$$

Differentiating gives

$$g' = 1 - \frac{N}{x}$$

The saddle-point equation $g_0' = 0$ gives

$$x_0 = N \quad \text{and} \quad f(x_0) = N^N e^{-N}$$

Further, $g_0'' = 1/N$, whereupon we find for the saddle-point approximation,

$$N! = \sqrt{\frac{2\pi}{\frac{1}{N}}} N^N e^{-N} = \sqrt{2\pi N}\, N^N e^{-N} \tag{14.22}$$

Taking logs gives the desired result

$$\ln(N!) = N \ln N - N + \ln\sqrt{2\pi N} \tag{14.21}$$

The very large numbers dealt with in statistical mechanics often justify a still cruder version of (14.21), obtained by dropping the $\ln\sqrt{2\pi N}$ term. The result, which we refer to as the *super* Stirling approximation, is

$$\ln(N!) \simeq N \ln N - N \quad \diamond \tag{14.23}$$

•••••• ••••••

14.4.1 Check the precision of the method by comparing the Stirling and super Stirling approximations with the exact value for $N = 1$, 2, and 10. Show that the percent of error decreases with increasing N by evaluating the ratios $(N!)_{\text{Stir}}/(N!)_{\text{exact}}$ and $(N!)_{\text{S.Stir}}/(N!)_{\text{exact}}$ for $N = 1$, 2, and 10.

14.4.2 Use the saddle-point method to evaluate

$$\int_0^\infty \exp\left[-\left(x^2 + \frac{a^2}{x^2}\right)\right] dx$$

The exact value is $\sqrt{\pi} e^{-2a}/2$.

14.4.3 Use the saddle-point method to show that

$$\int_0^\infty x^3 \exp[-(2x^2 + x^4)]\, dx \simeq \frac{\sqrt{\pi}}{8} \exp\left(-\frac{5}{4}\right)$$

•••••• ••••••

Example 14.4

It often happens that the equation which determines the saddle point is transcendental, that is, it is not an algebraic equation. To illustrate the methods which may be used in this case we evaluate

$$I = \int_0^\infty \frac{x^3\, dx}{e^x - 1}$$

an integral which arises in the theory of blackbody radiation. The saddle-point approximation for I is

$$I_{sp} = \sqrt{\frac{2\pi}{g_0''}}\frac{x_0^3}{e^{x_0}-1}$$

where

$$g(x) = -\ln\left(\frac{x^3}{e^x-1}\right) = -3\ln x + \ln(e^x-1)$$

The saddle-point equation is

$$g_0' = -\frac{3}{x_0} + \frac{e^{x_0}}{e^{x_0}-1} = 0$$

This may be rewritten as

$$x_0 = 3[1-e^{-x_0}]$$

This is a transcendental equation. By consulting tables of e^{-x} and exercising patience one can establish that $x_0 \simeq 2.82$. An iteration technique quickly leads to the same result. Inspection of the above equation suggests that x_0 is close to 3, that is, $1-\exp(-x_0)$ is nearly 1 when x_0 is close to 3 ($e^{-3} \simeq 0.05$). If we set $x_0 = 3$ on the right side, the saddle-point equation gives a better approximation for x_0

$$x_0 = 3(1-e^{-3}) = 2.85$$

This value of x_0 is iterated—reinserted into the exponential. This gives a still better approximation for x_0

$$x_0 = 3(1-e^{-2.85}) = 2.83$$

One more iteration gives the quoted value

$$x_0 = 3(1-e^{-2.83}) = 2.82$$

Note how quickly the iterations converge. Extreme precision in locating the saddle point is seldom warranted since the saddle-point integral is itself an approximation. In this instance the saddle-point value is 6.38, within 2% of the exact value, $\pi^4/15 \simeq 6.49$. ◇

•••••• ••••••

14.4.4 The number of photons per unit volume (photon number density) in blackbody radiation is given by

$$\frac{N}{V} = \frac{1}{\pi^2}\left(\frac{kT}{\hbar c}\right)^3 \int_0^\infty \frac{x^2\,dx}{e^x-1}$$

The integral may be expressed in terms of the Riemann zeta function

$$\int_0^\infty \frac{x^2\,dx}{e^x - 1} = 2\zeta(3) = 2.404$$

Evaluate the integral approximately using the saddle-point method. The saddle-point equation is transcendental. Use an iteration scheme to evaluate x_0.

•••••• ••••••

REFERENCES

Brief discussions of combinatorial problems ("number of ways") will be found in:

H. Margenau and G. M. Murphy, "The Mathematics of Physics and Chemistry," 2nd ed., Chapter 12. Van Nostrand-Reinhold, Princeton, New Jersey (1961).

and

B. V. Gnedenko and A. Y. Khinchin, "An Elementary Introduction to the Theory of Probability," Chapter 5. Dover, New York (1961).

By far the "quickest" derivation of the super Stirling approximation takes advantage of the identity $\ln(N!) - \ln(N-1)! = \ln N$. See:

P. A. H. Wyatt, Elementary Statistical Mechanics without Stirling's Approximation, *J. Chem. Educ.* **39**, 27 (1962).

Chapter 15

Classical Statistical Mechanics

15.1 THE RATIONALE OF STATISTICAL MECHANICS

People who wager on the outcomes of horse races answer to many names—bettors, losers, and so on. With varying degrees of success these individuals aspire to *predict* the outcome of each race. The principles which guide the deliberations of the more proficient bettors are remindful of the rationale of statistical mechanics.

The bettor collects facts regarding each horse and its jockey, the condition of the track, and so on. Every relevant scrap of information is considered. The information is always incomplete, that is, inadequate to allow a unique determination of the outcome of a race. On the basis of incomplete information he predicts the most likely or most probable outcome of a race.

The pronouncements of statistical mechanics are also of a probabilistic nature. Statistical mechanics is based on a mathematical scheme which takes account of all available data, such as the total number of particles in a system, its volume, the temperature of its surroundings, and so on. These data are always incomplete—they constrain, but do not uniquely fix the microscopic structure of the system. Statistical mechanics singles out the most probable configuration of the system, taking care to see that its choice is consistent with the constraints.

The bettor relies on personal experience as well as the assembled facts. Statistical mechanics relies on the experience of nature. Its rules are plausible and its predictions in accord with experiment. Of great significance is the fact that statistical mechanics permits a derivation of the laws of thermodynamics.

Briefly, the rules of statistical mechanics—to be illustrated and amplified in the following sections—are as follows:

I. Consider all distinguishable microscopic configurations of the system consistent with the constraints. In general, many different microscopic configurations result in the same *distribution.*

II. Identify the distribution corresponding to the largest number of microscopic configurations. Name this distribution *most probable.*

III. When making predictions as to the behavior of the system assume that the most probable distribution prevails. That is, the pressure, temperature, and other observable properties are assumed to be those corresponding to the most probable distribution.

We now show how adherence to these rules leads to a not unexpected prediction: The most probable configuration for a gas free of external forces is one of uniform spatial density. Consider our favorite system—a gas of N atoms. We mentally partition the container into C rooms or cells of equal volume. The cells are labeled $1, 2, \ldots, C$ according to their locations. The number of atoms occupying the ith cell is denoted by N_i. The only information we have regarding the gas is that it is composed of a definite number of atoms. There are N atoms and we know the value of N, but this is all we know about the gas. In particular, we proclaim our complete ignorance as to what the energy of the gas might be. A further implication of our limited knowledge is the absence of any net external force on the gas. A net external force would define a preferred direction in space and perhaps preferred positions in the container.

Our task is to determine the most probable distribution of the atoms. In the present context a distribution is defined by specifying the value of N_i for all C cells. According to our definition of most probable this means that we seek the distribution which can be formed in the largest number of ways. The search is subject to just one constraint, namely, that all candidates for the most probable distribution be comprised of exactly N atoms.

What we hope to find is that the most probable distribution is the one for which each cell contains the same number of atoms. Why? Because such a distribution makes the number density uniform throughout the container volume and a uniform density is one of the requirements for thermodynamic equilibrium. With N atoms in each of C cells we would have N/C atoms in each cell.

$$N_i = \frac{N}{C}, \qquad i = 1, 2, \ldots, C \tag{15.1}$$

The quantity to be maximized is the number of ways of partitioning N objects into C groups with N_i in the ith group. This number is

$$W\{N_i\} = \frac{N!}{N_1!\,N_2!\cdots N_C!} = \frac{N!}{\prod_{i=1}^{C} N_i!} \tag{15.2}$$

This can be argued quite readily, or extracted from (14.8) by setting $Q_i = 1$ (each floor becomes 1 cell), and with C in place of the index M. The value of W for the uniform density distribution is

$$\frac{N!}{\left[\left(\frac{N}{C}\right)!\right]^{C}} \equiv W_{\text{mp}} \tag{15.3}$$

Direct comparison shows that W_{mp} is larger than the value of $W\{N_i\}$ for any other distribution. For example, the distribution which has $(N/C)-1$ atoms in one cell, $(N/C)+1$ in another, and N/C in each of the remaining $C-2$ cells has a value of $W\{N_i\}$ given by

$$\frac{N!}{\left(\frac{N}{C}-1\right)!\left(\frac{N}{C}+1\right)!\left[\left(\frac{N}{C}\right)!\right]^{C-2}} \equiv W'$$

Comparing W' and W_{mp} shows

$$\frac{W'}{W_{\text{mp}}} = \frac{\frac{N}{C}}{\frac{N}{C}+1} < 1$$

Thus W' is less than W_{mp} so that the uniform distribution is more probable than an adjacent slightly nonuniform distribution.[†] Similarly, the distribution which has $(N/C)-2$ atoms in one cell, $(N/C)+2$ in another, and N/C in each of the other cells has a value of $W\{N_i\}$ which is smaller than W_{mp} by the factor

$$\frac{\left(\frac{N}{C}-1\right)\left(\frac{N}{C}\right)}{\left(\frac{N}{C}+1\right)\left(\frac{N}{C}+2\right)} < 1$$

······ ······

15.1.1 Consider the case of $N = 2$ atoms distributed over $C = 2$ cells of equal volume. Enumerate all distributions and verify that the uniform density distribution is most probable.

······ ······

[†] Two distributions are said to be *adjacent* when their cell populations differ only slightly.

Verifying that any particular distribution that has a value of $W\{N_i\}$ less than W_{mp} does not rigorously establish the uniform distribution as most probable. It might be just a local maximum. Some "distant" distribution might have a still larger value of $W\{N_i\}$. It is somewhat like being on a mountain top, surrounded by clouds. You are certain that no adjacent points are higher, but still loftier peaks might be hidden some distance away.

The student unconvinced by the foregoing arguments will find the following demonstration more rigorous. The derivation is not presented solely in the interest of rigor. It serves mainly to introduce the ideas and methods used in Section 15.7 where the all-important *Boltzmann distribution* is deduced. As a prelude to these proceedings we establish some idea of the sizes of N and C.

Consider 1 liter of gas under standard conditions. This represents about 1/20 of a mole or $N \simeq 3 \times 10^{22}$ atoms. The C volume elements are to be small on a macroscopic scale. An optical microscope is able to resolve lengths as small as 10^{-4} cm, so that a cell volume of 10^{-12} cm^3 would be on the borderline of visibility. Let us take the cell volumes to be 10^{-14} cm^3, thereby ensuring their submacroscopic size. The characteristic atomic volume is roughly a 1 Å cube (10^{-24} cm^3) so that the cell volumes are definitely large by microscopic standards. The number of cells is

$$C = \frac{\text{container volume}}{\text{cell volume}} = \frac{10^{+3}}{10^{-14}} = 10^{17} \text{ cells}$$

Thus $N = 3 \times 10^{22}$ and $C = 10^{17}$ are large numbers, as is the average number of atoms per cell, $N/C = 3 \times 10^5$.

We now demonstrate that $W\{N_i\}$ is a maximum for the uniform distribution. The one assumption made at the outset is that

$$N_i \gg 1, \qquad i = 1, 2, \ldots, C$$

that is, we assume large cell populations. This appears quite reasonable because both N and N/C are large numbers.

When the cell populations are very large $W\{N_i\}$ changes only slightly from one distribution to an adjacent one. Consequently $W\{N_i\}$ should be essentially the same for the most probable distribution and its most adjacent neighbors. The situation is suggested by Figure 15.1 which plots $W\{N_i\}$ versus $\{N_i\}$. Since $\{N_i\}$ denotes a set of numbers, such a plot is only symbolic. Near its maximum, the interchange of one atom between two cells does not appreciably alter $W\{N_i\}$. Reasoning conversely we can use this fact to locate the maximum of $W\{N_i\}$. We write

$$W\{N_i\} = W\{N_i'\} \tag{15.4}$$

where $\{N_i\}$ is the distribution of atoms in the most probable distribution and

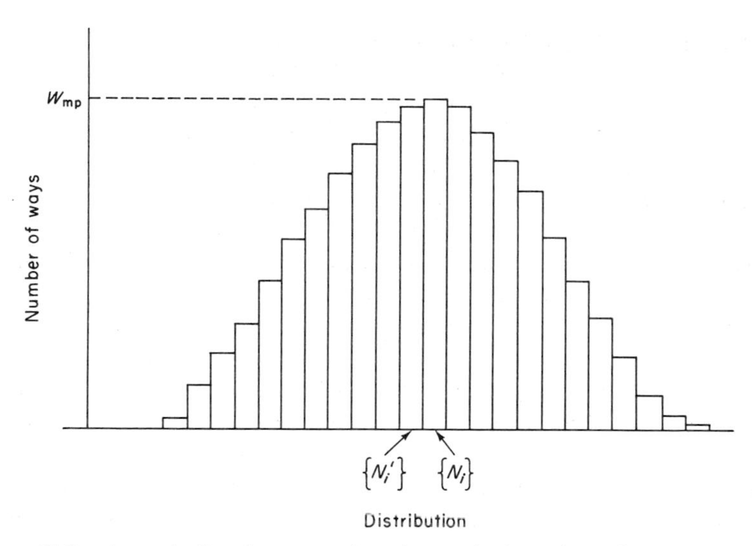

Figure 15.1. A symbolic plot suggesting the method used to determine the most probable distribution. The requirement that $W\{N_i\}$ have the same value for two adjacent distributions locates the peak.

$\{N_i'\}$ is the distribution for an adjacent distribution.† Note that $\{N_i'\}$ is immediately adjacent to $\{N_i\}$; the two distributions differ only in the population of two cells, the kth and sth. Specifically,

$$\{N_i\} = (N_1, N_2, \ldots, N_k, \ldots, N_s, \ldots, N_C) \tag{15.5}$$

and

$$\{N_i'\} = (N_1, N_2, \ldots, N_k - 1, \ldots, N_s + 1, \ldots, N_C) \tag{15.6}$$

From the way in which they have been constructed it is apparent that $\{N_i\}$ and $\{N_i'\}$ correspond to the same total number of atoms. Thus if $\{N_i\}$ is a distribution of N atoms, so also is $\{N_i'\}$

$$\sum_{i=1}^{C} N_i = \sum_{i=1}^{C} N_i' = N \tag{15.7}$$

We emphasize that k and s are arbitrary; the kth cell can be any one of the C and the sth can be any different cell. Inserting these values of N_i and N_i' into the maximization relation (15.4) gives

$$\frac{N!}{N_1!\,N_2!\cdots N_k!\cdots N_s!\cdots N_C!} = \frac{N!}{N_1!\,N_2!\cdots(N_k-1)!\cdots(N_s+1)!\cdots N_C!} \tag{15.4a}$$

† One does essentially the same thing in an ordinary maxima–minima problem. Thus, for example, requiring that $f(x) = f(x+dx)$ locates the maxima and/or minima of $f(x)$—the points at which $df/dx = 0$.

Canceling common factors and noting that

$$\frac{(N_s+1)!}{N_s!} = N_s + 1, \qquad \frac{N_k!}{(N_k-1)!} = N_k$$

reduces (15.4a) to

$$N_s + 1 = N_k \tag{15.8}$$

Because the cell populations are large we can replace N_s+1 by N_s with the result that (15.8) becomes

$$N_s = N_k \tag{15.9}$$

However, k and s are arbitrary; (15.9) tells us that for the most probable distribution the populations of *any* two cells we might choose must be the same. This means that all cells must contain the same numbers of atoms. Thus we have followed a different path to a determination that the uniform distribution is most probable. A parallel path is followed to derive the famous Boltzmann distribution law.

In the next section we introduce the important concept of phase space. This will permit us to move from the cells of ordinary coordinate space used in this section to the cells of phase space with which statistical mechanics deals.

15.2 PHASE SPACES

In Chapter 11 we introduced the notion of velocity space by analogy with coordinate space. With the abstraction of velocity space now firmly entrenched it is not difficult to visualize a combined position–velocity space. Such a space has six dimensions. A point in the space is specified by six quantities

$$x, \quad y, \quad z, \quad v_x, \quad v_y, \quad v_z$$

The combined position–velocity space is one type of phase space. Geometrically, a point in phase space fixes the phase of particle motion; where it is and how it is moving. It is not possible to plot or visualize points in six dimensions. Nevertheless, phase space provides an elegant framework for the description of particle behavior. Moreover, phase space is essential for the logical development of many segments of statistical mechanics. Phase space is both economical and essential.

The motion of a particle describes a real trajectory in coordinate space. With each point along the path of the particle we associate a position $\mathbf{r}$ and a velocity $\mathbf{v}$. For each point on the real path there corresponds a representative point in the six-dimensional phase space $(\mathbf{r}, \mathbf{v})$. As the particle moves, its representative point also changes and we can speak of a trajectory in phase space.

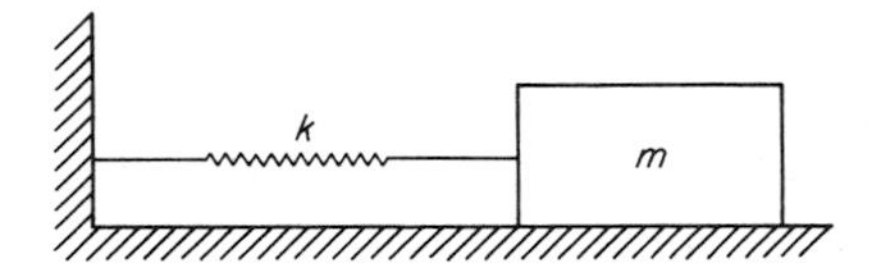

Figure 15.2. One-dimensional simple harmonic oscillator. The symbol k refers to the spring constant, not the Boltzmann constant.

The particle moves in a real three-dimensional space; its representative point moves in a six-dimensional phase space.

Phase-space trajectories can be pictured in two dimensions (on these pages!) if we deal with a system undergoing one-dimensional motion. The phase space then has two dimensions with, say, x and v_x denoting the coordinates of the representative point. As an example, consider the one-dimensional simple harmonic oscillator (SHO) of Figure 15.2. The motion of the oscillator can be described by

$$x(t) = A \cos \omega t, \qquad \omega^2 = \frac{k}{m} \tag{15.10}$$

$$v_x(t) = -\omega A \sin \omega t \tag{15.11}$$

From these equations we find

$$\frac{x^2}{A^2} + \frac{v_x{}^2}{A^2\omega^2} = 1 \tag{15.12}$$

This, in turn, shows that the phase-space trajectory of the representative point is an ellipse, as depicted in Figure 15.3. The arrows on the ellipse indicate the sense in which the representative point traverses the phase-space trajectory.

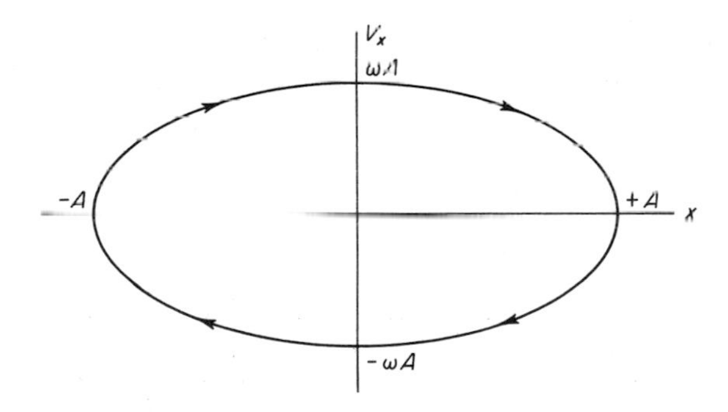

Figure 15.3. Phase-space trajectory of representative point for the simple harmonic oscillator. The representative point circulates in clockwise fashion around the ellipse.

•••••• ••••••

15.2.1 Explain why the phase-space trajectory of Figure 15.3 is traced clockwise rather than counterclockwise. [*Hint:* Start at $t = 0$ and consider the signs of x and v_x as time elapses.]

15.2.2 A particle undergoes one-dimensional motion, being subject to a constant acceleration g. The initial velocity is v_0. Determine $x(t)$ and $v_x(t)$. Eliminate t to obtain an equation relating x and v_x. Plot the resulting phase-space trajectory for $g = \frac{1}{2}$ and $v_0 = -1$. Locate the initial point ($t = 0$) and indicate the subsequent path of the representative point along the trajectory. The initial position of the particle is $x(t = 0) = 0$.

15.2.3 A rubber ball bounces in such a fashion that it rebounds to a height of 9/10 that to which it rose on the preceding bounce. *Sketch* its phase-space trajectory, assuming that its velocity changes discontinuously at impact, that is, that it remains in contact with the wall for zero time.

•••••• ••••••

Particles with coordinates in d^3r and velocities in d^3v have representative points which cluster closely in phase space. It is natural to introduce a phase-space volume element. We refer to the six-dimensional element

$$d^3r\, d^3v = dx\, dy\, dz\, dv_x\, dv_y\, dv_z \tag{15.13}$$

as the phase-space volume element.

For systems in thermodynamic equilibrium the density must be uniform. It is then unnecessary to deal with small spatial volume elements—the observable properties of the system are the same at all points. This permits us to deal with a large volume element—the total volume of the system. The corresponding phase-space volume element is then

$$d^3r\, d^3v = V\, dv_x\, dv_y\, dv_z \tag{15.14}$$

For reasons which must be left unstated at this point, a slightly different version of phase space is generally employed in statistical mechanics. Points in phase space are specified in terms of position and linear momentum rather than in terms of position and velocity. In many situations the difference between momentum and velocity is merely a factor of m, the particle mass. Thus

$$\mathbf{p} = m\mathbf{v}, \qquad dp_x = m\, dv_x$$

Such a simple relation between momentum and velocity does not always hold. For example, a photon traveling at speed c has zero rest mass. However, the photon momentum is not zero. The magnitude of the photon momentum follows from the relation

$$\varepsilon^2 = (pc)^2 + (mc^2)^2 \tag{15.15}$$

where ε is the energy, p is the momentum, and m is the rest mass. The rest mass of the photon is zero so that

$$\varepsilon = pc \tag{15.16}$$

In situations where we deal with the position–momentum phase space, a representative point is labeled by the six variables

$$x, \quad y, \quad z, \quad p_x, \quad p_y, \quad p_z$$

Velocity space and the accompanying volume element d^3v are replaced by momentum space and the volume element d^3p. In Cartesian momentum space

$$\mathbf{p} = (p_x, p_y, p_z), \qquad d^3p = dp_x dp_y dp_z \tag{15.17}$$

A description in terms of spherical momentum variables is often desired. The momentum can be specified by giving its magnitude p and direction (θ, φ). Volume elements needed for spherical momentum space are

$$d^3p = p^2\, dp \sin\theta\, d\theta\, d\varphi \qquad \text{and} \qquad d^3p = 4\pi p^2\, dp \tag{15.18}$$

The volume element in the position–momentum phase space is

$$d^3r\, d^3p = \begin{cases} \text{volume of position–momentum phase space} \\ \text{containing representative points of particles} \\ \text{with positions in } d^3r \text{ and momenta in } d^3p \end{cases}$$

Having identified phase space as the proper setting for statistical mechanics we proceed to develop the theory in some detail.

15.3 PHASE SPACES AND HOTELS: AN ANALOGY

In Chapter 14 we introduced some of the mathematical ideas needed to develop statistical mechanics. The treatment there was animated in terms of hotel guests, room and floor assignments, and so on. Our attention shifts from hotel guests to atoms and molecules. Certain ideas are most clearly presented with reference to a specific physical system. For this role we choose our favorite of long standing, a gas of N atoms confined within a volume V. The gas acts only as a familiar frame of reference; the results are of wide applicability within the framework of classical statistical mechanics. Our analysis is restricted to systems in thermodynamic equilibrium. Thus in the case of the gas, its pressure, density, and temperature are spatially uniform and do not change with time.

The results established in the preceding chapter enable us to form helpful correspondences between hotels and their guests and phase space and its occupants. The atoms are analogous to the hotel guests, while the complete phase space corresponds to the hotel. To construct the analogs of hotel rooms we mentally partition the phase space into a great many cells. Each phase-space cell is a suitable home for one or more atoms. Our discussion of hotels was concerned with both room and floor designations. Some care is required to establish the phase-space counterpart of a hotel floor assignment. With each phase-space cell there is associated a definite energy—the energy of an atom in that cell.† An important assumption is implicit in assigning a definite energy to each segment of phase space. To say that an atom has a certain energy because it is in a particular region of phase space ignores the effects of other atoms. The proximity of other atoms is partially responsible for determining the total energy of any given atom. Thus our basic approach here is limited to situations in which mutual interactions can be ignored. We deal with an independent-particle model. Such a model proves adequate in a number of cases. For example, in Chapter 11 such a model led us to the ideal gas law. Statistical mechanics copes with the problem of interacting particles by introducing more sophisticated phase spaces. Happily, these more powerful approaches are conceptual extensions of the ideas we now develop in connection with the independent-particle model.

Let us imagine a line along which the phase-space cell energies are plotted. We refer to this energy-axis pattern as the energy spectrum for the system. The nature of the physical system determines the details of an energy spectrum. However, certain characteristics are common to all energy spectra. The most important feature is that large numbers of phase-space cells have nearly equal energies. We may express this idea in a more quantitative way through the relation

$$\frac{\Delta\varepsilon}{\varepsilon} \ll 1$$

where ε is a representative cell energy and $\Delta\varepsilon$ a typical energy difference between adjacent cells. The high density of cells along the energy axis is very important. When the energy spacing of cells is small by comparison with the energies themselves we may legitimately group several cells and assign each the same energy. This grouping enables us to deal with wholesale numbers of atoms having nearly equal energies. We can pursue this idea through the hotel analogy. In any large hotel there are many rooms which rent for the same price even though the characteristics are not exactly the same for each room. Some rooms renting for $15.00 a day will be slightly larger than others, some will have newer furniture, and so on. It would be very cumbersome to have rents of $15.05, $14.98, $15.12, $15.07, ..., even though these figures might represent

† Once again we economize on words by stating that the atom is in a phase-space cell, when in fact it is the representative point of the atom.

the relative merits of the rooms. An average figure of $15.00 is far more expedient. To complete the analogy between phase space and hotels we must group rooms of the same rent on the same floor. We play the same game in statistical physics, grouping phase-space cells with nearly equal energies and assigning each an average energy. A group of cells with the same energy corresponds to a floor of rooms which rent for the same rate.

In summary, we have the following analogies between phase space and hotels: The complete phase space corresponds to a hotel. Individual phase-space cells are analogs of hotel rooms. Rooms which command equal rent are grouped on the same floor. Phase-space cells whose energies are nearly the same are considered in groups, all cells in the same group being assigned equal energy.

15.4 MICROSTATES AND MACROSTATES

The concepts of *microstates* and *macrostates* of a physical system are of great importance in statistical mechanics. Both microstates and macrostates are defined with reference to the populations of groups of phase-space cells.

Let Q_i denote the number of phase-space cells in the ith group and let ε_i be the corresponding cell energy. Hotels generally have some rooms occupied and others empty. Phase space is very similar. There will be both occupied and empty phase cells. Let N_i denote the total number of atoms which reside in the Q_i cells. Further, let M denote the total number of groups of cells. The energy axis is thereby partitioned into M segments, the ith segment comprising Q_i cells which contain a total of N_i atoms. A *macrostate* of the system is defined by specifying the populations of the M groups of cells. Thus a macrostate is determined by a specification of the numbers $N_1, N_2, N_3, \ldots, N_M$, that is, by the numbers of atoms in each group of cells. Evidently a *macrostate* is a *distribution* in phase space. Note in particular that a macrostate specifies only how many atoms, and not which atoms occupy each group of cells. A *microstate* is defined by specifying which phase-space cell each atom occupies. As such, a microstate corresponds to one way of realizing a particular distribution of hotel guests. Table 15.1 summarizes the analogies between hotel guests and the occupants of phase-space cells.

We have previously stressed the point that the measurable properties of the gas characterize the average behavior of many atoms. The measurable properties depend on the numbers of atoms occupying the various phase-space cells. The observable properties do not depend on which atoms these might be. A microstate indicates "which ones" as well as "how many" and thus supplies more information than is needed to determine the measurable properties of the system. A less detailed description of the gas, but one adequate to explain its

TABLE 15.1

Guest–Atom Analogies

Guest	Atom
N guests reside in a hotel with M floors, there being Q_i rooms on the ith floor. All rooms on the same floor rent at the same rate.	N atoms are located in M groups of phase-space cells, there being Q_i cells in the ith group. Each cell in a group corresponds to the same energy.
Specifying the numbers of guests $N_1, N_2, \ldots, N_M$ on each floor defines a *distribution*.	Specifying the numbers of atoms $N_1, N_2, \ldots, N_M$ in each group defines a *macrostate*.
A *way* of realizing a particular distribution is defined by identifying which room each guest occupies.	A *microstate* corresponding to a particular macrostate is defined by identifying which cell each atom occupies.

measurable properties, is provided by the macrostate. If the measurable properties depend only on the macrostate of the gas it might seem that the concept of a microstate is superfluous. However, this is not the case—we need microstates. In particular, microstates are essential to the development of the notion of thermodynamic probability.

15.5 THERMODYNAMIC PROBABILITY

Visualize some microstate of the gas. This microstate corresponds to some macrostate of the gas. In general, a great many different microstates correspond to the same macrostate. A macrostate is a distribution of the atoms, while each microstate giving rise to it constitutes a distinct way of achieving the distribution. The number of different microstates which result in the same macrostate is readily deduced. In fact, the direct analogies between atoms and hotel guests let us recognize $W\{N_i\}$ as the number of microstates giving rise to the same macrostate

$$W\{N_i\} = N! \prod_{i=1}^{M} \frac{Q_i^{N_i}}{N_i!} \tag{14.8}$$

Here the N objects are atoms, distributed over M groups of phase-space cells, such that the ith group of Q_i cells contains N_i atoms.

Microstates consistent with all of the macroscopic constraints imposed on the system by its surroundings are referred to as *accessible*. In the case of our gas, for example, microstates which have one or more atoms outside the container are conceivable but not accessible. Similarly, if we specify the total energy of the gas to be 16 J, the accessible microstates are restricted to be those which result in this energy. Other microstates are conceivable, but not accessible under the constraints imposed.

The macroscopic constraints imposed by the surroundings of a system determine its thermodynamic state. In general, the thermodynamic state of a system admits an enormous number of different macrostates. That is, a great many different macrostates can result in the same thermodynamic state. As we have just seen, there are also numerous microstates for each macrostate, $W\{N_i\}$ of them. Figure 15.4 suggests the fragmenting association of thermodynamic state, macrostates, and microstates.

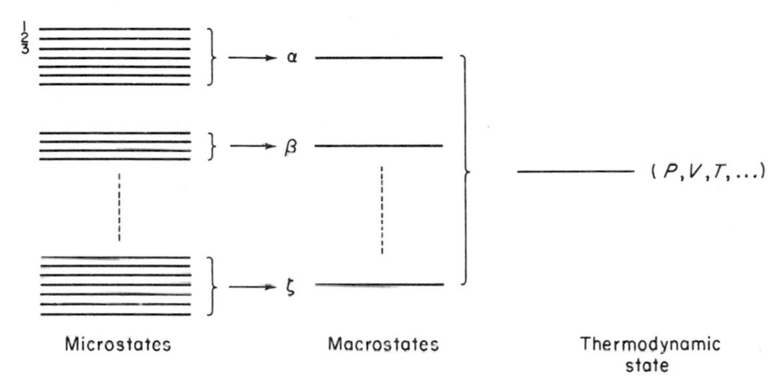

Figure 15.4. A great many accessible microstates correspond to the same macrostate. Similarly, a vast number of macrostates give rise to the same thermodynamic state.

At this point we introduce a crucial statistical hypothesis—a *postulate* of statistical mechanics.

All accessible microstates are equiprobable

The basis for advancing the postulate is quite simple; we have no reason—no evidence—which favors one accessible microstate over any other. The postulate is little more than Laplace's principle of insufficient reason. Assuming that all accessible microstates are equiprobable it follows that the probability of finding a system in a particular macrostate is directly proportional to $W\{N_i\}$, the number of microstates which result in that macrostate. In statistical mechanics $W\{N_i\}$ is generally referred to as the *thermodynamic probability* of the macrostate.

$$\begin{array}{c}\text{thermodynamic probability of finding a system in}\\ \text{the macrostate specified by the distribution } \{N_i\}\end{array} \equiv W\{N_i\} = N!\prod_{i=1}^{M}\frac{Q_i^{N_i}}{N_i!} \tag{15.19}$$

Although $W\{N_i\}$ is not a probability in the usual sense, it does convey the same information as a properly normalized probability.[†]

15.6 THE MOST PROBABLE MACROSTATE AND THERMODYNAMIC EQUILIBRIUM

Knowing which macrostate a system occupies allows one to compute mean values and thus to predict values for its measurable properties. Our analysis will reveal only the most probable macrostate, the macrostate for which the thermodynamic probability is a maximum. However, the system cannot always be in the most probable macrostate. The motions of the atoms carry the system through a succession of microstates. Not all of the microstates through which the system passes can correspond to the most probable macrostate. Such a feat would mean that external (macroscopic) constraints were acting to control the motions of the atoms. It would mean that Maxwell demons are possible. The fact that no such control is possible forms the basis for assuming that all accessible microstates are equiprobable. We conclude that the macrostate of the system changes with time. On the other hand the fact that we deal with a system in thermodynamic equilibrium means that the measurable properties remain virtually constant.[‡] This pair of observations leads to an important conclusion: The measurable properties of the system are nearly the same for a vast majority of the macrostates. The plausibility of this judgement is strengthened by considering its antithesis. If the measurable properties varied markedly with changes in the macrostate, a system could never reach thermodynamic equilibrium. Changes in the macrostate occur billions upon billions of times each second. The system would be unable to settle down into an equilibrium state if the observable properties changed appreciably with changes in the macrostate. The realization that the system continually changes macrostates, coupled with the existence of thermodynamic equilibrium leads us to believe that the measurable properties of the system are essentially the same for a sizable majority of all macrostates. Not every macrostate, because we can observe fluctuations. The fluctuations arise whenever the system ventures into a series of markedly offbeat macrostates. Because the most probable macrostate is representative of virtually all macrostates, statistical mechanics directs us to assume that

---

[†] Here is another situation where specifying the number of ways is equally as informative as giving the probability.

[‡] The measurements may reveal fluctuations, the variations about equilibrium values. We can picture a fluctuation as a succession of nonequilibrium macrostates.

The most probable macrostate corresponds to the thermodynamic equilibrium state.

By corresponds we mean that the macroscopic properties—pressure, temperature, internal energy, and so on—have the same values for the thermodynamic equilibrium state and the most probable macrostate.

15.7 THE BOLTZMANN DISTRIBUTION

We now seek the most probable macrostate for an isolated system in thermodynamic equilibrium. An isolated system can exchange neither particles nor energy with its surroundings. Therefore, the total number of particles N and the internal energy U of the system are constants. Macrostates aspiring to gain the status of most probable must specify a distribution of N atoms with a total energy equal to U.

The thermodynamic probability $W\{N_i\}$ as expressed by (15.21) is a function of the M variables $N_1, N_2, N_3, \ldots, N_M$. The most probable macrostate is determined by an extension of the approach used in Section 15.2. Let $\{N_i\}$ denote the most probable macrostate†

$$\{N_i\} = (N_1, N_2, \ldots, N_k, \ldots, N_s, \ldots, N_M) \tag{15.20}$$

and let $\{N_i'\}$ denote the macrostate immediately adjacent to $\{N_i\}$. Specifically, $\{N_i'\}$ is to differ from $\{N_i\}$ only in the populations of two groups of cells, the kth and sth:

$$\{N_i'\} = (N_1, N_2, \ldots, N_k - 1, \ldots, N_s + 1, \ldots, N_M) \tag{15.21}$$

We emphasize that k and s are arbitrary; the kth group of cells can be any one of the M and the sth any one of the remaining groups. The adjacent macrostate $\{N_i'\}$ can evidently be formed from the most probable macrostate $\{N_i\}$ by the transfer of one atom from the kth to the sth group of cells. This guarantees that both macrostates correspond to the same number of atoms

$$\sum_{i-1}^{M} N_i = \sum_{i=1}^{M} N_i' = N \tag{15.22}$$

Let W and W' denote the thermodynamic probabilities for $\{N_i\}$ and $\{N_i'\}$, respectively. The most probable macrostate is located by requiring that it have the same thermodynamic probability as its immediately adjacent neighbors

$$W = W' \tag{15.23}$$

---

†Henceforth, $\{N_i\}$ will be referred to as a macrostate instead of a distribution.

The idea behind (15.23) is that the macrostate having the largest thermodynamic probability differs in the most minimal way from adjacent macrostates. The differences are so slight that the most probable and immediately adjacent macrostates have essentially equal thermodynamic probabilities. Reasoning conversely, we can use this equality, as expressed by (15.23), to identify the most probable macrostate.

If the N_i atoms in the ith group of cells each have energy ε_i, their combined energy is $N_i\varepsilon_i$. The total internal energy of the N atoms distributed over M groups of cells is $\sum_{i=1}^{M} N_i\varepsilon_i$. In the competing macrostate $\{N_i'\}$ there are N_i' atoms in the ith group of cells so that the total internal energy is $\sum_{i=1}^{M} N_i'\varepsilon_i$. To satisfy the constraint

$$\text{total internal energy of gas} = U$$

subjects the macrostates $\{N_i\}$ and $\{N_i'\}$ to the constraint

$$U = \sum_{i=1}^{M} N_i\varepsilon_i = \sum_{i=1}^{M} N_i'\varepsilon_i \tag{15.24}$$

The distributions of atoms producing $\{N_i\}$ and $\{N_i'\}$ are given by (15.20) and (15.24) so that both macrostates correspond to the same total number of atoms. We now have a well-posed mathematical problem: Determine the form of the N_i which simultaneously satisfy (15.23) and (15.24). If we insert the explicit forms of $W\{N_i\}$ and $W\{N_i'\}$ given by (15.19), (15.23) becomes

$$W = N! \prod_{i=1}^{M} \frac{Q_i^{N_i}}{N_i!} = N! \prod_{i=1}^{M} \frac{Q_i^{N_i'}}{N_i'!} = W' \tag{15.23a}$$

Whereas (15.23a) is a relation involving products of functions of the N_i, (15.24) is a relation between sums of the N_i. For this reason it is mathematically more convenient to replace (15.23) by

$$\ln W = \ln W' \tag{15.25}$$

This is equivalent to requiring $W = W'$ because the macrostate for which $\ln W$ is a maximum is also the macrostate for which W is largest. Furthermore, $\ln W$ involves a sum of functions of the N_i, as does U.

•••••• ••••••

15.7.1 Show that the function

$$\varphi(x) = x^3 \exp(-x^2)$$

and $\ln \varphi(x)$ attain their maximum values at the same point.

•••••• ••••••

There is also a very strong physical reason for choosing to maximize $\ln W$ rather than W itself. As we shall discover, $\ln W$ is directly proportional to the entropy of a thermodynamic state, and thermodynamics has taught us that the state of maximum entropy is the equilibrium state. Thus by maximizing $\ln W$ we locate the state of largest entropy, the equilibrium state. In order to guarantee that both (15.24) and (15.25) are satisfied simultaneously we write

$$\ln W - \beta \sum_{i=1}^{M} N_i \varepsilon_i = \ln W' - \beta \sum_{i=1}^{M} N_i' \varepsilon_i \tag{15.26}$$

where $-\beta$ is a constant with the dimensions of (1/energy) introduced to render $-\beta U$ dimensionless. Thus both $\ln W$ and $-\beta U$ are dimensionless quantities, and (15.26) holds only when both (15.24) and (15.25) are satisfied. We will show later that β is the reciprocal of the characteristic thermal energy kT

$$\beta = \frac{1}{kT} \tag{15.27}$$

Inserting the explicit forms for W and W' into (15.26), and using the fact that

$$\ln\left(\prod_i X_i\right) = \sum_i \ln X_i \tag{15.28}$$

results in

$$\ln N! + \sum_{i=1}^{M}\left[\ln\left(\frac{Q_i^{N_i}}{N_i!}\right) - \beta N_i \varepsilon_i\right] = \ln N! + \sum_{i=1}^{M}\left[\ln\left(\frac{Q_i^{N_i'}}{N_i'!}\right) - \beta N_i' \varepsilon_i\right] \tag{15.29}$$

The most probable macrostate and the immediately adjacent macrostate differ only in the populations of the kth and sth groups of cells

$$N_k' = N_k - 1$$

$$N_s' = N_s + 1$$

For the other $M-2$ groups of cells

$$N_i' = N_i$$

Canceling identical factors in (15.29) leaves

$$\ln\left(\frac{Q_k^{N_k}}{N_k!}\right) - \beta N_k \varepsilon_k + \ln\left(\frac{Q_s^{N_s}}{N_s!}\right) - \beta N_s \varepsilon_s = \ln\left(\frac{Q_k^{N_k-1}}{(N_k-1)!}\right) - \beta (N_k - 1)\varepsilon_k$$
$$+ \ln\left(\frac{Q_s^{N_s+1}}{(N_s+1)!}\right) - \beta (N_s + 1)\varepsilon_s$$

Grouping k and s terms on opposite sides of the equation and noting that, for example,

$$\ln\left(\frac{Q_k^{N_k-1}}{(N_k-1)!}\right) - \ln\left(\frac{Q_k^{N_k}}{N_k!}\right) = \ln\left(\frac{N_k}{Q_k}\right)$$

results in

$$\ln\left(\frac{N_s+1}{Q_s}\right) + \beta\varepsilon_s = \ln\left(\frac{N_k}{Q_k}\right) + \beta\varepsilon_k$$

Assuming that the population of each group of cells is large allows us to replace N_s+1 by N_s, yielding

$$\ln\left(\frac{N_s}{Q_s}\right) + \beta\varepsilon_s = \ln\left(\frac{N_k}{Q_k}\right) + \beta\varepsilon_k \tag{15.30}$$

However, k and s are arbitrary; hence the quantity $\ln(N_i/Q_i) + \beta\varepsilon_i$ must be the same for all M groups of cells,

$$\ln\left(\frac{N_i}{Q_i}\right) + \beta\varepsilon_i = \text{constant},^\dagger \qquad i = 1, 2, \ldots, k, \ldots, s, \ldots, M$$

Calling the constant $\ln A$, where A is itself constant, we arrive at the distribution

$$N_i = AQ_i e^{-\beta\varepsilon_i}, \qquad \beta = \frac{1}{kT} \tag{15.31}$$

This is the famous *Boltzmann distribution*. It is the basis for a vast number of applications of statistical mechanics. The constant A may be expressed in terms of N and β by noting that the Boltzmann distribution satisfies the equation of constraint $N = \sum_{i=1}^{M} N_i$. Thus

$$N = \sum_{i=1}^{M} N_i = A \sum_{i=1}^{M} Q_i \exp(-\beta\varepsilon_i)$$

If we define

$$Z(\beta) \equiv \sum_{i=1}^{M} Q_i \exp(-\beta\varepsilon_i) \tag{15.32}$$

we have

$$A = \frac{N}{Z(\beta)}$$

† In general the constant will depend on the thermodynamic state of the system, its temperature, volume, and so on. It is a constant only in the sense that it has the same value for all M phase-space cells.

The populations may now be expressed as

$$N_i = \frac{NQ_i \exp\left(\frac{-\varepsilon_i}{kT}\right)}{Z(\beta)}, \qquad \beta = \frac{1}{kT} \tag{15.33}$$

Observe that N_i/N, the fraction of the atoms which have energy ε_i, is directly proportional to the exponential $\exp(-\varepsilon_i/kT)$. This all-important quantity governing the cell populations is known as the Boltzmann factor:

$$e^{-\varepsilon/kT} \Rightarrow \text{Boltzmann factor}$$

It is essential to understand how $e^{-\varepsilon/kT}$ behaves as ε and/or kT change. The Boltzmann factor reveals that there is a competition for the available internal energy. For a fixed temperature the Boltzmann factor decreases as the energy ε is increased. A modest change in the ratio ε/kT changes the Boltzmann factor dramatically. For example,

$$\varepsilon = 0, \qquad \exp\left(\frac{-\varepsilon}{kT}\right) = e^{-0} = 1$$

$$\varepsilon = kT, \qquad \exp\left(\frac{-\varepsilon}{kT}\right) = e^{-1} = 0.367$$

$$\varepsilon = 10kT, \qquad \exp\left(\frac{-\varepsilon}{kT}\right) = e^{-10} \simeq 5 \times 10^{-5}$$

$$\varepsilon = 30kT, \qquad \exp\left(\frac{-\varepsilon}{kT}\right) = e^{-30} \simeq 10^{-13}$$

The Boltzmann factor is a sharply decreasing function of energy. The majority of atoms have energies comparable to kT. Atoms with energies over $10kT$ are exceptional and constitute only a small fraction of all atoms. The quantity kT emerges as the characteristic thermal energy.

For a given energy, the Boltzmann factor decreases as T decreases. As a consequence a decrease in temperature shifts the population toward phase-space cells of lower energy.

We remark in passing that classical statistical mechanics (CSM) is often referred to as Boltzmann statistics, especially in situations where its predictions are contrasted with those of quantum statistical mechanics.

The quantity

$$Z(\beta) = \sum_{i=1}^{M} Q_i \exp(-\beta\varepsilon_i) \tag{15.32}$$

answers to several names, Zustandsumme and partition function being the most widely used. We use the latter.

Example 15.1 The Two-Level System

The so-called two-level system has only two available phase-space cells ($Q_1 = Q_2 = 1$). The energies associated with the two cells are

$$\varepsilon_1 = 0, \qquad \varepsilon_2 = +\varepsilon, \qquad \varepsilon > 0$$

With ε positive, ε_1 is the lower of the two energies. The partition function for the two level system is

$$Z(\beta) = \sum_{i=1}^{2} Q_i \exp(-\beta\varepsilon_i) = \exp(-\beta \cdot 0) + \exp(-\beta \cdot \varepsilon)$$

or

$$Z(\beta) = 1 + \exp(-\beta\varepsilon) \quad \diamond$$

•••••• ••••••

15.7.2 (a) With reference to Example 15.1 show that the fractional populations of the two cells are

$$\frac{N_1}{N} = \frac{1}{1+e^{-\beta\varepsilon}}, \qquad \frac{N_2}{N} = \frac{e^{-\beta\varepsilon}}{1+e^{-\beta\varepsilon}}$$

(b) With $\varepsilon = \text{constant} = 1$ eV, determine the population ratio N_2/N_1 for the three cases,

$$kT = 10 \text{ eV}, \qquad kT = 1 \text{ eV}, \qquad kT = 0.1 \text{ eV}$$

How do your results support the statement that "a decrease in temperature shifts the population toward phase-space cells of lower energy."

•••••• ••••••

15.8 THE PARTITION FUNCTION FOR AN IDEAL GAS

The parameter β is determined implicitly by (15.24) and (15.33), from which we find

$$\frac{U}{N} = \sum_{i=1}^{M} \frac{Q_i \varepsilon_i \exp(-\beta\varepsilon_i)}{Z(\beta)} \tag{15.34}$$

The right side of (15.34) can be expressed as a derivative of $\ln Z(\beta)$. Note first that

$$-\frac{\partial}{\partial\beta}\ln Z(\beta) = -\frac{1}{Z(\beta)}\cdot\frac{\partial Z}{\partial\beta}$$

From the definition of the partition function $Z(\beta)$ we have

$$\frac{\partial Z}{\partial \beta} = \frac{\partial}{\partial \beta}\sum_{i=1}^{M} Q_i \exp(-\beta\varepsilon_i) = -\sum_{i=1}^{M} Q_i \varepsilon_i \exp(-\beta\varepsilon_i)$$

so that

$$-\frac{\partial}{\partial \beta}\ln Z(\beta) = \sum_{i=1}^{M} \frac{Q_i \varepsilon_i \exp(-\beta\varepsilon_i)}{Z(\beta)} \tag{15.35}$$

Comparing (15.34) and (15.35) shows that

$$\frac{U}{N} = -\frac{\partial}{\partial \beta}\ln Z(\beta) \tag{15.36}$$

While this diabolical maneuver simplifies the form of the equation relating β and U/N it still does not provide an explicit equation of the form

$$\beta = \text{some explicit function of } \frac{U}{N}$$

We now show that $\beta = 1/kT$. The scheme is to evaluate $Z(\beta)$ for an ideal monatomic gas, which leads to

$$\frac{U}{N} = \frac{3}{2}\frac{1}{\beta} \tag{15.37}$$

We have previously shown that the internal energy of an ideal gas is related to its temperature by

$$\frac{U}{N} = \frac{3}{2}kT \tag{15.38}$$

from which it follows that $\beta = 1/kT$.

Although the notation suggests that $Z(\beta)$ depends on β alone this is most definitely not the case. In general the partition function depends on the full complement of thermodynamic variables. However, to avoid unwieldy notations we write simply $Z(\beta)$.

To evaluate $Z(\beta)$ we must establish the connection between the number of cells Q_i and the cell energy ε_i. This is done here for an ideal monatomic gas. Our original decision to consider groups of cells was prompted by the fact that many particles have nearly equal energies. It also means that there are many phase-space cells corresponding to a small range of energy. Suppose we consider the energy range from ε to $\varepsilon + d\varepsilon$, where $d\varepsilon$ is macroscopically small, but microscopically large. That is, there are many cells whose energies fall in the range from ε to $\varepsilon + d\varepsilon$. The high density of cells along the energy axis allows us to replace the summation over groups of cells by an integration over phase space. We make the replacement

$$Z(\beta) = \sum_{i=1}^{M} Q_i \exp(-\beta\varepsilon_i) = \int_{\substack{\text{phase}\\ \text{space}}} dQ(\varepsilon) \exp(-\beta\varepsilon) \tag{15.39}$$

where $dQ(\varepsilon)$ is the number of phase-space cells with energy in the range from ε to $\varepsilon + d\varepsilon$.

······ ······

15.8.1 Show how the sums

$$\sum_{n=0}^{N} n = \frac{N(N-1)}{2}$$

and

$$\sum_{n=0}^{N} n^2 = \frac{N(2N^2-3N+1)}{6}$$

are closely approximated by integrals when $N \gg 1$.

······ ······

For an ideal monatomic gas the energy of an atom is purely translational kinetic energy. The energy and speed are related by

$$\varepsilon = \tfrac{1}{2}mv^2 \tag{15.40}$$

Cells with energies in the range ε to $\varepsilon + d\varepsilon$ correspond to a particular range of speeds v to $v + dv$, and thus to a definite chunk of phase space. We denote the associated volume of phase space by $d^3v\, d^3r$. Let Ω denote the volume of each phase-space cell. The numerical value of Ω is of no consequence at the moment; it conveniently cancels out of most of the ensuing computations. We can express $dQ(\varepsilon)$ as

$$\begin{aligned} dQ(\varepsilon) &\equiv \text{number of phase cells in range } \varepsilon \text{ to } \varepsilon + d\varepsilon \\ &= \frac{\text{corresponding volume of phase space}}{\text{volume of one cell}} \\ &= \frac{d^3v\, d^3r}{\Omega} \end{aligned} \tag{15.41}$$

(15.39) becomes

$$Z(\beta) = \frac{1}{\Omega}\int_r d^3r \int_v \exp\left(\frac{-\beta mv^2}{2}\right) d^3v \tag{15.42}$$

The integrand is independent of position so that the integration over position space results in a factor V, the container volume, leaving

$$Z(\beta) = \frac{V}{\Omega}\int_v \exp\left(\frac{-\beta mv^2}{2}\right) d^3v \tag{15.43}$$

The spherical symmetry of the integrand allows us to integrate over spherical shells. Taking $d^3v = 4\pi v^2\, dv$ results in

$$4\pi\int_{v=0}^{\infty} \exp\left(\frac{-\beta mv^2}{2}\right) v^2\, dv = \left(\frac{2\pi}{m\beta}\right)^{3/2} \tag{15.44}$$

The partition function is then

$$Z(\beta) = \frac{V}{\Omega}\left(\frac{2\pi}{m\beta}\right)^{3/2} \equiv KV\beta^{-3/2} \tag{15.45}$$

Although the unit cell volume Ω appears in $Z(\beta)$ it drops out of the calculations which follow. The internal energy per atom now follows at once from (15.45)

$$\frac{U}{N} = -\frac{\partial}{\partial\beta}\left(-\frac{3}{2}\ln\beta - \ln V - \ln K\right) = \frac{3}{2}\frac{1}{\beta}$$

Thus as advertised previously

$$\frac{U}{N} = \frac{3}{2}\frac{1}{\beta} \tag{15.37}$$

However, the kinetic theory model of the ideal gas led to the result $U/N = (3/2)kT$, from which it follows that

$$\beta = \frac{1}{kT} \tag{15.27}$$

This establishes that $\beta = 1/kT$ for an ideal gas. However, $\beta = 1/kT$ in all cases where the Boltzmann distribution prevails. A proof of this statement is presented in Appendix B. The proof proceeds by showing that β has the same value for two systems in mutual thermodynamic equilibrium, regardless of their physical nature. Allowing one of the systems to be an ideal gas, together with the thermodynamic requirement that both systems have the same temperature establishes that $\beta = 1/kT$ for all systems.

••••••

15.8.2 A system with f degrees of freedom obeys the equipartition of energy theorem. Using the fact that

$$\frac{U}{N} = -\frac{\partial}{\partial\beta}\ln Z(\beta)$$

show that the partition function must be proportional to $T^{f/2}$.

••••••

We note in passing that, when energy is treated as a continuous variable, the discrete version of the Boltzmann distribution

$$N_i = \frac{NQ_i \exp(-\beta\varepsilon_i)}{Z(\beta)} \tag{15.33'}$$

is replaced by

$$dN(\varepsilon) = \frac{N\,dQ(\varepsilon)}{Z(\beta)}e^{-\beta\varepsilon} \tag{15.33a}$$

The quantity

$$\frac{dN(\varepsilon)}{dQ(\varepsilon)} = \frac{Ne^{-\beta\varepsilon}}{Z(\beta)} \tag{15.33b}$$

is the number of atoms per cell for phase-space cells corresponding to energies in the range ε to $\varepsilon + d\varepsilon$.

••••••

15.8.3 The expression for the partition function of an ideal monatomic gas given by (15.39) may also be written as

$$Z(\beta) = \int_{\varepsilon=0}^{\infty} e^{-\beta\varepsilon}\rho(\varepsilon)\,d\varepsilon$$

where $\rho(\varepsilon)$ gives the density of cells along the energy axis and is generally referred to as the density of states. The density of states is defined by the statement that $\rho(\varepsilon)\,d\varepsilon$ equals the number of phase-space cells with energies in the range ε to $\varepsilon + d\varepsilon$. Start from (15.43), use the relations $\varepsilon = \frac{1}{2}mv^2$ and $d^3v = 4\pi v^2\,dv$, and show that

$$Z(\beta) = \frac{4\pi V}{\Omega}\sqrt{\frac{2}{m^3}}\int_0^{\infty} e^{-\beta\varepsilon}\sqrt{\varepsilon}\,d\varepsilon$$

that is, that

$$\rho(\varepsilon) = \frac{2\pi V}{\Omega}\left(\frac{2}{m}\right)^{3/2}\sqrt{\varepsilon}$$

Evaluate the integral and verify that (15.45) again results. [*Hint:* It is necessary to form the differential of the relation $\varepsilon = \frac{1}{2}mv^2$ in order to relate $d\varepsilon$ and dv.]

•••••• ••••••

Example 15.2 Maxwellian Velocity Distribution Function

Chapter 12 promised a derivation of the Maxwellian velocity distribution law. With the results of Section 15.8 fresh at hand we can quickly fulfill that promise. From (15.33a) we have

$$dN(\varepsilon) = \frac{Ne^{-\beta\varepsilon}}{Z(\beta)}\,dQ(\varepsilon) \tag{15.33a}$$

as the number of atoms with energies in the range ε to $\varepsilon + d\varepsilon$. Next we consider a group of

$$dQ = \frac{V\,d^3v}{\Omega}$$

cells, each of whose energies differs only microscopically from $\varepsilon = \frac{1}{2}mv^2$.† Inserting the expression just obtained for $Z(\beta)$ gives (note that Ω cancels out)

$$dN = N\left(\frac{m}{2\pi kT}\right)^{3/2} \exp\left(\frac{-\beta mv^2}{2}\right) d^3v \tag{15.46}$$

The distribution function $f(v)$ is defined as the density of particles in velocity space

$$dN = f(v)\,d^3v \tag{12.3'}$$

Comparing (15.46) and (12.3′) shows that

$$f(v) = N\left(\frac{m}{2\pi kT}\right)^{3/2} \exp\left(\frac{-mv^2}{2kT}\right) \tag{12.5}$$

the promised Maxwellian distribution. ◇

15.9 THE STATISTICAL INTERPRETATION OF ENTROPY

Consider two systems, A and B, initially isolated and each separately in thermodynamic equilibrium at the same temperature T. We may suppose that the two systems are initially separated by a rigid adiabatic wall. Collectively A and B form a "super system" AB which is isolated and in thermal equilibrium.

---

† We have taken the coordinate space–volume element equal to the container volume ($d^3r = V$). Thus dQ refers to the total number of cells corresponding to velocities in d^3v.

Let $W(\text{A})$ and $W(\text{B})$ denote the thermodynamic probabilities for A and B, with $W(\text{A})$ the number of macrostates which correspond to the thermodynamic state of A; $W(\text{B})$ the number of macrostates which correspond to the thermodynamic state of B. With A and B isolated, the thermodynamic probability for the super system $W(\text{AB})$ is simply the product of $W(\text{A})$ and $W(\text{B})$

$$W(\text{AB}) = W(\text{A}) \cdot W(\text{B}) \tag{15.47}$$

This relation follows because A and B are isolated. They can independently exist in $W(\text{A})$ and $W(\text{B})$ macrostates so that $W(\text{A}) \cdot W(\text{B})$ macrostates are possible for the combined system.

Suppose that we now strip away the adiabatic coating from the wall separating A and B. No changes occur in the thermodynamic states of A or B because they have the same temperature. The state of the super system is likewise unchanged. It follows that the thermodynamic probability of the super system is still given by (15.47). The act of establishing thermal contact does not alter the number of macrostates open to A and B if they already have equal temperatures. Thus (15.47) holds whenever A and B are in mutual thermal equilibrium.

Next, we note that the extensive properties of the super system, its mass, internal energy, entropy, and so on, are additive. Thus

$$M(\text{AB}) = M(\text{A}) + M(\text{B}) \tag{15.48}$$

$$U(\text{AB}) = U(\text{A}) + U(\text{B}) \tag{15.49}$$

$$S(\text{AB}) = S(\text{A}) + S(\text{B}) \tag{15.50}$$

where M, U, and S denote mass, internal energy, and entropy, respectively. It is (15.50) which concerns us here. The entropy of a thermodynamic state is somehow related to the thermodynamic probability for that state. The functional form of the relation between entropy and thermodynamic probability must be such that both (15.47) and (15.50) are satisfied. It is readily verified that taking

$$S = C \ln W, \qquad C = \text{constant} \tag{15.51}$$

satisfies the prescribed conditions. Thus if we apply (15.51) to the super system and use (15.47), we find

$$S(\text{AB}) = C \ln W(\text{AB}) = C \ln W(\text{A}) + C \ln W(\text{B})$$

However, $C \ln W(\text{A}) = S(\text{A})$, and $C \ln W(\text{B}) = S(\text{B})$; it follows that

$$S(\text{AB}) = S(\text{A}) + S(\text{B}) \tag{15.50}$$

Thus a logarithmic relation leads to the required additive (extensive) property for entropy. The constant C is the same for all systems and must have the dimensions of entropy but is otherwise arbitrary at this point. We may fix its value by evaluating $\ln W$ for a specific system whose entropy is already known. The particular system which springs to mind is the ideal monatomic gas (I.G.) whose entropy is (for one mole)

$$S_{\text{I.G.}} = R(\ln V + \tfrac{3}{2}\ln T + \text{constant}) \tag{6.16'}$$

We must now evaluate $\ln W$, where

$$W = N! \prod_{i=1}^{M} \frac{Q_i^{N_i}}{N_i!} \tag{15.19'}$$

and where the most probable population of cells is given by the Boltzmann distribution

$$\frac{N_i}{Q_i} = \frac{N \exp(-\beta\varepsilon_i)}{Z(\beta)} \tag{15.33'}$$

Recall first that the logarithm of a product equals the sum of the logarithms of the factors. Thus

$$\ln W = \ln N! + \sum_{i=1}^{M} (N_i \ln Q_i - \ln N_i!)$$

Next we use the super Stirling approximation for $\ln N!$ and for each of the $\ln N_i!$. This gives

$$\ln N! = N \ln N - N$$

and

$$-\sum_{i=1}^{M} \ln N_i! = -\sum_{i=1}^{M} (N_i \ln N_i - N_i)$$

Taking account of the fact that $N = \sum_{i=1}^{M} N_i$, and that

$$\sum_{i=1}^{M} (N_i \ln Q_i - N_i \ln N_i) = -\sum_{i=1}^{M} N_i \ln \frac{N_i}{Q_i}$$

leaves us with

$$\ln W = N \ln N - \sum_{i=1}^{M} N_i \ln \frac{N_i}{Q_i}$$

For the most probable distribution (15.33)

$$\ln \frac{N_i}{Q_i} = \ln N - \ln Z(\beta) - \beta\varepsilon_i$$

Once again taking account of the fact that $N = \sum_{i=1}^{M} N_i$ shows that the $N \ln N$ term cancels, leaving

$$\ln W = N \ln Z(\beta) + \sum_{i=1}^{M} N_i \varepsilon_i \tag{15.52}$$

We recognize $\sum_{i=1}^{M} N_i \varepsilon_i$ as the internal energy U. Thus

$$\ln W = N \ln Z(\beta) + \beta U \tag{15.53}$$

It should be noted that this result for $\ln W$ is completely† general and is not restricted to the ideal gas. To evaluate $\ln W$ for the ideal gas we use (15.37) to find

$$\beta U = \tfrac{3}{2} N \qquad \text{(ideal gas)}$$

while (15.45) gives

$$N \ln Z(\beta) = N(\ln V + \tfrac{3}{2} \ln T + \text{constant})$$

Inserting these results into (15.53) and taking $N = N_0$ (since we deal with 1 mole of gas) gives

$$(C \ln W)_{\text{I.G.}} = C N_0 (\ln V + \tfrac{3}{2} \ln T + \text{constant}) \tag{15.51a}$$

Comparing this with (6.16′) shows that $CN_0 = R$ or

$$C = \frac{R}{N_0} = k = 1.38 \simeq 10^{-16} \ \frac{\text{erg}}{\text{K}} \tag{15.54}$$

Thus $C = k$, the Boltzmann constant, whereupon (15.51) becomes

$$S = k \ln W \tag{15.55}$$

This quantitative relation between entropy and thermodynamic probability unites two previous pronouncements concerning equilibrium, one thermodynamic the other statistical. To wit,

The state of maximum entropy corresponds to the thermodynamic equilibrium state;

and

The most probable macrostate corresponds to the thermodynamic equilibrium state.

The relation $S = k \ln W$ establishes the connection between entropy, a macroscopic quantity, and thermodynamic probability, a microscopic concept. In Chapter 6 we observed that an entropy increase was the result of processes which disordered or randomized a system. Our observations there helped to

† Provided we restrict ourselves to systems which are adequately described by classical physics.

establish a valuable but purely qualitative feeling for entropy. The key shortcoming was the lack of any quantitative definition of disorder. The relation $S = k \ln W$ provides the quantitative link between the concepts of entropy and disorder. Disorder is measured by $\ln W$. At least it is plausible and consistent to interpret $\ln W$ as a direct measure of disorder. For example, a highly-ordered distribution of atoms can be formed in relatively few ways, and therefore has a low thermodynamic probability, whereas the most probable distribution can be formed in the largest number of ways. The most probable distribution gives the greatest randomization, or mixing, consistent with the constraints.

•••••• ••••••

15.9.1 Show that

$$S(xy) = S(x) + S(y) \tag{15.50'}$$

has the solution

$$S(x) = C \ln x$$

by forming the partial derivatives of $S(xy)$ with respect to x and y. Proceed as follows: Observe that, with $z \equiv xy$, the chain rule leads to

$$\frac{\partial S(z)}{\partial x} = y\frac{\partial S(z)}{\partial z} \quad \text{and} \quad \frac{\partial S(z)}{\partial y} = x\frac{\partial S(z)}{\partial z}$$

so that

$$\frac{1}{y}\frac{\partial S(xy)}{\partial x} = \frac{1}{x}\frac{\partial S(xy)}{\partial y}$$

Having established this result, use (15.50) to show

$$h(x) \equiv x\frac{\partial S(x)}{\partial x} = y\frac{\partial S(y)}{\partial y} = h(y)$$

and remember that if $h(x) = h(y)$ for all values of two independent variables x and y, $h(x) = h(y) =$ constant.

•••••• ••••••

15.10 THERMODYNAMIC PROPERTIES: RELATIONSHIP TO THE PARTITION FUNCTION

In addition to supplying a conceptual link, (15.55) provides the basis for relating the partition function to the thermodynamic properties of a system. Using (15.53) in (15.55) gives a general relation between S, U, and $Z(\beta)$

$$S = k\beta U + kN \ln Z(\beta) \tag{15.56}$$

Multiplying through by T and noting that $kT\beta = 1$ results in

$$U - TS = -NkT \ln Z(\beta) \tag{15.57}$$

However, $U - TS$ is the Helmholtz free energy F, whence,

$$F = -NkT \ln Z(\beta) \tag{15.58}$$

We have previously established the relationship between $Z(\beta)$ and the internal energy U [compare (15.36)]. The Gibbs free energy G and the enthalpy H also may be expressed in terms of $Z(\beta)$.

······ ······

15.10.1 Verify that the Gibbs free energy G and the enthalpy H are related to the partition function by

$$\frac{G}{N} = kT\left[V\frac{\partial}{\partial V}\ln Z - \ln Z\right]$$

$$\frac{H}{N} = \frac{V}{\beta}\frac{\partial}{\partial V}\ln Z - \frac{\partial}{\partial \beta}\ln Z$$

······ ······

From thermodynamics it will be recalled that the equation of state emerges as one of the Maxwell relations

$$P = -\left(\frac{\partial F}{\partial V}\right)_T$$

Using (15.58) it follows that

$$P = \frac{NkT}{Z(\beta)}\left(\frac{\partial Z(\beta)}{\partial V}\right)_\beta \tag{15.59}$$

······ ······

15.10.2 Use (15.59) and the ideal gas partition function to obtain the equation of state

$$P = \frac{NkT}{V}$$

······ ······

The thermodynamic properties of a system, its internal energy, entropy, and so on, and its equation of state all follow from the partition function. To know $Z(\beta)$ is to know all at the thermodynamic level. There are enormous mathematical difficulties in evaluating $Z(\beta)$ for realistic models of specific physical systems. Failure to obtain agreement with experiment may be traced to some deficiency of the model because, as we have noted previously, our faith in the principles of statistical mechanics rests primarily on one supreme fact; the laws of thermodynamics may be derived from statistical mechanics. Of course the fun is in playing the game, that is, in applying statistical mechanics to specific physical systems. Proving that thermodynamics follows from statistical mechanics involves too much mathematical juggling to be classified as fun. The student interested in juggling should consult the references. The next chapter is devoted to the fun aspects of statistical mechanics.

15.11 SUMMARY

Several key concepts have been developed in this chapter. Among these are the notions of a microstate and a macrostate. A *microstate* specifies which atoms occupy each phase-space cell. A *macrostate* tells only how many atoms occupy each group of cells.

Our disbelief in Maxwell demons, that is, our refusal to accept the possibility that macroscopic constraints can control atomic motions, led to the postulate of equal *a priori* probability for all accessible microstates.

The *thermodynamic probability* was introduced, being defined as the number of microstates corresponding to the same macrostate. The most probable macrostate is the one having the largest thermodynamic probability.

The fact that thermodynamic equilibrium persists while a system moves through a succession of different macrostates was noted. This suggests that many different macrostates are equally valid representatives of the gross (macroscopic) behavior of a system. It proves mathematically feasible to determine the nature of the most probable macrostate. We therefore settle on it as *the* representative macrostate. That is, we assume that the thermodynamic equilibrium state has the same macroscopic properties as the most probable macrostate.

The most probable macrostate results when the members of the system have a Boltzmann distribution with respect to energy

$$N_i = \frac{NQ_i \exp\left(\dfrac{-\varepsilon_i}{kT}\right)}{Z(\beta)}$$

where $Z(\beta)$ is the partition function

$$Z(\beta) = \sum_{i=1}^{M} Q_i \exp(-\beta\varepsilon_i), \qquad \beta \equiv \frac{1}{kT}$$

The summation runs over groups of cells with different energies, Q_i being the number of cells with energy ε_i, and M denotes the total number of groups of cells.

The Boltzmann distribution reveals the "competition" for energy. The energy characteristic of the thermal motion of a particle is kT. Phase-space cells corresponding to energies much greater than kT are sparsely populated.

The entropy of a system is related to the thermodynamic probability by (15.55)

$$S = k \ln W$$

This relation reinforces our earlier observations which suggested that entropy measures the randomization or disorder of a system.

Finally, we have seen that the partition function $Z(\beta)$ is a key quantity, leading directly to the four thermodynamic potentials U, F, G, and H.

REFERENCES

Recommended for its clarity and wit is:

K. K. Darrow, Memorial to the Classical Statistics, *Bell Sys. Tech. J.* **22**, 108 (1943).

Most texts derive the Boltzmann distribution by using the method of Lagrangian multipliers. Methods intermediate between the one employed in Section 15.7 and the full Lagrangian treatment are given in the following three references:

R. Weinstock, On Introducing the Boltzmann Distribution, *Contemporary Phys.* **7**, 234 (1966).

D. Kleppner, Avoiding Lagrange Multipliers in Introductory Statistical Mechanics, *Am. J. Phys.* **36**, 843 (1968).

J. W. Lorimer, Elementary Statistical Mechanics without Lagrange Multipliers, *J. Chem. Educ.* **43**, 39 (1966).

A concise and coherent discussion of microstates, macrostates, and thermodynamic probability is given in:

F. W. Sears, "An Introduction to Thermodynamics, The Kinetic Theory of Gases, and Statistical Mechanics," 2nd ed., Chapter 14. Addison-Wesley, Reading, Massachusetts (1953).

Chapter 16

Applications of Statistical Mechanics

In this chapter the methods of statistical mechanics are applied to three rather diverse topics. At places the discussion is just a notch above the qualitative level. The aim of the chapter is threefold; to impress the student with the scope and power of statistical mechanics, to reinforce the ideas and formalism presented in the preceding chapter, and to pave the way for the presentation of quantum statistical mechanics in the next chapter.

In Section 16.1 the microscopic basis for the Clausius–Clapeyron equation is revealed. In Section 16.2 several aspects of chemical reactions are studied from the viewpoint of statistical mechanics. The chemist's "ten-degree rule"—which states that the rate of a reaction is doubled by a ten-degree increase in temperature—is shown to have a statistical mechanical basis. Finally, in Section 16.3 the statistical character of paramagnetism is studied and the origin of Curie's law uncovered.

16.1 MICROSCOPIC BASIS OF THE CLAUSIUS–CLAPEYRON EQUATION

The Clausius–Clapeyron equation was derived in an earlier chapter by considering a special sort of Carnot cycle. For a liquid–vapor transition, the slope of the vapor pressure versus temperature curve is given by the

Clausius–Clapeyron equation in the form

$$\frac{dP_v}{dT} = \frac{L_v}{T(V_v - V_l)} \simeq \frac{L_v}{TV_v} \tag{16.1}$$

where P_v is the vapor pressure, T is the absolute temperature, L_v is the molar latent heat of vaporization, and V_v and V_l are the molar volumes of vapor and liquid. Because of the large expansion in the liquid-to-vapor transition $V_v \gg V_l$, and the approximation in (16.1) is generally quite satisfactory. We now want to see inside the Clausius–Clapeyron equation—to understand it at the molecular level. Statistical mechanics can furnish such insight.

Consider a container, partially filled with liquid, and let N_v and N_l denote the numbers of molecules in equal volumes of vapor and liquid (see Figure 16.1). If the vapor is not too dense, the molecules in it will behave like those of

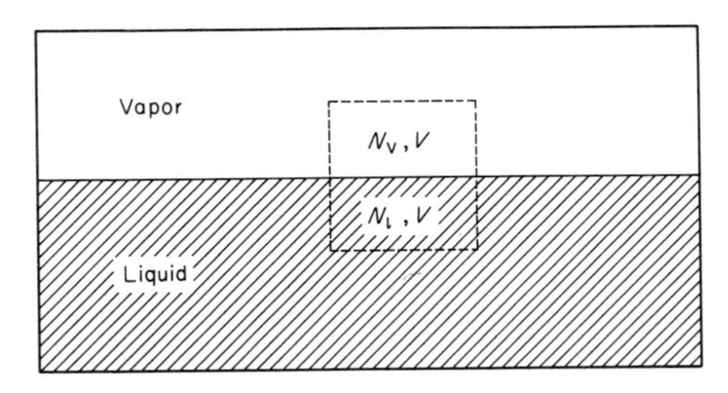

Figure 16.1. The liquid and vapor are in equilibrium at temperature T.

an ideal gas. The characteristic energy of vapor molecules is then $\frac{1}{2}kT$ per degree of freedom. The molecules in the liquid fall much lower on the energy scale. Strong intermolecular forces cause them to cling together in a configuration whose average energy per molecule is largely in the form of potential energy. This potential energy is negative and its magnitude is much greater than the mean kinetic energy of molecular motion. For these reasons we can write for the energies characterizing the vapor and liquid states as

$$\varepsilon_v \simeq kT$$

$$\varepsilon_l \simeq -E, \qquad E \gg kT$$

If we now regard the vapor and liquid as two different equilibrium states of a system characterized by energies ε_v and ε_l, the Boltzmann distribution law sets the ratio of populations at

$$\frac{N_v}{N_l} = \frac{\exp(-\varepsilon_v/kT)}{\exp(-\varepsilon_l/kT)} \tag{16.2}$$

Inserting the energies representative of liquid and vapor and noting that $E \gg kT$ produces

$$\frac{N_v}{N_l} \simeq e^{-E/kT} \tag{16.3}$$

The competition for thermal energy, as described by the Boltzmann distribution, is the origin of the exponential temperature dependence of the vapor pressure which we found in Chapter 5. Equation (16.3) leads directly to the Clausius–Clapeyron equation and in doing so adds support to the plausible notion that the latent heat of vaporization is the energy necessary to break the intermolecular bonds and thereby convert one mole of liquid into one mole of vapor. It must be confessed that (16.3) is not a rigorous application of the Boltzmann formula because we have used the average energies of two groups of particles whose individual energies are spread over fairly wide ranges. However, because the average energies differ so markedly, no essential error results if we consider the vapor and liquid as two different energy states of a single system. Multiplying both sides of (16.3) by $N_l kT/V$ gives

$$\frac{N_v kT}{V} = \frac{N_l kT}{V} e^{-E/kT} \tag{16.4}$$

If we recall the ideal gas equation of state in the form $P = NkT/V$, we recognize the left side as the vapor pressure. Thus

$$P_v = \frac{N_l kT}{V} e^{-E/kT} \tag{16.5}$$

The factor N_l/V is the number density for the liquid, a quantity which is virtually independent of temperature. To convert this into the Clausius–Clapeyron form we differentiate P_v with respect to T, ignoring any slight temperature dependence of the liquid density. The result is

$$\frac{dP_v}{dT} = \frac{P_v}{T}\left(1 + \frac{E}{kT}\right)$$

In view of the fact that $E \gg kT$ we can take this to be

$$\frac{dP_v}{dT} = \frac{P_v}{T}\frac{E}{kT}$$

Finally, if we note again that $P_v/kT = N_v/V$, we have

$$\frac{dP_v}{dT} = \frac{N_v E}{TV} \tag{16.6}$$

To compare this with the Clausius–Clapeyron equation, let N_v and V refer to 1 mole of vapor

$$N_v = N_0, \qquad \text{Avogadro's number}$$

$$V = V_v, \qquad \text{molar volume of vapor}$$

Then

$$\frac{dP_v}{dT} = \frac{N_0 E}{TV_v} \tag{16.7}$$

which concurs with the Clausius–Clapeyron equation (16.1) provided we identify $N_0 E$ with L_v, the latent heat. This is precisely how we should interpret $N_0 E$, for we originally argued that E was the difference between the average energies of a molecule in the vapor and liquid phase. Thus $N_0 E$ is the difference in energy per mole—the energy we must add to 1 mole of liquid to convert it into vapor at the same temperature. With

$$N_0 E = L_v$$

(16.7) assumes the form of the Clausius–Clapeyron equation.

•••••• ••••••

16.1.1 The relation $N_0 E = L_v$ enables us to check the validity of the assumption $E \gg kT$.

(a) Determine the ratio E/kT_B for the liquids listed in the accompanying tabulation. (T_B is the normal boiling point.) Note that L_v is expressed in calories per mole.

(b) Evaluate E and kT_B for water and hydrogen. Express the energies in electron volts.

Substance	L_v (cal/mole)	T_B (K)
Water	9717	373
Ethyl ether	6870	308
Ammonia	5560	240
Hydrogen sulfide	4490	213
Oxygen	1610	90
Hydrogen	216	20

•••••• ••••••

From (16.5) it is evident that the vapor pressure falls quite rapidly with temperature. From our inside vantage point we can interpret this as a direct consequence of the exponential Boltzmann factor, which favors states of low energy. In this instance it is the liquid state with its negative potential energy configurations which is favored over the vapor state in which the molecules have modest, but positive energies.

The Boltzmann factor is in evidence whenever there is competition for energy at the atomic-molecular level—it is the recurring theme of statistical physics.

16.2 CHEMICAL REACTIONS

In this section three aspects of chemical reactions are discussed from the statistical mechanical viewpoint. First the temperature dependence of chemical equilibrium constants is investigated. This is followed by a discussion of activation energies. Finally, an empirical rule of thumb—the ten-degree rule—is shown to have a statistical mechanical basis.

A Equilibrium Constants

We select as a model reaction the formation and dissociation of a diatomic molecule AB from atoms A and B

$$A + B \rightleftharpoons AB \tag{16.8}$$

The formation of an AB molecule occurs whenever A and B atoms collide under favorable conditions. Dissociation results when AB molecules acquire energy enough to tear themselves apart. Figure 16.2 suggests the energy level scheme for the model reaction. The energy E is the dissociation energy of the

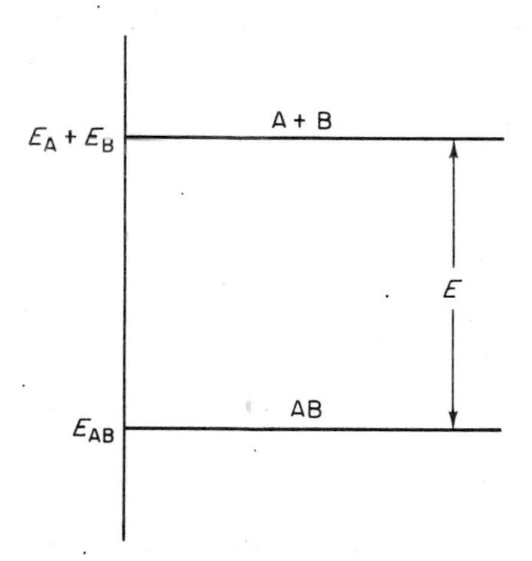

Figure 16.2. An energy level scheme for the reaction $A+B \rightleftharpoons AB$. Adding energy E to the molecule AB causes its dissociation into the atoms A and B.

molecule, the minimum energy which must be acquired by the molecule in order for it to break up into its component atoms A and B. The value of E is a characteristic of the particular reaction under study, not some universal constant. However, many such energies fall in the electron volt (eV) range

$$1 \text{ eV} = 1.6 \times 10^{-12} \text{ erg} = 3.8 \times 10^{-20} \text{ cal}$$

••••••

16.2.1 The Coulomb potential energy of two charges, q_1 and q_2, separated by a distance r is $q_1 q_2/r$. The characteristic atomic charge is the electronic charge, $e = 4.8 \times 10^{-10}$ esu, and the characteristic atomic dimension is 1 Å = 10^{-8} cm.

(a) Evaluate the corresponding characteristic Coulomb energy in electron volts and calories (1 erg = 1 esu^2/cm).

(b) Can you suggest why molecular dissociation energies are smaller than the corresponding atomic ionization energies? (To cite a typical example, it requires about 4.7 eV to dissociate the hydrogen molecule, but nearly 13.6 eV to ionize the hydrogen atom.)

••••••

Let n_A, n_B, n_{AB} denote the concentrations (number densities) of the reaction participants. If n_A were doubled, the probability of "A meets B" would be doubled and hence the rate of molecule-forming collisions would also be doubled. Likewise, doubling n_B doubles the chance for an encounter between A and B atoms thereby doubling the rate of AB formation. The rate of AB production is therefore proportional to the product of the concentrations $n_A n_B$. The rate of formation is also proportional to the mean relative speed at which A and B atoms encounter each other, which in turn, is proportional to the square root of the absolute temperature. However, there are other temperature-dependent factors affecting the rate of formation. These additional factors depend strongly on the nature of A and B as well as the temperature; they are not universal factors. If we lump together all of the temperature-dependent factors into a reaction constant $K_f(T)$ we can write the rate of formation as

$$R_f = K_f(T) n_A n_B \tag{16.9}$$

The rate of dissociation depends on three quantities: (a) the number of AB molecules present, (b) the probability that a molecule has acquired enough energy to dissociate, and (c) the probability that dissociation occurs, given that the needed energy is available. Only if an AB molecule has sufficient energy—at least E—can dissociation proceed. The competition for energy does not favor dissociation. The probability that a molecule has acquired the dissociation energy E is proportional to the Boltzmann factor $e^{-E/kT}$. The total number of molecules with energy enough to dissociate is therefore proportional to $n_{\mathrm{AB}} e^{-E/kT}$, and the rate of dissociation can be written as

$$R_{\mathrm{d}} = K_{\mathrm{d}}(T) n_{\mathrm{AB}} e^{-E/kT} \tag{16.10}$$

Our ignorance of the third factor cited above is buried in the dissociation constant $K_{\mathrm{d}}(T)$, which, like $K_{\mathrm{f}}(T)$, depends on the specific nature of the reaction.

Chemical equilibrium prevails when the rates of formation and dissociation are equal, $R_{\mathrm{f}} = R_{\mathrm{d}}$. From (16.8) and (16.10) it follows that, at equilibrium,

$$\frac{n_{\mathrm{A}} n_{\mathrm{B}}}{n_{\mathrm{AB}}} = \frac{K_{\mathrm{d}}(T)}{K_{\mathrm{f}}(T)} e^{-E/kT} \equiv K(T) \tag{16.11}$$

The form of the ratio $K_{\mathrm{d}}(T)/K_{\mathrm{f}}(T)$ depends strongly on the details of the chemical reaction. However, as we shall see, the crucial temperature dependence of the equilibrium constant $K(T)$ resides in the Boltzmann factor, which is universally present,

$$K(T) = \frac{K_{\mathrm{d}}(T)}{K_{\mathrm{f}}(T)} e^{-E/kT} \tag{16.12}$$

Let us summarize the development to this point; the equilibrium constant $K(T)$ relates the equilibrium concentrations n_{A}, n_{B}, and n_{AB} of the principals in the reaction

$$\mathrm{A} + \mathrm{B} \rightleftharpoons \mathrm{AB}$$

By equating the rates of molecular formation and dissociation at equilibrium we arrived at (16.12), showing that the equilibrium constant $K(T)$ has the form of a product. One factor, the ratio $K_{\mathrm{d}}(T)/K_{\mathrm{f}}(T)$, has a gradual or weak temperature dependence. The decisive importance of the universal Boltzmann factor $e^{-E/kT}$ remains to be demonstrated (compare, however, Problem 16.2.3).

16.2.2 In the molecular reaction

$$\mathrm{AB} + \mathrm{CD} \rightleftharpoons \mathrm{AC} + \mathrm{BD}$$

an amount of energy E is released (per reaction) when AB and CD combine to give AC and BD.

(a) Show that the equilibrium concentrations of the four constituents are related by an equation of the form

$$K(T) = \frac{n_{\mathrm{AB}} n_{\mathrm{CD}}}{n_{\mathrm{AC}} n_{\mathrm{BD}}}$$

(b) $K(T)$ contains an exponential Boltzmann factor. Explain its origin.

16.2.3 Consider the function

$$K(x) = xe^{-30/x}$$

over the interval from $x = 3$ to $x = 7$. Tabulate or plot $K(x)$ and the approximations

$$K_1(x) = \langle x \rangle e^{-30/x}$$

$$K_2(x) = xe^{-30/\langle x \rangle}$$

where $\langle x \rangle$ is the arithmetic average of x over the interval. Which approximation is the more faithful?

•••••• ••••••

B Activation Energies

With reference to our model reaction

$$A + B \rightleftharpoons AB$$

the formation of a stable molecule is said to be energetically possible, or favored. By forming a molecule, the atoms decrease their mutual potential energy. The energy released shows up largely as the heat of reaction. Frequently situations arise wherein an energetically possible reaction does not go or else proceeds at an extremely slow pace. The formation of water vapor according to the scheme

$$2H_2 + O_2 \rightleftharpoons 2H_2O$$

is a well-known example. If hydrogen and oxygen are mixed at ordinary temperatures, the above reaction proceeds so slowly as to be imperceptible. However, if an electric spark or a suitable catalyst is introduced, the reaction rate can jump to explosive proportions. Similar situations occur frequently in organic reactions. Often it is necessary for the reactants to form an intricate geometrical configuration—an *activated complex* as it is called—before they can reach a final product form. A catalyst speeds up the reaction by aiding the formation of activated complexes. For example, enzymes apparently serve only as templates in biological reactions. They help the large and clumsy biological molecules maneuver into activated complexes.

In order to form an activated complex the reactants must acquire a certain activation energy. The catalyst loans energy to the reactants and is repaid as the complex breaks up to form the reaction products. Mechanically the situation is analogous to the one shown in Figure 16.3a. In order to reach the position of lower potential energy the skier must first increase his potential energy by surmounting the hill. He accomplishes this by doing an amount of work mgH against the force of gravity. Energetically, the activated complex is

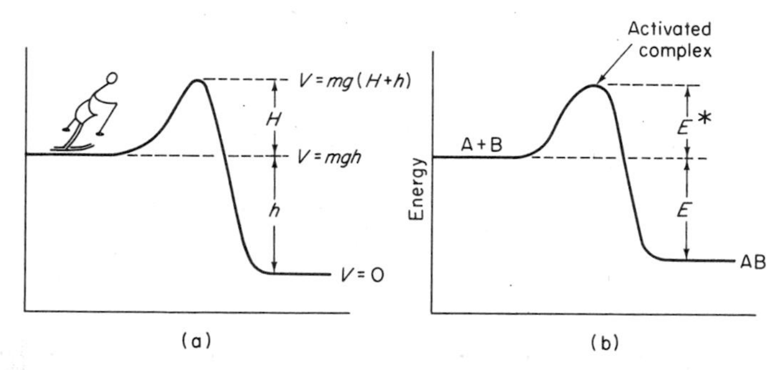

Figure 16.3. (a) To reach the lower level, the skier must first reach the hillcrest; this, in turn, requires that he increase his potential energy. (b) The activation energy E^* is the energy needed by A and B to form an activated complex.

a hilltop, as suggested by Figure 16.3b. The reacting atoms A and B must acquire an activation energy E^* in order to form a complex from which they can proceed to the final product state. In many cases the activation energy is much larger than the characteristic thermal energy kT. Only rarely then will the kinetic energies of the reacting atoms be as large as E^*, and the reaction rate will be small unless a catalyst lends a hand—and some energy.

Our earlier analysis is readily modified to accommodate the effects of an activation energy. The formation rate given by (16.9) must be altered by multiplying the right side by the probability that A and B possess the activation energy E^*. This probability is proportional to the Boltzmann factor $e^{-E^*/kT}$. Thus the rate of formation can be written as

$$R_f = K_f n_A n_B e^{-E^*/kT}$$

The rate of dissociation is given by (16.10), but with $E+E^*$ replacing E. For an AB molecule to break up it must first revert to the activated complex form and this demands an energy $E+E^*$

$$R_d = K_d n_{AB} e^{-(E+E^*)/kT}$$

Equating the rates of formation and dissociation shows that the equilibrium concentrations are still related by (16.11) since the common factor $e^{-E^*/kT}$ cancels. So, the existence of an activation energy has no bearing on the equilibrium concentrations of the constituents. It can dramatically affect the *rate* at which equilibrium is reached. For example, suppose that E^* is 1 eV and that T is 400 K. Then kT is roughly (1/30) eV and

$$e^{-E^*/kT} = e^{-30} \simeq 10^{-13}$$

Under these circumstances the rate of formation is 10^{-13} below what it would be in the absence of any activation energy demands—a dramatic reduction indeed.

••••••

16.2.4 Assume a Maxwellian distribution of atoms at temperature T and determine the fraction with kinetic energies in excess of $30kT$. [*Hint:* Refer to Section 12.5 and to Figure 12.6 in particular.]

••••••

In the final segment of this section we investigate how the reaction rate changes with temperature; in doing so we uncover the basis for the chemist's ten-degree rule.

C The Ten-Degree Rule

Chemists and biologists have long used a rule of thumb which states that the rate of a chemical reaction doubles with a 10 °C increase in temperature. The figure of 10 °C is a rough approximate value; the precise change in temperature required to double the rate of a reaction depends on the nature of the compounds involved and on the temperature itself. The basis for the rule stems from the fact that the characteristic activation energies associated with a great many chemical reactions are of roughly the same magnitude—namely, about one electron volt (1 eV) on a per-molecule basis.

In the chemical reaction

$$\mathrm{A} + \mathrm{B} \rightleftharpoons \mathrm{AB}$$

the rate at which A and B atoms unite to form AB molecules can be expressed as

$$R = K_{\mathrm{f}} n_{\mathrm{A}} n_{\mathrm{B}} e^{-E^*/kT}$$

wherein the decisive temperature dependence of R is contained in the exponential $e^{-E^*/kT}$. At a typical cooking temperature of 400 K, the characteristic activation energy is roughly 1 eV, and E^*/kT is in the neighborhood of 30. As we now demonstrate, a small change in temperature can change $e^{-E^*/kT}$ by a sizeable factor. So—we ignore the generally weak temperature dependence of K_{f} and focus on the exponential. If the temperature changes from T to $T+\delta T$, the exponent changes to

$$\frac{E^*}{kT\left(1 + \dfrac{\delta T}{T}\right)} \simeq \frac{E^*}{kT}\left(1 - \frac{\delta T}{T}\right)$$

The change in the exponent is seen to be

$$\frac{E^*}{kT}\frac{\delta T}{T} \simeq 30\frac{\delta T}{T}$$

Observe that a 1% change in T gives a 30% change in the exponent. The reaction rate at the temperature $T+\delta T$ may be expressed as

$$R(T+\delta T) = R(T)\exp\left(+\frac{E^*}{kT}\frac{\delta T}{T}\right) \tag{16.13}$$

It is evident how a small fractional change in the temperature can produce a sizable change in the exponent—and thence the reaction rate. The fractional change $\delta T/T$ is multiplied by a whopping $E^*/kT \simeq 30$. To make $R(T+\delta T)$ twice $R(T)$ the exponent in (16.13) must equal $\ln 2 = 0.69$. With $E^*/kT = 30$, and $T = 400$ K, the reaction rate doubles when $\delta T = 9\,°C$. Here then is the origin of the ten-degree rule. The exponential factor which governs the availability of activation energy is extremely sensitive to slight changes in temperature.

The ten-degree rule has some bearing on the existence of life. A drop in temperature reduces the rates of chemical reactions. Each reaction in the complicated schemes of biological oxidation and photosynthesis is similarly affected. As the temperature falls, the reaction rates diminish until life becomes dormant or ceases.

A healthy warm-blooded animal maintains a constant body temperature. A cold-blooded creature assumes the temperature of its surroundings. The built-in thermostats which human beings and other warm-blooded animals enjoy is of the greatest importance. Imagine what life would be like if man were cold-blooded. Going from a heated room at 72 °F to a frigid 0 °F surroundings would result in a 40 °C change of body temperature. The ten-degree rule advises that a 40° temperature drop depresses the biological reaction rates by a factor of 16. Something between death and monumental lethargy would surely follow.

A temperature rise increases the activation energy reservoir; a larger fraction of the molecules acquire the needed activation energy. However, two other effects of high temperature prove disastrous for life. As the temperature rises there is an increase in the kinetic energies of the long chain molecules which typify living systems (for example, the DNA molecule). The entire fabric of the living system jiggles and squirms more violently. Some of the molecules literally shake themselves to pieces. More important is the fact that the enzymes, the biological catalysts, gradually lose their shapes. Unable to provide a faithful template for the formation of activated complexes, their biological function ends, and life ceases.

•••••• ••••••

16.2.5 The ten-degree rule is only an approximate rule of thumb. The value of the temperature change needed to double the reaction rate depends on the values of E^* and T. Determine the approximate value of δT required to double the reaction rate when

(a) $E^* = 1$ eV, $T = 300$ K,
(b) $E^* = 2$ eV, $T = 400$ K.

16.2.6 How would measurements of the reaction rate and the temperature permit the activation energy to be determined? [*Hint:* Consider the shape of a $\ln R$ versus $1/kT$ plot of the data.]

•••••• ••••••

16.3 THE LANGEVIN THEORY OF PARAMAGNETISM

The response of a magnetic material to an applied magnetic field is measured by the magnetization $\mathcal{M}$, the magnetic dipole moment per unit volume. The total magnetic moment of a substance $\mathbf{M}$ is the vector sum of the moments of individual atoms or ions. An applied magnetic field tends to align individual moments while thermal agitation promotes random orientations. We anticipate that this competition will show itself in the form of an appropriate Boltzmann factor.

Let $\mathbf{H}$ denote the external magnetic field applied, and let $\mathcal{M} = \mathbf{M}/V$ denote the resulting magnetization. The magnetic properties are conveniently classified in terms of the susceptibility χ

$$\mathcal{M} = \chi \mathbf{H} \tag{16.14}$$

The susceptibility of a *diamagnetic* material is negative and temperature independent. The atoms of a diamagnetic substance do not have permanent magnetic moments. The applied field induces a magnetic moment in accord with Lenz's law. That is, the response of the atomic electrons produces a current whose magnetic field opposes the applied field. Diamagnetic effects are swamped by *paramagnetism* in materials whose atoms or molecules have permanent magnetic dipole moments. The dipoles tend to align in a way which augments the applied field. The susceptibility of a paramagnetic material is positive and temperature-dependent. At room temperature χ is roughly 10^{-2}. The paramagnetic susceptibility varies approximately as $1/T$ at ordinary temperatures. Pierre Curie parlayed this fact into what is now known as Curie's law

$$\chi = \frac{C}{T} \tag{16.15}$$

where C, called the Curie constant, is not a universal constant, but is different for each paramagnetic material. Curie's analysis was strictly empirical and his expression for χ is inadequate at a sufficiently low temperature.

In 1905, P. Langevin advanced a theory of paramagnetism. It is essentially the Langevin theory which we present here. A limiting form of the Langevin theory leads to Curie's law. Our goal is to understand the microscopic basis of the Langevin theory, to derive Curie's law from it, and to speculate as to why it requires modification at low temperatures.

The particles responsible for the paramagnetic behavior of a material may be atoms, or molecules, or even ions. Each such particle has a permanent magnetic dipole moment $\boldsymbol{\mu}$. As a result, each particle behaves like a tiny bar magnet. Because each particle (atom, molecule, ion) for a given material is identical, the size of the dipole moment μ is the same for all. The atomic structure of the particle determines the size of μ. When an external magnetic field is applied the magnetic dipoles experience a torque

$$\boldsymbol{\tau} = \boldsymbol{\mu} \times \mathbf{H} \tag{16.16}$$

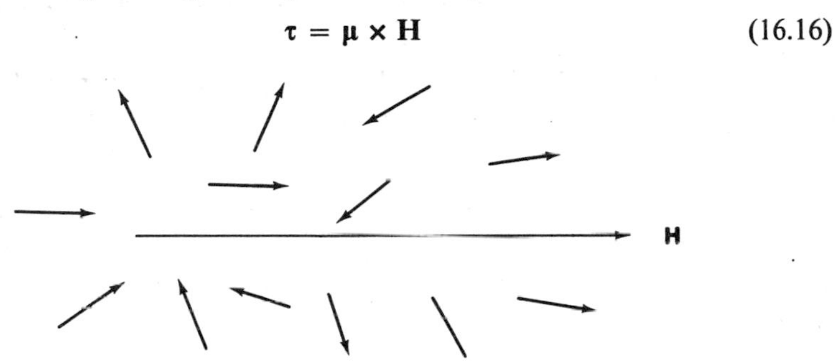

Figure 16.4. The z component of the magnetic dipole moment varies. Orientations with $\boldsymbol{\mu}$ parallel to $\mathbf{H}$ are favored as they correspond to the lowest energy.

This torque tends to rotate the dipole moments into a stable, torque-free position, which is to say, it tends to line up the magnetic moments in the direction of the applied field. Such ordering is only partially successful because of thermal agitation. Thermal motions tend to randomize the orientation of the moments. In the absence of an external field such randomization is complete and the vector sum of the magnetic moments is zero.† So, we have competing effects. Thermal energy tends to randomize; the external field tends to order. Whenever there is a competition for energy the Boltzmann factor appears.

To calculate the magnetization $\mathscr{M}$ and thence the susceptibility χ, we proceed as follows: Take the z axis along the direction of $\mathbf{H}$. The potential energy of a permanent magnetic dipole $\boldsymbol{\mu}$ in the field $\mathbf{H}$ is

$$\varepsilon = -\boldsymbol{\mu} \cdot \mathbf{H} = -\mu_z H \tag{16.17}$$

Although μ is the same for each of the particles the z component of $\boldsymbol{\mu}$ will vary (see Figure 16.4). From (16.17) we see that the favored orientation—the one

† Because individual magnetic moments frequently have their orientation altered it is more precise to say that the vector sum fluctuates about an average value of zero.

of lowest energy—has **μ** and **H** parallel

$$\varepsilon_{\text{min}} = -\mu H$$

whereas the configuration of maximum energy has **μ** antiparallel to **H**,

$$\varepsilon_{\text{max}} = +\mu H$$

The probability that the z component of the dipole moment lies in the range μ_z to $\mu_z + d\mu_z$ is proportional to $d\mu_z$ and to the Boltzmann factor

$$\exp\left(\frac{-\varepsilon}{kT}\right) = \exp\left(\frac{+\mu_z H}{kT}\right)$$

Thus

$$dP = A\exp(\alpha\mu_z)\,d\mu_z, \qquad \alpha = \frac{H}{kT} \tag{16.18}$$

••••••

16.3.1 Show that

$$A = \frac{\alpha}{e^{\mu\alpha} - e^{-\mu\alpha}}$$

follows from the requirement that $\int_{-\mu}^{+\mu} dP = 1$. What is the basis for imposing this requirement?

••••••

The mean value of μ_z is

$$\langle\mu_z\rangle = \int_{-\mu}^{+\mu} \mu_z\,dP$$

The result may be expressed as

$$\langle\mu_z\rangle = \mu\left[\coth(\mu\alpha) - \frac{1}{\mu\alpha}\right] \tag{16.19}$$

where

$$\coth(x) = \frac{e^x + e^{-x}}{e^x - e^{-x}} \tag{16.20}$$

is the hyperbolic cotangent function. The total magnetic moment of a material containing N such dipoles is $N\langle\mu_z\rangle$. The magnetization is just the dipole moment per unit volume $N\langle\mu_z\rangle/V$. Thus

$$\mathscr{M} = \frac{N\mu}{V}\left(\coth(\mu\alpha) - \frac{1}{\mu\alpha}\right) \tag{16.21}$$

The function

$$\mathscr{L}(x) \equiv \coth(x) - \frac{1}{x} \tag{16.22}$$

is termed the *Langevin function*. Under ordinary conditions, the variable

$$x = \mu\alpha = \frac{\mu H}{kT} \tag{16.23}$$

is generally much smaller than unity. For example, the magnetic moment has a magnitude of roughly one Bohr magneton† μ_B

$$\mu_B \equiv \frac{e\hbar}{2mc} = 0.927 \times 10^{-20} \ \frac{\text{erg}}{\text{G}} \tag{16.24}$$

With $H = 1000$ G and $T = 300$ K,

$$\frac{\mu_B H}{kT} \simeq 2 \times 10^{-4}$$

Note that x is the ratio of two characteristic energies; μH is the energy associated with the magnetic ordering, and kT is the characteristic thermal energy. The fact that $x = \mu H/kT$ is small at ordinary temperatures means that the magnetic ordering is weak or incomplete.

To obtain Curie's law we make use of the fact that x is small and use an approximate form of the Langevin function (see Problem 16.3.2)

$$\mathscr{L}(x) \simeq \frac{x}{3}, \qquad x \ll 1$$

Using this approximation the Langevin function in (16.21) gives

$$\mathscr{M} = \frac{N\mu^2 H}{3VkT} \tag{16.25}$$

Since $\mathscr{M} = \chi H$ defines the susceptibility we have

$$\chi = \frac{N\mu^2}{3VkT} \tag{16.26}$$

which is Curie's law, the Curie constant being given by

$$C = \frac{N\mu^2}{3Vk} \tag{16.27}$$

† The Bohr magneton is the characteristic atomic unit of magnetic dipole moment. In (16.24), e and m denote the electron charge and mass, c is the speed of light, and $\hbar$ is Planck's constant divided by 2π.

The values of χ are generally reported on a molar basis. For 1 mole we take $N = N_0$, Avogadro's number. Then V is the volume of 1 mole, and may be expressed in terms of the mass density ρ and molecular weight M through the relation $\rho = M/V$. The molar value of χ predicted by the Langevin analysis is

$$\chi_{\text{mol}} = \frac{\rho N_0 \mu^2}{3MkT} \tag{16.28}$$

••••• •••••

16.3.2 Show that the Langevin function

$$\mathscr{L}(x) = \coth(x) - \frac{1}{x}$$

has the limiting forms

(a) $\mathscr{L}(x) \simeq (1/3)x, \quad x \ll 1,$
(b) $\mathscr{L}(x) \simeq 1 - (1/x), \quad x \gg 1.$

For part (a) it is recommended that the power series expansions

$$e^{\pm x} = 1 \pm x + \tfrac{1}{2}x^2 \pm \tfrac{1}{6}x^3 + \cdots$$

be employed to find the approximate form of

$$\coth(x) = \frac{e^x + e^{-x}}{e^x - e^{-x}}$$

for small x.

16.3.3 Praseodymium is a rare-earth element of atomic weight 140.9 and density 6.78 gm/cm^3. Its magnetic susceptibility is 5.01×10^{-3} cgs units/mole, at a temperature of 293 K. Determine the magnetic moment of the Praseodymium atom. Express your result in units of the Bohr magneton μ_B.

••••• •••••

Langevin's classical theory has been superceded by Brillouin's quantum theory of paramagnetism. The physical pictures are much the same for both the classical and quantum theories. Experimentally, very low temperatures and very strong magnetic fields are needed to distinguish between the predictions of the Langevin and Brillouin theories. A strong field and a low temperature cooperate to produce a sizable value of the parameter $\mu H/kT$. When μH becomes comparable to kT, the effects of quantization become discernible and Brillouin's quantum theory of paramagnetism is required to obtain agreement with experiment.

••••• •••••

16.3.4 Quantum effects become important in paramagnetic materials when the parameter $\mu H/kT$ reaches (or exceeds) values near unity. Assuming that μ is roughly 1 Bohr magneton, what magnetic field strengths are needed to study quantum paramagnetism at

(a) liquid helium temperature (4 K)?
(b) liquid nitrogen temperature (80 K)?
(c) room temperature (300 K)?

Are the required magnetic fields presently attainable?

••••• •••••

REFERENCES

The thermodynamic and kinetic bases for the Clausius–Clapeyron equation are compared in:

R. P. Feynman, R. B. Leighton, and M. Sands, "The Feynman Lectures on Physics," Volume 1, Lecture 45. Addison-Wesley, Reading, Massachusetts (1963).

The Langevin theory of paramagnetism is developed in:

F. W. Sears, "An Introduction to Thermodynamics, The Kinetic Theory of Gases, and Statistical Mechanics," 2nd ed., Chapter 15. Addison-Wesley, Reading, Massachusetts (1953).

Chapter 17

Quantum Physics and Quantum Statistics

17.1 INTRODUCTION

The microscopic world is the realm of quantum phenomena. At the microscopic level the fundamental properties of matter are found in packages which are always integral multiples of some basic unit. For example, electric charge is *quantized*—it comes in discrete quantities. The smallest unit of electric charge has the value†

$$e = 4.803 \times 10^{-10} \text{ esu}$$

This is the charge carried by a proton, for example. A charge of equal magnitude but opposite sign resides on the electron. All distributions of electric charge are collections of these elementary quanta of electric charge.

The discrete nature of mass is also a familiar but perhaps unappreciated aspect of quantization. At the macroscopic level mass distributions appear continuous. On the microscopic level we know that matter is granular—we believe in atoms. Each atom has a characteristic mass, as does each of its component parts, the electrons, protons, and neutrons. So, when we speak of a quantity as being quantized, we mean that it is observed in nature only in integral multiples of some basic unit.

† In the mks system, $e = 1.609 \times 10^{-19}$ C.

Modern atomic and nuclear physics have revealed several other quantal features of nature. Of special significance for statistical mechanics is the quantum nature of angular momentum and energy. To develop a feeling for certain aspects of quantum theory we trace its early history in the next section. The student who wishes to recapture a sense of the excitement, the controversy, and the revitalization of physics which followed Max Planck's introduction of the quantum of action will find the articles by Martin J. Klein (1966) and K. K. Darrow (1952) most enlightening.

17.2 THE ORIGINS OF QUANTUM THEORY

Toward the end of the 19th century, several inadequacies of classical physics became apparent. Among these shortcomings was the failure of the equipartition of energy theorem to correctly describe the temperature variations in the heat capacity of diatomic gases. Nor did the specific heats of good electrical conductors follow the equipartition theorem. For some inexplicable reason the mobile electrons were able to transport energy and electric charge without making a detectable contribution to the specific heat. Gibbs' paradox concerning the entropy of mixing was another warning that all was not right with classical statistical physics.

Quantum theory was born, rather hesitantly, with Max Planck's investigation of still another riddle of nature—the blackbody radiation problem. In general, electromagnetic radiation reaching a material surface is partially absorbed and partially reflected. A blackbody is one which completely absorbs any electromagnetic radiation reaching its surface—a perfect absorber. Blackbody radiation may be studied experimentally using an electric furnace whose inner walls can be held at a constant temperature. A small hole in one wall will permit a tiny fraction of the thermal radiation to escape and be sampled. Any light which travels inward through the hole will almost certainly be absorbed before it has a chance to escape, whether or not the furnace walls are black. It is thus the hole in the furnace wall which functions as a blackbody. In principle, a hole in the side of a shoebox, a doorway leading to a darkened room, or any similar cavity opening would serve equally well. Because the radiation trapped in an isothermal cavity is identical with that emitted by a blackbody, the terminology cavity radiation is often used.

Under steady-state conditions, a blackbody emits and absorbs radiant energy at the same rates. The absorbed and emitted radiations may have quite different characteristics. The character of the radiation reaching the blackbody is determined by its sources. It may be monochromatic or a blend of different wavelengths. The flux of radiation emitted by a blackbody depends exclusively on one quantity—the temperature of the blackbody. The rate at which energy is radiated per unit area (total energy flux) is termed the *surface flux* and is designated by the symbol F. For a blackbody, F is proportional to the fourth power of its absolute temperature

$$F = \sigma T^4 \tag{17.1}$$

This equation is known as the *Stefan–Boltzmann Law*; the constant σ is Stefan's constant

$$\sigma = 5.67 \times 10^{-5} \text{ erg/cm}^2\text{-sec-K}^4 \tag{17.2}$$

Stefan formulated this relation in 1879 on the basis of experimental data. In 1884, Boltzmann used a novel Carnot cycle and classical thermodynamics to prove that F was proportional to T^4, although his analysis could not predict the value of σ.

The spectrum of a blackbody is continuous, that is, all frequencies are present. If a blackbody is scanned by a detector sensitive to a narrow range of frequencies, the surface flux F can be analyzed to give what is called the monochromatic emissive power, $e(\nu)$. The monochromatic emissive power is defined by the statement that $e(\nu)\,d\nu$ is the portion of the flux composed of radiation in the frequency range from ν to $\nu + d\nu$. Integrating $e(\nu)\,d\nu$ over all frequencies gives F

$$F = \int_0^\infty e(\nu)\,d\nu \tag{17.3}$$

In Figure 17.1, $e(\nu)$ is plotted versus ν for several different blackbody temperatures. When Planck began his theoretical work on the blackbody spectrum,

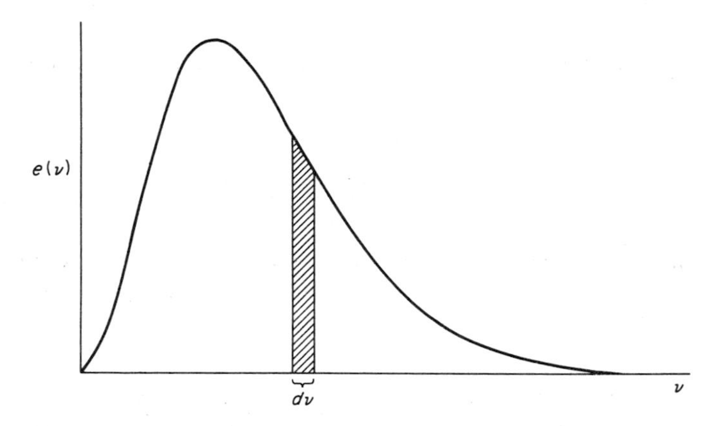

Figure 17.1. The monochromatic emissive power $e(\nu)$ for a blackbody. The hatched area $e(\nu)\,d\nu$ equals the intensity contributed by radiation in the range of frequency from ν to $\nu + d\nu$.

only the high-frequency end of the spectrum had been investigated experimentally. The problem confronting Planck, and others, was to formulate a theory which predicted the shape of the blackbody spectrum—a theory which predicted the functional form of $e(\nu)$.

In 1887 Hertz generated and detected electromagnetic waves, thereby confirming Maxwell's theory of light. Planck envisioned the surfaces of a radiating body as containing innumerable electrical oscillators. Each Hertzian oscillator had a characteristic frequency. An oscillator could absorb radiation of the same frequency it emitted. Thermal equilibrium in a cavity was reached when the rate of emission by each oscillator matched its rate of absorption.

Initially, Planck sought to derive the semiempirical relation for $e(\nu)$ advanced by Wien in 1896. Wien's expression for $e(\nu)$ is

$$e(\nu) = A\nu^3 e^{-B\nu/T} \tag{17.4}$$

where A and B are constants and T is the blackbody temperature. The available experimental data were in close agreement with Wien's formula. Planck assumed that Wien's expression for $e(\nu)$ was the correct one and managed to "derive" it, but only by a process of working backward from the answer—a trick mastered by virtually every student! Early in 1900, experiments by Rubens and Kurlbaum, two of Planck's colleagues, and by Lummer and Pringsheim, extended the measurements of $e(\nu)$ into the low frequency range (infrared). These measurements were in clear disagreement with Wien's formula.

In October, 1900, Planck presented a revised form of $e(\nu)$ which agreed with all data then available—and all accumulated since. Planck had no theory but advanced his formula as the simplest modification of Wien's formula which was in agreement with the data and consistent with thermodynamic considerations. The expression for $e(\nu)$, which proved to be the correct one, and for which Planck as yet had no theory, is

$$e(\nu) = \frac{A\nu^3}{e^{B\nu/T} - 1} \tag{17.5}$$

••••••

17.2.1 Show that for sufficiently high frequencies Planck's form of $e(\nu)$ reduces to Wien's expression (17.4).

••••••

Once again convinced that he had the right answer, Planck plunged ahead with the task of deriving this result. To do so he adopted Boltzmann's statistical approach. This was painful for Planck. He had long adhered to the belief that the laws of thermodynamics were of a causal nature—that they could be derived from the known laws of mechanics and electromagnetism without introducing assumptions of a statistical nature. Boltzmann was able to derive the laws of thermodynamics only by adding a statistical postulate; namely, that the observed properties of a system should correspond to the most probable value, or an appropriate average over a distribution of values. During the course of the analysis Planck found it mathematically advantageous to assign discrete energies to the oscillators. The energies open to oscillators of a given frequency were permitted to be multiples of some basic unit ε. Thus the energy of an oscillator of frequency ν was restricted to be

$$0, \varepsilon, 2\varepsilon, 3\varepsilon, \ldots, n\varepsilon, \ldots, \qquad n = \text{integer}$$

This trick had been employed previously by Boltzmann. Its use reflects the fact that it is conceptually easier to perform addition than it is to integrate. However, when the development reaches the point where mean values must be computed, summations are replaced by integrations, and so on. Our evaluation of the ideal gas partition function in Section 15.8 is an example of such a switch from a discrete to a continuous variable.

To his astonishment, Planck found that he could derive (17.5) only by retaining a finite value for ε. Furthermore, the form of ε was determined. The characteristic energy had to be directly proportional to the oscillator frequency ν. Thus for the first time there was written the now-famous relation

$$\varepsilon = h\nu \tag{17.6}$$

Planck had derived the blackbody spectrum by quantizing energy—by restricting the oscillator energies to multiples of the energy quantum $h\nu$. The constant h is called *Planck's constant* and has the value

$$h = 6.624 \times 10^{-27} \text{ erg-sec}$$

•••••• ••••••

17.2.2 Determine the dimensions of h. What other physical quantity has the same dimensions?

•••••• ••••••

Incredible as it may seem, Planck initially did not believe in energy quantization! He was convinced that quantization was a mathematical artifice—a neat trick which led to the correct blackbody spectrum. Planck spent the next 10 years retreating from quantization—trying to find an alternate route to the blackbody radiation law.

One of the major reasons for his refusal to embrace energy quantization was its consequences for electromagnetic theory. If the state of an oscillator is quantized, then it can emit and absorb only discrete quanta of energy. This in turn suggests that the energy of the electromagnetic field is quantized. Planck was not about to scrap the immensely successful theories of Maxwell and Huygens for the sake of the blackbody radiation law.

By 1910, physicists were more or less polarized on the quantization issue. The conservative element, led by Sir James Jeans, maintained that classical physics was "OK" as it stood. Jeans argued that blackbody radiation was not in thermodynamic equilibrium with the radiating surface and so could not be expected to follow the equipartition theorem. The radical elements, championed by Einstein, eagerly accepted the idea of energy quantization. Einstein argued that light quanta were a necessary consequence of Planck's theory. If an oscillator absorbs and emits electromagnetic energy in packets of size $h\nu$, then it would seem that the electromagnetic field is merely a collection of energy parcels—or photons as they are now called. In 1905, Einstein used the photon theory of light to explain the photoelectric effect, an achievement which later earned him the Nobel Prize.

In 1913, Niels Bohr advanced his quantum theory of the hydrogen atom. Both energy and angular momentum are quantized in the Bohr theory. Bohr's theory was subsequently modified and eventually replaced (in 1925–1926) by the less mechanistic modern quantum theory.

Today we have a relativistic quantum theory, thanks to the pioneering work of Dirac and Pauli some forty years ago. Quantum electrodynamics, a quantized electromagnetic theory—inconceivable to Planck—has been with us for 25 years, largely through the efforts of Bethe, Feynmann, Schwinger, and Tomonaga.

Quantum theory has had a profound effect on statistical physics. The following sections sketch the basic modifications wrought by quantum theory.

17.3 ANGULAR MOMENTUM AND STATISTICS

In classical mechanics it proves extremely fruitful to introduce the concept of angular momentum. Its importance stems in part from the fact that the angular momentum of a torque-free system is conserved; it is a constant quantity. For example, the conservation of angular momentum is the basis for Kepler's second law of planetary motion. However, the angular momentum of a planet is composed of two parts. Take Earth as a typical example. Our planet has orbital angular momentum by virtue of its yearly motion around the Sun. In addition it has a spin angular momentum by virtue of its daily rotation. For Earth and the other members of our solar system the spin angular momentum is always small by comparison with the orbital angular momentum.

In his 1913 theory, Bohr envisioned the hydrogen atom as a miniature solar system with the proton serving as a sun and the electron revolving about it in a planetary orbit. A basic postulate of Bohr's theory was that the orbital angular momentum of the electron was quantized, that is, it could take on only certain discrete values given by

$$L_{\text{orb}} = n\hbar, \qquad n = 1, 2, 3, \ldots$$

The quantity $\hbar$ (read "h cross") is Planck's constant divided by 2π

$$\hbar = \frac{h}{2\pi} = 1.054 \times 10^{-27} \text{ erg-sec} \tag{17.7}$$

The Bohr theory did not survive—it was replaced by a more satisfactory quantum theory. However, the quantization of orbital angular momentum forms a key part of the newer theory as well.

In 1925 Uhlenbeck and Goudsmit postulated a quantum mechanical analog of the planetary spin angular momentum. They noted that certain discrepancies between theoretical and experimental spectroscopy could be removed by assigning to the electron an intrinsic angular momentum over and above any possible orbital angular momentum. In particular, they showed that theory and experiment would be brought into accord if the electron had an intrinsic angular momentum of $\frac{1}{2}\hbar$. The terminology spin angular momentum was adopted to describe this new property because one could envision an electron not only moving about the atomic nucleus in a planetary-like orbit but also spinning rapidly about an axis of its own as does the Earth. The picture of an electron spinning rapidly about its own axis is a picturesque analogy, but a misleading one. The angular speed of rotation required to produce an angular momentum of $\frac{1}{2}\hbar$ is so large that it would give the surface of the electron a speed in excess of the speed of light. So, electrons have a spin angular momentum just because they are electrons, not because they are spinning spheres. Nevertheless, the name spin angular momentum has persisted. Uhlenbeck and Goudsmit introduced the spin angular momentum of the electron in a purely *ad hoc* manner. They offered no theoretical basis for the existence of the electron spin momentum. It remained for P. A. M. Dirac to show, in 1928, that the spin of $\frac{1}{2}\hbar$ emerged automatically by requiring that the quantum theory of the electron be in accord with the principles of the special theory of relativity.

The other two primary constituents of matter, the neutron and proton, also have an intrinsic (spin) angular momentum of $\frac{1}{2}\hbar$. The electron, proton, and neutron are often referred to as "spin $\frac{1}{2}$" particles.

Particles with spin other than $\frac{1}{2}$ exist. For example, the α particle (the nucleus of a helium atom) is a spin zero particle; it has no intrinsic spin angular momentum. The pion (pi meson) is another spin zero particle. The photon is a spin 1 particle; it possesses an intrinsic angular momentum of $\hbar$. Gravitons are spin 2 particles.[†] The spins of many different atomic nuclei have been measured. Spin values of 0, $\frac{1}{2}$, 1, $\frac{3}{2}$, 2, $\frac{5}{2}$, and so on, have been detected.

The quantum unit of spin angular momentum is $\frac{1}{2}\hbar$. The spin angular momentum of every particle can be expressed as an integral multiple of $\frac{1}{2}\hbar$ (including zero). Nature has seen fit to distinguish rather sharply between particles of integral spin and those of odd half-integral spin. Entities with integral spin are termed *bosons*; those with odd half-integral spin are called *fermions*. Thus, electrons, protons, and neutrons are fermions while photons, gravitons, and α particles are bosons.

It would carry us too far afield to delve further into the foundations of the fermion–boson scheme of classifying particles. It suffices to state that one set of quantum mechanical rules applies to a system of fermions and a different set is applicable to a collection of bosons. There is a carryover to statistical physics—there are two versions of quantum statistical mechanics. The so-called Bose–Einstein statistics version applies to systems composed of bosons. For a collection of fermions the appropriate statistical mechanics is called Fermi–Dirac statistics. Fermi–Dirac statistics is developed in the next section. Bose–Einstein statistics is treated in Section 17.5.

The word "statistics" which follows Bose–Einstein and Fermi–Dirac is a truncation of statistical mechanics. It is not meant to imply that different statistical methods are involved. Rather, two different quantum mechanical constraints lead to different particle distributions via the same statistical arguments.

17.4 FERMI–DIRAC STATISTICS

Consider a system of N identical fermions, and mentally chop its phase space into a great many cells of equal volume. With each cell there is associated a definite energy—the energy of a particle in that cell. As we noted in Chapter 15, an important assumption is implicit in assigning a definite energy to each segment of phase space. Assigning a particle a certain energy because it is in a particular region of phase space ignores the effects of other particles and thereby restricts us to situations in which mutual interactions can be ignored. We deal with an independent particle model. Even when mutual interactions are not ignorable the present approach can be generalized in a fruitful fashion.

[†] The quanta associated with gravitation are termed *gravitons*. The weakness of gravity has prevented their detection as yet, but theory dictates that their spin angular momentum be $2\hbar$.

The Boltzmann artifice of treating the energies as discrete becomes a necessity in QSM[†]—cell energies are quantized. In many instances the energy spectrum is very compact; there are many phase-space cells whose energies differ only slightly. We therefore consider groups of cells on a wholesale basis and assign the same energy to each. In QSM the distinction between fermions and bosons arises with respect to the capacity of the phase-space cells. Each phase-space cell has a maximum capacity of one fermion. Fermions are terribly introverted—they demand privacy.[‡] Thus if we look at the phase space of a fermion system we see some cells with one occupant while the remaining cells are empty. The introverted nature of fermions is a manifestation of the *Pauli exclusion principle* which decrees that identical fermions cannot occupy the same quantum state. To define precisely what is meant by a quantum state would take us far afield of our purpose here. In effect each phase-space cell corresponds to a distinct quantum state. In this connection, we remark that quantum physics forces an enlargement of the number of variables needed to specify the address of a phase-space cell. In addition to the position and linear momentum variables we must specify the value of a spin variable. Geometrically the spin angular momentum variable is represented as a vector, with different spin states corresponding to different orientations of the vector. For a particle of spin S there are $2S+1$ permitted orientations, that is, $2S+1$ distinct spin states. For electrons and other spin $\frac{1}{2}$ particles there are two possible spin states. The two permitted orientations of the spin angular momentum vector are antiparallel and the states are usually referred to as the spin up and spin down states.

If fermions are introverts, bosons are extroverts for there is no limit to the number of bosons which a given cell can accommodate. In a boson system some phase-space cells might be empty, others might contain a single boson. Still others might house two, three, and so on. It is a peculiarity of Bose–Einstein statistics that each phase-space cell must be prepared to accommodate even an infinite number of boson guests.

There is one other essential difference between CSM and QSM. In developing CSM we carefully numbered the atoms. We took great care to count the number of different ways of arranging the atoms. Alas, atoms do not obey the laws of classical mechanics. They do not, like pool balls, bear different numbers and

[†] Henceforth we use QSM to signify quantum statistical mechanics and CSM to stand for classical statistical mechanics.

[‡] We *do not* wish to suggest that the fermions have a will or that they are conscious. Such a suggestion was advanced recently. See A. A. Cochran, Relationships Between Quantum Physics and Biology, *Foundations Phys.* **1**, 235 (1971). Needless to say, Cochran's views did not go unchallenged. See J. F. Woodward *et al.*, Photon Consciousness: Fact or Fancy? *Foundations Phys.* **2**, 241 (1972).

colors. Atoms of the same element are indistinguishable.† QSM formally recognizes the meaning of identical particles. In CSM the microstates depicted in Figure 17.2 are distinct. In QSM they are identical. There is no way to distinguish the two atoms, hence we must remove the numbers. With the numbers removed, the two microstates of Figure 17.2 are identical. So, when it comes time to count microstates in QSM we must take care not to indulge in double-entry bookkeeping. We must count only distinct microstates.

Figure 17.2. Two microstates: With the atoms numbered the microstates are distinguishable. Removing the numbers from the atoms renders the microstates identical.

Consider the ith group of phase-space cells, Q_i in number. Each of these cells is characterized by the same energy ε_i. Let N_i denote the number of fermions distributed over the Q_i cells. The number of distinct microstates associated with a particular distribution of fermions is given by

$$W_i = \frac{Q_i!}{N_i!(Q_i - N_i)!} \tag{17.8}$$

To trace the origin of this assertion we imagine ourselves building up the W_i microstates by tossing fermions into phase-space cells one at a time. We first pretend that the fermions are distinguishable, that is, that they bear numbers $1, 2, 3, \ldots, N_i$. We can correct for the indistinguishability as a final step. We start with Q_i empty cells. The first fermion can be placed in any of the Q_i cells. The second fermion can be placed in any of the remaining $Q_i - 1$ empty cells. The one occupied cell is off limits for a second fermion—remember, they are introverts. Collectively there are $Q_i(Q_i - 1)$ ways of distributing the first two fermions. For each of these $Q_i(Q_i - 1)$ ways a third fermion can take its pick of the $Q_i - 2$ empty cells. Thus there are $Q_i(Q_i - 1)(Q_i - 2)$ ways of distributing three fermions. Pushing on, the N_ith fermion finds only $(Q_i - N_i + 1)$ empty cells from which to choose. The total number of ways of distributing the N_i fermions, assuming them to be distinguishable, is

$$Q_i(Q_i - 1)(Q_i - 2) \cdots (Q_i - N_i + 1) = \frac{Q_i!}{(Q_i - N_i)!}$$

•••••

17.4.1 Establish the above equality.

•••••

† This presumes that they are not different isotopes of the element.

We correct for the indistinguishability of the fermions by dividing by $N_i!$ That is, there are $N_i!$ rearrangements of N_i distinguishable particles, all of which become identical when the particles are rendered indistinguishable. Thus the $Q_i!/(Q_i - N_i)!$ ways enumerated above counts each distinct distribution $N_i!$ times, whence,

$$W_i = \frac{Q_i!}{N_i!(Q_i - N_i)!} \tag{17.8}$$

is the number of microstates corresponding to a distribution of N_i fermions over Q_i cells. If there are M groups of cells, the grand total of possible microstates is

$$W\{N_i\} = \prod_{i=1}^{M} W_i = \prod_{i=1}^{M} \frac{Q_i!}{N_i(Q_i - N_i)!} \tag{17.9}$$

••••• •••••

17.4.2 Explain the logic which dictates that $W = \prod_{i=1}^{M} W_i$. Why is the total number of microstates not merely the sum of the W_i?

••••• •••••

The procedure for finding the most probable distribution follows the lines established in CSM. We maximize $\ln W$ subject to the constraints that

$$\sum_{i=1}^{M} N_i = N \tag{17.10}$$

and

$$\sum_{i=1}^{M} N_i \varepsilon_i = U \tag{17.11}$$

To this end we introduce the most probable distribution

$$\{N_i\} = N_1, N_2, \ldots, N_k, \ldots, N_s, \ldots, N_M$$

and an adjacent distribution

$$\{N_i'\} = N_1, N_2, \ldots, N_k - 1, \ldots, N_s + 1, \ldots, N_M$$

This automatically guarantees that both distributions correspond to the same total number of particles. The most probable distribution follows by requiring that

$$\ln W\{N\} = \ln W\{N_i'\} \tag{17.12}$$

and that, simultaneously,

$$U = \sum_{i=1}^{M} N_i \varepsilon_i = \sum_{i=1}^{M} N_i' \varepsilon_i$$

As in CSM we can write

$$\ln W\{N_i\} - \beta \sum_{i=1}^{M} N_i \varepsilon_i = \ln W\{N_i'\} - \beta \sum_{i=1}^{M} N_i' \varepsilon_i$$

$\ln W'$ is given by a similar expression. Inserting the explicit forms for W_i and W_i' gives

$$\sum_{i=1}^{M} \ln \frac{Q_i!}{N_i!(Q_i - N_i)!} - \beta N_i \varepsilon_i = \sum_{i=1}^{M} \ln \frac{Q_i!}{N_i'!(Q_i - N_i')!} - \beta N_i' \varepsilon_i$$

As only the kth and sth groups of cells differ in composition there is a wholesale cancellation of terms on either side, leaving

$$\ln\left(\frac{Q_k - N_k + 1}{N_k}\right) - \beta\varepsilon_k = \ln\left(\frac{Q_s - N_s}{N_s + 1}\right) - \beta\varepsilon_s$$

It is legitimate to replace $Q_k - N_k + 1$ by $Q_k - N_k$ and $N_s + 1$ by N_s, leaving

$$\ln\left(\frac{Q_k}{N_k} - 1\right) - \beta\varepsilon_k = \ln\left(\frac{Q_s}{N_s} - 1\right) - \beta\varepsilon_s$$

However, k and s are arbitrary so that $\ln((Q_i/N_i) - 1) - \beta\varepsilon_i$ must have a common value for all M groups of cells. We denote this value as $-\beta\mu$, thereby obtaining

$$\ln\left(\frac{Q_i}{N_i} - 1\right) - \beta\varepsilon_i = -\beta\mu, \qquad i = 1, 2, \ldots, M$$

Solving for N_i results in

$$N_i = \frac{Q_i}{\exp[\beta(\varepsilon_i - \mu)] + 1} \tag{17.13}$$

This is the sought-for Fermi–Dirac distribution. The quantity μ is the chemical potential while β proves to be $1/kT$ as before. Mathematically β and μ enter as quantities which guarantee satisfaction of the conditions

$$N = \sum N_i$$

and

$$U = \sum N_i \varepsilon_i$$

The basic relation (17.13) allows a most fruitful interpretation. The quantity

$$\frac{N_i}{Q_i} = \frac{1}{\exp[\beta(\varepsilon_i - \mu)] + 1} \tag{17.13a}$$

is the fraction of the Q_i cells of energy ε_i which are occupied. Statistically we can interpret N_i/Q_i as the probability that any particular cell of energy ε_i is occupied by a fermion. In many situations the index i need not be displayed explicitly. We shall refer to

$$\frac{N}{Q} = \frac{1}{e^{\beta(\varepsilon-\mu)} + 1} \tag{17.14}$$

as the *occupation probability*. In situations where the discrete nature of the energy spectrum is of no consequence, we can use the technique of converting sums to integrals and write the occupation probability as

$$\frac{dN}{dQ} = \frac{1}{e^{\beta(\varepsilon-\mu)} + 1} \tag{17.14a}$$

where dN is the expected number of fermions in the dQ phase-space cells characterized by energy ε.

One of the earliest applications of Fermi–Dirac statistics was made by Arnold Sommerfeld in 1928 to the problem of determining the heat capacity of electrical conductors. A conducting solid is viewed as a collection of three-dimensional simple harmonic oscillators, with each atom possessing six degrees of freedom. The empirical law of Dulong–Petit states that the molar heat capacity of a solid is 6 cal/mole-K. This can be understood on the basis of energy equipartition. The equipartition theorem assigns a molar heat capacity of $\frac{1}{2}R$ to each degree of freedom. With six degrees of freedom for the three-dimensional simple harmonic oscillator (see Figure 11.4)

$$C_v = 3R \simeq 6 \text{ cal/mole-K}$$

The paradox arises because many of the solids which have $C_v \simeq 6$ cal/mole-K are also good electrical conductors (Ag, Au, Cu, Na, and so on). To explain their high electrical conductivity it is necessary to hypothesize that many of the atoms forming the crystalline lattice have sacrificed an electron. It is the directed motions of these electrons which constitute an electric current. Early models could explain the observed electrical conductivity by assuming that virtually every atom contributed one electron. Such an electron gas should also contribute to the heat capacity. Heat absorbed by the solid should be shared by the electrons and by the lattice of ions and atoms. It was assumed that the electrons behaved like a monatomic gas. The equipartition theorem predicted that they should have a molar heat capacity of $\frac{3}{2}R$. Virtually every atom donates an electron, leaving an ionic lattice. Each lattice ion (or atom) contributes six degrees of freedom. With each atom contributing one electron the total molar heat capacity should be $\frac{3}{2}R$ for the electrons plus $3R$ for the lattice, giving

$$C_v = \tfrac{9}{2}R$$

Experimentally, however, $C_v \simeq 3R$; the electrons apparently do not "play ball." Their contribution of $\frac{3}{2}R$ to the heat capacity is missing. Yet, the electrons seem to behave normally insofar as the electrical conductivity is concerned.

The apparent paradox evaporates once Fermi–Dirac statistics are brought to bear, and for this reason: Electrons distribute themselves over phase space according to the Fermi–Dirac expression (17.14), and not according to the classical Boltzmann distribution law (15.31). As a consequence of their Fermi–Dirac distribution, only a tiny fraction of the electrons are free to increase their energy and thereby share in the partitioning of energy. The electron degrees of freedom are suppressed and the ionic lattice receives virtually all of the absorbed thermal energy. What prevents the electron gas from absorbing thermal energy? Briefly, it is the Pauli exclusion principle. The exclusion principle forbids electrons from occupying the same phase-space cell. Since most of the electrons are unable to find a vacant phase-space cell of higher energy they are unable to partake of the absorbed energy. The small fraction that find empty phase-space cells nearby are able to absorb energy. Because they are relatively few in number their contribution to the heat capacity of the solid is ordinarily much smaller than that made by the lattice. We must study the Fermi–Dirac distribution in depth to see just how the exclusion principle operates to insulate the electron gas.

For openers, let us estimate the value of the chemical potential μ appropriate to the electron gas in a good electrical conductor. As it turns out, a typical value for μ is a few electron volts. This is extremely large by comparison to kT, which is about 1/40 eV at room temperature. In general, the value of μ depends on the concentration of electrons and the temperature. As a very good first approximation we can take $T = 0$ for the purposes of estimating μ. It has become customary to refer to the value of μ at absolute zero as the Fermi energy ε_F

$$\mu(T=0) \equiv \varepsilon_F$$

In many instances it is permissible to use ε_F in place of μ even at room temperature because the chemical potential varies only weakly with the temperature.

With $T = 0$, the parameter $\beta = 1/kT$ is infinite and

$$\exp[\beta(\varepsilon - \varepsilon_F)] + 1 = \begin{cases} 1, & \varepsilon < \varepsilon_F \\ \infty, & \varepsilon > \varepsilon_F \end{cases}$$

The occupation probability is given by

$$\frac{N}{Q} = \frac{1}{\exp[\beta(\varepsilon - \varepsilon_F)] + 1} = \begin{cases} 1, & \varepsilon < \varepsilon_F \\ 0, & \varepsilon > \varepsilon_F \end{cases}$$

Figure 17.3 is a plot of N/Q versus ε at $T = 0$. The figure simply reveals that electrons fill the cells of lowest energy consistent with the Pauli exclusion principle.

The value of the Fermi energy is determined by the condition that the total number of electrons equal the total number of phase-space cells whose energy does not exceed ε_F. In QSM it is preferable to deal with a phase space which combines position and momentum rather than position and velocity. Because the Fermi distribution is isotropic in momentum space we employ spherical momentum space variables. Cells corresponding to momenta in the spherical shell defined by p and $p+dp$ occupy a phase-space volume of

$$8\pi p^2\,dp\,V$$

A factor of 2 enters because of the electron spin. Thus for the momentum p to $p+dp$ there is a phase-space volume $4\pi p^2\,dp\,V$ open to electrons with spin up and an equal volume open to electrons with spin down. The volume of the unit cell is h^3, where h is Planck's constant. Dividing the phase-space volume by the volume per cell gives the number of cells

$$dQ = \frac{8\pi p^2\,dp\,V}{h^3} \tag{17.15}$$

The electron energy and momentum are related by

$$\varepsilon = \frac{p^2}{2m}$$

so that the radius of the momentum-space sphere corresponding to the energy ε_F is given by

$$p_F = \sqrt{2m\varepsilon_F}$$

The quantity p_F is called the *Fermi momentum*. The surface of the momentum-

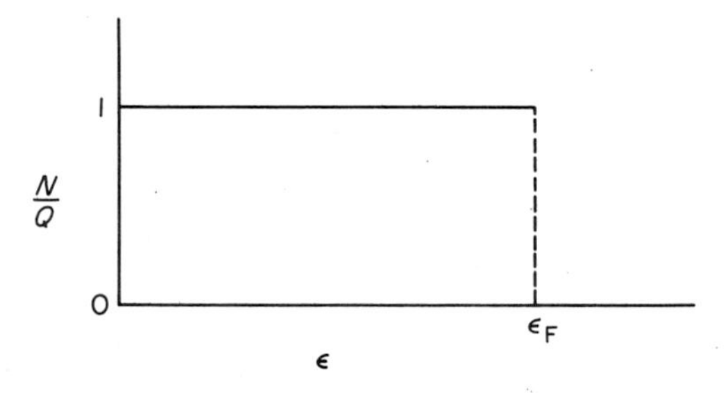

Figure 17.3. The occupation probability N/Q at absolute zero. Phase-space cells of energy greater than the Fermi energy ε_F are empty; those of energy up to ε_F are completely filled.

space sphere of radius p_F is referred to as the *Fermi surface*. At absolute zero, all electrons are beneath the Fermi surface. The total number of occupied cells is

$$Q = \frac{8\pi V}{h^3}\int_0^{p_F} p^2\,dp = \frac{8\pi V}{3h^3}(2m\varepsilon_F)^{3/2}$$

Since the N electrons completely fill these cells we set $Q = N$ and obtain for the Fermi energy

$$\varepsilon_F = \frac{h^2}{2m}\left(\frac{3n}{8\pi}\right)^{2/3} \tag{17.16}$$

where m is the electron mass and $n = N/V$ is the electron concentration (number density). For a good conductor, n is likely to exceed 10^{22} electrons/cm^3, which places ε_F in the electron volt range. For future reference, we remark that the Fermi momentum is independent of the mass of the particle. The value of p_F is fixed by the number density n. A convenient expression for p_F is

$$p_F = \hbar(3\pi^2 n)^{1/3}, \qquad \hbar = \frac{h}{2\pi} \tag{17.17}$$

•••••• ••••••

17.4.3 The density of metallic sodium is 0.971 gm/cm^3 and its atomic weight is 23.0. Assume that each sodium atom donates one electron and compute the Fermi energy of the electron gas.

•••••• ••••••

Example 17.1 Average Energy versus Fermi Energy

Even if the electron gas is at absolute zero, the mean kinetic energy of the electrons is quite large—another consequence of the Pauli exclusion principle. To show this, we compute the mean kinetic energy of the electron gas at $T = 0$.

All phase-space cells are occupied for $\varepsilon < \varepsilon_F$ and all are empty for $\varepsilon > \varepsilon_F$. We can therefore write

$$\langle\varepsilon\rangle = \frac{1}{N}\int_{\substack{\text{all}\\\text{cells}}} \varepsilon\,dN$$

where

$$dN = \begin{cases} dQ, & \varepsilon < \varepsilon_F \\ 0, & \varepsilon > \varepsilon_F \end{cases}$$

dQ is the number of phase-space cells for energies in the range ε to $\varepsilon + d\varepsilon$. If we write

$$dQ = \frac{8\pi}{h^3}Vp^2\,dp$$

and then use $\varepsilon = p^2/2m$ and its differential $d\varepsilon = p\,dp/m$, we find

$$dQ = \frac{4\pi V}{h^3}(2m)^{3/2}\sqrt{\varepsilon}\,d\varepsilon$$

Observing that $N = \int dN = \int_0^{\varepsilon_F} dQ$ we obtain

$$\langle\varepsilon\rangle = \frac{\int_0^{\varepsilon_F} \varepsilon^{3/2}\,d\varepsilon}{\int_0^{\varepsilon_F} \varepsilon^{1/2}\,d\varepsilon} = \frac{3}{5}\varepsilon_F$$

For good electrically conducting solids ε_F is typically a few electron volts. The corresponding electron speed is about 1000 times the speed of sound. The exclusion principle really keeps them jumping—even at absolute zero! ◇

•••••• ••••••

17.4.4 (a) Show that the volume of a spherical shell in momentum space may be expressed in terms of the energy ε by

$$4\pi p^2\,dp = 4\pi\sqrt{2}\,m^{3/2}\sqrt{\varepsilon}\,d\varepsilon$$

provided $\varepsilon = p^2/2m$.

(b) Supply reasoning to establish that the equation relating μ to the number density n and temperature is

$$n = \frac{4\pi(2m)^{3/2}}{h^3}\int_0^\infty \frac{\sqrt{\varepsilon}\,d\varepsilon}{e^{\beta(\varepsilon-\mu)}+1}$$

(c) Show that (17.16) results as a special case of the above equation when $\beta = \infty$ $(T = 0)$.

•••••• ••••••

Plots of the occupation probability N/Q versus ε for several values of ε_F/kT are shown in Figure 17.4. Where ε differs from ε_F by more than a few times kT, the exponent $\beta(\varepsilon-\varepsilon_F)$ has a large magnitude and the exponential is either very large or very small by comparison with unity, that is,

$$\exp[\beta(\varepsilon-\varepsilon_F)]\begin{cases} \simeq \infty, & \varepsilon - \varepsilon_F \gg kT \\ = 1, & \varepsilon = \varepsilon_F \\ \simeq 0, & \varepsilon_F - \varepsilon \gg kT \end{cases} \qquad \beta = \frac{1}{kT}$$

Thus for energies much below ε_F, N/Q is nearly one; N/Q drops sharply to $\frac{1}{2}$ at $\varepsilon = \varepsilon_F$ and tails off exponentially to zero when ε exceeds ε_F. We can describe the situation at $T > 0$ as one in which a small fraction of the electrons have been excited into phase-space cells above the Fermi surface. Most of the

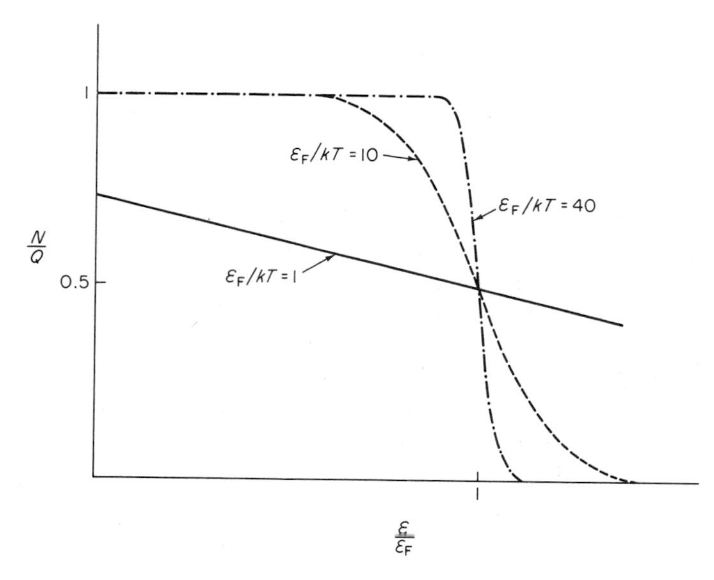

Figure 17.4. The Fermi–Dirac occupation probability for several values of ε_F/kT.

electrons are buried beneath the Fermi surface. They are unable to share in energy absorbed by the material because there is no phase space available to them—no vacant phase-space cells above them on the energy scale. Only the small fraction of the electrons in the vicinity of the Fermi surface can respond to external forces. Thus the electron gas in a metal is more or less insulated from the rest of the solid. With this picture in mind we can estimate the molar heat capacity C_v of the electron gas, which is considerably smaller than the value of (3/2) R predicted by the equipartition theorem.

It is not difficult to write down an exact expression for C_v. Evaluating the resulting integral is a bit tricky. Instead of seeking the exact result, let us determine C_v approximately. Electrons near the Fermi surface behave classically, that is, they move about freely like the atoms of an ideal gas. The molar heat capacity of the electron gas (e.g.) should then be (3/2) R times the fraction of the electrons near the Fermi surface

$$(C_v)_{\text{e.g.}} = \tfrac{3}{2}R(\text{fraction of the electrons near the Fermi surface}) \quad (17.18)$$

A span of energies of order kT above and below ε_F defines the region where there are electrons *and* empty phase-space cells. For $\varepsilon \gg \varepsilon_F + kT$ there are empty cells but virtually no electrons. For $\varepsilon \ll \varepsilon_F - kT$ there are plenty of electrons but no vacant cells.

Occupied energy states span an energy range of roughly ε_F, so that the range of $2kT$ over which vacancies occur encompasses a fraction of order $2kT/\varepsilon_F$ of all electrons. Taking $2kT/\varepsilon_F$ for the fraction of electrons near the Fermi surface in (17.18) gives

$$(C_v)_{\text{e.g.}} \simeq 3\left(\frac{kT}{\varepsilon_F}\right)R \ll \frac{3}{2}R \tag{17.19}$$

A sophisticated calculation using the exact but unwieldy form of N/Q gives a value of $\pi^2/2 \simeq 4.93$ in place of the 3 in (17.19). This result for the electron gas heat capacity resolves the apparent paradox. At ordinary temperatures the electron gas contribution to C_v is undetectably small by comparison to that of the lattice. The electron gas contribution to the heat capacity of metals has been measured in experiments conducted at extremely low temperatures ($\sim$1 K). At low temperatures quantum effects force the lattice heat capacity toward zero. At very low temperatures, the lattice heat capacity is proportional to T^3. As T approaches zero, the electron gas heat capacity—which by (17.19) decreases only linearly with T—dominates the lattice contribution.

Further applications of the Fermi–Dirac statistics are presented in the next chapter.

•••••• ••••••

17.4.5 (a) Show that a two-term Taylor series for the occupation probability about $\varepsilon = \varepsilon_F$ gives

$$\{\exp[\beta(\varepsilon-\varepsilon_F)]+1\}^{-1} = \tfrac{1}{2} - \tfrac{1}{4}\beta(\varepsilon-\varepsilon_F)$$

Verify that the right side equals 1 at $\varepsilon = \varepsilon_F - 2kT$ and 0 at $\varepsilon = \varepsilon_F + 2kT$.

(b) What is the average occupation probability over the range of energies from $\varepsilon_F - 2kT$ to $\varepsilon_F + 2kT$?

(c) Can you reconcile the results of (a) and (b) with the text statement that the fraction of electrons near the Fermi surface is $2kT/\varepsilon_F$?

•••••• ••••••

17.5 BOSE–EINSTEIN STATISTICS

The phase-space distribution of bosons is determined by precisely the same methods used for fermions. The major difference is that each phase-space cell can accommodate any number of bosons. There is no exclusion principle for bosons.

Consider N_i bosons distributed over Q_i phase-space cells. The Q_i cells correspond to the same energy ε_i. Regard the bosons as distinguishable at first, then correct at the end of the game by dividing by $N_i!$. To identify the Q_i cells we use $Q_i - 1$ walls. To count the number of ways W_i of distributing the N_i

Figure 17.5. Two ways of distributing 4 bosons over 3 cells. The bosons are supposed distinguishable at this stage and so are numbered.

bosons over the Q_i cells we imagine $N_i + Q_i - 1$ *positions* along a line. We place either a boson or a wall at each position. Figure 17.5 illustrates two ways of distributing $N_i = 4$ bosons over $Q_i = 3$ cells. There are $N_i + Q_i - 1 = 6$ positions; four occupied by bosons, two by cell walls.

Start dropping in bosons—one at a time. When all N_i bosons have been positioned, the remaining $Q_i - 1$ positions are called *walls*. The first boson can be placed at any one of the $N_i + Q_i - 1$ positions, the second at any one of the remaining $N_i + Q_i - 2$, and so on. The N_ith boson can be placed at any of the Q_i positions, leaving $Q_i - 1$ positions for cell walls. Because we regard the particles as distinguishable at this stage, the total number of ways of distributing them is the product

$$(N_i + Q_i - 1)(N_i + Q_i - 2)(N_i + Q_i - 3) \cdots (Q_i) = \frac{(N_i + Q_i - 1)!}{(Q_i - 1)!}$$

To correct for the indistinguishability we divide by $N_i!$ giving

$$W_i = \frac{(N_i + Q_i - 1)!}{N_i!(Q_i - 1)!} \tag{17.20}$$

The total thermodynamic probability W is the product of the W_i for all groups of phase-space cells

$$W = \prod_{i=1}^{M} W_i \tag{17.9'}$$

It is left as an exercise for the student (see Problem 17.5.1) to show that the most probable distribution has the form

$$N_i = \frac{Q_i}{\exp[\beta(\varepsilon_i - \mu)] - 1} \tag{17.21}$$

This is the Bose–Einstein distribution. By an extension of the arguments presented in Appendix B it can be shown that $\beta = 1/kT$. As before, μ is the chemical potential; its value depends on the nature of the bosons, their number density, and the temperature.

In the next chapter we apply the Bose–Einstein statistics to the problem of blackbody radiation and indicate how the theory has recently been used to support arguments for the "big bang" theory of the universe.

••••••

17.5.1 Use (17.20) in (17.9′) and follow the procedures used with Fermi–Dirac statistics to derive the Bose–Einstein distribution (17.21). The equations of constraint are (17.10) and (17.11).

••••••

17.6 QSM OR CSM?

The Bose–Einstein and Fermi–Dirac distributions differ in form only in that there is a -1 in the denominator of (17.21) and a $+1$ in the Fermi–Dirac denominator (17.13). Under circumstances where

$$e^{-\beta\mu} \gg 1 \tag{17.22}$$

both Fermi–Dirac and Bose–Einstein statistics emulate the classical Boltzmann statistics. Thus to the extent that the ± 1 is ignorable by comparison with the exponential, N_i is given by the Boltzmann distribution of CSM

$$N_i = \frac{Q_i}{\exp[\beta(\varepsilon_i - \mu)] \pm 1} \simeq \exp(\beta\mu)\, Q_i \exp(-\beta\varepsilon_i) \tag{17.23}$$

The right side is identical with the form of N_i given in (15.33), wherein $N/Z(\beta)$ is the quantity designated as $e^{\beta\mu}$ in (17.23).

At ordinary temperatures and densities (17.22) holds for most systems. Quantum statistical effects are then not in evidence and one can say that QSM reduces to CSM. Quantum effects become evident at low temperatures and/or high densities—conditions which tend to make $e^{-\beta\mu} < 1$.

When $e^{-\beta\mu} \ll 1$, that is, when $\mu \gg kT$, the distribution is said to be degenerate. As the ratio μ/kT increases the distribution degenerates, in this sense; all phase-space cell populations approach values appropriate to the distribution at absolute zero. The metallic electron gas studied in Section 17.4 is degenerate, its distribution differing from the $T = 0$ form only in the vicinity of the Fermi surface.

••••••

17.6.1 Explain why $\mu/kT \gg 1$ is equivalent to the condition $e^{-\beta\mu} \ll 1$.

••••••

To determine whether or not a distribution is degenerate one must evaluate the chemical potential μ and then compute $e^{-\beta\mu}$. In general the equation determining the ratio μ/kT is transcendental (compare Problem 17.4.4). Fortunately, it is often not necessary to know the precise value of $e^{-\beta\mu}$, it being sufficient to know whether it is large or small in comparison with unity. A "quick-and-dirty" test uses (17.23) to evaluate $e^{-\beta\mu}$. That is, one assumes that the Boltzmann distribution gives a satisfactory description and then checks for consistency. Failure to meet the test warns of quantum degeneracy. We first evaluate the nondegenerate expression for $e^{-\beta\mu}$ and then illustrate the technique. From (17.23) the total number of particles follows as

$$N = \sum_i N_i = \exp(\beta\mu) \sum_i Q_i \exp(-\beta\varepsilon_i)$$

The summation is converted to an integration over energy

$$\sum_i Q_i \exp(-\beta\varepsilon_i) = \int dQ \exp(-\beta\varepsilon)$$

Taking $dQ = 4\pi V p^2\, dp/h^3$ and $\varepsilon = p^2/2m$ results in

$$e^{-\beta\mu} = \left(\frac{2\pi mk}{h^2}\right)^{3/2} \frac{T^{3/2}}{N/V} \tag{17.24}$$

This form of $e^{-\beta\mu}$ points up our earlier remark that low temperatures and high densities tend to make $e^{-\beta\mu} \ll 1$ and promote a quantum degeneracy. The result (17.24) can be expressed in a form inviting an intuitive and easily remembered interpretation, namely,

$$e^{-\beta\mu} = \frac{(2\pi mkT)^{3/2}\, V/N}{h^3} \tag{17.25}$$

The quantity $(2\pi mkT)^{1/2}$ has the dimensions of momentum and is very nearly equal to the rms momentum. The quantity $(2\pi mkT)^{3/2}$ is the characteristic momentum space volume per particle. The quantity V/N is the spatial volume per particle. Their product, the numerator of (17.25), is the representative phase-space volume per particle. The factor h^3 in the denominator of (17.25) is the volume of one phase-space cell. It follows from (17.25) that the quantity $e^{-\beta\mu}$ measures the average number of phase-space cells allotted to each particle. The condition validating the use of classical statistics is $e^{-\beta\mu} \gg 1$, or

$$\frac{\text{volume of phase space per particle}}{\text{volume of one phase-space cell}}$$

$$= \text{number of phase-space cells per particle} \gg 1$$

Classical statistics therefore apply under conditions which produce a sparse population of phase-space cells. Each particle has—on the average—a large number of phase-space cells to roam about in. Conversely we expect quantum degeneracy to appear when the population per phase-space cell approaches unity. We next present examples illustrating this point.

Example 17.2 Degenerate or Nondegenerate?

We consider in turn the cases of (a) nitrogen molecules at normal atmospheric conditions, (b) liquid helium at the normal boiling point, and (c) electrons under typical laboratory plasma conditions. For each of the three we evaluate

$$e^{-\beta\mu} = \frac{(2\pi mkT)^{3/2}\, V/N}{h^3}$$

$e^{-\beta\mu} \gg 1$ implies that classical statistics is adequate, while $e^{-\beta\mu} \lesssim 1$ warns that quantum degeneracy may invalidate considerations based on CSM.

(a) For N_2 molecules at 1 atm and $T \simeq 300$ K we have (in cgs units)

$$m \simeq 5 \times 10^{-23}, \qquad \frac{V}{N} \simeq 4 \times 10^{-20}, \qquad (2\pi mkT)^{3/2} \simeq 6 \times 10^{-53}$$

$$h^3 \simeq 2 \times 10^{-79}, \qquad e^{-\beta\mu} \simeq 10^7$$

It follows that the air we breathe is adequately described by CSM.

(b) For the atoms of liquid helium at the normal boiling point we have $m \simeq 6 \times 10^{-24}$ and $T \simeq 4.2$ K. The volume per atom may be determined from the mass density of 0.12 gm/cm^3,

$$\frac{V}{N} = \frac{\text{mass per atom}}{\text{mass per unit volume}} \simeq 5 \times 10^{-23}$$

These figures give $e^{-\beta\mu} \simeq 1$. It is therefore not surprising to find that liquid helium exhibits quantum degeneracy effects near its normal boiling point. Indeed, helium exhibits superfluid properties below a temperature of 2.17 K.

(c) As a final example, consider electrons under conditions typical of a laboratory plasma; $m \simeq 10^{-27}$, $N/V \simeq 10^{12}$, $T \simeq 10^3$ K. One finds $e^{-\beta\mu} \simeq 10^8$ which ensures that quantum statistical effects are negligible. ✧

It must be pointed out that our considerations have been based entirely on the independent particle model. Our conclusions turn out to be reliable over a much wider range of conditions than might be expected. After all, the effects of quantum degeneracy appear at high densities and low temperatures—conditions where interactions between particles would seem paramount and where thermal motions are minimal.

In the case of fermions our good fortune may be traced to the Pauli exclusion principle. Regardless of how strongly the fermions interact, the exclusion principle limits the population to one per phase-space cell. Fermions which are free to respond to external forces (those near the Fermi surface) have kinetic energies far in excess of the thermal energy. Their high kinetic energy enables them to ignore interparticle forces and imitate the motion of nearly free particles.

•••••• ••••••

17.6.2 The Fermi energy ε_F of a degenerate gas of free electrons is given by (17.16). As was shown in Example 17.1, the mean kinetic energy of such electrons is $(3/5)\varepsilon_F$. The Coulomb potential energy of two electrons separated by a distance r is e^2/r. Taking $(V/N)^{1/3} = n^{-1/3}$ as a measure of the mean distance between charges suggests that $e^2 n^{1/3}$ is a measure of the mean potential energy of an electron. Provided $(3/5)\varepsilon_F/e^2 n^{1/3} \gg 1$ we expect that the electrons will behave like free particles. This inequality will prevail at sufficiently large n because the potential energy is proportional to $n^{1/3}$ while the kinetic energy is proportional to $n^{2/3}$.

(a) Show that $(3/5)\varepsilon_F/e^2 n^{1/3} = (6\pi^2/5)(3/8\pi)^{2/3} a_0 n^{1/3}$, where $a_0 = \hbar^2/me^2 = 5.29 \times 10^{-9}$ cm is the Bohr radius.

(b) Evaluate the ratio for the typical metallic density $n = 10^{22}/\text{cm}^3$.

•••••• ••••••

The existence of two versions of quantum statistics may cause the student to wonder out loud about the range of validity of thermodynamics. We have demonstrated how QSM reduces to CSM at low densities and high temperatures. Is it necessary to revise the laws of thermodynamics to encompass quantum phenomena? The answer is "no"; the old laws will do nicely. Distinctions between QSM and CSM are felt through different expressions used to evaluate thermodynamic properties. For example, the internal energy of a noninteracting gas of bosons or fermions can be expressed by the general relation

$$U = \sum \varepsilon_i N_i$$

where N_i is the number of particles with energy ε_i; N_i depends on T, V, and perhaps other thermodynamic variables. For fermions, N_i is given by (17.13); for bosons N_i is given by (17.21). Despite the fact that the form of $U(T, V, \ldots)$ is different for fermions and bosons, the first law of thermodynamics still reads $dQ = dU + dW$. Likewise, the vast arrays of thermodynamic formulas, such as Maxwell's relations, remain valid regardless of whether one deals with fermions, bosons, or mixtures thereof.

REFERENCES

The story of Planck's struggle with quantum theory is delightfully told in:

M. J. Klein, Thermodynamics and Quanta in Planck's Work, *Phys. Today* **19**, 23 (1966).

Another view of the birth of quantum theory is given in:

K. K. Darrow, The Quantum Theory, *Sci. Am.* **186** (March, 1952). (Also reprinted as *Sci. Am. Offprint* #205.)

The hotel analogy was inspired by a modern classic:

J. D. Trimmer, Hotel Management in Ergodia, *Phys. Today* **15**, 28 (November, 1962).

For alternate derivations of the QSM distributions see:

R. D. Cowan, Derivation of the Energy—Distribution Formulas in Quantum Statistics, *Am. J. Phys.* **25**, 463 (1957)

and

K. K. Darrow, The New Statistical Mechanics, *Bell Sys. Tech. J.* **22**, 362 (1943).

For a careful discussion of the statistical interpretation of the entropy and the revelation which shows h^3 to be the volume of a single phase-space cell consult:

K. K. Darrow, Entropy, *Bell Sys. Tech. J.* **21**, 51 (1942).

Elementary discussions of quantum statistics can be found in several texts. Two which circumvent the quantum mechanical wave function approach and include applications are:

F. W. Sears, "An Introduction to Thermodynamics, The Kinetic Theory of Gases, and Statistical Mechanics," Chapter 16. Addison-Wesley, Reading, Massachusetts (1953).

and

W. Kauzmann, "Thermodynamics and Statistics," Chapter 4. Benjamin, New York (1967).

Chapter 18

Applications of Quantum Statistical Mechanics

This chapter is the quantum mechanical counterpart of Chapter 16. The treatment of the various topics is uneven; at points it is necessary to substitute description for quantum mechanics.

In Section 18.1 we take a detailed look at the theory of blackbody radiation. Recently this oldest of quantum theories has attracted attention in connection with experiments which test the merits of various cosmological hypotheses—notably the "big bang" and "continuous creation" theories. In Section 18.2 we describe the three basic mechanisms by which atoms interact with radiation and then show how these processes are capable of producing both the blackbody spectrum and the spectacular laser action.

Section 18.3 provides an introduction to the physics of neutron stars and white dwarfs—two of the three possible endpoints of stellar evolution. Sections 18.4 and 18.5 revert to the frigid world of dilution refrigerators and superconductivity. The chapter closes with the revelation that there is an upper limit for temperature—an ultimate temperature which cannot be exceeded regardless of how much energy is fed into a physical system.

18.1 BLACKBODY RADIATION, BIG BANG, AND THE PRIMEVAL FIREBALL

Electromagnetic radiation exhibits particle-like properties. A stream of radiation may be treated as a photon gas. The integral spin of a photon brands it a boson. Within the framework of nonrelativistic quantum theory the photons do not interact with each other. Accordingly, they constitute a gas of independent bosons and their phase-space distribution is given by (17.24). The energy ε of a photon is related to the frequency ν of its source by Planck's famous relation

$$\varepsilon = h\nu$$

The linear momentum of a photon follows from the relation furnished by the special theory of relativity. For a particle of rest mass M, the linear momentum p and energy E are related by (c is the speed of light)

$$E^2 = p^2c^2 + (Mc^2)^2$$

For entities such as the photon, whose rest mass is zero, this reduces to $E = pc$. Thus the linear momentum of a photon p is given by

$$p = \frac{\varepsilon}{c} = \frac{h\nu}{c} \tag{18.1}$$

•••••• ••••••

18.1.1 The special theory of relativity shows that the linear momentum of a particle of rest mass M and speed v has the magnitude

$$p = \frac{Mv}{\left(1 - \dfrac{v^2}{c^2}\right)^{1/2}}$$

Use this to prove that any particle with a nonzero linear momentum which has zero rest mass (that is, which does not exist as an object at rest) necessarily travels at the speed of light c.

•••••• ••••••

The blackbody spectrum is continuous—all frequencies are present. This situation is handled by writing the Bose–Einstein distribution (17.24) as

$$dN = \frac{dQ}{e^{\beta(\varepsilon-\mu)} - 1} \tag{18.2}$$

where dQ is the number of phase-space cells with energies in the range ε to $\varepsilon + d\varepsilon$ and dN is the corresponding number of photons. Because of the relation $\varepsilon = pc$, we can readily determine the range of momenta corresponding to any range of energies. The number of phase-space cells dQ equals the volume of phase space corresponding to the prescribed range of momenta, divided by the volume of one cell

$$dQ = \frac{(\text{volume of phase space corresponding to momenta in } p \text{ to } p + dp)}{(\text{volume of one phase-space cell})}$$

The spatial distribution of blackbody photons in a cavity is uniform and isotropic. If V denotes the total volume occupied by the photon gas, we can write for dQ

$$dQ = 2\frac{4\pi p^2\,dp\,V}{h^3} \tag{18.3}$$

We are permitted to use V as the spatial volume element because of the uniformity. The isotropy of the radiation allows us to employ a spherical momentum space volume element $4\pi p^2\,dp$. The volume of one phase-space cell is h^3. The factor 2 displayed so prominently in (18.3) arises from the fact that the spin angular momentum vector of a photon must be parallel or antiparallel to its linear momentum vector. For the prescribed range of p there are $4\pi p^2\,dp/h^3$ cells open to photons having their spin vector parallel to their linear momentum. An equal number of cells can accommodate photons with antiparallel spin and linear momentum vectors. Using (18.1) we can express dQ as

$$dQ = \frac{8\pi V}{c^3}\nu^2\,d\nu \tag{18.4}$$

The chemical potential of a photon gas (ph.g.) is zero,

$$\mu_{\text{ph.g.}} = 0$$

This is related to the fact that the absorption and emission of radiation require an overall conservation of energy, but not photon conservation. For example, an ultraviolet photon could be absorbed and several infrared photons emitted. The chemical potential was introduced as a factor which guaranteed that all candidates for most probable distribution were composed of the same number of particles. Because the total number of photons need not be conserved, the constraint $N = \sum_i^M N_i$ cannot be imposed on the photon gas, with the result that $\mu = 0$. With these results for μ and dQ we can write

$$dN = \frac{8\pi V/c^3}{e^{\beta h\nu}-1}\nu^2\,d\nu \tag{18.5}$$

The energy of each of the dN photons is $\varepsilon = h\nu$, so that the blackbody energy contained in the frequency range from ν to $\nu+d\nu$ is

$$dU = h\nu\,dN = \frac{8\pi hV/c^3}{e^{\beta h\nu}-1}\nu^3\,d\nu \tag{18.6}$$

The spectral energy density $u(\nu)$ is defined such that $u(\nu)\,d\nu$ is the energy density (energy per unit volume) in the frequency range ν to $\nu+d\nu$.

Evidently,

$$u(\nu)\,d\nu = \frac{dU}{V} \tag{18.7}$$

giving

$$u(\nu)\,d\nu = \frac{8\pi h}{c^3}\frac{\nu^3\,d\nu}{e^{\beta h\nu}-1} \tag{18.8}$$

The total energy density $u = U/V$ is obtained by integrating $u(\nu)\,d\nu$ over all frequencies.

$$u = \frac{8\pi h}{c^3}\int_0^\infty \frac{\nu^3\,d\nu}{e^{\beta h\nu}-1} \tag{18.9}$$

Setting $x = \beta h\nu$ and recalling that $\beta = 1/kT$, gives

$$u = \left(\frac{8\pi k^4}{h^3c^3}\right)T^4\int_0^\infty \frac{x^3\,dx}{e^x-1}$$

The value of the integral is $\pi^4/15$ (see Example 14.4). The blackbody energy density is proportional to T^4

$$u = aT^4 \tag{18.10}$$

The blackbody energy density constant is

$$a = \frac{8\pi^5k^4}{15h^3c^3} = 7.56 \times 10^{-15} \quad \frac{\text{erg}}{\text{cm}^3-\text{K}^4}$$

••••••

18.1.2 The energy density of starlight is approximately 1 eV/cm^3. What is the corresponding blackbody temperature?

••••••

As shown below in Example 18.1, the energy density u and the surface flux are related by

$$F = \frac{c}{4}u \tag{18.11}$$

where c is the speed of light. This result leads to the Stefan–Boltzmann Law (17.1), with $\sigma = ca/4$

$$F = \sigma T^4 \tag{17.1}$$

$$\sigma = \frac{2\pi^5k^4}{15h^3c^2} = 5.67 \times 10^{-5} \quad \frac{\text{erg}}{\text{sec-cm}^2\text{-K}^4}$$

The differential form of (18.11) also holds, the monochromatic emissive power $e(\nu)$ and the spectral energy density being related by

$$e(\nu) = \frac{c}{4}u(\nu) \tag{18.12}$$

Inserting (18.8) for $u(\nu)$ gives

$$e(\nu) = \frac{2\pi h}{c^2}\frac{\nu^3}{e^{\beta h\nu}-1} \tag{18.13}$$

which is Planck's form (17.4), with

$$A = \frac{2\pi h}{c^2}, \qquad B = \frac{h}{k}$$

Figure 18.1 plots $e(\nu)$ versus ν for several different blackbody temperatures.

•••••• ••••••

18.1.3 Show that the number density of blackbody photons is given by

$$n = 8\pi\left(\frac{kT}{hc}\right)^3\int_0^\infty \frac{x^2\,dx}{e^x-1}$$

The value of the integral is 2.404 (see Problem 14.4.4).

•••••• ••••••

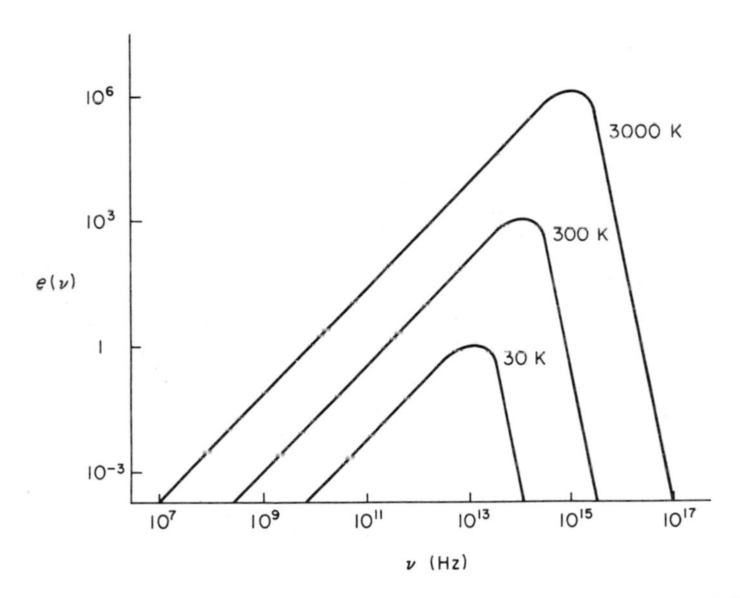

Figure 18.1. The blackbody spectrum for temperatures of 30 K, 300 K, and 3000 K. Both scales are logarithmic, giving the spectrum an appearance different from the version depicted in Figure 17.1. The vertical scale indicates only relative values of $e(\nu)$.

Example 18.1 Blackbody Surface Flux and Energy Density

Theoretical considerations favor the use of the spectral energy density $u(\nu)$. Experimentally, one measures the emissive power $e(\nu)$. The relation between the two is given by (18.12). To derive this, recall the result of Chapter 13, where we showed that the flux of any kinetic quantity $\Phi(\mathbf{v})$ is given by

$$\mathbf{S}_\Phi = \frac{1}{V}\int_{\substack{\text{all}\\\text{veloc.}}} \Phi(\mathbf{v})\,\mathbf{v}f(\mathbf{v})\; d^3v$$

To adapt this to the photon gas we replace $f(\mathbf{v})\,d^3v$ by dN, the number of photons with momenta in a prescribed range. An integration over all phase space replaces the integration over velocities. Since we want the energy flux, we take $\Phi = \varepsilon = h\nu$. Each photon moves at the same speed c and the integration over phase space boils down to integrations over frequency and direction. Let the positive z axis define the outward normal to the radiating blackbody surface. (See Figure 13.3 for the appropriate geometry.) With this choice of coordinates, only the z component of the energy flux will be nonzero, its value being the surface flux F. The z component of the photon velocity is $c\cos\theta$ so that

$$F = \frac{1}{V}\int h\nu c\cos\theta\, dN(\nu,\theta)$$

For dN we write

$$dN = \frac{dQ(\nu,\theta)}{e^{\beta h\nu}-1}$$

with

$$dQ(\nu,\theta) = \frac{2V}{c^2}\nu^2\, d\nu \sin\theta\, d\theta\, d\varphi \tag{18.14}$$

in place of (18.3). The factor $\sin\theta\, d\theta\, d\varphi$ cannot be replaced by 4π here, since photons leave the surface only over the range $\theta = 0$ to $\theta = \pi/2$ (compare Figure 13.3). The integration over phase space gives

$$F = \left\{\frac{2h}{c^3}\int_0^\infty \frac{\nu^3\, d\nu}{e^{\beta h\nu}-1}\right\}\left\{c\int_0^{2\pi} d\varphi \int_0^{\pi/2} \cos\theta\sin\theta\, d\theta\right\}$$

The second factor produces

$$c\int_0^{2\pi} d\varphi \int_0^{\pi/2} \cos\theta\sin\theta\, d\theta = 4\pi\left(\frac{c}{4}\right)$$

so that

$$F = \frac{c}{4}\left\{\frac{8\pi h}{c^3}\int_0^\infty \frac{\nu^3\, d\nu}{e^{\beta h\nu}-1}\right\} \tag{18.15}$$

Comparison of this with (18.9) reveals the desired result

$$F = \frac{c}{4} u \tag{18.11}$$

Further, since

$$F = \int_0^\infty \varepsilon(\nu)\, d\nu, \qquad u = \int_0^\infty u(\nu)\, d\nu$$

it is obvious from (18.15) that the differential relation

$$\varepsilon(\nu) = \frac{c}{4} u(\nu) \tag{18.12}$$

also holds.

It is often desirable to express the energy density in terms of wavelength λ rather than frequency. This is accomplished by introducing a quantity $\Psi(\lambda)$ such that $\Psi(\lambda)\, d\lambda$ is the energy per unit volume contained in the wavelength range from λ to $\lambda + d\lambda$. We write

$$\Psi(\lambda)\, d\lambda = u(\nu)\, d\nu$$

and use the relations†

$$\nu = \frac{c}{\lambda}, \qquad d\nu = \frac{c\, d\lambda}{\lambda^2}$$

to obtain

$$\Psi(\lambda) = \frac{8\pi ch}{\lambda^5} \frac{1}{e^{\beta hc/\lambda} - 1} \tag{18.16}$$

•••••• ••••••

18.1.4 (a) Derive Wien's displacement law, which states that the wavelength λ_{max} for which $\Psi(\lambda)$ is a maximum is inversely proportional to the blackbody temperature. Wien advanced the displacement law in 1896, as the empirical relation

$$\lambda_{max} T = 0.294 \quad \text{cm-K}$$

In the course of the solution you will encounter a transcendental equation which may be put in the form

$$(5 - y) = 5e^{-y}$$

† We denote by $d\nu$ and $d\lambda$ the magnitudes of the frequency and wavelength ranges; the formal act of differentiating $\nu = c/\lambda$ gives $d\nu = -c\, d\lambda/\lambda^2$, the minus sign indicating that changes in ν and λ are of opposite sign.

Inspection reveals that the solution is nearly $y = 5$. Taking $y = 5$ in the exponential, but not in the factor $(5-y)$, gives a better approximation. Show that

$$\lambda_{max} T = \frac{ch}{4.97k}$$

and use the tabulated values of c, h, and k to evaluate the right side.

(b) It is a perhaps surprising fact that the frequency ν_{max} at which $u(\nu)$ has its maximum value does not correspond to the wavelength λ_{max}. That is

$$\nu_{max} \lambda_{max} \neq c$$

Determine ν_{max} and verify this fact.

18.1.5 A typical room temperature is near 300 K. Would you expect radiation of the corresponding λ_{max} to be visible? Explain, without benefit of numerical calculation. The human eye is most sensitive to wavelengths near 5600 Å. What blackbody temperature corresponds to a λ_{max} of 5600 Å? The Sun emits like a blackbody with a temperature of about 5800 K (over a wide range of wavelengths, including the visible spectrum). What is the corresponding λ_{max}? Can you think of any reason for the near coincidence of the two wavelengths?

•••••• ••••••

It was the theory of blackbody radiation which prompted the birth of quantum theory. Recently, experimental measurements of blackbody radiation have helped to settle a question of truly cosmic proportions—the early history of our universe. Until recently two cosmological theories were foremost. The big bang theory was advanced and artfully exposited by Gamow. The steady-state theory was shaped by Bondi, Gold, and Hoyle.

The big bang theory holds that our universe evolved explosively from a superdense state. In its earliest stages the universe was little more than an expanding fireball, inside of which matter was completely dominated by radiant energy—blackbody radiation.

The steady-state theory contends that the large scale observable properties of the universe do not change with time. Observation has shown the universe to be expanding. The mass density is among the observable properties of the universe. For the density to remain constant while the universe expands, there must be a continuous creation of matter. Matter must be created out of nothing to maintain a constant ratio of mass to volume.

Both theories were equally adept at explaining the meager experimental data until recent blackbody radiation measurements ruled in favor of the big bang theory. We can discuss this fascinating topic only briefly. An authoritative account of the recent experiments is given by Peebles and Wilkinson. A thorough discussion of the experiments and their interpretations is presented by Peebles in his monograph "Physical Cosmology." Brief discussions of the theories, written by Gamow and Hoyle, are included in "The Universe" (Newman, 1957).

By detecting remnants of the primeval fireball, the experiments rule against the steady-state theory. The steady-state cosmology does not contain a plausible mechanism for producing the observed blackbody radiation. The universe envisioned by the steady-state theory was never in a dense state—it must have always been the way we find it today.

The experimental evidence for the primeval fireball, and thus for the big bang theory, is most impressive. The fireball remnant completely fills the universe. In a sense the universe is the granddaddy of all radiation cavities. One of the characteristics of blackbody radiation is its isotropy—the intensity of the radiation is the same in all directions. The observations have confirmed the isotropy of the fireball radiation. There is a slight anisotropy in the observed radiation. This has been attributed to the Doppler effect resulting from the motion of the Earth about the center of our Milky Way galaxy. An even more crucial test is whether or not the spectrum of the radiation fits that of a blackbody. It is here that the experiments appear to clinch the case for the fireball hypothesis. Gamow's 1948 theory of the big bang predicted a blackbody temperature of about 10 K. Subsequent theoretical estimates placed the temperature in the 1–30 K range. Measurements of the fireball radiation have been made at wavelengths ranging from a few millimeters to several centimeters. The measured intensities conform nicely with the emissive power $e(\nu)$ of a 2.7 K blackbody.

Several direct measurements fall in the so-called radio window of our atmosphere. At wavelengths longer than about 10 cm electromagnetic noise from our galaxy obscures the fireball signal. Radiation produced in our own atmosphere and radiation scattered by interstellar dust grains hamper measurements at shorter wavelengths. However, an indirect method has given a fourth point on the fireball spectrum (see Figure 18.2) at a wavelength of 2.6 mm. Because the fireball radiation pervades the universe, the vast interstellar gas clouds (along with everything else in the universe) are swimming about in a 2.7 K ocean of radiation. Among the constituents of these gas clouds is the diatomic cyanogen molecule (CN). These molecules are capable of absorbing radiation of many wavelengths, one of which is 2.6 mm. When a CN molecule absorbs a photon corresponding to the 2.6 mm wavelength its (quantized) energy is increased slightly. In the jargon of quantum theory, the molecule is raised from the ground state to its first excited state. A definite fraction of the CN molecules will be in this excited state. The fraction depends directly on the everpresent Boltzmann factor $e^{-\varepsilon/kT}$ and thus on the temperature. What

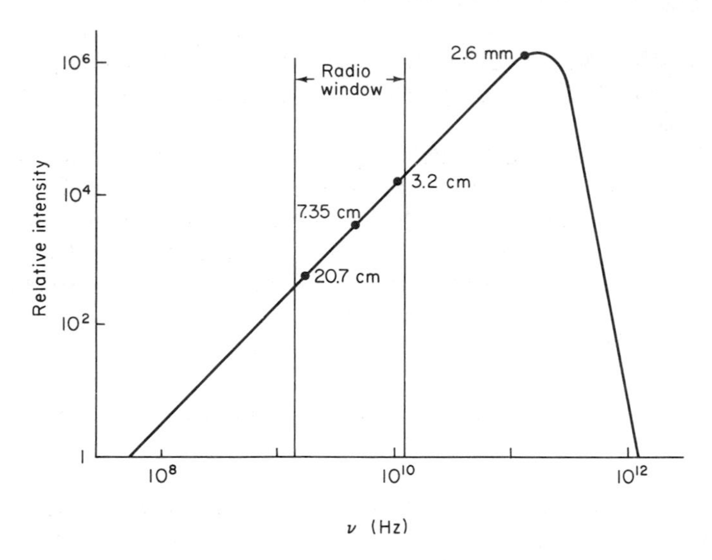

Figure 18.2. The relative intensity for a 2.7 K blackbody plotted versus frequency. Both scales are logarithmic. The relative intensity may be interpreted as the ratio $e(\nu)/e(\nu_0)$, where ν_0 is chosen for convenience. The experimental data points are labeled according to wavelength. The vertical lines define the radio window.

temperature?—the temperature of the CN molecule surroundings, which in this case is the 2.7 K radiation bath. The CN molecules produce certain *absorption* lines in stellar spectra, that is, they intercept certain wavelengths which would otherwise reach the astronomer's photographic plates. The wavelengths absorbed by CN molecules in the ground state are slightly different than those absorbed by molecules in the excited state. By measuring the relative intensities of the absorption lines astronomers can infer the fraction of CN molecules in the excited state. This in turn allows the temperature of the radiation bath to be computed. The experimental results are consistent with a fireball temperature of 2.7 K.

•••••• ••••••

18.1.6 The equation of state of a blackbody photon gas is

$$PV = \tfrac{1}{3}U$$

(a) Use this in conjunction with the result of (18.10), $U = aT^4V$, to show that the entropy of such a gas is

$$S = \frac{4}{3}\frac{U}{T}$$

[*Hint:* Start with $dS = dQ/T$, and use the first law of thermodynamics. In an expanding universe, the density of ordinary matter decreases inversely as the volume ($\rho = m/V$). The corresponding energy density (mc^2/V) also decreases proportional to V^{-1}.]

(b) Show that the blackbody energy density decreases as the inverse 4/3 power of V. [*Hint:* Decide what type of thermodynamic expansion is involved. This result is of importance in understanding the big bang theory. Initially the blackbody energy density far outweighed that of the matter. However, the more rapid decline with volume of the former eventually (about 250 million years after the expansion began) gave ordinary matter the upper hand, and it became possible for galaxy formation to begin.]

18.1.7 The intensity of a typical flashlight beam is roughly 1 W/cm^2. How does this compare with the present fireball energy flux σT^4?

18.1.8 The estimated average mass density in the universe is about 7×10^{-31} gm/cm^3. Cosmological models based on Einstein's general relativity theory require a density of at least 2×10^{-29} gm/cm^3 to give a closed universe. On the basis of present observations the observed expansion of the universe should increase indefinitely. According to Einstein's famous mass–energy relation, $E = mc^2$, any form of energy has a certain mass equivalent. The energy density associated with the primeval fireball is therefore equivalent to a certain mass density. Compute this equivalent mass density. Is it a significant fraction of the value 7×10^{-31} gm/cm^3?

18.1.9 (a) Explain how you could estimate the temperature of the Sun using only your eyes and a thermometer. [You are not permitted to know the Earth–Sun distance! You are permitted to know the angular diameter of the Sun ($\simeq \frac{1}{2}$ degree).]

(b) Make such an estimate. [*Hint:* Assume that the Earth is in radiative equilibrium with the Sun, and that both are blackbodies. Note that the effective absorbing area of the Earth is only $\frac{1}{4}$ of its radiating area.]

•••••• ••••••

18.2 STIMULATED EMISSION: THE LASER

Consider a collection of identical atoms capable of absorbing and emitting radiation of frequency ν. The atoms are situated in a blackbody radiation field with which they exchange photons. Each atom has open to it a discrete set of energy states, the quantum energy levels. An atom can undergo a transition—change its energy state—by absorbing or emitting energy in the form of photons. Absorption of a photon increases the energy of an atom. Photon emission decreases the energy of the atom. Such emissions and absorptions by the atoms represent gains and losses for the blackbody radiation field. At equilibrium, the rates of absorption and emission are equal. It was Niels Bohr who first pictured the emission of radiation as the transition of an atom from a state of higher energy ε_2 to a state of lower energy ε_1. The energy difference $\varepsilon_2 - \varepsilon_1$ is carried away by the emitted photon, it being assumed that the energy and frequency of the photon are related by Planck's relation

$$\varepsilon_{\text{ph}} = h\nu$$

Thus the Bohr theory assumes

$$\varepsilon_2 - \varepsilon_1 = h\nu$$

Einstein recognized three radiative processes: induced absorption, induced emission, and spontaneous emission. Figure 18.3 illustrates the action of the three processes. Einstein discovered the balance existing between these processes by requiring that the energy density of the blackbody radiation field be given by Planck's formula. Figure 18.3a suggests the action of induced absorption. Atoms of energy ε_1 find themselves "bathing" in a stream of

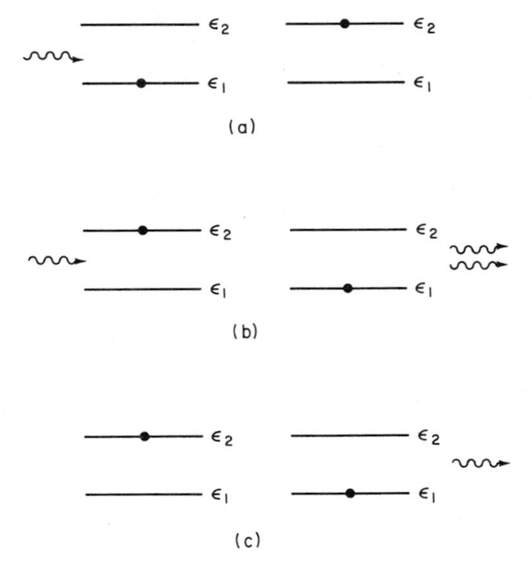

Figure 18.3. The three basic mechanisms by which atoms interact with photons. (a) Absorption: A photon of energy $\varepsilon_2 - \varepsilon_1$ is absorbed by an atom in the level of energy ε_1, raising it to the level of energy ε_2. (b) Induced emission: A photon of energy $\varepsilon_2 - \varepsilon_1$ encounters an atom in the upper level. The incident photon triggers the birth of its "twin." The two photons travel off together, leaving the atom in the lower level. (c) Spontaneous emission: an atom in the upper level spontaneously emits a photon of energy $\varepsilon_2 - \varepsilon_1$. The atom is left in the lower level.

photons. Photons of energy $h\nu = \varepsilon_2 - \varepsilon_1$ can be absorbed, thereby raising the absorbing atoms to the higher energy level. The inverse of induced absorption is induced emission. A stream of photons of energy $h\nu$ may coax atoms of energy ε_2 to emit such photons. As Figure 18.3b suggests each such transition increases the number of photons by one and leaves the atom in the lower energy state.

The rate of induced absorption (emission) is proportional to the number of atoms in the state of energy ε_1 (ε_2) and the density of photons corresponding to frequency ν. This photon density in turn is proportional to the spectral energy density $u(\nu)$. The rates for induced absorption (emission) may therefore be written

$$\frac{\text{number of induced absorptions}}{\text{sec}} = B_{12} N_1 u(\nu)$$

$$\frac{\text{number of induced emissions}}{\text{sec}} = B_{21} N_2 u(\nu)$$

where N_1 and N_2 are the numbers of atoms with energies ε_1 and ε_2 and B_{12} and B_{21} are the Einstein coefficients for induced absorption and emission. The emission of radiation can also proceed spontaneously (Figure 18.3c). Even without the coaxing radiation of frequency ν, an atom of energy ε_2 can "decide" to emit a photon and drop to the state of lower energy. The rate of spontaneous emission is unaffected by the surroundings. It depends only on the internal structure of the atom and the number of atoms in the state of energy ε_2. The rate of spontaneous emissions may therefore be written

$$\frac{\text{number of spontaneous emissions}}{\text{sec}} = A_{21} N_2$$

where A_{21} is called the Einstein coefficient for spontaneous emission. For equilibrium to prevail, the rates of emission and absorption must balance. Adding the rates of spontaneous and induced emission and equating the sum with the rate of absorption yields

$$B_{21} N_2 u(\nu) + A_{21} N_2 = B_{12} N_1 u(\nu)$$

Solving this for $u(\nu)$ gives

$$u(\nu) = \frac{\dfrac{A_{21}}{B_{21}}}{\dfrac{B_{12}}{B_{21}} \dfrac{N_1}{N_2} - 1}$$

With the atoms in thermal equilibrium the ratio of their populations is given by the Boltzmann factor. Since $\varepsilon_2 - \varepsilon_1 = h\nu$ we find

$$\frac{N_1}{N_2} = \frac{\exp(-\beta\varepsilon_1)}{\exp(-\beta\varepsilon_2)} = \exp(\beta h\nu) \tag{18.17}$$

so that

$$u(\nu) = \frac{\dfrac{A_{21}}{B_{21}}}{\dfrac{B_{12}}{B_{21}} e^{\beta h\nu} - 1} \tag{18.18}$$

If this is to agree with the Planck formula (18.8),

$$u(\nu) = \frac{\dfrac{8\pi h\nu^3}{c^3}}{e^{\beta h\nu} - 1} \tag{18.8$'$}$$

we must have

$$B_{12} = B_{21} \tag{18.19}$$

$$A_{21} = \frac{8\pi h\nu^3}{c^3} B_{21} \tag{18.20}$$

It is still necessary to compute one of the coefficients via quantum theory. In fact, quantum theory renders it possible to compute all three Einstein coefficients which may be inserted into (18.18) to produce the Planck distribution.

The competition between emissions and absorptions results in the Planck distribution only when the participating atoms are distributed according to the Boltzmann factor (18.17). The Boltzmann formula describes an equilibrium distribution of atoms. When the atoms are not in thermodynamic equilibrium the intensity of the radiation may depart sharply from the Planck form. Such nonequilibrium situations prevail in lasers.[†] A laser produces an intense, nearly monochromatic, and highly directional beam of radiation. More importantly, laser light is *coherent*. The light waves emitted by individual atoms are locked in step—they are in phase.

To achieve laser action it is necessary to create a nonequilibrium distribution of atoms in which stimulated emission greatly outweighs absorption and spontaneous emission. With reference to Figure 18.3, consider the results which follow if we start with the higher energy level ε_2 populated much more heavily than the lower energy level ε_1. A photon released by spontaneous emission liberates a second photon via stimulated emission. This pair of photons in

[†] The word LASER is an acronym standing for Light Amplification by Stimulated Emission of Radiation.

turn stimulates further emissions. An avalanche of photons quickly builds up.

The important feature of stimulated emission is that the resulting photons travel off "hand-in-hand." That is, the photon released travels off in the same direction as the photon which triggered its emission. From a wave viewpoint we describe this by saying that the incident wave prompts the atom to radiate a wave of the same frequency, traveling in the same direction. The resultant superposition of the two waves differs from the incident wave in only one essential respect—its amplitude is greater. Each stimulated emission strengthens the wave without altering its frequency or direction.

The growing flux of photons is depleted by absorptions. If the two energy levels are populated in accordance with the Boltzmann formula, absorptions match emissions and no photon avalanche develops. When the population of the upper level is greater than that of the lower level, stimulated emission wins out over absorption and laser action becomes possible.

The Boltzmann population ratio may be upset in a variety of ways, two of which will be discussed shortly. The resulting distribution of atoms is often called an inverted population. Producing an inverted population is referred to as pumping since one destroys the Boltzmann distribution by raising atoms from the lower to the higher energy level.

The two essential components of a laser are the active medium and an optical resonator. The active medium contains the atoms or molecules which emit the laser radiation. In addition the active medium may contain other atoms which assist in producing the required population inversion. The optical resonator is simply a pair of mirrors positioned at either end of the active medium. The mirrors enable the laser light to bounce back and forth through the active medium. Each trip through the active medium strengthens the laser beam. One of the mirrors is coated in such a way that about 1% of the incident light is transmitted. This slight leakage through the mirror constitutes the output of the laser.

The first laser action was achieved by T. H. Maiman in 1960. The active medium in Maiman's laser was a ruby crystal. Ruby is aluminum oxide in which a tiny fraction of the aluminum atoms have been replaced by chromium ions. The captivating color of a ruby stems from the fact that the chromium ions absorb visible light strongly except at the red and blue ends of the spectrum. Pumping was performed optically using a xenon flash tube. Figure 18.4 shows a simplified version of the energy levels and transitions for the ruby laser. The flash pumps chromium ions from the ground state 1 into a band of excited states 3. The upper laser level 2 is populated by nonradiative transitions, in which the chromium ions transfer energy to the crystal lattice instead of emitting a photon. If the xenon flash is sufficiently intense, the ground state population falls below that of level 2, whereupon stimulated emission occurs.

The ruby crystal was a cylinder 4 cm in length and $\frac{1}{2}$ cm in diameter. The ends

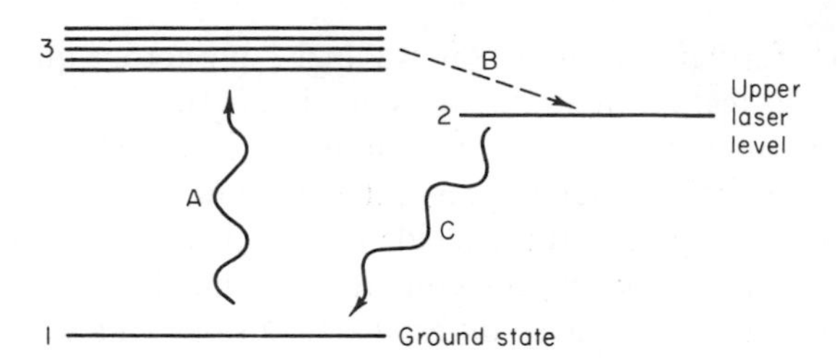

Figure 18.4. Energy levels and transitions for the ruby laser. A, Pump transition $1 \to 3$; B, nonradiative transition $3 \to 2$; C, laser transition $2 \to 1$. (See text for details.)

of the ruby were cut parallel, polished optically flat, and partially silvered, thus forming an optical resonator. Maiman's laser produced an intense pulse of coherent red light at a wavelength of 6943 Å. The pulse lasted about $\frac{1}{2}$ msec. Before moving on to describe the helium–neon laser we must understand those features of the ruby energy levels which promote laser action.

An important consideration in selecting an active medium is the scale of spontaneous emission *lifetimes* of the energy levels involved. Spontaneous emission is the optical counterpart of radioactive decay. While it is not possible to predict when a given radioactive nucleus will decay nor when a given excited atomic state will emit spontaneously, it is possible to establish average lifetimes against such events. For excited atoms typical lifetimes are on the order of 10^{-8} sec. Unusually long lifetimes of a few milliseconds are not uncommon. Such relatively long-lived states are referred to as *metastable*. In ruby the upper laser level 2 is metastable with a lifetime of about 3 msec. The long lifetime against spontaneous emission lets the population of the level build up, helping establish a population inversion. Also helpful is the fact that the nonradiative transitions $3 \to 2$ occur more rapidly than the radiative transitions which return the chromium ion to its ground state $3 \to 1$. Thus the $1 \to 3$ pumping transition populates level 3, which empties into level 2 more rapidly than the inverse $3 \to 1$ transitions can return the ions to the ground state. The fraction of the ions pumped out of the ground state increases with the intensity of the flash. A sufficiently large flash intensity produces a population inversion and thereby triggers laser action.

A prime drawback of the three-level operation typified by the ruby laser is that the laser transition terminates on the ground state level. A significant fraction of the ions must be pumped from the ground state to achieve an inverted population. A more efficient scheme is offered by four-level operation (see Figure 18.5). The fourth level replaces the ground state as the terminus of the laser transition. Prior to pumping, both laser levels are essentially empty by comparison to the ground state, because of the exponential Boltzmann factor. A weak but selective pumping operation suffices to establish an inverted

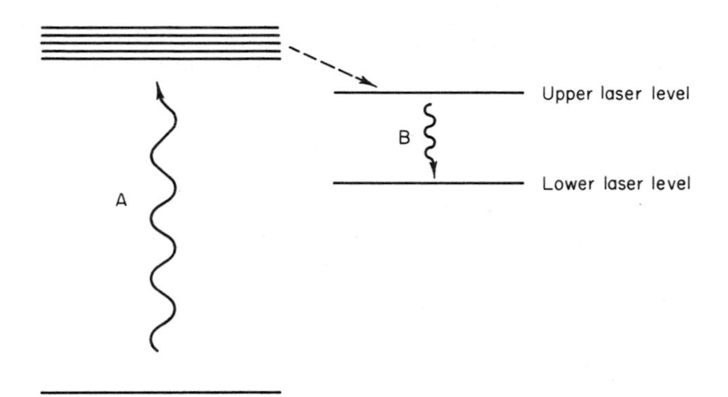

Figure 18.5. A four-level scheme for achieving an inverted population. A, Pump transition; B, laser transition.

population. The pumping action must be effective in populating the upper laser level, but not the lower level. In practice this selectivity is achieved in a variety of ways.

The high powers needed to achieve an inverted population in the three-level ruby laser restricted it to pulsed operation. The four-level scheme makes it feasible to operate a continuous wave (CW) laser. The helium–neon laser is a versatile CW device. It gives the appearance of being a neon sign placed between two mirrors. A popular version of the helium–neon laser operates in the visible spectrum at a wavelength of 6328 Å. A mixture of gaseous helium and neon constitute the active medium. The laser transitions between two levels of the neon atom produce the red 6328 Å radiation. The helium plays a role in the pumping scheme. Figure 18.6 shows the pertinent helium and neon

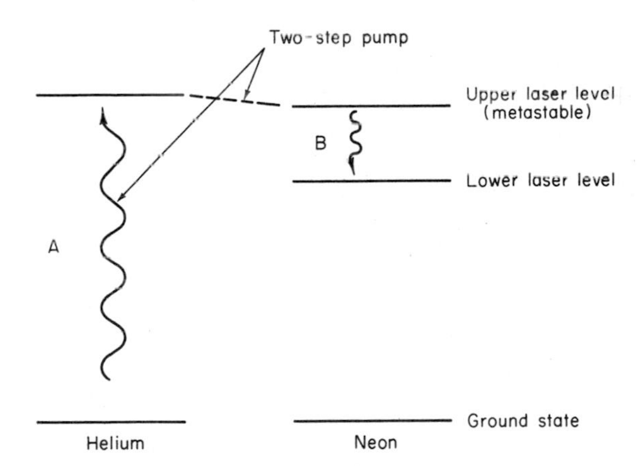

Figure 18.6. Energy levels involved in producing an inverted population in the He–Ne laser. A, Electrical discharge excitation; B, laser transition.

energy levels. An electrical discharge provides energetic electrons which pepper the helium atoms, boosting them into excited states. The energy of the excited state shown for the helium atom is a few hundredths of an electron volt greater than the energy needed to excite the upper laser level of the neon atom. A collision between an excited helium atom and a ground state neon atom can result in a transfer of energy to the neon atom. The neon atom is raised to the upper laser level and the helium atom returns to its ground state. The energy transfer is favored by the very small difference in the excitation energies. The small excess of energy is parceled out as translational kinetic energy between the colliding atoms. This collisional process is selective. The excitation energy of the helium atoms is nearly two electron volts higher than the excitation energy of the lower level of the laser transition. This large difference makes it very unlikely that an excited helium atom will transfer its energy to a neon atom and thereby populate the terminal state of the laser transition.

Of course, collisions between the swift electrons and neon atoms in the ground state will populate both laser levels. However, such collisions are not selective. They do not tend to produce a population inversion. The key to the pumping scheme is the tiny energy difference between the two excited states of helium and neon which favors the upper laser level.

The laser output is highly directional. This results from the fact that the amplitude of the wave builds up via repeated trips through the active medium. Only waves starting out more or less parallel to the axis of the optical resonator will be trapped and have the opportunity to grow in amplitude. Waves starting in other directions are refracted or reflected out of the active medium. The slight divergence of the laser beam is caused by diffraction as the escaping light passes through the mirror. The high directionality of laser light was demonstrated spectacularly in an experiment which beamed laser light to the Moon. The light was reflected back to Earth by an array of corner cubes left by the Apollo 11 astronauts. Optical components expanded the laser beam to give it a diameter of 10 ft when it left Earth. After traveling the nearly 250,000 miles to the Moon its diameter was approximately 1 mile. In traveling to the Moon the beam diameter grew by only 1 ft every 50 miles.

Many different types of lasers have been developed. Technical advances are so swift that any listing of laser applications would quickly become obsolete. Early solid-state lasers, such as the ruby laser, were much more powerful than the gas lasers simply because of the greater density of the active medium in a solid. Gas lasers offered the advantage of continuous operation in contrast to the pulsed action of the solid-state lasers. The hectic pace at which laser research progresses is rapidly eliminating these early drawbacks. There are now CW solid-state lasers. The recently developed carbon dioxide gas laser develops several kilowatts of power—enough to cut through a $\frac{1}{4}$-in. steel plate in a few seconds. Whereas the early lasers operated at a fixed wavelength, recent innovations have made it possible to tune a laser—to select any wavelength within a definite range. Chemical lasers offer another promising area for development. In a conventional laser the active medium draws its energy from an external source. This source may be optical, as in the case of the xenon flash tube, or it may be electrical, as it is with the helium–neon laser. In a chemical laser the active medium draws its energy from a chemical reaction. In fact, the chemical reactions take place inside the active medium. The chemical laser is self-pumping. It is somewhat like having a flashlight in which the bulb and batteries are a one-piece unit.

18.3 WHITE DWARFS, NEUTRON STARS, AND PULSARS

Stars are born when gravitational contraction causes the density and temperature of a protostar to reach levels where thermonuclear reactions are ignited. Initially, the nuclear fuel is hydrogen. Hydrogen burning produces helium and releases energy. Part of this energy remains in the star as the kinetic energy of thermal motion while the rest escapes as radiation, giving the star its luminosity. The pressures associated with the thermal motions and the flow of radiation counterbalance the compressive gravitational force, thereby enabling the star to maintain a stable configuration.

When a nuclear fuel is depleted, gravitational contraction begins anew. Part of the gravitational energy release is used to maintain the stellar luminosity. The remainder serves to raise the temperature, perhaps to a point where the ashes of earlier reactions ignite. The intricacies of stellar burning need not concern us here. After a lifetime on the order of billions of years the stellar furnace runs out of nuclear fuel.

Once a star has exhausted its accessible nuclear fuel the inevitable gravitational contraction resumes. According to currently accepted views this contraction terminates in one of three possible configurations—a white dwarf, a neutron star, or a black hole. The mass of the star as it begins to collapse is the most important quantity in determining which of these stellar corpses results. This need not be the final mass of the star since gravitational collapse may trigger a supernova which blasts a portion of the stellar mass outward into space.

The Sun is likely to die as a white dwarf. Its final gravitational collapse will cause the density to increase to the point where the electrons become degenerate. As degeneracy mounts the electron pressure increases sharply. This rise in pressure slows the gravitational collapse and eventually causes the contraction to subside. A stable configuration results in which the gravitational compression is balanced by the pressure of the degenerate electron gas. There is an upper limit to the mass which can be supported by the electron degeneracy pressure. This limiting value is about 1.2 times the mass of the Sun† and is known as the *Chandrasekhar limit.*

If the mass of the collapsing star is sufficiently great, gravitation can overwhelm the electron degeneracy pressure and a neutron star may result. The density becomes so great in the interior of a neutron star that atomic nuclei dissolve. A mixture forms in which about 99% of the particles are neutrons, hence the name of the star. The remaining 1% is made up of protons and electrons along with a smattering of more exotic particles. The neutrons are strongly degenerate and it is their pressure which resists the crush of gravity and sustains a stable configuration.

Radio astronomers recently discovered the so-called pulsars. Pulsars are characterized by the periodic emission of intense bursts of electromagnetic energy. The presently accepted thinking is that pulsars are rotating neutron stars which generated intense magnetic fields during gravitational collapse. The pulsar magnetic fields are a million times stronger than any yet produced in a laboratory. This field renders the pulsar atmosphere anisotropic, permitting charged particles and electromagnetic radiation to escape only along certain special directions. Two models, one as plausible and picturesque as the other, have attracted the most attention. One model has the rotation of the star and the channeling action of its magnetic field producing a narrow funnel-shaped flow of radiation (see Figure 18.7). The rotating beams of radiation sweep through space like an airport beacon and periodically illuminate the radio telescopes.

The second model proposes that the radiation is emitted by electrons which leave the stellar surface and are accelerated as they slide along the rotating magnetic field lines. Neither the particle nor the field can rotate at a speed in excess of the speed of light. As the particles approach the speed of light their acceleration results in an intense and highly directional beam of radiation. The beam originates well away from the stellar surface, near the velocity-of-light circle. The velocity-of-light circle is centered at the star and has a radius given by

$$R = \frac{2\pi c}{P}$$

where P is the rotational period of the star and c is the speed of light. Most pulsars have rotation periods on the order of 1 sec. We will see later that this was an essential clue to their identification as rotating neutron stars.

The third possible endpoint of stellar evolution is a black hole. A black hole presumably results when the collapsing mass is so great that not even the

---

† The mass of the Sun (2×10^{33} gm) is a convenient stellar mass unit and is designated as $1M_{\odot}$. The Chandrasekhar limit is approximately $1.2M_{\odot}$.

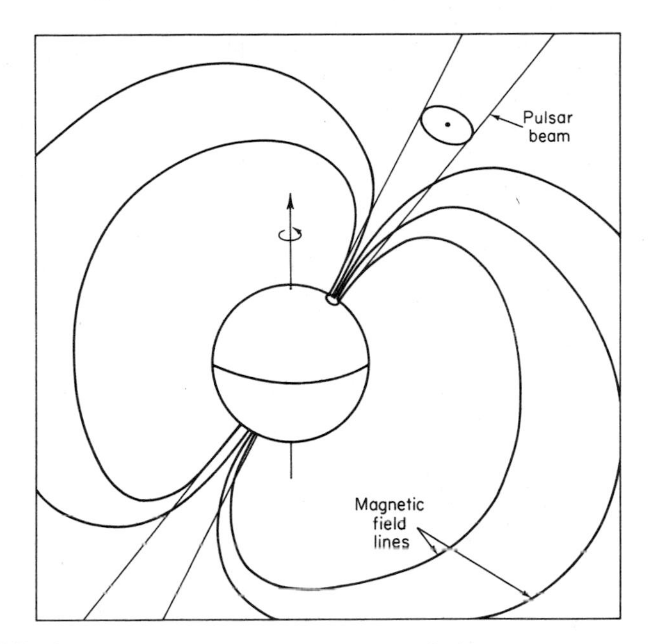

Figure 18.7. The pulsar magnetic field channels radiation, allowing intense beams to escape from the magnetic polar regions. The axis of rotation does not coincide with the magnetic axis. As a result the beams sweep through space like gigantic searchlights.

neutron degeneracy pressure can check its course. The black hole gains its name from the fact that not even light can escape its gravitational pull.

The black hole is presently a speculation, based on the general theory of relativity. None has been detected and it is not yet clear how astronomers might best search for them. Neutron stars have only recently been observed as pulsars. Of the endpoint trio, white dwarfs have the oldest pedigrees. Their discovery early in this century was a major surprise to astrophysicists and an understanding of their structure had to await the invention of quantum statistical mechanics.

Our discussion is limited to white dwarfs and neutron stars.[†] These objects are alike in that their stability against gravitational collapse is sustained by the pressure of degenerate fermions. This pressure is a direct manifestation of the Pauli exclusion principle.

In the development which follows we obtain estimates for the mass, radius, and magnetic field strength of degenerate stars—white dwarfs and neutron stars. The analysis is the sort physicists are most fond of; it amounts to a

[†] References at the end of this chapter include articles that discuss gravitational collapse and black holes.

sophisticated dimensional analysis. The technique is not geared for numerical precision. It can tell us, for example, that the mass of a white dwarf is approximately $1M_{\odot}$ but it can pin down the precise value only to within a factor of 5 or so.

Because of the inherent crudeness of the method we do not hesitate to take minor liberties with numerical quantities. For example, the total number of nucleons (protons or neutrons) in the Sun is 1.2×10^{57}. It is completely within the spirit of our procedures to round this number to 10^{57}. The solar nucleon number is designated by $N_{\odot}$

$$N_{\odot} = 10^{57}$$

Another dimensionless number which enters our considerations is $\hbar c/Gm_n^2$, where G is the gravitational constant and m_n is the nucleon mass. The numerical value of this quantity is 1.6×10^{38}, which we round to 10^{38}. A windfall of this minor fudging is the discovery that

$$N_{\odot}^{2/3} \simeq \frac{\hbar c}{Gm_n^2} \simeq 10^{38}$$

We make frequent use of this near equality in our analysis.

The gravitational potential energy of a star decreases as it collapses. The potential energy sacrificed shows up as an increase in kinetic energy and in other forms of energy, notably radiation. Energy is conserved; gravitational energy is converted into kinetic energy and radiation.

The kinetic energy increase is easily understood. Objects falling under the influence of gravity speed up. However, only about $\frac{1}{2}$ of the gravitational energy is converted into kinetic energy. Much of the remainder escapes from the star as radiation.

If a star of mass M collapses to some final radius R (small by comparison to its initial radius), the gravitational energy released is of order GM^2/R. This figure disregards a numerical factor of order unity whose value depends on how the mass is distributed throughout the star. In order of magnitude then, the kinetic energy increase of the particles is also GM^2/R. Further, the kinetic energy gained during collapse is substantially greater than the initial kinetic energy. We may therefore set

$$\text{kinetic energy of particles in final configuration} = \frac{GM^2}{R}$$

Let N denote the number of degenerate particles whose pressure supports the star against gravitational compression. As degenerate matter these particles are packed as tightly as the exclusion principle permits. The average kinetic energy of such particles is close to $\frac{3}{5}\varepsilon_F$, where ε_F is the Fermi energy (compare Example 17.1). We ignore the factor $\frac{3}{5}$ and adopt $N\varepsilon_F$ as the kinetic energy of the degenerate matter. Equating $N\varepsilon_F$ with the expression for the kinetic energy deduced above gives

$$\frac{GM^2}{R} = N\varepsilon_F \tag{18.21}$$

To make use of this relation we must express the Fermi energy in terms of R and N. Let p_F denote the Fermi momentum. The representative volume of momentum space per particle is $\frac{4}{3}\pi p_F{}^3$; the average spatial volume per particle is $\frac{4}{3}\pi R^3/N$. The product of these two gives the phase-space volume per particle. Changing the numerical factor slightly for later convenience we can write

$$(2\pi p_F)^3 \frac{R^3}{N} = (\text{phase-space volume per fermion})$$

The degenerate fermions fill phase space as densely as the exclusion principle permits, namely one per phase-space cell. This means that the phase-space volume per particle equals h^3, the volume of a single phase-space cell,

$$(2\pi p_F)^3 \frac{R^3}{N} = h^3 \tag{18.22}$$

The Fermi momentum follows as

$$p_F = \hbar \left(\frac{N}{R^3}\right)^{1/3} \tag{18.23}$$

Observe that the Fermi momentum is independent of the mass of the fermions; it is completely determined by their number density.

By eliminating p_F and ε_F between (18.21) and (18.23) we obtain a relation between R and N. Since N is directly proportional to the stellar mass this result also establishes the mass–radius relation for the star.

The relation between ε_F and p_F for nonrelativistic fermions is $\varepsilon_F = p_F{}^2/2m$. However, in white dwarfs the electrons move at relativistic speeds—they have kinetic energies comparable to their rest mass energy mc^2. The special theory of relativity shows that the kinetic energy of a particle is given by

$$\sqrt{(pc)^2 + e_0{}^2} - e_0$$

where p is the linear momentum and $e_0 = mc^2$ is the rest mass energy.

•••••• ••••••

18.3.1 (a) Use the binomial expansion to show that $\sqrt{(pc)^2 + e_0{}^2} - e_0 \simeq p^2/2m$, in the nonrelativistic limit where $pc \ll e_0 = mc^2$.

(b) What is the ultrarelativistic limit for the kinetic energy ($pc \gg e_0$)?

•••••• ••••••

Since we have identified ε_F as the kinetic energy we write

$$\varepsilon_F = \sqrt{(p_F c)^2 + e_0^2} - e_0 \tag{18.24}$$

Inserting this into (18.21) gives

$$\frac{GM^2}{R} = N[\sqrt{(p_F c)^2 + e_0^2} - e_0] \tag{18.25}$$

Minor algebraic manipulation converts this into

$$\left(\frac{GM^2}{RN}\right)^2 + 2e_0\left(\frac{GM^2}{RN}\right) = (p_F c)^2 \tag{18.26}$$

Finally, we use the result of (18.23) to eliminate p_F, obtaining

$$\left(\frac{GM^2}{RN}\right)^2 + 2e_0\left(\frac{GM^2}{RN}\right) = (\hbar c)^2\left(\frac{N}{R^3}\right)^{2/3} \tag{18.27}$$

Once we specify the relation between M and N, (18.27) gives us the sought-for relation between M and R.

For a neutron star, where the neutrons supply the degeneracy pressure and the mass we have

$$M = Nm_n \tag{18.28}$$

where $m_n = 1.67 \times 10^{-24}$ gm is the nucleon mass. In a white dwarf, where electrons supply the degeneracy pressure, (18.28) does not apply because the number of nucleons is not equal to the number of electrons. Let μ_e denote the number of nucleons per electron. In place of (18.28) we have

$$M = \mu_e Nm_n \tag{18.29}$$

In a white dwarf it is expected that $\mu_e \simeq 2$. This is the value of μ_e appropriate for a star which has burned its hydrogen into helium.†

In keeping with the spirit of our method we ignore the μ_e factor and use (18.28) throughout. Problem 18.3.3 provides an opportunity to demonstrate that our method yields the correct μ_e dependence.

•••••• ••••••

18.3.2 Explain why $\mu_e = 2$ for a pure helium star.

•••••• ••••••

† The temperature of a stellar interior is so high that all matter is completely ionized. Thus helium is not present as a neutral atom, but as an α particle and 2 electrons. Likewise, the hydrogen exists as free protons and electrons.

If we insert (18.28) into (18.27), multiply through by $R^2/(Gm_n^{\ 2})^2$, and recall that $N_\odot^{2/3} \simeq \hbar c/Gm_n^{\ 2}$, we obtain

$$N^2 + 2N^{4/3}N_\odot^{2/3}\left(\frac{R/N^{1/3}}{\lambda_c}\right) = N^{2/3}N_\odot^{4/3} \tag{18.30}$$

The quantity

$$\lambda_c = \frac{\hbar}{mc} \tag{18.31}$$

is known as the *Compton wavelength* wherein m is the mass of the degenerate fermion which supplies the pressure. For white dwarfs m is the electron mass; for neutron stars m is the nucleon mass. Numerically,

$$\begin{aligned} \lambda_c &= 3 \times 10^{-11} \text{ cm}, \qquad \text{white dwarf} \\ \lambda_c &= 2 \times 10^{-14} \text{ cm}, \qquad \text{neutron star} \end{aligned} \tag{18.32}$$

Solving (18.30) for $(R/N^{1/3})$ gives the mass–radius relation

$$\left(\frac{R}{N^{1/3}}\right) = \frac{1}{2}\lambda_c\left(\frac{1}{x} - x\right) \tag{18.33}$$

where

$$x = \left(\frac{N}{N_\odot}\right)^{2/3} = \left(\frac{M}{M_\odot}\right)^{2/3} \tag{18.34}$$

The last equality follows from (18.28) and the fact that $M_\odot = N_\odot m_n$ is the mass of the Sun. The quantity $((1/x)-x)$ is negative for values of x greater than one. Since the radius cannot be negative we conclude that x must not exceed unity

$$x = \left(\frac{M}{M_\odot}\right)^{2/3} \leqslant 1 \tag{18.35}$$

Equation (18.35) shows that there is an upper limit to the mass of a degenerate star, namely,

$$M \leqslant M_\odot \tag{18.36}$$

The upper limit of $M_\odot$ corresponds to $x = 1$ and $R = 0$. The idea of a zero radius star is of course not realistic. Nevertheless, the existence of an upper limit has an important implication. It means simply that normal stars of mass substantially greater than $M_\odot$ cannot die as degenerate stars unless they first manage to shed mass. Observations reveal that a sizable fraction of the stars have masses above $M_\odot$ and this poses a question as to their ultimate fate.

The original analysis by B. S. Chandrasekhar gave an upper limit of $5.76 M_\odot/\mu_e^2$ for white dwarfs. More refined calculations lowered the Chandrasekhar limit to approximately $5 M_\odot/\mu_e^2$. For the expected value of $\mu_e \simeq 2$ the Chandrasekhar limit is about $1.2 M_\odot$. Refined calculations, which account for the interaction between nucleons, show that neutron star configurations can exist over the range from about $0.25 M_\odot$ to near $2 M_\odot$. Thus the scale of one solar mass suggested by our rudimentary analysis is not far off the mark.

•••••• ••••••

18.3.3 Show that the limiting mass for a white dwarf emerges as $M_\odot/\mu_e^2$ when (18.29) is used in place of (18.28).

•••••• ••••••

For modest values of x, (18.33) shows that the scale for the stellar radius is

$$R \simeq \lambda_c N^{1/3} \tag{18.37}$$

For a star with $M \simeq M_\odot$ we have $N \simeq N_\odot$. Inserting $N_\odot^{1/3} \simeq 10^{19}$ for $N^{1/3}$ and noting (18.32) we find†

$$\begin{aligned} R_{\text{W.D.}} &\simeq 3 \times 10^8 \text{ cm} \simeq 5 \times 10^{-3} R_\odot \\ R_{\text{N.S.}} &\simeq 2 \text{ km} \end{aligned} \tag{18.38}$$

The white dwarf (W.D.) figure is comparable to the radius of the Earth (6×10^8 cm). Precise calculations for white dwarfs indicate somewhat larger radii, the exact value depending on the stellar mass. Calculations for neutron stars (N.S.) predict a radius of approximately 13 km, independent of the stellar mass.

As yet no experimental determinations of mass or radius have been obtained for neutron stars. Over 200 white dwarfs have been identified, but mass and radius determinations have been made for only eight. Data for three white dwarfs are

$$\begin{aligned} &\text{40 Eri B } (R = 1.6 \times 10^{-2} R_\odot; \quad M = 0.44 M_\odot) \\ &\text{Sirius B } (R = 2 \times 10^{-2} R_\odot; \quad M = 0.98 M_\odot) \\ &\text{Procyon B } (R = 10^{-2} R_\odot; \quad M = 0.65 M_\odot) \end{aligned}$$

The data for all eight stars are consistent with our results. Although meager, they reassure us that our primitive approach defines reliable mass and radius scales.

---

† $R_\odot$ is the radius of the Sun 7×10^{10} cm.

••••• •••••

18.3.4 Estimate the average mass density of a white dwarf and a neutron star. Compare the latter with the density of an atomic nucleus ($\rho_{\text{nuc}} \simeq 10^{15}$ gm/cm^3). At your computed densities, what would be the weights (on Earth) of 1 cm^3 of such matter?

••••• •••••

The magnetic field at the surface of a neutron star is expected to be on the order of 10^{12} G. By comparison, the strongest fields produced by superconducting magnets are less than 10^6 G. Such intense stellar magnetic fields presumably arise from flux trapping. The ionized stellar material is a nearly perfect electrical conductor. As a consequence of this the magnetic lines of force are swept inward with the collapse. The magnetic flux threading the star is thereby trapped—it remains constant during the collapse. The magnetic flux Φ is proportional to the product of the magnetic field H and the cross-sectional area of the star πR^2,

$$\Phi \propto \pi R^2 H$$

It follows from the constancy of Φ that

$$R_i^2 H_i = R_f^2 H_f \tag{18.39}$$

where i and f refer to the initial and final configurations. If a star the size of the Sun ($R_i \simeq 10^{11}$ cm), with a field of order 100 G collapses to a neutron star configuration ($R_f \sim 10^6$ cm), the resulting magnetic field strength is 10^{12} G. Field strengths of this size are consistent with the observed rates of pulsar slow-down, the barely perceptible increases of pulsar periods.

A similar multiplication of the magnetic field strength should accompany the formation of a white dwarf. Recent observations of white dwarf spectra have indicated field strengths in the 10^4–10^5 G range.

••••• •••••

18.3.5 The magnetic field at the surface of the Sun is roughly 1 G. The strength of the field in the solar interior is not known with any degree of certainty. However, the field is stronger in the interior than at the surface. Estimate the field strength which would result if the Sun collapsed to a white dwarf of radius $0.01 R_\odot$.

••••• •••••

It was noted earlier that most pulsars have a period near 1 sec. The range of pulsar periods was an important clue in their identification as rotating neutron stars. After the discovery of pulsars in 1967 both neutron stars and white dwarfs were suggested as possible sources of the radiation. Both rotation and vibration were advanced as the mechanisms responsible for the clock-like regularity of pulsar signals. White dwarfs and vibrating neutron stars were shown to lack one or more of the necessary characteristics, leaving the rotating neutron star as the only likely candidate.

The rotating white dwarf was eliminated on dynamic considerations. The gravitational acceleration at the surface of a star of radius R and mass M is $g = GM/R^2$. The minimum period of rotation (P_{min}) of such a star is fixed by Newton's second law

$$g = R\omega^2, \qquad \omega = \frac{2\pi}{P_{\text{min}}}$$

Periods less than P_{min} would result in the star tearing itself to shreds. Gravity would no longer be able to supply the necessary centripetal acceleration. From $g = R\omega^2$ we find

$$P_{\text{min}} = 2\pi \frac{R^{3/2}}{\sqrt{GM}} \tag{18.40}$$

For a neutron star one finds

$$P_{\text{min}} \simeq 10^{-3} \text{ sec}$$

while for a white dwarf

$$P_{\text{min}} \simeq 10 \text{ sec}$$

The observed pulsar periods range from 33 msec to about 3 sec, thereby eliminating the rotating white dwarf. A white dwarf is simply too big to rotate once each second.

Vibrating white dwarfs and neutron stars also may be ruled out as likely seats of pulsar activity. The fundamental period of pulsation of a star is proportional to the quotient of the stellar radius and the characteristic soundspeed (V_s)

$$P \sim \frac{R}{V_s} \tag{18.41}$$

The vibrations propagate as pressure variations. Hence the soundspeed in (18.41) is the speed characteristic of the degenerate particles which supply the stellar pressure.

In a neutron star the pressure is so enormous that the soundspeed approaches a fundamental limit—the speed of light. With $R \simeq 10^6$ cm and $V_s \simeq 10^{10}$ cm/sec the fundamental pulsation period of a neutron star is

$$P_{\text{N.S.}} \simeq 10^{-4} \text{ sec}$$

The somewhat less rigid structure of white dwarfs propagates sound at speeds of order 10^8 cm/sec. With $R \simeq 10^8$ cm, the fundamental pulsation period for white dwarfs is

$$P_{\text{W.D.}} \simeq 1 \text{ sec}$$

The pulsar located in the Crab nebula has a period of 1/30 sec. The Vela pulsar period is less than 1/10 sec. These pulsars cannot be vibrating white dwarfs. Conversely, pulsating neutron stars are ruled out because their fundamental period is much too low.

The remaining pulsar candidate—the rotating neutron star—is left as the only serious contender. Indeed, pulsar models based on the neutron star configuration have successfully explained many of the observations.

Problems associated with the structure of neutron stars provide a fertile field for applications of statistical physics. We conclude our discussion of degenerate stars by showing how one arrives at the conclusion that a neutron star is a blend of 99% neutrons and 1% electrons and protons.

The free neutron is an unstable particle and decays into a proton, electron, and antineutrino,

$$\text{n} \rightarrow \text{p} + \text{e}^- + \tilde{\nu} \tag{18.42}$$

The rest mass energy of a neutron exceeds that of the products. The energy difference shows up as the kinetic energy of the decay products. If an electron and proton collide with sufficient energy, the inverse reaction can occur

$$\text{p} + \text{e}^- \rightarrow \text{n} + \nu \tag{18.43}$$

In a neutron star interior the Fermi energy of the electrons is sufficiently great that this inverse reaction is energetically possible. An equilibrium is established, the rate of neutron decay being balanced by the rate of neutron production.

The condition for chemical equilibrium was established in Chapter 8 (see Example 8.3). The balance between neutron formation and decay is subject to the same conditions. In particular, (8.47) applies to (18.42). Equation (8.47) describes the balance of chemical potentials. For (18.42) it becomes

$$\mu_{\text{n}} = \mu_{\text{p}} + \mu_{\text{e}} + \mu_{\tilde{\nu}} \tag{18.44}$$

The chemical potential for the antineutrinos is zero. For the other species there is only an insignificant difference between the chemical potential and the Fermi energy. Thus (18.44) may be replaced by

$$\varepsilon_{\text{n}} = \varepsilon_{\text{p}} + \varepsilon_{\text{e}} \tag{18.45}$$

Electrical neutrality demands equal numbers of electrons and protons. As was demonstrated earlier, the Fermi momentum depends on the number density but not on the mass of the fermion. The electrons and protons therefore have equal Fermi momenta. By virtue of their smaller masses the electrons acquire a much higher Fermi energy than do the protons. Ignoring ε_p in favor of ε_e in (18.45) gives

$$\varepsilon_n \simeq \varepsilon_e \tag{18.45a}$$

The neutrons move with nonrelativistic energies, whereas the electrons are ultrarelativistic (see Problem 18.3.6). We have

$$\frac{p_n^2}{2m_n} = \varepsilon_n, \qquad p_e c = \varepsilon_e$$

where p_n and p_e are the Fermi momenta. The Fermi momentum follows from the "pack-'em-in-as-tightly-as-Pauli-allows" principle,

$$p_F = \hbar(3\pi^2 n)^{1/3} \tag{17.17'}$$

where n is the number density. The equilibrium condition (18.45a) then becomes

$$\frac{\hbar^2}{2m_n}(3\pi^2 n_n)^{2/3} = \hbar c(3\pi^2 n_e)^{1/3}$$

which converts to

$$\frac{n_e}{n_n} = \frac{3\pi^2}{8}\left(\frac{\hbar}{m_n c}\right)^3 n_n \simeq 10^{-41} n_n \tag{18.46}$$

The mass density in a neutron star is on the order of 10^{15} gm/cm^3 which means the neutron number density is about 10^{39} cm^{-3}. Thus $n_e/n_n \simeq 10^{-2}$; neutrons are roughly 100 times more plentiful than electrons and protons.

•••••• ••••••

18.3.6 (a) Take $n_e = 10^{37}$ cm^{-3} and compute the electron Fermi energy (assume they are ultrarelativistic). How does your result compare with the electron rest mass energy $mc^2 = 0.51$ MeV? Was it justifiable to treat the electrons as ultrarelativistic?

(b) Determine the neutron number density and mass density corresponding to $n_e = 10^{37}$ cm^{-3}.

•••••• ••••••

18.4 ^{3}He–^{4}He DILUTION REFRIGERATORS

During the past decade a new type of ultra low temperature refrigerator has been developed. It utilizes a mixture of helium isotopes, ^{3}He and ^{4}He. It is referred to as a *dilution refrigerator* because cooling is achieved by diluting the mixture. There are, in fact, two ways of cooling via dilution. Interestingly enough, the two dilution schemes are analogous to two classic methods of cooling—namely, adiabatic expansion and forced evaporation. To help expose the analogies we first review these two classic coolers. Following this review we catalog the relevant properties of ^3He and ^4He, and mixtures of the two. The two methods of cooling via dilution are then discussed in some detail. Commercially available dilution refrigerators can maintain continuous temperatures below 0.03 K.

The thermodynamic basis of the adiabatic expansion method is this: A gas expands adiabatically and performs work against its surroundings. The gas imports no heat, hence the work is done at the expense of its internal energy. An internal energy decrease is accompanied by a temperature drop. At the microscopic level one pictures molecules striking the expanding walls of their container. They transfer energy to the moving walls. The decrease in their kinetic energy registers macroscopically as a temperature drop.

Cooling by forced evaporation works like this: Suppose one starts with a liquid in equilibrium with its vapor, and then begins to pump away the vapor. This upsets the thermodynamic equilibrium. Liquid evaporates in an effort to restore the equilibrium, but evaporation requires heat—the latent heat of vaporization. The vapor carries off this latent heat, which is furnished by the liquid. So, in trying to maintain thermodynamic equilibrium with its vapor, the liquid rejects heat and is cooled. Microscopically, we know that temperature measures the average kinetic energy of the molecules. To overcome the attractive forces of its neighbors in the liquid and escape to the vapor, a molecule must have a kinetic energy considerably above the average. To put it briefly, only the fast molecules can escape. Their exodus lowers the average kinetic energy—and thus the temperature—of those which remain in the liquid.

There are inherent limitations on the classic coolers which prevent them from reaching temperatures below about 0.3 K. Adiabatic expansion machines suffer from the obvious difficulty of lubricating moving parts at liquid helium temperatures. Furthermore, the fractional change in temperature $\Delta T/T$ resulting from an adiabatic expansion decreases with the temperature. These drawbacks limit adiabatic expansion coolers to temperatures above about 5 K.

The lowest temperature which can be achieved by forced evaporation is limited by the rate at which the vapor can be pumped away. The pumping speed is proportional to the pressure head, the pressure difference between the vapor and the pump. For practical purposes the pressure head equals the vapor pressure, but the vapor pressure is proportional to $\exp(-L_v/RT)$, where L_v is the latent heat of vaporization. Thus as T decreases the vapor pressure falls exponentially, and with it the pumping speed. Microscopically it is a case of having fewer and fewer energetic molecules capable of escaping to the vapor. The vapor becomes tenuous. Pumping it away carries off little heat and the rate of cooling diminishes. A lower limit of about 0.3 K can be reached by pumping on ^{3}He, the rare isotope of helium.

An understanding of the principles of dilution refrigeration requires a familiarity with the properties of the stable helium isotopes, ^{3}He and ^{4}He, and mixtures of the two. The ^{4}He atoms are bosons; the ^{3}He atoms are fermions. This very important distinction stems from the fact that the ^{3}He nucleus contains an odd number of fermions, one neutron and two protons. The ^{4}He nucleus contains an even number of fermions, two neutrons and two protons. Both ^{3}He and ^{4}He remain liquids under their saturated vapor pressure at the lowest temperatures yet achieved. All available experimental evidence indicates they would remain liquids even at absolute zero. Solid phases can be produced by raising the pressure to near 30 atm.

^{4}He exhibits superfluid behavior below the lambda point $T_\lambda = 2.172$ K. Below 0.5 K, ^{4}He has essentially zero viscosity, zero heat capacity, zero entropy, and zero pressure. This parade of near zeroes may be traced to the momentum distribution of ^{4}He atoms below T_λ. As the temperature falls below T_λ the ^{4}He atoms begin to condense in the state of zero momentum. As bosons, there is no exclusion principle operating to prevent such a pile-up of ^{4}He atoms in the same momentum space cell. The fraction of the ^{4}He atoms in the zero momentum state increases sharply from zero, at T_λ, to one at absolute zero. In mixtures of ^{3}He and ^{4}He, the ^{4}He behaves like a mechanical vacuum; its near-zero pressure allows it to play a neutral role in so far as the mechanical aspects are concerned. The ^{4}He is essential to the dilution refrigerator for two reasons; first, it takes up space—it dilutes the ^{3}He. Of greater importance is the fact that pure ^{4}He has a greater affinity for a ^{3}He atom than does pure ^{3}He. We will touch on this point in some detail after describing the properties of ^{3}He–^{4}He mixtures.

Mixtures of ^{3}He and ^{4}He are tantalizing. Above 0.87 K the composition can be varied at will. (The ^{3}He boils away above 3.2 K.) At temperatures below 0.87 K there can be a phase separation. The mixture separates into two distinct components. The precise temperature of the phase separation depends on the relative concentrations of ^{3}He and ^{4}He. Both phases are liquids and in fact both are mixtures of ^{3}He and ^{4}He. The two phases are distinguished by the relative concentrations of ^{3}He and ^{4}He. The phase separation was discovered in 1956 by G. Walters and W. Fairbank. Figure 18.8 shows the phase diagram for ^{3}He–^{4}He mixtures. The solid line in the figure locates the phase separation boundary. The quantity X measured along the abscissa is the ^{3}He fraction

$$X \equiv \frac{N_3}{N_3 + N_4}$$

where N_3 and N_4 are the numbers of ^{3}He and ^{4}He atoms. At any temperature above 0.87 K a horizontal line on the phase diagram does not intercept the

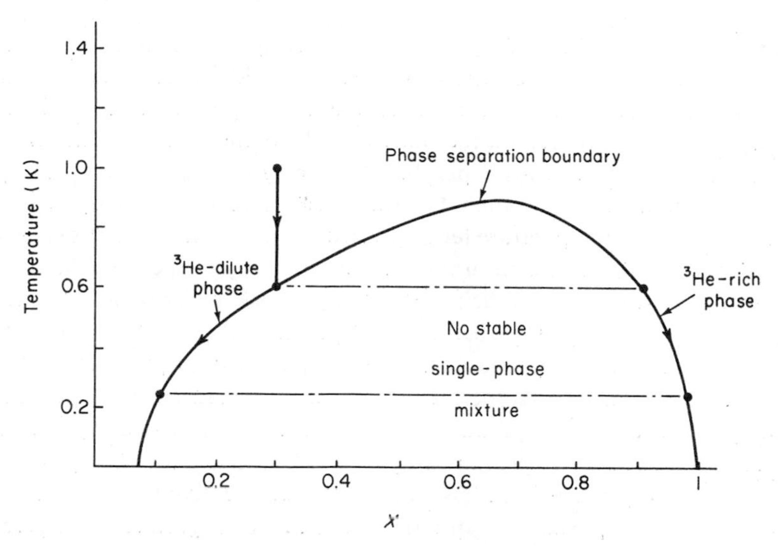

Figure 18.8. Phase diagram for ^3He ^4He mixtures, where X denotes the fraction of ^3He atoms in the mixture(s).

phase separation boundary, indicating that any mixing ratio of ^3He–^4He forms a stable single-phase solution. At any temperature below 0.87 K a horizontal line on the diagram intersects the phase separation boundary at two points. These two points fix the values of X for the two phases. The region beneath the phase separation boundary does not correspond to any stable thermodynamic state of ^3He–^4He mixtures.

The arrows in Figure 18.8 indicate the sequence of events which accompany the progressive cooling of a ^3He–^4He mixture. The system starts at a temperature of 1 K as a single-phase mixture in which 30% of the atoms are ^3He. As the mixture is cooled, nothing spectacular happens until the temperature reaches about 0.60 K. At this point there is a phase separation. The mixture separates into two distinct liquid layers. One layer is greatly enriched in ^3He, being nearly 90% ^3He. This ^3He-rich phase is less dense and so floats on the dilute phase. As the temperature is lowered still further, the ^3He-rich phase grows in size and becomes richer in ^3He, while the dilute phase is progressively impoverished.

Below a temperature of about 0.2 K the ^3He-rich phase is essentially pure ^3He while the dilute phase—the bottom layer—is about 6% ^3He and 94% ^4He. As the diagram reveals there will be about 6% ^3He in the dilute phase even at absolute zero. This unique property of ^3He–^4He mixtures was not confirmed until nearly 10 years after the discovery of the phase separation. It was expected that the ^3He-rich phase would behave as indicated in the diagram. The 6%

solubility of ^{3}He in the dilute phase was a welcome surprise: it is the key to dilution refrigeration. Research workers in the field felt that the ^{3}He-dilute phase would become pure ^{4}He as the temperature approached absolute zero. The arguments were based on the third law of thermodynamics and the tried and sometimes deceptive principle that "believing is seeing." The pre-experimental argument went like this: The third law decrees that the entropy tend toward zero as the temperature tends toward zero. Furthermore, a decrease in entropy means an increase in order. An obvious way to increase the order is via a phase separation, with the pure ^{3}He and pure ^{4}He forming two different layers. So went the argument before the experimental evidence argued otherwise. The argument is weak because the entropies of both pure ^{3}He and pure ^{4}He tend to zero as the temperatures approach absolute zero. The benefits accruing from a phase separation are of secondary importance. The configuration which actually prevails hinges on the question of energy. In other words, all mixing ratios have essentially the same entropy—nearly zero. It therefore becomes primarily a question of which configuration will give the lowest free energy.

We stated earlier that it is energetically favorable for ^{3}He atoms to mix with ^{4}He atoms. They are more tightly bound that way. Figure 18.9 compares answers to the questions "What energy levels are open to ^{3}He atoms added to pure ^{3}He?" and "What energy levels are open to ^{3}He atoms added to pure ^{4}He?" The figure indicates conditions at absolute zero and therefore ignores purely thermal energies. In part (a) the first ^{3}He atom added to pure ^{4}He is bound with an energy E_{34}. Subsequently added ^{3}He atoms are not as tightly bound. The Pauli exclusion principle forces the ^{3}He fermions into states of higher energy. In part (b) the first ^{3}He added to pure ^{3}He is bound with an energy E_{33}. The fact that $E_{34} > E_{33}$ means that ^{3}He will dissolve in ^{4}He rather than separating as a separate phase. There is a limit to the amount of ^{3}He which will dissolve in ^{4}He (for $T < 0.87$ K) because the Pauli principle forces each succeeding ^{3}He atom into a level of energy greater than E_{34}. Eventually it no longer becomes energetically favorable for the ^{3}He to dissolve in the ^{4}He. At temperatures below about 0.1 K the limit is near 6% ^{3}He.

It is instructive to see how theory and experiment are combined to predict the limiting value of 6% for the solubility of ^{3}He in the dilute phase. Thermodynamically, phase equilibrium requires equality of the chemical potentials of the two phases. From a microscopic viewpoint phase equilibrium results when a ^{3}He atom does not find it energetically more favorable to join one phase than the other. This interpretation permits us to write the condition for phase equilibrium as

energy of ^{3}He atom added to the ^{3}He-rich phase
= energy of ^{3}He atom added to dilute solution of ^{3}He–^{4}He

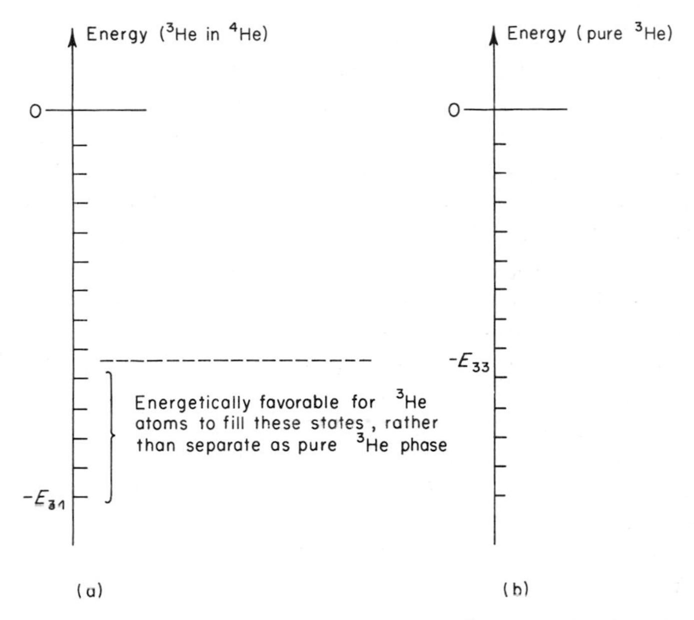

Figure 18.9. Schematic of the energy levels open to (a) ^{3}He atoms in a ^{3}He–^{4}He mixture and in (b) pure ^{3}He at absolute zero. A single ^{3}He atom is bound more tightly in ^{4}He than in pure ^{3}He. The Pauli exclusion principle forces additional ^{3}He atoms into states of higher energy, but these states are of lower energy than the states open in pure ^{3}He. Thus ^{3}He will dissolve in ^{4}He rather than separate as a second phase.

This is expressed symbolically as

$$\mu_{33} = \mu_{34}$$

Because we restrict ourselves to temperatures near absolute zero where the purely thermal energies are ignorable, expressions for μ_{33} and μ_{34} follow at once from Figure 18.9. At sufficiently low temperatures the ^{3}He-rich phase is nearly pure ^{3}He. This fact allows us to write

$$\mu_{33} \simeq -E_{33} \tag{18.47}$$

where E_{33} is the energy required to remove a ^{3}He atom from pure ^{3}He. Experimentally, E_{33} is the latent heat (per atom) of pure ^{3}He. Figure 18.9 reveals the energy of a ^{3}He atom in the dilute phase as a sum of kinetic and potential energies

$$\mu_{34} = -E_{34} + \varepsilon_F \tag{18.48}$$

The negative term, $-E_{34}$, is the potential energy with which the ^{3}He atom is bound to its neighbors in the liquid. The positive term, ε_F, is the kinetic energy forced upon the atom by the Pauli exclusion principle, and this equals the Fermi energy of the added ^{3}He atom [compare Eq. (17.16)]

$$\varepsilon_F = \frac{\hbar^2}{2m}(3\pi^2 n_3)^{2/3} \tag{18.49}$$

The mass m in (18.49) is not the mass of a ^{3}He atom. The effective mass of the ^{3}He is increased by virtue of the interaction with neighboring atoms. Experiments which measure the speed at which temperature differences propagate through ^{3}He–^{4}He mixtures show that m is equal to 2.5 times the mass of a ^{3}He atom. For a dilute mixture the ^{3}He fraction is related to the number density of the ^{3}He by

$$X \equiv \frac{N_3}{N_4+N_3} \simeq \frac{N_3}{N_4} = \frac{n_3}{n_4}$$

where n_4 is the number density of ^{4}He. Thus the Fermi energy term is proportional to $X^{2/3}$

$$\varepsilon_F = \frac{\hbar^2}{5m_3}(3\pi^2 n_4)^{2/3} X^{2/3} \tag{18.50}$$

With these results the condition for phase equilibrium becomes

$$-E_{33} = -E_{34} + \frac{\hbar^2}{5m_3}(3\pi^2 n_4)^{2/3} X^{2/3} \tag{18.47a}$$

The energy E_{34} depends on X. This dependence is determined from measurements of the temperature at which the phase separation occurs. The experiments conducted by Edwards and his co-workers show that $E_{34}(X) - E_{33}$ increases linearly with X. Over the experimental range of X their results may be represented by

$$E_{34}(X) - E_{33} \simeq 2.7(1+14X) \times 10^{-17} \text{ erg}$$

The values of n_4 and m_3 are

$$n_4 \simeq 2.2 \times 10^{22} \text{ cm}^{-3}, \qquad m_3 \simeq 5 \times 10^{-24} \text{ gm}$$

The limiting value of X emerges as the solution of

$$1 + 14X = 12X^{2/3} \tag{18.51}$$

As the student may verify, the solution is $X \simeq 0.06$; at absolute zero the dilute phase will contain about 6% ^{3}He.

••••• •••••

18.4.1 Verify that the solution of (18.51) is $X \simeq 0.06$ by locating the intersection of the graphs of $1+14X$ and $12X^{2/3}$.

••••• •••••

Earlier in this section we noted that the adiabatic expansion of a gas is one of the classic methods of cooling. The analogous way of cooling ^{3}He is by diluting it with ^{4}He. At temperatures below 0.5 K the ^{4}He has a negligible heat capacity. Dilution therefore enlarges the volume occupied by the ^{3}He without any appreciable heat transfer, that is, it expands the ^{3}He adiabatically. A dilution refrigerator based on this type of adiabatic expansion has been operated by deBruyn and his colleagues.† This type of refrigerator is suited only to single-cycle operation; it is a one-shot process. We turn next to the description of a dilution refrigerator which may be operated continuously. This second version of a dilution refrigerator is analogous to the forced evaporation type of classic cooler.

The cooling occurs in the so-called mixing chamber, where a layer of nearly pure ^{3}He floats on the ^{3}He-dilute mixture. At 0.2 K, for example, the upper layer is about 99% ^{3}He; the lower layer is roughly 8% ^{3}He. The ^{3}He is withdrawn from the dilute solution, circulated through heat exchangers, and so on, and eventually returned to the upper layer. As ^{3}He is removed from the bottom layer, ^{3}He atoms move downward from the upper layer in an effort to maintain the phase equilibrium. In moving across the phase boundary the ^{3}He carries away latent heat. This heat eventually leaves the mixing chamber by way of the ^{3}He circulation. This mechanism operates just like forced evaporation. The upper layer of ^{3}He acts like the liquid being evaporated. The ^{3}He in the dilute lower layer plays the role of a vapor. It is pumped away and carries with it the latent heat. So, if you take a schematic diagram of an evaporation cooler, turn it upside down, label the liquid "pure ^{3}He," and label the vapor "dilute ^{3}He–^{4}He mixture," you end up with a schematic of the mixing chamber segment of a dilution refrigerator.

There is a strong analogy between latent heats as well. In both cases the atoms which move across the phase boundary are those with much higher than average energy. The hot atoms are pumped away, leaving the cooler ones behind. One of the virtues of the dilution refrigerator is that the dilute ^{3}He phase—which corresponds to the vapor—never falls below a concentration of 6%. By contrast, the vapor pressure of the classic cooler vanishes exponentially as T tends toward absolute zero. In short, ^{3}He is self-refrigerating; classic coolers are not.

† A schematic diagram of this device is shown as Fig. 19 of the article by J. C. Wheatley, Dilute Solutions of ^{3}He in ^{4}He at Low Temperatures, *Am. J. Phys.* **36**, 181 (1968).

18.5 SUPERCONDUCTIVITY

In 1908 Heike Kammerlingh Onnes succeeded in liquifying helium. Three years later he discovered that mercury loses all electrical resistance below a temperature of 4.17 K. This phenomenon is called *superconductivity*. Since the initial discovery by Onnes, several thousand compounds and over one-third of the elements have been found to exhibit superconductivity. The behavior of the electrical resistivity ρ as a function of temperature is shown in Figure 18.10

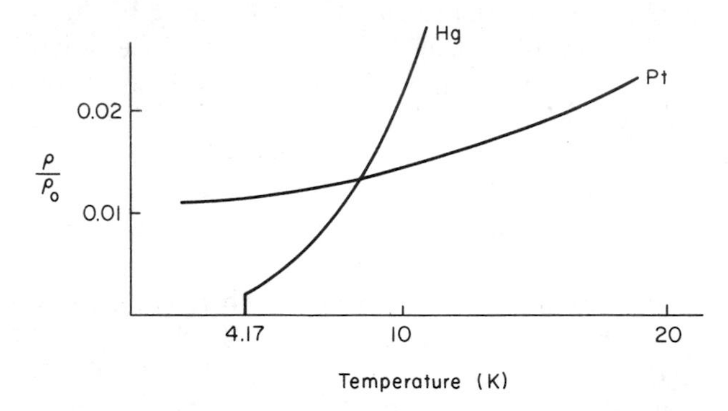

Figure 18.10. Resistivity versus temperature for mercury and platinum. The quantity ρ_0 is the resistivity at 0 °C. Mercury becomes superconducting at its transition temperature of 4.17 K.

for mercury and for platinum, a nonsuperconductor. The quantity ρ_0 denotes the resistivity at 0 °C. Platinum was one of the first materials investigated by Onnes. He attributed the leveling off of ρ/ρ_0 to impurities. He then turned to mercury which can be purified by vacuum distillation. Onnes expected that the resistivity of a pure metal would fall toward zero as the temperature approached absolute zero. The abrupt† disappearance of electrical resistance at 4.17 K was completely unexpected. Onnes subsequently found that tin and lead also become superconducting at sufficiently low temperatures. The resistivity of a good conductor at 0 °C is on the order of 10^{-5} ohm-cm. At liquid helium temperatures it has dropped by a factor of 100 or so to, say, 10^{-7} ohm-cm. Onnes showed that the resistivity of superconducting lead is less than 10^{-16} ohm-cm by observing the persistent current induced in a lead ring. The Onnes ring experiment consists of three steps:

(a) A lead ring is placed in the field of a magnet and then cooled below its superconducting transition temperature by bathing it in liquid helium.

† The transition from the normal to the superconducting state takes place over an interval of less than 0.1 K.

(b) The magnet is removed, inducing a "super" current in the ring. Since the current experiences no resistance it persists indefinitely.

(c) The persistent current induced in the ring creates a magnetic field which may be detected, for example, by observing the deflection of a compass needle. Observations over a period of a few hours established that the current did not diminish by as much as 1%, the precision of the measurement. Recent observations of persistent currents over a period of months have established $\rho < 10^{-26}$ ohm-cm.

The experimental data are consistent with the presumption that $\rho = 0$ for a superconductor. This means that there can be no electric field E inside a superconductor. This conclusion follows from Ohm's law which specifies the relation between E, ρ, and the electric current density J,

$$E = \rho J$$

Since $\rho = 0$ and J is finite, E must vanish.†

The temperature at which a material becomes superconducting is termed the *transition temperature*. The value of the transition temperature depends upon whether or not an external magnetic field is present. Transition temperatures for several superconductors are listed in Table 18.1. The temperatures are zero magnetic field values. The effects of an applied magnetic field will be discussed shortly.

TABLE 18.1

Transition Temperatures for Type I and Type II Superconductors

Type I	T_0 (K)	Type II	T_0 (K)
Lead (Pb)	7.19	Nb_3Sn	18.05
Mercury (Hg)	4.17	V_3Si	17.0
Tin (Sn)	3.72	Niobium (Nb)	8.70
Indium (In)	3.40	Vanadium (V)	4.89
Aluminum (Al)	1.20	Tantalum (Ta)	4.38

There appear to be two distinct types of superconducting materials. Type I, or soft superconductors include elements like Hg, Pb, Bi, In, Al, and Sn: they are mechanically soft and malleable, thus the terminology. By contrast, type II superconductors which include niobium (Nb, formerly called Columbium) and tantalum are hard and brittle, a troublesome property when trying to draw such materials into fine wires. The type II superconductors are a breed by

† Strictly speaking these conclusions are only correct for direct currents and static fields. In very-high-frequency fields the electrical resistivity of a superconductor does not vanish.

themselves and only in recent years have they been subjected to intensive study. The reasons for the great interest in type II superconductors will be discussed later.

In addition to the disappearance of electrical resistance, type I superconductors display several other distinctive features. We consider five of these. In the course of the discussion we develop the thermodynamics of superconductivity and comment on some aspects of the microscopic theory of superconductivity. The section concludes with remarks about type II superconductors and the quantization of magnetic flux.

A Meissner Effect

In 1933 Meissner and Ochsenfeld performed an experiment which revealed that a superconductor is not only a perfect conductor, but a perfect diamagnet as well. They observed the following: When a tin cylinder was (a) cooled below its superconducting transition temperature and then (b) subjected to a magnetic field, the lines of magnetic flux did not penetrate the sample. This is the behavior expected for a perfect conductor. As the magnetic field is turned on currents are induced tending to prevent any change in the magnetic flux through the sample (Lenz' Law). In the absence of any resistance this current assumes whatever value is necessary to prevent flux penetration.

The unexpected occurred when Meissner and Ochsenfeld reversed the order of (a) and (b). The magnetic field was first applied to the sample in its normal, that is, nonsuperconducting, state. The magnetic properties of the normal state are very weak, so that the magnetic flux lines fully penetrated the sample. The material was then cooled below its transition temperature. If zero resistivity were the only attribute of superconductivity, the flux threading the specimen would have remained unchanged as the transition temperature was passed. What Meissner and Ochsenfeld observed was something quite different. They found that the magnetic flux was completely expelled from the sample as it became superconducting. This perfect diamagnetism means that the magnetic induction B inside a superconductor is zero.† Thus $E = 0$ for a perfect conductor and $B = 0$ for a perfect diamagnet. The superconducting state exhibits both of these extreme properties.

The importance of the perfect diamagnetic behavior of a superconductor cannot be overemphasized. If superconductivity were solely a matter of zero resistivity no thermodynamic treatment of the phenomenon would be possible. Thermodynamically, the normal $\leftrightarrow$ superconducting transition qualifies as a

† As we point out in Section C an applied magnetic field does penetrate the surface skin layers to a depth of about 10^{-5} cm.

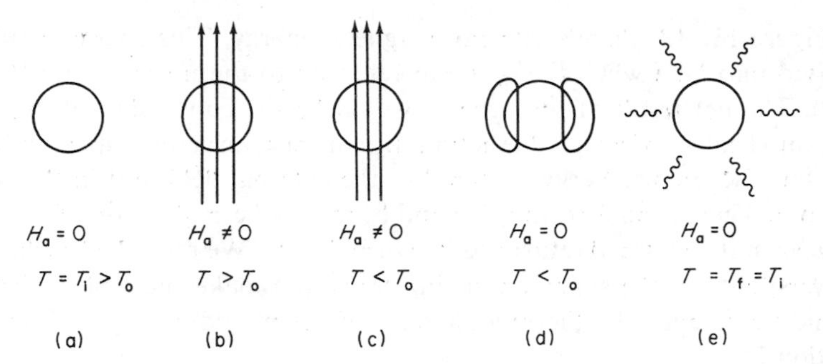

Figure 18.11. (a) If a superconductor were only a perfect conductor the $s \rightarrow n$ transition would be irreversible ($H_a = 0$, $T = T_i > T_0$). (b) The field is applied ($H_a \neq 0$, $T > T_0$) and (c) the sample cooled below its transition temperature ($H_a \neq 0$, $T < T_0$) at which point flux lines still thread the sample. (d) When the external field is removed flux is trapped, storing magnetic energy in the material ($H_a = 0$, $T < T_0$). (e) This energy is transferred to the surroundings as heat in returning the sample to its initial state ($H_a = 0$, $T = T_f = T_i$). The net result is the conversion of magnetic energy into heat—an irreversible action.

reversible process. In fact, it is a phase transition. However, if a superconductor were not a perfect diamagnet, the transition would be an irreversible process.

The concept of a reversible process was developed at length in Chapter 4. Figure 18.11a–e illustrates how irreversibility would arise if a superconductor exhibited zero resistivity but were not a perfect diamagnet. Figures 18.12a–e show how perfect conductivity and diamagnetism combine to achieve reversibility in a superconductor. The key point is that a perfect conductor would trap

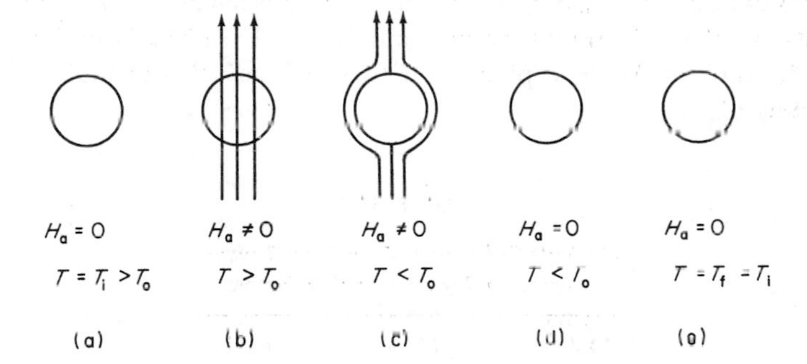

Figure 18.12. (a) The sequence of operations depicted in Figure 18.11 results in a reversible process for an actual superconductor ($H_a = 0$, $T = T_i > T_0$). (b) Flux penetrates the sample when the temperature exceeds the transition temperature T_0 ($H_a \neq 0$, $T > T_0$). (c) When the sample is cooled below T_0, flux is expelled ($H_a \neq 0$, $T < T_0$). (d, e) The system returns reversibly to its initial state because no magnetic energy has been stored in the superconducting material [(d) $H_a = 0$, $T < T_0$; (e) $H_a = 0$, $T = T_f = T_i$)].

flux (Figure 18.11d) thereby storing magnetic energy. This energy would be converted into heat when the material returned to the normal state (Figure 18.11e). The net result of the process would be the conversion of magnetic energy into heat—an irreversible action. By contrast, an actual superconductor expels flux. Magnetic energy is stored in the external field, not in the superconductor. There is no flux trapping and hence no thermalization of magnetic energy when the material returns to its normal state. We remarked earlier that the reversibility of the superconducting transition makes possible a thermodynamic treatment. The thermodynamics of superconductivity is developed in Section D.

B Critical Magnetic Field

In 1914 Onnes discovered that a superconductor reverts to its normal state if subjected to a magnetic field of sufficient strength. The critical field strength at which superconductivity is quenched depends on the temperature. Experimentally the relation between the critical field and temperature is closely approximated by the parabolic form

$$H_c = H_0\left[1 - \left(\frac{T}{T_0}\right)^2\right] \tag{18.52}$$

H_c denotes the value of the applied field strength for which the superconducting $\leftrightarrow$ normal transition occurs at the temperature T. The zero-field transition temperature is T_0 and H_0 is the critical field at absolute zero. The values of T_0 and H_0 are characteristics of the particular superconductor. Table 18.2 lists a few representative values. A plot of the relation between H_c and T, Figure 18.13, amounts to a phase diagram for a superconductor. The phase boundary defined by (18.52) divides the H–T plane into two regions—the normal and superconducting phases.

TABLE 18.2

Critical Fields and Transition Temperatures for Type I Superconductors

Material	H_0 (G)	T_0 (K)
Al	100	1.20
In	275	3.40
Sn	300	3.72
Hg	420	4.17
Pb	800	7.19

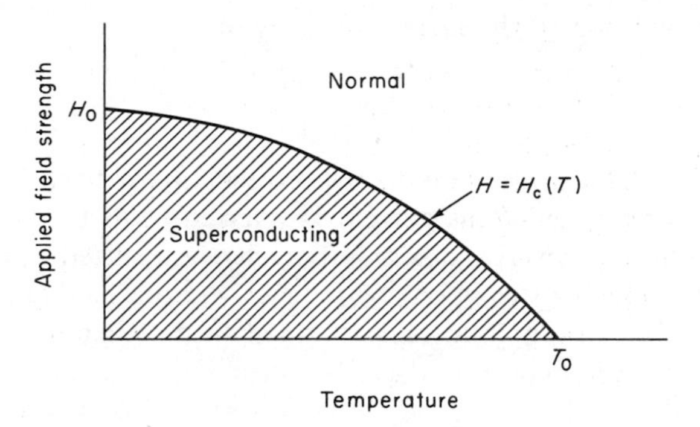

Figure 18.13. The phase boundary $H = H_c(T)$ divides the H–T plane into two regions. Experimentally the equation defining the boundary has very nearly the parabolic form of Eq. (18.52).

It was also observed that superconductivity vanished whenever the current carried by a superconductor exceeded some critical value (dependent on material)—even in the absence of an external magnetic field set up by the current itself. Experiment showed this to be the case. Recall Ampere's Law

$$\oint \mathbf{H} \cdot d\mathbf{l} = \frac{4\pi}{c} I_{\text{enc}}$$

Referring to the accompanying diagram shows

$$H(r) 2\pi r = \frac{4\pi}{c} I, \qquad r > a$$

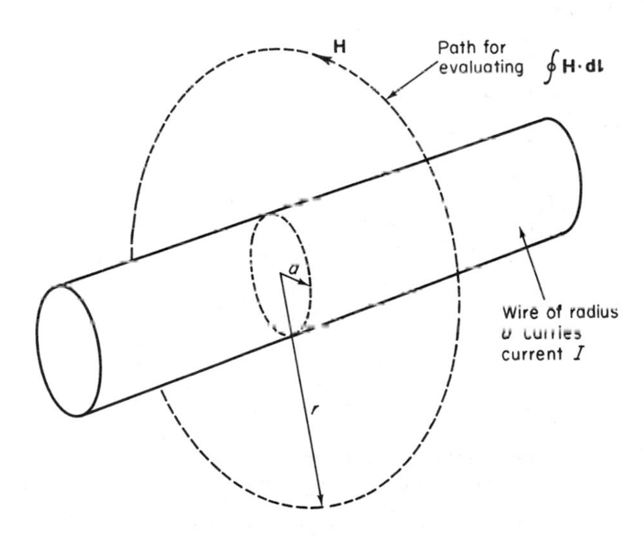

For $r = a$ we obtain H at the surface of the wire

$$H_{\text{sur}} = \frac{2I}{ca} \tag{18.53}$$

When the current I is large enough, H_{sur} exceeds the critical field and superconductivity is quenched. This current is referred to as the *critical current.* When Onnes first discovered superconductivity he realized at once its great potential for producing strong magnetic fields without having to pay for the Joule heating. His disappointment was equally keen when he found that the superconducting current produced a self-destructing magnetic field. The recent exploitation of type II superconductors has seen Onnes' dream turn into reality.

C Penetration Depth

A superconductor does not completely expel a magnetic field. Experiments show that an applied field penetrates the surface, but that the field decreases exponentially with depth (see Figure 18.14). The characteristic penetration depth λ is about 10^{-5} cm. Thus any applied magnetic field is confined to the surface skin layers of a superconductor. It follows from Ampere's Law that any supercurrents must flow in the skin of the material, that is, from

$$H(r) = \frac{2I_{\text{enc}}}{cr}$$

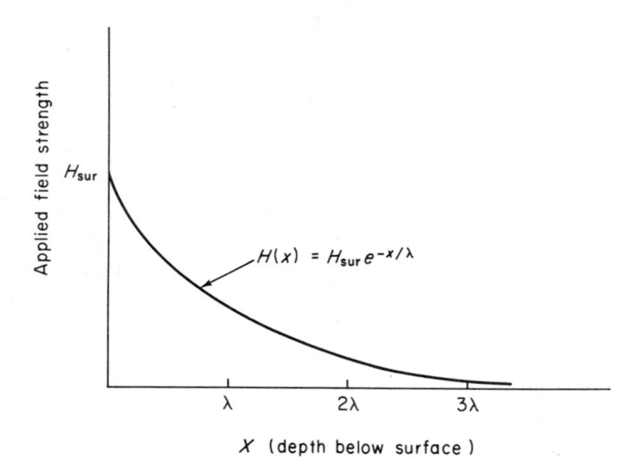

Figure 18.14. An applied magnetic field decreases exponentially with depth below the surface of a material in the superconducting state. The penetration depth λ is in the 10^{-6}–10^{-5} cm range.

it follows that, if H decreases exponentially beneath the surface, so also must the enclosed current. In passing we note that there is no corresponding penetration depth for a static electric field. This was demonstrated by H. London who measured the capacitance of a capacitor having superconducting plates.

D Phase Transition

Thermodynamically, the normal $\leftrightarrow$ superconducting transition qualifies as a phase change. In the presence of an applied magnetic field a latent heat is rejected as a substance goes from the normal to superconducting phase. There is no latent heat in the absence of a magnetic field. Further, the specific heat of a superconductor undergoes a discontinuous change at the transition temperature (see Figure 18.15). The details of these effects follow from the thermodynamics of superconductivity which we now develop.

In Chapter 4 we showed that the magnetic work done by a system could be expressed in the form

$$dW' = -H\,dM \tag{4.15}$$

The total thermodynamic work is the sum of the mechanical work and the magnetic work

$$dW = P\,dV - H\,dM \tag{18.54}$$

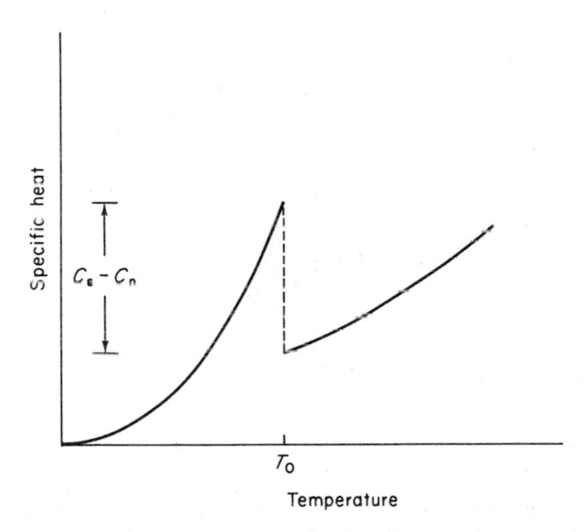

Figure 18.15. A superconductor exhibits a specific heat discontinuity at the transition temperature T_0.

The mechanical work is generally negligible compared to the magnetic work.[†] It is helpful to carry both terms because the analogs

$$P \leftrightarrow H, \qquad dV \leftrightarrow -dM \tag{18.55}$$

permit an immediate generalization of the equations for the thermodynamic potentials and their differentials. For example, consider the Gibbs free energy G

$$G(T,P) = U - TS + PV \tag{8.12'}$$

and its differential

$$dG = -S\,dT + V\,dP \tag{8.13'}$$

An appeal to (18.55) shows that we can include magnetic effects by writing

$$dG = -S\,dT + V\,dP - M\,dH \tag{18.56}$$

By regarding G as a function of the independent variables T, P, and H we obtain the generalization of (8.14)

$$\begin{aligned} S &= -\left(\frac{\partial G}{\partial T}\right)_{P,H} \\ V &= \left(\frac{\partial G}{\partial P}\right)_{T,H} \\ M &= -\left(\frac{\partial G}{\partial H}\right)_{T,P} \end{aligned} \tag{18.57}$$

The Gibbs free energy is

$$G = U - TS + PV - \int_0^H M(H)\,dH \tag{18.58}$$

This form of G ignores any H dependence of U and S. The first three terms on the right side are essentially independent of H. The magnetic field effects are incorporated in the last term. Ordinarily, the magnetization M is related to the magnetic induction B and the applied magnetic field H by

$$B = H + \frac{4\pi M}{V} \tag{4.11}$$

[†] Moderate changes in pressure have little effect on the properties of superconducting materials. However, a number of elements (for example, Te, Ge, Se), which are not superconducting at ordinary pressures, do make the transition at pressures of many thousands of atmospheres.

If we apply this to a superconductor wherein $B = 0$, we have

$$M = -\frac{VH}{4\pi} \tag{18.59}$$

The last term in (18.58) then gives

$$-\int_0^H M(H)\,dH = \frac{V}{4\pi}\int_0^H H\,dH = \frac{VH^2}{8\pi} \tag{18.60}$$

With this result we can write the Gibbs free energy of the superconducting phase as

$$G_s(T,P,H) = G_s(T,P,0) + \frac{VH^2}{8\pi} \tag{18.61}$$

where $G_s(T,P,0)$ comprises the field-independent terms of (18.58). The normal phase of a superconductor displays only a very weak dependence on a magnetic field. It is so weak in fact, that we can ignore the magnetic contribution to $G_n(T,P)$, the Gibbs free energy of the normal phase.

Equilibrium between the normal and superconducting phases—described pictorially in Figure 18.13—requires equality of the Gibbs free energies of the normal and superconducting phases. We ignore the pressure as a variable and concentrate our attention on the H, T dependence. The phase equilibrium condition reads

$$G_n(T) = G_s(T,0) + \frac{VH_c^2}{8\pi} \tag{18.62}$$

This equation is the thermodynamic relation defining the n–s phase boundary in Figure 18.13.

The relationship between the entropies of the normal and superconducting phases follows from

$$S = -\left(\frac{\partial G}{\partial T}\right)_{P,H} \tag{18.57a}$$

namely,[†]

$$S_n = S_s - \frac{VH_c}{4\pi}\frac{dH_c}{dT} \tag{18.63}$$

Experimentally, dH_c/dT is negative for all superconductors. It follows that $S_s < S_n$; the entropy of the superconducting phase is less than that of the

[†] Note the distinction between H the applied field and H_c the field which defines the phase boundary. The former is an independent thermodynamic variable. The latter is a temperature-dependent quantity [compare Eq. (18.52)].

normal phase. This in turn implies that the superconducting state is one of greater order than the normal state. The origin of the increased order in the superconducting phase lies in an attractive interaction between electrons. The nature of this attractive electron–electron interaction is discussed later in our outline of the theory of superconductivity.

The latent heat absorbed by a specimen going from the superconducting to normal phase is

$$L_s = T(S_n - S_s) = -T\frac{VH_c}{4\pi}\frac{dH_c}{dT} \tag{18.64}$$

Note that the latent heat vanishes in the absence of an external field. The observed values of latent heats are in complete accord with this result, thereby confirming the reversible character of the transition.

•••••• ••••••

18.5.1 In (18.63) we showed that $S_s < S_n$. Stability of an equilibrium state also requires that $F_s < F_n$, where F denotes the Helmholtz free energy. Make use of the analogs (18.55) to obtain an expression for F which includes a magnetic term. Show that the difference of F_n and F_s at the phase boundary is given by

$$F_n - F_s = \frac{VH_c^2}{8\pi}$$

•••••• ••••••

By combining $dQ = T\,dS$ and the definition of the specific heat $dQ = C\,dT$ we find

$$C = T\left(\frac{dS}{dT}\right)$$

Applying this to (18.63) we obtain the specific heat† discontinuity

$$C_s - C_n = \frac{TV}{4\pi}\left[H_c\left(\frac{d^2H_c}{dT^2}\right) + \left(\frac{dH_c}{dT}\right)^2\right] \tag{18.65}$$

Unlike the latent heat, the discontinuity in the specific heat is nonzero even in the absence of an applied field. Experimentally, the discontinuity is about twice C_n in the absence of an applied field ($H_c = 0$). The heat capacities become equal when $H_c \simeq \frac{2}{3}H_0$. In stronger fields C_n exceeds C_s (see Problem 18.5.2.)

---

† Equation (18.65) gives the total heat capacity of a sample of volume V. To obtain the specific heat capacity it is necessary only to replace V by the reciprocal of the mass density.

••••• •••••

18.5.2 (a) Prove that L_s is positive (that is, heat is absorbed in the s → n transition) for the parabolic form of $H_c(T)$ given by (18.52).

(b) Using (18.52) show that

$$C_s - C_n = \frac{VH_0{}^2}{2\pi T_0} x(3x^2 - 1)$$

where $x = T/T_0$.

(c) Show that $C_s = C_n$ when $H_c = \frac{2}{3}H_0$.

(d) Demonstrate that a superconductor which follows (18.52) obeys the third law of thermodynamics. [*Hint:* Show that the entropy difference $S_s - S_n$ vanishes at absolute zero.]

••••• •••••

Before moving on to consider the fifth and final notable property of superconductors, a pedagogic confession is in order. It may be misleading to infer from $B = 0$ that $M = -VH/4\pi$, as was done earlier. The implication of this relation is that $H \neq 0$ even though $B = 0$. The actual state inside a superconductor has $M = 0$, $H = 0$, as well as $B = 0$. The two descriptions turn out to be equivalent because the former description ignores the induced super currents. These currents flow in the skin and shield the interior from the applied field. The advantage of using $M = -VH/4\pi$ is that it allows us to employ the conventional expression for magnetic work $-H\,dM$. This in turn leads to the magnetic version of the Gibbs free energy, the condition for phase equilibrium, and so on. The student may rest assured that the same results follow—if less swiftly—from a more proper analysis.

E Isotope Effect

The transition temperature T_0 is not the same for different isotopes of certain superconducting elements. In 1950 Serin and his co-workers and, independently, Maxwell, observed the zero-field transition temperature of several isotopes of mercury. They found that the transition temperature decreased with increasing mass number M. Their measurements indicated that†

$$M^{1/2} T_0 = \text{constant}$$

In one of the few instances in the brief history of superconductivity where theory preceded experiment, Fröhlich (1950) suggested a mechanism which

† Experimentally, Serin and co-workers found $M^{\alpha} T_0 = \text{constant}$, with $\alpha = 0.504$; Maxwell found $\alpha = 0.486$. The theory advanced by Fröhlich implied $\alpha = \frac{1}{2}$. Subsequent studies revealed superconducting elements with values of α differing significantly from $\frac{1}{2}$.

might give rise to superconductivity. His was not a complete theory of superconductivity but it did predict the isotope effect.

Our presentation of the theory of superconductivity is necessarily superficial. Quantum mechanics is indispensible in any treatment which would furnish a deep understanding of superconductivity. Considering that the development of a satisfactory microscopic theory defied a generation of theoretical physicists, we offer no apologies for our abbreviated remarks.

At first thought it seems that the occurrence of superconductivity is a miracle. This impression is prompted by a calculation of the energy difference between the normal and superconducting phases. This difference is extremely small—on the order of 10^{-7} eV/atom. The figure of 10^{-7} eV/atom can be argued as follows: The energy difference between the normal and superconducting phases is of order $VH_0^2/8\pi$—the energy increase for a superconducting-to-normal transition near absolute zero. The energy difference per atom is $H_0^2/8\pi n$, where n is the number of atoms per unit volume. Taking $H_0 = 300$ G and $n \simeq 10^{22}$ cm^{-3} leads to the 10^{-7} eV/atom value.

•••••• ••••••

18.5.3 Verify the estimate of 10^{-7} eV/atom for mercury and aluminum. Consult Table 18.2 for H_0 values.

•••••• ••••••

By contrast the binding energy which glues the atoms into a lattice is on the order of a few electron volts per atom. Similarly, at 1 K the random thermal energy of vibration associated with the atomic lattice is on the order of 10^{-4} eV/atom. Thus it would seem that superconductivity is almost a miracle—requiring a delicate balance between several big effects before it can be observed. In a sense this is the case—there are many effects which cancel because they are unchanged by the transition. For example, the metallic lattice of atoms and ions is not changed by the superconducting transition. This fact was established long ago by Keesom and Onnes through X-ray diffraction. The particles directly responsible for the superconducting transition are the valence electrons. These are the electrons which, in a normal metallic state, are more or less free to move in response to an external electric field. However, the valence electrons interact with each other; they interact with defects in the lattice, with vibrations of the lattice, and with impurities randomly distributed throughout the material. It is necessary to isolate the interaction which precipitates the superconducting transition and ignore the rest—on the grounds that these are the same for both phases. Singling out the essential interaction was a crucial step in the development of a microscopic theory of superconductivity.

It was here that Fröhlich's work and the experimental confirmation of the isotope effect attained their great significance. The interaction giving rise to the isotope effect is called the *electron–phonon interaction.* It gives rise to an *attractive* force between electrons. This attraction fosters the increased order, that is, the lower entropy, characteristic of the superconducting phase. The attractive interaction springs from a polarization of the lattice. Each electron distorts the lattice in its immediate vicinity. The distortion of the lattice at one point is felt throughout the solid—the disturbance propagating as quantized sound waves, or phonons, as they are called. Other electrons feel the effects of such a distortion because the configuration of the lattice in their immediate vicinity is altered by the phonons. Thus the electrons interact with each other via the phonons.

The attractive electron–electron interaction, mediated by the phonons, is opposed by the repulsive Coulomb force. The effective Coulomb energy between electrons separated by a distance r is not e^2/r because of *shielding*. A shielding of the Coulomb interaction results because, on the average, there is a net positive charge in the vicinity of an electron and a net negative charge surrounding a positive ion. Each charge continually strives to surround itself with particles of opposite charge. The result may be described as one in which each charge is dressed in a screening cloud of charge of the opposite sign. The potential energy of two dressed electrons is reduced from its Coulomb value by an exponential factor $\exp(-r/r_s)$. The quantity r_s is a screening length, roughly equal to the mean distance between electrons. Thus two electrons separated by a distance of a few times r_s repel each other rather weakly. Dynamically they are hidden from each other by their screening clouds. The screening of the repulsive Coulomb forces makes it possible for the attractive electron–phonon interaction to compete effectively and give a net attraction between two electrons.

In 1956 L. N. Cooper showed how an attractive electron–electron interaction would promote the formation of electron pairs. The energetically most favorable configuration of such a pair is one in which the two electrons have antiparallel spins and equal but opposite linear momenta. Thus the combined spin angular momentum of the pair is zero, as is its total linear momentum. When a substance becomes superconducting, the attractive interaction between electrons causes a condensation of these Cooper pairs, as they are now called. At absolute zero all of the electrons form Cooper pairs. The number of Cooper pairs decreases with increasing temperature, falling to zero at the transition point.

The attractive electron–electron interaction is intrinsically weak, but has a long range—roughly 10^{-4} cm. This is more than a thousand times greater than the typical screening length r_s which defines the range of the repulsive Coulomb force. The range of the attractive electron–electron interaction is usually referred to as the *coherence length*. The coherence length may also be thought of as the characteristic dimension of a Cooper pair.

Somewhat less energetically favorable is the formation of Cooper pairs which have zero spin angular momentum but nonzero linear momentum. Such pairs travel without resistance—their motions produce the super current. In a current-carrying superconductor the full assembly of Cooper pairs is locked in step. Their collective motions result in a quantum system of macroscopic size. The only way to introduce resistance is to disrupt the entire ensemble of pairs, that is, one must destroy the superconducting state.†

In 1957 a microscopic theory of superconductivity was formulated by Bardeen, Cooper, and Schrieffer. The BCS theory assumes that the superconducting state forms as a strongly correlated assemblage of Cooper pairs. Bardeen, Cooper, and Schrieffer showed that such a state is energetically favored over one formed from normal electrons whenever the attractive electron–phonon interaction dominates the repulsive Coulomb force between electrons. The BCS theory exhibits a Meissner effect, the phase change and specific heat discontinuity, the isotope effect, and the other electrodynamic properties of superconductors. Furthermore, the theory predicted new effects which have since been verified experimentally. It seems safe to say that although the theory of (type I) superconductivity has not achieved its final form, the mechanism by which the phenomenon operates has been identified.

In view of the successes of the BCS theory this last remark may seem too modest a claim. There remain, however, many aspects of superconductivity which are not understood. For example, a superconductor with a transition temperature of 25 K–30 K would revolutionize the technologies of power transmission and communication. At present, an alloy of Nb, Al, and Ge has the highest transition temperature, 21 K. The search for these relatively high-temperature superconductors has been sparked by B. T. Matthias. Despite its many virtues, the BCS theory is powerless to suggest how one might engineer higher transition temperatures. Matthias has discovered empirical rules which suggest combinations of elements likely to have a high transition temperature. Unfortunately, many of the most promising compounds are chemically unstable. Type II superconductivity is another area where the BCS theory is not directly applicable. The distinctive characteristic of type II superconductors is the absence of a sharp Meissner effect. Magnetic flux is excluded until a critical field H_{c1} is reached. When the magnetic field is increased beyond H_{c1}, there is a gradual penetration of flux. There is a second critical field H_{c2}, greater than H_{c1}, at which flux penetration becomes complete. Above H_{c2} no

† Again we remark that this is strictly true only for direct current. At any temperature above absolute zero there will be some normal electrons. In dc operation the super current short circuits the normal electron flow. In a high-frequency electric field the inertia of the super electrons limits their current, enabling a finite resistive current of normal electrons to develop.

vestige of superconductivity remains. The range of field strengths defined by

$$H_{c1} < H < H_{c2}$$

is called the *mixed state* since portions of the sample are superconducting while others are normal. The flux penetration takes the form of flux lines which lead magnetic field lines through the sample. The flux lines arrange themselves in a triangular lattice. For fields only slightly greater than H_{c1} most of the sample is in the superconducting state. On and near the flux lines the material is normal. As the applied field strength is raised the density of flux lines increases. At the upper critical field H_{c2} the density of flux lines reaches a point where the surrounding normal regions overlap, leaving no superconducting paths through the material.

The upper critical field is generally much larger than H_{c1} or typical H_0 values for type I materials. For Nb_3Sn, $H_{c2} = 88{,}000$ G, and materials with H_{c2} as large as 400,000 G are known. Other high-field superconductors will doubtless be discovered. High-field superconductors have been used to fabricate extremely powerful electromagnets. Power is consumed only in the cryogenic system which maintains the coils in the superconducting state.

The BCS theory is not sufficiently general to deal with the mixed state of type II superconductors. Historically, a phenomenological theory encompassing the mixed state was developed by Landau and Ginzburg in 1950. The Landau–Ginzburg equations were solved by the Russian physicist A. A. Abrikosov in 1957. Abrikosov's solution predicted the triangular array of flux lines characterizing the mixed state. Furthermore, his calculations indicated that the flux which threaded the specimen was quantized. Each flux line was associated with a single flux quantum. The quantum of flux is the *fluxoid*

$$\Phi = \frac{hc}{2e} = 2.07 \times 10^{-7} \text{ G-cm}^2 \tag{18.66}$$

The existence of such flux quanta was verified experimentally in 1961 by W. Fairbank and Deaver and by Doll and Nabauer. More recently, the triangular flux line pattern has been studied directly using a novel photographic technique (see References).

The world of superconductivity shivers in the shadow of absolute zero, the unattainable minimum for temperatures. In the next and final section we discover that there is also a maximum temperature—the ultimate temperature.

18.6 THE ULTIMATE TEMPERATURE

Traditionally a textbook is not a proper place for speculative ideas. Nevertheless, in this final section of this final chapter we wish to bring the student to one of the many places along the frontiers of physical science.

Current research unites two areas, seemingly poles apart. These are the fields of high-energy physics and thermodynamics. The partnership has led to an exciting discovery: There is an *ultimate temperature*—an upper limit for temperature which cannot be exceeded regardless of how much energy is poured into a system. The value of this ultimate temperature is slightly over 10^{12} K. This temperature is established by fitting data from high energy scattering experiments to a theoretical model. Later in this section we show how the ultimate temperature may be deduced from theoretical considerations.

The origin of an ultimate temperature can be made plausible by consideration of the following thought experiment: Take 1 mole of molecular hydrogen confined in a fixed volume and add energy. Keep adding energy and note the temperature at which significant events occur. Further, note the behavior of the heat capacity C_v. A running record of the experiment might look something like that in the accompanying tabulation.

Temperature (K)	Remarks
300–600	Heat capacity nearly constant $C_v \simeq \frac{5}{2}R$
1000	Dissociation begins, C_v rises
5000	Dissociation virtually complete. System now consists of 2 moles of atomic hydrogen $C_v \simeq 3R$
10^4	Ionization of atoms noticeable, C_v rising
10^5	Ionization virtually complete. The system now consists of 2 moles of protons and 2 moles of electrons, C_v has leveled off at a value near $6R$
10^5–10^8	C_v remains nearly constant
10^9	Collisions of particles produce electron–positron pairs, C_v rising
10^{11}	Proton–proton collisions produce pions, C_v rising
10^{12}	Pions abundant. Energetic proton–proton collisions produce proton–antiproton pairs, C_v rising
$\simeq 1.2 \times 10^{12}$	Temperature has reached the ultimate limit. Addition of still more energy merely increases the number and variety of particles, C_v rising

How are we to interpret these results? An important point to note about the "data" is that increases in the heat capacity accompany events which increase the total number of particles (dissociation, ionization, pair-production, and so on). On the other hand C_v remained constant over temperature ranges where the number of particles did not change appreciably.

Since we prescribed that the experiment proceeds at constant volume there is no $P\,dV$ work; the energy added to the system shows up as an increase of internal energy

$$\text{energy added} = \Delta U$$

The definition of C_v is

$$C_v = \left(\frac{\partial U}{\partial T}\right)_V$$

Over a temperature range where C_v remains constant we may conclude that all of the energy added goes toward increasing the kinetic energies of the particles and this registers as a temperature increase

$$\text{energy added} = \Delta U = C_v \, \Delta T, \qquad C_v = \text{constant}$$

An increase of C_v with T shows that energy is being used for purposes beyond the increase of kinetic energy. These other uses are obvious; it takes energy to pull apart a hydrogen molecule. Energy is required to tear the electron from a hydrogen atom. In the process of pair-production, energy is converted from one form (kinetic) into another (mass).

The change of C_v from (5/2) R to $3R$ brought about by molecular dissociation simply reflects the increase in the number of particles. The gas goes from 1 mole of diatomic molecules (each having 5 degrees of freedom) to 2 moles of monatomic atoms (each having 3 degrees of freedom). The values of (5/2) R and $3R$ follow from the equipartition prescription of $\frac{1}{2}R$ per degree of freedom per mole. Ionization of the atomic hydrogen again doubles the number of particles. With protons and electrons alike possessing 3 degrees of freedom the heat capacity is again doubled.

Energy used to break atomic bonds is not available to increase the temperature. Processes which soak up energy by creating additional particles act as thermostats—they act to prevent an increase in temperature. Let E denote the energy characteristic of the particle-producing reaction. For example, in the dissociation of molecular hydrogen, $E = 4.48$ eV. For the creation of an electron–positron pair, $E = 2m_e c^2 \simeq 1$ MeV, the rest mass energy of the pair. At temperature T the average kinetic energy per particle is on the order of kT. The relation

$$kT \simeq \tfrac{1}{10}E \tag{18.67}$$

furnishes a *rough estimate* of the ignition temperature of the reaction. This rule-of-thumb follows from two considerations. The first point is an obvious one; particle-producing reactions are a result of collisions in which kinetic energy is converted into other forms. The second point is that a small but significant fraction of the particles will have kinetic energies at least 10 times as great as the average energy. Collisions between these above average particles initiate the reaction.

The relation (18.67) gives only a rough estimate—it ignores the fact that the pressure also influences the reaction rate. The greater the pressure, the more likely it is for particles to interact.

••••• •••••

18.6.1 Using the rule-of-thumb (18.67), evaluate ignition temperatures for the following reactions:

(a) dissociation of H_2,
(b) ionization of H ($E_{\text{ion}} = 13.6$ eV),
(c) electron–positron pair production.

How do your values compare with the temperatures listed in the thought experiment?

••••• •••••

When the temperature reaches the 10^8–10^9 K range, the production of electron–positron pairs becomes an important process. Electron–proton collisions and photon–proton collisions provide two mechanisms for electron–positron pair production.

At temperatures above 10^{11} K proton–proton collisions begin to produce *pions*.† The pions are the least massive members of the family of particles known as *hadrons*. Hadron is a generic name assigned to particles which interact through the powerful, short-range, nuclear force. In addition to the pions, the hadron family includes the neutron, the proton, and other more massive particles.

Experimentally, the number of hadrons of mass less than m is a rapidly increasing function of m. Theoretical considerations suggest that the number of hadrons with mass less than m increases exponentially for large m. This exponential increase in the number of mass states is countered by the Boltzmann factor $e^{-\beta\varepsilon}$ which governs their occupation probability. The ultimate temperature is a consequence of these two competing exponential factors.As the ultimate temperature is approached, energy added to the system is used to produce particles rather than to increase the kinetic energies of the existing particles. The thermostat action of particle production becomes complete—there can be no further temperature increase.

We have so far envisioned the system as a collection of particles which started out as hydrogen molecules and evolved into a "super-zoo" of hadrons as we added energy. On the other hand, a value for the ultimate temperature T_u is established by fitting the theory to experimental data on proton–proton collisions. In matching theory and experiment one assumes that two colliding protons momentarily form a system which achieves internal thermodynamic equilibrium. The two protons lose their identities and become a fireball—a

† There are three varieties of pions, π^0, π^+, and π^-. The superscript denotes the electric charge of the pion. The neutral pion (π^0) has a rest mass energy of 135 MeV. The charged pions $\pi^{\pm}$ have rest mass energies of 139.6 MeV.

system of many hadrons which distribute themselves over the available energy states in accordance with the laws of statistical mechanics.

After a lifetime of about 10^{-23} sec the fireball breaks up. In some instances two protons emerge, in which the case the event is described as an elastic scattering. However, as the kinetic energies of the colliding protons is increased the fireball debris becomes more varied and is likely to include other hadrons. Pion production is a frequent result. Two reactions which produce pions are

$$\mathrm{p} + \mathrm{p} \rightarrow \mathrm{p} + \mathrm{n} + \pi^{+}$$

and

$$\mathrm{p} + \mathrm{p} \rightarrow \mathrm{p} + \mathrm{p} + \pi^{0}$$

The production of proton–antiproton pairs becomes possible when the kinetic energies of the colliding protons reach the GeV range (1 GeV = 10^9 eV),

$$\mathrm{p} + \mathrm{p} \rightarrow \mathrm{p} + \mathrm{p} + \mathrm{p} + \tilde{\mathrm{p}}$$

The collision products are restricted only by the various conservation laws. For example, the conservation of electric charge prohibits the creation of a proton, but allows the production of proton–antiproton pairs.

•••••

18.6.2 The rest mass energy of a proton is 938 MeV. Two protons with equal kinetic energies collide head on, and produce a proton–antiproton pair

$$\mathrm{p} + \mathrm{p} \rightarrow \mathrm{p} + \mathrm{p} + \mathrm{p} + \tilde{\mathrm{p}}$$

What is the minimum kinetic energy required of the protons? Use (18.67) to estimate the ignition temperature for this creation.

•••••

At high energies the exponential character of the hadron mass spectrum makes itself felt. As the fireball temperature approaches the ultimate value it becomes more efficient to produce particles than to raise the temperature. The experimental evidence for thermodynamic equilibrium at an ultimate temperature is drawn from data on the momentum distribution of fireball fragments. If the fireball hadrons are in mutual thermodynamic equilibrium, they will have an isotropic distribution of momenta (in the rest frame of the fireball). One can measure the momentum distribution of the fragments in the plane perpendicular to the direction of motion of the fireball. Theory predicts that the distribution function for the perpendicular component of the momentum ($p_\perp$) is proportional to $\exp(-p_\perp c/kT_\mathrm{u})$, where T_u is the ultimate temperature. The fireball reactions include elastic scattering as well as the more exotic particle production schemes. Measurements of the momentum distribution of elastically scattered protons is consistent with the expected form provided one takes $kT_u \simeq 160$ MeV or

$$T_u \simeq 1.2 \times 10^{12}\,\text{K} \tag{18.68}$$

Can we understand why T_u has this particular value? The answer is "yes," and the argument is as follows: We observe, for the final time, that kT measures the average kinetic energy. We also note that the pion is the least massive hadron. The average kinetic energy cannot be raised above $m_\pi c^2$, the rest mass energy of the pion. Why? Because collisions very effectively convert kinetic energy into mass. It is not possible to increase the kinetic energies when pion-producing collisions soak up energy added to the system. The production of pions acts as a thermostat—as a governor for the temperature. We are therefore led to write

$$kT_u \simeq m_\pi c^2 \tag{18.69}$$

as the condition establishing the value of the ultimate temperature. With $m_\pi c^2 = 135$ MeV we find from (18.69) that $T_u \simeq 10^{12}$ K.

In brief we can say that the particular value of the ultimate temperature stems from the fact that particles with kinetic energies in excess of $m_\pi c^2$ will interact and produce pions. The existence of an upper limit for temperature seems destined to become the fourth law of thermodynamics.

REFERENCES

A survey of optical methods of temperature measurement, which draws heavily on the blackbody theory, is given in:

G. A. Hornbeck, Optical Methods of Temperature Measurement, *Appl. Optics* **5**, 179 (1966).

The essays by G. Gamow (Big Bang) and F. Hoyle (Continuous Creation) are included in:

J. R. Newman (ed.), "The Universe." Simon and Schuster, New York (1957).

A highly readable account of the remnant radiation is presented in:

P. J. E. Peebles and D. T. Wilkinson, The Primeval Fireball, *Sci. Am.* **216**, 28 (June 1967).

One measure of the scientific and technological impact of the laser is the number of Scientific American articles devoted to it! Included among these are the following:

A. L. Schawlow, Optical Masers, *Sci. Am.* **204**, 52 (June, 1961).

A. L. Schawlow, Advances in Optical Masers, *Sci. Am.* **209**, 34 (July 1963).

G. C. Pimentel, Chemical Lasers, *Sci. Am.* **215**, 32 (April 1966).

C. K. N. Patel, High Power Carbon Dioxide Lasers, *Sci. Am.* **219**, 23 (Aug. 1968).

A. L. Schawlow, Laser Light, *Sci. Am.* **219**, 120 (Sept. 1968).

D. R. Herriott, Applications of Laser Light, *Sci. Am.* **219**, 140 (Sept. 1968).

J. E. Faller and E. J. Wampler, The Lunar Laser Reflector, *Sci. Am.* **222**, 38 (March 1970).

M. J. Lubin and A. P. Fraas, Fusion by Laser, *Sci. Am.* **224**, 21 (June 1971).

The concept of a negative absolute temperature is discussed in:

M. W. Zemansky, "Temperatures Very High and Very Low." Van Nostrand-Reinhold, Princeton, New Jersey (1964).

The experiment which led to consideration of a negative temperature is described in the brief letter:

E. M. Purcell and R. V. Pound, A Nuclear Spin System at Negative Temperature, *Phys. Rev.* **81**, 279 (1951).

Subsequent studies of negative temperatures include:

N. F. Ramsey, Thermodynamics and Statistical Mechanics at Negative Absolute Temperatures, *Phys. Rev.* **103**, 20 (1956).

and

M. J. Klein, Negative Absolute Temperatures, *Phys. Rev.* **104**, 589 (1956).

The use of dimensionless ratios in astrophysics is discussed at length in:

E. E. Salpeter, Dimensionless Ratios and Stellar Structure, R. Marshak (ed.), "Perspectives in Modern Physics." Wiley (Interscience), New York (1966).

The black hole concept is explained in:

R. Ruffini and J. A. Wheeler, Introducing the Black Hole, *Phys. Today* **24**, 30 (Jan. 1971)

and

R. Penrose, Black Holes, *Sci. Am.* **226**, 38 (May 1972).

An incisive look at gravitational collapse is presented in:

K. S. Thorne, Gravitational Collapse, *Sci. Am.* **217**, 88 (Nov. 1967).

A broad view of cosmological aspects of gravitational collapse and "missing mass" is presented in:

J. A. Wheeler, Our Universe: The Known and the Unknown, *Am. Sci.* **56** (1), **1** (1968).

A broader view of cosmological questions is presented in the excellent monograph:

D. W. Sciama, "Modern Cosmology." Cambridge Univ. Press, London and New York (1971).

An informative and well-written article describing the structure of degenerate stars is:

M. Ruderman, Solid Stars, *Sci. Am.* **224**, 24 (Feb. 1971).

Additional arguments favoring the rotating neutron star model for pulsars is presented in:

J. P. Ostriker, The Nature of Pulsars, *Sci. Am.* **224**, 48 (Jan. 1971).

The phase separation in ^3He–^4He mixtures was detected using a magnetic resonance technique:

G. K. Walters and W. M. Fairbank, Phase Separation in He^3–He^4 Solutions. *Phys. Rev.* **103**, 262 (1956).

An excellent treatment of the theory and design of dilution refrigerators is presented in:

D. S. Betts, Helium Isotope Refrigeration, *Contemp. Phys.* **9**, 97 (1968).

A nontechnical discussion of dilution refrigeration and other quantum coolers is given in:

O. V. Lounasmaa, New Methods for Approaching Absolute Zero, *Sci. Am.* **221**, 26 (Dec. 1969).

The dependence of E_{34} on the ^3He fraction X is reported in:

D. O. Edwards, D. F. Brewer, P. Seligman, M. Skertic, and M. Yaqub, Solubility of He^3 in Liquid He^4 At 0 °K. *Phys. Rev. Letters* **15**, 773 (1965).

A comprehensive study of ^3He–^4He mixtures is presented in:

J. C. Wheatley, Dilute Solutions of ^3He in ^4He at Low Temperatures, *Am. J. Phys.* **36**, 181 (1968).

A selection of reprints, which includes Onnes' report of the discovery of superconductivity, has been published for the American Association of Physics Teachers. It is entitled:

"Superconductivity, Selected Reprints." Am. Inst. Phys., New York.

A splendid survey of all experimental aspects of superconductivity—up to 1955—is given in:

B. Serin, Superconductivity: Experimental Part, *Handbuch Phys.* **15**, 210 (1956).

A brilliant review of the theory of superconductivity is offered in:

H. Fröhlich, The Theory of the Superconductive State, *Reports Progress Phys.* **24**, 1 (1961).

Most parts of this review can be read with profit by an advanced student or an instructor. Sections 1 and 6 should be required reading for all aspiring scientists.

An inexpensive cryogenic system and an experiment suited to the measurement of transition temperatures above 2 K is described in:

E. P. Stillwell, Jr., R. L. Gardner, and H. T. Littlejohn, Experiments in Cryogenics for the Undergraduate, *Am. J. Phys.* **35**, 502 (1967).

The effects of ultra-high pressure on the superconducting properties of materials is surveyed in:

N. B. Brandt and N. I. Ginzburg, Superconductivity at High Pressure, *Sci. Am.* **224**, 83 (April 1971).

A recent review of progress toward the 30 K transition temperature is given in:

B. T. Matthias, The Search for High-Temperature Superconductors, *Phys. Today* **24**, 21 (Aug. 1971).

Type II superconductors are discussed in:

A. C. Rose-Innes, The New Superconductors, *Contemp. Phys.* **7**, 135 (1965).

Independent experiments confirming the existence of flux quantization were reported in:

B. S. Deaver, Jr., and W. M. Fairbank, Experimental Evidence For Quantized Flux in Superconducting Cylinders. *Phys. Rev. Letters* **7**, 43 (1961).
R. Doll and M. Nabauer, Experimental Proof of Magnetic Flux Quantization in A Superconducting Ring. *Phys. Rev. Letters* **7**, 51 (1961).

The quantized vortex pattern in superconductors has been studied directly using a novel photographic technique. See:

U. Essmann and H. Trauble, The Magnetic Structure of Superconductors, *Sci. Am.* **224**, 75 (March 1971).

See also:

P. J. Collins and J. E. Gordon, An Undergraduate Experiment Demonstrating Flux Quantization and Superconductivity, *Am. J. Phys.* **37**, 293 (1969).

Introductory level books dealing with elementary particles which may prove helpful are the following:

C. E. Swartz, "The Fundamental Particles." Addison-Wesley, Reading, Massachusetts (1965).

Chapter 6 focuses on the conservation laws.

K. W. Ford, "The World of Elementary Particles." Ginn Blaisdell, Boston, Massachusetts (1963).

Chapter 4 deals with the conservation laws.

Appendix A

Gaussian Integrals

The function $e^{-\alpha x^2}$ occurs in many areas of statistical analysis. We refer to $e^{-\alpha x^2}$ as a *Gaussian function*. The Maxwellian distribution function

$$f(v) = \text{constant} \cdot \exp\left(\frac{-\frac{1}{2}mv^2}{kT}\right)$$

is a Gaussian function. In kinetic theory it is frequently necessary to evaluate integrals involving the Gaussian function. Specifically, one needs to evaluate integrals having the form

$$G_n(\alpha) = \int_0^\infty x^n e^{-\alpha x^2}\, dx \tag{A.1}$$

The $G_n(\alpha)$ are termed *Gaussian integrals*. Note that although $G_n(\alpha)$ is defined by a definite integral its value depends on the parameter α. We can treat $G_n(\alpha)$ as a function of α. In particular we can *differentiate* $G_n(\alpha)$ with respect to α. By differentiating $G_n(\alpha)$ we obtain what is known as a *recurrence relation*. The recurrence relation for Gaussian integrals lets us compute $G_{n+2}(\alpha)$ from $G_n(\alpha)$, via differentiation. In kinetic theory one is generally interested in cases where n is an integer. If we know $G_0(\alpha)$, the recurrence relation readily yields $G_2(\alpha)$, which in turn gives $G_4(\alpha)$, and so on. From $G_1(\alpha)$ we can compute $G_3(\alpha)$, $G_5(\alpha)$, and so on. Thus only $G_0(\alpha)$ and $G_1(\alpha)$ need be evaluated directly; the recurrence relation takes over from there.

To derive the recurrence relation we differentiate $G_n(\alpha)$ with respect to α. The integral defining $G_n(\alpha)$ is a definite integral—its limits are constants, independent of α. Consequently the integration over x in (A.1) and a differentiation with respect to α may be interchanged. Thus

$$\frac{dG_n}{d\alpha} = \int_0^\infty x^n \frac{d}{d\alpha}(e^{-\alpha x^2})\,dx$$

Remembering that we differentiate with respect to α,

$$\frac{d}{d\alpha}(e^{-\alpha x^2}) = -x^2 e^{-\alpha x^2}$$

so that

$$\frac{dG_n}{d\alpha} = -\int_0^\infty x^{n+2} e^{-\alpha x^2}\,dx$$

The right side is $-G_{n+2}(\alpha)$ so that the recurrence relation is

$$G_{n+2}(\alpha) = -\frac{dG_n}{d\alpha} \tag{A.2}$$

As remarked earlier, only $G_0(\alpha)$ and $G_1(\alpha)$ need be evaluated directly. Consider first

$$G_1(\alpha) = \int_0^\infty x e^{-\alpha x^2}\,dx \tag{A.3}$$

The substitution

$$\alpha x^2 = z, \qquad 2\alpha x\,dx = dz$$

converts the right side to

$$G_1(\alpha) = \frac{1}{2\alpha}\int_0^\infty e^{-z}\,dz \tag{A.4}$$

However,

$$\int_0^\infty e^{-z}\,dz = -e^{-z}\Big]_0^\infty = 1$$

so that (A.4) gives

$$G_1(\alpha) = \tfrac{1}{2}\alpha^{-1} \tag{A.5}$$

Using the recurrence relation (A.2) we find

$$
\begin{aligned}
G_3(\alpha) &= -\frac{dG_1}{d\alpha} = \frac{1}{2}\alpha^{-2} \\
G_5(\alpha) &= -\frac{dG_3}{d\alpha} = \alpha^{-3} \\
G_7(\alpha) &= -\frac{dG_5}{d\alpha} = 3\alpha^{-4}
\end{aligned} \tag{A.6}
$$

The evaluation of $G_0(\alpha)$ is a bit more tricky. In fact, we compute $[G_0(\alpha)]^2$. The symmetry of the integrand allows us to write

$$G_0(\alpha) = \int_0^\infty e^{-\alpha x^2}\,dx = \tfrac{1}{2}\int_{-\infty}^{+\infty} e^{-\alpha x^2}\,dx$$

Furthermore, x is a dummy variable of integration so that we may write

$$G_0(\alpha) = \tfrac{1}{2}\int_{-\infty}^{+\infty} e^{-\alpha x^2}\,dx = \tfrac{1}{2}\int_{-\infty}^{+\infty} e^{-\alpha y^2}\,dy$$

and thus

$$[G_0(\alpha)]^2 = \tfrac{1}{4}\int_{-\infty}^{+\infty} e^{-\alpha x^2}\,dx \int_{-\infty}^{+\infty} e^{-\alpha y^2}\,dy$$

Grouping factors

$$[G_0(\alpha)]^2 = \tfrac{1}{4}\iint_{-\infty}^{\infty} e^{-\alpha(x^2+y^2)}\,dx\,dy \tag{A.7}$$

we can regard (A.7) as an integration over the entire x–y plane, with $dx\,dy$ denoting the differential element of area. By switching to polar coordinates we can immediately evaluate $[G_0(\alpha)]^2$. From Figure A.1 we see that $x^2 + y^2 = r^2$.

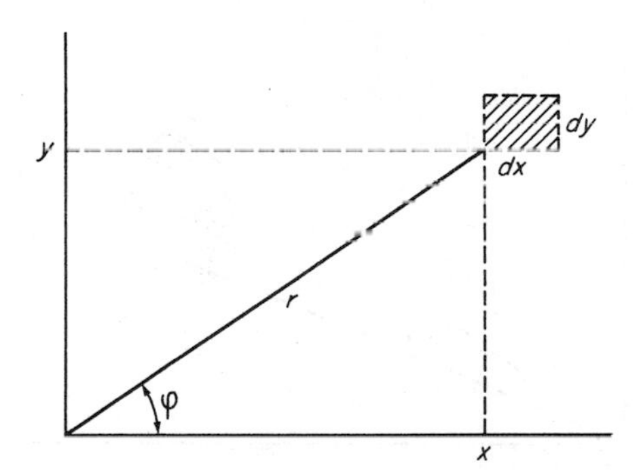

Figure A.1. Cartesian and polar coordinates. The hatched area $= dx\,dy$.

Thus

$$e^{-\alpha(x^2+y^2)} = e^{-\alpha r^2}$$

The integration

$$\iint\limits_{-\infty}^{+\infty} \cdots \, dx \, dy$$

is replaced by

$$\int_{\varphi=0}^{2\pi} \int_{r=0}^{\infty} \cdots \, r \, dr \, d\varphi$$

that is, $r\,dr\,d\varphi$ is the differential element of area in polar coordinates (see Figure A.2). The limits on φ and r are such as to cover the entire x–y plane.

We can now write

$$[G_0(\alpha)]^2 = \tfrac{1}{4} \int_0^{2\pi} d\varphi \int_0^{\infty} e^{-\alpha r^2} r \, dr \tag{A.8}$$

The φ integration gives 2π while

$$\int_0^{\infty} e^{-\alpha r^2} r \, dr = G_1(\alpha) = \tfrac{1}{2}\alpha^{-1}$$

Thus

$$[G_0(\alpha)]^2 = \frac{\pi}{4}\alpha^{-1}$$

which gives

$$G_0(\alpha) = \frac{\sqrt{\pi}}{2}\alpha^{-\frac{1}{2}} \tag{A.9}$$

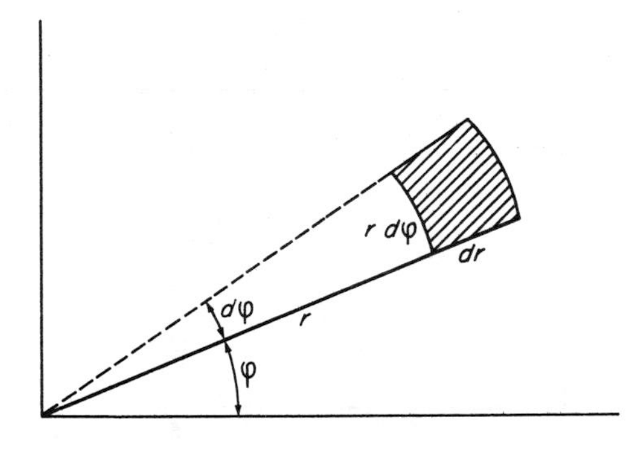

Figure A.2. Polar coordinate element of area. The hatched area $= r\,dr\,d\varphi$.

Using the recurrence relation we can generate $G_2(\alpha)$, $G_4(\alpha)$, and so on.

$$G_2(\alpha) = -\frac{dG_0}{d\alpha} = \frac{\sqrt{\pi}}{4}\alpha^{-3/2}$$

$$G_4(\alpha) = -\frac{dG_2}{d\alpha} = \frac{3\sqrt{\pi}}{8}\alpha^{-5/2}$$

$$G_6(\alpha) = -\frac{dG_4}{d\alpha} = \frac{15\sqrt{\pi}}{16}\alpha^{-7/2}$$

For ready reference the first eight Gaussian integrals are listed in Table A.1.

TABLE A.1

Gaussian Integrals

n	$G_n(\alpha) = \int_0^\infty x^n e^{-\alpha x^2}\,dx$	n	$G_n(\alpha) = \int_0^\infty x^n e^{-\alpha x^2}\,dx$
0	$\frac{1}{2}\sqrt{\pi}\,\alpha^{-1/2}$	4	$\frac{3}{8}\sqrt{\pi}\,\alpha^{-5/2}$
1	$\frac{1}{2}\alpha^{-1}$	5	α^{-3}
2	$\frac{1}{4}\sqrt{\pi}\,\alpha^{-3/2}$	6	$\frac{15}{16}\sqrt{\pi}\,\alpha^{-7/2}$
3	$\frac{1}{2}\alpha^{-2}$	7	$3\alpha^{-4}$

Appendix B

The Universal Character of β

In Chapter 15 we obtained a general relation for $\ln W$, with W being the thermodynamic probability

$$\ln W = N \ln Z(\beta) + \beta U \tag{15.53}$$

The parameter β originally entered as a convenience factor which rendered βU dimensionless. Considerations involving the internal energy of an ideal gas showed that

$$\beta = \frac{1}{kT} \tag{B.1}$$

at least for the ideal gas. We now establish the universal character of β, that is, we show that $\beta = 1/kT$ for all systems. The proof employs the "super system" device used in Section 15.9 to deduce the relationship between entropy and thermodynamic probability.

Let two systems, A and B, be in mutual thermodynamic equilibrium at a temperature T. Collectively A and B constitute a super system AB which is also in an equilibrium state at the temperature T. The populations of the phase-space cells of A and B are given by the Boltzmann distribution (15.33)

$$N_{i\mathrm{A}} = \frac{N_{\mathrm{A}} Q_{i\mathrm{A}}}{Z_{\mathrm{A}}(\beta_{\mathrm{A}})} \exp(-\beta_{\mathrm{A}} \varepsilon_{i\mathrm{A}}) \tag{B.2}$$

$$N_{k\mathrm{B}} = \frac{N_{\mathrm{B}} Q_{k\mathrm{B}}}{Z_{\mathrm{B}}(\beta_{\mathrm{B}})} \exp(-\beta_{\mathrm{B}} \varepsilon_{k\mathrm{B}}) \tag{B.3}$$

To establish the universal character of β we must show that $\beta_A = \beta_B$. Since A and B are arbitrary this step would show only that β has the same value for all systems. We have already established that $\beta = 1/kT$ for an ideal gas. Thus by demonstrating that $\beta_A = \beta_B$, then allowing B to be an ideal gas, we can show $\beta_A = 1/kT$, where A denotes any system. Because of the relation

$$W(AB) = W(A) \cdot W(B) \tag{15.47}$$

we have

$$W(AB) = \frac{N_A!}{\prod_{i=1}^{M_A} N_{iA}!} \frac{N_B!}{\prod_{k=1}^{M_B} N_{kB}!}$$

If one carries out the same maximization procedure for $\ln W(AB)$, namely,

$$\ln W(AB) - \beta_s U_{AB}(\beta_s) = \ln W'(AB) - \beta_s U'_{AB}(\beta_s)$$

the result is

$$N_{iA} N_{kB} = \frac{N_A Q_{iA} \exp(-\beta_s \varepsilon_{iA})}{Z_A(\beta_s)} \frac{N_B Q_{kB} \exp(-\beta_s \varepsilon_{kB})}{Z_B(\beta_s)} \tag{B.4}$$

From (B.2) and (B.3) we have

$$N_{iA} N_{kB} = \frac{N_A Q_{iA} \exp(-\beta_A \varepsilon_{iA})}{Z_A(\beta_A)} \frac{N_B Q_{kB} \exp(-\beta_B \varepsilon_{kB})}{Z_B(\beta_B)} \tag{B.5}$$

Comparing (B.4) and (B.5) leads to

$$\frac{Z_A(\beta_A)}{Z_A(\beta_s)} \exp[-(\beta_s - \beta_A)\varepsilon_{iA}] = \frac{Z_B(\beta_s)}{Z_B(\beta_B)} \exp[-(\beta_B - \beta_s)\varepsilon_{kB}] \tag{B.6}$$

On the left side stands a function referring to A alone; on the right side there stands a quantity referring to B alone. The only possible way for the equality to prevail is if both sides equal the same constant. This in turn is possible only if $\beta_s = \beta_A$; otherwise the exponential factor

$$\exp[-(\beta_s - \beta_A)\varepsilon_{iA}]$$

would be different for each different value of ε_{iA}. Likewise, the right side of (B.6) is a constant only if $\beta_s = \beta_B$. Thus we have

$$\beta_A = \beta_s, \qquad \beta_B = \beta_s$$

and so

$$\beta_A = \beta_B$$

Hence β has the same value for all systems at the same temperature. We have seen that $\beta = 1/kT$ for an ideal gas. It must therefore have this same value for all systems.

Subject Index

C
D
E 9
F 0
G 1
H 2
I 3
J 4

II

The Federal Reserve System

The Federal Reserve System has been given the responsibility for monetary controls. Monetary controls are the older, more orthodox method of achieving economic stability—the real goal of the monetary policy of the Federal Reserve System. However, the goal of the System is more than just stability; it is stability at a high level of employment. In addition, the goal includes growth. The Federal Reserve System's goals, then, are price stability at a high level of employment together with economic growth. To achieve these goals, the System has certain tools which work by influencing the supply of credit and the interest rate. The principal tools are open market operations, control of the discount rate and changes in reserve requirements. Changes in the supply of credit and the interest rate however must have an impact upon demand for monetary policy to be effective. That is, monetary controls are, for the most part, indirect in that they can influence aggregate demand only indirectly.

How successful the Federal Reserve has been in achieving its goals has always been subject to some debate. The article by Deane Carson suggests that the System has grown too complex over the years and that it should be streamlined. One way of doing this might be to make membership for national banks voluntary as it now is for state banks. Another recommendation for change is set forth in Harry G. Johnson's testimony in hearings before the congressional subcommittee on domestic finance. He questions the independence of the Federal Reserve System, and suggests that national economic goals might be better served if the monetary authority were not independent of the executive and legislative branches of government.

G. L. Bach considers the same question of the independence of the monetary authority in reviewing the Report of the Commission on Money and Credit. He examines possible changes in the degree of independence of the Federal Reserve System, as well as possible changes in its organization in terms of the System's goals and capabilities as well as in the context of the economic and political environment within which it operates.

The remaining articles in this section look at two of the principal tools of the Federal Reserve System. Robert Roosa explains how transactions are made for the System open market account. In the final article, the discount mechanism is considered in terms of both its role in implementing monetary policy and its use in coordination with other monetary tools.

6

Is the Federal Reserve System really necessary?

*Deane Carson**

Since 1964 marks the golden anniversary of the Federal Reserve System, the title of this essay may appear somewhat uncharitable to those who have come to think of the Federal Reserve in terms only slightly less affectionate than those accorded to the Old Lady of Threadneedle Street.[1] I hasten to assure the reader that my heresy, if that is what it is, involves principally the word *System*; that is to say, I shall examine the need for a Federal Reserve *System* as it is presently constituted, quite apart from the generally acknowledged need for central bank monetary policy. While this task might be thought properly to lie within the province of the political scientist, I shall show that, on the contrary, there are many important economic aspects involved in such an inquiry.[2]

Central to the analysis which follows is the proposition that the

Reprinted with permission from *The Journal of Finance,* Vol. XIX, No. 4 (December, 1964), pp. 652-661.

* Senior Economist, Office of the Comptroller of the Currency, and Associate Professor of Economics (on leave), Brown University. Views expressed in this paper are those of the author and do not necessarily reflect those of the Office of the Comptroller of the Currency. [*Editors' note:* Mr. Carson is now at Northwestern University.]

[1] If the timing of this critique seems somewhat uncharitable, I must fall back on an amusing precedent: Allan Sproul, one of the more astute central bankers in recent times, advocated abolition of the Office of the Comptroller of the Currency in his contribution to a book of essays sponsored by the Comptroller to mark the Centennial of the National Banking System. See his "The Federal Reserve System—Working Partner of the National Banking System for Half a Century" in Deane Carson (ed.), *Banking and Monetary Studies: In Commemoration of the Centennial of the National Banking System* (Homewood, Ill.: Richard D. Irwin Inc., 1963), p. 77.

[2] This has been recognized by Representative Wright Patman who has marked the Federal Reserve's fiftieth milestone in his own inimitable fashion; namely, by a thoroughgoing investigation of the structure of the Federal Reserve System, with an overriding emphasis upon its "independence" and mix of public and private powers.

success of the essential function of the central bank, monetary management, is independent of the structural arrangements that characterize its organization. This is to say, central bank policies can be executed within a variety of organizational structures, both internal as well as external vis à vis the commercial banking system. The Federal Reserve System *qua System* is but one of a number of such structural arrangements within which a monetary policy can be carried on.

Unfortunately, this fact is little appreciated. A fair sampling of money and banking textbooks, while explicitly silent on the point, leave one to infer that in some unique sense the existing system is a necessary adjunct to the pursuit of successful monetary management. After a chapter or two on the structure of the Federal Reserve System, the student is successively introduced to *functions* and to *policy*.

Out of this, or perhaps independent from this, have developed a mythology and a basic fallacy. The mythology has many aspects: it is generally believed that "member banks" are necessary to the conduct of monetary policy; it is generally believed not only that legal reserve requirements are necessary to the conduct of monetary policy but also that these reserves have to be held at the central bank; it is widely if certainly not universally believed that the Federal Reserve Banks serve many useful functions that could not and are not performed by private institutions, such as discounting, clearing of checks, and provision of vaults for the safekeeping of securities; and, without exhausting the mythology, a rather substantial sentiment exists to the effect that the whole pyramid of varying *authority*—the two hundred and sixty-one Directors of Federal Reserve Banks and their branches, the twelve-man Open Market Committee, the seven-man Board, and the twelve-man Federal Advisory Council—somehow formulates a monetary policy superior to that which could be conjured up by a single Governor of the calibre of Montague Norman or Benjamin Strong.

The fallacy that all this has fostered is simply this: monetary policy, being an extremely complex matter, requires a very complex *System* to make it operative, and the resources that we now allocate to monetary management are required to maintain a viable central banking function in relation to the goals we have assigned to the Federal Reserve.[3] In opposition to this I would advance the proposition that a simple central banking structure is most conducive to successful monetary management, other things equal, and that we can reduce both its internal and its external costs by adopting certain basic reforms.[4]

[3] The recent Patman inquiry (Hearings *op. cit.*) dwelt at some length on this matter although, unfortunately, its shots were so scattered that the essential allocation problem was submerged.

[4] Lack of space prohibits discussion of all such reforms. One in particular deserves separate treatment and cannot be included in this paper; namely, the need to transfer the supervisory functions now performed by the Federal Reserve to some

Basically, my proposals involve two such reforms which, while perhaps not interdependent at first glance, are closely related in fact. I propose, first, that membership in the Federal Reserve be placed on a completely voluntary basis; and, second, that compulsory legal reserve requirements be abolished. These, together with their corollary structural changes, are discussed in turn below.

I. The Case for Voluntary Membership

At the present time, state-chartered banks may elect to become members of the Federal Reserve System; banks chartered by Federal authority must become members as a matter of law. This distinction between banks according to the source of charter was initially imposed on the grounds that the purposes of the Federal Reserve Act could only be carried out if a substantial fraction of the cash reserves of commercial banks were mobilized in the Federal Reserve District Banks, and if a substantial number of banks had access to the discounting privileges afforded by these regional arms of the Federal Reserve System. Fears that compulsory membership for all commercial banks would compromise the rights of the several states, together with the easy expediency of subjecting Federally chartered banks (which were already subject to Federal control) to captive membership in the System, were responsible for the distinction between banks as written into the Federal Reserve Act.

I shall demonstrate in this section that voluntary membership (1) would not, as some have alleged, destroy the effectiveness of monetary management, and (2) would reduce the discrimination against (particularly) smaller Federally-chartered banks that are now captive members. Initially, we assume that the second part of the suggested reform is not adopted, that is to say, member banks continue to be subject to compulsory legal reserve requirements which must be held with the District Banks. This assumption is dropped in Section II of the paper.

Our initial task is to estimate the probable results of legislation providing for voluntary membership in the Federal Reserve. Such an estimate is based upon the assumption that National banks of any given size would elect to remain in the System in the same proportion that State-chartered banks of that size are presently members. Since we have data at hand on the assets of member National banks, and member State banks in various size groups, estimates can easily be generated. Tables 1 and 2 provide the basic data for these estimates. By summing the totals

---

other agency. This has been suggested by at least one present member of the Board, J. L. Robertson, and is reportedly looked upon with favor by others. In any case, the complexity of monetary management would seem to argue for single-minded attention of the Board.

for various classes of banks in Table 1, we observe that insured bank assets totaled $310.8 billions at the end of December, 1963. Next, summing the totals of column 7 in each Table, we find that if all insured commercial banks were accorded the right to forgo System membership, something like $98.1 billion of commercial bank assets would be "outside" the Federal Reserve. This represents 31.5 percent of total assets.

The effectiveness of monetary policy depends to some extent on the pervasiveness of its impact and possibly but not clearly upon the percentage of bank institutions that have access to the discount window.[5]

Table 1. Number and Assets of Insured Commercial Banks, by Size, December 1963 (dollar amounts in millions)

1	2	3	4	5	6	7
Deposit Size (millions of dollars)	National		State-Member		Insured Nonmember	
	No. Banks	Assets	No. Banks	Assets	No. Banks	Assets
Less than 1.0	132	$ 123	24	$ 22	630	$ 535
1.0 to 1.9	388	702	131	224	1,665	2,766
2.0 to 4.9	1,316	5,100	465	1,758	2,563	9,228
5.0 to 9.9	1,145	9,082	328	2,530	1,282	9,760
10.0 to 24.9	935	16,037	277	4,647	688	11,314
25.0 to 49.9	329	12,739	104	4,068	144	5,434
50.0 to 99.9	167	13,257	68	5,459	48	3,573
100.0 to 499.9	164	41,052	64	15,170	30	6,102
500 and over	39	72,143	27	57,337	1	677
Total	4,615	$170,233	1,488	$91,215	7,051	$49,390

Any correlation between policy effectiveness and *number* of member banks, however, must certainly be weak, since the impact of scarce or ample funds would not appear to depend upon the presence of Federal Reserve stock in the portfolio of any particular bank. Furthermore, our highly developed system of correspondent banking relationships insures that monetary policy changes will be transmitted to the entire banking structure. I would certainly argue, in any case, that the effectiveness of monetary policy with 68.5 percent of commercial bank assets covered will be no less than when 90 or 100 percent coverage obtains.[6] Since the reasons for this are covered in the following section, they need not be considered here.

[5] Under present law the Federal Reserve banks can technically make advances to non-member banks under 12 U.S.C. 347c.

[6] As a matter of fact, the middle 1920's are often considered years of effective monetary policy; at that time approximately 69 percent of commercial bank assets were covered.

Table 2. Estimate of Assets of Nonmember National Banks if Membership Were Optional (dollar amounts in millions)

1	2	3	4	5	6	7
	Assets of					
Deposit Size (millions of dollars)	Assets of Insured State Banks	Insured Non-member Banks	Insured Non-member Banks as Percent of Insured State Banks	Assets of National Banks	Assets of National Banks that would be non-members (column 4 times column 5)	Cumulative Nonmember Assets
Less than 1	$ 557	$ 535	96.1	$ 123	$ 118	$ 118
1.0 to 1.9	2,990	2,766	92.5	702	649	767
2.0 to 4.9	10,986	9,228	84.0	5,100	4,284	5,051
5.0 to 9.9	12,290	9,760	79.4	9,082	7,212	12,263
10.0 to 24.9	15,961	11,314	70.9	16,037	11,368	23,531
25.0 to 49.9	9,502	5,433	57.2	12,739	7,285	30,816
50.0 to 99.9	9,032	3,573	39.6	13,257	5,244	36,060
100.0 to 499.9	21,272	6,102	28.7	41,052	11,776	47,836
500 and over	58,014	677	1.2	72,143	842	48,678

An alternative to the voluntary membership proposal discussed above would provide for compulsory membership of all insured commercial banks above a given size. The cutoff asset size that has been occasionally mentioned is $10,000,000. Under this proposal, obviously, larger non-member State banks would be required to join, while all National banks under the cutoff size would be afforded the choice now open to State-chartered banks. For the latest available data (end of 1963) I have calculated that this cut-off point would reduce "covered" assets by only approximately $6.2 billion under the extreme assumption that all National banks with less than $10 million total assets elect to forgo Federal Reserve membership. At the same time, voluntary membership would be extended to approximately 77 percent of all insured commercial banks, from the present 66 percent.[7]

On its face, this proposal would seem to be a superior alternative to completely voluntary membership. And indeed, it probably is a more satisfactory basis for discrimination than that found in the present law. On the other hand, its superiority to complete voluntarism can only be defended on the grounds that effective monetary policy requires that a large proportion of the reserves of the commercial banks be held in the form of compulsory balances at the Reserve Banks. More precisely, it requires the finding of a positive correlation between effectiveness of monetary policy and the percent of total bank reserves held within the System. Again this is properly a matter for consideration in Section II and is therefore postponed for the moment.

There are, however, clear advantages to the completely voluntary membership proposal. Certainly the most important of these is that it would enable all insured banks to choose between public and private suppliers of banking services to banks. In this connection it is worthy of note that large private banks, as correspondents, now provide a very wide range of such services on terms that are clearly superior to similar services provided by the Federal Reserve Banks. Among the more important of the latter are check-clearing arrangements, temporary loan accommodation, credit and operations analysis, and provision of economic information. Small National banks find it convenient to utilize these privately supplied services, against which they must carry correspondent balances, in spite of the fact that they must also carry legal reserves with the District Banks. In effect, compulsory membership imposes a discriminatory burden on these banks in the form of double cash balances.

Table 3 demonstrates the extent to which Federal Reserve membership leads to this result. It indicates a consistent pattern of higher cash holdings to total assets for member banks than for nonmember banks. This is not due to lower reserve requirements for State-chartered banks; indeed,

[7] Table 1.

Table 3. Cash and Balances with Banks as a Percentage of Total Assets of National, State Member, and State Non-Member Banks in Selected Areas[a] June 29, 1963*

State or Area	National Banks	State-chartered Member Banks	State-chartered non-Member Banks
United States	17.6	18.4	12.5
Alaska	12.9	—	13.0
Connecticut	17.5	18.3	10.4
District of Columbia	18.3	16.4	14.5
Hawaii	17.1	—	12.4
Idaho	11.8	12.4	11.7
Kansas	18.6	17.9	14.1
Maine	13.8	12.9	9.2
Mississippi	18.6	18.5	16.7
New Hampshire	17.6	—	6.2[1]
New Jersey	12.9	12.5	10.5
New Mexico	17.7	18.5	15.0
North Dakota	12.6	—	9.4[2]
South Carolina	20.3	16.4	15.2
South Dakota	13.4	13.6	11.3
Vermont	11.4	—	6.8
West Virginia	17.2	19.2	12.7
Wyoming	15.2	17.2	14.0

* Source: *FDIC Assets, Liabilities and Capital Accounts of Commercial and Mutual Savings Banks March 18 and June 29, 1963.*

[a] States were selected to exclude all those in which banks subject to Reserve City legal reserve requirements were in operation.

[1] Includes 20 banks, 1 of which was a member bank.

[2] Includes 115 banks, 2 of which were member banks.

of the selected states, nine have substantially higher reserve requirements than those currently imposed by the Federal Reserve,[8] six states impose legal reserve requirements that are substantially the same as System requirements,[9] and only two states[10] have reserve requirements that are substantially less than the Federal Reserve's 12.5 percent and 4 percent requirements against demand deposits and time deposits respectively.

The clear implication of these comparisons is that membership in

[8] Wyoming (20 and 10 DD and TD); Alaska (20 and 8); Idaho (15 percent of all deposits); Kansas (12½-20 and 5); West Virginia (15 and 5); South Dakota (12-20 depending on size but one-third may be held in bonds); New Hampshire (15 and 15); Vermont (30 and 8); and Mississippi (15-25 and 7-10).

[9] New Mexico and New Jersey (12 and 4); Hawaii, Connecticut and Maine (12 and 5); and District of Columbia (12½ and 4).

[10] North Dakota (10 and 5); and South Carolina (7 and 3).

the Federal Reserve leads banks to hold a higher proportion of their assets in cash than is considered necessary by banks that are not in the System. From this we deduce that compulsory membership of National banks, where it is due to a "locked-in" effect,[11] discriminates without economic justification against banks holding Federal charters. In effect, captive banks, particularly the smaller National banks, maintain sterile cash reserves required by law for which they receive few compensating benefits; in order to carry on their business, they also must carry correspondent balances which do bear a return in the form of needed services. Non-member banks, which may and almost invariably do make their legal reserves serve double-duty as service-generating correspondent balances, are placed in a position of competitive advantage.

While the inequity of present membership requirements would be somewhat modified if compulsory membership were adopted, discrimination would not be eliminated. Indeed, while discrimination by charter would be avoided, total inequity might well increase. Under compulsory membership all banks that find privately produced bank services to banks superior to those provided by the Federal Reserve would be deprived of the choice now accorded to State banks. Since it is principally the larger banks that find Federal Reserve membership attractive, such a plan would tend to discriminate against small banks in general rather than against a particular segment of this group.

II. The Need for Reserve Requirements and Reserve Balances at the Federal Reserve Banks

Desired cash holdings of the banking system limit the marginal expansion of bank deposits and, to the extent that they are influenced by legal reserve requirements, it can be said that the latter serve as a fulcrum for credit control. More precisely, however, the monetary control mechanism operates *via* changes in the level of total reserves relative to desired cash holdings of the banking system. I shall contend in this section that the necessity for legal reserve requirements and minimum cash balances at Federal Reserve banks is a function of the particular objectives of Federal Reserve policy; I shall further argue that the locus of the banking system's cash reserves is of little significance with respect to either the structure of the Federal Reserve System, or its effectiveness as a central bank.

[11] All National banks, of course, could escape the burdens of membership by changing to State charters. The costs of this, however, are quite high in many cases. When a bank changes its charter it must also change its name, entailing considerable out-of-pocket expenses and loss of "good will." It is not reasonable to impose this cost in order to reduce other costs that have no economic justification, and where a reasonable alternative remedy is at hand.

A. The functions of reserve requirements and a proposal

Reserve requirement changes are a substitute for open market operations.[12] An initial justification for the existence of legal reserve requirements is, therefore, that their levels can be changed, and with them monetary and credit expansion potentials. It is not within the scope of this discussion to weigh the merits of changes in reserve requirements *versus* changes in the open market portfolio of the central bank. In a zero percent reserve requirement banking system, however, it must be recognized that the substitute, imperfect as it now is from the standpoint of effectuating monetary control, would no longer exist.

It can be argued, therefore, that some future situation might arise that would call for the raising of reserve requirements, even though the Federal Reserve Board has not seen a need to do so since February 1951, thirteen years and several business expansions ago.[13] I recognize this possibility as a defect in the plan, but a defect which could be easily remedied through congressional action, given the compelling circumstances that would give rise to the need.

Quite apart from the above, a great deal of emphasis has been given to the *level* of legal reserve requirements as a base which limits the potential expansion of money and credit. Arithmetical exercises in standard textbooks "prove" that the height of reserve requirements determines the maximum expansion potential of any given amount of excess reserves, subject to assumptions that are usually specified.[14] It is not at all clear that this fact is relevant to the functionality of legal reserve requirements. In the first place, banks individually and in the aggregate would hold some level of desired cash reserves against deposits in the absence of legal requirements,[15] thus providing the "base" for monetary and credit expansion (or contraction).

In the second place, since the levels of reserve requirements have been progressively lowered (with few reversals) in the postwar period without appreciably affecting the performance of monetary policy, the question can be raised as to why they are at all necessary in the present context—that is, as a limitation on the potential expansion of money and credit.

[12] Cf. Joseph Aschheim, *Techniques of Monetary Control* (Baltimore: The Johns Hopkins Press, 1961), Chapter II.

[13] On November 26, 1960, the Board raised country bank reserve requirements from 11 to 12 percent, while simultaneously permitting the calculation of vault cash in the reserve base. This increase was a technical adjustment to the inclusion of vault cash and therefore does not count as a monetary policy action.

[14] Zero desired excess reserves, and no change in cash in circulation.

[15] For example, state chartered banks in Illinois are not subject to reserve requirements, yet they keep something in the order of 12 percent of their deposits in cash.

Cash reserves can be controlled by open market operations, and the tone of the market observed by the simple device of central bank hypothecation of the market's desired level of bank cash reserves. Given continuation of reporting requirements, the device of "shadow reserve requirements"[16] suggested here would enable the central bank to observe "excess reserves," "free reserves" and "net borrowed reserves" as indicators of money market conditions without the necessity of formal requirements.

The plan would work in the following way: suppose the Federal Reserve Board were to announce that it considered X percent of deposits (details aside) an appropriate level of cash reserves for the commercial banks (or some segment of the banking system).[17] Periodic reports to the Federal Reserve on actual cash holdings and deposits would give the monetary authorities precisely the same "feel of the market" that they now require to conduct defensive open market operations to offset very short-term disturbances in the money market.

It is of course a debatable question whether offsetting these changes is an appropriate objective of monetary control in the pursuit of longer range goals of full employment, price level stability, and economic expansion. Many would argue that day-to-day fluctuations in cash reserve need not interfere with the achievement of an appropriate level of change in the money supply which, after all, is the most important means of realizing the goals. Beyond this, it has been argued persuasively that free reserves are a misleading guide for monetary management.[18]

B. Slippage effects of the zero reserve requirement proposal

The proposal set out in skeleton form above[19] raises a very obvious question: will the abolition of reserve requirements increase the slippage that now exists between policy actions and policy results? Contrary to one's first inclination to answer affirmatively, it is not at all certain that this should be the case.

We are not concerned with slippages in general, but rather with one segment of the total lag between policy actions and their ultimate effects upon income and prices. This segment is the initial one, that

[16] I am indebted to Sherman Shapiro for coining this phrase to describe the mechanism.

[17] It is not necessary to make such an announcement to generate the statistical indicators. However, an announced level of appropriate reserves would benefit portfolio managers and managers of reserve positions in that it would remove one source of uncertainty as to central bank policy that would exist if the announcement were not made.

[18] Cf. A. James Meigs, *Free Reserves and The Money Supply* (Chicago: The University of Chicago Press, 1962).

[19] Rather than extend this essay unduly by discussing the details of the proposal (transition problems, the eligibility of various cash assets for reserve computation, and other technicalities), I choose to leave these to future discussion.

which spans the sequence between a change in total cash reserves of the commercial banks and the employment of these reserves in loans and investments.

While this is basically an empirical question, intuition leads to the belief that if banks individually and collectively are in equilibrium (in the sense that their cash-to-deposit ratios are at the desired level), changes in cash reserves occasioned by open market operations will elicit responses quickly and in the right direction. If the Federal Reserve purchases securities (presumably, but not necessarily with Federal Reserve notes), the banks will find actual cash in excess of desired cash, and will take steps (loans, investments) to return to equilibrium.

On the other hand, sales of securities by the central bank will push the banks into equilibrium in the opposite direction. If the Federal Reserve retains its discount window, the deficit banks could choose between "borrowing" from themselves and borrowing from the Federal Reserve Bank. As Sprinkel has pointed out, the discount window is itself an institutionally sanctioned source of slippage[20]; I would suggest that its usefulness would depart with the demise of legal reserve requirements.

In effect each bank would have its own discount window; but we know that banks eschew borrowing as sin, and there is no reason to believe that this attitude would change just because the lender was the bank itself. I suspect that loan and investment officers would keep an even sharper eye on the actual cash ratio than they now do on the free reserve position. Temporary departures from desired equilibrium would occasion furrowed brows in the Board room and charges to the operating officers to "get the cash ratio back where it is supposed to be."

Over the monetary cycle the banks might well change their levels of desired cash reserves relative to deposits in a way that would counteract monetary policy. But this is hardly a peculiar defect of the zero reserve requirement proposal, since in effect precisely the same thing occurs with existing legal reserve requirements.

III. Conclusions

I have presented the case for voluntary membership in the Federal Reserve and a system of zero required cash reserves. The Federal Reserve System has evolved in the past half century into a vast and cumbersome machine; a quasi-private organization, its regional staffs have grown far out of proportion to their importance in conducting monetary policy. The tourist business in Maine may indeed be an important area of economic inquiry, but it is difficult to see its connection with the goals

[20] Beryl Sprinkel, "Monetary Growth as An Economic Predictor," *Journal of Finance*, September 1959, p. 342.

of monetary control. The district Federal Reserve Banks engage in such irrelevancies simply because of the archaic notion of membership in the Federal Reserve System. Catering to the banks to induce them to retain membership diverts a good deal of the attention of our monetary authorities from the main business at hand. Voluntary membership would go far toward a solution to this problem.

Reserve requirements are unnecessary to the effective conduct of monetary policy. They impose a tax on member banks that might well be levied in another way, if the revenue is needed or a need exists for penalizing this particular industry. Since they serve no liquidity purpose, it is extremely difficult to justify their existence.

7

Should there be an independent monetary authority?

Harry G. Johnson

Mr. Johnson. Thank you, Mr. Chairman. I want to begin with the argument for an independent monetary authority which is, I think, the key——

The Chairman. Will you identify yourself, please, for the record first?

Mr. Johnson. I am Prof. Harry Johnson, of the University of Chicago.

I want to begin with the argument for an independent monetary authority, which is the crux of the issues facing the committee, in my judgment.

The argument for an independent monetary authority has two facets to it. One is the political argument that an independent monetary authority is desirable to prevent Government from being able to indulge in its natural propensity to resort to inflation. The other, which is less explicitly political, is that a stable monetary environment is essential to the proper functioning of a predominantly free enterprise society, and that an independent monetary authority is essential to maintain such a monetary environment.

The first argument seems to me utterly unacceptable in a democratic country. Indirectly, it is an argument for establishing the monetary authority as a fourth branch of the Constitution, charged with the function of forcing the Legislature and the Executive to follow conservative economic policies involving the balancing of the budget and restraint on

Reprinted from *The Federal Reserve System After Fifty Years,* Hearings before the Subcommittee on Domestic Finance, Committee on Banking and Currency, House of Representatives, 88th Cong., 2d Session (Washington, D.C., U.S. Government Printing Office, 1964), pp. 970-973. Professor Harry G. Johnson is at the University of Chicago.

Government expenditures. In other words, it involves the establishment of a special position in Government for the owners of one form of property—owners of money and of assets fixed in terms of money—a position which is inconsistent with the principles of democratic equality and the presumption of democracy that the purpose of government is to serve the social good.

Turning to the second argument, granted that a stable monetary environment is desirable, the question arises whether an independent monetary authority as presently understood is sufficient to provide such stability. The argument that it is assumes that, if free of control by the Executive and Legislature, the monetary authority will govern monetary policy in the light of the longrun best interests of the economy, and will conduct its policy flexibility and efficiency in the short run. This assumption is not consistent with the historical evidence of the behavior of monetary authorities; the evidence is rather that central banks have done little if anything to restrain inflationary policies in wartime—and war and its aftermath have been the almost exclusive source of serious inflation in the major countries in the 20th century—while in peacetime they have displayed a pronounced tendency to allow deflationary policies on the average. Moreover—I refer here particularly to the behavior of the United States and Canadian central banks in the past decade—in the short-run conduct of policy they have tended to overreact to changes in the economy and to reverse their policy with a substantial delay, thereby contributing to the economic instability that their policies are intended to combat.

These defects are in my judgment inherent in the conception, constitution, and operating responsibilities and methods of an independent monetary authority, and are unlikely to be modified greatly by gradual improvement of the techniques of central banking on the basis of accumulated experience and research. For one thing, freedom of a central bank from direct political control does not suffice to render it insensitive to contemporary political opinion. On the contrary, its position as the one agency of economy policy formation outside the normal political structure both exposes it to subtle and sustained political pressures and forces it to become a political animal on its own behalf, devoting considerable effort either to justifying its policies by reference to popularity-esteemed objectives or to denying responsibility for economic conditions and passing the buck on to the Executive or the Legislature, the result being to obfuscate the policy choices that have to be made. Secondly, the position of the central bank as controller of the money supply inevitably must bias the monetary authority—except in times of national emergency such as war—toward emphasizing the pursuit of objectives connected with the value of money—resistance to domestic inflation, and preservation of the international value of the currency—to the underemphasis or neglect of other objectives such as high employment and economic growth. Thirdly, the methods of monetary management, which involve the central bank concentrating its attention on money market conditions and interest rates, and on member bank reserve positions and lending, rather than on the performance of the economy in general, are extremely conducive to the behavior pattern of overreaction and delayed correction of error already mentioned.

Because it concentrates on money market and banking phenomena, rather than the effects of its policies on the quantity of money and economic activity, and because the effect of monetary policy on the economy operates with a substantial lag, the central bank is extremely likely to push its policy too far and too fast before it realizes that the policy has taken effect and begins to consider moderating it; and because the realization of effectiveness comes late, it is likely to reverse its policy too sharply. In addition, the fact that the central bank stands in a special relation to its Government and domestic economy fosters the existence of an international fellow-club-member relationship among central banks, a relationship congenial to the formation and propagation of policy fads in central banking. It is only on the basis of fads in central banking opinion, I believe, that one can understand the emergence of the fear of runaway inflation as a dominant motif in central bank policy statements in 1957-58 and the belief at that time in the need to reduce bank liquidity by debt-funding, or the widespread belief that the dollar would soon be devalued that emerged in 1958-59 and persisted thereafter in spite of reiterated statement of the U.S. determination not to devalue.

Recognition of the undemocratic nature of the political argument for an independent monetary authority, together with scholarly documentation of the inadequacies of the historical performance of the Federal Reserve System, has led a number of economists—including my distinguished colleague, Milton Friedman, who will appear before this committee at a later date—to recommend that the goal of providing a stable monetary environment should be implemented, not by entrusting discretionary monetary management to an independent monetary authority, but by legislating that the monetary agency be required to increase the quantity of money at a fixed rate determined from historical experience. I have a certain sympathy with this recommendation, as an alternative to discretionary management as it has been conducted in the past, but there are, in my opinion, some overriding objections to it. In the first place, the proposal is essentially a component of a much broader program for transforming the country into a working model of an ideal competitive system, which system it is assumed would require no deliberate economic management; since the majority of public opinion seems in fact committed to the belief that economic management can improve on unfettered competition, adoption of the proposal would entail accepting a self-denying ordinance in a crucial area of policy, an inconsistency which I doubt would prove acceptable for long. In the second place, the proposal depends on the empirical assumption that the demand for money depends primarily on income and is relatively insensitive to changes in interest rates, an assumed fact concerning which the results of empirical research are in substantial conflict; if the demand for money is not a stable function of income only, the proposal might lead to more instability than discretionary management. Thirdly, the proposal abstracts from the complications of international competition. To be feasible, the proposal would have to be accompanied by the adoption of floating exchange rates, and even in that case might aggravate instability associated with international movements of capital in response to interest-rate differentials between countries; alternatively, it would have to be accomplished by policies of direct intervention in international trade and payments inconsistent with the efficient operation of a competitive economy.

My own view is that the pursuit of monetary stability through the separation of monetary management from other economic policy, and its placement under either an independent authority or a strict rule of increase, is an illusory solution to the problem. Instead, I believe that monetary policy should be brought under the control of the Executive and legislature in the same way as other aspects of economic policy, with the administration bearing the ultimate responsibility for monetary policy as part of economic policy in general. In making this recommendation, I must admit that there is a danger of monetary mismanagement in the pursuit of political objectives; but I consider it preferable for such mismanagement to be a clear responsibility of the administration, and accountable to the electorate.

I would also point to a danger emphasized by the British economist, Sir Roy Harrod, at the time of the nationalization of the Bank of England, and confirmed, in my judgment, to some extent by subsequent British experience; namely, that bringing the monetary authority within the fold of Government may give more rather than less weight in policymaking to its definitions of, and opinions on, policy problems. In this connection, though, I would like to point out that the monetary authority can only too easily be cast as a scapegoat to conceal the unwillingness of public opinion and the administration to recognize and resolve policy conflicts. In particular, in this country in recent years the fundamental policy problem has been the conflict between equilibrium in the balance of payments and a satisfactory level of domestic activity, imposed by the overvaluation of the dollar relative to the major European currencies; and I do not believe that an administration armed with complete control of monetary policy, but committed to preserving the international value of the dollar, would have conducted a monetary policy very different from what the Federal Reserve has in fact conducted. It might, of course, have taken the bold step of raising foreign loans on the order of $15 to $25 billion to tide over the years of waiting for European prices to inflate up to the American level, but there is no evidence that the administration has been prepared to contemplate such a policy. Given the commitment to a fixed exchange rate, domestic monetary policy must necessarily be subordinated to the balance-of-payments position; the burden of achieving a satisfactory level of employment and activity must be borne by fiscal policy rather than monetary policy, which, in recent circumstances, has meant a substantial tax cut; and it is the reluctance of public opinion, the Congress, and the administration to resort to a tax cut, rather than the policy of the Federal Reserve System, that is ultimately responsible for the unsatisfactory levels of employment and activity that have characterized the economy during the recent years.

While I believe that the monetary authority should be made part of the regular machinery of governmental economic policy making and policy execution, I am not too hopeful about the possibility that this change would result in a significant improvement in the efficiency of monetary policy as an instrument of shortrun economic stabilization. My reasons for skepticism stem from the analysis of the influence of the monetary authority's position and responsibilities in the economy on its methods of conducting monetary policy that I have already sketched. This analysis leads me to a conclusion basically similar to that of the proponents of a fixed rule of monetary expansion: that the monetary policy instrument is not well adapted to the pursuit of shortrun stabilization policy, and that it should instead be devoted, so far as possible, to the goal of providing a stable longrun monetary environment. I differ from the advocates of an expansion rule, however, in recommending that this goal should be established as a priority objective of discretionary monetary policy, operating as one of a group of instruments of economic policy rather than legislated as a rigid obligation on the monetary authority. I have developed the analysis underlying this recommendation in a lengthy document prepared for the Canadian Royal Commission on Banking and Finance, which is attached to this statement.

The Chairman. It may be inserted in the record at this point.

8

Economics, politics, and the Fed

G. L. Bach

The economics and politics of modern money and credit, with special emphasis on the Federal Reserve System, was the main assignment of the Commission on Money and Credit.[1] The need for such an examination is clear: for since the Federal Reserve was established a half century ago, much has changed—the role of government has mushroomed; the national debt has grown from virtually nothing to nearly $300 billion; the United States has become the world's undisputed economic leader, the one massive international creditor and international banker for the noncommunist world.

One looks in vain in the Report presented by the Commission for penetrating new analyses or perspectives on the problems wrought by these changes, or for new policy approaches or machinery in the field of money and credit. Above all, the Commission has not dealt convincingly with the No. 1 dilemma of modern monetary policy—how to encourage high-level employment and rapid growth while repressing inflation. But, notwithstanding these shortcomings, the Commission's recommendations should not be written off by thoughtful businessmen. For while they contain little that is new, they are sober and responsible, and they do raise important questions about the way monetary policy works and how it might work better.

What the Commission has to say about the Federal Reserve, money, and monetary policy can be broadly lumped under four main heads: (a) the goals of Federal Reserve policy; (b) what monetary policy can

Reprinted with permission from the HARVARD BUSINESS REVIEW, January-February, 1962. Pp. 81-91. G. L. Bach was Dean of the Graduate School of Industrial Administration, Carnegie Institute of Technology. He is now Professor of Economics and Policy, Stanford University, Graduate School of Business.

[1] *Money and Credit, Their Influence on Jobs, Prices, and Growth: The Report of the Commission on Money and Credit* (Englewood Cliffs, New Jersey, Prentice-Hall, Inc., 1961).

do; (c) Federal Reserve independence; and (d) Federal Reserve organization and operations.

Economic Goals

The Commission begins its Report with a chapter about "National Economic Goals." Its position can be summarized as follows:

1. The Federal Reserve's goals should be part of, and substantially the same as, the goals of general national economic policy.
2. Reasonable price stability, sustained high levels of employment and production, and an adequate rate of economic growth should be the three main objectives of monetary policy—and these must be achieved within a framework of economic freedom and primary reliance on the market mechanism, while recognizing the need for adequate national security measures.
3. These three goals are roughly of equal importance, and they can be attained simultaneously within the general framework indicated above.

The first two propositions are eminently sensible and widely accepted. But the third is the rub. Are high employment, rapid growth, and stable prices of roughly equal importance? Can they all be attained together? If something must give, what should it be?

Here the Commission, with its objectivity, its generous budget, and its large group of economic researchers, could have done a vital job which apparently went undone. There is strong reason to suppose that we may face painful choices in the future, especially between keeping prices stable and unemployment low. If so, sound policy decisions require a careful weighing of the costs of trading one objective off for the other.

For monetary policy this is especially crucial, because avoidance of inflation has been traditionally the No. 1 goal of central bankers. Maintenance of high-level employment and output has generally been secondary; and, at least until recently, little direct attention has been paid to the encouragement of economic growth. But many, perhaps most, economists today argue that high-level employment and output of wanted goods and services are the fundamental goals—that large-scale unemployment of men and machines is *the* major economic waste. They say that moderate inflation, while undesirable, involves mainly a redistribution of financial claims rather than a diminution of real output of goods and services. In short, it is more important to maximize total output than to quarrel over who gets how much of a smaller total.

Choice among goals

The Commission puts avoidance of inflation first in its discussion. But it devotes less than a page to the basic case for reasonable price-level stability. Its statement is careful, avoiding the unbased claims of disaster from moderate inflation that are made by so many. Indeed, it is so cautious that the case for elevating stable prices to a parity with high-level employment and adequate economic growth seems inadequately demonstrated. The Commission refers to others' claims about the costs of creeping inflation—that inflation slows down the rate of economic growth, produces serious inequities, reduces productive efficiency, generates waste in the allocation of resources, and leads toward runaway inflation and economic collapse. Yet existing economic research on the effects of moderate inflation fails to support these fears convincingly. In fact, creeping inflation in America seems to have much less sweeping results than many had anticipated, for either the better or the worse.

These comments are not intended to imply that inflation is unobjectionable. Rather, I wish to stress how little has been accomplished—and, unfortunately, how little the Commission has added—in careful, dispassionate weighing of the social gains and costs involved in giving up some of one national goal in order to achieve more of another.

The Commission's treatment of the basic case for an "adequate" rate of economic growth is similarly brief. Here, too, one is reluctant to criticize a statement of economic goals so much more balanced and careful than most. But the need for such objective, comparative analysis is a basic one—indeed, the foundation for all rational choice as to the best monetary policy. More attention to these long-range fundamental issues and less day-to-day operating advice on monetary and fiscal policy would have made the Report a more valuable document.

Employment vs. stable prices

Can we have both high employment and stable prices? The Commission says *yes*, but its discussion leaves me uneasy. *If* we face inflation based on excess income claims, then there is a dilemma that cannot be wished away. That is, if labor unions, business, agriculture, and all the others demand larger incomes in the market place than can be satisfied at high-employment national output, then their higher wage and price demands generate inflation—higher prices, wages, profits, and money incomes generally, rather than more real output. Such inflation cannot go far without more money to support the higher spending and prices; therefore, the Federal Reserve can stop it merely by sitting tight and providing no new bank reserves. But the price of doing so is high—growing unemployment, with falling output and sales, at the higher prices. Then we must choose; stable prices or higher employment and rapid growth.

The Commission hopes that excess-income-claims inflation will not arise before unemployment has been substantially eliminated, so that if inflation does resume, restrictive monetary policy can be used safely.

The possible dilemma worries it, however, and it mentions briefly the need for nonmonetary reforms to make the goals compatible. It suggests steps to increase labor mobility, to offer better information on job openings, to present fewer restrictions on job entry, and the like. I suspect that such steps can indeed help. Without them, monetary expansion to fight unemployment is likely to generate inflation long before unemployment is brought down to acceptable levels. Thus, such nonmonetary reforms may hold a big key to the effective use of monetary policy—but, I suspect equally, not a complete key in a world of excess-income-claims inflation. More about this later.

The Commission also proposes modernization of the Congressional mandate to the Federal Reserve, to spell out the goals it urges. The present law is little changed from 1913, when the Fed's main purposes were "to furnish an elastic currency, to afford means of rediscounting commercial paper, and to establish a more effective supervision of banking." These are now routine. The Commission would substitute a mandate parallel to that established for the administrative agencies of the government in the Employment Act of 1946, with reasonable price stability added as a goal, but without specifying any target rate of economic growth. It is hard to disagree with the modernization recommendation, if we agree on what our national goals are.

Role of Monetary Policy

What can monetary policy do toward achieving these goals? A good deal, the Commission says, but probably not as much as fiscal policy; and monetary policy needs a lot of help in preventing inflation in a world of strong unions, of administered prices, and of restrictions on job mobility.

By and large, the Commission's views are orthodox on money and credit. The Report calls for strong use of the traditional monetary policy weapons, and finds that, in general, the Fed has done a good job in using them. It wants primary emphasis on open market operations (as does just about everybody else); doubts the wisdom of frequent changes in reserve requirements; thinks the rediscount rate bears about the right (minor) amount of emphasis now; and opposes "bills only." On selective controls, it is generally negative except for margin requirements in the stock market; but it could not agree on the control of consumer credit, where about half the members were for reestablishing Federal Reserve stand-by controls, and half against.

There are some technical recommendations offered, notably that reserve requirement differentials on demand deposits be eliminated and reserve requirements be removed on savings and time deposits, as part of a general plea for more equal treatment of commercial banks and other financial intermediaries. Finally, to spread the impact of federal monetary policy and to make its burden more equitable, the Commission recommends that all insured banks be required to become members of the Federal Reserve System, subject to the System's reserve requirements and supervisory standards.

The Commission's Report ranges over the entire area of monetary policy, with a tone that is generally optimistic. A number of points are especially worthy of note.

Lags and uncertainties

Behind the scenes, a battle rages among economic researches over the precise ways that monetary controls affect private spending and the performance of the economy. There is general agreement that the Federal Reserve can control the quantity of money (demand deposits and currency) within reasonably narrow limits by controlling the volume of excess reserves held by the commercial banks, although it sometimes faces substantial lags in getting the results it wants. But how much and how fast any changes in the supply of money affect national production, employment, and prices is another matter.

One set of research results suggests that the supply of money does indeed dominate the level of economic activity, but that there is a highly uncertain lag—varying generally from six to eighteen months—between Federal Reserve action and the resulting change in economic activity. If this is so, it throws grave doubts on the efficiency of Federal Reserve attempts to mitigate booms and depressions. With such lags current policy changes may have their effect only when the need is past, and may indeed make things worse.

Another main line of research agrees on the importance of money, but suggests that the lag probably is not over six months or so for the main impact of Federal Reserve policy. This, with a little luck in forecasting, is short enough to provide real hope for helpful results. The right timing of monetary ease and restraint is crucial. Alas, the Commission throws no new light on such a fundamental issue.

Another basic, unsettled question is the extent to which money and credit conditions affect the level of business investment in plant, equipment, and inventories. Still another question is their impact on residential construction. Another is how much they influence the spending of state and local governmental units. Investment spending by these three groups plays a central role in business fluctuations. Standard monetary doctrine holds that tighter money checks such investment significantly and promptly, that easy money stimulates it. But modern research leaves the truth in the shadows and, again, the Commission throws no new light.

Controlling "new" inflation

Most of what the Commission says about using tight money to check inflation could have been written a generation ago. For traditional "demand pull" inflation, where inflation comes from excessive total demand, a gradual squeeze on spending by limiting the growth of bank credit, or even contracting it, is the accepted remedy. And there is no reason to suppose that it will not work, if the medicine is applied judiciously. True, an overdose of tight money could send the patient into a relapse of unemployment and recession, but this is the "art" of central banking—to fit the monetary remedy smoothly to the need for it.

The nasty problem arises when the inflation comes from excess income claims—when wages and prices push up in some sectors even though substantial unemployment remains. As was indicated above, monetary restraint can still check the inflation—*if* we are willing to pay the price. Federal Reserve action holding down bank lending and limiting the money supply will sooner or later check rising prices. But, in doing so, it will hold total spending below the level needed to produce high employment and rapid growth at the higher wages and prices set by some economic units, even in the face of unemployment and weak markets elsewhere in the economy. To pin the blame on the trouble-making groups who push up wages and prices too fast is right—but it is not very helpful to the monetary authorities, who can only control the supply of money.

No one knows how important this kind of inflationary pressure will be in the future. The Commission hopes, wishfully, I suspect, that it will not be very important. It emphasizes that other measures can help too—better information on job openings, higher labor mobility, elimination of both union and management barriers to job shifting, antitrust action to encourage price flexibility, and elimination of government policies which support rigid prices. It calls for business and labor statesmanship to avoid unreasonable income demands, and turns its back firmly on direct wage and price controls. All things considered, the Commission suspects that attempts to get unemployment below 4% (or so) of the labor force by expansionary monetary-fiscal policy will bring some inflation. Most economists, I believe, would consider this optimistic indeed; the figure may well be nearer 5% or 6%, and possibly even higher. And this means 4 million or 5 million people unemployed, with a corresponding waste of potential output and an inevitable drag on economic growth.

Businessmen and bankers, therefore, who have a simple answer to the inflation problem—just tell the Federal Reserve to stand firm with tight money—would do well to face up to this dilemma. The Commission has certainly not overemphasized it. The answer is far from easy; neither the Commission nor anyone else has yet come up with a clear answer. This is the greatest challenge to monetary policy today.

International liquidity

After many years, the balance of international payments has returned as a major constraint on the free use of monetary policy to fight domestic unemployment and stimulate economic growth. Although we still have the lion's share (nearly half) of the world's monetary gold, over the past decade we have annually owed foreigners more than they owed us. And over the last few years they have chosen to take a significant part of this excess in gold. (Previously they had taken it largely in short-term dollar balances—deposit accounts at U.S. banks and short-term U.S. investments.) In 1960 temporary speculation developed against the dollar, and fears of a major withdrawal of foreign short-term balances were widely expressed. Our gold stock has declined about $5 billion over the past few years, to about $17 billion, of which some $11 billion is now earmarked by legislation as "backing" for Federal Reserve notes and deposits at the Reserve Banks.

The Commission recognizes the increased danger of short-term capital flights in today's edgy world. Such flights could put great pressure on any nation's currency by depleting its gold reserves. Wisely, the Commission refuses to succumb to panic. It urges strengthening of the International Monetary Fund, as a source of temporary additional reserves for currencies under temporary pressure. It wisely rejects restrictionist trade measures as a device for increasing our favorable trade balance. Turning to monetary policy, it significantly rejects domestic monetary restraint that would induce unemployment and slower growth as a means of holding down prices and costs here, to improve our trade balance. Finally, to free the Fed from excessive pressures of temporary gold drains, it recommends Congressional elimination of the 25% gold backing now required by law for each dollar of Federal Reserve notes and deposits. This would make all of the nation's gold stock available to meet possible international drains.

If this country is to be the world's major banker, and the dollar (along with gold) is to be the world's major monetary reserve unit, we need to reassess our monetary arrangements. To freeze a large part of our gold stock means that in the face of short-run international liquidity crises, our monetary authorities must either subordinate basic domestic policies to preserving the gold stock, or risk both domestic and international alarm if "crisis" escape clauses are invoked or if Congress is asked to change the gold cover law under pressure. The gold arrangement proposed by the Commission is the one long followed by most major nations. The threat of an international confidence crisis against the dollar would surely be lessened if the world knew that our full gold stock, instead of only one third of it, was available to meet temporary drains.

Thus, I suspect, it is doubtful that we can much longer afford the rigidity imposed by present legislation. There is now no close relationship between gold flows and the supply of money, nor should there be. It is unlikely that freeing the entire gold stock for international use would make either the monetary authorities or the Congress any less responsible in trying to meet their total economic responsibilities. In a world that values gold as a reserve, it is important that the United States maintain a sizable gold stock, although there is no one "right" amount or ratio for all times and conditions. The monetary authorities must balance multiple domestic and international goals under rapidly changing conditions. The main advantage of the present law is that the 25% gold cover requirement acts as a bright red warning light if gold is withdrawn. But flashing red lights create panic as well as warn; and, to shift my image, the 25% gold requirement ties one hand behind the authorities' backs and forces them to do their job the hard way.

A much bigger question is whether the United States should lead the way toward establishment of a greatly expanded International Monetary Fund, or new supranational central reserve bank. The Commission is not convinced that we need such a drastic step yet. But in the small, highly interconnected world of today—with increasingly volatile international balances—an improved international liquidity and payments mechanism may soon be essential. Here again, the Commission could usefully have gone deeper. More effective meshing of domestic and international requirements has become again one of the biggest jobs faced by the monetary authorities.

Economic growth

Over the past decade Americans, used to being first in everything, have had some uncomfortable facts to face. In economic growth, we are not only not leading the race, but we are back with the also-rans. Both the communists and our Western allies seem to be outdoing us by substantial margins. Could monetary policy help more?

The Commission's Report is sobering. There are no panaceas, though somewhat easier money might have helped a little. The money supply needs to grow fast nough to support a growing total expenditure, but not at any set rate like the much-discussed 3% per year proposed by those who doubt the ability of the Federal Reserve to open and close the money spigot at the right times on a discretionary basis. The Commission seems almost to come to the position that when growth is too slow because of underemployment (as over the past several years), this is evidence of inadequate growth in the money stock. But it doesn't quite say so.

Many other observers are more critical, arguing that too-slow monetary growth, due to fear of inflation, has significantly slowed U.S. economic growth since 1955. The old dilemma of how to restrain inflationary pressures (real or feared) and simultaneously to expand employment and growth is found everywhere throughout the volume.

The monetary authorities need the finesse of a tightrope walker. Too little money will hold back spending, investment, and employment—wasting output and slowing economic growth below otherwise attainable levels. Too much money encourages inflation, without appreciably speeding economic growth. What we need is just the right amount of money, in between underemployment and inflation. Once the Federal Reserve has approximated that and thus minimized the waste of recession, there is little more than monetary policy can do. Easier money—especially lower long-term interest rates—might safely help stimulate investment if we could count on a budget surplus to restrain inflationary total spending, but this is a nice balance indeed to hope for. As a practical matter, it is largely up to fiscal policy, revision of the tax system, more investment in education and scientific research, and a step-up in just plain hard work to push the growth rate up still further.

Through this maze of unsettled but, alas, fundamental disputes underlying monetary policy, the Commission has trodden a careful, middle-of-the-road path. The public or private administrator can never wait for absolute truth. He must act on the best information available to him. The Commission's views represent a fair picture of well-informed professional opinion. Perhaps this is all we could ask.

It is only fair to recognize that the economics profession has not been able to provide the guidance needed through its own research over many years. But, as on the problem of economic goals, the Commission had vast opportunity and large resources. A more fundamental study of the role of money and the foundations of monetary policy would have been of greater long-run value, both in providing a firmer basis for current policy and in showing the gaps in knowledge we need to fill to make better policy decisions in the future.

Issue of Independence

In a temperate analysis of governmental operations and Federal Reserve responsibilities, the Commission draws these conclusions:

1. The President must bear the central responsibility for governmental economic policy recommendations and execution.

2. Federal Reserve responsibilities for national economic policy are closely intertwined with those of other government agencies, especially the Executive Office of the President and the Treasury.

3. Federal Reserve independence is now adequately protected, and Federal Reserve influence could be increased by closer participation in governmental policy determination.

4. To the end of closer and more informal working relationships between the Federal Reserve and the White House, the Federal Reserve Board chairman and vice chairman should be designated by the President from among the Board's membership, with four-year terms coterminous with the President's.

5. To improve efficiency and attract more able members, the Federal Reserve Board should be reduced from seven to five members, and all major Federal Reserve monetary powers should be centered in the Board.

6. To improve national economic policy formulation and coordination, the President should establish a cabinet-level "Advisory Board on Economic Growth and Stability," including the chairman of the Federal Reserve Board.

These proposals have been widely criticized by conservatives on the ground that they would undermine the independence of the Federal Reserve. The critics suggest that the "liberals" on the Commission somehow outflanked the "conservatives" in bringing about this stab in the back for financial soundness (a neat trick if indeed it occurred, since two thirds of the twenty Commission members were highly successful businessmen and banks, only two were labor leaders, and the other five were independent professional men). On such a vital issue of monetary arrangements as this, it is well to take a closer look.

Case for independence

Stated bluntly, the traditional argument for Federal Reserve independence is that, if independent, the Fed will stand against inflation and financial irresponsibility in the government. History tells of many treasuries which have turned to money issue to pay their bills when taxes were inadequate. The modern world's major inflations have all come with large governmental deficits, covered by the issue of new money (currency or bank deposits). While legislatures vote the expenditures, treasuries must pay the bills. Thus, it is argued that treasuries have a predictable inflationary bias, however well-intentioned their secretaries may be. Against this bias, central bankers are alleged to be basically conservative; they can be counted on to look out for the stability of the monetary unit.

Another variant is based on the presumption that the entire political process is inherently inflationary. It is always easier for Congress to spend money than to raise taxes; "politicians" are inherently financially irresponsible. Thus, an independent Federal Reserve is needed to call a halt to the overspending tendencies of the politicians, and to the tendency of the politicians to plump too readily for good times for the economy as a whole, even though these good times may generate some inflation.

Lastly, there is an argument that the President, the politician par excellence, is not to be trusted on financial matters, and that an independent Federal Reserve is needed to see that he does not go too far with expansionary, inflationary economic policies.

Meaning of independence

These arguments suggest that we need to examine the meaning of the term "independence." Independence from whom? A Federal Reserve independent of the U.S. Treasury rests squarely on the realistic assessment of history. Treasuries *have* been inflationary in their biases, and we therefore need a powerful agency in governmental economic circles to stand against these inflationary biases when they threaten the soundness of our economic structure.

But Federal Reserve independence from the Congress is hardly meaningful in our governmental system. Congress established the Federal Reserve. It can change it any time it wishes, or call it to account for any of its actions. Federal Reserve officials readily acknowledge their responsibility to Congress—though the Fed need not go to Congress for appropriations to conduct its affairs and though, in practice, Congress, happily, is reluctant to intervene directly in Federal Reserve policy making.

The really difficult question is this: Should, or can, the Federal Reserve be independent from the President? The Constitution clearly allots to the federal government the power to create money and regulate the value thereof. In our society, where bank deposits comprise some 80% of our total money supply and currency only 20%, control over the supply of bank credit *is* control over the volume of money. Federal Reserve officials have consistently recognized the basically governmental nature of their function, though they value the close relationships they have with private bankers.

Furthermore, control over the money supply of the nation is a vital operating responsibility. Monetary policy is inextricably intermingled with fiscal policy and debt management policy, if the nation's economic goals are to be achieved effectively. The President must ultimately be responsible for recommendation and execution of the nation's basic economic policy. This logic leads clearly to the conclusion that the Federal Reserve must work closely with other agencies under the general responsibility of the President for executing national economic policy.

To give an independent Federal Reserve the power to negate the basic policies arrived at by the executive and legislative branches of the federal government would be intolerable for any administration, Republican or Democratic. But independence, looked at practically, is a matter of degree, not of black and white. The real question, thus, is the terms on which the Federal Reserve participates in governmental policy making and execution.

Need for cooperation

To be most effective, the Federal Reserve needs to be in a position to work closely with the other major government agencies responsible for national economic policy—especially the Treasury, the Budget Bureau, and the Council of Economic Advisers. No Federal Reserve chairman has ever claimed that the Board should disregard the debt management problems of the Treasury, or that the government's financial needs should be given no weight.

On the contrary, all major Federal Reserve officials have agreed on the need for close working relationships with the Treasury on monetary, fiscal, and debt policy. The times when the Federal Reserve has been least effective have been the times when it has been most isolated from the President and from effective, coequal working relationships with the Secretary of the Treasury and other high-level government officials. This was substantially the case throughout the much-discussed decade of the 1940's when the Federal Reserve was most subservient to Treasury debt-management needs. Secretaries Morgenthau and Snyder were close personal confidants of Presidents Roosevelt and Truman; but Federal Reserve officials seldom saw either President.

An effective Federal Reserve voice for the stable money point of view can best be assured if the Fed is an active, continuous participant in the day-to-day process of governmental economic policy formation. Seldom indeed does a central bank undertake a major war with the Congress and the Administration in a showdown on economic policy. Federal Reserve participation in policy making will generally be a more effective device for presenting the sound money point of view than will spectacular defiance of the government's policies. Extreme independence is, unfortunately, likely to mean splendid isolation from the decisions that matter.

On balance

The need is for recognized Federal Reserve independence from the Treasury *and* for coequal voice with other major agencies in the economic policy councils of the government. In other words, the need is to maintain a strong and substantially independent voice for a stable-money point of view without placing Federal Reserve officials in an untenably isolated position, where to use their independence involves major intragovernmental conflict and divided national economic policy. Budgetary and monetary matters call for the best efforts of wise men. But we must not fall into the trap of supposing that all wisdom will reside in appointed Federal Reserve officials, rather than in other government officials appointed by the same President and approved by the same Senate. The President, the Secretary of the Treasury, and other high governmental officials also seek to advance the national welfare, as they see it. How best to mesh the judgments and responsibilities of these various public officials is the problem, not simply to set up an independent nongovernmental board with a legal (but seldom practical) power to say *no* to the U.S. Government.

Recommended changes

To improve the coordination of over-all economic policy and to increase the influence of the Federal Reserve while maintaining its special quasi-independent status, the Commission recommends primarily two modest changes:

1. The President should establish a cabinet-level "Advisory Board on Economic Growth and Stability" which would include the chairman of the Federal Reserve Board.
2. The term of office of the chairman (and vice chairman) of the Federal Reserve Board should be made coterminous with that of the President, to eliminate the possibility that a Federal Reserve chairman would be personally unacceptable to a President.

A new President could (as now) immediately appoint one new Board member, and could name him chairman; or he could name a new chairman from among existing Board members. The staggered-term membership of the Board would remain unchanged, except that it would be reduced from seven to five members. While further centralization of System authority in the Board would increase somewhat the President's power over the Fed, overlapping ten-year terms would go far to protect the stability and independence of the Board members from short-run political pressures.

These two recommendations might help substantially to assure effective working relationhips between the Fed, the Presidency, and the rest of the administrative branch of the government. To insist that a new President accept a Federal Reserve chairman to whom he objected strongly would probably serve little purpose, and would be more likely to decrease the effectiveness of the Fed than to increase it. As a practical matter, the chairman must represent the System in its most important contacts with the President, as well as with the Treasury and in most cases with Congress. Making the chairmanship coterminous with the President's term, though it might have little importance in most instances, makes practical administrative sense. It is significant that both William M. Martin, the present chairman of the Fed, and Marriner S. Eccles, chairman for longer than any other man and the individual who was most responsible for the restored independence of the Fed in 1951, concur in the recommendation to make the chairmanship coterminous with the President's term.

Appointment by the President of an "Advisory Board on Economic Growth and Stability" would be one device for assuring closer coordination among the governmental agencies (including the Fed) responsible for national economic policy. Whether such a special advisory board would be effective would depend heavily on whether the President wanted to use it. Some such device is obviously necessary. The Commission wisely avoids a recommendation to make such an advisory board mandatory by legislation, while stressing the importance of coordinated national policy formation in which the Federal Reserve has a strong voice.

Critics have labeled these recommendations a stab in the back for Federal Reserve independence. This appears to be a serious exaggeration. They reflect operating realities, and are modest proposals indeed when viewed in the light of the experience of most other nations, where central banks have been completely subordinated to treasuries or to governments.

Organization and Operations

Since the Commission apparently feels that the Federal Reserve has done a good job on monetary policy, we might expect few recommendations for change in Federal Reserve organization and operation. The Commission believes, however, that the Federal Reserve could do its job more effectively if its organization were modernized and if its controls were extended to all insured banks. It does not suggest that structural changes are of overriding importance, but that they would be useful and in keeping with modern needs and mores.

In essence, beyond the recommendations concerning the chairmanship of the Board, the Commission recommends that:

1. The Federal Reserve Board should, as noted, be reduced in size from seven to five members, with staggered ten-year terms.
2. Special occupational and geographical qualifications for Board members should be eliminated, and replaced by statutory stipulation that members be positively qualified by experience or education, competence, independence, and objectivity.
3. All major policy powers (over open market operations, Reserve requirements, and discount rates) should be vested in the Federal Reserve Board. The separate Federal Open Market Committee would be abolished, but the Board would be required to consult regularly with the twelve Reserve Bank presidents in determining its policies. Discount rate changes would no longer be inaugurated separately at the twelve regional Reserve Banks.

4. Technical ownership of the Federal Reserve Banks by member banks should be eliminated through retirement of the present capital stock; instead, membership should be evidenced by a special nonearning certificate for each member.

5. All insured banks should be required to become members of the Federal Reserve System.

The basic purpose of these recommendations is to centralize the policy-making functions of the Federal Reserve System in one governmental body (the Federal Reserve Board), unmistakably responsible to the public rather than to the commercial banks; to increase the efficiency of Federal Reserve operations by streamlining the present complex organizational structure; and to extend the direct impact of monetary controls to substantially all commercial banks in the country.

The present complex Federal Reserve organization reflects the regional needs of a half century ago, modified here and expanded there as the focus shifted to national monetary policy and as new instruments developed. It looks terrible on paper. But, most observers agree, it works well on the whole. Given these facts, is the Commission right that some changes ought to be made? To answer this question thoroughly would take a small book. Only a few major issues can be noted here.

Focus of responsibility

Few businessmen would tolerate in their own firms the complex organization and overlapping responsibilities for major policy that exists in the Fed. Open market operations, reserve requirements, and discount rates—all have identical general policy goals and need to be completely coordinated. To have a different group responsible for each invites delay and confusion. The day when discount rates needed to be set separately to meet differing regional needs is long past. Monetary policy is national policy, and is recognized as such by all concerned. Information on regional developments is indeed valuable in forming monetary policy, but this could be arranged readily without diluting and diffusing responsibility for monetary policy. So runs the argument for modernization.

But there are counter arguments. The main one is that, while this may well be true in principle, the present arrangement works well. Why change it? In fact, all 19 major Federal Reserve officials (7 Board members plus the 12 Reserve Bank presidents) consult together on all major policy issues, and in effect make policy together. Policy responsibility is thus not scattered. Instead, Federal Reserve policy is made in the best tradition of wide representation and careful consideration by a large group of responsible men. While monetary policy should not, of course, be regional in nature, the present system of regional banks draws presidents and board members of high ability into the System, where they could not be pulled without the attraction of policy responsibility. In policy deliberations, it is thus argued, Reserve Bank presidents both reflect regional interests and bring monetary judgments and insights which add significantly to those found in the Washington Board.

Conclusions? Much depends on this last argument. Over much of Reserve Bank history, there is little evidence that Reserve Bank presidents (with the exception of the New York president) have added much to policy making. The last decade has seen the appointment of a number of Reserve Bank presidents of especially high ability, men whose competence in monetary economics and in practical banking compares favorably with the best of the Washington Board. It is argued that such men could not be drawn to membership on the Washington Board. The facts are not clear, either on this or on the contribution now made by the Reserve Bank presidents to policy formation. A priori, the case for a simpler organization is strong. And some argue that the national Board is more attractive to top-quality men. The Commission has a strong point, but not all the evidence is in to permit an unequivocal conclusion.

A smaller board

Suppose we agree that policy-making responsibility should be centered in one group, be it the Federal Reserve Board or the Federal Open Market Committee. How big should this group be? The Commission believes that it should be smaller than the present 12-man Open Market Committee, and much smaller than the *de facto* 19-man group which now makes Federal Reserve policy. The Commission plumps for a 5-man board because it feels this would be more efficient, less cumbersome, and less given to delay and indecision.

Few businessmen or students of organization believe that a 19-man, or even a 12-man, committee is small enough to do an effective job of running an organization and making day-to-day decisions on intricate major policy issues. A large decision-making group is needed when many separate interests must be represented. But sound monetary policy formation does not rest on a compromise of conflicting regional or occupational group interests represented on the Board. Excellent regional information is needed, but the information providers do not have to be policy makers. Except for regional differences, it is not clear that the 12 Reserve Bank presidents represent very different interests.

Or a large group is justified if additional members add significantly to the decision process. Both widespread experience and a priori reasoning cast doubt on the marginal gain from additional members after a committee totals a half dozen or so, unless the additional man is of especially high ability or holds quite different views from the others. In the Federal Reserve case, there seems little reason to suppose that going beyond the half dozen or so ablest men in the System is justified on either count.

We Americans traditionally distrust concentration of power in government; we value the combined judgment of a number of men. But the case for 19 decision-makers is hard to defend. The Commission's figure of 5 appears reasonable, and the smaller the group the better will be the chance of getting first-class men to serve on it. If a mixture of Reserve Bank presidents and Washington officials is desired, a small decision-making group could still be obtained by combing, say, 3 Board members with 2 Reserve Bank presidents on a rotating basis.

History suggests that, as a practical matter, System leadership has usually been highly concentrated in a few hands, notably Marriner S. Eccles' for many years and before that in those of Benjamin Strong (long time president of the New York Fed). Realistically, the chairman must represent the System in its most important contacts with the President, the Treasury, and Congress. The Federal Reserve is, in fact, a policy-making and operating agency, and it inescapably will have one or a very few men who carry most of the burden. The old judicial parallel, with the Reserve Board termed the "Supreme Court of Finance," is not a realistic analogy. Courts apply common and statute law under an elaborate set of judicial precedents and safeguards.

As was noted above, the Fed sails on seas virtually uncharted by Congress and with heavy day-to-day operating responsibilities for our monetary mechanism. Hence, the Board is more like the Secretary of the Treasury than like the Supreme Court in its basic role (though it does have some commission-type regulatory duties). This fact further weakens the case for a large policy-making board. Chairman Martin plays the role of cooperative leader superbly, but even today there is some question whether System policy would be much different if he were a single governor, or if policy-making power were centered in a small board as the Commission recommends.

Reserve Bank ownership

Technical ownership of the Federal Reserve Banks by the commercial banks flies in the face of the basic Constitutional provision that the federal government shall "coin money [and] regulate the value thereof." Surely the Federal Reserve authorities in regulating our money must be responsible, not to the bankers, but to the people of the United States, just as are the Secretary of the Treasury and other governmental officials. Perceptive bankers are the first to agree.

It is clear that both Board members in Washington and the presidents of the Reserve Banks, in fact, view themselves as public officials, sworn to advance the welfare of the people, rather than as representatives of the bankers. Why, then, bother to change the situation, even though the present arrangement is admittedly a vestige of the thinking of a half century ago? The main answer is that our national monetary authorities must, like Caesar's wife, be above suspicion and reproach. Even though commercial bank ownership of Reserve Bank stock clearly does not now mean control by the bank over national monetary policy, it opens a suspicion that such improper influence might be exerted.

Balancing the arguments

The logical case for the Commission's recommendations on Federal Reserve structure is a good one. But as a practical matter, history throws doubt on their importance. Moreover, the cost of these changes would be great—in bitter argument and in deep wounds within the System. It seems unlikely today that the gains would be worth the price. My guess is that before another decade goes by, Congress will want to take a hard look at Federal Reserve organization, possibly under the pressure of new financial needs generated by changing international or national conditions. If so, the Commission's recommendations will deserve a careful look, as part of a more thorough study of both alternative organizational arrangements and the lessons of monetary history.

Conclusion

Now that the Commission's proposals concerning money and credit are out, what do they amount to in total? What will be their impact on monetary thinking and on legislation and administration?

In sum, the Commission—a group of 20 able, conscientious men—have found that our monetary and credit institutions serve us well, that monetary policy can play a significant role in helping to provide stable economic growth without inflation, and that no sweeping changes are needed in our monetary arrangements. These are encouraging conclusions, though one can wish that the Commission had dug deeper in a search for new insights into our monetary problems.

The Commission suggested no real solutions for our toughest monetary problems, notably how to achieve rapid growth, high employment, and stable prices together, and how to compromise effectively domestic and international monetary requirements. It is easy to criticize the findings. But critics need to face up to the hard job of themselves producing realistic, integrated proposals for improving monetary arrangements and policy. A half century after the first National Monetary Commission led to establishment of the Federal Reserve, finding Commission consensus on so much of modern monetary policy is encouraging. Businessmen and other citizens, along with government officials, will do well to listen to the Commission's analyses and advice. But for them, and especially for professional economists, the challenge to understand better and to devise more imaginative monetary and credit policies remains.

9

Transactions for the System Open Market Account

Robert Roosa

Trading Methods and Objectives; Coordination with Transactions for Other Official Accounts

Once the Account Management has decided, on the basis of all presently available information, that action should be taken in order to further the current objectives of the Federal Open Market Committee, the choice as to procedure lies mainly between outright transactions and repurchase agreements in Government securities. On a very broad basis, of course, outright transactions are used when there is good reason to believe that the change to be made in the reserve base may be of some duration; and repurchase agreements are used when a need for funds is most likely to be quite temporary, soon to be followed by a need to reverse the effect of the operation through the absorption of funds. But it would be difficult indeed to catalogue all outright transactions under the first set of conditions, or to catalogue all repurchase agreements under the second. Further description of the methods and procedures inherent in each type of operation may help to suggest some of the peculiar qualities of each that make the choice of one or the other particularly appropriate in various circumstances. Following the discussion of outright transactions and of repurchase agreements in Government securities, there is a third part of this present section devoted to operations in bankers' acceptances; and then a fourth, indicating the variety of ways in which the Account Management may effect coordination between the Federal Reserve

Reprinted with permission from *Federal Reserve Operations in the Money and Government Securities Markets* published by the Federal Reserve Bank of New York, by Robert V. Roosa, 1956, pp. 80-83. Mr. Roosa is now with Brown Brothers, Harriman and Company, New York.

System's objectives, in its own Government securities transactions, and transactions that also come to the Trading Desk for execution from various Treasury trust funds and investment accounts, from foreign central banks and governments, and from some of the smaller member banks in the Second Federal Reserve District, as well as from a few other miscellaneous accounts.

The execution of outright transactions for the System Open Market Account

In recent years virtually all purchases or sales in the market for System Account on an outright basis have been made in Treasury bills. As already explained, any dealer who consistently demonstrates his readiness to make markets is welcome to compete for transactions with the Account, provided the firm meets reasonable standards of credit-worthiness and provides the financial statements needed in reaching a determination to that effect.

Once the Account Management decides, for example, to purchase Treasury bills in a magnitude of 50, or 75, or 100 million dollars, its approach to the market is governed by two overriding considerations. First, it must trade with all dealers on a freely competitive basis. This means, when buying, that it will take bills at the lowest offered prices (highest yields), up to the point that the intended total of purchases is reached. Second, the Account Management must give some weight, if the size and variety of dealers' offerings are great enough to permit any leeway, to the need for maintaining a maturity distribution of the portfolio that will best contribute to the practical administration of the Account over the months ahead. This means that when the Account holds no bills of a particular weekly issue, perhaps because it had let the preceding issue run off at maturity in order to absorb reserves at that earlier time, some preference would be given to restoring a moderate holding of that particular maturity, all other things being equal. There are other similar operating considerations that have to be taken into account.

The usual procedure, once the Vice President-Manager has reached his decision, is to brief all available members of the trading staff on the language to be used in contacting dealers, and to assign each person to two, three, or four dealers as his responsibility. On instructions from the Manager, Trading and Markets, all traders then begin simultaneously contacting dealers to ask for bids or offerings, as the case may be, specifying particular maturities where that is appropriate. Under present arrangements, it is usually possible for the traders to reach all dealers and note their bids or offers within an elapsed time of three to five minutes. It is necessary to act quickly so that no dealer can be placed in a privileged position either by obtaining knowledge of a Federal

Reserve System operation in advance of other dealers or by being able to bid or offer after the initial impact on prices becomes apparent (which might be a particular advantage, in the event of Federal Reserve purchases, for example, if the purchases had an immediate upward influence on Treasury bill prices).[1] As each trader completes his contacts with dealers, all of the bids or offers he receives are promptly assembled on a single worksheet, or "blanket," which the Manager, Trading and Markets, and any other officers currently working with him, can then use in order to select those bids or offers which, in conformity with the two principles just mentioned, will fulfill the System's buying or selling objective.

The bids or offers received from the dealers are requested on a "firm" basis for twenty minutes or thereabouts, as a general rule, and by the elapse of that time the selections will have been completed and each trader can begin return calls to the dealers indicating which, if any, of their bids or offers are to be taken. Generally, this entire operation, known colloquially as a "go-around," is completed within thirty minutes. The Account Management usually conducts operations in the manner described, unless unsettled conditions are so pervasive in the market that the broadside effect of a full-scale "go-around" threatens to be unduly disruptive. Under those circumstances, the Account Management may achieve its objective in terms of absorbing or releasing reserves, with less pronounced impact on market psychology, by simply taking advantage of bids or offers that are volunteered by the dealers during the course of the routine conversations continually in process between the traders on the Trading Desk and those at various dealer firms. In no event, however, would purchases or sales be made at prices out of line with those currently prevailing in the market, as checked and cross-checked by all of the traders in their continuing conversations with the market.

Outright transactions are normally executed either for "cash" (that is same-day) or for "regular" (next-day) delivery and payment. Due to mechanical problems of physical delivery, it is ordinarily not practicable to initiate negotiations with respect to cash transactions after 12 o'clock noon, although in special circumstances, particularly when the Federal Reserve is a seller, transactions may at times be executed as late as 1:00 p.m. One advantage or disadvantage of cash transactions, depending on the circumstances, which also applies in the use of repurchase agreements to be described below, is that they provide or withdraw Federal Reserve funds to or from the market immediately. There are times when, in the nature of a developing situation, a prompt release or withdrawal of

[1] System action does exert an influence on prices of bills, and other Government securities as well, but fortunately the market is no longer as sensitive to this particular procedure as formerly.

funds is particularly desirable. Transactions for regular delivery may, of course, be executed at any time until the 3:30 p.m. closing hour. Thus, it is broadly correct to generalize, as far as outright transactions are concerned, that any action intended to influence reserves today must be decided upon, and execution must begin, before or close to noon. Any reserve effect which can suitably be exerted on the following day may be brought about through regular delivery transactions at any time during the trading day.

Because of the difficulties of assuring timely physical deliveries, particularly when dealers may have to withdraw securities already pledged on loans at other institutions in order to effect delivery, or must await deliveries of the securities from customers, the Account Management often finds on its "go-arounds" that dealers are not prepared to offer enough securities on a cash basis to meet the Federal Reserve System's objective. Conversely, when the System is on the selling side, dealers may also be somewhat reluctant to take as many securities as the Account Management may have in mind because of the difficulties of subsequently making arrangements late in the day for additional loan facilities, both in order to be able to pay for the securities when picking them up (in Federal funds) and in order to carry the securities in position after receiving them. By comparison, when the Federal Reserve policy objective indicates a need for buying, the repurchase agreement facility usually provides somewhat greater flexibility. Securities that a dealer may still be in need of financing ordinarily will be more easily accessible, right down to the last minute at which the dealer closes his own books. Repurchase agreements can physically be negotiated as late as 1:30 p.m., and timely deliveries subsequently effected. Although it is rarely practicable from the Federal Reserve System's own point of view to operate that late, there are some occasions when the Account Management is glad to take advantage of this avenue for placing funds immediately in the market at a relatively late hour in the trading day. Of course, the greater flexibility of operations in repurchase agreements is dependent upon the dealers having securities in position that they are willing to finance through a repurchase agreement with the Reserve Bank at the rate of interest the Reserve Bank is charging on that day.

10

The discount mechanism and monetary policy

The primary objectives of the Federal Reserve System are to provide monetary conditions that will facilitate economic growth, a high level of employment, and price stability. The System's ability to contribute toward these objectives rests, in large measure, upon affecting the money supply. To this end, the Federal Reserve depends chiefly on its ability to affect member bank reserves which in turn affect the quantity of bank credit and money. The proximate objective then is to influence, at any given time, the volume, costs, and availability of bank reserves in such manner as to promote the primary objectives noted above.

In order to regulate member bank reserves the Federal Reserve authorities have three major instruments—open market operations, discount rate changes, and changes in reserve requirements. During the period June-August the System made net open market purchases of about $730 million, lowered the discount rate at most Federal Reserve Banks in two steps from 4 per cent to 3 per cent, and made adjustments in reserve requirements and vault cash which were expected to make available about $600 million of reserves. As a result, total reserves of member banks expanded and the cost of borrowing additional reserves from the Reserve Banks was reduced.[1] Of all these actions taken by the System, the change in the discount rate probably received the most attention, while the change in reserve requirements was noted mainly in official and banking circles and the public remained relatively unaware of the direction of open market operations. This concern with movements in the discount rate arises because such changes are viewed by many as an indication of a change in Federal Reserve policy. This article attempts to

Reprinted with permission from "The Discount Mechanism and Monetary Policy," *Monthly Review*, Federal Reserve Bank of St. Louis, September 1960, pp. 5-9.

[1] See "Recent Financial Developments" in this month's *Review*.

analyze the role of discount policy in relation to overall monetary actions and thereby place changes in the discount rate in perspective.

Member banks obtain additional reserves from time to time by borrowing from their Reserve Bank. Typically, a member bank will borrow from a Federal Reserve Bank in order to avoid a temporary reserve deficiency arising from an unexpected drain in its deposits. The individual bank has several means by which it can obtain additional reserves. It may dispose of assets (usually a short-term marketable security such as Treasury bills), borrow in the Federal funds market, borrow from a correspondent bank, or borrow from the Reserve Bank.[2] The decision as to which form the adjustment will take is influenced to a large extent by policies and actions of the System.

The primary function of the Federal Reserve System under the original act of 1913 was to provide for a more elastic currency. As originally conceived the discount mechanism was to be the major instrument of this policy. Member banks were permitted to discount notes, drafts, and bills of exchange of relatively short maturity arising out of actual commercial and agricultural transactions. Three years later Government securities were added to the list of eligible paper. During the 1930's, when member banks had large excess reserves, the discount mechanism assumed primarily a standby significance.

During the war and until the Treasury-Federal Reserve "accord" in 1951, member banks made their short-run reserve adjustments chiefly by buying or selling Government securities rather than borrowing from the Federal Reserve. In this period banks held large quantities of these securities, the prices of which were supported by the Federal Reserve.

After the accord, member banks once again began to rely more frequently on the discount mechanism to make short-run reserve adjustments. During this period the amount of outstanding borrowing rather closely paralled fluctuations in the level of economic activity. Member bank borrowing reached peak levels in the months of December 1952, April 1956, and August 1959. Borrowings were lowest in this period during the months of July 1954 and July 1958. These dates correspond roughly to the peaks and troughs of business cycles since 1951.

Discount Policy

Discount policy at any time consists primarily of two aspects: administration of the "discount window" and setting the discount rate.

[2] The Federal funds market consists of the borrowing and lending, primarily by member banks, of deposit balances at the Federal Reserve Banks. For a more complete discussion of this institution see "The Federal Funds Market," in the April 1960 issue of this *Review*.

Table 1. Member Bank Borrowing

Selected Months (Monthly averages of daily figures. In millions of dollars.)	
April 1951	161
December 1952	1,593
July 1954	64
April 1956	1,060
July 1958	109
August 1959	1,007
August 1960	293

Administration of the discount window

The twelve Reserve Banks administer the function of lending to member banks in their respective districts as well as setting the rate which is subject to approval by the Federal Reserve Board. The principles used by each Reserve Bank in judging an application for a loan are set forth in Regulation A of the Board of Governors which reads in part as follows:

> Federal Reserve credit is generally extended on a short-term basis to a member bank in order to enable it to adjust its asset position when necessary because of developments such as a sudden withdrawal of deposits or seasonal requirements for credit beyond those which can reasonably be met by use of the bank's own resources. . . . Under ordinary conditions, the continuous use of Federal Reserve credit by a member bank over a considerable period of time is not regarded as appropriate.
>
> In considering a request for credit accommodation, each Federal Reserve Bank gives due regard to the purpose of the credit and to its probable effects upon the maintenance of sound credit conditions, both as to the individual institution and the economy generally. It keeps informed of and takes into account the general character and amount of the loans and investments of the member banks. It considers whether the bank is borrowing principally for the purpose of obtaining a tax advantage or profiting from rate differentials and whether the bank is extending an undue amount of credit for the speculative carrying of or trading in securities, real estate, or commodities, or otherwise.

Administration of the discount privilege does not change with shifts in monetary policy. The Reserve Banks are aided in their enforcement of Regulation A by the traditional reluctance of some commercial banks to remain indebted.

The discount rate

The discount rate is the interest rate charged by the Reserve Banks on loans to member banks. This then becomes the cost of obtaining additional reserves through such borrowing. As brought out above a member bank has several alternatives in adjusting to short-run changes in its reserve position. The decision as to which method is adopted is determined in large part by the relative cost. The relative cost is frequently determined by (1) the loss or gain realized on the sale of a short-term earning asset, and (2) the relation btween the discount rate and other short-term money market rates.

Insofar as an individual member bank is concerned, an adjustment in its reserve position through any of the alternative methods stated above solves the bank's immediate problem. From the standpoint of monetary policy the type of reserve adjustment is important. Adjustments in reserves which are made through transactions in the Federal funds market or in Treasury bills represent merely a transfer of funds. No reserves are created or destroyed in this process. On the other hand, borrowing or the repaying of borrowing from the Reserve Banks increases or diminishes total reserves of the banking system. As we have seen, this is the variable upon which the System operates in order to affect bank credit.

Open market operations have been in recent years the primary means through which the Federal Reserve exercises control over member bank reserves. Changes in borrowing from the Reserve Banks, which are at the initiative of the individual member banks, may offset temporarily the effects of open market operations. However, since the discount rate has an influence on the decisions of member banks either to borrow from the System or make their temporary reserve adjustments in some other way, the relationship of the discount rate to other market rates may tend to cause member bank borrowing in the aggregate to supplement, rather than offset, open market policies.

The following set of examples are designed to show how discount policy combined with open market operations function first in a period when the Federal Reserve is attempting to exercise credit restraint, and then again in a period of credit ease. It will be assumed in both cases that the banking system is initially in a state of equilibrium. That is, total bank credit is at a desired level and excess reserves and borrowings from the Reserve Banks are at levels which the member banks consider satisfactory. In addition we will assume that the economy is operating at a relatively high level and experiencing growth with prices about stable.

Example 1—Credit restraint. Assume for the moment that the demand for bank credit increases and that interest rates and prices are tending to creep up, and that the System would decide that supplying additional reserves via open market operations to meet this credit demand would be inflationary. As the demand for credit increases, member banks will seek additional reserves by selling securities or borrowing from the Federal Reserve. As short-term interest rates rise relative to the discount rate, banks will find it more desirable to adjust their reserve positions by increased borrowings from the System rather than by selling Treasury bills. As a result commercial bank credit would expand—new reserves being supplied by the System in the form of additional loans to member banks.

In order to slow up the rate of increase in member bank reserves, the System might reduce its open market purchases or allow market forces (gold outflows, cash drainage, or a rise in Treasury balances) to pinch reserves: Commercial banks would now find that in order to avoid reserve deficiencies it becomes necessary to further increase their indebtedness to the Federal Reserve. Although the intent of the System is to restrain the credit expansion, it will permit the use of the discount window to cushion the shock of reserve stringency for individual banks. In this sense the discount mechanism will act as a safety valve.

It should be recognized that the increase in member bank borrowing will offset initially the objectives of open market policy oriented toward restraint. To make this borrowing more costly and reduce the incentive to use the discount window the System may raise the discount rate. It may be noted that the Federal Reserve policy of restraint is already underway. The rise in the discount rate is not a signal initiating a change in policy as much as it is a move to strengthen a policy already in effect.

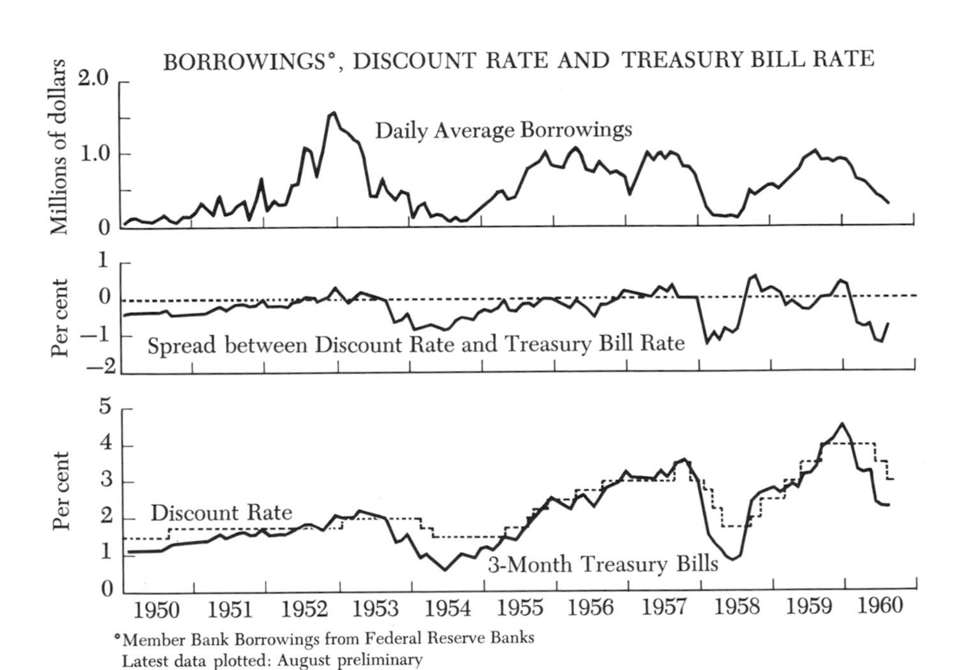

*Member Bank Borrowings from Federal Reserve Banks
Latest data plotted: August preliminary

Example 2—Credit ease. Starting once again from our assumed initial position, let us consider what happens if the demands for credit begin to slow up or contract, market interest rates are falling, and the possibility of a weakening in economic activity appears. The banking system will probably find itself with more than the desired level of borrowing and may begin to repay borrowings from the Federal Reserve. The Federal Reserve with a view to encouraging full use of resources may supply reserves through open market purchases of securities. As interest rates continue to fall relative to the discount rate, banks would have an added incentive to repay their debt to the Reserve Banks rather than use their reserves for lending or investing. Thus, despite Federal Reserve action to increase reserves, the decline in outstanding borrowing may actually reduce reserves and total bank credit. In order to reduce the incentive to make further adjustments in reserves through repayment to the System the discount rate might be lowered. Here, again, the change in the discount rate cannot be considered as signalling a change in Federal Reserve policy, but rather a move designed to reinforce open market operations. The marginal advantage of the new-found reserves now rests with expanding investments or loans. Thus, the reserves made available to banks as their loans decline will be used to expand investment holdings when the discount rate is lowered relative to other money market rates. If investments increase more than loans decline, total bank credit and the money supply would tend to rise.

As has been pointed out in the examples above, discount rate changes can be used to keep changes in member bank borrowing from adversely affecting open market operations. Appropriate discount rate policy may be used to supplement open market operations as well as providing a safety valve. In practice, a major difficulty in implementing discount policy is to maintain the discount rate in proper relation to other short-term rates, primarily Treasury bills.

The Federal Reserve may not find it feasible to adjust the discount rate to maintain the desired relationship between it and other rates, for reasons relating to Treasury financing, sudden changes in short-term interest rates, and public reaction to discount rate changes. Thus, it is possible that with infrequent discount rate changes, the movement of other rates can alter the effectiveness of a given discount rate. In periods of boom with rising interest rates, a constant or "lagging" discount rate would provide the same incentive to member bank borrowing as a reduction in the discount rate with other rates unchanged. Similarly, in periods when credit policy is oriented toward ease a discount rate which lags behind the fall in market rates increases the "effective" cost of member bank borrowing, thus inducing a decline in member bank borrowings (reflected in a greater reluctance to borrow and a stronger incentive to use excess reserves to repay outstanding borrowings). In such situations the discount rate tends to reduce the effectiveness of the open market operations designed to encourage credit expansion.

In light of the above analysis, many of the discount rate changes made by the Federal Reserve System may be considered as "technical adjustments" to market rates reflecting the efforts of monetary authorities to establish a relationship between the discount rate and other market rates appropriate for the effective accomplishment of the objectives of open market operations.

In recent years, among the numerous proposals providing for some modification in the operations and policies of the Federal Reserve System, are those relating to the discount mechanism. Several of the better known proposals are listed below.

Alternative Methods

1. One alternative which has been advocated would eliminate all discretion associated with discount rate policy. This plan involves tying the discount rate automatically to a particular short-term money market rate. The Treasury bill rate usually is recommended for this purpose. This procedure would eliminate the possibility of the discount rate becoming out of line with the Treasury bill rate and would still retain fully the safety valve advantages of discounting. Since 1956 the Canadian central bank has followed this procedure by setting its discount rate each week at ¼ of one per cent above the latest average tender rate for Canadian Treasury bills.

There are three primary arguments against such a technique. First, no one market rate is really "ideal" as a guide and if it happens that the bill rate becomes out of line with other short-term rates this would automatically place the discount rate out of line also. Second, there is no general agreement even among the advocates of this plan as to the frequency with which the discount rate should be changed. Many argue that weekly changes generate too much uncertainty. If a longer period is adopted, such as a month, lags in the discount rate as against other rates become an increasing problem. Third, fixing the discount rate in a set relationship with the bill rate eliminates the possibility of actively changing the discount rate to contribute to economic stability by supplementing open market operations.

2. Another alternative is an adaptation of the method described above which would eliminate the last criticism. This plan would tie the discount rate to the bill rate but would allow the spread between the two to vary with changes in monetary policy. The discount rate may be placed below or above the Treasury bill rate, depending upon the degree of ease or restraint the System wished to follow. The spread between the two rates would not vary with changes in the bill rate but would be changed only in response to a change in Federal Reserve policy. The advantages

claimed for this procedure are that technical adjustments would be made automatically and any change in the spread would be associated with a definite change in monetary policy. This would eliminate any doubt as to whether a given change in the discount rate represents a change in policy or is merely a technical adjustment to changing market conditions.

3. There is some support for eliminating the discount mechanism entirely. This would leave open market operations as the primary tool of monetary policy. Member banks would then have to make their reserve adjustments through carrying larger idle cash balances, the Federal funds market, other forms of interbank borrowing, the securities market, or changing their loan policy. Otherwise, they would be subject to penalties. Since banks would not be permitted to borrow, there would be no changes in member bank borrowing which might frustrate the economic stabilization policies of the System. The plan, however, would eliminate the "safety-valve" feature of discounting during periods when the Federal Reserve is attempting to curb credit expansion through open market sales. During such periods reserve adjustments might become extremely costly and subject individual banks to severe penalties.

A modified version of this approach would be to maintain a relatively high discount rate of about 2 or 3 percentage points above the current Treasury bill rate. This would usually discourage borrowing. In an emergency there would still be a safety valve with only a modest penalty.

III

Monetary Theory and Monetary Policy

Monetary policy is concerned with the establishment of guides to the actions of the monetary authority in its use of monetary controls. It is expected that monetary policy will be directed toward achieving the goals of the system, such as economic stability and growth. To a considerable degree, monetary policy is based upon the explanations of how the monetary system works that are to be found in monetary theory. Monetary theory, however, not only takes many different forms but also changes as our knowledge expands. Recent thinking on the monetary system is reviewed and related to policy by Milton Friedman in the first article. In particular, he points out the difference between what he terms "credit policy" and monetary policy and considers the theoretical bases of both.

The nature and degree of the impact of monetary policy upon the price level, as well as upon the economy generally, has been subject to disagreement and has led to criticism of the monetary authority and its policies. George Horwich addresses himself to this question by examining in a theoretical perspective the relationship between tight money, monetary restraint, and the price system. Criticisms of monetary policy have not been confined to the question of its effect on the price system. It has been argued that it is not only ineffective, but also discriminatory and even contrary to social policy. These questions are examined critically by James R. Schlesinger in "Monetary Policy and Its Critics."

Specific aspects of monetary policy and the tools used to implement them are examined in some detail in the next two articles. A. Dale Tussing looks at credit availability both as a method of monetary control and as an aspect of theory. *The Morgan Guaranty Survey* reviews the open market operations of the Federal Reserve System in terms of their role in implementing monetary policy.

Both theoretically and practically, monetary policy has international as well as domestic significance. The principal features—and defects—of present international monetary arrangements are examined by Warren L. Smith, who also suggests possible alternatives. Similarly, James Tobin discusses the "dollar crisis" and the adjustments that have been made to counteract it. In the final article Milton Friedman places the international payments system in a perspective broader than that of current problems alone and suggests a possibly permanent solution to the balance of payments problem.

11

Post war trends in monetary theory and policy

Milton Friedman

The post war period has seen a dramatic change in the views of academic students of economics about monetary theory and of governmental officials about monetary policy. At the end of the war most professional economists and most governmental officials concerned with economic policy took it for granted that money did not matter, that it was a subject of minor importance. Since then there has been something of a counter-revolution in both theory and policy.

In theory, the direction of change has been toward the earlier attitudes associated with the quantity theory of money, but with a different emphasis, derived from the Keynesian analysis, on the role of money as an asset rather than as a medium of exchange. In the field of policy, the direction of change has been away from what we might call "credit policy," i.e., policy which emphasizes rates of interest and availability of credit, and toward monetary policy, i.e., policy which is concerned with the quantity of money. The emphasis has been away from qualitative controls and toward quantitative controls. And, finally, in the field of policy there has been renewed attention to the problem of relating internal stability to external stability. In examining these changes I shall outline briefly what the situation was at the end of the war; I shall then discuss in more detail the changes in theory that I have just sketched, and finally analyze the changes in policy.

Reprinted with permission from *The National Banking Review*, Vol. II (September, 1964), pp. 1-10.

Milton Friedman is Paul S. Russell Distinguished Service Professor of Economics at the University of Chicago.

This paper is adapted from a talk given in Athens in January 1963, under the auspices of the Center for Economic Research.

I. The Post War Situation

Economic thought at the end of the war was greatly affected by the Keynesian revolution which occurred in the 1930's. Keynes himself was much less extreme in rejecting the importance of money than were some of his later disciples. Keynes stressed the particular problem of underemployment equilibrium. He argued that under such circumstances one might run into something he called absolute liquidity preference. His analysis concentrated on the relation between money, on the one hand, and bonds or other fixed interest securities, on the other. He argued that bonds were the closest substitute for money, and that in the first instance one could regard people as choosing between holding their wealth in the form of money or holding it in the form of bonds. The cost of holding wealth in the form of money was the interest that could otherwise be received on bonds. The higher the rate of interest, the less money people would want to hold and vice versa. But, Keynes said, there exists some rate of interest so low that if the rate were forced still lower nobody would hold any bonds.

At that interest rate, liquidity preference is absolute. At that rate of interest, if more money were introduced into the economy people would try to get rid of the money by buying bonds. This, however, would tend to lower the rate of interest. But even the slightest decline in the rate of interest would lead people to hold money instead. So, said Keynes, under such circumstances, with the interest rate so low that people were indifferent whether they held money or bonds, no matter what quantity of the one they held or what quantity of the other, changes in the stock of money would have no effect on anything. If the quantity of money were increased by buying bonds, for example, the only effect would be that people would substitute money for bonds. If the quantity of money were decreased by selling bonds, then the opposite effect would occur.

Keynes did not of course deny the validity of the famous quantity equation, $MV = PT$. That is an identity which is a question of arithmetic not of theory. What he said, in effect, was that in conditions of under-employment, V (velocity) is a very unstable, passive magnitude. If M (quantity of money) increases, V will go down and the product will not change. If M decreases, V will go up and the product will not change. I emphasize this point in order to make clear that the question at issue is an empirical question and not a theoretical question. There was never any dispute on a purely theoretical level in this respect between Keynes and the quantity theorists.

Keynes himself felt that such a position of unstable velocity would occur only under conditions of under-employment equilibrium. He said that under conditions of inflation the quantity theory comes into its own. But some of his disciples went much farther. They argued that even under conditions less extreme than those of absolute liquidity preference, changes in the stock of money would not have any significant effect. It is true, they said, that under such circumstances changes in the stock of money would lead to changes in interest rates. But, changes in interest rates, they argued, would have little effect on real flows of spending: the amount of money people want to invest in projects is determined by considerations other than the rate of the interest they have to pay; in technical language, the demand for investment is highly inelastic with respect to the interest rate. And consequently, they argued that, even under conditions of full employment or of inflation, changes in the quantity of money are of minor importance. An increase in M would tend to lower the interest rate a little, but this in turn would have very slight effect in expanding investment. And hence, they argued, one would find again that V of the MV equation fluctuated widely, tending to offset changes in M.

The general presumption among most economists at the end of the war was that the postwar problem was going to be depression and unemployment. The problem was going to be to stimulate sufficient investment and sufficient consumption to prevent substantial unemployment. The appropriate monetary policy in their view was very simple. The monetary authorities should keep money plentiful so as to keep interest rates low. Of course, interest rates according to this view did not make much difference, but insofar as they had any effect it would be in the direction of expanding investment slightly and hence contributing to the investment that would be urgently needed to offset deficiencies of demand. Nearly two decades have elapsed since then, and it is hard now to remember how widespread these views were and how strongly they were held by people in responsible positions, as well as by economists in general. For example, in 1945, E. A. Goldenweiser who at the time was the Director of Research of the Federal Reserve Board's Division of Research and Statistics wrote:

> This country will have to adjust itself to a 2½ per cent interest rate as the return on safe, long-time money, because the time has come when returns on pioneering capital can no longer be unlimited as they were in the past.[1]

This whole approach was shattered by the brute evidence of experience. In the first place, and most important, the problem of the post-

[1] "Postwar Problems and Policies," *Federal Reserve Bulletin*, February 1945, p. 117.

war world turned out to be inflation and not deflation. Country after country which adopted an easy money policy because of the views I just described discovered that it was faced with rising prices. Equally important, no country succeeded in stopping inflation without taking measures which had the effect of controlling the quantity of money. Italy stopped inflation in 1947. How? By measures designed to hold down the quantity of money. The experience was repeated in Germany after the monetary reform in 1948; in the U.S., after the Federal Reserve-Treasury Accord in 1951; in Britain, when it restored orthodox monetary policy in 1951 to keep prices down; in Greece; and in France, a recent (1960) addition to the list. Those countries which continued to follow low interest rate policies or continued to increase the quantity of money rapidly, continued to suffer inflation, whatever other measures they took.

Though this experience was in many ways the most important single factor that produced a radical change in attitudes toward money, it was reinforced by several other factors. One was the developments which were proceeding in the world of economic theory in the analysis and re-examination of the body of doctrine which had emerged out of the Keynesian revolution. The most important element here was the emphasis on the role of real cash balances in affecting flows of expenditures, first pointed out by Haberler and then by Pigou in several articles which received more attention. An essential element of the Keynesian approach has been the view that only substitution between money and bonds is important, that real goods or real expenditures are not an important substitute for cash balances, and that when cash balances are larger than people desire to hold, they alter solely their desired holdings of other securities. The intellectual importance of the forces brought to the fore by Haberler and Pigou was the emphasis they placed on the possibility of substitution between cash on the one hand and real flows of expenditures on the other. This contributed to a re-emphasis on the role of money.

Another development that had the same effect, in a negative way, was the disillusionment with fiscal policy. The counterpart of the Keynesian disregard for money was the emphasis placed on fiscal policy as the key element in controlling the level of aggregate demand. In the U.S. in particular, governmental expenditures have proved to be the most unstable element in the economy in the postwar years, and they have been unstable in a way that has tended to increase fluctuations rather than to decrease them. It has proved to be extremely hard to change expenditures and receipts in advance in such a way as to offset other forces making for fluctuations. This led to re-emphasis on monetary policy as a more flexible instrument which could be used in a sensitive way.

II. Developments in Monetary Theory

Let me turn now to the developments in monetary theory that have followed this postwar experience and the re-emphasis on money as an important economic magnitude. One development has been that many economists who continue to use the Keynesian apparatus have revised their empirical presumptions. These economists now say that liquidity preference is seldom absolute, that there is some elasticity in the demand for cash balances, and that if there are changes in the stock of money there will be changes in interest rates. They say also that investment is not completely insensitive to interest rates, that when borrowing becomes more expensive, the amount spent on investment is reduced, and conversely. This view goes along with the attitude that, while money is more important than these economists used to think it was, monetary policy still can influence income only indirectly. A change in the stock of money may affect the interest rate, the interest rate may affect investment, the change in investment may affect income, but it is only by this indirect route, the argument runs, that monetary changes have an effect on economic change.

This is purely a semantic question of how one wants to describe the channels of influence. The crucial issue is the empirical one of whether in fact the links between money and income are more stable and more regular than the links between investment and income. And it is on this empirical issue that the postwar evidence spoke very strongly and led to a re-examination of the role of money.

A more fundamental and more basic development in monetary theory has been the reformulation of the quantity theory of money in a way much influenced by the Keynesian liquidity preference analysis. That analysis emphasizes money as an asset that can be compared with other assets; its emphasis is on what is called "portfolio analysis," analysis of the structure of peoples' balance sheets, of the kinds of assets they want to hold. This emphasis looks at monetary theory as part of capital theory, or the theory of wealth. This is a rather different emphasis than that derived from earlier approaches, particularly that of Irving Fisher, which put major emphasis on transactions and on money as a mechanical medium of exchange somehow connected with the transactions process.

The emphasis on money as an asset has gone in two different directions. On the one hand, it has led to emphasis on *near moneys,* as an alternative source of liquidity. One example is the work of Gurley and Shaw and their analysis of financial intermediaries as providing money substitutes. Another example, in its most extreme form, is in the

Radcliffe Committee report which attempts to widen the concept of money to make it synonymous with the concept of liquidity, itself an undefined term which covers the universe. My own view is that this particular trail toward widening the range of reference of the concept of money is a false trail. It will peter out and will not in fact be followed. The reaction which the Radcliffe Committee analysis has received among academic economists and others seems to suggest that my opinion is widely shared.

The other direction in which the emphasis on money as an asset has led is toward the development of a theory of the demand for money along the same lines as the theory of the demand for other assets and for commodities and services. In such a theory, one asks what determines the amount of cash balances that people want to hold. Here it is essential to distinguish between cash balances in two senses: nominal cash balances, the nominal quantity of money as defined in terms of monetary units such as drachmas, dollars, and so forth; and real cash balances, the real stock of money as defined in terms of command over goods and services.

The essential feature of the quantity theory of money in both its older versions and its more recent and modern version is the assertion that what really matters to people is not the number of things called drachmas or dollars they hold but the real stock of money they have, the command which those pieces of paper give them over goods and services. In talking about the demand for money, one must ask what determines the command over goods and services that people want to keep in the form of money. For example, take a very simple definition of money as consisting only of currency, of the pieces of paper we carry in our pockets. We must then ask what determines whether the amount that people hold is on the average equal to a little over six weeks' income, as it is in Greece, or a little over four weeks' income, as it is in the U.S., or five weeks' income, as it is in Turkey. Thus, when we talk about the demand for money, we must be talking about the demand for real balances in the sense of command over goods and services, and not about nominal balances.

In the theory of demand as it has been developed, the key variables include *first*, wealth or some counterpart of wealth, for example, income or, preferably, something like permanent income which is a better index of wealth than measured income. Because the problem is one of a balance sheet, the first restriction is that there is a certain total amount of wealth which must be held in the form of money, or bonds, or other securities, or houses, or automobiles, or other physical goods, or in the form of human earning capacity. Hence, income or wealth acts as a restraint in determining the demand for money in exactly the same way that the total income people have operates to determine their demand for shoes or hats or coats by setting a limit to aggregate expenditures. The *second* set of variables that is important is the rates of return on substitute forms of holding money. Here, the most important thing that has happened has been a tendency to move away from the division of assets into money and bonds that Keynes emphasized, into a more pluralistic division of wealth, not only into bonds but also into equities and real assets. The relevant variables therefore are the expected rate of return on bonds, the expected rate of return on equities and the expected rate of return on real property, and each of these may of course be multiplied by considering different specific assets of each type. A major component of the expected rate of return on real property is the rate of change in prices. It is of primary importance when there is extensive inflation or deflation.

I should like to stress the significance of the emphasis on money as one among many assets, not only for the kinds of variables that people consider as affecting the demand for money, but also for the process of adjustment. According to the earlier view of money as primarily a medium of exchange, as something which is used to facilitate transactions between people, it was fairly natural to think of a short link between changes in the stock of money and changes in expenditure and to think of the effects of changes in the stock of money as occurring very promptly. On the other hand, according to the more recent emphasis, money is something more basic than a medium of transactions; it is something which enables people to separate the act of purchase from the act of sale. From this point of view, the role of money is to serve as a temporary abode of purchasing power. It is this view that is fostered by considering money as an asset or as part of wealth.

Looked at in this way, it is plausible that there will be a more indirect and complicated process of adjustment to a change in the stock of money than looked at the other way. Moreover, it seems plausible that it will take a much longer time for the adjustment to be completed. Suppose there is a change in the stock of money. This is a change in the balance sheet. It takes time for people to readjust their balance sheets. The first thing people will do is to try to purchase other assets. As they make those purchases, they change the prices of those assets. As they change the prices of those assets, there is a tendency for the effect to spread further. The ripples spread out as they do on a lake. But as prices of assets change, the *relative* price of assets, on the one hand, and flows, on the other hand, also change. And now people may adjust their portfolios not only by exchanging assets but by using current income to add to, or current expenditures to subtract from, certain of their assets and liabilities. In consequence, I think that this reformulation of monetary theory with its emphasis on monetary theory as a branch of the theory of wealth has very important implications for the process of adjustment and for the problem of time lags.

III. Developments in Monetary Policy

Policy does not always have a close relation to theory. The world of the academic halls and the world of policy makers often seem to move on two wholly different levels with little contact between them. The developments in post-war monetary policy have not been the same throughout the world. However, the makers of monetary policy in different countries have been in closer and more systematic touch with one another than the monetary theorists. As a result, I think one can speak to some extent of general trends in policy without necessarily referring to the country.

As I indicated earlier, I think two features dominate and characterize the trends in postwar monetary policy. The first is the shift of emphasis away from credit policy and toward monetary policy. I think this is a distinction of first rate importance, and yet one which is much neglected. Therefore let me say a word about the meaning of this distinction. When I refer to credit policy, I mean the effect of the actions of monetary authorities on rates of interest, terms of lending, the ease with which people can borrow, and conditions in the credit markets. When I refer to monetary policy, I mean the effect of the actions of monetary authorities on the stock of money—on the number of pieces of paper in people's pockets, or the quantity of deposits on the books of banks.

Policy makers, and central bankers in particular, have for centuries concentrated on credit policy and paid little attention to monetary policy. The Keynesian analysis, emphasizing interest rates as opposed to the stock of money, is only the latest rationalization of that concentration. The most important earlier rationalization was the so-called real bills doctrine. The belief is still common among central bankers today that, if credit were somehow issued in relation to productive business activities, then the quantity of money could be left to itself. This notion of the real bills doctrine goes back hundreds of years; it is endemic with central bankers today. It understandably derives from their close connection with commercial banking, but it is basically fallacious.

The emphasis on credit policy was closely linked with the emphasis at the end of the war on qualitative controls. If what matters is who borrows and at what rate, then it is quite natural to be concerned with controlling the specific use of credit and the specific application of it.

In the U.S., for example, emphasis on credit policy was linked with emphasis on margin controls on the stock market, and with controls over real estate credit and installment credit. In Britain, it was linked with controls over hire purchase credit. In each of these cases, there was a qualitative policy concerned with credit conditions. The failure of the easy money policy and of these techniques of qualitative control promoted a shift both toward less emphasis on controlling specific rates of return and toward more emphasis on controlling the total quantity of money.

The distinction that I am making between credit and monetary policy may seem like a purely academic one of no great practical importance. Nothing could be farther from the truth. Let me cite the most striking example that I know; namely, U. S. experience in the great depression from 1929 to 1933. Throughout that period the Federal Reserve System was never concerned with the quantity of money. It did not in fact publish monthly figures of the quantity of money until the 1940's. Indeed, the first mention in Federal Reserve literature of the quantity of money as a criterion of policy was in the 1950's. Prior to that time there was much emphasis upon easy or tight money, by which was meant low or high interest rates. There was much emphasis on the availability of loans, but there was no emphasis and no concern with the quantity of money.

If there had been concern with the quantity of money *as such*, we could not have had the great depression of 1929-33 in the form in which we had it. If the Federal Reserve System had been concerned with monetary policy in the sense in which I have just defined it, it literally would have been impossible for the System to have allowed the quantity of money in the U.S. to decline from 1929 to 1933 by a third, the largest decline in the history of the U.S. in that length of time. In reading many of the internal papers of the Federal Reserve Board during that period, the communications between the various governors of the Federal Reserve Banks and the Board of Governors, and so forth, I have been struck with the lack of any quantitative criterion of policy. There are vague expressions about letting the market forces operate. There are comments about "easy" money or "tight" money but no indication of precisely how a determination is to be made whether money is "easy" or "tight." This distinction between emphasis on credit policy and emphasis on monetary policy is a distinction of great importance in the monetary history of the U.S., and I think also in the monetary history of other countries.

The failure of the easy money policy was reinforced by another factor which promoted a shift in policy away from qualitative measures involving control of particular forms of credit, and toward quantitative measures involving concern with changes in the stock of money. This other factor was a reduction of exchange controls and quantitative restrictions on international trade, as in the postwar period one country after another began to improve its international position. There was a move toward convertibility in international payments. This shift toward convertibility led to a reduction of emphasis on qualitative direct controls and toward increased emphasis on general measures that would affect the course of events through altering the conditions under which people engaged in trade. In turn, this led to a final development in monetary policy—the renewed concern about the relation between internal monetary policy and external policy, the problem of the balance of payments. In this area we have had, most surprisingly of all I think, a return to an earlier era of something approximating a gold standard.

In the immediate postwar period, concern with the balance of payments tended to be centered in the countries of Western Europe that were having a so-called dollar shortage. Those countries were at that time facing the problem of recurrent drains of their international reserves. They were in the position of having somehow to restrain their residents from converting their local currencies into foreign currencies. Those were also the countries that emerged from the war with fairly extensive exchange controls and direct restrictions on trade. And thus in the first years after the war the solution to this problem took the form of direct control rather than of monetary policy.

At that time the U.S. was in a very different position. It was gaining gold and it was able to take the position that it could conduct its monetary policy entirely in terms of internal conditions and need pay no attention to the effects that its policies had abroad. Of course, that was not what happened. There is little doubt that during the immediate postwar period the ease in the U.S. gold position contributed toward a greater readiness to accept inflation than would otherwise have prevailed, so that the ease in the international balance produced a relatively easier monetary policy than we otherwise would have had. But once the U.S. started selling gold on net instead of buying gold on net, to use a more accurate term than the term "losing gold," the situation changed drastically and the U.S. itself became much more concerned with the effect of monetary policy and much more driven toward a pre-World War I gold standard approach.

In recent years, the concern with the international balance of payments has given rise to greater co-operation among central banks. They have tried to develop techniques which will assure that any temporary drains on the reserves of one country will be matched by offsetting movements by central banks in the other countries. Despite the immense amount of good will and of human ingenuity that has gone into this effort to avoid payments difficulties through central bank co-operation, I must confess that I regard the tendency as an exceedingly dangerous one. The danger is that the arrangements developed will provide an effective system for smoothing minor difficulties but only at the cost of permitting them to develop into major ones.

I am much struck by the analogy between what is now happening in this respect and what happened in the U.S. between 1919 and 1939. The U.S. in that earlier period developed a monetary system which turned out to be an effective device for smoothing minor difficulties. The system was highly successful in helping to make the years from 1922 to 1929 relatively stable. But this stability was purchased at the cost of major difficulties from 1920 to 1921, from 1929 to 1933, and again from 1937 to 1938. I very much fear that the same results may emerge from present trends toward international co-operation among central banks, because these measures do not go to the root of the problem of international adjustment.

In international financial arrangements, as in personal finances, the problem of having enough liquid assets to meet temporary drains must be sharply distinguished from the adjustment to changed circumstances. The central bank arrangements look only to providing liquidity for temporary drains. More fundamental adjustment to changed circumstances can come only through either: (1) domestic monetary and fiscal policy directed toward holding down or reducing domestic prices relative to foreign prices when the country is experiencing a deficit, or toward permitting domestic prices to rise relative to foreign prices when the country is experiencing a surplus; (2) changes in exchange rates to achieve a similar alteration in the relative level of domestic and foreign prices when expressed in the same currency; or (3) direct measures designed to alter the flows of receipts or expenditures, such as changes in tariffs, subsidies, and quotas, direct or indirect control of capital movements, restrictions on foreign aid or other governmental expenditures, extending ultimately to that full panoply of foreign exchange controls that strangled Western Europe after the war and remains today one of our most unfortunate gifts to many underdeveloped countries.

The great danger is that central bank co-operation and other means to enlarge liquidity, by providing palliatives that can at best smooth over temporary imbalances, will encourage countries to postpone undertaking such fundamental adjustments to changed circumstances. The consequence will be to allow minor imbalances to accumulate into major ones; to convert situations that could have been corrected by gradual and minor monetary tightness or ease, or by small movements in exchange rates, into situations that would require major changes in monetary policy or exchange rates. The consequence is likely to be not only international financial crises, but also the encouragement of the use of the third method of adjustment, direct controls. Paradoxically, most economists and most policy makers would agree that it is the worst of the three; yet it is the one that has most regularly been resorted to in the postwar period.

These developments in monetary policy are much more difficult to pin down precisely than the developments in monetary theory, as may be expected from the fact that monetary policy is and must be much more a matter of opportunism, of day-to-day adjustment, of meeting the particular problems of the time. The theorist can sit in his ivory tower and make sure that his structure is coherent and consistent. This is, I must say, an advantage of the theorist and a great disadvantage of the policy maker, and not the other way around. But I think it is clear that we are likely to see in the future still further developments in monetary policy.

There is almost invariably a long cultural lag before developments in theory manifest themselves in policy. If you were to look at what is being proposed today in domestic policy in the U.S., you would say that my analysis of changes in the field of monetary theory must be a figment of my imagination. The policy proposals that are being made in the U.S. today are all reflections of the ideas of the late 1930's, or at the latest of the early 1940's. That is natural and widespread. The people who make policy, who are involved in policy formation, are inevitably people who got their training and their education and their attitudes some 20 or more years earlier. This is a special case of a much more general phenomenon. I am sure all are aware of that famous book by A. V. Dicey on *Law and Public Opinion in the 19th Century*, the main thesis of which is precisely that trends in ideas take about 20 years before they are effective in the world of action. What is happening in the U.S. today is a dramatic illustration of his thesis. And so I expect that monetary policy will in the course of the next 20 years show some radical changes as a result of the changes I have described in monetary theory.

12

Tight money, monetary restraint, and the price level *

George Horwich†

I. Introduction

In recent years a substantial body of public opinion has opposed monetary restraint on the ground that higher interest rates, designed to curb inflation, raise prices by raising business costs. The net effect of monetary policy in controlling inflation would thus be negligible, if not actually perverse. This point of view has been particularly prevalent in official policy statements and public pronouncements by members of the Democratic party.[1] However, it is not limited to the latter group, for

Reprinted with permission from *The Journal of Finance,* Vol. XXI, No. 1 (March, 1966), pp. 15-28.

* The author has benefited from comments and suggestions by M. June Flanders, A. P. Lerner, and the editorial reader, none of whom is responsible for any remaining flaws in the analysis.

† Professor of Economics, Purdue University.

[1] As might be expected, these pronouncements were more in evidence before the party came to power in 1961. See, for example, the statements of Democratic senators in Hearings before the Committee on Finance, United States Senate (85th Cong., 1st Sess.), *Investigation of the Financial Condition of the United States,* June-August 1957, pp. 68, 323, 342, 345, 390-91, 733, 739, 785, 952, 1093, 1402-03. In the summer of 1959, Senator William Proxmire, of the Senate Banking and Currency Committee, circulated a questionnaire to a large number of economists in which the opening question was:

> What evidence is there to support or refute a conclusion that "tight money" serves to limit inflation enough to off-set the rising costs of borrowing and the higher total cost of everything that is paid for "on time?" (I am seeking evidence to determine the extent to which credit restraint counteracts over-all inflationary pressure, on the one hand, and the extent to which it channels the price increases into the time payment segment of the economy on the other.)

A more recent statement of Senator Proxmire's views is contained in Hearings before even Republicans, as well as nonpartisan monetary officials, have accepted the argument in principle.

The stock reply of the proponents of monetary policy is simply that interest is a small and unimportant element in total business costs.[2] But this is a fatal concession. For the rate of interest, narrowly defined (as it usually is) as the return on borrowings, is only one of many varieties of rent paid on the existing stock of capital. The relevant economic variable is in fact all property income, which constitutes as much as a third of the national income. As a first approximation, the various components of the property return, including bond interest, stock dividends, and building rentals, would be expected to rise and fall together over time.

The belief that higher interest as a cost might contribute to inflation has not been seriously discussed by contemporary economists, in spite of the widespread popularity of the argument. The failure to offer an analysis is due, I think, to our own incomplete integration of the theoretical role of interest as a production cost, as a return to the claimants of capital, and as a variable in monetary policy.

This paper outlines a framework within which the impact of interest on prices may be analyzed. There are two main building blocks in the approach, both Keynesian. One is a concept of monetary restraint, which derives from the *Treatise.* It is essentially Wicksellian: interest rates are fundamentally determined by internal forces, rather than by the monetary authority. The other building block is the assumed nature of the return to capital, which is advanced in the *General Theory.* This is the marginal efficiency relationship, which links interest movements closely to changes in the productivity of real capital. The Keynesian monetary and capital structure is set forth in Sections II and III, respec-

---

the Joint Economic Committee (87th Cong., 1st Sess.), *Review of Report of the Commission on Money and Credit,* August 1961, p. 256. Representatives of organized labor take a similar position. See Hearings before the Joint Economic Committee (85th Cong., 2nd Sess.), *The Relationship of Prices to Economic Stability and Growth: Commentaries,* October 1958, pp. 33 and 248.

The opinion among Democratic senators (and presidents) that higher interest rates are inflationary has an historical precedent in the writings of Thomas Tooke and the "Banking School." See, for example, Albert Feaveryear, *The Pound Sterling,* Oxford: The Clarendon Press, 1962, 2nd edition (edited and revised by Victor Morgan), p. 266, who quotes Tooke: 'A high rate of interest had no effect upon prices, except perhaps to raise them, in the long period, by raising the cost of production." This "heretical" view was supported by William Newmarch's investigation into the circulation of bills of exchange for the period 1830-1853. He found no evidence that commercial credit was limited at all by rising interest rates. See Tooke and Newmarch, *History of Prices,* New York: Adelphi Co., 1857, Vol. VI. For a survey of this whole controversy, see J. R. T. Hughes, *Fluctuations in Trade, Industry and Finance,* Oxford: The Clarendon Press, 1960, pp. 228-36.

[2] See the remarks by Treasury Secretary Humphrey, Hearings before the Committee on Finance, *op. cit.,* pp. 390-91; Under Secretary of the Treasury Burgess, *ibid.,* p. 733; and Chairman Martin of the Board of Governors, *ibid.,* pp. 1268-69, 1402.

tively. They are combined in Section IV, which examines the impact of tight money on the price level. Section V considers an alternative concept of monetary restraint, in which the monetary authority assumes a more independent role in setting interest rates. Section VI is a summary and conclusion. An Appendix offers a theoretical foundation for the determination of the price and output of capital and consumption goods.[*]

II. Monetary Restraint: The Keynes-Wicksell Framework

"Tight money" is the appellation generally applied to an economy experiencing rising interest rates and rising income.[3] In the Wicksellian-Keynesian view, interest rates rise because of an excess demand for loanable funds caused by an increase in desired investment or a decrease in desired saving. This constitutes an increase in the natural rate of interest, which the market rate follows with a lag. Simultaneously money income rises, according to Wicksell, because the quantity of money supplied rises with the rate of interest (the money supply is "elastic"); or, according to Keynes, the quantity of money demanded falls with the rise of the rate of interest. In either case, there is a higher equilibrium interest rate, which is associated with a higher equilibrium level of prices and income.

In this context monetary restraint, which raises interest rates, accelerates an adjustment that would occur in any case. In the language of Wicksell, the monetary authority raises the market rate to the level of the higher natural rate.[4] The nature of the disturbance and the role of monetary policy is shown with the aid of Figure 1.

In the left diagram are the real flow functions, saving (S), which is the complement of consumption, and investment demand (I). Desired investment increases with lower interest rates, while saving decreases. Unlike the investment-interest relation, the dependency of saving on

[* *Editors' note:* The Appendix referred to here and in notes 9, 11, 14, 19, 21, and 22 has not been included.]

[3] Rising interest rates are rarely accompanied by falling income. Hendershott and Murphy ("The Monetary Cycle and The Business Cycle: The Flow of Funds Re-Examined," *National Banking Review*, I, June 1964, p. 535) were able to find only one quarter in the entire period 1952-62 during which interest rose while economic activity declined.

[4] This view of monetary policy is advanced by Keynes in *A Treatise on Money*, London: Macmillan and Co., Ltd., 1950, Vol. I, p. 273, and Vol. II, pp. 351-52 and 362. See also D. H. Robertson, who bases similar policy prescriptions on a more purely Wicksellian model, in which the inflation is due to elasticity of the money supply: "Industrial Fluctuation and the Natural Rate of Interest," *Economic Journal*, XLIV, December 1934, pp. 650-56 (reprinted as Ch. V in *Essays in Monetary Theory*, Staples Press, Ltd., 1940). For recent treatments, see M. J. Bailey, *National Income and The Price Level*, New York: McGraw-Hill Book Co., Inc., 1962, pp. 156-57; and G. Horwich, *Money, Capital, and Prices*, Homewood: Richard D. Irwin, Inc., 1964, pp. 450-51.

the rate of interest is not essential to our argument. In the right diagram are the stock or existing-asset functions, the demand (L) and the supply of real balances (M/P), which are the complements of the demand and supply, respectively, of all non-monetary assets (typically financial claims or securities). The demand for real balances (liquidity preference) is a downward function of the rate of interest. The supply of real balances is assumed, for simplicity, to be insensitive to the interest rate. Both saving and the demand for money also depend on the level of real income, which we assume, for the present, is constant at the full employment level. Each pair of schedules in the diagram is in equilibrium at a common rate of interest r′, the natural rate. The equilibrium quantity of real balances (at point A) is the ratio of M′, which is the nominal stock of money, to P′, the index of all prices.

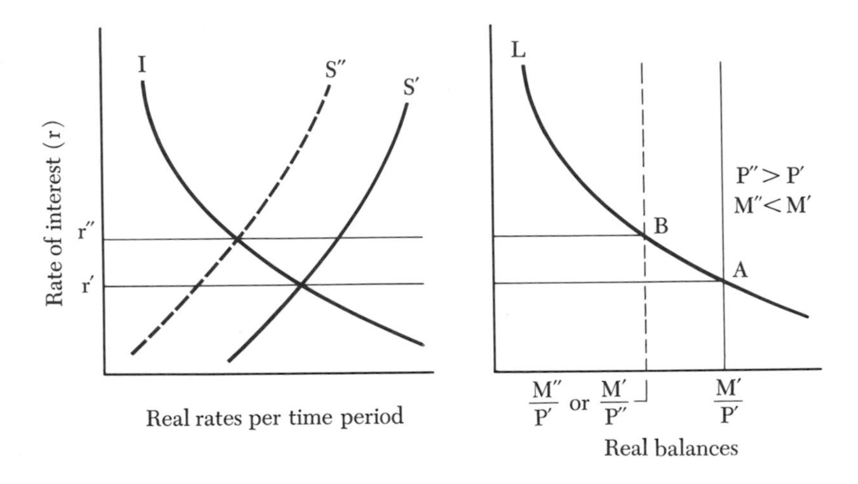

Figure 1

Now suppose that a higher natural (i.e., equilibrium) rate of interest r″ is created by a decrease in the saving schedule from S′ to S″. At r″ asset-holders want a lower quantity of real balances and, by implication, a greater quantity of nonmonetary assets or securities. These desires are measured by the movement from A to B along the liquidity preference schedule. In the absence of outside interference, the reduction in real balances will be obtained through an internally generated inflationary process.[5] Immediately following the decrease of saving, we have at r′, the initial—and, momentarily, still prevailing—rate of interest, $I > S''$, an excess of ex ante investment over desired saving. This is the spur to both higher prices and the higher market rate of interest. Given the level of

[5] Cf. Horwich, *op. cit.*, Ch. IV and pp. 277-79.

real income and stock of money M′, I > S″ persists until the market rate is r″ and prices are P″. The higher price level establishes the quantity of real balances desired at B, M′/P″.[6]

However, from the viewpoint of asset-holders, the new equilibrium at B could also be reached by reducing the quantity of money, the price level remaining constant. More precisely, there is a quantity M″ < M′ such that M″/P′ = M′/P″. Restrictive monetary policy accordingly takes the form of an open-market sale conducted by the central bank, which reduces the quantity of money to M″ and raises the market rate of interest to r″. This establishes swiftly and directly the new equilibrium interest rate and real balances, both of which are otherwise achieved internally through a prolonged inflationary adjustment.

The goal of a tight monetary policy is thus the avoidance of inflation, accomplished by "raising" interest rates along a course they would naturally follow. The criticism of this policy, referred to in Section I, is that the higher interest rate nevertheless contributes to inflation by raising the interest costs of production. We are now ready to analyze this argument by considering in more detail the real forces that give rise to an increase in the equilibrium (natural) rate of interest. We assume that the monetary authority is completely successful in offsetting inflationary pressures due to a reduction in the quantity of real balances demanded. The only disturbances to the system are thus in the rate of interest, the rate of saving and investment, and any other variables that are necessary concomitants of these three.

III. The Marginal Efficiency of Investment

The role of interest as a cost of production depends upon the nature of the investment demand function. The most widely accepted investment theory is the Keynesian marginal efficiency concept, as elaborated by Lerner and Robertson.[7] In this view investment demand (ex ante investment), a downward function of interest, is derived from a short-run inverse relation between the return to (the marginal efficiency of) investment and the rate of ex post investment. This latter relationship is

[6] If there are any unemployed resources or excess capacity in the economy, the adjustment will raise output along with prices. The real saving schedule and the demand for real balances both shift to the right. This will lower the natural rate somewhat below r″ and raise the equilibrium level of real balances above M′/P″. The adjustment thus terminates earlier and prices rise less than if output were constant.

[7] See Keynes, *The General Theory of Employment, Interest and Money,* New York: Harcourt, Brace and Co., 1936, Ch. 11; A. P. Lerner, *The Economics of Control,* New York: The Macmillan Co., 1944, Chs. 21 and 25, and "On Some Recent Developments in Capital Theory," *American Economic Review,* LV, May 1965, pp. 284-95; D. H. Robertson, *Lectures on Economic Principles,* Vol. II, London: Staples Press, Ltd., 1958, pp. 61-68; and Horwich, *op. cit.,* pp. 40-43.

represented by the MEI line in Figure 2. The entire line is based upon a given gross marginal product of capital, which is effectively a constant in any one- or two-year period.[8] Thus the downward slope of the line is due not to changes in the gross product, but rather to variations in the net product remaining after subtraction of variable depreciation charges. The latter in turn depend upon the marginal cost and supply price of new capital goods, which in the short run vary directly with their rate of output. The greater the rate of ex post investment, the greater therefore are the output and supply price of capital. Given that the capital

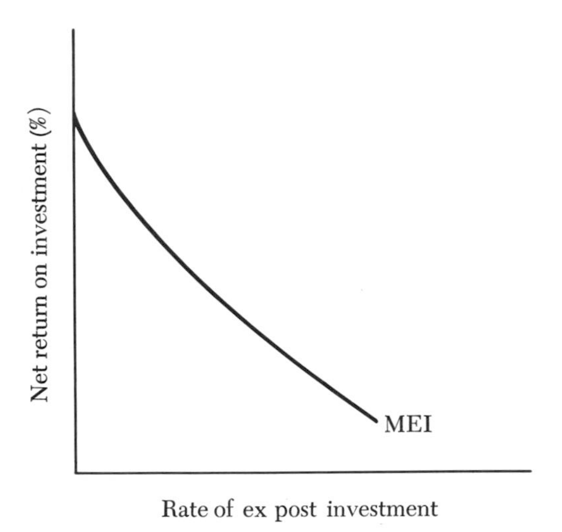

Figure 2

stock is valued at replacement, rather than historical cost, the greater are the depreciation allowances and the smaller is the net available return to the claimants of the capital stock. The rate of return, defined as the ratio of net earnings to the supply price of capital, thus falls both because of the decrease in the numerator and the increase in the denominator.[9]

Assuming that firms are rational and competitive, they will equate the percentage return on investment to the market rate of interest. That is, at any rate of interest, investment expenditures will be an amount, which, if realized ex post, will result in a return to investment just equal to the interest rate. Given accurate knowledge of the return correspond-

[8] In any individual firm, changes in capital may, of course, create substantial changes in its marginal product in any period of time. But the marginal product of capital, *averaged* over all firms in the economy, is not likely to move very much in the short run.

[9] A detailed two-sector account of the derivation of the marginal efficiency and investment demand schedules is presented in the Appendix, Section A.

ing to each rate of ex post investment, ex ante investment (I)—a function of interest—thus coincides exactly with the MEI schedule.[10]

In utilizing the Keynesian investment theory, we assume, for simplicity, that there exists only one variety of capital, the return on which is equated to "the" rate of interest. We assume also that the price of the capital stock and the direction of investment is such that the return on all existing capital units and all new investment projects is the same. Finally, we assume that all claims to capital held in the form of fixed-value financial instruments are renegotiated continuously in order to equate the coupon rate to the market rate of interest. This assumption maximizes the possible impact of higher interest rates on current business costs and the general price level.

IV. Two Disturbances

Within the framework of Sections II and III let us analyze a rise in the equilibrium rate of interest due to an autonomous (i) decrease in saving, and (ii) increase in investment.

A. Saving

A decrease in saving entails a leftward shift in the saving schedule and an upward movement along the I function. The monetary authority immediately carries out an open-market sale, which reduces the nominal supply of money and raises the market rate to the higher natural level (see Figure 1). Resources flow into consumption industries, ex ante saving and investment are maintained in constant equality, and the price level, up to this point, is unchanged.

The movement along the investment demand schedule coincides with an upward real movement along the MEI line. This implies that the increase in the rate of interest is accompanied by an increase in the net return to capital. The return rises as ex post investment and the marginal cost and supply price of capital-goods output all decline to lower equilibrium levels. All firms employing capital find that they may reduce their current depreciation charges, thereby financing the higher coupon rates created by the higher rate of interest. Thus, while interest costs are higher, there is for every firm an added revenue available to meet this cost. Under these circumstances, there are no forces on the side of cost and production tending to alter the average price of commodities. (Any

[10] The assumption of competitive behavior among firms is not essential to the analysis. In the absence of perfect competition, the ex post marginal efficiency schedule will lie above the ex ante investment demand function. The vertical distance between the schedules will be monopoly profit. This can be handled by the subsequent analysis (Sections IV and V) by simply requiring that the two schedules shift together simultaneously in the same direction.

change in the price level due to the lower price of capital goods is offset by a correspondingly higher price of consumption goods, which results when consumption demand increases at the expense of saving.[11])

Given the Keynesian "interest-equalization" monetary policy and the capital-investment decision process, there are thus neither monetary nor nonmonetary sources of inflation. The only qualification to this hinges on the possibility that the transfer of resources from investment to consumption industries might be facilitated by a given increase or rate of increase in the price level. If this were true, the Keynesian monetary action would contribute to a degree of immobility, unemployment, and temporary reduction of aggregate output. Prices, as a result, would rise. One can only speculate how this would compare quantitatively or—from a public policy viewpoint—how it should be evaluated, relative to the *permanent* inflation due to an internally stimulated movement along the liquidity preference schedule. Notice, moreover, that inflation attributable to resource immobility in this case cannot be characterized as resulting from higher interest "costs." Rather, the inflation is due to immobility originating in a decrease in demand and *revenues* in investment-goods industries.

B. Investment

An independent increase in investment results from an increase in the gross marginal product of capital. This implies that aggregate output y has also increased. The monetary disturbance is now somewhat more complex and is described with reference to Figure 3.

The initial schedules, drawn with solid lines, all meet at r′, the natural rate. In the left diagram the predisturbance functions are I′ and S(y′), which is saving based on output y′. The equilibrium is at point A. In the right diagram the beginning equilibrium is at point C, at which real balances are M′/P′, the quantity determined by r′ along the initial schedule L(y′). The investment curve increases to I″ and is accompanied by a simultaneous increase in output to y″.[12] This causes three simultaneous shifts: (i) real saving moves to the right to S(y″), establishing a new saving-investment equilibrium at point B; (ii) the demand for real balances increases to L(y″); (iii) given price flexibility, the increase in L lowers the general price level from P′ to P″, raising the supply of real balances to M′/P″. Assuming that the demand and supply of real

[11] See the Appendix, Section B, for a fuller description of the impact of a shift in saving on both the investment-goods and consumption-goods sectors.

[12] The increase in the marginal efficiency schedule may reflect an anticipated future increase, rather than a de facto current increase in output and the productivity of capital. In this event firms may have to resort to additional finance, including new money and the dishoarding of existing balances, in order to service new borrowings or security issues. However, as a source of inflation, this can only be a temporary, reversible phenomenon.

balances increase synchronously, the equilibrium in this market moves horizontally to point D. Now, if left to its own devices, the economy will establish a new money-market equilibrium at point E, at which the interest rate, determined by $S(y'')$ and I'', is $r'' > r'$; real balances are M'/P'''; and, since the money stock is constant throughout, the price level is higher at $P''' > P'$.[13]

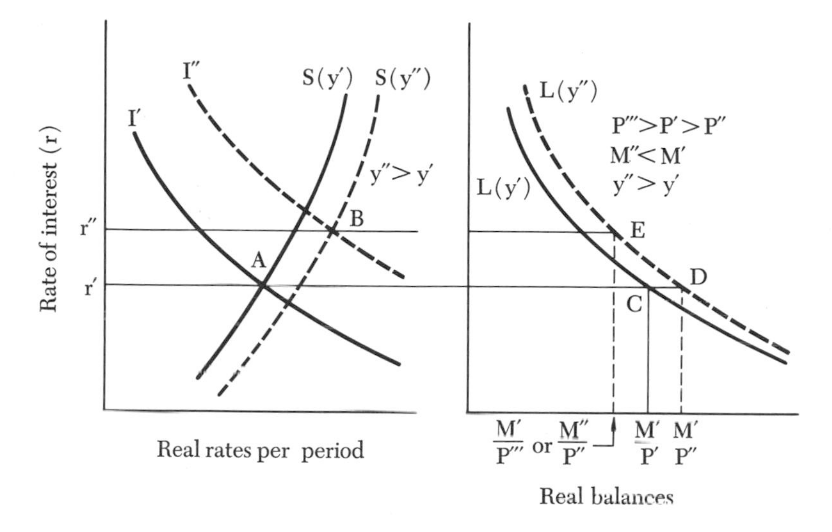

Figure 3

The monetary authority can again perform its catalytic role of raising the market rate to the natural rate. If it assumes full responsibility for doing so, it will conduct an instantaneous open-market sale, moving the equilibrium from D to E by reducing the money supply to $M'' < M'$ and real balances to $M''/P'' = /M'P'''$. However, at point D prices are $P'' < P'$, and at E the economy is thus left with a net reduction in the price level due to the increase in output. Since we are only interested in evaluating the role of monetary policy in raising interest rates, we shall assume for analytical purposes that the lower price level is allowed to remain. We assume that prices are perfectly flexible downward in response to the rise of output, which, for the moment, is constant at y''. Under these "pure" circumstances, does the higher interest rate, viewed as a *cost* phenomenon, promise to raise prices?

Let the productivity increase, which gave rise to the increase in investment, be distributed uniformly among all firms. There are thus

[13] It is possible (though unlikely) for point E to lie directly above or to the right of point C, implying that the final price level is equal to or less than the beginning one. This could be the result of a greater income coefficient in saving or in the demand for money, shifting either schedule farther to the right.

throughout the economy greater revenues with which to meet higher interest payments. The increased interest rate has no independent influence on costs, output, or prices, which remain at the levels described in the preceding paragraphs. Now suppose, more realistically, that some firms fail to share in the greater productivity of capital. The latter typically stems from an increase in the labor force or an innovation, neither of which will be experienced by every producer. Nevertheless, the greater interest charges on existing capital must be met by all firms, and those not sharing in the productivity rise will undergo an increase in costs unmatched by greater revenues. These firms must retrench, releasing resources. If the resources are mobile and find employment elsewhere, then aggregate output is constant and the price level again is not directly affected by the higher interest rate. However, if in the short run the necessary reallocation creates unemployment and a net reduction of output, prices will tend to rise. In the actual context of an increase in capital productivity, unemployment may simply prevent output from rising and prices from falling by their maximum amounts. It is in this sense that an increase in investment and the rate of interest—"tight money"—is a force tending to raise costs and also prices.[14]

A further possible source of inflation is again the required transfer of resources between consumption and investment industries. In the present disturbance the movement along $S(y'')$ from r' to r'' is an increase in the proportion of income saved; resources must accordingly move from consumption to investment-goods output.[15] If this transfer is in any way dependent upon inflation, then the Keynesian monetary policy will delay it. Output will fall below y'' and prices will rise above P''.

V. Monetary Restraint: An Alternative View

Let us now modify the analysis by assuming that monetary restriction does not coincide with an increase in the natural rate of interest. The monetary authority reduces money and raises interest in order to

[14] See Appendix, Section C, for the two-sector account of an increase in investment.

[15] The increase in saving at r' due to the shift of the schedule from $S(y')$ to $S(y'')$ will also require a movement of resources from consumption to investment industries, provided that (i) the spontaneous increase in output from y' to y'' is distributed between consumption and investment goods in the predisturbance ratio; (ii) the marginal rate of saving is greater than the average rate. If, for example, the marginal saving rate were less than the average, the de facto increase in investment goods due to the more productive capital stock would exceed the increase desired by savers, and resources would have to leave investment industries. However, (ii) is in fact widely accepted on both theoretical and empirical grounds. (i) is difficult to justify. I know of no commonly accepted production functions or types of technological change that guarantee this result.

offset inflationary disturbances which do not themselves operate through, or entail a change in, the rate of interest. An example of such a disturbance is a direct movement from cash balances in goods—i.e., a leftward shift in the demand schedule for money—or money creation via a fiscal deficit which does not raise bank reserves.[16] The impact of tight money against such a background is discussed with reference to Figure 4. For simplicity, we describe the disturbance and the monetary response separately, though in fact they may coincide.

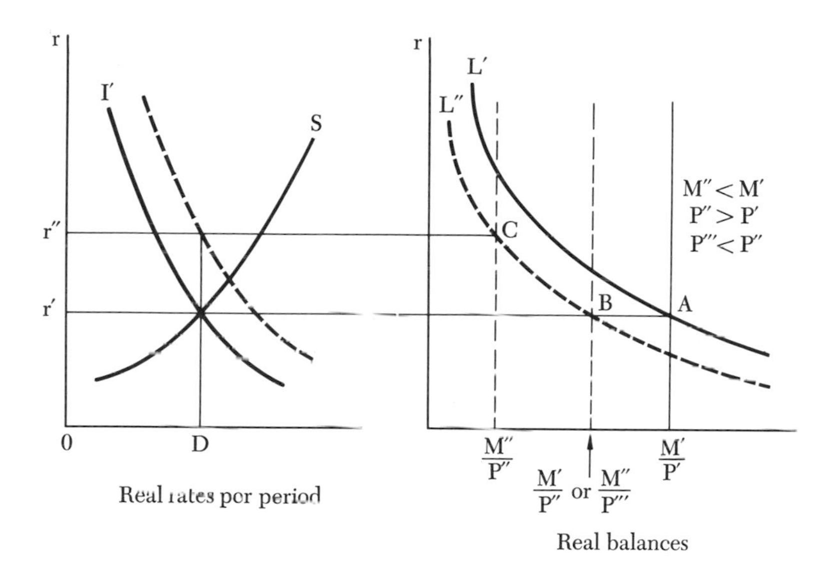

Figure 4

The system initially is in stock-flow equilibrium at interest rate r′ and real balances M′/P′. The existing-asset equilibrium is at point A. Suppose that the inflationary disturbance to which the monetary authority reacts is a once-for-all decrease in the demand-for-money schedule relative to goods. In the diagram L shifts leftward from L′ to L″. Assuming a fixed level of output, commodity prices rise directly to, say, P″.[17] The supply of real balances falls to M′/P″ and the existing-asset equilibrium moves horizontally from A to B.

[16] This is the kind of disturbance (though generally it is the opposite case of hoarding) that Robertson typically is concerned with. See, e.g., *Banking Policy and the Price Level*, London: Staples Press, Ltd., 1949, pp. 53-54; and "Saving and Hoarding," *Economic Journal*, XLIII, September 1933, pp. 401-2 (reprinted as Ch. IV in *Essays in Monetary Theory, op. cit.*).

[17] The increase in the demand for goods is assumed to be across the board, with no effect on relative prices or the rate of interest. Thus real saving and investment remain unchanged at their initial level of equality.

The central bank now strives, through open-market action, to offset the higher price level. It conducts an instantaneous open-market sale, which reduces M to M'' and raises r to r''. This entails an instantaneous movement in the stock equilibrium from B to C along L''. The price level remains, for the moment, unchanged at P''. But at the higher market rate r'', ex ante saving exceeds investment. This causes both prices and interest to fall over time via a stock-flow dynamic adjustment (output is still constant).[18] The final equilibrium is at point B, at which the market rate is again r', prices are $P''' < P''$, and real balances are $M''/P''' = M'/P''$. If the bank's objective were to stabilize the price level, then M'' would be a quantity such that $P''' = P'$.

Tight money is thus now characterized by a market rate of interest whose average level is temporarily raised. The rate rises first in response to an open-market sale, and then falls through the internal adaptive mechanism. This is in contrast to the Keynesian-Wicksellian framework in which tight money entailed an interest rate rising monotonically from one equilibrium level to another. The role of the central bank was to facilitate and accelerate that movement. In both cases the monetary policy serves to curtail simultaneously occurring inflationary pressures. In the Keynesian case the inflation was a built-in response to the movement of the rate of interest. In the present action the inflation is an autonomous disturbance which the bank-induced change in money and interest endeavors to offset.

Once again, in order to evaluate any impact on the price level of the higher interest rate as a cost phenomenon, we must turn to the underlying real variables. We have seen that the capacity of firms to absorb higher interest costs depends on the net return to capital, which is a function of the level of ex post investment. In Section IV monetary tightness, as imposed by the central bank, was accompanied by constant equality between desired saving and investment, to which ex post investment corresponded. But now desired saving and investment are unequal throughout the period of higher interest rates. Neither the disturbances nor the adjustment process indicate in themselves what the level of ex post investment will be. However, we will consider several possibilities lying between the blades of the ex ante saving-investment scissors and generalize from there. We continue to assume, for the present, that total output and employment are constant.

Perhaps the simplest assumption is that ex ante and ex post investment are equal. Beginning at the initial equilibrium yield r', ex post investment moves up along the ex ante investment demand schedule I to the height of the tight money rate r'', and then down again to the unchanged equilibrium yield r'. On our assumptions, this coincides with an identical movement along the schedule of the marginal efficiency of

[18] See Horwich, *op. cit.*, pp. 179-187.

investment. The return to investment thus moves exactly, and finances the production costs associated, with changes in the market rate of interest. The return first rises as resources move out of capital production, lowering the relative price of investment goods and the necessary depreciation allowances held against capital. Then the return falls to its original level as the sequence is reversed. This analysis parallels that of the decrease in saving (Section IV-A). In both cases the variable interest costs are accompanied by equal variations in net revenues for all firms. Profits, output, and prices are unaffected by the adjustment process.[19]

Perhaps a more realistic asumption is that resources are completely immobile. This would tend to be true the briefer is the time interval of the tight money period. Ex post investment would then remain at the predisturbance level, OD in Figure 4, which is above the ex ante levels indicated by the I schedule. It appears thus that the return to investment will remain equal to r′, below the prevailing interest rates. But the only way in which unwanted capital goods can be sold to firms is by lowering their price. In the face of falling demand, this implies that the entire investment-goods supply schedule drops, reflecting a reduction in the compensation paid to resources in that industry. The supply price of capital output falls—both absolutely and relatively—and the MEI schedule (based on a given investment-goods supply function) rises initially to the location of the dashed (unlabeled) curve in Figure 4. In brief, resources stay on in the investment industry by accepting lower real returns, thereby raising the profitability of each investment-output level; the ex ante I schedule follows the movement of the MEI line.[20] As the market rate of interest falls (owing to the continuing excess of ex ante S over ex ante I), investment demand rises along the dashed curve. This permits resource payments and the supply price of capital to increase by whatever amount is required to maintain OD as the profitable investment. Thus the movement to equilibrium is one in which the I and MEI schedules gradually shift downward to their original positions. The return to investment is always at least as great as the market rate of interest. Hence, there is no independent upward pressure on costs and prices originating in the higher rate of interest.[21]

Any tendency for ex post investment to exceed the ex ante amount would be analyzed in the manner of the preceding paragraph. The marginal efficiency schedule would shift along the locus of marginal efficiency-ex post investment points, financing the higher interest rates of the adjust-

[19] See the description in the Appendix, Section D.

[20] The upward shift in the I schedule reduces the magnitude of the saving-investment gap and consequently the fall over time of the rate of interest and the speed of the deflationary adjustment. For a similar point in his criticism of Keynes' *Treatise*, see D. H. Robertson, "Mr. Keynes' Theory of Money," *Economic Journal*, XLI, September 1931, p. 401.

[21] Cf. also Appendix, Section E.

ment period. As shown in the Appendix, the precise location of the ex post points, and thus the MEI schedule, depends on the relative shifts of the marginal cost schedules of investment and consumption industries. These shifts determine the ex post levels of output and the relative prices of the two commodities, which together determine the return to real capital.

The assumption of fixed total output and employment in a deflationary process is apt to be quite unrealistic, owing to cost and price rigidities. However, the present monetary policy is designed to offset an inflationary disturbance, and the *net* movement of the price level over time may not actually be great. The total demand for output would thereby be relatively stable, even though interest rises and stimulates an excess of saving over investment, as described.[22] But even with a stable price level, the adjustment will require a temporary reallocation of resources, except in the limiting case of constant output in each sector, noted above. Given less-than-perfect mobility between sectors, unemployment and a decline in aggregate output would thus generally occur. This would reduce aggregate saving and raise the natural rate of interest, terminating the deflationary adjustment at a higher price level and market rate of interest.

The development of unemployment and reduced output does not itself alter the basic analysis regarding the ability of firms to finance interest on capital. Whatever the degree of employment or unemployment, this ability depends on the location of the marginal efficiency schedule, which the ex ante I function will tend to follow. This is equivalent to saying that firms will not commit themselves to the payment of interest on new (and old) capital above what the return on investment promises to be.[23] And this has nothing to do, per se, with the degree of unemployment or the constancy of output. Conceivably, even though resources are idled, the movement of output and prices in the two sectors might be consistent with a movement along the predisturbance MEI schedule and with equality between ex ante and ex post investment,

[22] The inflationary disturbance was assumed in n.17 to have no effect on interest or real saving and investment. Hence these variables are influenced only by the monetary contraction. Given more or less stability of the general price level, the relevant aspects of the sectoral price changes described in the Appendix, Sections D and E, are thus the relative, rather than the absolute movements.

[23] We have been assuming that firms know what the marginal efficiency of various investment levels is, and that the ex ante I schedule invariably coincides with the MEI function. In Section IV, where the schedules undergo a once-for-all shift, and monetary policy contributes an element of stability to the adjustment process, this is not an unreasonable assumption. However, in the present context, where MEI may shift constantly, entrepreneurial error as to the return on investment is a more serious possibility. (Keynes himself stressed the underlying role of expectations in the marginal efficiency concept; cf. *The General Theory, op. cit.*, pp. 138-144.) Thus, if firms in the aggregate overestimate the return, offering to pay more interest than the de facto marginal efficiency on the investment justifies, I is above MEI. Costs, including interest, will exceed revenues, and output will be reduced, provoking unemployment and nonmonetary inflation.

as previously described. But unemployment ought certainly to increase the range of possibilities with respect to the shift of MEI. To the cases already considered, we might wish to add the possibility that the movement of variables will drive MEI and I *below* their predisturbance location.

VI. Summary and Conclusion

Sections II-IV advanced a model of money and capital in which (i) policy increases in market interest rates are merely a response to prior independent increases in the natural rate of interest, and (ii) higher interest cost—or, more generally, higher rent on the existing capital stock—is financed by an increase in the net productivity of capital. If tight money (defined as simultaneous rising interest rates and inflationary pressures) is due to a decrease of saving, the greater capital productivity is provided by an allowable reduction in depreciation charges. All firms share in this productivity rise, which finances the higher interest costs. Restrictive monetary policy, aimed at raising interest rates to the natural level, tends to stabilize the price level, unless the movement of resources from investment to consumption is delayed by price stability. For then unemployment and a reduction of aggregate output might furnish a nonmonetary source of inflation.

If tight money is caused by an independent increase in investment, the gross product of capital and the level of output rise directly. Interest-equalizing monetary policy is thus consistent with a decline in prices. However, the more bountiful capital stock is not likely to be distributed evenly throughout the economy. Resource reallocation within the given industrial structure (and also, possibly, between consumption and investment industries) is required. Unemployment may again result, reducing output and raising prices. Since capital productivity has increased, output may simply rise, and prices fall, less than otherwise. But just as in the decrease of saving, the monetary authority in this case is not responsible for the necessary movements of the interest rate or of resources. These would occur in any event. Only if, in curtailing the monetary inflation, the authority removes what happens to be a stimulus to resource mobility, is it guilty in any degree of causing unemployment and nonmonetary inflation.

In Section V tight money is caused by monetary contraction designed to offset inflationary disturbances which do not involve a movement in the market or natural rate of interest. The monetary authority is thus solely responsible for the higher interest rate. However, the rise in interest is only temporary, since the forces which counteract the rise in prices tend at the same time to return the interest rate to its original level. During the deflationary—or better, anti-inflationary—adjustment, ex ante saving exceeds ex ante investment. Ex post investment is thereby indeterminate and may take on a wide range of values. It is shown that any particular level of ex-post investment, entailing the purchase of new capital goods, can only be effected by raising the return on investment to the level of the higher interest rate. This is accomplished by appropriate shifts in the marginal cost and supply price of both investment and consumption output. The interest costs are thus always covered by productivity changes. But the adjustment very likely will require some reallocation of resources; imperfect mobility will again cause unemployment, reduced output, and a nonmonetary source of inflation.

The tendency of tight money (defined most generally as a tendency to rising income and rising, or higher average, interest rates), in the face of monetary restraint, to raise the price level thus depends on the ease with which associated resource movements are accomplished. Only in one case can the necessary reallocation be said in some sense to result from an increase in interest "costs." That is the case of an independent increase in the investment demand schedule. The increased costs are those of a limited number of firms whose share in the greater productivity of capital is below average, and who will, accordingly, lose resources. In all other cases of tight money the resource movements, while induced by the interest rise, are not directly attributable to an increase in interest costs as such, but rather to a decline of revenues in affected industries.

The claim that higher interest rates raise prices by raising costs is thus, within our analytical framework, without much substance. Whether the natural rate is constant or increasing, greater interest payments tend to be financed by greater net productivity of real capital. The single case in which higher interest rates raise costs—that of an increase in investment demand—is one in which the overall cost structure and price level are falling because of a more productive capital stock. The real thrust of the interest rate in this case is thus only to limit a deflationary movement, not literally to create inflation.

13

Monetary policy and its critics

James R. Schlesinger

In assessing the import of the criticisms of monetary policy which seemingly have flowed in an unending stream during the last three decades, it is wise to recognize at the outset that some people simply do not *like* monetary policy. This antipathy arises from nonlogical policy judgments concerning the *suitability* of monetary restraints, but it is reflected in a continuously evolving set of charges concerning the *operation* of monetary policy. Whenever one argument appears defective, these critics readily turn to another—one which may or may not be consistent with what was previously espoused. It is this underlying emotional response which helps to explain the quality of changeableness that has characterized the debate over monetary policy.

Much of the antagonism to monetary policy stems from its reliance on the price mechanism. A rationing process that operates through the market seems inhuman to many people, and the results are regarded as unfair or harmful. In addition, the impact of monetary policy on spending decisions is so subtle that many doubt that it is there at all. Influencing total spending by bringing about changes in the value, the volume, or the composition of the financial assets of the community strikes some observers as a mechanism too weak or too indirect to be relied upon. Thus there are doubts about the effectiveness of monetary policy as well as about its appropriateness, and, of course, legitimate doubts along these lines inevitably are seized upon by self-interested groups to attain monetary conditions more satisfactory from their point of view. Still, the genuine misgivings do raise certain issues of public policy which ought to be considered explicitly.

Reprinted from *The Journal of Political Economy*, Vol. LXVIII (December, 1960), pp. 601-616 by permission of The University of Chicago Press. James R. Schlesinger was at the University of Virginia; he is now with the RAND Corporation.

A substantial portion of the recent debate has been concerned not with the monetary mechanism itself but with a particular monetary policy. For example, it has been argued that the level of demand consistent with a stable price level is somewhat lower than that necessary to achieve full employment. Much of the present criticism of Federal Reserve policies consists of assertions that the System has chosen the wrong monetary goal—preventing cost inflation—and that its attempt to influence aggregate supply conditions through its control over the money stock is foolhardy. Such criticisms do not imply that the critic necessarily distrusts or disapproves of the use of monetary controls; they are simply disagreements over details of monetary policy. To cite one prominent example, Sumner Slichter was a vigorous critic of Federal Reserve policies, yet nowhere in his writings is there any indication that he felt any doubt that some degree of monetary restraint is necessary for the proper functioning of the economy. Other observers, especially quantity theorists, have argued that the Federal Reserve is at fault for permitting rises in the price level. By the very nature of their position, such critics cannot be taken to believe that monetary controls are either unnecessary or inappropriate.

In this paper we shall not be concerned with such surface disputes over goals but with the much more fundamental criticisms of those who argue that monetary policy is, for one reason or another, *inherently defective.* Put into three general categories, these charges maintain that monetary policy is (1) ineffective, (2) discriminatory, and (3) contrary to sound social policy. Each will be considered in turn.

I. The Question of Effectiveness

Although it is almost a truism that to question the effectiveness of monetary policy is potentially the most devastating of the criticisms, the neoclassical economists never appeared to entertain such doubts. From Wicksell to Keynes (of the *Treatise*) it was generally believed that monetary policy, by lowering interest rates and, concurrently, the supply prices of capital goods, could induce investment demand sufficient to maintain the constancy of the price level or, to stress a more modern consideration, sufficient to achieve the utilization of all factors of production. Conversely, it was believed that a rise in interest rates would serve to deter enough marginal borrowers to contain total spending within the limits of total supply at the prevailing price level. The investment demand scheduled was assumed to be sufficiently elastic so that correct monetary policy, in the long run if not in the business cycle, would insure thc absorption of full-employment savings. Yet this Age of Faith was soon to be followed by an Age of Despair.

A. The 1930's

The deep and seemingly unshakable depression of the thirties simultaneously dragged monetary policy down from the position of honor that it had occupied and raised doubts as to whether it had any influence on spending at all. Skeptics questioned whether so minor an item in the total cost picture as a small rise in the interest rate was sufficient to alter the spending decisions of borrowers. For short-lived investments such as those in machinery and equipment, the period of investment was too brief for changes in the interest rate to have any substantial influence on costs. (Such considerations were reinforced by corporate rules of thumb which required every piece of machinery to pay for itself in some arbitrary time period.) On the other hand, it was argued that, for long-lived investments, where it is obvious that even a small change in the rate of interest will have a substantial impact on costs, the risk allowance was so large that variations in the cost of borrowing would be swallowed up by the allowance for risk. Consequently, investment demand could be considered insensitive to interest-rate variation. Of course, the abler critics, such as the late Sir Hubert Henderson, who was the guiding figure in the iconoclastic Oxford studies of the price mechanism, did recognize that there were certain long-lived, relatively riskless investments—in housing, public utilities, and public investments—which were extremely sensitive to interest-rate changes. They argued, however, that population growth in most Western nations had either slowed down or ceased entirely and that it was toward population growth that the interest-sensitive categories of long-lived investment were oriented.[1] The fact that this area of investment activity had shrunk further reduced the impact of interest-rate changes.

Empirical studies during the period tended to confirm this skeptical appraisal of monetary policy. Investigators,[2] using either questionnaires or case studies, reached the conclusion that perhaps half the firms studied paid no attention whatsoever to interest rates and that only a small minority considered them to be significant. Of course, it is necessary to make allowance for the period in which these studies were made; nevertheless, the number of relevant issues which the investigators *failed* to consider is noteworthy:

[1] H. D. Henderson, "The Significance of the Rate of Interest," *Oxford Economic Papers*, No. 1 (January, 1938), reprinted in *Oxford Studies in the Price Mechanism*, ed. T. Wilson and P. W. S. Andrews (New York: Oxford University Press, 1951), pp. 20-22.

[2] See the articles by J. E. Meade and P. W. S. Andrews reprinted in *Oxford Studies in the Price Mechanism*, pp. 27-30, 51-66; also J. F. Ebersole, "The Influence of Interest Rates upon Entrepreneurial Decisions in Business—a Case Study," *Harvard Business Review*, XVII, No. 1 (Autumn, 1938), 35-39.

1. Since the bulk of investment activity is concentrated in the largest firms, is not the percentage-of-firms criterion misleading? Were the minority of firms that borrowed heavily and invested heavily the ones that were sensitive to interest-rate changes, as seems likely?

2. Does not the responsiveness of firms to monetary policy vary with the time and the economic climate? Consequently, will conclusions drawn in a depressed period characterized by excess capacity be applicable under other conditions—particularly periods of expansion?

3. If interest rates do not control the investment decision, do they influence the *timing* of the expenditures which follow from that decision?

4. Do interest rates affect corporate dividend policy (that is, corporate savings), thus providing non-credit sources of expenditures?

5. Are expenditures influenced by credit conditions other than interest rates or by the general tone of the money market of which the interest rate is simply a symptom?

6. Are businessmen actually able to appraise the determinants of their own decisions; is it not likely that they are constitutionally far more alert to positive inducements such as sales than to (negative) inhibitors like interest rates?

7. Finally—and perhaps most important of all—do not investigations of this sort, which make inquiries of individual businessmen and then argue from the specific to the general, ignore the *indirect* influence of interest rates on spending decisions? Cannot an all-round process of expansion be generated from slender beginnings? If even a few businesses are induced by cheaper credit to expand outlays, may not other concerns also be persuaded to increase expenditures as their sales rise?

To raise questions of this sort is to underscore the conceptual defects of the empirical investigations. Nevertheless, at the time, these studies did tend to confirm the new analytical presupposition that investment decisions were insensitive to changes in the interest rate. Moreover, difficulties posed by the inelasticity of the investment-demand schedule were compounded by the Keynesian view of liquidity preference, which hinted that monetary policy had little effect on interest rates anyway. Even with limited demand for investment funds, the long-term interest rate, it was believed, would not fall below some positive level, say 2 per cent, because the threat of capital loss at lower interest rates was so great that the public would absorb in cash balances more and more money without bidding up bond prices or lowering interest rates. Buttressed by such conceptions, dominant opinion in the late thirties in the government and in academic circles held that monetary policy was ineffective.

Nevertheless, it remains a distinct possibility that the sensitivity of investment to interest-rate changes and the strength of liquidity preference may vary substantially with changes in the over-all economic climate. If we recall the influences bearing on the effectiveness of monetary policy, it seems plausible to argue that the observers of the thirties erred in generalizing from conditions prevailing in the deepest depression ever experienced. When national income has at one point fallen by almost 50 per cent, when many industries are operating at 20-30 per cent of capacity, when new investment is deterred by excess capacity, when no new markets are foreseen and business confidence has ebbed, low interest rates are unlikely to have much stimulative effect, no matter how low they fall. Such conditions may well be described by an inelastic investment-demand schedule.

In addition, consider the strong desire for liquidity then prevailing, the willingness of the public and the banks to hoard rather than to commit funds (the theory of credit expansion precludes excess reserves). In the period after 1929 there was a run to liquidity. By 1933 the banking system had reached a state of collapse. Many banks were forced to shut their doors because of illiquidity at the same time that the financial community was still being blamed for the speculative excesses of the twenties. Is it surprising that both the public and the banking community exhibited under these conditions a strong liquidity preference, reflected in excess reserves and in the astonishing gap between long-term and short-term interest rates?

Moreover, the economic difficulties were reinforced by political conditions that were hardly conducive to business confidence. Monetary policy attempts to influence business decisions at the margin; yet such considerations become insignificant when the social system appears to be in chaos. Labor conditions were unsettled. Businessmen, who were widely used as scapegoats, were apprehensive. Neither the path to an effective monetary policy nor the path to recovery lies in the direction of alarming those who make investment decisions. Plainly, it would be unwise to regard monetary policy as *generally ineffective* on the basis either of the analysis or of the conditions of the thirties.

B. The 1940's

Financial developments of the forties provided an institutional rationalization for the skepticism concerning monetary policy, yet at the same time gave rise to the inflationary pressures which eventually were to strike the wartime chains from monetary policy. At the end of World War II, however, the heritage of control, associated with fear of the consequences of the use of the traditional weapons provoked by the enormously expanded public debt, reinforced the antimonetary attitudes of the thirties.[3] Rising interest rates, it was argued, would not inhibit

[3] Cf. Lawrence H. Seltzer, "Is a Rise in Interest Rates Desirable or Inevitable?" *American Economic Review,* Vol. XXXV, No. 5 (December, 1945). One interesting aspect of Seltzer's article is that it contains an early expression of the belief that monetary policy remains utterly useless up to the point that it becomes potentially

spending, yet would increase the cost of debt service to the taxpayer—failure to reckon the costs of inflation, of course, tended to lead to undue stress on the cost of debt service. Perhaps more important was the belief that the bulk of the debt was infirmly held and that rising interest rates would provoke a panic, in the course of which a substantial part of the debt, perhaps including even savings bonds, would be jettisoned. If, eventually, the Federal Reserve were forced to intervene to pick up the pieces, why not prevent such a cataclysm by an initial policy of support to the government securities market? Periodic Treasury refunding operations which would require Federal Reserve support were held to reinforce such considerations. These arguments were so widely accepted that for a time even the Federal Reserve System readily acquiesced in its own Babylonian captivity.[4]

In view of the excess liquidity which characterized the postwar period, it is certainly arguable that monetary controls would have had little immediate effect, even though, as a general proposition, monetary policy is more effective in coping with inflationary pressures than with depression. There may always be an interest high enough or monetary pressure severe enough to check investment demand, but, in conditions already characterized by monetary redundancy, it may not be practicable to bring this about. When anything that is bought can be sold, when inflation will justify any investment, when markets appear overwhelmingly promising, money expenditures will rise as the circulation of money increases, even though the supply be held constant. Some inflation is inevitable as a phase of the process of reducing excess liquidity. Yet, even if monetary restraint could not have dissipated the inflation potential in 1945-46, there is clearly no long-run case against monetary control. Moreover, events have demonstrated that fears of the collapse of the government securities market, understandable as they may have been in light of our lack of experience in handling so large a debt, have been excessive. The market for governments is normally stable in the sense that when some holders wish to sell securities, purchasers other than the central bank stand ready to buy. There is no *cumulative unloading* of securities but rather a *transfer* of securities among holders without the intervention of the Federal Reserve system. Nevertheless, the "loose-cargo" argument did cling curiously to life. As the years passed, the Federal Reserve

---

disastrous and that there are no intermediate effects. This view has been modified by more recent writers, but (for other reasons) it is still with us. Today it is argued that sizable interest-rate changes may precipitate a depression, but that, up to the point that it becomes dangerous, monetary policy is ineffective. Based on the presupposition that even a small rise in the interest rate would provoke a panic in the government securities market, Seltzer's formulation was far more coherent than the more recent one.

[4] See the 32d, 33d, 34th, 35th, and 36th *Annual Report of the Board of Governors of the Federal Reserve System* (1945-49).

System became increasingly restive under conditions such that its open-market operations accentuated, first, the inflation of 1946-48, then the recession of 1949, and then the renewed inflation of 1950-51. Yet down to the Accord of 1951 and later, a substantial body of academic opinion regarded monetary restraint as unnecessary, monetary policy as ineffective, and general credit controls as obsolete.[5]

C. The 1950's

With the revival of monetary policy, far greater stress was placed upon the *availability* of credit as opposed to its *cost* than would have been deemed appropriate in neoclassical thought. But interest rates move sluggishly. Debt instruments are imperfect substitutes for each other. Save in the open market, lenders are subject to a sense of restraint. Consequently, it seems clear that it is more than the price of credit itself that limits borrowing. The willingness of lenders to lend is an important consideration; lenders may prefer to curtail requests for credit without increasing rates.[6] Borrowers themselves may become reluctant to borrow when they do not feel assured about their long-run liquidity position, even if they are undeterred by the cost of borrowing per se. It is the availability of credit that is most important, and rising interest rates may merely be symptomatic of the several forces at work during periods of monetary stringency.

In response to the renewed emphasis upon monetary policy, a third type of criticism has developed, drawing on the older arguments, yet transposing or inverting the elements contained therein.[7] Basically, it is

[5] Consider the comments of the various contributors to the "Symposium on Monetary Policy," *Review of Economics and Statistics*, Vol. XXXIII, No. 3 (August, 1951), at the time of the revival of monetary policy in this country and also those of the contributors to "Monetary Policy: A Symposium," *Bulletin of the Oxford University Institute of Statistics* Vol. XIV, Nos. 4, 5, and 8 (April, May, and August, 1952), when British monetary policy was revived.

[6] The new emphasis on the lender as opposed to the borrower was stressed by Robert V. Roosa as a part of what came to be called "the availability doctrine" (see "Interest Rates and the Central Bank," in *Money, Trade, and Economic Growth: In Honor of John Henry Williams* [New York: Macmillan Co., 1951]; also I. O. Scott, Jr., "The Availability Doctrine: Development and Implications," *Canadian Journal of Economics and Political Science*, Vol. XXIII, No. 4 [November, 1957]). One aspect of the doctrine upon which Roosa laid some stress was the belief that a small rise in the rate of interest might bring about a curtailment of lending on the part of conservative financial institutions. Such a rise might generate caution in disposing of liquid assets like bills and at the same time lock these institutions into their portfolios of long-term governments on account of the reluctance to take capital losses. From the vantage point of the late fifties, it appears that this restraining influence was a phase of the transition from the kept markets of the forties to the free markets of the fifties. In recent years, financial institutions have not been at all reluctant to dispose of government securities in the face of rising interest rates.

[7] The most vigorous exponent is W. L. Smith. See his "On the Effectiveness of Monetary Policy," *American Economic Review*, Vol. XLVI, No. 4 (September, 1956);

contended that the nation's financial machinery, through the lubricating medium of the government debt, can effectively and automatically mobilize idle balances to maintain monetary expenditures whenever pressure is applied. Rising interest rates, which are a consequence of monetary restraint, supply the incentive and the mechanism through which such idle balances are mobilized. During boom periods, banks are subjected to pressure to expand business loans. As they attempt to sell bonds interest rates rise, and this increase induces those who held idle or excessive balances at lower interest rates to purchase securities. The sale of securities by the banks frees reserves and permits the expansion of business loans. This process of replacing investments by business loans is considered to be inflationary, even though the liabilities side of the banks' balance sheets is left unaffected. The exchange of assets permits the activation of the money supply. As velocity increases, so do monetary expenditures, even though the money stock is held constant. Thus "mere" control of the money supply will not seriously limit expenditures in the short run; monetary policy is ineffective.

It is interesting to compare the ingredients of the current critique with those of the older arguments. First, it is believed that rising interest rates in themselves have little deterrent effect on expenditures—this is, of course, a necessary element in any questioning of the effectiveness of monetary policy. Second, Keynes's notion of liquidity preference has, more or less, been turned on its head. Initially designed to demonstrate that increases in the money stock would not serve to lower interest rates in depression and therefore were an ineffective stimulus, it is now used to demonstrate that rising interest rates are the means through which monetary hoards are mobilized and consequently that control over the monetary stock in inflation is an ineffective restraint. Third, the government debt is seen not as an incubus making interest-rate variation risky but as the lubricating element in the financial structure. It is a point of historical irony that some of those who support the new criticism previously held the "loose-cargo" view of the government debt. Nevertheless, it would be folly—despite some rather abrupt changes in the positions of the critics—not to recognize that the new indictment is the most profound criticism of monetary policy yet devised, not at all dependent on the peculiarities of deep depression or the vagaries of wartime finance. There

---

his "Monetary Policy and the Structure of Markets," in *The Relationship of Prices to Economic Stability and Growth: Compendium of Papers Submitted by Panelists Appearing before the Joint Economic Committee* (Washington, D.C., 1958), pp. 493-98; and his "Some Unsettled Issues in Monetary Policy," in *United States Monetary Policy* (Durham, N.C.: American Assembly, Duke University, 1959), pp. 14-30; see also W. W. Heller, "CED's Stabilizing Budget Policy after Ten Years," *American Economic Review*, XLVII, No. 3 (September, 1957), 646-49; and L. S. Ritter, "Income Velocity and Monetary Policy," *American Economic Review*, Vol. XLIX, No. 1 (March, 1959).

is, no doubt, some element of truth in the argument. Increasing velocity, by activating idle balances, does reduce the *immediate* effectiveness of monetary policy. But there is some limit to the increase of velocity, so that in the intermediate period,[8] at least, monetary restraint does imply the ability to limit money expenditures. The problem is how rapidly. If there is a *substantial* lag before monetary restraints take hold, then, by their nature, monetary controls may be a weak tool to *rely* upon in dealing with *short-run* fluctuations.

Even in the short run, one should recognize the limits of the argument. First, investment demand is probably more sensitive to interest rates than the critics admit, and the declining liquidity associated with the growing pressure for bank loans undoubtedly plays some role in deterring expenditures. Second, the possibility of *substantial* loan expansion through the sale of securities may have been in part a temporary manifestation associated with the high proportion of governments in commercial bank portfolios after the war. The higher the ratio of loans to total bank credit becomes, the more limited is the possibility of further expansion through exchange. Third, the effect of the long-term decline in the securities ratio on the willingness of banks to dispose of investments may be reinforced by a debt-management policy which during recession (unlike the Treasury policy in 1954) prevents the excessive accumulation of highly liquid, short-term items which serve as the basis of loan expansion on the return of prosperity. Finally, it can be argued that increasing velocity is a part of the mechanism of restraint.[9] The Federal Reserve does not wish to close off spending from borrowed sums but merely to encourage reconsideration of spending decisions. Unless velocity is perfectly elastic in its response to monetary pressure, some restraint will occur. Although the argument based on loan expansion and variable velocity is the most reasonable of the criticisms of the effectiveness of monetary policy, it should be emphasized that the argument does not imply any doubt concerning the necessity for monetary control; it asserts only that monetary policy should not be *exclusively* relied upon for dealing

[8] In the short run, velocity might rise because of the activation of idle balances. In the long run, in principle at least, velocity might rise because of an adjustment in the community's methods of completing financial transactions (frequency of receipts and expenditures, etc.); this is particularly pertinent in the age of the credit card. It is in an intermediate period that monetary controls can take hold, that is, after idle balances have been exhausted but before the habits of the community have time to change.

[9] Since velocity is regarded as a constant, increases in velocity could not be expected to be *part* of the mechanism of restraint in a rudimentary quantity theory such as the one based on the Fisherine equation. Milton Friedman has argued, however, that in a more sophisticated version of the quantity theory one would not expect velocity to be a constant but rather to vary with interest rates (*Studies in the Quantity Theory of Money* [Chicago: University of Chicago Press, 1956], pp. 12-13). Without considering one's self a quantity theorist, one can surely accept this position with regard to the effect of restraint on velocity.

with short-term fluctuations. This position is perfectly consistent with advocating monetary restraint in boom times for the purpose of alarming potential spenders about their future liquidity positions. Monetary policy, although it should be employed, may not be *sufficiently* effective by itself; therefore, it should be strengthened and supplemented by other devices.

II. The Question of Discrimination

The broad charge of discrimination implies that monetary restraint is unfair because of its disproportionate impact on certain categories of borrowers who either lose access to funds or else become subject to *exceptionally* onerous terms of borrowing.[10] The ordinary indictment implies that monetary restraint affects the *allocation of resources* in a way that drastically and inappropriately affects the interests of certain categories of borrowers and at the same time is potentially damaging to the national economy. Plainly, this charge is wholly inconsistent with the preceding argument, for if monetary policy can affect the allocation of resources and the volume of expenditures, it cannot at the same time be *ineffective*. Yet resource allocation is inextricably meshed with the *distribution of income*, so that sometimes the charge becomes the assertion that monetary restraint unfairly alters the income distribution. This is a traditional political refrain among legislators whose constituents include large number of farmers or small businessmen—the charge having overtones of "the people" versus "the interests." It is essentially a protest against higher interest rates which ignores both the change in the demand for borrowed funds and the possible costs of inflation. To be valid, it has to be assumed that those who need borrowed funds are somehow more deserving than those who supply them. Some of the current cries of discrimination do involve such a notion—consumer credit is almost a pure case in point—yet, for the most part, charges of discrimination are concerned with the effect on resource allocation.

It must be recognized at the outset that monetary policy cannot fail to affect different citizens differently. Since it affects the availability of credit, the impact of monetary policy must be "disproportionate" in that it is asymmetrical in its consequences. Such inherent asymmetry may be traced to two causes: (1) Some institutions are more dependent on bor-

[10] The most prominent proponent of this position is Leon H. Keyserling, see, *inter alia,* his statement, *January 1957 Economic Report of the President: Hearings before the Joint Economic Committee* (Washington: Government Printing Office, 1957). It has also been indorsed by J. K. Galbraith and S. E. Harris, see Galbraith's statement, *January 1958 Economic Report of the President: Hearings before the Joint Economic Committee* (Washington: Government Printing Office, 1958), and the joint communication, "The Failure of Monetary Policy," which is included. This position has also been reiterated perennially by countless representatives of affected groups.

rowed funds than are others; of those dependent on borrowed funds, some are especially dependent on bank credit, whereas others have access to other sources—that is, security markets. (2) Within the camp of borrowers, the strength of the demand will vary among the several groups. Those whose demand is not so inelastic will be unwilling to pay as high a price as will others, and consequently, as interest rates rise, their share of the funds will fall. Now surely it would be trivial if those who charge discrimination had these kinds of disparities in mind. It is hardy logical to charge discrimination simply because the cost of borrowing has risen, even though those who are forced to pay higher rates are likely to be resentful. Nor would it appear logical to charge discrimination because those who use borrowed funds more than others are more damaged by increased competition (that is, demand) for such funds or a reduction in supply. (Is it contended that the available funds be allotted by some sort of parity system based on historical norms?) Nor would it appear logical to charge discrimination when those whose demand is less intense receive a smaller share of the available funds.[11] (Would it then be fairer or less discriminatory for those whose demand is less intense to share equally with those whose demand is more intense?)

Plainly, those who argue that monetary policy is discriminatory must have something more than these banalities in mind. After all, in a market *not characterized by discrimination,* one may anticipate that, with the tightening of credit, interest rates will rise most and borrowings fall least in money submarkets in which demand is most inelastic and that interest rates will rise least but borrowings will fall most in money submarkets in which demand is most elastic.[12] Any results other than these would constitute a prima facie case of discrimination. Critics of monetary policy, however, must have in mind situations in which the market does not behave in this normal way—that is, situations in which some borrowers are faced with a more-than-to-be-anticipated increase in rates or decrease in funds. Complaints of this type are put forward by, or on behalf of, five categories of borrowers—homebuilders, municipalities and state governments, small businesses, consumer borrowers, and affected industries. Each type of complaint will be examined in turn.[13]

[11] If this were not so, the charge of discrimination could be raised in behalf of any and all borrowers. When money becomes tight, some borrowers will pay substantially higher rates; all others will discover substantial reductions in their volume of borrowings. Since all pay higher rates or suffer from a reduction in funds, all may charge discrimination, according to the above logic.

[12] Those categories of borrowers who are unwilling to pay higher rates will probably not have their borrowings fall to zero because of the desire for the diversification of holdings on the part of lenders.

[13] In the discussion which follows I am very much indebted to the investigations of Harmon H. Haymes ("An Investigation of the Alleged Discriminatory Impact of General Credit Controls in Periods of Monetary Stringency" [unpublished Ph.D. dissertation, University of Virginia, 1959]).

A. Homebuilding

In periods of tight money, new housing starts sometimes show an extraordinarily rapid decline. For example, from 1955 to 1957—years of intensifying boom and rising interest rates—housing starts actually dropped from 1.3 million to 1.0 million per year, although housing starts had risen by some 20 per cent in the preceding period of recession. Such changes may in large degree reflect the elasticity of demand for housing credit—housing is a long-lived asset, so that a small rise in the interest rate means a sharp rise in the supply price. Nevertheless, other factors may help to account for the decline—in particular, the entire fall may be attributed to the decrease in Veterans Administration and Federal Housing Authority mortgages. While government-underwritten new housing starts fell by 400 thousand units, those with conventional mortgages rose moderately by some 70 thousand.[14] Congress has set a maximum rate on mortgages, beyond which the Veterans Administration and Federal Housing Authority cannot insure. Is the sharp drop in government-underwritten mortgages a type of discrimination attributable to monetary policy? When credit grows tight, what other result could be expected than that those who are restrained by legal restrictions from bidding emphatically will be eliminated from the market? Clearly, it is the law that discriminates, implying that home-builders should not be encouraged in an unwise inclination to sign a mortgage bearing (at that time) more than 4¾ per cent interest. In any event, the charge that it is monetary restraint which is discriminatory should be dropped.

B. Local government borrowing

During the boom, 1955-57, complaints were heard from both municipalities and state governments that they were unable to raise the funds that they needed. Despite rising demand for funds, new-security issues of state and municipal governments fell from $7.0 billion in 1954 to $5.4 billion in 1956. Yet, during the mid-fifties, interest rates rose pro-

[14] The *Federal Reserve Bulletin* provides the following data on new housing starts:

	New Housing Starts (Thousands)	Mortgages Gov't-Underwritten	Conventional
1955	1,329	670	659
1957	1,042	313	729

Since the value of construction as opposed to the number of starts fell relatively little, it may be argued that the effect of the law was to encourage the building of fewer but bigger houses.

portionately more rapidly for municipal bonds than for corporate bonds or for United States government bonds. To put the issue another way, the premium which municipals had previously enjoyed over competing bonds disappeared during the same period. Is this not a case of discrimination, with both volume falling and interest rates rising sharply? The fall in volume is largely explicable in terms of the attitudes of local governments. Many municipalities have ordinances prohibiting the paying of more than, say, 4 per cent on debt issues; when market rates rise, they are unable to borrow and are eliminated from the market. The same results follow in other municipalities because of the unwillingness of officials to pay more than some stipulated rate. Apparently, the citizens in the localities have examined the intensity of their demand for local improvements and their willingness to pay taxes and have correspondingly limited their bids for funds. Consequently, during periods of tight money, the volume of borrowing falls. The reluctance to pay higher taxes may be shortsighted, to be sure, but the fall in volume cannot be charged to the discriminatory impact of monetary policy.[15]

The proportionally more rapid rise of interest rates on municipal bonds may be attributed to another structural characteristic of the market —the response to the tax-exemption privilege. As long as demand is small enough to be satisfied by the supply of funds from those who benefit sufficiently to pay a premium for tax-exempt returns, interest rates will remain low. When, however, the demand rises (as it has in the fifties) or the supply of funds from those willing to pay a premium drops, interest rates will rise suddenly to a competitive level, since the municipalities will have to pay enough at the margin to fulfil their requirements. Such results can hardly be charged to the discriminatory impact of monetary policy but rather to the evaporation of a tax feature designed to aid (discriminate in favor of) local governments.[16]

C. Small business

During periods of credit restraint the accusation is invariably made that tight money discriminates against small business vis-à-vis large business. Of all the charges of discrimination, it is here that the evidence seems best to substantiate the charge. "True" discrimination—a failure of ordinary market forces—occurs when certain categories of borrowers

[15] The niggardliness of local governments has come in for considerable criticism lately. The underlying notion that the citizen-consumer is foolish or misguided, although difficult to express publicly, is really somewhat different from the charge that there has been discrimination.

[16] This does not eliminate the equity problem posed by the fact that certain well-to-do people have been obtaining high tax-free returns while the municipalities have not been obtaining any compensating advantage in the terms of borrowing. The justification that municipalities could get cheaper financing without dispensing much in the way of ill-gotten gains seems to have worn thin in recent years.

lose access to credit, although they are willing to pay competitive rates. Large business concerns usually maintain substantial lines of credit which may normally be unused. In boom periods they will draw upon these lines of credit; so that banks may be forced to reject loan requests by small businesses willing to pay market rates or higher. In the main this appears to be a problem of small businesses located in larger communities where banks are likely to have "big business" customers. Of course, during boom periods small businesses can and do turn to other sources of credit—open-book accounts, factors, etc. Such substitutes are inferior to bank credit and frequently far more costly. Much bank credit is siphoned to small business through the accounts receivable of large business borrowers. Small businesses do get by, but this does not solve the problem posed by the fact that our financial system discourages small business, while it is public policy to encourage it. Action by the Small Business Administration and the recently formed investment companies under the Small Business Investment Act seems desirable in counteracting such tendencies.

D. Consumer credit

Sometimes it is argued that tight money discriminates against users of consumer credit. In fact, consumer credit rises rapidly during periods of boom (tight money) and tends to contract mildly in periods of recession, so that the charge of discrimination must relate to its allegedly unfair impact on income distribution rather than resource allocation. A leading congressional critic is fond of rhetorically informing Federal Reserve officials that they are forcing the ordinary man to pay more for his house, for his car, and for any other purchase on time. With respect to consumer credit, at least, this charge seems to be inaccurate. Interest rates on consumer credit are normally very high and rigid, some of them pressing against state maximums; the demand for this credit is highly inelastic. Thus, during prosperity, rates rise little, if at all. Because of the relatively high rates, lenders, including banks, are always ready to satisfy any demand for consumer credit. In prosperity the rise of other rates reduces the relative attractiveness of consumer credit, but not enough to divert funds to other uses. In a market in which tight money barely affects the volume of funds lent and has only a slight effect on interest rates, the charge of discrimination would appear to be at its wildest.

E. Affected industries

As might easily be anticipated, industries especially sensitive to monetary restraint are likely to see themselves as victims of discrimination. This is especially true of the construction and railroad industries. The late Robert R. Young, for example, was one of industry's most persistent, if not most perceptive, critics of tight money, ready to explain to interested congressional committees how many more boxcars the New York Central could have bought, had interest rates not been so unwarrantedly high.[17] On occasion similar criticisms are heard from the electric-power, natural-gas-transmission, and automobile industries as—for other reasons—from the farming and small-business sectors of the economy. Most of such complaints use the word "discrimination" in the sense that a general control bears down more heavily on those sectors of the economy which are sensitive to it than on those sections which are not, rather than that there are discriminatory standards for the several industries. It should be understood that such complaints do bring to the surface grave policy issues. How fair is it, for example, to help stabilize the general economy by forcing a particular industry like housing through wider fluctuations, using monetary policy reinforced by Veterans Administration and Federal Housing Authority controls? But, for the most part, the policy issues raised by the affected industries verge upon the third general criticism of monetary policy—that its results are contrary to sound social policy.

III. The Question of Social Policy

Even if monetary policy is effective, even if it is non-discriminatory, it may be argued that its consequences are in conflict with either long-run welfare considerations or the national interest. In particular, restrictive monetary policy operates by cutting down on investment activity and thus militates against economic growth.[18] Monetary policy may be all too effective in reducing what may be referred to as "social investment" and for this reason may be less attractive than certain forms of fiscal policy. By its nature, monetary policy is particularly effective in deterring the long-lived, relatively riskless investment that can be considered es-

[17] In his statement (*Monetary Policy, 1955-56: Hearings before the Subcommittee on Economic Stabilization of the Joint Economic Committee* [Washington: Government Printing Office, 1957], p. 54), he comments: "We have slowed down the scheduling of our building of boxcars just because we cannot afford to pay 5½ per cent for money when the Interstate Commerce Commission gives us a 3-per cent return. It is just that simple. . . . If the figures were reversed, we would start building; if we paid 3 per cent for money and we were allowed to earn 5 per cent, we would cure the boxcar shortage overnight."

Young was, no doubt, correct in his last assessment. The problems of the regulated industries, in which investment tends to be long-lived and riskless, cannot simply be laid at the door of monetary policy, however.

[18] Keyserling and others have argued in these terms but have failed to recognize the implications of the argument, in that some alternative restraint on spending must be employed. Of those economists concerned with the growth issue, Arthur Smithies has been most forthright and consistent in urging not only fiscal restraints but the use of selective controls to restrict consumption (see his "Uses of Selective Credit Controls," *United States Monetary Policy* [New York: American Assembly, Columbia University, 1958], esp. pp. 73-81).

pecially conducive to progress (business investment in plant, electric-power production, transportation, including pipelines, and educational facilities) or conducive to social health (housing, municipal and other public services). Even if monetary policy is not discriminatory in the technical sense, it works by curtailing those expenditures which are in the public interest. The affected categories of borrowers (in effect, those charging discrimination) do not have to be protected against discrimination; they should nevertheless be encouraged in the long-run interests of the society.

Reliance on monetary policy, so the argument runs, means that national resources are wasted on additional consumption goods or on "silly" investment in neon lighting and amusement parks rather than being devoted to "worthwhile" purposes. This view of consumer expenditures as superfluous is plainly at the opposite pole from the one maintaining that the defect of monetary policy is that it discriminates against the poor man who must use consumer credit to lift his standard of living, since the chief defect alleged in this case is that credit resources are "wasted" on consumption goods and other fripperies. It may readily be understood how the appeal of an argument stating that our social values are wrong and that we are not sacrificing enough for economic growth has been reinforced by concern over the menacing posture of the Soviet Union.

It is difficult not to have some immediate sympathy for this position. Our scale of social priorities may indeed be askew. Many are affronted by the current "boom psychology" and its accompanying orgy of materialism. Surely we ought not to permit credit to be wasted on consumer goods when it might be used to create additional productive capacity. Still, if this orgy of materialism is so vicious, what is the purpose of additional economic progress? Why should consumers not buy automobiles, now, so that plants may be built to produce more automobiles in the future? Why should resources be diverted from washing machines to electric-power facilities at the present time, so that future generations of consumers have electric power for their washing machines? Viewed in this way, the criterion of more investment for more rapid growth becomes somewhat less compelling. Even in regard to social services, particularly those provided by municipalities, it seems perfectly apparent that the higher interest costs of new schools, new sewerage systems, and the like could be met without inducing a lower interest rate, if citizens saw a compelling social reason. The taxpayers have decided, rightly or wrongly, that they do not wish to bear the cost. Stern Galbraithian denunciations of "the unseemly economics of opulence" notwithstanding, the American people seem to like more, shiny, tasteless consumer goods. In monetary policy as elsewhere, responsibility for unsound social standards should be placed where it belongs and not attributed to the conventional wisdom of economists.[19]

[19] Economists as a group can no more be accused of discouraging the public from paying for public services than they can of urging the public to litter the national

Still, it may be that our scale of social priorities is distorted and that reliance on monetary policy tends to aggravate such perversions. Assuming that the American people could be converted to this way of thinking, what kinds of alternative policies seem appropriate? What remedies can be suggested by those who argue that restrictive monetary policy is inconsistent with a desirable level of social investment? Disregarding those who explicitly or implicitly argue that there are no limits to the nation's resources and all that need be done to increase production is to increase demand, it seems plain that if restrictive monetary policies are de-emphasized, some substitute method of restricting total expenditures must be employed. The likeliest choice is a more restrictive fiscal policy. If the protests against the social implications of restrictive monetary policy are to be anything more than the futile whine that it is unpleasant to have the nation's aspirations limited by its resources, those who make such criticisms must in all consistency demand more rigorous fiscal restraints, particularly those which bear down heavily on consumption. Failing in this, if the argument is to make any sense at all, it represents a plea for direct controls over investment activities (and other activities?), and, until now, the American people have given no indication that they would permit such powers to be exercised in peacetime. If one abandons general controls, there is no alternative save direct controls to rapid inflation. Selective controls, particularly on consumer spending, may ease the problem somewhat, but selective controls have in the past revealed administrative, political, and economic weaknesses that have inevitably led to their breakdown. If the argument that monetary restraint leads to undesirable social consequences is accepted, the nation must be assured that some alternative form of control will, in fact, be substituted for such restraint before it can be abandoned.

IV. Implications for Monetary Control

It should be clear that the various strands of criticism cannot be woven together to form a well-meshed case against monetary policy. Each line of argument is discrete, and sometimes inconsistent with other lines. Plainly, if monetary policy is ineffective, if it has no impact on total expenditures and resource-allocation, then it cannot be undesirable because it brings about an allocation of resources which is contrary to sound social policy or because it squeezes particular sectors of the economy in a discriminatory manner. Monetary policy can hardly be defective *both*

parks. (Despite the entreaties of billboards and broadcasts, the public seems to regard littering as a constitutional right, not to be compromised by the penalty of paying for the picking-up of the beer cans they have been unwilling to refrain from discarding. "The fault, dear Brutus, lies not in our stars but in ourselves that we are underlings.") Cf. J. K. Galbraith, *The Affluent Society* (New York: Houghton Mifflin Co., 1958), p. 253.

because it discriminates against consumer credit *and* because it permits credit that could be used "productively" to be diverted into frivolous consumption. Much of the criticism of monetary policy reflects the desire of the critic to substitute his own judgment for what he regards to be the defective results of the market process. Since individual judgments vary widely, clearly the critics are likely to disapprove as vehemently of each other's diagnoses and prescriptions as they are of monetary restraint itself.

Any single criticism may quite reasonably be defended. Most of the more perceptive critics do confine themselves to one line of attack—that is, ineffectiveness or discrimination. Others, with much less logic, attempt simultaneously to maintain several contradictory lines of criticism. Such attitudes can only be ascribed to the emotional, rather than the critical, faculties. Some of the inconsistencies may be attributed to the fact that the defects of monetary policy have varied with economic circumstances over the past three decades. Since the arguments have emerged fortuitously, they could not be expected to form a logical whole.[20] At one time monetary policy might be ineffective, at another it might be effective but discriminatory, etc. Nevertheless, the simultaneous employment of contradictory arguments can hardly be defended, and the rapidity with which new arguments emerged as the debate shifted over the years can be attributed to the antagonism felt by many professional and lay observers toward monetary policy.

[20] The views of Seymour E. Harris, who has frequently been a penetrating critic of monetary policy, may be one example. Over the years they have undergone various metamorphoses. In *Twenty Years of Federal Reserve Policy* (Cambridge, Mass.: Harvard University Press, 1933), he argued that monetary policy failed in the twenties because the Federal Reserve officials invariably were timid in the face of political pressure. To this theme he returned after a Keynesian interlude (*The New Economics* [New York: Alfred A. Knopf, 1948], pp. 50-51) in his comments in the "Symposium on Monetary Policy" (*op. cit.*, pp. 179-84, 198-200), arguing that it is the lack of courage on the part of the central bank rather than the weakness of its weapons that frustrates monetary policy. At various points in the argument he makes the following observations: "The problem [of monetary policy] has certainly not been one of impotency of weapons. . . . The Federal Reserve . . . is surely in a position to deny the economy the money without which a large inflation could not be carried on" (p. 183), and "monetary restraints are the easiest approach to inflation control—much less painful than more taxes or less public expenditures. . . . [T]heir atrophy is the result not of ignorance, but of the determination not to fight inflation which prevails in the country" (p. 180). In the *January 1959 Economic Report of the President: Hearings before the Joint Economic Committee* (Washington: Government Printing Office, 1959), he readily admits, though with mixed feelings, that this generation of Federal Reserve officials has not yielded to political pressure and has in his view been altogether too courageous in defending its convictions. All this is understandable. But in the communication with Galbraith ("The Failure of Monetary Policy," *loc cit.*), he argues that monetary policy simultaneously is ineffective, discriminatory, and dangerous. How the view that tight money is ineffective can be reconciled with his views of 1951 is not made clear, nor is it explained to the reader how monetary policy can be so ineffective in dealing with inflationary pressure, yet curtail demand sufficiently to bring on a depression.

Monetary policy is surviving the debate, albeit somewhat scarred in places. Plainly, the Arcadian view of monetary policy of the twenties has departed. Various institutional changes—the rise of liquidity, the declining importance of bank credit in the spending decisions of large corporations, the problems of debt management, and the removal of certain spending decisions from the market—have reduced somewhat the immediate effectiveness of monetary policy. But the major attacks on monetary policy have also been blunted. The technical charge of discrimination is, for the most part, fallacious, save in the case of small versus large businesses, and even here the importance should not be exaggerated. With regard to effectiveness, the extreme, depression-born doubts that monetary policy could have little influence on spending decisions have disappeared, along with the war-born refusal to use credit policy for any purpose other than maintenance of the interest-rate pattern on the government debt. What remains is a reasoned critique of monetary restraint, not so much a case *against* monetary control as a case *for* recognizing its limitations and its defects and for searching for alternative policies and tools. This critique consists of two parts:

1. In boom periods, monetary controls may "take hold" only after an operational lag of substantial duration. The integration of the financial community in association with the widespread holding of government securities permits rapid mobilization of idle balances during periods of pressure. A restructuring of the assets of the banking system may permit rising money expenditures through rising velocity, though the money supply is held constant. But this *does not imply that monetary control is unnecessary*. Control over the money stock is essential in the long run. Even in the short run, rising interest rates, rising velocity, and falling liquidity are all parts of the mechanism of restraint. Even if it operates slowly, monetary control operates in the right direction. At worst, all that this argument implies is that the nation should not rely upon monetary restraint as its sole instrument for combating short-term fluctuations. Monetary control cannot be dispensed with, but the search should continue for other instruments of general control.

2. Monetary restraint, as it becomes effective, may lead to results which we would not prefer on other grounds—national interests, economic progress, wefare considerations, etc. But this does not imply that monetary restraint in itself is undesirable, it merely hints that the results of alternative policies might be better. It imposes upon the critics the obligation of proposing and of obtaining public acceptance of alternative instruments of control. The use of alternative instruments may permit the alleviation of monetary restraint; it will never permit dispensing with monetary control.

From the standpoint of the more dramatic charges, the above critique is very modest indeed, representing a plea for de-emphasizing monetary policy, while accepting the necessity of monetary control. Needless to say, it would not be accepted by all economists, particularly those who feel that maximum welfare is invariably obtained through the market process. For the latter, monetary policy has additional advantages in that it reflects the current savings decisions by individuals in a way that fiscal restraints cannot. Yet, supposing the critique is accepted, it is clear even then that there is a residue of monetary policy which must be used. Until such time that fiscal restraints and other alternative controls have been perfected to the point that over-all demand can be precisely controlled, there will be minimal need for monetary policy at the fringes. To argue to the contrary is to imply that fiscal policy is more flexible, less crude, than we have experienced it to be in fact—a Beveridge Plan type of utopianism. In democracies, particularly those in which authority is divided, fiscal controls have proved to be incapable of achieving a delicate adjustment of demand. Since a free economy is prone to periodic surges of spending, inevitably, in the quest for stability, a minimal use of monetary policy cannot be avoided.

Much of the public criticism of monetary policy arises from restlessness under any form of restraint, a restlessness that reflects a natural and inevitable human distress at the fact that resources are limited. Economists should try to counteract such tendencies. In order to achieve maximum impact on public policy, it is necessary that economists occasionally coalesce on fundamentals. With the exception of a few ultramoderns,[21] virtually all economists do agree that monetary policy must be used to some extent in the attempt to stabilize the economy. True, under some conditions monetary policy may not be effective, particularly in the short run; in any given case, however, the only way to learn how effective it is, is to use it. In discussing the need for co-ordinating monetary policy and fiscal policy, economists have come to recognize that neither of these instruments is necessarily either immediately or precisely effective; that is the nature of instruments which seek to influence *voluntary* spending decisions on the part of the public. But both instruments have their roles to play; neither can be disregarded.

[21] It is somewhat ironical that the recent arguments that monetary policy is ineffective in the short run but potentially dangerous in the long run *because of the existence of lags* (as in the Galbraith-Harris communication cited in nn. 11 and 20) runs exactly parallel to criticisms directed against discretionary fiscal policy (see Milton Friedman, "A Monetary and Fiscal Framework for Economic Stability," *American Economic Review*, XXXVIII, No. 3 [June, 1948], esp. 254-58). Friedman has consistently adhered to the logically impeccable position that lags constitute an argument against all forms of discretionary authority. But many of those who currently criticize monetary policy on the basis of lags can hardly be described as skeptics about discretionary *fiscal* policy. To me, it appears true that lags do complicate the work of using either the fiscal or the monetary instrument and place limits on the effectiveness of both. This represents, however, an argument for greater flexibility rather than the abandonment of either instrument. That is tantamount to throwing out the baby with the bath water.

14

Can monetary policy influence the availability of credit?*

A. Dale Tussing†

Over the past decade and a half, it has become increasingly accepted that orthodox central bank policy exercises an influence over something called "the availability of credit." In the 1950's, this idea was actively promoted by the Federal Reserve, first to support their position that the post-war interest-rate "pegging" program should be discontinued and flexible monetary policy revived,[1] and then later to buttress the official Federal Reserve position of the middle and late 'fifties that interest rates were set by market forces and not by design of the Federal Reserve—monetary policy then being seen as primarily a matter of regulating the volume of commercial bank assets through controlling the amount of reserves owned by the banks.[2]

With the "revival of monetary policy" the "pegging" controversy is a thing of the past; and concern over interest rates (and a Federal Reserve

Reprinted with permission from *The Journal of Finance*, Vol. XXI, No. 1 (March, 1966), pp. 1-13.

* The author is indebted to Professor Melvin A. Eggers of Syracuse University, whose advice was invaluable through every stage of the preparation of the manuscript. In addition, the manuscript was read by Professors Omer Carey and Gary Sorenson of Washington State University, who offered useful comments. The *Journal's* referee also made helpful suggestions, both editorial and substantive, which improved the finished manuscript.

† Assistant Professor of Economics, Washington State University. [*Editors' note:* He is now Assistant Professor of Economics, Syracuse University.]

[*Editors' note:* Reference numbers are for the list at the end of this article.]

[1] The view that the "availability doctrine" of the 1950's was originally offered primarily as an argument to end the pegging program has appeared in many places, e.g., Modigliani [17, p. 97], Kareken [15, p. 322].

[2] This official view was associated with the practice which became known as the "bills only" policy. For a defense of the practice, see W. W. Riefler [19], and Young and Yager [33].

confession of control over them), has been forced by the worsening U.S. international payments situation. Thus the factors leading to official central bank advocacy of an "availability-of-credit" interpretation of the influence of monetary policy have all but disappeared. Moreover, the empirical foundations of Robert V. Roosa's "availability doctrine," on which much of the Federal Reserve's position had been based, have been effectively undermined.[3]

Nonetheless, the notion persists that monetary policy, besides (or instead of) influencing aggregate demand *via* the quantity of money and/or interest rates, does so through influencing the availability of credit. In a survey of contemporary monetary thinking, H. G. Johnson in 1962 reviewed the availability doctrine, noting the theoretical and empirical shortcomings which had been discovered in it during its brief life. "Nevertheless," Johnson added "the doctrine and discussion of it have helped to popularize the concept of 'availability of credit' as one of the main variables on which monetary policy operates" [12, p. 371].

Unfortunately, it has never been very clear what is meant by the availability of credit, especially as a variable subject to the influence of monetary policy, separate and distinguishable from the quantity of money and the interest rate, and nonetheless significant in its influence on aggregate effective demand. The virtually complete rejection of the availability doctrine has complicated, rather than simplified, the situation; we are left with a "popularized concept" (to use Johnson's words somewhat out of context) without either a definition or a theory.[4]

This paper has the modest objectives of setting forth a definition of the availability of credit and indicating, in general terms, what must be done to make it a variable "on which monetary policy operates." The definition must (1) be consistent with the most common doctrines offered in the name of availability, (2) show that the availability of credit is a variable distinguishable from the stock of money and the interest rate, and (3) relate the availability of credit to effective aggregate demand.

[3] See below, footnote 10.

[4] One element of the availability doctrine of the 1950's which has stood the test of time has been the concept of nonprice rationing of credit—an idea, however, neither original with nor rigorously developed in the doctrine. Recent theorizing points to the development of a theory of the bank as a firm, with particular attention to the portfolio selection and credit rationing problems See, e.g., Friemer and Gordon [4] and Hodgman [10]. Unlike the availability doctrine of the 1950's, the developing credit rationing theory does not concern itself, except in a minor way, with implications for the effectiveness of monetary policy. The main thrust of the availability doctrine was that there were credit effects of tight-money policies which supplemented or substituted for interest-rate and stock-of-money effects in limiting aggregate demand. It is these aspects of the doctrine on which this paper concentrates. Credit-rationing theory, which is not treated in this article, is concerned primarily with the question, "Should a rational profit-maximizing banker or even a rational utility-maximizing banker practice credit rationing, or should the amount he is willing to lend be an increasing function of the interest rate the borrower is willing to pay [4, p. 397]?"

Though this objective can be called "modest," inasmuch as it is clearly only a preliminary step in the formulation of a rigorous availability-of-credit theory, it is nonetheless an objective neither sought nor obtained by any of the economists, central bankers, and others who have contributed to the formulation or the criticism of the availability concept over the past fifteen years. This is especially true with respect to the third point, above: no one has bothered to indicate why, if control over the availability of credit is possible, it makes any difference.

I. Two Availability Approaches

The two approaches to monetary policy most commonly offered in the name of credit availability are (a) the availability doctrine and (b) the credit-creation doctrine. The former is associated with the names of Robert V. Roosa and others then at the Federal Reserve Bank of New York,[5] though in one sense it can be said that the availability doctrine has no real author at all. There is no internal or external evidence that those to whom the doctrine is ordinarily attributed intended to propose it as a new theory of credit control; and neither Roosa's article nor the subsequent "official" statements of the doctrine by the Board of Governors[6] was sufficiently integrated, precise, or operational that it would normally be accepted as a theory. The most (or only) clear-cut exposition of the doctrine was in professional economists' journal accounts of it ([11] [14] [22] [23] [24] [28]); but these economists cannot be credited with (or blamed for) proposing the doctrine, as their purpose was commonly to bury the doctrine, not to praise it. Moreover, and more oddly yet, the economists' journal accounts of the doctrine frequently contained elements found neither in the Roosa article nor in the official Board of Governors' statement.[7]

[5] At the time of the publication of Roosa's important article [21] he was vice president, Federal Reserve Bank of New York. The article appeared in a volume dedicated to John Henry Williams, also an F.R.B.N.Y. vice president; and Roosa gives Williams major credit, "largely through an oral medium, without benefit of a published written record" [21, p. 276] for development of the availability approach. In the same volume, Allan Sproul, president of the F.R.B.N.Y., offers strikingly similar views on monetary policy [26].

[6] The official Federal Reserve statement first appeared in Congressional testimony [31, pp. 368-383]. The statement was explicitly an official Federal Reserve statement on its theory of how monetary policy operates.

[7] For instance, Tobin [28] reviews what he calls the "new theory of monetary control" developed "under the leadership of Robert V. Roosa." His review of that new theory depends on the official Federal Reserve version; indeed, the article cited is a review of the published record of the Congressional hearings in which that version first appeared. Yet Tobin includes a theoretical point—substitution of government for private debt in institutional portfolios, motivated by rising yields on the former and rigid yields on the latter, and an accompanying pattern of credit rationing—not explicitly found in that version. In doing so, he does not cite any particular page or state-

The authorship of the credit-creation doctrine is only slightly less clouded than that of the availability doctrine. This second view is the commonly held one that central bank control over the assets of commercial banks is important because of the ability of banks to "create credit": only commercial banks, goes the doctrine, can be a source of expansional instability, because only they, of all lenders, can lend without there being a prior voluntary deposit by (presumably) a saver, an ability derived from commercial banks' money-creating ability. While many economists will agree that this doctrine has become an orthodox one, it is difficult to cite clear-cut statements of it. It is rather more often the underlying premise of published statements on monetary policy than it is their substance.

Fortunately (for purposes of this paper), two events of the past decade did elicit such substantive statements. One was the attempt of the Federal Reserve, already referred to, to play down the role of monetary policy in influencing interest rates, and correspondingly to play up its influence on commercial bank landing capacity. The other was the appearance of the "Gurley-Shaw thesis," the responses to which ordinarily relied heavily on the credit-creation doctrine [2] [3] [25]. It was not the Gurley-Shaw theory, as such, which elicited the strongest objections, but instead the two authors' conceptual framework, in which commercial banks were seen to play a role not fundamentally different from that played by other deposit-account financial intermediaries [6]. That commercial banks are fundamentally different from other intermediaries is, of course, of the essence in the credit-creation doctrine.

Both doctrines will be reviewed briefly, with their shortcomings noted.

The availability doctrine

The propositions of the availability doctrine are many, some of them resting on complex assumptions concerning expectational patterns. It is fair to say, however, that the doctrine's central proposition was the so-called "lock-in" effect: the argument that a tightening of monetary policy, with rising interest rates, would induce institutional holders of government securities to hold onto them, and perhaps even to increase them, thereby preventing them from financing new private loans through the sale of securities. This effect was to proceed from a number of causes: rising long-term rates, by reducing the current market values of long-term securities, would, it was thought, inhibit sales of these securities, through institutional lenders' unwillingness to realize capital losses; rising rates

---

ment in the official version; he cites only P. Samuelson's testimony criticizing such a theoretical point [30, pp. 696 ff]. Samuelson, in turn, does not attribute the theoretical point to either the Roosa or the Federal Reserve versions, nor in fact to anyone. In Kareken's critique of the doctrine [14] the same point is included, but Kareken relies for support only on Tobin's earlier article. In his answer to Kareken, Hodgman [11] includes the same point, but cites only Kareken.

would also reduce the liquidity of lenders' portfolios, and encourage them to maintain or increase their short-term holdings; and rising rates on marketable securities would not be matched, at least at first, by increases in loan rates, because the latter, as "administered prices," adjust to new equilibrium levels only with some lag [21, pp. 289f] [31, pp. 371f, 382f].[8] It is not really worth while exploring this reasoning in detail, since the lock-in effect has been shown to be theoretically defective,[9] and experience of the 1950's shows the lock-in effect to be empirically false, especially with respect to commercial banks, who have engaged in heavy sales of government securities in each tight-money period, replenishing them during the subsequent monetary ease of recessions.[10]

Proposals are made from time to time to enforce a lock-in effect through security-reserve requirements and similar devices [1] [2]. It is thus not idle to raise the question of what difference it would make if the doctrine on this point were both valid and true. Certain inferences can be drawn, but the doctrine itself was never explicit on the question. "General tightening of credit," begins the official Federal Reserve version of the doctrine, "restricts most directly the amount of spending which can be done with borrowed funds" [31, p. 369]. The theory implied by the proposition of the lock-in effect is that where banks change the *composition* of their assets, reducing security holdings and increasing loans, a contribution is made to aggregate effective demand. Why central bankers would argue that this is so is puzzling, in light of orthodox banking theory, for in engaging in such transactions commercial banks lose the special and unique characteristics which distinguish them from other institutional lenders. If they finance loan expansion through security sales, they do not create demand deposits; functionally, such transactions are little different from the operations of, say, savings and loan associations, which must finance lending operations through sale of share accounts.

Did the proponents of the doctrine, virtually all of them central

[8] Not all the above propositions were stated in such a clear-cut manner in the original sources; but the present writer is confident that all were originally, or later became, part of the availability doctrine, as that doctrine was understood by economists.

[9] In addition to the works already cited, see Robertson [20], Lindbeck [16], and the more recent study by Kane and Malkiel [13], which shows strong rational economic motives for bankers to shift from securities into loans in periods of high loan demand and tight money, even where the return on the former exceeds that on the latter.

[10] Between October, 1954, and October, 1956, commercial bank holdings of U.S. government securities declined by approximately $13 billion, remaining at their low level until November, 1957 (close to a monetary-policy turning point), at which time they recovered by $10 billion within the space of seven months Between January, 1959, and April, 1960, the banks disposed of approximately $10 billion worth of government securities. During both 1954-1957 and 1958-1960 tight-money periods, commercial bank business loans rose at a rate exceeding their post-Accord trend value. Life insurance companies' holdings show a similar pattern. Mutual savings banks and mutual savings and loan associations showed less regular behavior, but were clearly not "locked in" with respect to government security holdings.

bankers, join with and even lead those who have held nonmonetary financial intermediaries to be potentially destabilizing in their day-to-day activities? Apparently so, for the availability doctrine also included, as subsidiary arguments, the following two propositions:

1. Conventional monetary policy also has an impact on nonbank lenders. Open-market purchases and sales, when dealings are with nonbank lenders, increase and reduce, respectively, the funds available to these institutions for lending [21, p. 280]. The lock-in effect is also explicitly extended to nonbank lenders [31, p. 372].

2. Underwriters also respond to tightening credit conditions, discouraging new private security issues during periods of rising rates in order to avoid market losses on their own holdings [31, p. 373].

What these two points must be taken to mean, considered together with the lock-in effect, is that conventional monetary policy restricts the ability of potential borrowers to make financial arrangements, through intermediaries and market institutions, with households and businesses which, cash in hand, are ready to acquire interest-earning assets. Regardless of the empirical truth or falsity of the doctrine, it represented a radical revision in the official *modus operandi* of monetary policy.

The credit-creation doctrine

Commercial banks, according to this second availability approach to the effects of monetary policy, "create credit": that is, by virtue of their ability to pay for new assets with newly issued demand deposit liabilities (i.e., their ability to "create money"), commercial banks can lend (or acquire marketable financial assets) with no prior act of voluntary saving on the part of any other economic unit. This means that commercial banks can permit an excess of planned investment, for instance, over planned saving.

Most adherents to this view overstate their case. Not content to argue that commercial banks, in the sense defined, can create credit, they often argue (a) that commercial banks *always* create credit when they acquire additional assets, and (b) that *only* commercial banks can create credit, so that only commercial banks can lend to borrowers or acquire additional financial assets without a prior "act of saving" on the part of some other economic unit. Both these extreme statements of the doctrine are false. Commercial banks can also finance the purchase of "old" securities by creating demand deposits, thus creating money without creating credit. Other lenders, including households, can lend by running down money balances, thus in effect "creating credit" with no concurrent *ex ante* saving.[11]

[11] For an example of the overstated version of the credit-creation doctrine, see W. W. Riefler's statement [5, vol. 1, p. 301] to the Radcliffe Committee, in which

The fact that the doctrine is usually overstated does not mean that it is false. It is true that commercial bank finance *can*, and probably frequently does, account for an excess of *ex post* over *ex ante* saving, though the relationship is far from one of identity.

Even when stated in this more modest way, however, the credit-creation doctrine is unacceptable as a theory of the effect of monetary policy on aggregate demand. Clearly, all it tells us is that bank finance *can permit* an excess of *ex post* over *ex ante* saving. It does not explain why there must be such an excess. It does not, in short, present us with a theory of aggregate demand. By contrast, the quantity theory of money, for instance, argues that an increase in the stock of money *induces* more demand; the interest-rate approach argues that increases in the money supply relative to the demand for money reduce interest rates, thereby increasing investment demand and, perhaps, consumer demand (by reducing the propensity to save). But the credit-creation doctrine is not a theory of demand. It only says that *if* there is excess aggregate demand that excess can be financed through the banking system (and in extreme form, only through the banking system). This point provides a useful clue to the meaning of credit availability, and will be dealt with further momentarily.

II. The Framework for an Availability Theory

One matter on which virtually all widely accepted macroeconomic theories of instability find consensus is that the problem is in essence a financial one. Thus modern quantity-of-money theory stresses the search, by household and firm, for an optimum balance of assets, both real and

(since Riefler was testifying in behalf of the Federal Reserve) the doctrine was made official Federal Reserve theory. For similar statements, see also Aschheim [2, p. 125], Culbertson [3, p. 121], and Thomas [27, p. 273]. Gurley and Shaw [7, pp. 135f] show that the error in this version of the doctrine derives from an erroneous definition of saving, which identifies saving with the payment of money for a non-money asset, an act which may signify, but is not the same as, *ex ante* saving. A number of writers make a careful though erroneous distinction between money acquisition and saving, and an equally careful association between acquisition of non-money assets and saving. Thus Thomas [27, p. 273] writes that for stability, "the bulk of credit needs must be met out of saving, not by the creation of additional money," thus implying a clear-cut distinction between the two. And Culbertson [3, p. 122] writes, ". . . The usual banking theory would call for social accounting definitions representing creation of loan funds as involved in both of two types of actions; that of the banking system in acquiring additional debt by creating demand deposits (and the government in adding to currency in circulation) and that of other economic units in increasing their net holdings of nonmonetary debt." The latter type of action is described as an "act of saving." In fairness, it should be noted that those advocating the credit-creation view described may be asserting (but not demonstrating) an empirical proposition rather than a doctrine: that only commercial bank finance *does* (rather than *can*) account for an excess of *ex post* over *ex ante* saving.

financial, and argues that changes in the stock of money alter this balance, a correction being sought through acquisition of additional real assets. Thus income-expenditure theory stresses the investment-saving relationship (a financial relationship), and further usually stresses the role of changes in the money stock (*via* interest-rate changes) in influencing investment behavior. And thus, too, the availability theses, as seen, stress financial institutions and processes themselves. Indeed, the common characteristic which separates these macroeconomic theories from those structural and distributive theories emphasizing costs, wages, profit margins, etc., is that the former are financial theories, even though they may not be expressed in financial terms. The framework employed here is explicitly financial.

The framework is borrowed from the initial joint work of J. G. Gurley and E. S. Shaw [6]. The fundamental behavior units of the economy are "spending units"—households, business firms, individual government units, and miscellaneous other units such as churches, universities, and trade unions. These units are classified according to their "budget"[12] positions: those having balanced budgets, i.e., whose ordinary receipts, received as transfers, factor income (for households), sales receipts (for business firms), or taxes (for government units) exactly meet their transfer payments, factor payments, payments for product, and payment of taxes; those having deficit budgets, whose payments exceed their receipts; and those having surplus budgets, whose receipts exceed their payments. For simplicity, a closed economy is assumed, though no particular difficulty would be created by including the rest of the world as a surplus or deficit sector.

It is only a truism that the total net surpluses of surplus-budget spending units must equal the total net deficits of deficit-budget units. This is the same as saying that, in a closed economy, the economy as a whole is a balanced budget unit—that factor payments equal factor receipts, product sales equal product purchases, tax payments equal tax receipts, etc.

While there is an *ex post* identity between net deficits of deficit-budget spending units and net surpluses of surplus unit, there is no necessary reason for *ex ante* equality. Different spending units may respond to different inducements, and have conflicting intentions. The possibility for conflicting intentions is at the heart of conventional aggregate demand theory of instability, since that type of theory is, either explicity of implicity, a theory of the relative sizes of aggregate *ex ante* deficits and *ex ante* surpluses. This emphasis is most explicit in Keynesian income-expenditure theory, where concentration is on planned deficits of the

[12] "Budget," a word borrowed from the Gurley-Shaw nomenclature, is probably a misnomer, since it implies a planned or *ex ante* character, a connotation which is not here intended unless specifically so noted.

private business sector (investment) and planned surpluses of the household sector (saving).

Expressing the stabilization problem in explicitly financial terms underlines the fact that there are two distinct but interrelated types of questions with which theory and policy legitimately may become concerned:

1. There are behavioral questions concerning the determinants of planned deficits and planned surpluses, especially with respect to the inducements to invest and to save. Modern macroeconomic theory of instability has consisted in large part of a search for dependable behavior relationships which would permit an explanation of what changes in independent variables induce spending units to desire changed deficits or surpluses. Thus quantity-of-money doctrine can be described as concentrating on the motives of spending units to increase their intended expenditures, relative to their saving.[13] An increase in the quantity of money purportedly induces an increase in desired deficits, then, relative to surpluses. Similarly, income-expenditures theory ordinarily revolves around the influence of the interest rate on motives to invest (and, in some cases, on motives to save). In any event, there is typically a theory relating demand behavior to some variable subject to the influence of monetary policy.

2. There are also questions—again, behavioral questions—concerning the institutional processes through which destabilizing changes in deficits and/or surpluses can be achieved; i.e., there are questions concerning whether *ex ante* deficits in excess of *ex ante* surpluses can in fact be financed. Availability-of-credit theories concentrate on this second type of question. This is a point worth emphasizing: Neither the availability doctrine nor the credit-creation doctrine deals with the motives or behavior of ultimate lender (surplus-budget spending unit, or "saver") or ultimate borrower (investor or other deficit-budget spending unit). The concentration of both these doctrines is on financial processes and on institutions *relating* ultimate borrower and ultimate lender. This point is explicit in the availability doctrine, where much is made of the point that monetary policy operates through its effects on lenders, meaning in all cases institutional lenders and not households or other "ultimate" lenders. Though not explicit in the credit-creation doctrine, the point nonetheless applies there as well; as noted earlier, there is nothing in that doctrine to explain *why* borrowers want to borrow. The doctrine concentrates on their ability to do so, in the absence of "prior saving," when they borrow from commercial banks.

[13] Though a modern quantity theorist might not put it that way. Perhaps preferable would be a statement that an increase in the stock of money induces spending units to increase their demand for real as opposed to financial assets, in order to bring into an optimal arrangement the flow of benefits from all types of assets.

Hence availability-of-credit theories, in general, are those theories which deal exclusively with the financial mechanism rather than with the variables which influence demand. Influencing the availability of credit has to do with influencing deficit units' *access* to financing, as distinct from influencing their desire to borrow or the desire of savers to save.

The best definition of the availability of credit, then, is probably the "supply side" of the loanable funds equation: saving (*ex ante*), plus reduction of money balances by money-holders, plus new money creation.[14] But it is not the definition of availability of credit which is the significant characteristic of availability-of-credit theory distinguishing it from other types of theory: it is instead the new type of theory's concentration on the financial process, rather than the determinants of demand behavior by spending units, that marks it as a new approach.

This distinction between demand-behavior theories on the one hand, and availability-of-credit theories on the other, makes clear that a number of theories commonly grouped under the latter heading actually belong under the former. For instance, theories purporting to show an influence of monetary policy on the desire to borrow, *via* changes in such non-interest lending terms as compensating balances and loan discounts, are not really "availability" theories at all; they are essentially extensions of the interest-rate approach. Under them, the effectiveness of monetary policy depends on the response ("elasticity") of potential borrowers to changes in non-interest terms.[15] Similarly, as has already been noted by Musgrave [18, p. 166], the concept of credit availability does not apply when imperfection in credit markets "arises because of heterogeneity of product, so factors of risk and diversification enter," so long as the volume of lending in each market (determined by the lender, according to the characteristics of the borrower) depends on the interest-elasticity of loan demand in *that* market. The Gurley-Shaw thesis has achieved note as an assault upon orthodoxy because it argues for a major role in the stability problem for financial intermediaries other than commercial banks; but Gurley and Shaw assign this major role to these institutions not because

[14] But availability-of-credit theories are not otherwise related to the loanable funds theory. The latter is a theory of interest rate determination, and belongs under the "demand behavior" heading (number one, above), rather than availability of credit.

[15] See J. Guttentag [8], who proposes that the expression "availability of credit" be used to describe non-interest lending terms. He makes clear that his use of the term "availability" is quite different from that of Roosa, *et al.* though Guttentag was then an economist for the Federal Reserve Bank of New York) when he writes (p. 222): "One possible source of confusion between availability and supply is the emphasis given in the literature to the operations of lenders. It is frequently stated that the older monetary theory emphasized the borrower whereas the availability doctrine emphasizes the lender. Of course the fact that the lenders administer terms does not mean that the basic forces underlying changes in availability originate on the supply side of the market. On the contrary, it would appear that for the most part changes in availability originate on the demand side."

of their role in financing deficits, but because their liabilities can substitute for money for asset purposes, thus lowering interest rates, other things being equal, and inducing more investment. Even the Gurley-Shaw thesis, then, despite its concern with present-day financing institutions, must be classified as a demand-behavior and not an availability-of-credit theory. This is a point missed by most Gurley-Shaw critics, who concern themselves, as already noted, with commercial banks' ability to create credit.

Conventional aggregate-demand theory, where concern is exclusively with the determinants of demand (i.e., of *ex ante* budget positions), ordinarily leaves the methods of financing various budget positions an unexplored matter. Their tacit assumption must be that, whatever the economy's financial institutions, they are inevitably adequate to the task of financing whatever deficits and surpluses they are called upon to finance. The financing institutions have a wholly passive role, responding to decisions made by ultimate lenders and ultimate borrowers. Financial processes can be taken for granted. Acceptance of this type of theory implies employing stabilization policy directed at influencing the demand behavior of spending units (including the possibility, when the policy-maker is a spending unit, as is the Federal government, of it adjusting its own budget position in accordance with stabilization objectives).[16] This presumption on the part of conventional theory that financial processes may be taken for granted may help explain why such theory, which is at heart financial theory, can so often be presented in entirely nonfinancial terms[17]—in terms of "leakages," "injections," etc.

Where conventional theory has concentrated on the determinants of *ex ante* budget positions and taken adequate financing for granted, availability-of-credit theories have concentrated on financing processes and taken the existence of excess demand (excess of *ex ante* deficits over *ex ante* surpluses) for granted. Thus there is no availability-of-credit theory of instability, but only an availability-of-credit approach to stabilization policy. Instability derives from the existence of excess aggregate demand; the objective of policy, under this approach, is to frustrate, rather than to dissuade, those who intend to incur deficit positions.

The search for an availability theory on which policy could be based would seem likely to stem from two environmental conditions: (1) the existence of excess aggregate demand; and (2) an inability to deal directly, within the institutional constraints imposed on policy, with the determinants of demand. For instance, it might be held that planned

[16] Even though this type of theory can "take for granted" financial processes, an important role may nonetheless be given by them to the behavior of financial institutions, insofar as it influences *ex ante* budget positions, without altering the exclusively demand-behavior nature of the theories.

[17] Even, oddly enough, in money and banking textbooks. Usually the "theory section" is the only non-financial section.

deficits are sensitive to interest-rate changes, but that a significant reduction in planned deficits could only be achieved by means of an interest-rate increase exceeding institutional maxima imposed by convention, by the existence of a large public debt whose market value is sensitive to interest rate changes, or by the possibility that marked changes in the money supply and in commercial-bank-owned reserves that would be incident to an interest-rate increase might endanger the viability of parts of the financial system.

The fact that availability-of-credit theories can take for granted the existence of excessive aggregate demand may be the feature of these theories which explains why they are incompletely formulated, though a reading of the availability literature suggests that its authors simply did not recognize the important distinction between influencing ultimate demand and influencing the financial process.

III. Prolegomenon to any Future Availability Theory

It is not the purpose of this paper to present a full-fledged theory of credit availability. Indeed, it is not certain at all whether this is possible. The purpose is to indicate the meaning of credit availability as "one of the main variables on which monetary policy operates." It is possible here only to suggest some elements any such theory will have to contain, elements policy-makers will have to take immediate account of, if they do indeed operate on the assumption that they are influencing the availability of credit.

It adds little to knowledge to point out that *ex ante* deficits in excess of *ex ante* surpluses can be financed through money creation plus attempts at reduction of money balances by money-holders. To control the availability of credit sufficiently so that no excess demand can be financed would seem to require influencing both these two types of financing operations. Thus the credit-creation doctrine, in the more modest interpretation of it set forth above, must be a part of an overall availability-of-credit theory. Unfortunately, it is only a part, and a small part at that. The volume of money-creation in the United States each year (even assuming all of it to be associated with credit-creation) is small by comparison with the volume of money-transferring financial transactions, a great many (but an uncountable number) of which do not involve *ex ante* surpluses but merely the substitution of income-earning assets (both real and financial) for money assets. All the indicators we have—income velocity of money, volume of trade credit, commercial bank sales of government securities, etc.—suggest that in boom periods, money goes "where the action is," i.e., to deficit-budget spending units.

There is no doubt about the ability of the monetary authorities, using conventional techniques, to control money-creation. There is considerably more doubt about their ability to control the transfer of money from money-holders to deficit units (either directly or through intermediaries) in exchange for newly created non-money financial instruments. The availability doctrine is best interpreted, to be consistent with the view taken in this paper of the availability of credit, as an allegation that the volume of such transactions is limited as a byproduct of the exercise of conventional monetary techniques. If, for instance, commercial banks were to find money-holders willing to buy government securities, the lending capacity of banks would be enhanced, and banks, without "creating credit" in the conventional sense could nonetheless finance deficits unmatched by a corresponding volume of *ex ante* surpluses. But if the banks were to be locked in, some potential borrowers would be frustrated.

As noted, this major proposition of the availability doctrine has proved to be empirically false. Of course, this does not remove the problem the lock-in effect was intended to solve. Indeed, it points up the fact that monetary policy may have perverse effects on the availability of credit. For instance, it has often been noted that the same tight-money policies of the central bank which limit the ability of commercial banks to "create credit" may, through the inducement of higher interest rates, bring money-holders to desire to hold interest-earning assets instead of money, thereby permitting the achievement of deficits in excess of *ex ante* surpluses. This is not done solely through institutional lenders' sale of government securities; in the United States, there is a vast, complex, and highly efficient financial system well geared to satisfying the needs of deficit-budget units, whatever the circumstance.

When the matter of controlling the availability of credit is defined in terms of "clogging up" this financial process, rather than attempting to act upon motives to save, borrow, spend, etc., it becomes clear that many proposals have already been made to strengthen the hand of the Federal Reserve in controlling credit availability. The old secondary-security-reserve proposal, originally offered to "insulate" the public debt from interest-rate changes so as to permit resumption, in the immediate post-war period, of flexible monetary policy, has recently been revived as a means to limit credit availability [2, pp. 33ff]. A similar proposal has been to require commercial banks to hold, in place of long-term coupon-bearing government securities a special, nonmarketable issue [1]. There have been several proposals for limiting the assets of non-commercial-bank financial intermediaries, usually through a reserve requirement [9]. One proposal which deserves more attention than it has received is for the payment of a variable interest rate on demand deposits, "to tighten the Federal Reserve's control over the opportunity cost that bank depositors charge against any alternative investment of funds" [29, p. 278].[18]

Before any proposal be taken seriously, one thing should be carefully noted. An excess of *ex ante* deficits over *ex ante* surpluses is a disequilibrium situation. If the excess of deficits simply cannot be financed, the disequilibrium exists in financial markets. If they can be financed, the disequilibrium is transferred into product markets in the form of excess aggregate demand for output and, under capacity-output conditions, inflation. Conventional policy has as its objective preventing both types of disequilibrium by acting on the motives of spending units, encouraging smaller deficits and/or larger surpluses. To restrict the availability of credit, as defined here, means to prevent product market disequilibria by *maintaining* a disequilibrium condition in credit markets; as noted already, the objective is to frustrate, rather than to dissuade, those planning deficits.

Since an attempt to restrict the availability of credit means an attempt to maintain a disequilibrium, it shares some of the disabilities and disadvantages of direct controls. One suspects that, regardless of the techniques employed, ways will be found around the particular controls and techniques used. The objective of maintaining indefinitely a disequilibrium situation seems extremely ambitious; but that and nothing less is what is meant by limiting the availability of credit.

REFERENCES

1. Alhadeff, David A. "Credit Controls and Financial Intermediaries," *American Economic Review,* September, 1960, 655-671.
2. Aschheim, Joseph. *Techniques of Monetary Control,* Baltimore, 1961.
3. Culbertson, J. M. "Intermediaries and Monetary Theory: A Criticism of the Gurley-Shaw Theory," *American Economic Review,* March, 1958, 119-132.
4. Freimer, Marshall, and Gordon, M. J. "Why Bankers Ration Credit," *Quarterly Journal of Economics,* August, 1965, 397-416.
5. Great Britain, Committee on the Working of the Monetary System. *Principal Memoranda of Evidence,* London, 1960.
6. Gurley, J. G. and Shaw, E. S. "Financial Aspects of Economic Development," *American Economic Review,* September, 1955, 515-538.
7. ———. "Intermediaries and Monetary Theory: Reply," *American Economic Review,* March, 1958, 132-138.

[18] Commercial banks are now prohibited from paying interest on demand deposits. The rationale for this restriction has to do with supervisory control over bank performance rather than with monetary policy *per se;* and it appears that economists concerned with bank performance are in increasing numbers endorsing repeal of this prohibition. See, *inter alia,* J. G. Gurley [32, pp. 285f].

8. Guttentag, J. "Credit Availability, Interest Rates, and Monetary Policy," *Southern Economic Journal,* January, 1960, 219-228.
9. Henderson, J. M. "Monetary Reserves and Credit Control," *American Economic Review,* June, 1960, 348-369.
10. Hodgman, Donald R. "Credit Risk and Credit Rationing," *Quarterly Journal of Economics,* May, 1960, 258-278.
11. ———. "In Defense of the Availability Doctrine: A Comment," *Review of Economics and Statistics,* February, 1959, 70-73.
12. Johnson, H. G. "Monetary Theory and Policy," *American Economic Review,* June, 1963, 335-384.
13. Kane, E. J. and Malkiel, B. G. "Bank Portfolio Allocation, Deposit Variability, and the Availability Doctrine," *Quarterly Journal of Economics,* February, 1965, 113-134.
14. Kareken, J. H. "Lenders' Preferences, Credit Rationing, and the Effectiveness of Monetary Policy," *Review of Economics and Statistics,* August, 1957, 292-302.
15. ———. "Post-Accord Monetary Developments in the United States," *Quarterly Review, Banca Nazionale del Lavoro,* September, 1957, 322-351.
16. Lindbeck, Assar. *The "New" Theory of Credit Control in the United States, rev. ed.,* Stockholm, 1962.
17. Modigliani, Franco. "The Monetary Mechanism and Its Interaction with Real Phenomena," *Review of Economics and Statistics,* Supplement February, 1963, 79-107.
18. Musgrave, R. A. "Monetary-Debt Policy Revisited," *Public Policy,* 1954, 155-176.
19. Riefler, W. W. "Open-Market Operations in Long-Term Securities," *Federal Reserve Bulletin,* November, 1958, 1260-1274.
20. Robertson, D. H. "More Notes on the Rate of Interest," *Review of Economic Studies,* February, 1954, 136-146.
21. Roosa, Robert V. "Interest Rates and the Central Bank," *Money, Trade and Economic Growth: in Honor of John Henry Williams,* New York, 1951, 270-295.
22. Scott, Ira O. "The Availability Doctrine: Development and Implications," *Canadian Journal of Economics and Political Science,* November, 1957, 532-539.
23. ———. "The Availability Doctrine: Theoretical Underpinnings," *Review of Economic Studies,* October, 1957, 41-48.
24. Smith, Warren. "On the Effectiveness of Monetary Policy," *American Economic Review,* September, 1956, 588-606.
25. Solomon, Ezra. "Financial Institutions in the Saving-Investment Process," *Conference on Savings and Residential Financing, 1959 Proceedings,* Chicago, 1959, 29-41.
26. Sproul, Allan. "Changing Concepts of Central Banking," *Money, Trade and Economic Growth: in Honor of John Henry Williams,* New York, 1951, 296-325.
27. Thomas, Woodlief, "How Much Can be Expected of Monetary Policy?" *Review of Economics and Statistics,* August, 1960, 272-276.

28. Tobin, James. "Monetary Policy and the Management of the Public Debt," *Review of Economics and Statistics,* May, 1953, 118-127.
29. ———. "Towards Improving the Efficiency of the Monetary Mechanism," *Review of Economics and Statistics,* August, 1960, 276-279.
30. U. S. Congress, Joint Committee on the Economic Report. *Monetary Policy and the Management of the Public Debt, Their Role in Achieving Price Stability and High-Level Employment, Hearings* before the Subcommittee on General Credit Control and Debt Management, 82nd Congress, 2nd Session, Washington, 1952.
31. ———. *Monetary Policy and the Management of the Public Debt, Their Role in Achieving Price Stability and High-Level Employment, Replies to Questions and Other Material* for the Use of the Subcommittee on General Credit Control and Debt Management, 82nd Congress, 2nd Session, Washington, 1952.
32. U. S. Congress, Joint Economic Committee, *Review of the Report of the Commission on Money and Credit, Hearings,* 87th Congress, 1st Session, 1961.
33. Young, R. A. and Yager, C. A. "The Economics of Bills Preferably," *Quarterly Journal of Economics,* August, 1960, 341-373.

15

Opening the books on monetary policy

Every third Tuesday, or thereabouts, there convenes in Washington at the headquarters of the Board of Governors of the Federal Reserve System a group which represents a uniquely American contribution to the art of central banking. This is the Federal Open Market Committee: twelve men responsible for administering the Federal Reserve's principal monetary weapon—the power to buy and sell securities in the open market, subject only to the Federal Reserve Act's broad guideline that operations shall be conducted "with a view to accommodating commerce and business and with regard to their bearing upon the general credit situation of the country."

The FOMC is the heart of the modern Federal Reserve System. In effect, the Committee's open market authority gives it day-to-day control of the main valve that regulates the nation's money supply. To greatly oversimplify a process of vast complexity, a net buying stance by the FOMC over a period of time sets the stage for monetary expansion by creating new reserves for the banking system; a net selling stance shrinks reserves and tends toward contraction of the money supply.

Furthermore, in a Federal Reserve System where power is divided by statute between the Board of Governors in Washington and the twelve regional Federal Reserve Banks, the Open Market Committee—on which both the Board and the Banks are represented—has become the natural focal point for virtually all decisions on credit policy. These include actions on matters over which the Committee itself has no formal authority—for example, changes in the discount rate, reserve requirements, and regulations governing the payment by commercial banks of interest on time and savings deposits.

Despite its central position, the Committee has had relatively low visibility. Its meetings, logically, are held under conditions of tight security (officially, Federal Reserve employees are not even supposed to con-

Reprinted with permission from the March 1965 issue of *The Morgan Guaranty Survey*, a monthly publication of the Morgan Guaranty Trust Company of New York.

firm when meetings are taking place). When the Committee makes a change in monetary policy, there is no announcement. Changes ultimately are confirmed in the "record of policy actions" that the Board of Governors by law must publish in its annual report; but, by the time this document appears, the effect of any policy change normally has become apparent in the market. The account given, moreover, tends to be terse in style, with little inkling of the give-and-take of debate leading to policy decisions.

Reticence is, for good reasons, the tradition of central bankers everywhere. Members of the Open Market Committee, arguing against more timely disclosure of the details of deliberation, have held it would inhibit them in discussion and would corrode their freedom from partisan political pressure. Also, only by avoiding public commitment to a given policy line can the authorities maintain their own flexibility of action.

The minutes go public

The Committee's concept of the need for permanent secrecy has undergone a gradual change, however, and an increasing amount of information has been made public—after suitable aging. In the Board's 1963 annual report, for example, the FOMC summary and related material occupy 143 pages, just less than half the total bulk of the document. A decade earlier the comparable material filled only eight pages.

The most significant move thus far toward greater disclosure came last year when the Board and the Committee decided to transfer to the public files of the National Archives the actual minutes of Open Market Committee deliberations from 1936—just after the FOMC had been cast in substantially its present form—through 1960.[1] No formal commitment has been made, but Federal Reserve officials have indicated that minutes for subsequent periods probably will be released regularly in the future, with a lag of some three to four years.

The decision to lay bare the verbatim minutes has opened up some 8,700 pages of hitherto largely unavailable monetary history. Most of it is concerned with analysis of business conditions, and with the recurring task of producing a consensus within the Committee on the monetary policy appropriate to those conditions. This material, keenly attuned to the nuances of change in the economic outlook, provides a rich mine for the economic historian.

What mainly emerges from the massive 25-year record, however, is the Committee's own development as a unique organism in central banking, the decision center where national monetary policy is made by

[1] Available on microfilm from Exhibits and Publications Division. National Archives and Records Service, Washington, D.C. $55.

a central banking system which is distinctly federal in character. Much of the story of how the Committee's role has been forged falls into three chapters, each centering on one of the three "great debates" that marked the Committee's deliberations during the 1950's:

1. The battle with the U.S. Treasury, ending in 1951 in the "accord" that allowed monetary policy to become an effective contracyclical force;

2. The controversy over whether System open market operations should be confined to the very shortest-term Treasury securities ("bills only") or spread through all maturity ranges;

3. The sometimes stormy relationship which existed during the early 1950's between the Board of Governors and the Federal Reserve Bank of New York (which as agent executes all of the Committee's transactions).

The meat of the recently opened minutes, in other words, is in the last decade of the quarter-century span which they cover. The Committee's earliest roots actually go back to 1922, when the regional Reserve Banks first got together in informal meetings to coordinate their purchases and sales of government securities.

In those days the Reserve Banks were in the securities market as investors, seeking income with which to meet operating expenses. Their market transactions assumed significant proportions early in 1922 as the Banks sought to offset a sharp drop in income from member bank borrowings. Rather by accident, it was discovered that their purchases of securities had an impact on the banking system's reserves—and potentially on money supply—and furthermore that this impact quickly passed beyond the regional confines of any given Reserve District.

The coordination that began in 1922 continued on an informal basis until the Banking Act of 1933 gave the Open Market Committee official status. Revisions in 1935 set the composition of the group in essentially its present form. The seven members of the Board of Governors became members of the Committee, and the remaining five seats were made subject to rotation among the twelve Reserve Banks on an annual basis. This arrangement still prevails, except that in 1942 the place of the Federal Reserve Bank of New York on the Committee was made permanent, leaving only four places to rotate among the eleven other regional Banks. However, all twelve Reserve Bank presidents, senior economists from the Banks, and various members of the staff of the Board of Governors regularly participate in FOMC meetings (*box on page 202*).

By law, a Reserve Bank holding a place on the Open Market Committee must designate as its representative either its president or its first vice president (appointments of both these officials to their Reserve Bank positions are subject to Board approval). By tradition, the Chairman of the Board of Governors has always served as Chairman of the FOMC,

Around the Table at 2001 Constitution Ave., N.W.

When the Federal Open Market Committee gathers around the 30-foot mahogany table in the Board of Governors' room at 10 o'clock of a Tuesday morning, the first item is a vote on the minutes of the preceding meeting.

Next comes the report of the Special Manager of the System Open Market Account in charge of foreign currency operations. This has been a feature of the meeting only since early in 1962, when the System began operating in the foreign exchange markets to defend the dollar from danger of speculative attack. After discussion, the Committee votes to authorize continued exchange operations.

The meeting then turns to its principal business, formal discussion of economic conditions and monetary policy. This opens with a review by the Manager of the Open Market Account of transactions in the domestic market since the last meeting. The Committee is asked to "approve, ratify, and confirm" what has been done.

Next there are detailed presentations by three members of the staff of the Board of Governors: the first on the general domestic business outlook, the second on domestic financial developments, the third on the balance of payments and international finance. Then the presidents or first vice presidents of the Reserve Banks (only five are now members of the Committee, but all twelve attend) and the seven Governors comment in turn. The Vice Chairman of the Committee (always the President of the Federal Reserve Bank of New York or the First Vice President as his alternate) talks first. The order following him switches from clockwise around the table at one meeting to counterclockwise at the next, then back to clockwise, etc. Either way, the go-around ends with the Vice Chairman of the Board of Governors and, finally, the Chairman, who distills a consensus from the comments around the table. The end-product of this discussion at each meeting is a vote either to renew or to amend the Committee's formal directive to the Federal Reserve Bank of New York.

From meeting to meeting, the directive guides the Manager of the System Account in the conduct of open market operations. It is broad enough to let him react flexibly to changing conditions in the market place. Since actual changes in the wording of the directive are relatively infrequent, the Manager also adjusts his reading of it to the consensus in which the Chairman sums up each meeting's discussion.

Thus the consensus is of focal importance in the Committee's work. In it, typically, the Chairman states what he regards as the common ground among the opinions—sometimes rather widely divergent—that have been presented. His own view may be stated alongside this synthesis, or blended into it.

Not uncommonly, further discussion ensues and exceptions are registered to one part or another of the summation before the formal vote on the directive is taken. At least through 1964, however, no FOMC vote on open market policy (as shown in the Board of Governors' annual reports) has ever gone against the judgment expressed in Chairman Martin's consensus.

and the President of the Federal Reserve Bank of New York has always served as Vice Chairman.

The 1935 legislation requires that the Committee meet "at least four times each year," and from its formation in March 1936 until June 1955 it usually met no oftener than that. To carry on its operations during the intervening periods, it created a five-man executive committee (three members of the Board and two Reserve Bank presidents), which met at irregular intervals at the call of the chair. In 1955 the executive committee was abolished and the present pattern of meetings of the full Committee every third week was established.

I. The "Accord"

Aside from the actual founding of the Federal Reserve System by passage of the Federal Reserve Act in 1913, it can be argued that the most important event in the System's history thus far has been the understanding reached with the Treasury in 1951 after a period of bitter controversy with that agency. The crux of the dispute was the Treasury's insistence—supported for a while by President Truman—that the Federal Reserve should continue its practice, carried on throughout World War II, of supporting the prices of Treasury securities (and thus holding down their yields) through its open market operations.

The minutes of the Open Market Committee provide a valuable record of how and why the Reserve-Treasury controversy developed. The definitive account of the whole episode is likely to prove to be the one written for the November 1964 issue of the *Monthly Review* of the Federal Reserve Bank of New York by Mr. Allan Sproul, who as President of the New York Bank and Vice Chairman of the Open Market Committee was a principal in the events. But the minutes make worth-while additions to the perspective given by his narrative.

They make clear, for instance, the essentially passive role which open market operations played in national economic policy for more than a decade preceding the accord. During the latter 1930's, excess reserves in the banking system were so great—because of gold imports from Europe at a time when private credit demands were slack—that securities purchases and sales by the central bank made little practical difference. Mr. Marriner S. Eccles, as Chairman of the Board of Governors and of the FOMC, told the Committee on December 13, 1939:

> Open market operations during recent months were not directed toward changing or influencing money rates. The large volume of excess reserves has removed the System so far from contact with the money market that the System's purchases and sales of securities have had little effect on the market and practically no effect on the aggregate volume of available funds. In these cir-

cumstances, transactions for the System open market account should continue to be in the direction of maintaining stability in the Government securities market, which in turn reflects itself in the entire capital market.

As early as the previous April, the FOMC had authorized the Federal Reserve Bank of New York to cushion the market impact if there should be an outbreak of hostilities in Europe. In taking on market stabilization as a continuing assignment, the Committee presumably realized, it was risking a possible conflict at some future point with its responsibility to restrain threatened inflation. Concern on that score, however, was to recede before the more compelling urgency of financing the rapidly developing U.S. defense effort. On September 27, 1941, the Committee showed itself willing to accept a course proposed by its economist, Mr. E. A. Goldenweiser, in these terms: "To adopt a policy under which a pattern of interest rates would be agreed upon from time to time and the System would be pledged to support that pattern for a definite period."

Direct involvement of the United States in World War II came less than three months later, and the policy of supporting a "pattern of interest rates"—that is, "pegging" the prices of Treasury securities across the whole maturity range by standing ready to buy whenever sellers appeared—was set for the duration. It was the Open Market Committee's continued adherence to this policy *after* the war that subordinated monetary policy, and hence the Committee itself, to the financing needs of the Treasury. The minutes show that Committee meetings became little more than perfunctory discussions of what rate and what maturity to recommend to the Treasury for its next financing.

As the postwar economy gathered momentum, the Federal Reserve System's role as a residual buyer in the government securities market made the System, in Mr. Eccles' classic phrase, an "engine of inflation." As it bought and bought to maintain the market prices of Treasury paper, the System paid for its purchases by crediting the accounts of member banks. Thus it poured reserves into the banking system—the next thing to printing money.

The start of the Korean War in June 1950, and the inflationary surge that followed, brought to a head the concern Reserve officials were feeling over their policy of accommodating the Treasury. President Sproul, in the meeting of August 18, 1950, told the Open Market Committee he thought it was time to tighten up the state of the credit markets:

"The question today is what we are going to do in our sphere of primary responsibility, not what we are going to recommend to the Treasury that it do in its primary sphere. It is not a question of the long-term bond issue or of refunding the September-October maturities, but what are we going to do about making further reserve funds available to the banking system in a dangerously inflationary situation. . . . We can't do the whole job with credit measures but in view of our responsibility and the national program I think that general credit measures should now be used. . . . This time I think we should act on the basis of our unwillingness to continue to supply reserves to the market by supporting the existing rate structure and should advise the Treasury that this is what we intend to do . . ."

The Committee was flexing its muscles, although it was not yet prepared to let the credit markets find their own level of interest rates. That same day the Board approved an increase in the discount rate from 1½% to 1¾%. The announcement of the change said: ". . . the Board of Governors of the Federal Reserve System and the Federal Open Market Committee are prepared to use all the means at their command to restrain further expansion of bank credit *consistent with the policy of maintaining orderly conditions in the Government securities market*" (italics supplied).

The minutes show that Mr. Thomas B. McCabe, Chairman of the Board of Governors, and Mr. Sproul had an appointment to see Treasury Secretary John W. Snyder at 4:30 that afternoon, to inform him of the action before it was publicly announced. After their meeting the Treasury announced a $13.5 billion refunding on terms that in effect ignored the Board's decision on the discount rate. The offering was a catastrophic flop in the market, and the Federal Reserve had no choice but to bail it out. Foreseeing the possibility of such a development, the Committee had authorized the Open Market Account to purchase the new securities "to the extent necessary to assure that the issue would not fail . . ." The Account ended up buying more than 80% of the offering.

The FOMC minutes show that, on September 28, 1950, Messrs. McCabe and Sproul met with Secretary Snyder to seek "his counsel as to how the expansion [in bank credit] could be curbed." In particular, they wanted the Treasury's view on "an increase in reserve requirements, and the possibility of restricting bank reserves by being reluctant buyers of securities and allowing a moderate further increase in short-term rates." Secretary Snyder's rejoinder, as reported in the minutes of the same meeting, was to doubt the effectiveness of these instruments of general monetary control and to suggest that the FOMC might "get the new President of the [American Bankers] association and other officials to put on a strong campaign to restrain credit."

Letter from the White House

Despite opposition from the Administration, the FOMC continued to assert its independence through the rest of 1950. In early December, following press reports of "open speculation as to whether the Federal Reserve is again undercutting the (Treasury) financing," President Truman sent Chairman McCabe the following letter, which appears in the FOMC minutes for January 31, 1951:

"I am enclosing you the article to which I referred in my telephone conversation with you Friday night.

"It seems to me that this situation is a very dangerous one and that the Federal Reserve Board should make it perfectly plain to the open market committee and to the New York Bankers that the peg is stabilized.

"I have succeeded in getting the Treasury to appreciate the fact that we should have our obligations financed on longer terms than has been the case generally but, if the Federal Reserve Board is going to pull the rug from under the Treasury on that, we certainly are faced with a most serious situation, because we are going to have an immense amount of Federal financing in the next six months.

"I hope the Board will realize its responsibilities and not allow the bottom to drop from under our securities. If that happens, that is exactly what Mr. Stalin wants."

In his reply to the President, Chairman McCabe made it plain that the reports of Federal Reserve intention to "undercut" the Treasury were inaccurate; at the same time, he reiterated the Federal Reserve's view that "the creation of additional bank reserves in a period like this only adds more fuel to the fire of inflation."

The matter passed the limits of tolerance early in 1951. Conferences involving the White House, the Treasury, and the Federal Reserve only succeeded in deepening misunderstanding. Twice the Administration released public statements reporting Federal Reserve willingness to support the then existing pattern of interest rates. Mr. Sproul recalls in his recent article that "this was at variance with what the Federal Open Market Committee believed had been said and done . . ."

Leading members of Congress began to call for a closing of the rift between the Reserve and the Treasury. A fresh attempt to settle their differences was launched early in February, with the Administration represented this time principally by Mr. William McC. Martin, Jr., then an Assistant Secretary of the Treasury. In the course of these discussions, it soon developed that the gap between the two agencies was not so wide as had earlier appeared and that concessions were possible on both sides. By early March, details of a formal settlement had been worked out—the Federal Reserve had adequate freedom to start moving toward an independent monetary policy, and the Treasury had adequate assurance that the long-term government bond market would not be allowed to collapse. Shortly after, Mr. Martin was named by President Truman to be Chairman of the Board of Governors of the Federal Reserve System (and therefore chairman of the Open Market Committee as well).[2]

[2] Mr. Martin, who has held both chairmanships longer than any other man, is the second William McChesney Martin to serve on the FOMC. His father, head of the Federal Reserve Bank of St. Louis in the 1930's, was a member of the present FOMC in 1937 and 1939, and an alternate member in 1936, 1938, and 1940.

II. "Bills Only"

It was only through the enhanced freedom won by the Federal Reserve in the accord that monetary policy was able to come to full flower in the United States. Yet, in a sense, the accord was only the first in a series of events that have helped shape the FOMC and form that body's conception of its own role. For, having gained its freedom, the Committee had to decide what to do with it, and in this it was breaking new ground.

At the meeting held on May 17, 1951, the first at which Mr. Martin presided as Chairman, a study project was proposed. The minutes relate:

> [Chairman Martin] suggested that the Committee authorize him to appoint a committee to consist of himself as chairman and not less than two or more than four other members of the Federal Open Market Committee to make a study of the scope and adequacy of the Government securities market . . . Chairman Martin emphasized that his suggestion for a study of this type was in no sense a criticism of the operations of the System account. In response to questions, he indicated that there was a need for a broader market for Government securities, that perhaps there would be value in studying the British Government securities market, that the time may come when the Federal Open Market Committee might find it necessary to change the procedure whereby it did business with only a small number of qualified dealers, and that he hoped the study proposed would make some worthwhile suggestions along these and other lines.

It was a full year later, May 1952, when the study group—with Governor Abbot L. Mills, Jr. of the Federal Reserve Board and President Malcolm Bryan of the Federal Reserve Bank of Atlanta as members along with Chairman Martin—began its work. Mr. Robert H. Craft, then Vice President and Treasurer of Guaranty Trust Company of New York, was retained as a consultant to direct the study.

Initially, it was contemplated that the job would take about 90 days. But it soon was apparent that this schedule could not be met. The study panel—which came to be known as the Ad Hoc Subcommittee on the Government Securities Market—developed an unexpectedly large volume of material in conversations with government securities dealers. More important, controversy was stirring within the Federal Reserve System over the conduct of the study. The minutes of the Open Market Committee meeting on June 6 relate sharp criticism by Mr. Sproul of the proposed method of operation, which would have prevented the New York Bank from seeing the full record of testimony:

> The argument . . . apparently was that some dealers might not talk as freely as they otherwise would in the fear that there might be some retaliation

against them by the New York Bank. This, Mr. Sproul said, was offensive to him as chief executive officer of the New York Bank and to Mr. [Robert G.] Rouse as Manager of the System Account.

As part of the dialogue which had been touched off, the New York Bank prepared an analysis of the techniques used in conducting open market operations. This was circulated within the System but has never been released publicly. Reference is made to it in a letter, reproduced in the FOMC minutes, which was addressed to the presidents of all the Reserve Banks by Chairman Martin in late August 1952:

"In addition, . . . we have received, as you have, the extensive report of the New York Bank entitled 'Federal Reserve Open Market Operations.' This has deserved and required the most careful review and discussion, tasks on which the Committee is now engaged.

"It has become evident, in hearings, discussions, and from the New York report, that the issues involved in the Committee's terms of reference are of the most fundamental and far-reaching character. They involve not only the most complicated problems of technique and organization but profound problems of a more theoretical and philosophical nature . . ."

The Subcommittee filed its report on November 12, 1952. It took as an initial imperative the need for a government securities market possessing "depth, breadth, and resiliency"—where there are orders in volume both above and below the market, where the orders come from widely divergent investor groups, and where new orders pour promptly into the market to take advantage of sharp and unexpected fluctuations in prices. Without such a market, the Subcommittee argued, the FOMC would be hampered in turning its policies into action. To impart the desired characteristics to the market, it recommended that the FOMC should confine its operations to the very shortest-term government securities, preferably Treasury bills, except in the extraordinary instance when intervention elsewhere in the maturity range was required to correct "a disorderly situation in the Government securities market."

Mr. Craft's argument on behalf of this policy is reported in the minutes of the FOMC meeting held on March 4-5, 1953:

Even today many of the more sophisticated people in the Government securities business were still not convinced that the Federal Open Market Committee had abandoned the theory that the Government securities market must continue to be controlled within limits. This was illustrated, he said, by the fact that purchases by the Federal Reserve Bank of New York of Government securities for Treasury trust accounts might be the cause of rumors that the Open Market Committee was "back in the market."

Mr. Craft emphasized what he conceived to be the advantages of confining transactions for the System account normally to Treasury bills as a means of permitting greater flexibility in open market account operations, with a minimum of disturbance to prices and yields on longer-term securities. He said that would permit the market (a) to reflect the natural forces of demand and supply, and (b) to furnish a signal of the effectiveness of credit policy aimed primarily at the volume and availability of bank reserves . . .

Mr. Craft also brought out the view that it was desirable to limit intervention by the System outside the bill market to periods when it seemed desirable to correct disorderly conditions in the market. It was his view that this was desirable so as to avoid imposing on the market any particular pattern of prices and yields, and he felt that assurances along the lines recommended in the subcommittee's report should be given by making known to the dealers the "ground rules" which would govern System operations in the market . . .

The President and the Chairman

The minutes for the same meeting record the case made against "bills only" by Mr. Sproul:

The principal reason why the Government securities market did not have depth, breadth, and resiliency at all times is now due to uncertainties regarding general credit policy and the Treasury's debt management program, rather than because of any concern that the Federal Reserve might intervene in the market. . . . What the market wants to know, he said, is whether interest rates and, therefore, security prices are going up or down; this is tied in with the whole question of credit policy.

With respect to the proposal for confining open market operations to the short-term sector of the market, Mr. Sproul said that there might be times when the System would wish to intervene in other than the short-term area in order to get direct effects on the availability and cost of credit in the capital market or the mortgage market, as a means of effectuating credit policy. . . . And quite apart from what the Committee might decide as a matter of current policy on the suggestion that operations be confined to the short-term area, Mr. Sproul said that public assurance as to the continuance of this policy could not be given to the market, as proposed, without misinterpretation and misunderstanding and without seeming to bind future open market committees, which could not be bound by statements made by predecessor committees. . . . Mr. Sproul stated that at the present time he thought it was desirable to operate only in the short-term sector of the market as far as that was possible, but that he could not say what would be desirable next year or two years from now. . . .

Among points made by Chairman Martin in support of his Subcommittee's recommendations was the one in this excerpt from the minutes:

Chairman Martin said that the idea that the Open Market Committee should carry on operations having to do with the supply of reserves by operating in the long-term market was entirely inconsistent with having a good Government securities market, that a dealer could not be expected to stay in the business if he felt that the Federal Reserve in its judgment would attempt to effectuate credit policy by intervening in the long-term market. He said that he was not interested in the Government securities dealer per se but that he was greatly interested in the Government securities market, that over a period of time there must be a reasonably good Government securities market in order that the Committee might effectuate its credit policies, and that while general credit policies which might be adopted by the Committee would affect prices and yields on Government securities, the additional uncertainties that might be caused by the threat of Committee operations on a large scale in long-term government securities might destroy the market.

While specifically differing on this point, and steadfastly opposing any commitment to stick to a given sector of the market, the President of the New York Bank found a limited area of agreement with the subcommittee:

Mr. Sproul thought dealers could and would stay in business even though the possibility of Federal Reserve intervention in the long-term market continued. . . . He was talking about preserving freedom of action for the Committee in the future. He thought the Committee could say, in season and out, that its purpose and policy now is to effectuate credit policies through supplying or absorbing reserves and not to support any pattern of rates or prices or yields in the Government securities market, but he did not think the Committee could give any other assurance which would be worth while in terms of its effect in the market or in terms of what the Committee might or might not do at some future time.

Despite the sharpness of the debate and the width of the gulf that separated Mr. Sproul and the Ad Hoc Subcommittee, the minutes of the March 4-5 session record that the vote to approve the Subcommittee's recommendations (insofar as they related to operating policies) was unanimous. Mr. Sproul's assent, however, was accompanied by his stated assumption that the Committee's consensus referred to the "present situation in the market and present open market policy, and not to a permanent philosophy with respect to nonintervention in the long-term market." The question of a public anouncement of "ground rules" was shelved for further study.

After 1953, "bills only" did not again play a significant part in Committee discussions for some seven years. It remained a sore point between the New York Bank and the Board of Governors, however, and sharp skirmishes over it occasionally flared in the FOMC. A vote on renewing the policy came up annually; almost every year there was a sole dissent by the President of the New York Bank (Mr. Sproul until his retirement in 1956; thereafter his successor, Mr. Alfred Hayes).

The policy remained in effect until February 1961, when a combination of balance-of-payments considerations and the state of the domestic economy led the Committee to vote a departure from it. Details of the

Committee discussion that led to this decision presumably won't be on view until release of the next batch of FOMC minutes. A foreshadowing of the move, however, appears in the record of the meeting of October 25, 1960, when the Committee voted to extend its operations to securities of up to fifteen months' maturity (previously the one-year Treasury bill had been the limit). Mr. Hayes, citing balance-of-payments considerations, argued against "further overt measures of monetary ease." He said that it was going to be necessary to give the banking system a "substantial injection of reserves," and that this created a problem because any further decline in market yields on Treasury bills was undesirable from an international standpoint. He suggested an approach:

". . . it might be well to broaden our open market purchases to include short-term securities other than bills, even though the available supply of such securities is probably not very great. . . .

"Beyond this, I think we should remain especially alert to developments in the long-term capital markets. It may be that the present heavy atmosphere will clear up once the current American Telephone and Telegraph financing is out of the way, particularly if the Treasury refunding is kept out of the long-term area. But if congestion still remains, this may prove to be one of those comparatively rare occasions when the Committee should give serious study to the possibility of limited operations in the longer end of the market to clear the air and encourage a lower rate level."

III. Washington and New York

While debate over "bills only" in the Committee was not heavy after March 1953, the minutes of a number of meetings make clear that positions taken on the policy had created tension between the Board of Governors and the New York Bank. This was not only because the Board's staff had played a major role in developing the "bills only" concept. It also reflected the fact that execution of the policy fell to the trading desk of the New York Bank, whose President was in essential disagreement with the policy.

At an executive session of the full FOMC held on the afternoon of March 2, 1955, Chairman Martin drew the issue in sharp focus. Speaking of the relationship between the Open Market Committee and the Manager of the Open Market Account, he said:

". . . we have a structure and procedure that is neither fish, nor flesh, nor fowl. . . . We find ourselves, as a Committee, in the position of managing a $25 billion bond account. We are directly responsible in the public eye for the operations of that account in a far greater degree than the Board of Governors or the Board of Directors of any individual Reserve Bank is for the operations of that Federal Reserve Bank. In my thinking—and again I want to say I eliminate all personalities—that responsibility is shared by all of the members of the Open Market Committee with respect both to the management of the account and the manager of the account. I do not think our present arrangements permit us to discharge that responsibility adequately. . . . I am clear that the manager of the account should not be reporting to the Board of Directors of the New York Bank either directly or through the President of the New York Bank. . . .

"I am not putting this in terms of abuse or of fraudulent operations, but I do not consider it proper that the manager of the account, because he is also a Vice President of the New York Bank, should be in a position where he might have to resist pressure to report operations that the directors of that Bank may wish him to report to them. . . . What we are really doing here is talking about the evolution of the place of the manager of the Open Market Account in the System. That account has become much more important than it was when it was set up a good many years ago. It is now becoming vital to our operations that we are and appear to be absolutely clean-cut in the exercise of our responsibility, the shared responsibility of all of us. We should be clear that nobody have any inhibitions with respect to moving at any time to ask for the removal of the manager of the account, if he feels that request is his duty at the time."

Mr. Martin then made a formal motion that "As a matter of formal status it is the sense of the Federal Open Market Committee that the Manager of the System Open Market Account be made more directly responsible to the Open Market Committee as a whole."

In reply, Mr. Sproul proposed "to discuss this whole matter, and in more specific terms than might be implied by the motion which has been placed before us. Its importance lies in what is not explicit in the motion." He continued:

"There is and can be no question of deviation from the authority of the Federal Open Market Committee unless the Federal Reserve Bank of New York is incompetent or unworthy of trust. If what is now contemplated is to remove the Federal Reserve Bank of New York from the chain of command, it means placing all of our confidence in one man, instead of in an institution and a man. That one man would report directly to the Committee, presumably to the Chairman or the Vice Chairman.

"The result would be either (a) to vest more discretionary authority in the Manager of the System Open Market Account or (b) to place upon the Chairman the dual role in the Committee now attributed by the ad hoc subcommittee to the President of the Federal Reserve Bank of New York or (c) to have the job of continuous and detailed administrative surveillance of the Manager of the Account devolve upon some member of the staff of the Chairman or [*sic*] the Board of Governors, or (d) to leave the President of the Federal Reserve Bank of New York about where he is, but without institutional responsibility and integration. . . .

"Make no mistake, the change in our organizational structure, which is here suggested, would be taken as a sign of no confidence in the competence and trustworthiness of the Federal Reserve Bank of New York."

The executive session concluded by authorizing appointment of a six-man committee "to bring back concrete proposals for perfecting the structural and operating organization that will best implement the policies of the Federal Open Market Committee." Mr. Martin appointed the committee, but it never rendered a report and the issue was allowed to die. One change, however, did grow out of the Board-Bank debate: before, the Manager of the Open Market Account had been selected by the directors of the Federal Reserve Bank of New York and approved by the Open Market Committee; this sequence was reversed, so that now the selection is at the initiative of the FOMC, subject to approval of the New York Bank.

Enlarging the role

What the Chairman had sought in the executive session was an overt further affirmation of Committee authority, at some symbolic cost to the position of the principal regional Bank—and inferentially to the positions of all twelve. He did not get the specific thing he had gone after, but he did open the way to considerable subsequent enlargement of the Committee's role. Mr. Martin had commented at the March 2, 1955 meeting: "If my concept is correct as to the 'System', then it is important that we discuss here [in the FOMC] possible changes in margin requirements, discount rates, and reserve requirements, to consider how they may be integrated with open market operations."

The minutes of the Open Market Committee meeting on November 16, 1955, show the leadership role Mr. Martin saw for the Committee:

> Chairman Martin stated that before proceeding with discussion of open market operations, he wished to comment on a telegram he had sent to the President of each Federal Reserve Bank under date of November 9, 1955 suggesting that, without implying that action should be taken on the matter, there be a full review of the discount rate by the directors of the Reserve Bank at their next meeting. That wire, Chairman Martin said, was sent out in accordance with the general thought that the Federal Open Market Committee should be the focal point of discussions of all aspects of System credit policy. . . .

The consensus of the meeting strongly favored an increase. Two days later, six Reserve Banks announced that, with Board approval, they were raising the discount rate from 2¼% to 2½%. The others followed within a week.

Further insight into the Committee's evolving role—and an interesting Chairman's-eye view of the Federal Reserve System—emerge in the record of the March 3, 1959 meeting of the FOMC:

It was at a time like this, the Chairman said, that the System must face up to the complexities of its organization. While there might be some question whether the System was organized properly, it was necessary to work with what existed and not with what some would like the situation to be. As a general rule, Reserve Bank directors' meetings were spread over a period of weeks, and there would not be a conjunction unless the Board set the meeting dates for every Federal Reserve Bank. From time to time in the past, the System had relied on the New York Bank or some other Reserve Bank to act as a leader to pull the thing together, but at present the situation was difficult because there would be different judgments, as in fact there were around the table today.

In his own judgment, Chairman Martin said, the economic factors now called for an increase in the discount rate. . . . Where the matter would end up he did not know, but personally he would like to see those banks whose directors were to meet on Thursday of this week [New York, Philadelphia, Chicago] take action on the discount rate and dispose of the matter, for he felt that this would be better for the System, even realizing that many misinterpretations would be placed on the action. He would be glad if Mr. Hayes were to express his (Chairman Martin's) personal judgment to the New York Board of Directors for he would like them to know what he thought about the situation.

On March 5 the Federal Reserve Banks of New York, Chicago, Philadelphia, and Dallas raised their discount rates from 2½% to 3%, effective the following day.

A place in the process

Besides tracing the course by which the Open Market Committee has assumed a broader monetary role, the annals of the group's meetings also shows a delicate awareness—especially on the part of the Chairman—of the relationship that role bears to the economic policies of the national Administration. This has not meant acceptance by the Federal Reserve of a subservient place; rather, it has meant recognition that the Federal Reserve is part of, and not separate from, the process by which national economic policy is formulated.

A case in point appears in the minutes of a Committee meeting held July 7, 1959. At the time, the Eisenhower Administration was struggling to finance a massive budget deficit caused by the 1957-58 recession. The FOMC was pursuing a vigorous policy of credit restraint. The Administration had a bill pending before the Democratic-controlled Congress to lift the 4¼% interest rate ceiling on Treasury bonds of more than five years' maturity. This excerpt relates Chairman Martin's advice to the Committee:

From the minutes of the preceding Committee meeting, which he had been unable to attend, he noted that comments had been made by several persons to the effect that any errors should be on the side of restraint rather than ease. Errors this time, he suggested, ought to be on the side of ease rather than restraint, although he would not want to see policy changed at all. He felt that he would not be discharging his responsibility unless he indicated to the Committee that he felt the Government securities market was in a critical state. Here he was not talking about ghosts but about the actual situation, a situation complicated by the fact that every move on the part of the System was being watched closely at the present time.

The last thing he would want to suggest, the Chairman said, would be to ease System policy simply to obtain legislation to eliminate the interest rate ceiling on Treasury bonds and to increase the rates on savings bonds. He had been under personal pressure recently to indicate that the Federal Reserve might ease its policy but he had not given an inch. However, looking at the pressures over which no one has control, he felt it necessary to be careful that the System did not conduct itself in a way that might look like a deliberate action to nettle people. Interest rates were now a major issue politically. They might be made a national issue in 1960 or during the next few months, and at the moment he was not sure whether it would be possible to get a bill on the interest rate ceiling through Congress. Again he emphasized that he was not suggesting an easing of policy, but he thought the System must be careful not to overdo the matter of restraint at a time when the money market had a real bite in it. . . .

Chairman Martin then referred to the saying he had used on other occasions that the iron which breaks is not as strong as the steel which bends. . . .

IV. What May Come of It

On the whole, the U.S. approach to central banking and monetary policy comes off well under the scrutiny made possible by the unlocking of the Open Market Committee's record of deliberations over a quarter-century period. Mythologists will be disappointed at the total absence of "money mystique" in the talk that goes around the Committee table. Scholars may wish there were more dwelling on theoretical concepts. But anyone who troubles to read the record for any distance should be pleased at the amount of common sense that laces the pages.

Proponents of radical revision of the Federal Reserve System—for example, abolition of the Open Market Committee, as was recommended by the Commission on Money and Credit in 1961—will find in the minutes little if any support for their case. The CMC proposal, which would turn the Committee's functions over to the Board of Governors, is embodied in legislation introduced in the present session of Congress by Representative Wright Patman.

In effect, those who would do away with the FOMC deny the contribution of the twelve Reserve Banks to this country's unique federal system of central banking; they deny the value of having a group within the central bank which can bring to the policy process an intimate, working knowledge of business and finance across the nation; they deny the "cardinal principle," as Mr. Sproul has called it, of central banking—namely, that some of those who determine policy should also live and work in the private money markets. In all this they go against the evidence of the 1936-60 minutes, which bear overwhelming witness to the soundness and the worth of the national-regional, public-private system as it has evolved in the U.S.

At the same time, study of the minutes points up that formality is significantly at variance with fact in some aspects of monetary management. Changes to bring prescription into line with practice would give official sanction to what logic already has installed.

For example, the statutory rotation of places on the Open Market Committee among eleven regional Reserve Banks—with four having seats at a time, and only New York having a permanent seat—understates the role of the regional Banks in making policy. It is clear from the Committee minutes that, in the actual give-and-take of FOMC meetings (*box on page 202*), the consensus which the Committee forges every three weeks as a preface to its policy vote has been the product not only of the twelve statutory members, but rather of the full group of nineteen (seven Governors and twelve presidents). It would be logical to recognize this situation and give voting power, as well as talking rights, to all the regional Banks at all times.

There is a mathematical snag in doing this while preserving a voting majority in the Committee for the Board of Governors, the publicly appointed members of the group. One solution would be to make all twelve Reserve Bank presidents permanent members of the FOMC, each with half a vote. The Board would still dominate, seven to six. This formula might be thought to entail a relative loss of position for the Federal Reserve Bank of New York, which now has both a permanent seat and a full vote. That institution's influence in the Committee and in the Reserve System, however, derives from factors more fundamental than the size of its vote. And the proposed change would strengthen and regularize the contribution which the regional Banks as a group make to the work of the Committee.

Another change suggested by the record concerns the location of the initiative for discount rate changes. Formally it now is lodged in the boards of the regional Banks, each numbering nine directors. The immediate impetus for a rate change, however, usually comes out of a meeting of the Committee; and the Governors, when considering a Reserve Bank proposal for a rate change, try to get the matter discussed by the Committee before deciding whether or not to approve. When the initiative actually comes from the Committee, the sheer job of communications that must be done in order to get action by the individual Reserve Banks—and under the most compelling need for secrecy—is formidable even in an electronic age. When it is desirable to move extremely fast—as was the case last November, when the rate was raised from 3½% to 4% in response to the posting of a 7% bank rate in London—the difficulty is compounded.

Giving the FOMC formal authority to initiate and announce changes in the discount rate would eliminate a fair amount of procedural fuss and bother merely by formalizing what in many cases is now done informally. The autonomy of the regional Banks would be preserved by continuing their right to propose discount rate changes for Board approval as at present, and also by allowing any of them to decline to follow the FOMC in a change if the directors felt conditions in their Reserve District were sufficiently different from those in the nation as a whole to justify such a departure.

Actually, the central role which the Open Market Committee now plays in the whole process of forming and executing monetary policy would be fully recognized only by bringing officially within its scope also such matters as reserve requirements, the administration of regulation Q, and the setting of margin requirements for the carrying of securities. All this suggests that the Committee's official title, connoting limitation to but one of the Reserve System's means of monetary management, is a gross understatement. Perhaps it too should be amended.

Whether or not specific changes in structure—or in name—come about as a result, the opening to public view of the formal record of monetary deliberations can hardly help being of benefit to the ongoing work of the money managers. To the extent that the record is examined, it will improve understanding and reinforce confidence. The only possible cause for regret is that disclosure waited so long. Hopefully the release of minutes after a reasonable period of aging—say three years—will become routine. Secrecy is indispensable to the makers of monetary policy for the near term, but in the long run they have nothing to hide.

16

Are there enough policy tools?

Warren L. Smith

I. Introduction

If I were asked to sketch the main features of the world economy as it exists today, with particular reference to the advanced industrial countries, I would describe the situation in the following way:

1. Each country insists on the free exercise of its sovereign authority to regulate its level of internal demand for the purpose of maintaining suitable economic conditions at home in terms of employment and the behavior of its internal price level. In addition, the idea of governmental responsibility for the attainment of a satisfactory rate of long-term economic growth is widely accepted, and some efforts are made to influence the composition of demand in favor of investment-type activities as a means of accelerating growth.

2. In many countries, fiscal policy is, as a practical matter, a relatively inflexible instrument, partly because of unsatisfactory administrative arrangements and in some cases also because of outmoded and unenlightened views about budget deficits. Monetary policy is administratively more flexible, but, in practice, freedom to use it for domestic purposes has become increasingly limited by balance-of-payments considerations.

3. Free international movement of goods and of capital as a means of achieving efficient use of resources is generally accepted as a goal, and substantial progress has been made in achieving it. In particular, since the advent of general currency convertibility in 1958, controls over the international flow of capital have been relaxed and investors have become increasingly inclined to shift funds internationally in response to differential changes in expected rates of return.

Reprinted by permission of the publisher and author. *The American Economic Review,* Vol. LV, No. 2 (May, 1965), pp. 208-220. Warren L. Smith is Professor of Economics, University of Michigan.

4. Subject to some important qualifications to be discussed below, trade is conducted under a system of fixed exchange parities, with actual exchange rates fluctuating only within very narrow limits around these parities.

5. Countries hold limited supplies of monetary reserves in the form of gold, foreign exchange, and lines of credit at the IMF. The reserves available and potentially obtainable set a limit—albeit a somewhat elastic one—on the cumulative size of a country's balance-of-payments deficit. For this reason, each country operates subject to a balance-of-payments constraint—not in the sense that the balance of payments must always be in equilibrium but in the sense that there is some limit on the size and duration of deficits that can be tolerated. There is no corresponding limit for surpluses.

Some of the features in the above list need to be spelled out a little more fully. First with regard to the goal of internal stability (item 1 above), countries are often said to seek the twin goals of "full employment" and "price stability." For some time, however, it has been getting increasingly clear that this way of describing the situation is quite out of touch with reality. Indeed, the concept of full employment, while perhaps useful as a slogan, is without precise meaning. A more accurate way to describe the situation is as follows: There is in each country a "trade-off" between employment (or unemployment) and price stability; that is, over a considerable range the more unemployment is reduced by policies to expand aggregate demand the higher is the price that must be paid in terms of inflation. This relation holds primarily because of the tendency of money-wage increases to outstrip increases in productivity even under conditions of substantial unemployment. The trade-off varies from country to country, depending on the organization, traditions, and aggressiveness of the labor movement, the price policies followed by industry, and so on, and from time to time depending on the attendant circumstances. The trade-off may be influenced by policy measures—wage-price guideposts, incomes policies, etc.—but I am not aware of any cases in which efforts to change it have been notably successful. Not only does the trade-off itself vary from country to country but so also do the relative weights attached to price stability and employment in the implicit social welfare functions that govern the behavior of the authorities responsible for economic policy in the various countries. As a consequence, to the extent that each country is left free to decide what combination of price inflation and employment to select from the many choices open to it, price trends may vary considerably from country to country.[1]

[1] Much lip service is paid to price stability as a goal of national economic policy but in reality it is an extremely unrealistic objective, since no one is in fact willing to pay the price in terms of unemployment required to achieve it. And since it is a practically unattainable goal for individual countries, it is obviously an unrealistic objective for the world as a whole.

The other feature that calls for further comment is exchange rate arrangements (item 4 above). Although, as indicated above, fixed exchange rates appear to be one of the generally accepted goals of economic policy, we do not now have a system of really fixed rates. Indeed, the most serious difficulty with the present international monetary system seems to lie in the area of exchange rate policy. The present arrangements, under which exchange rates are fixed within very narrow limits at any particular time but are subject to readjustment from time to time to correct "fundamental" disequilibria in national balances of payments, seem ideally calculated to encourage speculation. Since opportunities for the investment of capital, viewed broadly, do not ordinarily vary widely as between major countries, even a mild suspicion that a country may devalue its currency can cause a speculative outflow of capital from that country. And, as more and more investors become familiar with the possibilities of transferring capital internationally, it seems likely that the potential size of speculative capital flows will become even larger. The result is that most countries will entertain the possibility of devaluation only in the most dire emergency, but the threat is nevertheless sufficient to induce speculation. And there is always the possibility that speculation will exhaust a country's reserves and force the devaluation that the speculators are hoping for.

It does not strain reality very much to describe the world economy we seek to achieve as one in which (1) there is complete freedom in the international movement of goods and capital, (2) exchange rates are fixed, and (3) individual countries are free to use monetary and fiscal policy—primarily the former—to attain their domestic price and employment goals. And the system is subject to the constraint that each country possesses a limited supply of reserves with which to cover deficits in its balance of payments.

The trouble with this system is that it is basically inconsistent. There are three possible ways of correcting a deficit or surplus through adjustment of the current account: through the use of trade or exchange controls, through an adjustment of exchange rates, and through internal price and income changes. Since all of these violate the principles of the system, they are ruled out. Consequently, when a country experiences a deficit, there is no assurance that the deficit will be eliminated before its limited supply of reserves is used up. And the difficulty is further compounded by the nature of the prevailing exchange rate system under which a persistent deficit creates a fear of devaluation, possibly leading to a speculative outflow of capital and an unlimited self-generating expansion of the deficit.

Thus, if the underlying principles are adhered to, there is no mechanism that can be depended upon to eliminate a balance-of-payments disequilibrium brought about by such disruptive forces as changes in tastes or technology. Beyond that, even if the system is in equilibrium to begin with, the diverse price trends inherent in the independent economic policies of the member countries may themselves in due course produce disequilibrium.

Of course, the system has worked after a fashion—although the road has been pretty rocky, especially in view of the fact that only six years have elapsed since the restoration of convertibility in 1958. And it has survived only because its fundamental principles have been violated in various ways.

1. The underlying principle of free movement of goods and capital has been compromised through the imposition of trade restrictions for balance-of-payments reasons by Canada in 1962 and the United Kingdom in recent weeks. The United States has also persistently violated the principle by tying foreign aid, by discriminating in favor of American suppliers in its defense procurement policies, and by the recent enactment of the so-called "Interest Equalization Tax."

2. Minor use has been made of exchange rate adjustments in the Dutch and German revaluations of March, 1961. Such adjustments, however, probably do more harm than good by weakening confidence in the overall stability of exchange rates and encouraging speculation.

3. In practice, domestic monetary and fiscal policies have not been entirely unaffected by the balance-of-payments situation. In part, this is because, due to the less than perfect effectiveness of domestic monetary and fiscal tools, it has not always been feasible to offset completely the automatic corrective effects of deficits and surpluses on internal demand. But beyond that, deficit countries have found it necessary to adapt their domestic policies to the exigencies of the balance of payments—albeit reluctantly—when their international reserves have been seriously threatened. The leading example here is the United States, which has suffered from an unnecessarily high rate of unemployment and a resultant irrecoverable loss of output amounting perhaps to $150 billion in the last five years, partly as a result of its balance-of-payments deficit. This has not been due, in my opinion, however, to the fact that policies to expand aggregate demand have been held back by fear that they would worsen the trade balance. Rather it has been a result of the administrative inflexibility of fiscal policy (as a consequence, in part, of unenlightened views about budget deficits and growth of the public debt) combined with the fact that the need to avoid an accentuation of short-term capital outflows has acted as a significant constraint on monetary policy—the one flexible and acceptable instrument that might in the absence of the balance-of-payments constraint have been used to expand aggregate demand. No doubt similar considerations have to some extent operated in European countries to limit the use of restrictive monetary policies to check excessive inflation in the face of balance-of-payments

surpluses. But it is quite clear that the system has an inherent deflationary bias. The limited supply of reserves sets some upper bound on the size of a nation's cumulative deficit, whereas there is no equivalent upper bound to the size of a cumulative surplus and the associated expansion of monetary reserves.

Techniques of central bank cooperation through the use of currency "swaps" and intervention in foreign exchange markets to deal with minor speculative crises have been progressively developed and refined. To deal with more serious crises—the most dramatic of which, to date, is the sterling crisis of recent weeks—massive supplies of foreign exchange have been mobilized to support threatened currencies and combat the activities of speculators. Up to now, these efforts have been successful in fending off disaster, but the world lives in dread of a forced devaluation of sterling or the dollar, which would in all probability create a state of international financial chaos from which recovery would be extremely difficult. Moreover, the present situation gives the opinions of international currency speculators an entirely disproportionate weight in the determination of economic policy.

It seems clear that the present international economic arrangements are seriously defective. It is doubtful whether they can survive in their present form for very long. The question is: What changes should be made? In the extensive discussions of the international economy that have taken place in the last few years, much energy has been expended on the formulation of proposals for reform of the world's monetary institutions, and many ingenious schemes have been proposed. However, it is difficult for me to see how the mere establishment of an improved international banking arrangement for providing liquidity can be depended upon to yield a satisfactory solution to our problems as long as private international transactions are conducted by the use of national currencies whose exchange parities are felt to be subject to change. Even if gold were completely demonetized and official settlements between countries were carried out by the transfer of credits on the books of a reformed IMF, the problem would remain. Suppose, for example, that under such an arrangement the United States were to run a substantial balance-of-payments deficit. The dollar would decline in foreign exchange markets, and foreign central banks would have to buy dollars to prevent the exchange rate from moving outside the prescribed limits. The dollars would be deposited with the IMF for credit to the other country's account, and they would be debited to the U.S. account. If the latter account were to become exhausted, the United States would presumably have to arrange a loan from the IMF which would replenish its balance. As long as there was felt to be an effective limit to the credit line available, so that eventually devaluation (or direct controls) might have to be employed to correct the deficit, private investors would have essentially the same motive to speculate against the dollar as they have under the present system, and the speculation itself would help to exhaust the credit line. Since claims against the IMF would presumably be subject to an exchange guarantee, countries might be willing to extend sufficiently large credits through the IMF to make the effective supply of reserves available for meeting speculative runs larger than it is at present. But there is no assurance that the fundamental problem of speculation would be eliminated.

Schemes to reform the financial system and increase the supply of monetary reserves can undoubtedly be of some help. But the more fundamental need is to introduce some workable mechanism for restoring and maintaining balance-of-payments equilibrium. Can this be done more or less within the confines of the present system by making more flexible use of traditional policy instruments? Or are some fundamental reforms of the system required, and if so, what reforms would be appropriate? These are the questions to which I shall devote the remainder of my paper.

II. More Flexible Use of Monetary and Fiscal Measures

I believe a considerable improvement in economic performance could be achieved within the framework of the present international monetary system if fiscal policy could be rendered substantially more flexible than it now is. For example, in the United States some arrangement, such as that proposed by the Commission on Money and Credit in its 1961 report or that recommended in the January, 1962, *Annual Report of the Council of Economic Advisers,* giving the President discretionary authority to change personal income tax rates for countercyclical purposes would be very helpful.

The idea would be to develop a policy arrangement under which the Western industrial countries would agree to rely on flexible fiscal policy, implemented primarily through tax adjustments, to regulate internal demand to achieve domestic goals. Monetary policy would then be assigned the task of maintaining balance-of-payments equilibrium by establishing interest rates at levels which would induce a sufficient inflow or outflow of private capital to cover the deficit or surplus on current account (including government military and foreign aid transactions) that would occur at target levels of income and employment. To be workable, such an arrangement would require that tax policy be rendered much more flexible than at present, not only in the United States, but in other countries as well.

Machinery would need to be set up to provide careful international coordination of the monetary policies of participating countries. The objective would be to establish a matrix of interest rate differentials among countries which would be sufficient to achieve approximate overall equilibrium in the balance of payments of each country. Marginal adjustments would need to be made in interest rates from time to time to preserve equilibrium in the face of changes in underlying conditions. Strong efforts would be needed in order to avoid competitive increases in interest rates which would raise the general level of rates without contributing to the maintenance of payments equilibrium. It would be highly desirable that steps be taken to increase the freedom of capital movements—especially of long-term funds—in order to make capital flows adjust more sensitively to interest rate changes.

Under such a system, the mix of monetary and fiscal policies would be used to achieve internal and external equilibrium simultaneously. If, for example, it was necessary to raise interest rates for balance-of-payments reasons, any undesired restrictive effects on internal demand would be offset by a reduction in taxes. Fairly frequent adjustments in both monetary and fiscal policies would presumably be necessary. Since, at best, only an imperfect adjustment could be achieved, monetary reserves would, of course, continue to be needed to deal with temporary balance-of-payments deficits. But, hopefully, persistent large deficits resulting in heavy drains on reserves could be avoided more and more effectively as accumulating experience led to increased skill in the use of the available policy instruments and in the coordination of monetary policies.

One of the objections to such an arrangement is that the use of the monetary-fiscal mix as a means of dealing with the balance of payments precludes its use to regulate capital formation for economic growth. To enable the country to regulate capital formation and thereby influence growth in the face of the adjustments in interest rates that would be needed to maintain balance-of-payments equilibrium, a second flexible fiscal instrument could be introduced. The best possibility for this purpose would probably be an investment tax credit, along the lines of the 7 percent credit introduced in the United States in the Revenue Act of 1962. Provision could be made for flexible adjustments of the rate of tax credit when such adjustments were deemed desirable—as, for example, to offset the restrictive effect on investment of a rise in domestic interest rates called for by balance-of-payments considerations. It is in principle possible to have—within limits at least—any desired level of investment in combination with any desired level of interest rates through appropriate use of fiscal incentives to shift the marginal efficiency of investment schedule.

Would such an arrangement be feasible as a means of eliminating the contradictions in the present system? This I find difficult to judge. If price trends among participating countries diverged persistently leading to progressively larger current account deficits and surpluses to be covered by interest-induced capital flows, the arrangement would in due time prove to be unworkable. Some supplementary efforts to coordinate other policies to prevent this would therefore be necessary. The possibilities of success seem sufficient to warrant consideration. Of course, the political difficulties of obtaining greater flexibility of fiscal policy are undoubtedly substantial in some countries. But this problem will almost certainly have to be tackled anyway, because monetary policy is already sufficiently hamstrung by the balance-of-payments problem to make it an ineffective instrument of domestic policy in many countries. More use will have to be made of fiscal policy to achieve domestic goals even if no effort is made to achieve such far-reaching international coordination of policies as that described above.

This means of achieving balance-of-payments equilibrium would not, of course, be entirely optimal, because the equilibrating adjustments would occur entirely in the capital account. It should be remembered, however, that a balance-of-payments surplus is a form of national investment, and a deficit is a form of disinvestment. If a country ran a persistent surplus, it could, in principle at least, use tax adjustments to trim its rate of domestic investment so as to achieve the desired overall division of national resources between consumption and capital formation. Similarly, a country having a persistent deficit could offset its adverse growth effects by employing a policy mix that would encourage domestic investment. Some inefficiency would nevertheless be present, because if domestic investment was viewed as more conducive to growth than an equal amount of accumulation of foreign claims, there would be no way to shift resources between the two. The current account surplus or deficit would be primarily determined by the level of overall domestic demand and there would be no policy instrument available to change it.

III. Approaches to Fundamental Reform of the System

The arrangements outlined above would be a possible way of making the present system work more effectively by means of a much more flexible and sophisticated use of monetary and fiscal policies. If such an arrangement is unacceptable—or, after a trial, proved to be unworkable—there are as far as I can see only two approaches to fundamental reform: currency unification or the adoption of flexible exchange rates. Experience gives us strong reasons for believing that the first of these is definitely workable; however, it would involve a substantial cost that countries might well not be willing to pay. The second might involve a less serious cost but is less certain to work effectively. The two solutions could be combined in various ways.

A. Currency unification

One solution to the problem of lack of sufficient tools to achieve the desired goals would be through the establishment of a unified currency system with absolutely and permanently fixed exchange rates. In this way, the balance-of-payments problem could be eliminated entirely, but I shall argue that the price that would have to be paid to make such an arrangement acceptable and viable would be the surrender of sovereignty over monetary and fiscal policies by the nations involved to a central body.

Under a fixed exchange rate system, individual countries would from time to time experience structural balance-of-payments deficits resulting from changes in tastes, technology, and so on, and some corrective mechanism would be necessary to restore external equilibrium in these cases. Since prices in industrial countries are characteristically rigid in a downward direction, deflation, which would serve primarily to create unemployment of labor and capital, is both an economically inappropriate and politically unacceptable means of dealing with a balance-of-payments deficit. Under modern conditions, the maintenance of high levels of employment and capacity utilization requires that such changes in relatives prices as are needed to correct chronic balance-of-payments disequilibria be accomplished primarily through price increases in surplus countries rather than price declines in deficit countries. In addition, as indicated earlier in this paper, the procedures of wage determination are such in most countries as to produce some inflationary tendencies at acceptable levels of unemployment. Thus, with fixed exchange rates it is necessary that the international monetary system have a moderate inflationary bias built into it. However, under fixed exchange rate systems in the past the pressure on deficit countries to eliminate deficits has been much stronger than the pressure on surplus countries to eliminate surpluses, so that such arrangements have characteristically had a substantial deflationary bias. Unless some means could be found to eliminate this bias, it is doubtful whether a fixed exchange rate system would be able to survive in the modern world.

By means of a thoroughgoing currency unification, however, it would be possible, in principle, to devise a fixed exchange rate system which would not have the deflationary bias that has characterized past arrangements of this kind. The banks in each country might agree to accept for deposit at par checks drawn on banks in all of the other countries. Thus, exchange rates would be absolutely fixed (with no margin of fluctuation). To simplify the bookkeeping, it would be desirable to redefine the national units of account so as to permit all exchange rates to be set equal to unity. The banks of each country would give credit in that country's currency for all checks deposited, no matter where the checks originated or in what currency they were denominated. Each bank would send all checks denominated in other currencies to its central bank for collection. Settlements between central banks would be handled by reciprocal accounts, or, better yet, through an international clearing agency along the lines of our Federal Reserve Interdistrict Settlement Fund. All barriers impeding the free flow of capital among countries would be removed.

The key characteristic of such an arrangement is that the central bank of each country would have an unlimited credit line with the central banks of the other participating countries. The availability of unlimited credit would eliminate the deflationary bias that has ordinarily characterized fixed exchange rate systems. However, if each country was left free to pursue an independent monetary and fiscal policy, serious difficulties might arise as a result of differences in economic structure or policies among the various participating countries. Suppose, for example, that an important country chose to follow an inflationary domestic policy—perhaps because cost-push pressures on its price level made such a policy necessary for the achievement of the desired level of employment. The inflationary policy would tend to generate a balance-of-payments deficit, through which the inflationary pressure would be transmitted to other countries. These countries could take domestic action to offset the inflationary pressures, but this would mean an enlargement of the inflating country's balance-of-payments deficit, the counterpart of which would be a surplus in the consolidated balance of payments of the other countries. If the other countries did not wish to run surpluses in their balances of payments and thereby provide a flow of goods and services to meet the rapacious demands of the inflating country, their only recourse would be to inflate their own economies in pace. In other words, by following an inflationary policy, a single major country might be able to force the rest to choose between balance-of-payments surpluses in its favor and domestic inflation. Thus, a fixed exchange rate system which provided unlimited automatic credits to deficit countries while leaving participants free to pursue independent national monetary and fiscal policies would probably have an immoderate inflationary bias and would certainly be unacceptable.

The necessary condition for currency unification to be workable is that the participating countries give up their sovereign authority to conduct independent monetary and fiscal policies directed at internal price and employment goals. Such policies would have to be conducted by a centralized monetary and fiscal authority charged with responsibility for internal stability for the group of countries as a whole. This centralized authority, in order to carry out its responsibilities effectively, would need to have sole power to regulate the supply of money and credit and to levy certain taxes and control certain categories of government expenditures. Under such an arrangement, internal price and income changes and interest rate adjustments would take care of the balance-of-payments problems of the individual countries. The arrangement would be very similar to the internal monetary system of the United States, and balance-of-payments problems would presumably no longer be a matter of concern.

The price that would have to be paid for this arrangement would be the loss of sovereignty over economic policy by the individual participating countries; in this respect, their position would become similar to that of individual states or Federal Reserve districts in the United States. Thus, individual countries would no longer be able to choose their optimal levels of internal demand but would have to accept the levels that were associated with the overall policies judged by the central authorities to be appropriate for the group as a whole. In other words, some countries might find themselves in the position of depressed areas—a position very similar to that of a state like West Virginia in the U.S. federal system. The central government could, of course, alleviate localized distress by programs of expenditures and tax incentives aimed at the stimulation of production and employment in depressed areas, just as the federal government is able to do in the United States.

The above argument needs to be qualified a little. Actually, it would be possible for participating countries to engage to a limited extent in fiscal policy to stimulate or retard aggregate demand to influence their internal employment and price levels. However, monetary policy would have to be centralized, and countries would have to finance their deficits by selling their securities at interest rates that would make them acceptable to investors in the financial environment generated by the central monetary authorities. And, since they would not possess the power to create money, their securities would not be free of default risk; indeed, like the securities of our state governments, their rating would presumably depend on their financial condition. As a result, persistent deficit financing might be prohibitively difficult and expensive, especially for those countries in a depressed economic condition, for which it would be particularly important. In practice, it would no doubt be desirable for fiscal action to regulate demand to be conducted almost entirely by the central government, both because such action by the constituent members would prove to be difficult and costly and because there might otherwise be troublesome competition in economic policy and serious difficulties in the proper coordination of fiscal and monetary policy.

Since the power to create money is perhaps the fundamental element of national sovereignty, it would be difficult—although perhaps possible—for the participating countries to carry out independent military and foreign policies, especially those involving heavy expenditures overseas for national security and economic aid to underdeveloped countries. With monetary sovereignty eliminated, the other elements of national sovereignty would probably wither away, with the corresponding powers being shifted to the central government.

While currency unification might provide a satisfactory solution to the problems of economic policy, it would require, directly and indirectly, such a sweeping surrender of the accepted and widely revered prerogatives of national sovereignty that it is hard to believe that it would be acceptable at the present time to many countries. The United States would be especially unlikely to be willing to accept it, given our sense of responsibility as the political and economic leader of the free world.

B. A system of flexible exchange rates

A second fundamental reform that would eliminate the inconsistencies of the present system would be the adoption of a system of flexible exchange rates. This would, in principle, permit the participating countries to carry out independent monetary and fiscal policies directed at the maintenance of adequate levels of internal demand, with exchange rates adjusting in such a way as to maintain balance-of-payments equilibrium.

This is not the place for an extended discussion of the already hotly debated question of the merits of a system of flexible exchange rates. While I tend to be sympathetic to such an arrangement, I realize that it would be impossible to tell for sure how it would work until it had been tried. Moreover, the prevailing views of important officials and men of affairs are generally so hostile to the idea that I judge its general adoption to be impracticable.

C. A mixed system

The two solutions described above—currency unification and centralization of responsibility for monetary and fiscal policy on the one hand, and flexible exchange rates on the other—can be combined. According to the recent work on optimum currency areas, countries having close trading relations might properly be combined into blocks within which currencies would be unified.[2] It should be clearly recognized, in my opinion, that to be workable this would require the acceptance by the members of each bloc of a common centralized monetary and fiscal

[2] See especially R. I. McKinnon, "Optimum World Monetary Arrangements and the Dual Currency System," *Banca Nazionale del Lavoro Quar. Rev.*, Dec., 1963, pp. 366-96.

policy. Then flexible exchange rates could be employed between the blocs—which would constitute areas between which trading relations were more limited. As an example, the countries of Western Europe might constitute one bloc and the United States and the United Kingdom another, with a flexible exchange rate between the bloc currencies. A solution somewhat along these lines was suggested in the Brookings report on the U.S. balance of payments.[3]

On the face of it, this sounds like a reasonable solution. However, I find it difficult to believe that, even in such an economically interrelated area as Western Europe, individual countries would be willing to give up their historic sovereign power to control money, as would, in my judgment, be absolutely vital to the success of monetary unification. In many ways, monetary sovereignty lies at the very heart of national sovereignty in all fields, including foreign and military affairs. Moreover, in strictly economic affairs, the countries may differ very substantially in their trade-offs between price stability and employment, as well as in the weights they attach to these two competitive goals. This, too, might make them very cool toward accepting a group consensus with regard to monetary and fiscal policy.

IV. Concluding Comments

It seems to me that the present adjustable-peg exchange rate system is unworkable and has to be abandoned in favor of either firmly fixed rates or continuously flexible rates. I do not believe any of the proposed purely financial schemes for providing more reserves can be depended upon to shore up the present system and make it workable. The trouble is that with adjustable parities, the possible size of speculative runs is so vast—remember that the entire stock of private financial claims denominated in a particular currency is potentially available to finance a run on that currency—that unlimited, or virtually unlimited, supplies of reserves are needed to provide firm assurance that a speculative run could not succeed in forcing a devaluation.

What is vitally necessary is to introduce into the system some means of maintaining or restoring balance-of-payments equilibrium. Possibly this could be done by using monetary and fiscal policies in a flexible way to provide for a systematic offset of deficits and surpluses on current account with surpluses and deficits on private capital account—with some limited interim reliance on monetary reserves while the necessary adjustments were being brought about. This would have to be done, of course, without any use of exchange-parity adjustments so that over

[3] W. S. Salant *et al., The United States Balance of Payments in 1968* (Brookings Institution, 1963), pp. 258-62.

a period of time the system would become one of reliable *de facto* exchange rate stability.[4]

As I have indicated, the only other alternatives I can see are (1) currency unification combined with full unification of monetary and fiscal policies, (2) flexible exchange rates, or (3) some combination of the two. Of course, with suitable financial tinkering and *ad hoc* adjustments, the present system may survive for many years, even without a basic reorientation of monetary and fiscal policy. But its deflationary bias and its basic instability seem inherent and likely to constitute a continuing element of weakness in the world economy that may be especially dangerous in times of crisis.

[4] I suppose there are other possible ways of making the present system work, such as the introduction of exchange or trade controls to be employed under accepted rules to deal with balance-of-payments problems. However, I am convinced that such arrangements would prove in practice to be so cumbersome, unworkable, and subject to evasion and abuse that they would ultimately collapse under their own weight.

17

Europe and the dollar

James Tobin

The dollar crisis will no doubt be surmounted. "The dollar" will be saved. Its parity will be successfully maintained, and the world will be spared that ultimate and unmentionable calamity whose consequences are the more dreaded for never being described. The world monetary system will stay afloat, and its captains on both sides of the Atlantic will congratulate themselves on their seamanship in weathering the storm.

But the storm is in good part their own making. And if the financial ship has weathered it, it has done so only by jettisoning much of the valuable cargo it was supposed to deliver. Currency parities have been maintained, but full employment has not been. The economic growth of half the advanced noncommunist world has been hobbled, to the detriment of world trade in general and the exports of the developing countries in particular. Currencies have become technically more convertible but important and probably irreversible restrictions and discriminations on trade and capital movements have been introduced. Some government transactions of the highest priority for the foreign policy of the United States and the West have been curtailed. Others have been "tied" to a degree that impairs their efficiency and gives aid and comfort to the bizarre principle that practices which are disreputably illiberal when applied to private international transactions are acceptable when government money is involved.

These are the costs. Were, and are, all these hardships necessary? To what end have they been incurred?

They have been incurred in order to slow down and end the accumulations of dollar obligations in the hands of European central banks. It is fair to ask, therefore, whether these accumulations neces-

Reprinted by permission of the publishers from James Tobin THE REVIEW OF ECONOMICS AND STATISTICS Cambridge, Mass.: Harvard University Press, Copyright, 1964, by the President and Fellows of Harvard College. Vol. XLVI (May, 1964), pp. 123-126. James Tobin is Professor of Economics, Yale University.

sarily involved risks and costs serious enough for the countries concerned and for the world at large to justify the heavy costs of stopping them.

Which is easier? Which is less disruptive and less costly, now and in the long run? To stop the private or public transactions that lead one central bank to acquire another's currency? Or to compensate these transactions by official lending in the opposite direction? I do not suggest that the answer is always in favor of compensatory finance. But the issue always needs to be faced, and especially in the present case.

Several courses were open to European countries whose central banks had to purchase dollars in their exchange market in recent years. (a) They could have built up their dollar holdings quietly and gladly, as they did before 1959. (b) By exercising their right to buy gold at the United States Treasury, they could have forced devaluation of the dollar or suspension of gold payments. (c) They could have taken various measures to correct and reverse chronic European payments surpluses. (d) By occasional withdrawals of gold and by constant complaints they could have brought tremendous pressure for "discipline" upon the United States without forcing a change in the dollar parity.

European central banks and governments chose the fourth course, with token admixtures of the third. They have made world opinion, and American opinion, believe there is no other choice. Almost everyone agrees that the pressure of the balance of payments deficit upon the United States is inescapable arithmetic rather than the deliberate policy of foreign governments. Yet for almost ten years previously, United States deficits were no problem. Clearly it is a change in human attitude and public policy, not inexorable circumstance, which has compelled us to take "corrective" actions.

It is true that the concern of financial officials about "the dollar" was only an echo—and a subdued echo at that—of the fears, hopes, anxieties, and speculations that arose in private financial circles in the late 1950's. But financial officials do not have to follow the private exchange markets; they can lead instead. By an equivocal attitude toward private suspicions of the dollar, European officials kept pressure on the United States. Never did they firmly say that they would not force devaluation or suspension of gold payments. Instead, they succeeded in making the maintenance of gold-dollar convertibility at $35 per ounce a unilateral commitment of the United States, under three successive Administrations. Once a banker has solemnly assured the world and his depositors that he will never fail, he is at the mercy of those depositors capable of making him fail.

Memories are short, and gratitude is not a consideration respected in international relations, especially when money is involved. But the United States had and has considerable moral claim on European governments and central banks.

The present excess supply of dollars is in many respects an unwinding of the dollar shortage of the immediate postwar period. Capital left Europe because the continent was vulnerable to military attack, its governments were unstable, its industries were prostrate and uncompetitive, and its currencies were inconvertible. Capital has returned to Europe when events have overcome the special advantages which North America seemed to have in these respects. It is therefore relevant to recall the behavior of the United States when the shoe was on the other foot.

During the dollar shortage the United States: gave Western European countries (other than Greece, Turkey, and Spain) $32 billions of military and economic aid; lent them $11 billions additional (in spite of the default of European governments of debts connected with World War I); acquiesced in substantial devaluations of European currencies, without which European exports would still not be competitive; and acquiesced in exchange controls, capital controls, quantitative restrictions on imports, and discriminations against the United States and other non-European countries—by no means all of which are liquidated even now. After enabling Europe to overcome the dollar shortage, the United States has been expected to adjust to its reversal *without* the tools that Europe used in its turn. Rightly so, because many of these tools were illiberal expedients—the more reason for replacing them now with compensatory intergovernmental finance.

The United States has undertaken, at considerable cost in real resources and foreign exchange, to defend Western Europe against the Soviet Union. This is in theory a joint effort, but European governments do not even yet fulfill their modest commitments to NATO. While European political leaders solicit constant reassurance that United States military power will remain visibly in Europe, their finance ministers and central bankers complain about the inflow of dollars.

The United States has not only tolerated but encouraged the development of a European customs union which attracts American capital and discriminates against American exports (especially the products of industries, notably agriculture, where North America has a clear comparative advantage).

The United States has borne a disproportionate share of the burden of assistance to uncommitted and underdeveloped nations, in which European countries have a common political and, one might hope, humanitarian interest.

The United States has provided a reserve currency. In the late forties no other international and intergovernmental money was available except gold; and the supply of gold was not keeping up with the demand. United States deficits filled the gap with dollars. It is true that this gave the United States a favored position among countries.

Anyone who can print money can choose how new money will be first spent. The United States did not seek this privileged role; it arose by accidental evolution rather than conscious design. As it happens, the United States did not exploit it to live beyond our means, to make the American people more affluent. We used it rather for broad international purposes. No doubt in the long run the creation of new international money should be a privilege and responsibility more widely and symmetrically shared. But once the United States and the world are adjusted to the creation of international money via United States deficits, it is scarcely reasonable suddenly to ring a bell announcing that the world's financial experts have now decided that these deficits—past, present, and future—are pernicious.

The United States has not pushed its moral case before world public opinion. This is because many Americans believe, or prefer to believe, that balance of payments deficits, like venereal diseases, betray and punish the sins of those whom they afflict. Others regard them as simply matters of arithmetic and circumstance. Still others are afraid that making a moral argument will indicate to our all-powerful European creditors insufficient resolution to overcome the difficulties. On their side, the Europeans have neatly segregated the contexts. Their financial officials wash their hands of tariff and trade policies, agricultural protection, defense and aid appropriations, and their governments' budgets. Any European failings on these counts are facts of life to which the United States must adjust, rather than reasons for more patience or more credit.

By the narrowest of bankers' criteria—all moral claims aside—the United States is a good credit risk. Its balance sheet vis-à-vis the rest of the world, not to mention its internal productive strength, indicates the capacity to service a considerably increased external public debt. The United States has been confined to the types of credit that can be given on the books of central banks. European Parliaments cannot be asked to vote long-term loans to Uncle Sam, although the American people voted through the Congress to tax themselves to finance the Marshall Plan when Europe's credit rating was nil.

Meanwhile, European central banks are uneasy holding short-term dollar assets. They prefer gold. Why? Because they might some day force us to give them a capital gain on gold holdings. We compensate them with interest on their dollar holdings when they forego this speculative possibility. But bygones are bygones; and past interest earnings are irrelevant when future capital gains beckon. On its side, the United States has had nothing to lose and much to gain in guaranteeing to maintain the value of official dollar holdings. After stubbornly resisting this suggestion on obscure grounds of principle, the United States Treasury now belatedly and selectively guarantees value in foreign currency.

The only remaining reason to refuse the United States credit is that the United States, like any other deficit country, must be "disciplined." Disciplined to do what?

To stop an orgy of inflation? The United States has the best price record of any country, except Canada, since 1958—before there was a Balance of Payments Problem. The rates of unemployment and excess capacity during the period scarcely suggest that the government has been recklessly overheating the economy with fiscal and monetary fuel.

Nevertheless, many Europeans say that when they buy dollars they are importing inflation. It is hard to take this claim seriously. First of all, if acquisitions of dollars are inflationary so are acquisitions of gold, and Europe shows no signs of saturation with gold. Second, the classic mechanism of international transmission of inflation is certainly not operating. We have not inflated ourselves into an import surplus adding to aggregate demand in Europe. To the contrary, we have maintained a large and secularly growing export surplus. Third, although central bank purchases of foreign exchange have the same expansionary monetary effects at home as other open market purchases, it is not beyond the wit or experience of man to neutralize these effects by open market sales or other monetary actions. Fourth, United States farmers and coal producers, and Japanese light manufacturers, among others, stand ready to help European governments reduce their living costs and their payment surpluses at the same time. The truth is that Europe does not really want a solution at the expense of its balance of trade.

Perhaps we are to be disciplined to cut foreign aid. European governments do not attach the same importance as we do to aid programs, especially in the Western Hemisphere. Clearly we need a better understanding on development assistance and "burden sharing" among the advanced countries.

Should the United States be disciplined in order to cut off private exports of capital, by controls or by tight monetary policy or both? This has been a major and successful focus of European pressure. The United States authorities have responded by pushing up United States interest rates, more than a full point at the short end, and by proposing the Interest Equalization Tax. European pressure is motivated in part by nationalistic and protectionist aims—keep the rich Americans from buying up or competing with local industry. This may or may not be a worthy objective, but its worth is the same whether international payments are in balance or not.

Two other issues are involved. The first concerns capital markets and controls. Should the United States move toward poorer and more autarkic capital markets, or should the Europeans move toward more efficient and freer capital markets? Much of United States long-term capital movement to Europe does not represent a transfer of real saving. Instead it is a link in a double trans-Atlantic chain connecting the European saver and the European investor. The saver wants a liquid, safe, short-term asset. The investor needs long-term finance or equity capital and seeks it in the United States. Unfortunately, another link in the same chain is official European holding of short-term dollar obligations. But the Europeans themselves could, through institutional reforms, do a great deal to connect their savers and investors more directly and to reduce the spread between their long and short interest rates.

The second issue is the appropriate international level of interest rates. Evidently national rates must be more closely aligned to each other as international money and capital markets improve. But surely the low-rate country should not always do the aligning. This would impart a deflationary bias to the system. In principle, easy fiscal policy could overcome this bias, but only at the expense of investment and growth. In the present situation European countries are fighting inflation by tightening their money markets rather than their budgets. They are forcing the United States to fight unemployment with a tight money-easy budget mixture. If interest rates are raised whenever a country faces either inflation or balance of payments difficulties, while expansionary fiscal policy is the only measure ever used to combat deflation, a number of swings in business activity and in payments will move the world to a mixture of policies quite unfavorable to long-run growth.

In summary, the adjustments forced on the United States to correct its payments deficit have not served the world economy well. Neither were they essential. European countries have had at their disposal several measures which are desirable in their own right, not just as correctives to the present temporary imbalance in payments. To the extent that they are unprepared to take these measures, they should willingly extend compensatory finance. International financial policy is too important to leave to financiers. There are more important accounts to balance than the records of international transactions, and more important markets to equilibrate than those in foreign exchange.

18

Exchange rates—
How flexible should they be?

Milton Friedman

Chairman Douglas. The committee will come to order. . . .

We are happy to have as our first witness a former colleague of mine, a very distinguished economist, Prof. Milton Friedman, who has been perhaps the foremost advocate in this country of flexible exchange rates and, along with Professor Meade of England, one of the two leading advocates in the world of flexible exchange rates.

We are very glad to welcome you, Professor Friedman.

Mr. Friedman. Thank you, Professor Douglas—Senator Douglas.

Discussions of U.S. policy with respect to international payments tend to be dominated by our immediate balance-of-payments difficulties. I should like today to approach the question from a different, and I hope more constructive, direction. Let us begin by asking ourselves not merely how we can get out of our present difficulties but instead how we can fashion our international payments system so that it will best serve our needs for the long pull; how we can solve not merely this balance-of-payments problem but the balance-of-payments problem.

A shocking, and indeed, disgraceful feature of the present situation is the extent to which our frantic search for expedients to stave off balance-of-payments pressures has led us, on the one hand, to sacrifice major national objectives; and, on the other, to give enormous power to officials of foreign governments to affect what should be purely domestic matters.

Reprinted from *The United States Balance of Payments,* Hearings before the Joint Economic Committee, 88th Cong., 1st Session (Washington, D.C., U.S. Government Printing Office, 1963), pp. 451-468. Milton Friedman is Professor of Economics, University of Chicago.

Chairman Douglas. May I say, so far, so good. I enjoyed that 100 percent.

Representative Reuss. It might be wise to stop there.

Mr. Friedman. Foreign payments amount to only some 5 percent of our total national income. Yet they have become a major factor in nearly every national policy.

I believe that a system of floating exchange rates would solve the balance-of-payments problem for the United States far more effectively than our present arrangements. Such a system would use the flexibility and efficiency of the free market to harmonize our small foreign trade sector with both the rest of our massive economy and the rest of the world; it would reduce problems of foreign payments to their proper dimensions and remove them as a major consideration in governmental policy about domestic matters and as a major preoccupation in international political negotiations; it would foster our national objectives rather than be an obstacle to their attainment.

To indicate the basis for this conclusion, let us consider the national objective with which our payments system is most directly connected: the promotion of a healthy and balanced growth of world trade, carried on, so far as possible, by private individuals and private enterprises with minimum intervention by governments. This has been a major objective of our whole postwar international economic policy, most recently expressed in the Trade Expansion Act of 1962. Success would knit the free world more closely together, and, by fostering the international division of labor, raise standards of living throughout the world, including the United States.

Suppose that we succeed in negotiating far-reaching reciprocal reductions in tariffs and other trade barriers with the Common Market and other countries. To simplify exposition I shall hereafter refer only to tariffs, letting these stand for the whole range of barriers to trade, including even the so-called voluntary limitation of exports. Such reductions will expand trade in general but clearly will have different effects on different industries. The demand for the products of some will expand, for others contract. This is a phenomenon we are familiar with from our internal development. The capacity of our free enterprise system to adapt quickly and efficiently to such shifts, whether produced by changes in technology or tastes, has been a major source of our economic growth. The only additional element introduced by international trade is the fact that different currencies are involved, and this is where the payment mechanism comes in; its function is to keep this fact from being an additional source of disturbance.

An all-around lowering of tariffs would tend to increase both our expenditures and our receipts in foreign currencies. There is no way of knowing in advance which increase would tend to be the greater and hence no way of knowing whether the initial effect would be toward a surplus or deficit in our balance of payments. What is clear is that we cannot hope to succeed in the objective of expanding world trade unless we can readily adjust to either outcome.

Many people concerned with our payments deficits hope that since we are operating further from full capacity than Europe, we could supply a substantial increase in exports whereas they could not. Implicitly, this assumes that European countries are prepared to see their surplus turned into a deficit, thereby contributing to the reduction of the deficits we have recently been experiencing in our balance of payments. Perhaps this would be the initial effect of tariff changes. But if the achievement of such a result is to be sine qua non of tariff agreement, we cannot hope for any significant reduction in barriers. We could be confident that exports would expand more than imports only if the tariff changes were one sided indeed, with our trading partners making much greater reductions in tariffs than we make. Our major means of inducing other countries to reduce tariffs is to offer corresponding reductions in our tariff. More generally, there is little hope of continued and sizable liberalization of trade if liberalization is to be viewed simply as a device for correcting balance-of-payments difficulties. That way lies only backing and filling.

Suppose then that the initial effect is to increase our expenditures on imports more than our receipts from exports. How could we adjust to this outcome?

One method of adjustment is to draw on reserves or borrow from abroad to finance the excess increase in imports. The obvious objection to this method is that it is only a temporary device, and hence can be relied on only when the disturbance is temporary. But that is not the major objection. Even if we had very large reserves or could borrow large amounts from abroad, so that we could continue this expedient for many years, it is a most undesirable one. We can see why if we look at physical rather than financial magnitudes.

The physical counterpart to the financial deficit is a reduction of employment in industries competing with imports that is larger than the concurrent expansion of employment in export industries. So long as the financial deficit continues, the assumed tariff reductions create employment problems. But it is no part of the aim of tariff reductions to create unemployment at home or to promote employment abroad. The aim is a balanced expansion of trade, with exports riding along with imports and thereby providing employment opportunities to offset any reduction in employment resulting from increased imports.

Hence, simply drawing on reserves or borrowing abroad is a most unsatisfactory method of adjustment.

Another method of adjustment is to lower U.S. prices relative to foreign prices, since this would stimulate exports and discourage imports. If foreign countries are accommodating enough to engage in inflation, such a change in relative prices might require merely that the United States keep prices stable or even, that it simply keep them from rising as fast as foreign prices. But there is no necessity for foreign countries to be so accommodating, and we could hardly count on their being so accommodating. The use of this technique therefore involves a willingness to produce a decline in U.S. prices by tight monetary policy or tight fiscal policy or both. Given time, this method of adjustment would work. But in the interim, it would exact a heavy toll. It would be difficult or impossible to force down prices appreciably without producing a recession and considerable unemployment. To eliminate in the long run the unemployment resulting from the tariff changes, we should in the short run be creating cyclical unemployment. The cure might for a time be far worse than the disease.

This second method is therefore also most unsatisfactory. Yet these two methods—drawing on reserves and forcing down prices—are the only two methods available to us under our present international payment arrangements, which involve fixed exchange rates between the U.S. dollar and other currencies. Little wonder that we have so far made such disappointing progress toward the reduction of trade barriers, that our practice has differed so much from our preaching.

There is one other way and only one other way to adjust and that is by allowing (or forcing) the price of the U.S. dollar to fall in terms of other currencies. To a foreigner, U.S. goods can become cheaper in either of two ways—either because their prices in the United States fall in terms of dollars or because the foreigner has to give up fewer units of his own currency to acquire a dollar, which is to say, the price of the dollar falls. For example, suppose a particular U.S. car sells for $2,800 when a dollar costs 7 shillings, tuppence in British money (i.e., roughly £1=$2.80). The price of the car is then £1,000 in British money. It is all the same to an Englishman—or even a Scotsman—whether the price of the car falls to $2,500 while the price of a dollar remains 7 shillings, tuppence, or, alternatively, the price of the car remains $2,800, while the price of a dollar falls to 6 shillings, 5 pence (i.e., roughly £1=$3.11). In either case, the car costs the Englishman £900 rather than £1,000, which is what matters to him. Similarly, foreign goods can become more expensive to an American in either of two ways—either because the price in terms of foreign currency rises or because he has to give up more dollars to acquire a given amount of foreign currency.

Changes in exchange rates can therefore alter the relative price of U.S. and foreign goods in precisely the same way as can changes in internal prices in the United States and in foreign countries. And they can do so without requiring anything like the same internal adjustments.

If the initial effect of the tariff reductions would be to create a deficit at the former exchange rate (or enlarge an existing deficit or reduce an existing surplus) and thereby increase unemployment, this effect can be entirely avoided by a change in exchange rates which will produce a balanced expansion in imports and exports without interfering with domestic employment, domestic prices, or domestic monetary and fiscal policy. The pig can be roasted without burning down the house.

The situation is, of course, entirely symmetrical if the tariff changes should initially happen to expand our exports more than our imports. Under present circumstances, we would welcome such a result, and conceivably, if the matching deficit were experienced by countries currently running a surplus, they might permit it to occur without seeking to offset it. In that case, they and we would be using the first method of adjustment—changes in reserves or borrowing. But again, if we had started off from an even keel, this would be an undesirable method of adjustment. On our side, we should be sending out useful goods and receiving only foreign currencies in return. On the side of our partners, they would be using up reserves and tolerating the creation of unemployment.

The second method of adjusting to a surplus is to permit or force domestic prices to rise—which is of course what we did in part in the early postwar years when we were running large surpluses. Again, we should be forcing maladjustments on the whole economy to solve a problem arising from a small part of it—the 5 percent accounted for by foreign trade.

Again, these two methods are the only ones available under our present international payments arrangements, and neither is satisfactory.

The final method is to permit or force exchange rates to change—in this case, a rise in the price of the dollar in terms of foreign currencies. This solution is again specifically adapted to the specific problem of the balance of payments.

Changes in exchange rates can be produced in either of two general ways. One way is by a change in an official exchange rate; an official devaluation or appreciation from one fixed level which the Government is committed to support to another fixed level. This is the method used by Britain in its postwar devaluation and by Germany in 1961 when the mark was appreciated. This is also the main method contemplated by the IMF which permits member nations to change their exchange rates by 10 percent without approval by the Fund and by a larger amount after approval by the Fund. But this method has serious disadvantages. It makes a change in rates a matter of major moment, and hence there is a tendency to postpone any change as long as possible. Difficulties cumulate and a larger change is finally needed than would have been required if it could have been made promptly. By the time the change is made, everyone is aware that a change is pending and is certain about the direction of change. The result is to encourage flight from a currency, if it is going to be devalued, or to a currency, if it is going to be appreciated.

There is in any event little basis for determining precisely what the new rate should be. Speculative movements increase the difficulty of judging what the new rate should be, and introduce a systematic bias, making the change needed appear larger than it actually is. The result, particularly when devaluation occurs, is generally to lead officials to "play safe" by making an even larger change than the large change needed. The country is then left after the devaluation with a maladjustment precisely the opposite of that with which it started, and is thereby encouraged to follow policies it cannot sustain in the long run.

Even if all these difficulties could be avoided, this method of changing from one fixed rate to another has the disadvantage that it is necessarily discontinuous. Even if the new exchange rates are precisely correct when first established, they will not long remain correct.

A second and much better way in which changes in exchange rates can be produced is by permitting exchange rates to float, by allowing them to be determined from day to day in the market. This is the method which the United States used from 1862 to 1879, and again, in effect, from 1917 or so to about 1925, and again from 1933 to 1934. It is the method which Britain used from 1918 to 1925 and again from 1931 to 1939, and which Canada used for most of the interwar period and again from 1950 to May 1962. Under this method, exchange rates adjust themselves continuously, and market forces determine the magnitude of each change. There is no need for any official to decide by how much the rate should rise or fall. This is the method of the free market, the method that we adopt unquestioningly in a private enterprise economy for the bulk of goods and services. It is no less available for the price of one money in terms of another.

With a floating exchange rate, it is possible for Governments to intervene and try to affect the rate by buying or selling, as the British exchange equalization fund did rather successfully in the 1930's, or by combining buying and selling with public announcements of intentions, as Canada did so disastrously in early 1962. On the whole, it seems to me undesirable to have government intervene, because there is a strong tendency for government agencies to try to peg the rate rather than to stabilize it, because they have no special advantage over private speculators in stabilizing it, because they can make far bigger mistakes than private speculators risking their own money, and because there is a tendency for them to cover up their mistakes by changing the rules—as the Canadian case so strikingly illustrates—rather than by reversing course. But this is an issue on which there is much difference of opinion among economists who agree in favoring floating rules. Clearly, it is possible to have a successful floating rate along with governmental speculation.

The great objective of tearing down trade barriers, of promoting a worldwide expansion of trade, of giving citizens of all countries, and especially the underdeveloped countries, every opportunity to sell their products in open markets under equal terms and thereby every incentive to use their resources efficiently, of giving countries an alternative through free world trade to autarchy and central planning—this great objective can, I believe, be achieved best under a regime of floating rates. All countries, and not just the United States, can proceed to liberalize boldly and confidently only if they can have reasonable assurance that the resulting trade expansion will be balanced and will not interfere with major domestic objectives. Floating exchange rates, and so far as I can see, only floating exchange rates, provide this assurance. They do so because they are an automatic mechanism for protecting the domestic economy from the possibility that liberalization will produce a serious imbalance in international payments.

Despite their advantages, floating exchange rates have a bad press. Why is this so?

One reason is because a consequence of our present system that I have been citing as a serious disadvantage is often regarded as an advantage, namely, the extent to which the small foreign trade sector dominates national policy. Those who regard this as an advantage refer to it as the discipline of the gold standard. I would have much sympathy for this view if we had a real gold standard, so the discipline was imposed by impersonal forces which in turn reflected the realities of resources, tastes, and technology. But in fact we have today only a pseudo gold standard and the so-called discipline is imposed by government officials of other countries who are determining their own internal monetary policies and are either being forced to dance to our tune or calling the tune for us, depending primarily on accidental political developments. This is a discipline we can well do without. See my article entitled "Real and Pseudo Gold Standards" which I will present later for inclusion in the record.

Chairman Douglas. The article will be placed in the record at the end of your oral presentation.

Mr. Friedman. A possibly more important reason why floating exchange rates have a bad press, I believe, is a mistaken interpretation of experience with floating rates, arising out of a statistical fallacy that can be seen easily in a standard example. Arizona is clearly the worst place in the United States for a person with tuberculosis to go because the death rate from tuberculosis is higher in Arizona than in any other State. The fallacy in this case is obvious. It is less obvious in connection with exchange rates. Countries that have gotten into severe financial difficulties, for whatever reason, have had ultimately to change their exchange rates or let them change. No amount of exchange control and other restrictions on trade have enabled them to peg an exchange rate that was far out of line with economic realities. In consequence, floating rates have frequently been associated with financial and economic instability. It is easy to conclude, as many have, that floating exchange rates produce such instability.

This misreading of experience is reinforced by the general prejudice against speculation; which has led to the frequent assertion, typically on the basis of no evidence whatsoever, that speculation in exchange can be expected to be destabilizing and thereby to increase the instability in rates. Few who make this assertion even recognize that it is equivalent to asserting that speculators generally lose money.

Floating exchange rates need not be unstable exchange rates—any more than the prices of automobiles or of Government bonds, of coffee or of meals need gyrate wildly just because they are free to change from day to day. The Canadian exchange rate was free to change during more than a decade, yet it varied within narrow limits. The ultimate objective is a world in which exchange rates, while free to vary, are in fact highly stable because basic economic policies and conditions are stable. Instability of exchange rates is a symptom of instability in the underlying economic structure. Elimination of this symptom by administrative pegging of exchange rates cures none of the underlying difficulties and only makes adjustment to them more painful.

The confusion between stable exchange rates and pegged exchange rates helps to explain the frequent comment that floating exchange rates would introduce an additional element of uncertainty into foreign trade and thereby discourage its expansion. They introduce no additional element of uncertainty. If a floating rate would, for example, decline, then a pegged rate would be subject to pressure that the authorities would have to meet by internal deflation or exchange control in some form. The uncertainty about the rate would simply be replaced by uncertainty about internal prices or about the availability of exchange; and the latter uncertainties, being subject to administrative rather than market control, are likely to be the more erratic and unpredictable. Moreover, the trader can far more readily and cheaply protect himself against the danger of changes in exchange rates, through hedging operations in a forward market, than he can against the danger of changes in internal prices or exchange availability. Floating rates are therefore more favorable to private international trade than pegged rates.

Though I have discussed the problem of international payments in the context of trade liberalization, the discussion is directly applicable to the more general problem of adapting to any forces that make for balance-of-payments difficulties. Consider our present problem, of a deficit in the balance of trade plus long-term capital movements. How can we adjust to it? By one of the three methods outlined: first, drawing on reserves or borrowing; second, keeping U.S. prices from rising as rapidly as foreign prices or forcing them down; third, permitting or forcing exchange rates to alter. And, this time, by one more method: by imposing additional trade barriers or their equivalent, whether in the form of higher tariffs, or smaller import quotas, or extracting from other countries tighter "voluntary" quotas on their exports, or "tieing" foreign aid, or buying higher priced domestic goods or services to meet military needs, or imposing taxes on foreign borrowing, or imposing direct controls on investments by U.S. citizens abroad, or any one of the host of other devices for interfering with the private business of private individuals that have become so familiar to us since Hjalmar Schacht perfected the modern techniques of exchange control in 1934 to strengthen the Nazis for war and to despoil a large class of his fellow citizens.

Fortunately or unfortunately, even Congress cannot repeal the laws of arithmetic. Books must balance. We must use one of these four methods. Because we have been unwilling to select the only one that is currently fully consistent with both economic and political needs—namely, floating exchange rates—we have been driven, as if by an invisible hand, to employ all the others, and even then may not escape the need for explicit changes in exchange rates.

We affirm in loud and clear voices that we will not and must not erect trade barriers—yet is there any doubt about how far we have gone down the fourth route? After the host of measures already taken, the Secretary of the Treasury has openly stated to the Senate Finance Committee that if the so-called interest equalization tax—itself a concealed exchange control and concealed devaluation—is not passed, we shall have to resort to direct controls over foreign investment.

We affirm that we cannot drain our reserves further, yet short-term liabilities mount and our gold stock continues to decline.

We affirm that we cannot let balance-of-payments problems interfere with domestic prosperity, yet for at least some 4 years now we have followed a less expansive monetary policy than would have been healthy for our economy.

Chairman Douglas. We thank you for that, Professor Friedman.

Mr. Friedman. Even all together, these measures may only serve to postpone but not prevent open devaluation—if the experience of other countries is any guide. Whether they do, depends not on us but on others.

For our best hope of escaping our present difficulties is that foreign countries will inflate.

In the meantime, we adopt one expedient after another, borrowing here, making swap arrangements there, changing the form of loans to make the figures look good. Entirely aside from the ineffectiveness of most of these measures, they are politically degrading and demeaning. We are a great and wealthy Nation. We should be directing our own course, setting an example to the world, living up to our destiny. Instead, we send our officials hat in hand to make the rounds of foreign governments and central banks; we put foreign central banks in a position to determine whether or not we can meet our obligations and thus enable them to exert great influence on our policies; we are driven to niggling negotiations with Hong Kong and with Japan and for all I know, Monaco, to get them to limit voluntarily their exports. Is this posture suitable for the leader of the free world?

Chairman Douglas. I do not wish to interrupt you, but I would like to say that I think many visits to Monaco are for a different purpose. [Laughter.]

Go ahead.

Mr. Friedman. It is not the least of the virtues of floating exchange rates that we would again become masters in our own house. We could decide important issues on the proper ground. The military could concentrate on military effectiveness and not on saving foreign exchange, recipients of foreign aid could concentrate on how to get the most out of what we give them and not on how to spend it all in the United States; Congress could decide how much to spend on foreign aid on the basis of what we get for our money and what else we would use it for and not how it will affect the gold stock; the monetary authorities could concentrate on domestic prices and employment, not on how to induce foreigners to hold dollar balances in this country; the Treasury and the tax committees of Congress could devote their attention to the equity of the tax system and its effects on our efficiency, rather than on how to use tax gimmicks to discourage imports, subsidize exports, and discriminate against outflows of capital.

A system of floating exchange rates would render the problem of making outflows equal inflows unto the market where it belongs and not leave it to the clumsy and heavy hand of Government. It would leave Government free to concentrate on its proper functions.

In conclusion, a word about gold. Our commitment is to buy and sell gold for monetary use at a fixed price of $35 an ounce is, in practice, the mechanism whereby we maintain fixed rates of exchange between the dollar and other currencies—or, more precisely, whereby we leave all initiative for changes in such rates to other countries. This commitment should be terminated. The price of gold should be determined in the free market, with the U.S. Government committed neither to buying gold nor to selling gold at any fixed price. This is the appropriate counterpart of a policy of floating exchange rates. With respect to our existing stock of gold, we could simply keep it fixed, neither adding to it nor reducing it; alternatively, we could sell it off gradually at the market price or add to it gradually, thereby reducing or increasing our governmental stockpiles of this particular metal. In any event, we should simultaneously remove all present limitations on the ownership of gold and the trading in gold by American citizens. There is no reason why gold, like other commodities, should not be freely traded on a free market.

The Chairman. Thank you very much, Professor Friedman. Your paper, entitled "Real and Pseudo Gold Standards" will appear at this place in the record.

Real and Pseudo Gold Standards[1]

International monetary arrangements have held a consistently important place among the topics discussed at the meetings of our society. This is eminently fitting, since there is probably no other major facet of economic policy with respect to which liberals (in the sense of our society) reach such divergent conclusions from the same underlying principles.

One group, of which Philip Cortney is a distinguished member, favors a continuation of the formal linking of national currencies to gold, rigid exchange rates between different national currencies, a doubling or more than doubling of the official price of gold in terms of national currencies, and an abandonment of governmental measures designed to evade the discipline of gold. This group is apparently indifferent about whether gold circulates as coin; it is satisfied with a gold bullion standard.

A second group, represented by the Economists' National Committee on Monetary Policy, also favors a continuation of the formal linking of national currencies to gold together with rigid exchange rates between different national currencies. But it emphasizes the importance of gold coinage and of a widespread use of gold coin as money in national as well as international payments. Apparently, this group be-

[1] Paper written for the Mont Pelerin Society meetings in September 1961. [Reprinted with permission from *The Journal of Law and Economics* of the University of Chicago Law School, Vol. 4 (October, 1961), pp. 66-79.]

lieves there is no need for a change in present official prices of gold, or, at least, in the U.S. price.

A third group, of which I count myself a member, favors a separation of gold policy from exchange-rate policy. It favors the abandonment of rigid exchange rates between national currencies and the substitution of a system of floating exchange rates determined from day to day by private transactions without government intervention. With respect to gold, there are some differences, but most of us would currently favor the abandonment of any commitment by governments to buy and sell gold at fixed prices and of any fixed gold reserve requirements for the issue of national currency as well as the repeal of any restrictions on private dealings in gold.

I have stated and defended my own policy views elsewhere, at some length.[2] Hence, I would like to use this occasion instead to explore how it is that liberals can reach such radically different conclusions.

My thesis is that current proposals to link national currencies rigidly to gold whether at present or higher prices arise out of a confusion of two very different things: the use of gold as money, which I shall call a real gold standard; governmental fixing of the price of gold, whether national or international, which I shall call a pseudo gold standard. Though these have many surface features in common, they are at bottom fundamentally different—just as the near identity of prices charged by competitive sellers differs basically from the identity of prices charged by members of a price ring or cartel. A real gold standard is thoroughly consistent with liberal principles, and I, for one, am entirely in favor of measures promoting its development, as I believe, are most other liberal proponents of floating exchange rates. A pseudo gold standard is in direct conflict with liberal principles, as is suggested by the curious coalition of central bankers and central planners that has formed in support of it.

It is vitally important for the preservation and promotion of a free society, that we recognize the difference between a real and pseudo gold standard. War aside, nothing that has occurred in the past half century has, in my view, done more to weaken and undermine the public's faith in liberal principles than the pseudo gold standard that has intermittently prevailed and the actions that have been taken in its name. I believe that those of us who support it in the belief that it either is or will tend to be a real gold standard are mistakenly fostering trends the outcome of which they will be among the first to deplore.

This is a sweeping charge, so let me document it by a few exam-

[2] See, in particular, "The Case for Flexible Exchange Rates" and "Commodity-Reserve Currency," in my Essays in Positive Economics, pp. 157-203, 204-250 (1953), and "A Program for Monetary Policy, pp. 77-84 (1959).

ples which will incidentally illustrate the difference between a real and pseudo gold standard before turning to an explicit discussion of the difference. My examples are mostly for the United States, the country whose monetary history I have studied in most detail.

A. Examples of effects of a pseudo gold standard

1. U.S. monetary policy after World War I. Nearly half of the monetary expansion in the United States came after the end of the war, thanks to the acquiescence of the Federal Reserve System in the Treasury's desire to avoid a fall in the price of Government securities. This expansion, with its accompanying price inflation, led to an outflow of gold despite the great demand for U.S. goods from a war-ravaged world and despite the departure of most countries from any fixed parity between their currencies and either gold or the dollar. The outflow of gold finally overcame Treasury reluctance to see the price of Government securities fall. Beginning in late 1919, then more sharply in January 1920 and May 1920, the Federal Reserve System took vigorous deflationary steps that produced first, a slackening of the growth in the stock of money, and then a sharp decline. These brought in their train a collapse in wholesale prices and a severe economic contraction. The near-halving of wholesale prices in a 12-month period was by all odds the most rapid price decline every experienced in the United States before or since. It was not of course confined to the United States, but spread to all countries whose money was linked to the dollar either by having a fixed price in terms of gold or by central bank policies directed at maintaining rigid or nearly rigid exchange rates. Only those countries that were to experience hyperinflation escaped the price collapse.

Under a real gold standard, the large inflow of gold up to the entry of the United States into the war would have produced a price rise to the end of the war similar to that actually experienced. But neither the postwar rise nor the subsequent collapse would have occurred. Instead, there would have been an earlier and milder price decline as the belligerent nations returned to a peace-time economy. The postwar increase in the stock of money occurred only because the Reserve System had been given discretionary power to "manage" the stock of money, and the subsequent collapse occurred only because this power to manage the money had been accompanied by gold reserve requirements as one among several masters the System was instructed to serve.

Under a wholly fiduciary currency, with floating exchange rates, the initial postwar expansion might well have occurred much as it did, though the depreciating value of the dollar in terms of other currencies might have been a quicker and a more effective check than slowly declining gold reserves. But the subsequent collapse would almost surely not have occurred. And neither the initial price inflation nor the subsequent price collapse would have been communicated to the rest of the world.

The worldwide inflation and then collapse was at the time a severe blow to a belief in free trade at home and abroad, a blow whose severity we now underrate only because of the later catastrophe that overshadowed it. Either a real gold standard or a thoroughly fiduciary standard would have been preferable in its outcome to the pseudo gold standard.

2. U.S. monetary policy in the 1920's and Britain's return to gold. There is a widespread myth among gold standard advocates that the U.S. monetary policy during the 1920's paved the way for the great depression by being unduly inflationary. For example, Cortney writes, "the Federal Reserve Board succeeded in the 1920's in holding up the price level for a surprising length of time by an abnormal expansion of inflationary credit, but in so doing it helped produce the speculative boom."[3] Nothing could be farther from the truth. The U.S. monetary policy in the 1920's and especially in the late 1920's, judged in terms of either a real gold standard in the abstract or prior U.S. experience, was if anything unduly deflationary.

The sharp 1920-21 price decline had brought prices to a level much closer to the prewar level than to the postwar peak though they were still appreciably above the prewar level. Prices rose only moderately in the subsequent cyclical expansion which reached its peak in 1923. From then until 1929, wholesale prices actually fell, at a rate of roughly 1 percent a year.

As to gold, credit, and money, the Federal Reserve System sterilized much of the gold inflow, preventing the gold from raising the stock of money anything like as much as it would have done under a real gold standard. Far from the Reserve System engaging in an "abnormal expansion of inflationary credit," Federal Reserve credit outstanding in June 1929 was 33 percent lower than it had been in June 1921 and only 16 percent higher than in June 1923 although national income was nearly 25 percent higher in 1929 than in 1923 (in both money and real terms). From 1923 to 1929, to compare only peak years of business cycles and so avoid distortion from cyclical influence, the stock of money, defined to include currency, demand deposits, and commercial bank time deposits, rose at the annual rate of 4 percent per year, which is roughly the rate required to match expansion of output.

[3] Introduction to Charles Rist, *The Triumph of Gold*, p. 8 (1961).

On a narrower definition, excluding time deposits, the stock of money rose at the rate of only 2½ percent per year.[4]

The deflationary pressure was particularly strong during the great bull market in stocks, which happened to coincide with the first few years after Britain returned to gold. During the business cycle expansion from 1927 to 1929, wholesale prices actually fell a trifle; one must go back to 1891-93 to find another expansion during which prices fell and there has been none since. The stock of money was lower at the cyclical peak in August 1929 than it had been 16 months earlier. There is no other occasion from the time our monthly data began in 1907 to date when so long a period elapsed during a cyclical expansion without a rise in the stock of money. The only other periods of such length which show a decline have an end point in the course of severe contractions (1920-21, 1929-33, 1936-37).

So far as the United States alone was concerned, this monetary policy may have been admirable. I do not myself believe that the 1929-33 contraction was an inevitable result of the monetary policy of the 1920's or even owed much to it. What was wrong was the policy followed from 1929 to 1933, as I shall point out in a moment. But internationally, the policy was little short of catastrophic. Much has been made of Britain's mistake in returning to gold in 1925 at a parity that overvalued the pound. I do not doubt that this was a mistake—but only because the United States was maintaining a pseudo gold standard. Had the United States been maintaining a real gold standard, the stock of money would have risen more in the United States than it did, prices would have been stable or rising instead of declining, the United States would have gained less gold or lost some, and the pressure on the pound would have been enormously eased. As it was, by sterilizing gold, the United States forced the whole burden of adapting to gold movements on other countries. When, in addition, France adopted a pseudo gold standard at a parity that undervalued the franc and proceeded also to follow a gold sterilization policy, the combined effect was to make Britain's position untenable. The adverse consequences for faith in liberal principles of the deflationary policies adopted in Britain from 1925 to 1931 in the vain effort to maintain the reestablished parity are no less obvious than they were far reaching.

3. U.S. policy in 1931-33. U.S. monetary behavior in 1931-33 is in some ways a repetition of that from 1920 to 1921, but on a more catastrophic scale, in less fortunate circumstances, and with less justification. As we

[4] These statements are based on estimates of the stock of money from 1867 to date constructed by Anna J. Schwartz and me in connection with a study for the National Bureau of Economic Research. Hereafter, I will use the term "stock of money" as referring to the first of these two definitions.

have seen, in 1919 the Reserve System deviated from the policy that would have been dictated by a real gold standard. In 1920, when it saw its gold reserves declining rapidly, it shifted rules, over-reacted to the outflow, and brought on a drastic deflation. Similarly, from 1922 to 1929, the Reserve System sterilized gold and prevented it from exercising the influence on the money stock that it would have had under a real gold standard. And again in 1931, when Britain went off gold and the United States experienced an outflow of gold, the Reserve System shifted rules, over-reacted to the outflow, and catastrophically intensified a deflation already 2 years old.

The circumstances were less fortunate in 1931 than in 1920 in two different respects, one domestic and the other foreign, and both in some measure the Reserve System's own creation.

The domestic difference was that the deflationary action of 1920 came at the end of a period of expansion which was widely regarded as temporary and exceptional, and served to intensify without necessarily prolonging a recession that would probably have occurred anyway. The deflationary action of 1931 came after 2 years of severe contraction which had been showing some signs of terminating; probably served to nip in the bud a revival; and both greatly intensified and substantially prolonged the contraction, turning it into the most severe for nearly a century.

This difference was largely the Reserve System's creation because of its inept handling of the banking difficulties that started in the fall of 1930. Until that date, the contraction, while rather severe, had shown no signs of a liquidity crisis. Widespread bank failures culminating in the failure of the Bank of the United States in late 1930 changed the aspect of the contraction. This episode turned out to be the first of a series of liquidity crises, each characterized by bank failures and runs on banks by depositors anxious to convert deposits into currency, and each producing strong downward pressure on the stock of money. The Reserve System had been set up with the primary aim of dealing with precisely such crises. It failed to do so effectively but not because it lacked the power or the knowledge. At all times, it had ample power to provide the liquidity that the public and the banks desperately sought and the provision of which would have cut short the vicious chain reaction of bank failures. The System failed because accidents of personality and shifts of power within the System left it with no dominant personality who could avoid the usual outcome of committee control: the evasion of responsibility by inaction, postponement, and drift. More fundamentally yet, the failure reflected the adoption of a monetary system that gave great power to a small number of men and therefore was vulnerable to such accidents of personality and shifts of power. Had the liquidity crisis been cut short at its onset in 1930 and the

Bank of the United States kept from failing (as very likely would have occurred before the Federal Reserve System), the economy would probably have been vigorously expanding by September 1931 instead of being precariously balanced on the verge of another liquidity crisis.

The international difference in circumstances that was less fortunate in 1931 than in 1920 was the monetary situation in other countries. In many countries, monetary arrangements in 1920 were in a state of flux, so they could adapt with some rapidity. By 1931, a new pattern of international monetary arrangements had become established, in considerable measure under the patronage of the Federal Reserve Bank of New York, as well as the Banks of England and France. More serious and more directly to be laid at the Reserve System's door, its gold sterilization policy had, as we have seen, increased the problem of adjustment for many other countries and so left them more vulnerable to new difficulties. In the event the monetary world split in two, one part following Britain to form the sterling area; the other following the United States, in the gold bloc. The sterling area countries all reached bottom and began to expand in late 1931 or early 1932; most gold bloc countries experienced further deflation and did not reach bottom until 1933 or 1934.

The deflationary monetary actions had less justification in the fall of 1931 than in 1920 for two different reasons. First, in 1920, the Federal Reserve System was still in its infancy, untried and inexperienced. Set up under one set of conditions, it was operating under a drastically different set. It had no background of operation in peacetime, no experience on which to base judgments. By 1931, the System had more than a decade of experience and had developed a well-articulated body of doctrine, which underlay the gold sterilization policy and which called for its offsetting an outflow of gold rather than reinforcing its deflationary effect. Second, the gold situation was drastically different. By early 1920, the gold stock was declining rapidly and the Reserve System's gold reserve ratio was approaching its legal minimum. Prior to September 1931, the System had been gaining gold, the monetary gold stock was at an alltime high, and the System's gold reserve ratio was far above its legal minimum—a reflection of course of its not having operated in accordance with a real gold standard. The System had ample reserves to meet the gold outflow without difficulty and without resort to deflationary measures. And both its own earlier policy and the classical gold standard rules as enshrined by Bagehot called for its doing so: the gold outflow was strictly speculative and motivated by fear that the United States would go off gold; the outflow had no basis in any trade imbalance; it would have exhausted itself promptly if all demands had been met.

As it was, of course, the System behaved very differently. It reacted vigorously to the external drain as it had not to the internal drain by raising discount rates within a brief period more sharply than ever before or since. The result was a major intensification of the internal drain and an unprecedented liquidation of the commercial banking system. Whereas the stock of money had fallen 10 percent from August 1929 to August 1931, it fell a further 28 percent from August 1931 to March 1933. Commercial bank deposits had fallen 12 percent from August 1929 to August 1931; they fell a further 35 percent from August 1931 to March 1933. Never was there a more unnecessary monetary collapse or one which did more to undermine public acceptance of liberal principles.

Once again, either a real gold standard throughout the 1920's and 1930's or a consistent adherence to a fiduciary standard would have been vastly preferable to the actual pseudo gold standard under which gold inflows and minor gold outflows were offset and substantial actual or threatened gold outflows were overreacted to. And this pattern is no outmoded historical curiosity: witness the U.S. reaction to gold inflows in the early years after World War II and its recent reaction to gold outflows; witness the more recent German sterilization of gold inflows. The pseudo gold standard is very much a living menace.

4. U.S. nationalization of gold. After going off gold in March 1933, the United States reestablished a fixed official price of gold in January 1934, raising the price to $35 an ounce. Many current proponents of a rise in the official price of gold approve this action, regarding it as required to bring the value of the gold stock into line with an allegedly increased fiduciary circulation. Perhaps a rise in the price of gold was desirable in 1934 but it cannot be defended along these lines, at least for the United States itself. In 1933, the ratio of the value of the gold stock to the total stock of money was higher than it had been in 1913 or at any date between. If there be any valid argument for a rise in the price of gold along these lines, it is for 1929, not 1934.

Whatever may be the merits of the rise in the price of gold, there can be little doubt that the associated measures, which were taken in order that the rise in the price of gold should have the effect desired by the Roosevelt administration, represented a fundamental departure from liberal principles and established precedents that have returned to plague the free world. I refer, of course, to the nationalization of the gold stock, the prohibition of private possession of gold for monetary purposes, and the abrogation of gold clauses in public and private contracts.

In 1933 and early 1934, private holders of gold were required by law to turn over their gold to the Federal Government and were compensated at a price equal to the prior legal price, which was at the time very decidedly below the market price. To make this requirement effective, private ownership of gold within the United States was made illegal except for use in the arts. One can hardly imagine a measure more destructive of the principles of private property on which a free enterprise society rests. There is no difference in principle between this nationalization of gold at an artificially low price and Fidel Castro's nationalization of land and factories at an artificially low price. On what grounds of principle can the United States object to the one after having itself engaged in the other? Yet so great is the blindness of some supporters of free enterprise with respect to anything touching on gold that as recently as last year Henry Alexander, head of the Morgan Guaranty Trust Co., successor to J. P. Morgan & Co., proposed that the prohibition against the private ownership of gold by U.S. citizens be extended to cover gold held abroad. And his proposal was adopted by President Eisenhower with hardly a protest from the banking community.

Though rationalized in terms of "conserving" gold for monetary use prohibition of private ownership of gold was not enacted for any such monetary purpose, whether itself good or bad. The circulation of gold and gold certificates had raised no monetary problems either in the 1920's or during the monetary collapse from 1930 to 1933. Except for the final weeks just preceding the banking panic, the internal drain had not been for gold but for currency of any kind in preference to deposits. And the final gold drain was the consequence of the rumors, which proved correct, that Roosevelt planned to devalue. The nationalization of gold was enacted to enable the government to reap the whole of the "paper" profit from the rise in the price of gold—or perhaps, to prevent private individuals benefiting from the rise.

The abrogation of the gold clauses had a similar purpose. And this too was a measure destructive of the basic principles of free enterprise. Contracts entered into in good faith and with full knowledge of the part of both parties to them were declared invalid for the benefit of one of the parties!

This collection of measures constituted a further step away from a real gold standard to a pseudo gold standard. Gold became even more clearly a commodity whose price was fixed by governmental purchase and sale and rationing rather than money or even a form of money.

5. International Monetary Fund and postwar exchange policy. I agree fully with Professor Rist's criticisms of the International Monetary Fund and the arrangements it embodied.[5] These arrangements are precisely those of a pseudo gold standard: each country is required to specify a formal price of gold in terms of its own currency and hence, by implica-

[5] See Charles Rist, op. cit., supra note 3, pp. 188, 193.

tion, to specify official exchange rates between its currency and other currencies. It is forbidden to change these prices outside narrow limits except with permission. It commits itself to maintaining these exchange rates. But there is no requirement that gold serve as money; on the contrary, many of the IMF provisions are designed to prevent it from doing so.

The results have been anything but happy from a liberal viewpoint: widespread controls over exchange transactions, restrictions on international trade in the forms of quotas and direct controls as well as tariffs; yet repeated exchange crises and numerous changes in official exchange rates. No doubt, conditions are now far better than shortly after the war, but clearly in spite of the IMF and not because of it. And the danger of foreign exchange crises and accompanying interferences with trade is hardly over. In the past year, the United States moved toward direct interferences with trade to cope with a balance-of-payments problem; Germany appreciated; and Britain is now in difficulties.

B. The distinction between a real and a pseudo gold standard

Because of its succinctness and explicitness, Cortney's numbered list of prerequisites for the restoration of "monetary order by returning to an international gold standard" forms an excellent point of departure for exploring the difference between a real and a pseudo gold standard. His point (6) concludes "the price of gold will have to be raised to at least $70 an ounce." His point (7) is "Free markets for gold should be established in all the important countries, and trading in gold, its export and import should be absolutely free."[6] Here is the issue in a nutshell. Can one conceive of saying in one breath that worldwide free markets should be established in, say, tin, and in the next, that the price of tin should "be raised" to some specified figure? The essence of a free market is precisely that no one can "raise" or "fix" price. Price is at whatever level will clear the market and it varies from day to day as market conditions change. If we take Cortney's point (7) seriously, we cannot simultaneously take his point (6) seriously, and conversely.

Suppose we follow up the logic of this point (7) and suppose a free market to prevail in gold. There might then develop, as there has in the past, a real gold standard. People might voluntarily choose to use gold as money, which is to say, to express prices in units of gold, and to hold gold as a temporary abode of purchasing power permitting them to separate an act of barter into a sale of goods or services for money and the purchase of goods or services with money. The gold used as money might be called different things in different languages: "or" in French, "gold" in English; it might be measured in different units:

[6] Ibid., p. 37.

say in grams in France and ounces in the United States; special terms such as "napoleon" or "eagle" might develop to designate convenient amounts of gold for use in transactions, and these might differ in different countries. We might even have governments certifying to weight and fineness, as they now inspect scales in meat markets, or even coining "eagles," "double-eagles," and the like. Changes in nomenclature or in units of measure, say, the shift from ounces to grams, might be made by legislation, but these would clearly have no monetary or income or redistributive effects; they would be like changing the standard units for measuring gasoline from gallons to liters; not comparable to changing the price of gold from $35 an ounce to $70 an ounce.

If such a real gold standard developed, the price of commodities in terms of gold would, of course, vary from place to place according to transportation costs of both the commodities and of gold. Insofar as different countries used gold, and used different units, or coins of different size, the price of one kind of gold in terms of another would be free to vary in accordance with preferences by each country's citizens for one kind or the other. The range of variation would, of course, be limited by the cost of converting one kind of gold into another, just as the relative price of commodities is similarly limited.

Under such a real gold standard, private persons or governments might go into the business of offering storage facilities, and warehouse receipts might be found more convenient than the gold itself for transactions. Finally, private persons or governments might issue promises to pay gold either on demand or after a specific time interval which were not warehouse receipts but nonetheless were widely acceptable because of confidence that the promises would be redeemed. Such promises to pay would still not alter the basic character of the gold standard so long as the obligors were not retroactively relieved from fulfilling their promises, and this would be true even if such promises were not fulfilled from time to time, just as the default of dollar bond issues does not alter the monetary standard. But, of course, promises to pay that were in default or that were expected to be defaulted would not sell at face value, just as bonds in default trade at a discount. And, of course, this is what has happened when a system like that outlined has prevailed in practice (e.g., in much of the pre-Civil War period in the United States).

Such a system might, and I believe would raise grave social problems and foster pressure for governmental prohibition of, or control over, the issue of promises to pay gold on demand.[7] But that is beside my present point, which is that it would be a real gold standard, that under it there might be different national names for the money, but there would not be in any meaningful sense either national currencies

[7] See my "A Program for Monetary Stability," pp. 4-9 (1959).

or any possibility of a government legislating a change in the price of gold.

Side by side with such a standard, there could, of course, exist strictly national currencies. For example, in the United States from 1862 to 1879, greenbacks were such a national currency which circulated side by side with gold. Since there was a free market in gold, the price of gold in terms of greenbacks varied from day to day; i.e., in modern terminology, there was a floating rate of exchange between the two currencies. Since gold was in use as money in Britain and some other countries, its main use in the United States was for foreign transactions. Most prices in the United States were quoted in greenbacks but could be paid in gold valued at the market rate. However, the situation was reversed in California, where most prices were quoted in gold but could be paid in greenbacks at the market rate. No doubt, in this historical episode, the expectation that greenbacks would some day be made promises to pay gold had an effect on their value by expanding the demand for them. But this was not essential to the simultaneous coexistence of the two currencies, so long as their relative price was freely determined in the market, just as silver and gold, or copper and silver, have often simultaneously circulated at floating rates of exchange.

If a government abjured a national currency, it might still borrow from the community in the form of securities expressed in gold (or bearing gold clauses), some of which might be demand obligations and might be noninterest bearing. But it would thereby surrender everything that we now call monetary policy. The resources it could acquire by borrowing would depend on the interest it was willing to pay on interest-bearing securities and on the amount of non-interest-bearing demand securities the public was willing to acquire. It could not arbitrarily issue any amount of non-interest-bearing securities it wished without courting inability to meet its promises to pay gold and hence seeing its securities sink to a discount relative to gold. Of course, this limitation in governmental power is precisely what recommends a real gold standard to a liberal, but we must not make the mistake of supposing that we can get the substance by the mere adoption of the form of a nominal obeisance to gold.

The kind of gold standard we have just been describing is not the kind we have had since at least 1913 and certainly not since 1934. If the essence of a free market is that no one can "raise the price," the essence of a controlled market is that it involves restrictions of one kind of another on trade. When the Government fixes the price of wheat at a level above the market price, it inevitably both accumulates stocks and is driven to control output; i.e., to ration output among producers eager to produce more than the public is willing to buy at the controlled price. When the Government fixes the price of housing space at a level below the market price, it inevitably is driven to control occupancy; i.e., to ration space among purchasers eager to buy more than sellers are willing to make available at the controlled price. The controls on gold, like the related controls on foreign exchange, are a sure sign that the price is being pegged; that dollar, pound, etc., are not simply different names for different sized units of gold, but are national currencies. Insofar as the price of gold in these currencies and the price of one currency in terms of another are stable over considerable periods, it is not because of the ease of converting one quantity of gold into another, and not because conditions of demand and supply make for stable prices, but because they are pegged prices in rigged markets.

The price of $35 an ounce at which gold was supported by the United States after January 1934 was initially well above the market price, like the price at which wheat is currently being supported. The evidence is in both cases the same: a rapid expansion of output and the accumulation of enormous stockpiles. From 1933 to 1940, production in the United States rose from less than 2.6 to 6 million ounces; in the world, from 25 to 41 million ounces; the gold stock in the Treasury rose from 200 to 630 million ounces, or by 1¾ times as much as the total of world output during the intervening period. Had this pace of increase in output and stock continued, the gold purchase program might well have been limited in scope; perhaps, as the U.S. silver purchase program finally was, to domestic output alone.

But the war intervened, which stopped the inflow of gold and brought a major rise in the stock of money. The resultant rise in other prices with no change in the price of gold has altered the character of the fixed U.S. price. It is now probably below the market price (given the present monetary use of gold), like rents under rent control. The evidence is again in both cases the same; a reduction in production, a decline in stocks, and a problem of rationing demanders. The U.S. gold output is now less than in 1933, though world output still exceeds the level of that year. The U.S. gold stock has declined to roughly 500 million ounces, well below its wartime peak but still 2½ times its level when the present price for gold was established. The restriction on the ownership of gold abroad by U.S. citizens is a first, and feeble, step toward still tighter rationing of demanders. The gentlemen's agreement among central banks not to press for conversion of dollar balances into gold is a more far reaching, if still rather weak, additional step. The history of every attempt at Government price fixing suggests that if the pegged price is far below the market price for long, such attempts are doomed to fail.

Doubling the price of gold would no doubt reverse the situation and raise the pegged price again above the market price. Gold production and U.S. gold stocks would no doubt rise. But to what avail? Gold would still be simply a commodity whose price is supported; countries would continue with their separate monetary policies; fixed exchange rates would freeze the only market mechanism available under such circumstances to adjust international payments; foreign exchange crises would continue to succeed one another; and direct controls of one kind or another would remain the last resort, and one often appealed to, for resolving them.

This kind of pseudo gold standard violates fundamental liberal principles in two major respects. First, it involves price fixing by Government. It has always been a mystery to me how so many who oppose on principle Government price fixing of all other commodities can yet approve it for this one. Second, and no less important, it involves granting discretionary authority to a small number of men over matters of the greatest importance; to the central bankers or Treasury officials who must manage the pseudo gold standard. This means the rule of men instead of law, violating one of our fundamental political tenets. Here again, I have been amazed how so many who oppose on principle the grant of wide discretionary authority to governmental officials are anxious to see such authority granted to central bankers. True, central bankers have on the whole been "sound money" men with great sympathy for private enterprise. But since when have we liberals tempered our fear of concentrated power by trust in the particular men who happen at a particular moment to exercise it? Surely our cry has been very different —that benevolent or not, tyranny is tyranny and the only sure defense of freedom is the dispersal of power.

C. Conclusion

Let me close by offering a proposal, not for reconciling our views, but at least for possible agreement among us on one part of the gold problem. Can we not all agree with Mr. Cortney's point (7): The establishment of a thoroughly free market in gold, with no restrictions on the ownership, purchase, sale, import, or export of gold by private individuals? This means in particular, no restrictions on the price at which gold can be bought or sold in terms of any other commodity or financial instrument, including national currencies. It means, therefore, an end to governmental price fixing of gold in terms of national currencies.

The major problem in achieving such a reform is, as for the U.S. wheat program, the transitional one of what to do with accumulated Government stocks. In both cases, my own view is that the Government should immediately restore a free market, and should ultimately dispose of all of its stocks. However, it would probably be desirable for the Government to dispose of its stocks only gradually. For wheat, 5 years has always seemed to me a long-enough period so I have favored the Government committing itself to dispose one-fifth of its stocks in each of 5 years. This period seems reasonably satisfactory for gold as well, and hence my own proposal for the United States, and also other countries, would be that the Government should sell off its gold in the free market over the next 5 years. Perhaps the greater ratio of the accumulated stock to annual production for gold than for wheat makes a longer transitional period appropriate. This seems to be a matter of expediency not of principle.

A world-wide free market in gold might mean that the use of gold as money would become far more widespread than it is now. If so, governments might need to hold some gold as working cash balances. Beyond this, I see no reason why governments or international agencies should hold any gold. If individuals find warehouse certificates for gold more useful than literal gold, private enterprise can certainly provide the service of storing the gold. Why should gold storage and the issuance of warehouse certificates be a nationalized industry?

IV

Non-Bank Financial Institutions

Until recently non-bank financial institutions did not receive the attention they deserved from the academic community. Recently, however, they have been studied more closely; the vast size of these institutions makes them impossible to ignore. Billions of dollars of savings are channelled through these institutions and find their way into the capital markets. Many of these savings seek fixed obligations rather than equities and because of this some observers have become concerned that there would be a lack of savings seeking equity capital.

The traditional view has been that since these institutions do not create money, they did not need to be regulated to the same extent as commercial banks. John G. Gurley and Edward S. Shaw challenge this view. In their article they maintain that non-bank financial institutions compete with the monetary system. Funds flowing through these institutions are not subject to the same control as are the monies created by commercial banks and thus monetary controls are less effective. Warren L. Smith in his article takes the opposite and traditional view. He argues that financial intermediaries are sufficiently different from commercial banks that Federal Reserve control over them is not needed.

The last article in this section, by Orson H. Hart, examines whether or not life insurance companies are likely to increase their holdings of equities in the future. To be sure, there are legal limitations imposed by the insurance laws of the major states on the amount of equities life insurance companies may own. Aside from the legal restrictions, the article does not foresee any increase in the proportion of insurance companies' assets made up of equities, except possibly if the companies capture a larger proportion of the pension fund business.

19

Financial intermediaries and the saving-investment process*

John G. Gurley and Edward S. Shaw

It is fashionable these days to speak of the growing institutionalization of saving and investment. Rapid advances in recent years by pension funds, open-end investment companies, credit unions, and savings and loan associations, among others, have caught our eye. But the advance has been going on at least since the Civil War, and, as Raymond Goldsmith has recently shown, it was quite pronounced during the first three decades of this century. It is with these three decades that our paper is primarily concerned. Our method of analyzing financial data, however, requires explanation since it is based on unconventional theory. Accordingly, the first portions of the paper are largely theoretical. After that, we get down to brass tacks.

Deficits, Security Issues, and GNP

It is easy to imagine a world in which there is a high level of saving and investment, but in which there is an unfavorable climate for financial intermediaries. At the extreme, each of the economy's spending units—whether of the household, business, or government variety—would have a balanced budget on income and product account. For each

Reprinted with permission from *The Journal of Finance,* Vol. XI, No. 2 (May, 1956), pp. 257-266. John G. Gurley was at the University of Maryland and is now at Stanford University. Edward S. Shaw is at Stanford University.

* This paper [was] presented at a joint meeting of the American Finance Association and the American Statistical Association held in New York City on December 27, 1956. The program was under the chairmanship of Robert V. Roosa, Federal Reserve Bank of New York.

spending unit, current income would equal the sum of current and capital expenditures. There could still be saving and investment, but each spending unit's saving would be precisely matched by its investment in tangible assets. In a world of balanced budgets, security issues by spending units would be zero, or very close to zero.[1] The same would be true of the accumulation of financial assets. Consequently, this world would be a highly uncongenial one for financial intermediaries; the saving-investment process would grind away without them.

Financial intermediaries are likely to thrive best in a world of deficits and surpluses, in a world in which there is a significant division of labor between savers and investors. In the ideal world for financial intermediaries, all current and capital expenditures would be made by spending units that received no current income, and all current income would be received by spending units that spent nothing. One group of spending units would have a deficit equal to its expenditures, and the other group would have a surplus equal to its income. And, of course, the *ex post* deficit would necessarily be equal to the *ex post* surplus. In this setting, the deficit group would tend to issue securities equal to its deficit, and the other group would tend to accumulate financial assets equal to its surplus. Security issues and financial-asset accumulations, therefore, would tend to approximate GNP or the aggregate of expenditures. No more congenial world than this could exist for financial intermediaries.

Unfortunately for these intermediaries, our own economy has been much closer to the first than to the second world. With some exceptions during the past half-century, the annual security issues of spending units over complete cycles have averaged somewhat below 10 per cent of GNP in current prices. These issues include government securities, corporate and foreign bonds, common and preferred stock, farm and nonfarm mortgages, and consumer and other short-term debt. We shall call these primary security issues. Thus, at the turn of the century when GNP was around $20 billion, primary security issues ran a bit less than $2 billion per annum. In the late 1940's, with a GNP of approximately $250 billion, primary issues hovered around $20 billion per annum. Dividing the half-century into thirteen complete cycles, we find that the average annual ratio of primary issues to GNP was between 7 and 10 per cent in nine of the cycles. The exceptional cases included World War I, when the ratio reached 20 per cent, the 1930's, when the ratio fell to 3 or 4 per cent, and World War II, when it climbed to 25 per cent. However, if we consider longer phases, 1897-1914, 1915-32, and 1933-49, the ratio was between 9 and 10 per cent in each phase. There

[1] Securities might be issued by spending units to build up their financial assets or their holdings of existing real assets. However, in a world of balanced budgets, no spending unit would have a *net* accumulation of these assets, positive or negative.

is sufficient strength, then, in the link between borrowing and GNP to make the relationship useful for financial analysis. And while the ratio lies closer to zero than to 100 per cent, still it is high enough to permit financial intermediation to be a substantial business.

The Role of Financial Intermediaries

What is the business of financial intermediaries? They lend at one stratum of interest rates and borrow at a lower stratum. They relieve the market of some primary securities and substitute others—indirect securities or financial assets—whose qualities command a higher price. This margin between yields on primary and indirect securities is the intermediaries' compensation for the special services they supply.

The financial institutions that fit these specifications are savings and loan associations, insurance companies, mutual savings banks, Postal Savings banks, investment companies, common trust funds, pension funds, government lending agencies, and others. In addition, we count the monetary system, including commercial banks, as one among many intermediaries. It is a vitally important intermediary, in view of its functions and its size. But its elevated rank among intermediaries does not alter the principle that the monetary system, like other intermediaries, transmits loanable funds by issues of indirect financial assets to surplus units and purchases of primary securities from deficit units. The indirect financial assets, deposits and currency that it issues or creates, are, like the indirect financial assets issued or created by other intermediaries, substitutes for primary securities in the portfolios of spending units. We shall return to this point in a few moments.

Internal and External Finance of Expenditures

In a world of balanced budgets, each spending unit's current and capital expenditures would be financed entirely from its current income. Thus, aggregate expenditures in the economy would be self-financed or internally financed. Internal finance would be equal to GNP.

In a world of deficits and surpluses, some expenditures would be financed externally. The extent of such financing is measured by the sum of the deficits (or surpluses) run by spending units. If at a GNP of $400 billion, the sum of all spending units' deficits is $40 billion, then 10 per cent of GNP is financed externally and 90 per cent is financed internally.

External finance may take two forms: direct finance and indirect finance. The distinction is based on the changes that occur in the financial accounts of surplus units' balance sheets. The finance is indirect if the surplus units acquire claims on financial intermediaries.[2] It is direct if surplus units acquire claims on debtors that are not financial intermediaries.[3]

While the proportion of GNP that is externally financed has not changed much over the past half-century, the proportion that is indirectly financed has risen and, of course, the proportion that is directly financed has fallen. In short, a growing share of primary issues has been sold to financial intermediaries.[4] But the relative gainers have been the non-monetary intermediaries and the relative loser has been the monetary system. Now, if we look at these trends from the standpoint of surplus spenders, we have the following picture: the surplus units have accumulated financial assets in annual amounts that, over long periods, have been a fairly steady percentage of GNP. However, these accumulations have been relatively more and more in the form of indirect financial assets, and relatively less and less in the form of primary securities. Moreover, the accumulations of indirect financial assets have swung toward the non-monetary types and away from bank deposits and currency. Commercial banks and the monetary system have retrogressed relative to financial intermediaries generally.

A Reconsideration of Banking Theory

A traditional view of the monetary system is that it determines the supply of money: it determines its own size in terms of monetary debt and of the assets that are counterparts of this debt on the system's balance sheet. Other financial intermediaries transfer to investors any

[2] In our empirical work, we exclude from indirect finance some kinds of claims on intermediaries, such as accrued expenses or even stockholder equities, that are essentially like debt issues of non-financial spending units.

[3] It may help to illustrate these financing arrangements. Suppose that at a GNP of $400 billion the sum of all spending units' deficits is $40 billion. Suppose further that $40 billion of primary securities, such as corporate bonds and mortgages, are issued to cover the deficits. The primary securities may be sold directly to surplus spending units whose aggregate surplus will also be equal to $40 billion, looking at it *ex post*. In this case direct finance will take place, with surplus spenders acquiring various types of primary securities. Alternatively, if the primary securities are sold to financial intermediaries, surplus spenders will accumulate claims on these intermediaries, indirect financial assets instead of primary securities. In this event we say that the expenditures represented by the primary securities have been indirectly financed. If indirect finance occurs through commercial banks, surplus spenders accumulate bank deposits; if through savings and loan associations, they acquire savings and loan shares; if through life insurance companies, policyholder equities; and so on.

[4] This growth has not been steady. Indeed, it is shown later that there was retrogression in intermediation from 1898 to 1921. The share of issues going to intermediaries rose in the 1920's, rose further in the 1930's, and remained high in the 1940's.

part of this money supply that may be deposited with them by savers. Their size is determined by the public's choice of saving media.

As we see it, on the contrary, the monetary system is in some significant degree competitive with other financial intermediaries. The growth of these intermediaries in terms of indirect debt and of primary security portfolios is alternative to monetary growth and inhibits it. Their issues of indirect debt displace money, and the primary securities that they hold are in some large degree a loss of assets to the banks.

Bank deposits and currency are unique in one respect: they are means of payment, and holders of money balances have immediate access to the payments mechanism of the banking system. If money were in demand only for immediate spending or for holding in transactions balances, and if no other financial asset could be substituted as a means of payment or displace money in transactions balances, the monetary system would be a monopolistic supplier exempt from competition by other financial intermediaries.

But money is not in demand exclusively as a means of payment. It is in demand as a financial asset to hold. As a component of balances, money does encounter competition. Other financial assets can be accumulated preparatory to money payments, as a precaution against contingencies, or as an alternative to primary securities. For any level of money payments, various levels of money balances will do and, hence, various sizes of money supply and monetary system.

The more adequate the non-monetary financial assets are as substitutes for money in transactions, precautionary, speculative, and—as we shall see—diversification balances, the smaller may be the money supply for any designated level of national income. For any level of income, the money supply is indeterminate until one knows the degree of substitutability between money created by banks and financial assets created by other intermediaries. How big the monetary system is depends in part on the intensity of competition from savings banks, life insurance companies, pension funds, and other intermediaries.

Financial competition may inhibit the growth of the monetary system in a number of ways. Given the level of national income, a gain in attractiveness of, say, savings and loan shares vis-à-vis money balances must result in an excess supply of money. The monetary authority may choose to remove this excess. Then bank reserves, earning assets, money issues, and profits are contracted. This implies that, at any level of income, the competition of non-monetary intermediaries may displace money balances, shift primary securities from banks to their competitors, and reduce the monetary system's requirement for reserves. In a trend context, bank reserves cannot be permitted to grow as rapidly as otherwise they might, if non-monetary intermediaries become more attractive channels for transmission of loanable funds.

Suppose that excess money balances, resulting from a shift in spending units' demand away from money balances to alternative forms of indirect financial assets, are not destroyed by central bank action. They may be used to repay bank loans or to buy other securities from banks, the result being excess bank reserves. At the prevailing level of security prices, spending units have rejected money balances. But cannot banks force these balances out again, resuming control of the money supply? They can do so by accepting a reduced margin between the yield of primary securities they buy and the cost to them of deposits and currency they create. But this option is not peculiar to banks: other intermediaries can stimulate demand for their debt if they stand ready to accept a reduced markup on the securities they create and sell relative to the securities they buy. The banks can restore the money supply, but the cost is both a decline in their status relative to other financial intermediaries and a reduction in earnings.

The banks may choose to live with excess reserves rather than pay higher prices on primary securities or higher yields on their own debt issues. In this case, as in the previous two, a lower volume of reserves is needed to sustain a given level of national income. With their competitive situation improved, non-monetary intermediaries have stolen away from the banking system a share of responsibility for sustaining the flow of money payments. They hold a larger share of outstanding primary securities; they owe a larger share of indirect financial assets. They have reduced the size of the banking system at the given income level, both absolutely and relatively to their own size, and their gain is at the expense of bank profits.[5]

[5] We may mention a few additional issues in banking theory. As intermediaries, banks buy primary securities and issue, in payment for them, deposits and currency. As the payments mechanism, banks transfer title to means of payment on demand by customers. It has been pointed out before, especially by Henry Simons, that these two banking functions are at least incompatible. As managers of the payments mechanism, the banks cannot afford a shadow of insolvency. As intermediaries in a growing economy, the banks may rightly be tempted to wildcat. They must be solvent or the community will suffer; they must dare insolvency or the community will fail to realize its potentialities for growth.

All too often in American history energetic intermediation by banks has culminated in collapse of the payments mechanism. During some periods, especially cautious regard for solvency has resulted in collapse of bank intermediation. Each occasion that has demonstrated the incompatibility of the two principal banking functions has touched off a flood of financial reform. These reforms on balance have tended to emphasize bank solvency and the viability of the payments mechanism at the expense of bank participation in financial growth. They have by no means gone to the extreme that Simons proposed, of divorcing the two functions altogether, but they have tended in that direction rather than toward indorsement of wildcat banking. This bias in financial reform has improved the opportunities for non-monetary intermediaries. The relative retrogression in American banking seems to have resulted in part from regulatory suppression of the intermediary function.

Turning to another matter, it has seemed to be a distinctive, even magic, charac-

A Reconsideration of Interest Theory

It is clear from the foregoing remarks that this way of looking at financial intermediaries leads to a reconsideration of interest theory. Yields on primary securities, the terms of borrowing available to deficit spenders, are influenced not only by the amount of primary securities in the monetary system—that is, by the supply of money—but also by the amount of these securities in non-monetary intermediaries—that is, by the supply of indirect financial assets created by these intermediaries. Suppose that savings and loan shares become more attractive relative to bank deposits, resulting in an excess supply of money. Now, if we suppose that the monetary system chooses and manages to keep the money supply constant under these circumstances, the excess supply of money will cause yields on primary securities to fall. The activities of non-monetary financial intermediaries, then, can affect primary yields. The same money supply and national income are compatible with various interest rate levels, depending upon the size of non-monetary intermedi-

---

teristic of the monetary system that it can create money, erecting a "multiple expansion" of debt in the form of deposits and currency on a limited base of reserves. Other financial institutions, conventional doctrine tells us, are denied this creative or multiplicative faculty. They are merely middlemen or brokers, not manufacturers of credit. Our own view is different. There is no denying, of course, that the monetary system creates debt in the special form of money: the monetary system can borrow by issue of instruments that are means of payment. There is no denying, either, that non-monetary intermediaries cannot create this same form of debt. They would be monetary institutions if they could do so. It is granted, too, that non-monetary intermediaries receive money and pay it out, precisely as all of us do: they use the payments mechanism.

However, each kind of non-monetary intermediary can borrow, go into debt, issue its own characteristic obligations—in short, it can create credit, though not in monetary form. Moreover, the non-monetary intermediaries are less inhibited in their own style of credit creation than are the banks in creating money. Credit creation by non-monetary intermediaries is restricted by various qualitative rules. Aside from these, the main factor that limits credit creation is the profit calculus. Credit creation by banks also is subject to the profit condition. But the monetary system is subject not only to this restraint and to a complex of qualitative rules. It is committed to a policy restraint, of avoiding excessive expansion or contraction of credit for the community's welfare, that is not imposed explicitly on non-monetary intermediaries. It is also held in check by a system of reserve requirements. The legal reserve requirement on commercial banks is a "sharing ratio"; it apportions assets within the monetary system. The share of assets allocated to the commercial banks varies inversely with the reserve requirement. The proportion of the commercial banks' share to the share of the central bank and Treasury is the "multiple of expansion" for the commercial banking system. The "multiple of expansion" is a remarkable phenomenon not because of its inflationary implications but because it means that bank expansion is anchored, as other financial expansion is not, to a regulated base. If credit creation by banks is miraculous, creation of credit by other financial institutions is still more a cause for exclamation.

aries and upon the degree to which their issues are competitive with money.[6]

The analysis is only a bit more complicated when we allow for issues of primary securities and the growth of income. Let us take these one at a time. At any income level, some spending units will have deficits and others surpluses. During the income period, the deficit spenders will tend to issue primary securities in an amount equal to their aggregate deficits. Now, if the surplus spenders are willing to absorb all of the issues at current yields on these securities, there will be no tightening effect on security markets. Surplus spenders will accumulate financial assets, all in the form of primary securities, and financial intermediaries will purchase none of the issues.

But this is an unlikely outcome. Ordinarily, surplus spenders can be expected to reject some portion of the primary securities emerging at any level of income and demand indirect financial assets instead, unless their preference for the latter is suppressed by a fall in prices of primary securities and a corresponding rise in interest rates charged to deficit spenders. This incremental demand for indirect financial assets is in part a demand for portfolio diversification. The diversification demand exists because there is generally no feasible mixture of primary securities that provides adequately such distinctive qualities of indirect securities as stability of price and yield or divisibility. The incremental demand for indirect assets, however, reflects not only a negative response, a partial rejection of primary securities, but also a positive response, an attraction to the many services attached to indirect assets, such as insurance and pension services and convenience of accumulation. Part of the demand is linked to the flow of primary security issues, but another part is linked more closely to the level of income.

For these reasons, then, ordinarily some portion of the primary issues must be sold to financial intermediaries if present yields on these securities are to be defended. Assuming for the moment that the monetary system is the only financial intermediary, the increase in the money supply must be equal to the portion of primary issues that spending units choose not to accumulate at current yields. If the monetary system

[6] We can reach the same conclusion by looking at the supply of and the demand for primary securities. The shift in demand to savings and loan shares reduces spending units' demand for bank deposits by, say, an equivalent amount. Consequently, the demand by spending units for primary securities is unchanged at current yields. Also, there is no change in this demand by the monetary system, since we have assumed the money supply constant. However, there is an increase in demand for primary securities by savings and loan associations. So, for the economy as a whole, there is an excess demand for primary securities at current yields, which is the counterpart of the excess supply of money.

Downward pressure on primary yields is exerted as long as the indirect debt of non-monetary intermediaries is to some degree competitive with money and as long as the additional demand for primary securities by these intermediaries is roughly equivalent to their creation of indirect debt.

purchases less than this, spending units will accumulate the residual supply at rising interest rates to deficit spenders. The emergence of security issues and a diversification demand for money based on these issues means that the money supply must rise at a given income level to maintain current yields on primary securities.

Still retaining the assumption that the monetary system is the only financial intermediary, we now permit income to grow. As money income gains, spending units demand additions to their active or transactions balances of means of payment. An upward trend in money payments calls for an upward trend in balances too. The income effect also applies to contingency or precautionary balances. If spending units are increasingly prosperous in the present, they feel able to afford stronger defenses against the hazards of the future.[7]

The combination of the income and diversification effects simply means that, when income is rising, a larger share of the issues must be purchased by the monetary system to prevent a rise in primary yields. The system must supply money for both diversification and transactions, including contingency, balances.

We may now introduce non-monetary intermediaries. The growth of these intermediaries will ordinarily, to some extent, reduce the required growth of the monetary system. We have already presented the reasons for this, so it suffices to say that primary yields may be held steady under growth conditions even with a monetary system that is barely growing, provided other intermediaries take up the slack.

In summary, primary security issues depend on aggregate deficits, and the latter in turn are related to the income level. At any income level, the diversification effect of these issues means that financial intermediaries must grow to hold primary yields steady. If income is rising, too, there is an incremental demand for money and perhaps for other indirect assets for transactions and contingency balances, requiring additional intermediary growth. To the extent that the issues of non-monetary intermediaries are competitive with money balances of whatever type, the required growth of the monetary system is reduced by the expansion of other intermediaries.

[7] For periods longer than the Keynesian short run, it is hardly safe to assume that transactions and contingency demands for additional money balances are proportional to increments in the level of money income. They may be elastic to interest rates on such primary securities as Treasury bills and brokers' loans. For any increment in money income, they may rise with real income. As a larger share of national income involves market transactions, as population moves from farms to cities, as a dollar of income is generated with more or fewer dollars of intermediate payments, as credit practices change, as checks are collected more efficiently, or as deposits cease to bear interest and bear service charges instead, one expects the marginal ratio of active balances to income to vary. And incremental demand for contingency balances must be sensitive not only to income, and perhaps to interest rates, but to the evolution of emergency credit facilities, to job security and social security, to an array of circumstances that is largely irrelevant in short-period analysis.

20

Financial intermediaries and monetary controls*

Warren L. Smith

I. Introduction

It has been argued recently that the rapid expansion of financial intermediaries other than commercial banks has undermined the effectiveness of the traditional instruments of monetary control, since these instruments have relatively little—or at least a very indirect—effect on the growing nonbank part of the financial system. The remedy that is sometimes suggested is that the monetary controls of the Federal Reserve be extended to cover nonbank financial intermediaries. This thesis has been advanced rather forcefully by Gurley and Shaw, and it seems to have attracted a number of other adherents.[1]

In this paper, I shall undertake an analysis of the role of nonbank financial intermediaries in the financial structure of the economy, their relation to the commercial banking system, and the impact that their growth has had upon monetary policy. My conclusion is that while the growth of these institutions has weakened monetary controls to some extent, the effects do not appear to have been very great, and the main sources of difficulty for monetary policy are to be found elsewhere.

Reprinted by permission of the publishers from Warren L. Smith THE QUARTERLY JOURNAL OF ECONOMICS Cambridge, Mass.: Harvard University Press, Copyright, 1959, by the President and Fellows of Harvard College. Vol. LXXIII, (November, 1959), pp. 533-553. Warren L. Smith is Professor of Economics, University of Michigan.

* Some of the ideas developed in this paper were first advanced in a preliminary way in one section of my address entitled, "Some Unsettled Issues in Monetary Policy," which was delivered at the Second Duke American Assembly in March 1959, and later published in the summary proceedings of that Assembly, *American Monetary Policy* (Durham, N.C., 1959), pp. 14-30.

[1] See particularly J. G. Gurley and E. S. Shaw, "Financial Aspects of Economic Development," *American Economic Review*, XLV (Sept. 1955), 515-38.

II. Financial Intermediaries and Commercial Banks

The rapid growth of nonbank financial intermediaries between 1900 and 1952 is amply documented and discussed in the recent study by Raymond Goldsmith.[2] Tabulations of the Federal Reserve flow-of-funds studies show clearly that the rapid expansion of intermediaries has continued since 1952.[3] Between 1952 and 1957, financial asset holdings of intermediaries increased about three and one-half times as fast as similar holdings of commercial banks (59 per cent compared with 17 per cent). When holdings of Treasury obligations and of cash assets are eliminated, however, thus limiting the comparison to private loans and securities and securities of state and local governments, intermediaries increased their holdings only about twice as fast as did commercial banks (78 per cent compared with 43 per cent). On the other hand, the activities of savings departments of banks should be included with those of intermediaries, and, although no satisfactory basis for segregation of assets exists, the fact that demand deposits increased by only 7 per cent while outstanding intermediary claims (including time deposits in commercial banks) increased by 47 per cent indicates clearly that purely

[2] R. W. Goldsmith, *Financial Intermediaries in the American Economy Since 1900* (Princeton: Princeton University Press, 1958). For a panoramic picture of the growth of various types of financial institutions, see particularly Tables 10–12, pp. 73-78. Joseph Aschheim, in his recent article, "Commercial Banks and Financial Intermediaries: Fallacies and Policy Implications," *Journal of Political Economy*, LXVII (Feb. 1959), 59-71, contends, quite correctly, that Goldsmith's findings exaggerate the growth of nonbank financial intermediaries in the relevant sense and that the growth has been uneven. He shows (see his Table 1, p. 60) that commercial banks held a larger percentage of the assets of private financial institutions in 1952 than in 1939. However, it should be noted that since 1945 this percentage has undergone a continuous and fairly substantial decline. Moreover, as Aschheim notes and as I shall argue, savings banking operations of commercial banks should be included with the operations of nonbank financial intermediaries, and while, as he points out (Table 2, p. 61), commercial bank time deposits grew less rapidly than demand deposits between 1939 and 1952, time deposits have grown considerably more rapidly than demand deposits since 1945. In other words, when Goldsmith's data are adjusted to take account of Aschheim's legitimate criticisms, they still indicate that nonbank financial intermediaries have grown much more rapidly than commercial banks since the end of World War II.

[3] I shall use the term "commercial banks," or sometimes just "banks," to refer to those institutions which accept demand deposits and are responsible for the management of the nation's check-payment system. The term, "financial intermediaries," or sometimes just "intermediaries," will refer to all other institutions whose primary line of business is the sale of their claims to the public and who, by investment of the funds so obtained in primary securities issued by various classes of spending units, are able to increase the aggregate supply of loanable funds. I shall not bother to append the word "nonbank" when I refer to financial intermediaries other than commercial banks partly to minimize cumbersome terminology and partly because I shall argue that commercial banks are not intermediaries anyway.

commercial banking activities have been growing at a very much slower pace than those of financial intermediaries.[4]

Several writers have argued recently that financial intermediaries are very similar to commercial banks, since, being subject to fractional reserve requirements (largely self-imposed), they are able, like banks, to engage in a process of "multiple credit creation."[5] Admittedly they cannot create money (i.e., means of payment) as can banks, but some of them issue claims to the public which are very close substitutes for money—such as time deposits, savings and loan shares, etc.[6] And, in any case, their credit-creating activities expand the supply of loanable funds available to finance expenditures, even if there is not a parallel expansion of the money supply.[7]

I believe this argument is a considerable oversimplification which exaggerates the similarities between commercial banks and financial intermediaries. Commercial banks do have a special ability to expand credit for a reason that is simple but often overlooked. What is truly unique about commercial banks is the speed and automaticity of the

[4] These calculations are based on data taken from mimeographed tabulations obtained from the Board of Governors of the Federal Reserve System.

[5] This view is expressed in Gurley and Shaw, *op. cit.*, pp. 534-36; G. W. McKinley, "The Federal Home Loan Bank System and the Control of Credit," *Journal of Finance*, XII (Sept. 1957), 319-32; Donald Shelby, "Some Implications of the Growth of Financial Intermediaries," *Journal of Finance*, XIII (Dec. 1958), 527-41.

[6] McKinley (*op. cit.*, pp. 321-22) argues for an expanded definition of money to take in claims other than means of payment. See also C. F. Haywood, "A Comment on 'The Federal Home Loan Bank System and the Control of Credit,'" *Journal of Finance*, XIII (Dec. 1958), 542-44; and G. W. McKinley, "Reply," *ibid.*, pp. 545-46.

[7] Assuming that the public's holdings of various claims expand in proportion, the traditional "credit expansion multiplier" can easily be expanded to encompass the operations of intermediaries. If the public holds demand deposits, currency, and claims against intermediaries in the proportions $1 : c : t$, the "multiplier" would be as follows:

$$E = \frac{1 + c + t}{r_d + c + (r_t + r_d r_s)\, t} X,$$

where E = the combined expansion of financial claims (currency, demand deposits, and intermediary claims), or, alternatively, the combined acquisition of credit instruments by banks and intermediaries, X = excess reserves (deposits at central bank) available to support expansion, r_d = reserve requirement for demand deposits, r_t and r_s = reserve requirements for intermediary claims (held in central bank deposits and currency and in demand deposits in commercial banks, respectively). If, for example, $c = .25$, $t = 1.25$, $r_d = .20$, $r_t = .03$, and $r_s = .02$, the multiplier will be approximately 5 ($2.50 \div .4925$), and \$100 of excess reserves will permit expansion of financial claims and of earning assets of banks and intermediaries combined by about \$500, of which \$200 will be demand deposits, \$50 currency, and \$250 intermediary claims. This approach is employed by Shelby, *op. cit.*, who presents a multiplier expression somewhat similar to the one given here. See also E. S. Shaw, *Money, Income and Monetary Policy* (Homewood, Ill.: Richard D. Irwin, 1950), pp. 133-42.

Some who view the role of intermediaries in this way feel that their growth has not undermined monetary controls, contending that, while the discipline imposed by the central bank is not applied directly to intermediaries, the indirect effects are powerful and dependable. This is the opinion of McKinley, *op. cit.*, and of Shelby, *op. cit.*

process by which reserves lost by one bank when it makes loans are restored to the banking system. It makes no difference what disposition is made of the funds paid out by the borrower—whether the recipient decides to spend the money, save it and use the savings to buy a claim in a financial intermediary or to buy a primary security, or to hold it in the form of an idle deposit, the reserves lost by the bank making the loan are normally restored to the banking system quite promptly and mechanically due to the virtually universal practice of depositing a check in a commercial bank promptly after receipt.[8] That is, the restoration of reserves to the commercial banking system within a few days of the time they are lost through lending is a built-in feature of our payments mechanism, and it is for this reason that *their distinctive role as issuers of means of payment gives commercial banks a peculiar ability to expand credit.*

In a sense, financial intermediaries do participate in the expansion of credit, but their role is different from that of commercial banks and has a different significance. When a person deposits a check in a commercial bank, there is absolutely no presumption that he has performed an act of saving,[9] whereas, subject to two important qualifications to be taken up in the next section of this paper, the receipt of funds by an intermediary does imply an act of saving. I might add that it is the nature and magnitude of these qualifications that determines the importance of intermediaries as a destabilizing influence.

As a first approximation, intermediaries can be said to participate in the process of credit expansion to the extent that expansion of loans and creation of money by the commercial banking system leads to an increase in expenditures, income, and saving. A portion of the saving generated in the process of income expansion is channeled to intermediaries. These institutions, in turn, invest the savings entrusted to them in loans and securities, thus helping to finance another round of income expansion. As the multiplier process of income growth continues, at each round part of the additional saving that is generated takes the form of intermediary claims, permitting an equivalent expansion of in-

[8] This statement must be qualified to the extent that the recipient of the check decides to take the proceeds in currency or the intermediary in which the check is deposited uses the proceeds to add to its vault cash reserves. This qualification is not ordinarily very important, because, under normal circumstances, the marginal currency-to-demand-deposit ratios of both the public and intermediaries are quite low. Another qualification is necessary if the funds are deposited in a savings account in a member commercial bank, since a small amount of member bank reserves (5 per cent of the deposit ordinarily) is drained away from the commercial banking portion of the system.

[9] This point is made by J. M. Culbertson in "Intermediaries and Monetary Theory: A Criticism of the Gurley-Shaw Theory," *American Economic Review*, XLVIII (Mar. 1958), 122. The above statement should not be interpreted as a denial of the proposition that when equilibrium is reached through income and interest rate adjustments, an increment to the money supply becomes part of somebody's saving.

termediaries' earning assets. Thus, the role of intermediaries in the process of credit expansion differs from that of commercial banks in three important respects.

1. The time periods are entirely different. The time period involved in the process of commercial bank credit expansion is closely related to the operation of the payments mechanism—it is the average time that elapses between the receipt of excess reserves by one bank and the receipt by other banks of that portion of the excess reserves that is transmitted to them as a result of the first bank's lending. This period is surely quite short—a matter of a few days—and is determined in good part by rather mechanical and well-ingrained habits which result, under ordinary circumstances, in the deposit of a check in one's account very soon after receipt.[1] We may designate this time period as a *payment-turnover period.* On the other hand, the time period involved in expansion by intermediaries is an *income-turnover period,* i.e., the average period that elapses between successive receipts of income in the income expenditure process. Clearly the time period for commercial banks is very much shorter.[2]

2. Credit expansion by intermediaries is subject to much more important leakages than is expansion by banks. That is, the entire flow of funds goes through the commercial banks each time around[3] (i.e., each payment-turnover period), whereas only that portion of saving which is allotted to intermediaries (obviously quite a small fraction of the total income flow) goes through these institutions each time around (i.e., each income-turnover period).

3. The two processes differ greatly in their economic significance. Commercial bank credit creation makes funds available to finance expenditures in excess of the funds arising out of the current income flow. Intermediaries, to the extent that their activities are as described so far, merely collect a portion of current voluntary saving and serve the

[1] The time period is also determined partly by two other factors which are not quite so mechanical—the time that elapses between the receipt of excess reserves by a bank and the consequent expansion of the bank's loans and the time that elapses between the borrowing of funds through a bank loan and the expenditure of the loan proceeds. However, it seems safe to assume that these lags are ordinarily quite short, and two recent studies seem to corroborate this view. See George Horwich, "Elements of Timing and Response in the Balance Sheet of Banking, 1953-55," *Journal of Finance,* VII (May 1957), 238-55; R. P. Black, *An Analysis of the Impacts of the 1953 and 1954 Reductions in Federal Reserve Member Bank Reserve Requirements* (Unpublished Ph.D. thesis, University of Virginia, 1955).

[2] The income-turnover period is the reciprocal of the income velocity of active money. See Fritz Machlup, "Period Analysis and Multiplier Theory," this *Journal,* LIV (Nov. 1939), 1-27, reprinted in *Readings in Business Cycle Theory* (Philadelphia: Blakiston Co., 1944), pp. 203-34. It should be noted that the payment-turnover period is *not* the reciprocal of the transactions velocity of active money, however, but is a smaller, more mechanical, and probably more stable magnitude.

[3] This statement needs to be qualified to take account of possible currency drain. (See footnote 8, p. 277.)

function of making these funds available for the financing of current expenditures—i.e., they help to channel saving into investment in a broad sense. Thus, intermediaries are exactly what their name indicates. Commercial banks, on the other hand, are distinctly *not* intermediaries. That is, the decision to save a portion of current income and to hold the savings in the form of a demand deposit does not make any more funds available to the capital market than would have been available had the decision been made to spend instead, and does no more than restore to the commercial banking *system* the lending power that was lost when the original check was written to transmit income to the recipient.[4]

In view of these major differences between the two processes, it seems doubtful whether a "credit expansion" approach to the study of intermediaries is desirable or useful. The mixing together under one heading of two phenomena as different as these seems better calculated to add to confusion than to enlightenment.[5]

III. Possible Destabilizing Effects of Intermediaries

The above analysis overlooks the possibility of substitution between money and intermediary claims as asset prices and interest rates change. Consider a situation in which the monetary authority in the interest of checking inflation reduces the rate at which it is permitting the supply of bank reserves to increase in the face of a rise in the rate of growth of demand for loanable funds.[6] The growth of the supply of money and bank credit is slowed down promptly and interest rates begin to rise. According to the above analysis, intermediaries will be affected in due course to the extent that rising interest rates slow down the rate of growth of income and thus the flow of new saving into their claims. However, intermediaries may be involved in another more immediate way. As interest rates rise on primary securities, intermedi-

[4] Culbertson, *loc. cit.*

[5] For some kinds of problems—especially in connection with economic growth—it may be important to consider the total amount of credit, bank and nonbank, that can be supported by a given reserve base. The simple "credit expansion multiplier" of the type referred to in footnote 7, p. 276 above would be of little use for this purpose, however. The correct approach would require that the bank credit expansion process be embedded in a comprehensive model of the income determination mechanism in order to take account of the expansion of time deposits and similar claims resulting from the flow of a portion of saving into these instruments as income expanded. However, these are second-order effects, and I believe one is justified in overlooking them in connection with short-run analysis of stabilization policy on the ground that they develop rather slowly and can be offset, if necessary, by the monetary authority.

[6] My discussion will, for the most part, be confined to circumstances in which a restrictive monetary policy is being applied, although much of it would apply with signs reversed to a situation in which credit is being eased.

aries may raise the interest rates they offer to holders of their claims in order to attract an increased flow of funds. To the extent that their claims are close substitutes for money—such as savings deposits and savings and loan shares—the more attractive terms may induce holders of money to shift to such claims.[7] In the simplest case, if intermediaries hold their reserves in the form of deposits in commercial banks, this shift increases the lending power of intermediaries without reducing that of the banks, thus increasing the supply of loanable funds. Cash balances are dishoarded with intermediaries acting as agents in the process. Such dishoarding could happen in the absence of intermediaries, with holders of cash balances giving them up in exchange for primary securities. But intermediaries may facilitate it by offering claims to the public that are better substitutes for money than are primary securities. To the extent that this is the case, intermediaries serve to increase the elasticity of liquidity preference, thus to some extent undercutting the restrictive efforts of the monetary authorities.

Shifts between money and intermediary claims are only part of a much more general problem. The greater the supply of liquid assets which are close substitutes for money, the greater the elasticity of liquidity preference is likely to be. As credit conditions tighten, holders of Treasury bills or other short-term—or not so short-term—government securities may be able to sell them to finance expenditures or may be able to borrow from someone else who is able to raise the funds by selling them. Thus, banks may sell such securities to obtain funds for lending and so may intermediaries. Commercial paper or other high-grade private debt near to maturity may similarly serve as a vehicle for such transfers of funds. The importance of intermediaries as a destabilizer depends upon their ability to sell their own newly-issued claims or liquid assets from their portfolios in exchange for existing cash balances and thus to obtain funds for lending to supplement the inflow of current saving. We shall now consider these possibilities in some detail.

A. Substitutions between money and intermediary claims

The effects on the supply of loanable funds produced by shifts of one dollar between various categories of fixed value redeemable claims are given by the coefficients in Table 1,[8] in which d is the reserve requirement for demand deposits, t the reserve requirement for time de-

[7] This possibility is suggested by J. G. Gurley in his discussion of papers by Herbert Stein, J. W. Angell, and E. M. Bernstein, *American Economic Review, Papers and Proceedings,* XLVIII (May 1958), 105.

[8] Expressions rather similar to some of those given here are worked out in J. M. Lishan and W. R. Allen, "Time Deposits, the Banking System, and Inflation," *Southern Economic Journal,* XXV (Oct. 1958), 133-43.

Table I. Effects on Supply of Loanable Funds of Shifts between Categories of Claims

Nature of shift			Example
From	To	Coefficient	$d = .18; t = .04; c = .1$
Currency	Demand deposits	$\frac{(1-d)\,(1+c)}{d+c}$	\$3.22
Currency	Time deposits in commercial banks	$\frac{(1-t)\,(1+c)}{d+c}$	3.77
Currency	Other intermediary claims	$\frac{(1-td)\,(1+c)}{d+c}$	3.90
Demand deposits	Time deposits in commercial banks	$\frac{(d-t)\,(1+c)}{d+c}$	.55
Demand deposits	Other intermediary claims	$\frac{d\,(1-t)\,(1+c)}{d+c}$	.68
Time deposits in commercial banks	Other intermediary claims	$\frac{t\,(1-d)\,(1+c)}{d+c}$	.13

posits and other intermediary claims, and c the coefficient of cash drain (the public's desired marginal ratio of currency to demand deposits). Reserves for demand and time deposits at commercial banks are assumed to be held in the form of currency or deposits at the central bank, while reserves of other intermediaries are held in the form of deposits in commercial banks. These coefficients take account of secondary expansion of credit by the commercial banking system, but, for reasons set forth earlier, do not allow for any induced expansion by intermediaries.[9]

[9] Vault cash reserves of intermediaries other than savings departments of commercial banks appear to be so small that it is legitimate to disregard them. Between the end of 1952 and the end of 1957, the vault cash reserves of insured mutual savings banks increased by less than 0.2 per cent of the increase in deposits. (Federal Deposit Insurance Corporation, *Annual Reports,* 1953 and 1957.) Published data do not show separately the vault cash reserves of savings and loan associations, but a similar relationship seems likely. The expression for the effect of a shift from demand deposits to intermediary claims (the most important type of shift) in the general case is as follows:

$$\frac{(1-t)\,(d+c)-(1-d)\,[c\,(1-t\,(1-p))+pt]}{d+c}$$

where p is the proportion of reserves of intermediaries held in cash or central bank deposits. The coefficients in the table for shifts from demand deposits to time deposits

The last column of the table shows the effects of the various shifts in the case in which reserve requirements are 18 per cent for demand deposits, 4 per cent for time deposits and other intermediary claims, and there is a cash drain of 10 per cent.[1] Shifts involving currency have by far the strongest effects; however, while currency holdings show erratic variations and a seasonal pattern, there is no evidence of systematic shifts of the type we are considering.[2] But there are some indications of destabilizing shifts from demand to time deposits in commercial banks. Table II shows the changes in holdings of various types of fixed value redeemable claims by the consumer sector for the years 1953-57. Consumers substantially reduced their rate of accumulation of claims between 1954 and 1955; this reduction probably was chiefly due to the durable goods boom—particularly in automobiles—in 1955, which involved the use of large amounts of liquid assets for the making of down-payments. However, the most interesting year is 1957, when holdings of time deposits in commercial banks increased by $5.1 billion—an amount much larger than in any previous year and an increase of $3.0 billion over 1956—while holdings of money and United States savings bonds showed substantial declines.

Interest rates on time and savings deposits in commercial banks increased sharply in 1957, partly as a result of an increase in the permissible ceilings on such rates by the banking authorities at the beginning of the year.[3] Rates of return to savers in mutual savings banks and savings and loan associations are substantially higher than interest

---

and from demand deposits to other intermediary claims are the two special cases of this general coefficient, where $p = 1$ and $p = 0$, respectively.

In deriving the coefficient, for the effect of a shift from time deposits at commercial banks to other intermediary claims, it was assumed for simplicity that the reserve percentages were the same but that the reserves were held in different forms, as indicated above.

[1] When allowance is made for nonmember banks and for working reserves of vault cash, these reserve percentages probably approximate those now in effect. For the decade 1948-58, the increase in currency outside banks was about 11 per cent of the increase in demand deposits adjusted.

[2] In 1957, when, as suggested below, there is evidence that rising interest rates induced a shift of funds from demand to time deposits, currency outside banks remained constant while demand deposits adjusted fell by over a billion dollars. In his recent article, "The Demand for Currency Relative to the Total Money Supply," *Journal of Political Economy*, LXVI (Aug. 1958), 303-28, Phillip Cagan concludes that the ratio of currency to total money supply falls as the interest rate on deposits rises. Since Cagan includes time deposits in commercial and mutual savings banks in the money supply and since the ratio of currency to total money supply in this sense did decline in 1957, the results for that year are consistent with his analysis. But there is no sign of a tendency to shift from currency to time deposits.

[3] The Board of Governors of the Federal Reserve System and the Federal Deposit Insurance Corporation raised the maximum rate that could be paid on time and savings deposits by member banks and insured nonmember banks from 2½ to 3 per cent effective January 1, 1957. (*Federal Reserve Bulletin*, December 1956, p. 1301, and FDIC, *Annual Report*, 1956, pp. 83-84.)

Table II. Changes in Consumer Sector Holdings of Fixed Redeemable Claims, 1953–1957 (billions of dollars)

	Net Change in Holdings				
	1953	1954	1955	1956	1957
Time deposits in commercial banks	2.5	2.3	1.7	2.1	5.1
Time deposits in mutual savings banks and Postal Savings deposits	1.6	1.8	1.6	1.6	1.4
Savings and loan and credit union shares	3.9	4.8	5.2	5.3	5.3
U. S. Savings Bonds	.3	.5	.3	−.2	−1.9
Demand deposits and currency	−.2	1.7	−1.5	−.3	−1.8
Total	8.1	11.1	7.3	8.5	8.1

Source: Board of Governors of the Federal Reserve System.

rates on commercial bank time deposits, and these rates of return also increased in 1957, although apparently to a smaller extent than interest rates on commercial bank time deposits.[4] However, there is no evidence of any acceleration in the rate of accumulation of mutual savings bank time deposits or savings and loan shares in 1957. Holdings of these claims increased by about the same absolute amount in 1957 as in each of the previous three years, and the percentage increase was actually somewhat smaller.

The change in composition of consumer asset holdings between 1956 and 1957 suggests a shift of funds from demand deposits and perhaps United States savings bonds[5] to time deposits in commercial

[4] According to data compiled by the United States Savings and Loan League, the average return on savings and loan accounts was 3.1 per cent in 1956 and 3.3 per cent in 1957, the average interest rate on savings deposits in mutual savings banks was 2.8 per cent in 1956 and 3.0 per cent in 1957, and the average interest rate on time deposits in commercial banks was 1.6 per cent in 1956 and 1.8 per cent in 1957. (*Savings and Loan Fact Book*, 1958, Table 34, p. 58.) However, computations based on data contained in the accounting statements of all insured commercial banks indicates a considerably greater increase in the average interest rate on time and savings deposits in commercial banks—from 1.58 per cent in 1956 to 2.08 per cent in 1957. (FDIC, *Annual Report*, 1957, p. 41.) Thus, it appears that commercial bank time deposit interest rates rose considerably more sharply in 1957 than interest rates on mutual savings bank time deposits and returns on savings and loan shares, but that yields on commercial bank time deposits still remained substantially below yields on the other two classes of claims. Of course, there is a wide variation from one bank or association to another, in each of these groups.

[5] The yields to maturity of U.S. savings bonds, Series E and H, were increased from 3 to 3.25 per cent, effective February 1, 1957, thus making them reasonably competitive with yields obtainable from other savings media. However, the yields on savings bonds rise with the period of holding and are less than 3 per cent for the

banks. It also suggests that for many persons time deposits in commercial banks are closer substitutes for money than are savings and loan accounts or time deposits in mutual savings banks.[6] It may be that the sharp increases in time deposit interest rates which were made by many commercial banks in 1957 pushed these rates above the "threshold" levels at which many small savers for the first time became aware of the advantages of investing in such claims. In such cases, savers may have failed to "shop around" to see where they could get the highest returns and simply taken the easy course of shifting funds from checking accounts to savings accounts in commercial banks.[7] Other

---

first three years; thus, low rates during the early period reduce their initial attractiveness relative to time deposits. But, for the investor who already holds savings bonds, the yields from continued holding are in excess of 3.25 per cent, so there should be little incentive to redeem them prior to maturity in order to shift to time deposits. (*Treasury Bulletin*, May 1957, pp. A-1ff.) The reduction in the amount of savings bonds outstanding increased from $1.5 billion in 1956 to $3.8 billion in 1957, reflecting an increased disinvestment by all sectors of $2.3 billion. This was the result of a reduction of $0.9 billion in Treasury sales of new bonds, a reduction of $0.1 billion in redemptions of matured bonds, and an increase of $1.5 billion in redemptions of unmatured bonds. (*Treasury Bulletin*, Feb. 1959, pp. 43-46.) Separate data for consumer sector holdings are not available in this breakdown, but since yields from continued holding of unmatured bonds already in an investor's possession are relatively attractive, it seems likely that only a relatively small part of the increased rate of disinvestment can be attributed to increased yields available on time deposits.

[6] For indications that consumers prefer, other things equal, to hold savings deposits in commercial banks, see D. A. Alhadeff and C. P. Alhadeff, "The Struggle for Commercial Bank Savings," this *Journal*, LXXII (Feb. 1958), 1-22, esp. 4-6. See also John Lintner, *Mutual Savings Banks in the Savings and Mortgage Markets* (Harvard Graduate School of Business Administration, 1948), pp. 143-47.

[7] It is doubtful whether commercial banks benefit by raising interest rates to attract time deposits. The additional earning assets that a bank can acquire due to a shift of funds from demand to time deposits is equal to the difference between the reserve requirements on the two classes of deposits multiplied by the amount of funds shifted. (For the banking system there is further multiple expansion possible, which is taken into account in deriving the coefficient in Table 1.) The following expression gives the break-even yield (i) that must be earned on the additional funds invested as a function of the prevailing interest rate on time deposits (r), the reserve requirements on demand deposits (d) and on time deposits (t), and the interest elasticity of the public's demand for time deposits ($n_{Tr} > 0$), when all of the increase in time deposits comes at the expense of demand deposits at the same bank.

$$i = \frac{r}{d - t}\left(1 + \frac{1}{n_{Tr}}\right)$$

If $r = .02$, $d = .20$, $t = .05$, and n_{Tr} is infinite (so that the funds shift to time deposits without any need to raise the interest rate), the yield, i, on the new loans and investments would have to be 13.3 per cent in order to permit the bank to break even. If $n_{Tr} = 1$, the yield required to justify a small increase in the interest rate to attract more funds into time deposits would be 26.7 per cent, because the bank would incur additional cost due to the necessity of paying the higher interest rate on its existing time deposits. To the extent that the rise in time deposit interest rates would draw in funds from outside the bank, the break-even yield on the new loans and investments would be reduced, but it is at least doubtful whether there would be realizable gains in most cases. Furthermore, once savers become interest conscious, they may eventually come to realize that they can get higher returns at mutual savings banks or savings and loan associations.

more sophisticated savers may have decided that the higher interest rates made it worth while to manage their cash balances more carefully, and engage in more frequent shifts between demand and time deposits.[8]

Although consumer saving, asset, and debt adjustments involve so many dimensions that simple interpretations are rather dangerous, it does seem at least plausible that the rise in time deposit interest rates in 1957 caused a shift of funds from demand to time deposits at commercial banks which may have been large enough to weaken perceptibly the restrictive monetary policy which the Federal Reserve was attempting to apply during most of the year.[9]

The destabilizing effects of shifts between demand deposits and time deposits at commercial banks could be eliminated (or reduced) by equalizing (or narrowing the differentials between) reserve requirements on demand and time deposits. This could be accomplished partly by lowering reserve requirements on demand deposits, a step which might be desirable on other grounds as well.[1] However, it is far from certain that there will continue to be systematic destabilizing effects.

[8] This is an instance of the interest elasticity of demand for transactions balances. There can scarcely be any speculative liquidity preference involved, since there is no liquidity risk on holdings of fixed value redeemable claims of the type under consideration. See James Tobin, "The Interest Elasticity of the Transactions Demand for Cash," *Review of Economics and Statistics*, XXXVIII (Aug. 1956), 241-47; W. J. Baumol, "The Transactions Demand for Cash: An Inventory Theoretic Approach," this *Journal*, LXVI (Nov. 1952), 545-56, and "Marginalism and the Demand for Cash in Light of Operations Research Experience," *Review of Economics and Statistics*, XL (Aug. 1958), 209-14; A. H. Hansen, *Monetary Theory and Fiscal Policy* (New York: McGraw-Hill, 1949), pp. 66-68. It may be noted that corporations and state and local government units increased their holdings of time deposits in member banks rather substantially between mid-1957 and mid-1958. (*Federal Reserve Bulletin*, Nov. 1958, pp. 1277-78.) Since these increases do not show up in the Federal Reserve flow-of-funds accounts for 1957, they are apparently related (especially in the case of corporations) to the rebuilding of liquidity that occurred in 1958.

[9] With reserve requirements at the 1957 level of about 19 per cent for demand deposits and 4 per cent for time deposits and assuming no cash drain (since currency holdings did not change perceptibly), the supply of loanable funds would be increased by about 79 per cent of any shift from demand to time deposits (using the appropriate coefficient from Table 1). Thus, if the entire increase of $3.0 billion in the increment to consumer holdings of time deposits between 1956 and 1957 was due to switches from demand to time deposits, the supply of loanable funds was increased by about 79 per cent of this amount, or $2.4 billion. Some of the incremental increase may have been due to switches from U.S. savings bonds (see footnote 5, p. 283), but, assuming (as seems reasonable) that the Treasury obtained the funds to redeem the bonds by borrowing from nonbanks, the effects of this would be about the same as shifts from demand to time deposits.

[1] The additional reserves needed to permit the money supply to expand to meet the requirements of economic growth can be supplied either by open market purchases or by reduction of reserve requirements. Over a period of years, if the federal cash budget is approximately balanced, the resulting constancy of the supply of government securities, combined with rapid growth of private debt, could result in a scarcity of government securities which might necessitate considerable adjustments in financing practices and portfolio policies. If the Federal Reserve should rely on open market purchases to supply the reserves for growth, the problem would be accentuated. It would be minimized if reserve requirement reductions were used as a means of supplying reserves.

To the extent that the transfer of funds that apparently occurred in 1957 reflects a newly-awakened awareness of the advantages of time deposits on the part of small savers, as suggested above, the funds will probably not be shifted back into demand deposits when interest rates fall. And there are systematic destabilizing effects only to the extent that funds regularly shift from demand to time deposits when interest rates rise and from time to demand deposits when they fall.

The possible destabilizing effects of shift between demand deposits and other intermediary claims, such as mutual savings bank time deposits and savings and loan shares, could be dealt with by imposing appropriate legal reserve requirements on these institutions. There is, however, no evidence that such shifts have been important. It is true that holdings of these intermediary claims have increased much more rapidly than the money supply during the postwar period. But the growth has been steady and has shown no discernible tendency to speed up when interest rates have risen or to slow down when they have fallen. In percentage terms, the growth of these claims accelerated up to 1954 and has since been steadily but slowly decelerating.[2] There is certainly no indication of systematic cyclical shifts between money and these intermediary claims. In the absence of such systematic shifts, there is no reason why the fact that the public's holdings of these claims has been increasing faster than its holdings of money balances should create any particular problems for the monetary authorities.[3]

B. Portfolio shifts by intermediaries

A second type of substitution effect which may exert a destabilizing influence operates through changes in the composition of inter-

[2] The consumer sector's holdings of mutual savings bank time deposits and shares in savings and loan associations and credit unions increased by $5.0 billion or 13.2 per cent in 1952, $5.7 billion or 13.3 per cent in 1953, $6.8 billion or 14.0 per cent in 1954, $7.0 billion or 12.6 per cent in 1955, $7.1 billion or 11.4 per cent in 1956, and $7.0 billion or 10.1 per cent in 1957. (Data from Federal Reserve flow-of-funds tabulations.) The increase in 1958 appears, on the basis of such data as are now available, to have been somewhat larger than the increases of the previous four years.

[3] It should be noted that intermediary claims (including time deposits in commercial banks) are substitutes for other assets as well as money. Under some circumstances, a rise in interest rates on such claims might induce investors to substitute them for goods —i.e., increase their rate of saving—with clear deflationary effects. However, there is little evidence that, under most conditions, the level (as distinct from the composition) of saving is much influenced by interest rates. Intermediary claims might also be substituted for primary securities. The effects of this would depend on the exact nature of the transactions involved. If investors diverted part of their current flow of saving from the purchase of newly-issued primary securities to intermediary claims, there would be little effect, since the intermediaries would, in effect, purchase the securities which would otherwise have been bought by the investors.

mediaries' asset portfolios. When the central bank is following a restrictive policy and interest rates are rising, intermediaries may sell government securities out of their portfolios, and invest the proceeds in higher-yielding private securities, mortgages, and other loans.[4] To the extent that the rising yields on government securities induce holders of idle cash balances to part with such balances in exchange for securities offered by financial institutions, these institutions are enabled to obtain funds in excess of their current inflows from saving, which funds can be used for lending to finance increased income-generating expenditures.[5] Of course, to the extent that expenditures schedules are interest elastic, the rise in interest rates caused by the security sales will cause expenditures to be cut back with restrictive effects. But if the demand for securities on the part of holders of idle cash balances is quite interest elastic, while expenditures schedules are interest inelastic, this process of portfolio adjustment may constitute an important offset to the effectiveness of a restrictive policy. Destabilizing portfolio adjustments may, of course, be made by commercial banks and other investors as well as by intermediaries.[6]

During the 1955-57 period of credit restriction, transfers of government securities among various investor groups played an important part in the process of mobilizing funds to finance current expenditures, but sales of government securities by intermediaries were a relatively unimportant factor. In 1955 commercial banks liquidated $7.4 billion of government securities to augment the supply of funds available to finance private spending. These securities were absorbed by other investor groups, the largest buyers being nonfinancial corporations, ($4.0 billion) and households ($1.8 billion). In 1956 banks continued to liquidate governments but at a reduced rate, while corporations, finding outside funds harder to obtain and more costly now reversed the process

[4] Instruments other than government securities could serve as a vehicle for adjustments such as those described above. For example, insurance companies could sell seasoned corporate securities out of their portfolios and use the proceeds to buy up newly-issued corporate securities through private placements. The fact that, in periods when credit conditions tighten and interest rates rise, the yields on new issues often rise sharply relative to the yields on more or less equivalent outstanding securities might provide an incentive for such switches. And the existence of investors who do not have the facilities to enter the new issue market directly would provide the opportunity for such adjustments.

[5] The willingness of investors to release idle cash balances and take up securities is doubtless partly due to the interest elasticity of demand for transactions balances—corporate treasurers find it worth-while to economize on cash balances and invest and disinvest in Treasury bills more frequently as interest rates rise (see footnote 8, p. 285). But here it is also partly speculative liquidity preference, or perhaps willingness to take a greater liquidity risk in return for a higher yield. See the excellent article by James Tobin, "Liquidity Preference as Behavior Towards Risk," *Review of Economic Studies,* LXVII (Feb. 1958), 65-68.

[6] For a more extensive discussion, see my article, "On the Effectiveness of Monetary Policy," *American Economic Review,* XLVI (Sept. 1956), 588-606.

of the previous year and sold governments in the amount of $4.5 billion. The securities sold by banks and corporations were absorbed through purchases by other sectors, but this time the largest absorber of securities was the federal government itself which retired $4.2 billion out of the proceeds of a cash surplus. A similar pattern prevailed during the first half of 1957.

Intermediaries did sell government securities in the amounts of $700 million in 1955 and $600 million in 1956, but such sales amounted to $500 million in the recession-recovery year 1954 when credit conditions were relatively easy. In fact, they have been net sellers of governments in most years since World War II, but with the exception of the immediate postwar years when they were heavily overloaded with governments as a result of the period of war finance, their sales have been relatively small and have not shown a systematic destabilizing pattern.[7]

Table III assembled from data contained in the Federal Reserve flow-of-funds accounts, shows where the banking system and financial intermediaries obtained the funds to finance the very large extensions of credit to the private sector during the tight-credit period from the end of 1954 to the middle of 1957. For commercial banks, the creation of money (demand deposits and currency) does not account for any of the expansion, since the money supply actually declined by $1.2 billion. Nearly all of the funds needed to cover the $22.5 billion increase in private loans and security holdings came from two sources—liquidation of government securities ($14.5 billion) and the expansion of time deposits ($7.0 billion). While a considerable part of the growth of time deposits undoubtedly represented the normal flow of current saving into this investment medium, it is probable that, as explained above, some of it reflected dishoarding of idle cash balances resulting from switches from demand deposits to time deposits in response to rising interest rates. Moreover, many of the government securities sold by banks were probably taken up by holders of idle cash balances, which were thereby destroyed. The deposits were then recreated through private lending and were promptly injected into the current expenditure

[7] Some intermediaries, such as life insurance companies, have sold government securities consistently during the postwar period, thus gradually restoring their portfolios to a better balanced condition following the tremendous accumulation of government securities during the war. Others, including savings and loan associations, have been rather consistently accumulating government securities at a modest rate in recent years. Thus, purchases and sales of governments by the various classes of intermediaries have offset each other to a considerable extent, and all intermediaries taken together have been net sellers of governments in most recent years. It may be noted that in our earlier discussion of shifts between money and intermediary claims, it was necessary to consider only those types of intermediaries that issue fixed value redeemable claims which are close money substitutes. The problem we are now dealing with, on the other hand, requires a consideration of all intermediaries, since even those—notably life insurance companies—whose claims are not close substitutes for money may engage in portfolio shifts of a destabilizing character.

stream. Although it is risky to attempt any estimate of the dollar magnitudes, it seems likely that through these two channels commercial banks served to facilitate a large amount of dishoarding during this period.[8]

Table III. Sources of Funds Supplied to the Private Sector by Banks and the Monetary System and by Intermediaries, December 31, 1954 to June 30, 1957 (billions of dollars)

	Commercial Banks and the Monetary System[1]	Financial Intermediaries[2]
Total funds supplied to private sector[3]	22.5	47.9[4]
Sources of funds—total	22.5	47.9[4]
Increase in money supply[5]	−1.2	–
Issuance of fixed value redeemable claims[6]	7.0	20.0
Saving through life insurance[7]	–	8.0
Pension accumulations and retained earnings available for financial investment[8]	1.4	12.6
Issuance of credit and equity market instruments	.5	6.1[4]
Sales of Federal obligations	14.5	.8
Other sources, net, and statistical discrepancy	.3	.4

Source: Board of Governors of the Federal Reserve System.

1 Commercial banks, the Federal Reserve System, and Treasury monetary funds.

2 Mutual savings banks, savings and loan associations, credit unions, insurance companies (life and nonlife), self-administered pension and retirement plans, insurance activities of fraternal orders, finance companies, open-end investment companies, brokers and dealers, banks in U.S. possessions, and agencies of foreign banks in the U.S.

3 Includes state and local governments.

4 Net amount of funds obtained by intermediaries from commercial banking system eliminated.

5 Demand deposits adjusted plus currency outside banks.

6 Commercial and mutual savings bank time deposits and savings and loan and credit union shares.

7 Estimated saving by policyholders resulting from life insurance premium and benefit operations.

8 Includes excess of pension plan premiums and interest earnings over benefits and expenses and excess of retained earnings over capital expenditures.

8 It is obvious that these processes are self-limiting at some point, since the marginal rate of substitution of time deposits for demand deposits diminishes as the shifting process goes on and since commercial bank liquidity is worn down by the shifting of portfolios from government securities to loans. However, the experience of 1955-57 suggests that they may, at least under some conditions, continue for a considerable time.

The amount of funds supplied to the private sector by intermediaries was more than double that supplied by banks, and an examination of Table III suggests that most of the funds supplied by intermediaries represented flows of current saving. Liquidation of Treasury securities was a negligible factor, amounting to less than a billion dollars. Of the remaining funds, $26.7 billion came from saving through life insurance, the issuance of new securities (debt and equity) by intermediaries, and pension accumulations and retention of earnings. Surely most of these funds represent flows of current saving, although there may be some indirect dishoarding involved in the sale of $6.1 billion of new securities.[9] The remaining $20 billion was obtained through investment by the public in time deposits in mutual savings banks and in shares in savings and loan associations and credit unions. As indicated earlier, most of this probably represents the normal flow of current saving into such media.

Commercial banks obtained $14.5 billion or 64.4 per cent of the funds they advanced to the private sector through sales of Treasury securities, whereas the corresponding magnitudes for intermediaries were $0.8 billion and 1.7 per cent. This striking difference reflects the much greater flexibility of portfolio policies on the part of commercial banks, which specialize in providing the marginal fluctuating portion of the financial requirements of the private sector of the economy.[1] Intermediaries are more specialized in providing long-term funds for mort-

[9] The major portion represents sales of securities—including open market commercial paper—by sales finance and consumer finance companies. (To the extent that such securities were absorbed by commercial banks, they have been eliminated from the financial intermediaries column.) Some of this financing undoubtedly resulted in activation of existing cash balances. For example, sales finance companies obtained an increasing amount of funds by tapping the liquid reserves of nonfinancial corporations. Rising yields on finance company paper induced corporations to invest their surplus cash balances in such paper, and finance companies passed these funds on to consumers to finance automobile and other durable goods purchases. On the developing financial practices of finance companies during this period, see D. P. Jacobs, "Sources and Costs of Funds to Large Sales Finance Companies, in *Consumer Installment Credit* (Board of Governors of the Federal Reserve System, 1957), Part II, Vol. I, pp. 353-79.

[1] The peculiar Federal tax provisions applicable to commercial banks also tend to encourage frequent portfolio adjustments. Capital losses are deductible from ordinary income, while capital gains are subject to the lower rates of the capital gains tax. Thus, when interest rates rise, banks have an incentive to sell securities at losses and reinvest in securities which will provide capital gains if held to maturity, since the tax saving of the capital losses will exceed the extra taxes on the capital gain. These tax provisions do not provide a direct inducement to shift from government securities to private debt, since the benefits can be obtained by reinvesting in government securities similar to those sold (the wash sale rule prevents reinvesting in identical securities within thirty days of the sale). But the provisions undoubtedly encourage more frequent portfolio adjustments and thus indirectly promote shifting. For a good discussion of these tax provisions, see R. H. Parks, "Income and Tax Aspects of Commercial Bank Portfolio Operations in Treasury Securities," *National Tax Journal*, XI, (Mar. 1958), 21-34.

gage finance, fixed capital investment, and capital improvements by state and local governments. Even during moderate declines in business activity, such institutions can ordinarily find sufficient outlets in the private sector for most of the funds they take in, and they usually exploit these outlets in a steady predictable fashion by letting out such funds as they receive on a current basis at whatever the prevailing yields may be. Rarely do they invest much in short-term government securities or other highly liquid assets while awaiting more favorable conditions for committing funds at long term.

Although systematic changes in the composition of the portfolios of intermediaries could exert destabilizing effects which would seriously weaken the effectiveness of monetary controls, recent experience suggests that when business fluctuations are as moderate as they have been in recent years, commercial banks constitute a much more serious problem in this respect, than do intermediaries.[2] What is needed to make monetary policy more effective is not additional controls over the activities of intermediaries but rather some means by which the scope for destabilizing portfolio adjustments by commercial banks can be reduced.[3]

IV. Concluding Comments

The analysis presented in this paper suggests that financial intermediaries have not contributed very much to instability or to the ineffectiveness of monetary policy in recent years. It does appear that in

[2] A major decline in demand in a particular sector could cause problems for financial institutions specializing in the financing of that sector. For example, a sharp decline in residential construction could impose difficult adjustments on savings and loan associations, whose activities are almost entirely in mortgage finance. They might react by accumulating government securities which would later be liquidated and the funds shifted into mortgages. But this is only one possible type of adjustment. Another possibility is the development of a somewhat more flexible and diversified portfolio policy by savings and loan associations. Still another is that continued bidding for a reduced supply of new mortgages by these institutions would reduce interest rates on mortgages and induce other financial institutions, such as commercial banks and life insurance companies, with more flexible and diversified portfolio policies, to reduce their participation in mortgage financing, thus leaving a larger share of this market to the more specialized savings and loan associations.

[3] Policies which keep down the supply of short-term Treasury securities which commercial banks use to make such adjustments and which are likely to prove to be good substitutes for idle cash balances on the part of nonbank investors when interest rates rise should be helpful in this regard. The avoidance of unnecessary aggressive pursuit of policies of easing credit during declines in business activity would also be desirable, since this would minimize the accumulation of liquid assets by the banking system and thereby minimize the ability of banks to engage in inflationary shifts from government securities to loans in an ensuing period of credit restrictions. If such measures are not sufficient, some method of influencing directly the composition of bank portfolios—such as a secondary reserve requirement—might be considered.

1957 rising interest rates induced a shift of funds by the public from demand deposits to time deposits at commercial banks, which increased the supply of loanable funds due to the fact that reserve requirements are lower for time than for demand deposits. It is not yet clear that we can expect such shifts of funds to be an important systematic destabilizer. If they should prove to be a serious problem, however, they can be dealt with by equalizing the reserve requirements on demand and time deposits at commercial banks. There is no evidence of systematic destabilizing shifts between demand deposits and claims against other intermediaries, such as mutual savings banks and savings and loan associations. However, if such shifts should raise difficulties in the future, their destabilizing effects can be eliminated by the application of appropriate legal reserve requirements to these institutions.[4] This is a straightforward remedy which is entirely consistent with the traditional concepts of central banking, and we should be prepared to put it into effect if necessary. It should be noted, however, that if we impose the same effective reserve requirements on savings institutions as are applicable to demand deposits in order to eliminate destabilizing effects, we also take away the intermediary status of these institutions, since, under these circumstances, a deposit of current saving in a savings institution will reduce the lending power of the banking system as much as it will increase the lending power of the savings institution.

Of course, stabilization policy need not be directed only at the sectors or institutions primarily responsible for the instability. One principle that has been suggested for monetary controls is that they should be devised to touch the economy in its "sensitive spots"—i.e., those sectors or institutions that are particularly amenable to controls.[5] It may be that certain financial intermediaries will prove to be convenient pegs on which to hang an effective monetary policy—either because controls can be devised that can influence them effectively or because from an administrative point of view they provide a convenient point for the application of controls. An example of this is the possibility of regulating consumer credit by controlling the availability of funds to sales finance companies.[6] Controls over intermediaries might also be desirable

[4] A fully effective reform would also include permission for member banks to count vault cash as legal reserves, the elimination of the present three-fold classification of member banks for reserve requirement purposes, and the extension of the reserve requirement provisions of the Federal Reserve System to nonmember commercial banks. [*Editors' note:* Effective November 24, 1960, member banks were allowed to count their vault cash as reserves; effective July 28, 1962, the authority of the Board of Governors to classify cities was terminated, with the result that there is now a two-fold classification for reserve requirement purposes.]

[5] R. S. Sayers, *Central Banking after Bagehot* (Oxford: The Clarendon Press, 1957), pp. 32-33.

[6] See J. S. Farley, "Further Comment," in *Consumer Installment Credit, op. cit.*, Part II, Vol. II, pp. 151-56.

as a means of broadening the impact of monetary policy in order to make it more equitable.

Thus, there may be valid reasons for applying special controls to intermediaries—or to some particular class or classes of intermediaries. But I do not think such controls can be justified on the grounds that intermediaries are, in some meaningful sense, similar to commercial banks. Nor do I believe that there is yet sufficient evidence to provide a strong case for the extension of Federal Reserve controls to financial intermediaries on the grounds that activities of these institutions have been a serious destabilizer.

21

Life insurance companies and the equity capital markets

*Orson H. Hart**

My assignment today is to offer some comments on the prospective interest of life insurance companies in the equity capital markets, with particular reference to the impact of variable annuities and separate accounts. These are both significant technical developments that over the years probably will draw a fair amount of life company money into the common stock market. It may take some time, however, and perhaps quite a bit of time, for the full potential of these developments to be realized. Their application is confined to the pension field; they do not apply to investments originating with the companies' general insurance business—80 to 85 per cent of the assets of the companies today. Moreover, these are not the only new developments available to the life companies in the pension field. The income tax law, a substantial obstacle to life company participation during the 1950's, has been amended. Beyond this, the so-called investment-year method for calculating policyholders' dividends on new pension contracts is an important development that greatly enhances the competitive prospects of the life companies in the field of fixed annuities. We can be sure, I think, that the life companies are going to be more formidable competitors in the pension field than they were during the last decade, but whether the predominant investment route is through common stocks remains to be seen.

Reprinted with permission from *The Journal of Finance*, Vol. XX, No. 2 (May, 1965), pp. 358-367.

* New York Life Insurance Company.

Note: The comments in this paper are the views of the author, and do not necessarily reflect those of the New York Life Insurance Company. The author has had the benefit of the opinions of many knowledgeable people with the New York Life and other companies, but after they read the paper they concluded that they ought not to compete with me in my hour of glory and that the credit was all mine.

Do not overlook the possibility that over the next several years the general price level will be little changed, interest rates will remain relatively high, and the virtues of the bond and mortgage markets as outlets for pension money will be quite impressive both to independent trustees and to the insurance companies. At the present time, bonds and mortgages have rate advantages of considerably more than 1 per cent over common stocks.

Although the trend is towards equity investment of pension money today, uncertainty over the prospective course of stock prices could easily cause a shift from the equity to the bond and mortgage markets. I am not going to assume the role of a stock price forecaster at this meeting, but I do think that it would be a mistake to assume a continuation of the postwar trend of stock prices without a pretty careful examination of fundamental business and monetary conditions. Over say the next decade, should we expect stock prices to rise faster than earnings, or earnings faster than the growth of the over-all economy? What should we expect with regard to the trend in long-term interest rates? These are some of the questions that need to be pondered before we conclude what the trend of pension investments is likely to be.

Finally, we should not conclude that an increase in life company investing in common stocks as a result of a possible gain in the companies' share of the pension business necessarily means a net addition to the over-all demand for common stocks. It is the growth of the pension business, and the relative attractiveness of the bond and mortgage markets versus the stock market, that will determine the demand for common stocks from this source. I doubt if it makes much difference who invests the funds—the life companies, the commercial banks, or other independent trustees.

Before I go into more detail on these matters, let me try to dispose of another important question—Is there much likelihood that the companies, apart from their pension business, will move heavily into common stocks? Certainly common stocks are not excluded as an outlet for the general investment accounts of the companies, but I would think the odds are strongly against such a development.

For since the time of the Armstrong Investigation in 1905, and actually for pretty much the history of the life insurance business in this country, American life insurance companies have been predominantly fixed income investors, seeking certainty in investment return and safety of principal over potential appreciation through stock market profits. It is a partly mutual and partly stock business in which the big New York mutuals in effect are required by law to pass along 90 per cent or more of their earnings every year to their policyholders, thus exerting an enormous competitive pressure not only on the stock companies but on each other. The result is that it is established insurance philosophy that only those earnings are retained in the business as are essential for guaranteeing the financial integrity of the companies. No company could stay in business over the long run without meeting this competition, and you therefore will find few companies with surpluses of the size needed to support a major common stock program in its general investment account.

This, it seems to me, establishes the investment philosophy of the business, quite apart from the limitations imposed by the insurance laws of New York and other major insurance states. A life insurance company must credit interest to its policyholders' reserves at the rate of 2½ to 3 per cent per year. With reserves ranging from 80 to 85 per cent of assets and surplus 10 per cent or less, it is obvious that continuity of income has to be an essential aspect of the financing planning of all companies. Larger surpluses relative to reserves must be built to accommodate larger equity programs, and in view of the competition that exists in the business today it is obvious that unless the Commissioners ordered a limitation on dividends—a far cry from established life insurance policy—most companies would find this a long and difficult job.

The fact is there is little demand from our policyholders, other than employers setting up retirement systems, to invest their reserves in common stocks. Our policyholders may seek appreciation with their own funds by investing them in common stocks. They may even neglect their real insurance needs to provide funds for common stock investment in periods when common stock prices are rising. But when it comes to their life insurance they want what they have to be safe beyond question. Safety is the paramount consideration. Policyholders do not regard their insurance as a vehicle of possible profit through fortunate equity investment. Even in periods when the price level seems to be indicating a sustained rise they do not ask us to move into common stocks, but rather that we oppose easy money, government spending, or other developments they believe are responsible for the inflation.

Thus all equity, including preferred stock, common stock, and investment real estate, accounts for less than 10 per cent of the assets of the life companies today. Substantial increases in the proportion of equity investments held by life insurance companies, apart from their pension business, are not likely to occur in the years ahead. A half century or more of successful accommodation to the present investment philosophy would cause all managements to think twice before venturing strongly into common stocks, whatever the laws might permit, with their general investment accounts. This means that unless the companies make a very rapid penetration of the pension business, and do this mostly via the equity route, the proportion of life company money going into the common stock market is going to continue to be relatively small for some time to come. It is the pension business, not the general insurance business of the life companies, that is focusing attention on our role in the equity markets, and it is principally to the recent developments in this field that I am going to address myself this morning.

Postwar Conditions in the Pension Field

The postwar inability of the life insurance companies to compete in the pension field on an equal footing with independent trustees is generally attributed to two facts, that their earnings were subject to income taxation while those of the noninsured plans were not, and that the life insurance companies were required to invest their funds largely in bonds and mortgages while the independent trustees were free to invest in common stocks. The tax inequality has been substantially eliminated by the Life Insurance Company Income Tax Act of 1959, and I shall not refer to the tax problem hereafter. The investment limitations, however, have continued to hamper the companies. Increasingly the pension business, which has been expanding rapidly during most of the postwar period, has been going the equity route. Employers are impressed with the rising trend of stock prices, of the order of 15 per cent per year since 1950. Within a few years after the end of the war the life companies found themselves on the fringes of a business they had pioneered more than a quarter of a century before. Since 1945 noninsured plans have grown nearly twice as fast as insured plans.

Faced with this situation the companies have sought changes in the insurance laws and regulations to restore their competitive position. First was a change in the insurance laws to permit them to invest in equities for their pension accounts. With stock prices rising so substantially during the 1950's, it was either this or getting out of the pension business, so it seemed. And the companies did not intend to get out of the pension business.

The first organization to gain approval of such investment was the College Retirement Equities Fund, a separate corporation established by the Teachers Insurance and Annuity Association to sell variable annuities. This they achieved when the New York Legislature authorized the sale of variable annuities, but only by this one company, beginning in 1952.[1] The New York Legislature has not yet approved the sale of variable annuities by other New York companies, but in 1962 it authorized the sale of separate pension accounts. In the meantime, the New Jersey Legislature has authorized the sale of both variable annuities and separate pension accounts by New Jersey companies. The Securities and Exchange Commission has ruled that such sales are exempt from

[1] It should be noted that CREF's operations are limited to a highly specialized market, almost entirely university teachers and research personnel.

the Investment Company Act of 1940, provided they are confined to the employers' contributions under group contracts. Finally a large insurance company has secured approval of a registration statement under the Securities Act of 1933, permitting sales of variable annuities to small employers in amounts not to exceed $25 million in premiums.

However, in addition to changes in the insurance law regarding equity investment of pension reserves, the companies also sought a change in the insurance regulations that would improve the competitive position of fixed annuities in the group field. This was the so-called investment-year method for calculating investment return on group annuities. By this method a life insurance company may compute dividends on group pension accounts on the basis of the company's rate of return on new investments rather than on the average return received on the outstanding portfolio. Since virtually all companies at present have a higher return on new investment than on their entire portfolio, they can offer lower-cost fixed annuity programs by use of this method. New York State approved use of the investment-year method in 1961.

Thus a considerable amount of progress has been made, particularly during the past few years, in opening up equity investment of pension funds by life insurance companies and otherwise restoring our competitive position. If New York State should approve the sale of group variable annuities, as many think it will do in the next few years, the competitive position of the life insurance companies in the pension field will be very strong. For the companies will then be able to sell either variable or fixed guaranteed annuities, which the independent trustees are not able to offer. They also will be able to sell separate pension accounts, which under 100 per cent employer-financed plans could be entirely common stock programs. It would seem that they will be able to offer everything the independent trustees can offer plus variable and fixed guaranteed annuities. But whether their efforts are directed towards the equity route or the fixed annuity route will depend upon the relative marketability of the respective programs, and in the last analysis this will depend upon conditions in the capital market. Let us now consider what these programs are and how they compare with each other.

Variable Annuities, Separate Accounts, and Funded Retirement Systems

All funded retirement systems operate on the principal that regular payments will be made to a retirement fund during working years, in order to provide the money later on for retirement payments. In the case of fixed annuities both the rate of accumulation of the fund and the size of the retirement payments are determined by the size of the net premiums and the rate of interest earned on the fund and assumed in the contract. The net premiums and the retirement payments are fixed by contract and by plan; policy dividends, a possible variable, usually go to the policyholder (in most instances the employer), not to the annuitants. Thus a fixed annuity program, unless it is supplemented by additional employer contributions after retirement, usually reflects the conditions prevailing when the program was formulated. Marked changes in these conditions during the accumulation period or after retirement either benefit or harm the annuitant, as the case may be, since payments to him remain unchanged under all circumstances.

In the case of a variable annuity the net premiums may be fixed, but the funds are invested in common stocks and so the rate of accumulation varies with the trend in stock prices. If the trend is upwards, as has been the case pretty much since the end of the war, the retirement fund may greatly exceed the comparable fund accumulated through fixed interest investment, and the annuitant may benefit from the larger fund through larger retirement payments. Moreover, since the fund continues to be invested in common stocks after retirement, the retirement payments will continue to reflect the trend of stock prices and will be further enhanced if the trend remains upwards. Conversely, however, a declining stock price trend will have the effect of reducing the retirement fund, and if the trend continues downwards after retirement the annuitant may find his payments grievously impaired.

There are several variable funds in existence today and we can gain a little perspective by reviewing the experience of the best known of them, the College Retirement Equities Fund (CREF). It has been spectacularly successful. It commenced operations in 1952 when it priced its accumulation units at $10 apiece. The value is adjusted each month to reflect the trend in stock prices. As of September 30, the accumulation unit was valued at $33.89, and the annuity unit at $26.48.

The program is administered by the Teachers Insurance and Annuity Association. For the purpose of this paper, it is important to note that participants in the program are permitted to devote one-quarter, one-third, or one-half of their pension premium to CREF, but at least one-half must be used to provide for fixed retirement through the conventional TIAA retirement program. Thus the combined TIAA-CREF program is putting an increasing amount of its funds into the equity market. CREF is now growing, on a cost basis, roughly twice as fast as TIAA. Now part of this growth is due to the reinvestment of profits realized on the turnover of stocks. (Of course such reinvestment is not a source of new funds to the stock market.) Premiums and investment income, net of expenses, are probably the best measures of the investment flows. These measures suggest that the rapid growth of CREF is going to moderate and may level off somewhat below the TIAA rate, apart from capital gains.

Now why is this so? Partly because CREF is now a young organization holding a relatively small proportion of its reserves on behalf of retirees. As the organization matures this proportion will rise and payouts to retirees will increase, cutting down the rate of asset growth. Beyond this, however, the division of the premium dollar and the comparative advantage of interest yields over dividends yields both favor the growth of TIAA.

Approximately 85 per cent of the TIAA participants are also participating in CREF, and of those participating in CREF 85 per cent or more are making the maximum contribution of one-half the premium. This is obviously quite a testimonial to CREF, but the fact remains that considerably more than half the premium income is going to TIAA—about $95 million to TIAA as compared with $56 million to CREF in 1963. Beyond this, as you all know, interest yields have exceeded dividend yields for a good many years. TIAA's rate of investment return is more than 150 basis points above CREF's return on cost, and 250 basis points above its market return.

CREF's portfolio is now roughly 30 per cent of TIAA's assets when measured at cost, and more than 40 per cent when measured at market. Conceivably the growth in market values will enable CREF in time to surpass TIAA in assets, but I should not think this a likely development at least for some years to come. For the combined operation seems destined to put something less than half its funds, year in and year out, into the equity market, unless a major change is made in the allocation of the premiums between the two components of the system, or market yields greatly increase relative to bond and mortgage yields. Beyond this the proportion of new funds going into the stock market could be considerably reduced if annuitants elect, as they can do, to convert their accumulation units into fixed annuities at retirement.

Now as to separate accounts. These accounts permit a life company to set up a fund separate from its general investment account, and invest in this account the reserves of group pension policies. They differ from a variable annuity in that when employees retire, their shares of the reserve must be used to purchase fixed life annuities. Furthermore, the New York legislation, which is the most important because of the size of the companies domiciled in New York, permits only the employer's share of the contribution to be so invested, the employee's share going into a conventional annuity.

The New York legislation became effective in 1962, but more or less similar legislation was already in effect in the other principal insurance states in the East. By the end of 1963 a check of the annual statements of companies prominent in the pension business indicated that $20 million had been invested in separate accounts as of that date. A more recent check by a John Hancock actuary disclosed that there are now 119 contracts in force, with assets of $65 million and annualized premiums of $70 million.[2] The separate account approach thus is growing rapidly, but of course these accounts are in the early stages of their development and the current rate of growth can hardly be expected to be maintained for long.

The growth of common stock investment under separate accounts is going to be limited by two considerations. The first is that under the New York law when employees reach retirement their shares of the fund must be used to purchase single-premium life annuities, thus in effect transferring the investment from common stocks in the separate account to fixed interest obligations in the general account of the company. Of course this does not necessarily mean that large amounts of common stock will have to be sold and the funds reinvested in bonds and mortgages, but it does mean that the flow of funds into the separate account will be diminished to provide for the necessary purchase of fixed annuities.

How much of an effect this requirement will have on the growth of common stock investment of course will depend upon a great many factors, including the age distribution of the participants, the relative size of the contributions on behalf of the different age groups, and so forth. There is no question, however, but that the impact will be sizeable and the growth of the fund considerably less than it would be if the retirees were to retain variable instead of fixed annuities. In a fully matured pension program approximately one-third of the reserves are on behalf of retired participants. A separate account therefore might approach two-thirds of the amount it would reach under a variable annuity program.

In the second place, as noted above, only the employer's share of the premium may be contributed to the separate account. Unless the employer makes the entire contribution some proportion of the reserves, and perhaps a sizeable proportion, will have to be invested in the company's general account, that is, largely in fixed interest obligations of one kind or another. The trend at the present time is toward increasing employer responsibility for the cost of pensions, and so it may be that over the next ten or fifteen years an increasing proportion of the typical pension account will be invested in common stocks. I suspect, however, that many employers will want some reflection of fixed interest stability in their programs. My guess is that the rate of growth in separate equity accounts probably is going to approximate the rate of growth of the general investment account, once the newness wears off and a normal rate is established.

[2] J. Darrison Sillesky before the Society of Actuaries' Annual Meeting, White Sulphur Springs, W. Va., November 9–11, 1964.

There remains a third new approach to the pension business open to life insurance companies, namely, the investment-year method for computing dividends to policyholders on group annuities. This method has been approved by several states, including the important state of New York in 1961. Prior to this time dividends had to reflect the average rate of return on invested assets. Since this average reflected investments made years in the past when long-term interest rates were below 3 per cent, the insurance companies of course were greatly handicapped in competing with independent trustees for new accounts. For the independent trustees, by the nature of their operations, automatically offer the current rate of interest. Now a company can guarantee something in line with its average rate of investment return, but compute dividends on the basis of its return on new investments. This puts the insurance companies in a much better competitive position on group annuity contracts, and in effect offers employers a lower-cost fixed annuity program than was available heretofore.

Thus there are three new routes through which life companies may be competing for the growing pension business in the years ahead. One points to equity investment during both the accumulation and the retirement period. One points to equity investment during the accumulation period only, and one points to fixed income investment. All limit the equity investment in one way or another. One, the variable annuity, is not yet available to most companies. Do these developments foretell a substantial increase in life company common stock buying? As I have already indicated, I think it is going to take some time for the trends to develop, but in any event they depend upon what employers want, and employers wants are going to be shaped by the relative cost to them of the various alternatives.

Some Comments on the Future of the Pension Business

Anyone interested in what the future holds for the pension business, and this should include most of us, must ask what the course of events is going to be in the general economy. He must think in very long terms, for pensions are a lifetime undertaking. I am going to talk in terms of the next ten or fifteen years, but this is due entirely to the fact that at my time of life the next ten or fifteen years looks a lot more important than the next forty or fifty.

If the long-term trend in prices is slowly upwards, as many economists think, then some method for increasing retirement income must be of concern to most of us, and within the current limits of pension planning this would doubtless mean growing interest in variable annuities and separate equity accounts as opposed to a fixed annuity program. If, however, the next ten or fifteen years are not going to be a period of gradual inflation, and there are quite a few qualified people who think this is the most likely course of events, then the virtues of fixed annuities are going to be increasingly apparent and the growth of both variable and separate equity accounts will be less substantial. It is not easy to draw a quick and decisive conclusion about anything with as many unknowns as we have in this equation.

We live in unusual times now, different from any of the postwar periods that have occurred earlier in our history. After all our previous wars we have demobilized, and the Federal government has retired to something approaching its prewar role in the economy. The Federal government did not do this when the war ended in 1945. It did indeed reduce the scope of its activities drastically in 1946 and 1947, cutting its tax take accordingly. But in 1948 the Berlin crisis arose and in 1950, only five years after the peace, the United States in effect was at war again in Korea. If there had been no other changes in the postwar role of the Federal government, the maintenance of military preparedness in all its aspects makes this a period greatly different from any other postwar period in our history. One way of looking at it is to say that the war never ended, but that some peacetime conditions were superimposed upon the economy and obscured the wartime underpinning.

All of this, it seems to me, poses quite a different economic outlook from anything we have had in our previous history. The economy has had no real letdowns, a remarkable postwar experience. But this is far from predicting continuing inflation; I am not sure we have had much over the past several years. Notwithstanding the heavy demand of the military for goods and services, the productive efficiency of the American economy has performed much as it did in previous postwar periods. The shortages in due course have been made good, and the economy in effect has acclimated itself to its new military burdens. Today it is more stable than it was thirty years ago, but it is still production oriented and as its capital supplies grow it seems to me that the prospects for continued inflation diminish. In the absence of new and large-scale military entanglements, I am inclined to think that national economic policy is going to have to orient itself increasingly towards promoting full employment of resources.

This does not trouble me, because I think it means simply that we have a rich and productive system capable of providing a rising standard of living for everybody, provided we only can maintain a reasonably satisfactory level of employment of resources. With a flexible price and wage system this should not be impossible of accomplishment, though I am under no illusions as to the difficulties involved.

The likelihood is that variable annuities, separate accounts, and fixed annuity programs all will have their innings at one time or another in the years ahead. Unless, however, we mismanage our economic affairs I do not see why the future, looking to the next fifteen years or so, has to be an inflationary one. The disproportionate economic power of the labor unions and their ability to raise the cost of doing business has diminished over the years, partly because the labor unions have become more aware of the consequences of their actions, and partly because there has been a great increase in both the quantity and the efficiency of capital. We are moving into an era when with a little luck and more good judgment we may be able to add to our leisure—one of the principal objectives, it seems to me, of a civilized economy.

So when I think about the course of events in pensions, I have in mind the hopes of most of us and I may be confusing hope with reality. I am essentially an optimist, many of my long-term friends will be surprised to learn, and as an optimist I am looking for a fair degree of price stability and a plentiful supply of goods and services in the years ahead. And I am looking beyond this for a fiscal and monetary policy that is dedicated to such an economic climate. Under these circumstances interest rates are going to be at reasonably attractive levels, and the rise in stock prices is going to be much more modest than during the past fifteen years. The virtues of the bond and mortgage markets will become more apparent to all pension investors, insured and noninsured, and the trend towards equity investment will subside. But whether or not my hopes are realized one thing seems sure—the life insurance companies are back in the pension business, and with contracts suitable for inflationary, stable, or deflationary conditions, as the times may require.

V

Fiscal Policy and Debt Management

Fiscal policy as a conscious method of exerting some control over the economy is of much more recent development than monetary policy. For this reason, and perhaps because it rests on a Keynesian intellectual foundation, it has been more controversial than monetary policy. Most economists now recognize its importance although it continues to be a subject for disagreement and even argument, especially when political questions become involved. Since fiscal policy is concerned with the ways in which the federal government acquires and spends its funds, an understanding of federal budgets and their multiple purposes is essential to its study. Joseph Scherer discusses and explains the three basic federal budgets in the first article.

Fiscal policy can be used in association with monetary policy to attain such economic goals as stabilization and growth. Paul A. Samuelson explains how this can be accomplished when he reports on "The New Look in Tax and Fiscal Policy." Elements of fiscal policy, in particular the question of balancing the budget, are still subject to popular and political debate, however. Arthur Smithies examines the pros and cons of an annually balanced budget and suggests some possible changes in governmental policy that could make fiscal policy a more effective means of attaining economic goals. In the process of evaluating the balanced budget, much popular attention has been focussed on budget deficits. F. M. Bator suggests a change in emphasis that may permit a more open strategy of planning and operating fiscal policy.

Debt management, although technically not a part of either fiscal or monetary policy, is interwoven with both. An introduction to the subject, including its historical development and its present importance, is provided by William Laird's article. Thomas R. Beard examines in some detail the relationship of debt management to monetary policy. It is, for instance, quite possible that Treasury debt policies may negate the effects of the controls managed by the monetary authorities.

Warren Smith explains the significance of the debt and of debt management both to the economy and to monetary and fiscal policy. He evaluates the techniques of debt management as well as its effects and suggests possible improvements. A general overview of the national debt and its importance to the economy is given by Roger L. Bowlby. He suggests that all too often we either overemphasize or underemphasize the debt's importance, and he attempts to assess its real significance within the economy.

22

A primer on federal budgets

*Joseph Scherer**

The Federal budget is a multipurpose document. Its original purpose had been, and its main purpose continues to be, to provide a system of planning and control over Government activities by the executive and legislative branches. In this respect, it serves the same functions that a budget plan performs for an individual or a business. But, unlike the budget of any other single economic unit, the Federal budget because of its sheer size—some $90-120 billion per year, depending upon the particular budget concept used—exerts a potent influence on the nation's economy. This influence, moreover, is being increasingly directed, as a matter of deliberate policy, toward assisting the economy to attain, and sustain, high levels of employment and economic activity. Not only have these growth and stability goals been incorporated in legislation, as in the Employment Act of 1946, but there appears to be a growing consensus among the citizenry that it is appropriate and desirable for the Federal Government to pursue such goals.

In order to evaluate how the Federal Government carries out these housekeeping and policy purposes, it is necessary to examine budget data, totals as well as components. This is not easy, since the needs of analysts have led to the development of a number of concepts that at times appear to provide conflicting data. For example, different dollar magnitudes can be found for categories designated by the same general name. Thus, the data for fiscal 1964 (the year ended June 30, 1964) show the Federal deficit as $8.2 billion in the administrative budget, or $4.8 billion in the consolidated cash budget, or $3.9 billion in the national income account budget. Likewise, for the current year, fiscal 1965, the

Reprinted with permission of the Federal Reserve Bank of New York from its *Monthly Review* (April 1965), pp. 79-88. Joseph Scherer, Economist, Research Department, Federal Reserve Bank of New York.

* Economist Domestic Research Division.

Bureau of the Budget estimates the deficit will be $6.3 billion in the administrative budget, $4.0 billion in the cash budget, and $5.0 billion in the national income account budget. These three different deficit amounts are, of course, neither arbitrary nor unnecessary. Instead, they reflect an attempt to provide appropriate data for unraveling some exceedingly complicated economic and accounting relationships.

As a very abbreviated summary, each of the three widely used measures of the Federal budget has its own appropriate use. Yet each measure, singly, as well as all measures together, still leaves something to be desired in terms of providing a complete picture of the role of the Federal Government in the economy. For example, none of the budgets integrate complete information, on a current basis, about Government lending activities and guarantees of loans, although this information may be assembled from other sources.

The administrative budget provides data which are most useful to the Government itself for housekeeping and control purposes. Because of the detail given for individual agencies and the availability of detailed monthly data, the administrative budget may prove helpful to an analyst focusing on some narrow aspect of the Federal impact on the economy.

The consolidated cash budget provides the most comprehensive view of Federal expenditures and receipts. Changes in these flows have a direct impact on the Government's cash balances and constitute a major determinant of Treasury debt operations with the non-Federal sector.

Finally, the Federal sector in the national income accounts is often used for formulating and analyzing problems primarily in the framework of the national income and product accounts data.

It is the purpose of this article to delineate in broad terms the uses and limitations of the alternative budget series and also to indicate the typical sources where these data can be found. First, the budget process is briefly described. Then an explanation is given of the basic characteristics of each budget concept, of some of the interlocking relationships among the budgets, and of the way in which each budget serves different analytical or administrative purposes.

The Budget Process

The President's budget message, and its accompanying documents, usually delivered to Congress in the third week of January, present a comprehensive view of Federal spending and receipts for the current and the next fiscal year. (The Federal Government's fiscal year runs from July 1 of one year to June 30 of the following year and is identified by the year in which it ends.) Implementation of the tax and spending programs described in the budget is dependent upon legislation already in effect, as well as on new legislation still to be enacted. The new legislation does not come in a single package, but is introduced and considered by Congress in separately proposed and separately enacted bills. It will be useful to consider for a moment the general process by which a bill is enacted, and then to focus more specifically on what further steps are necessary before a particular agency can actually spend funds for a program.

Each new activity of the Federal Government (or extension of an old activity) must be authorized by a bill which has passed both houses of Congress and has been signed by the President.[1] Such bills are considered first by the appropriate legislative committee responsible for the subject area (true of both the House of Representatives and the Senate), which in turn typically refers the bills to subcommittees specializing in particular segments of the over-all area covered by the full committee. After the relevant legislative subcommittee and committee have approved the bill—including, if necessary, authorization to appropriate up to a given amount of money for the program—the bill is brought to a vote before the full membership of each branch of Congress.

For major legislation in the House of Representatives, the Rules Committee ordinarily acts as an intermediary to determine when legislation can reach the floor for consideration. Failure of the Rules Committee to bring out a bill produces complications since the bill cannot be voted on by the full chamber, except by a cumbersome procedure which is not often tried. If the bills for a particular program passed by the two houses differ in any respect, these differences must be resolved by a conference committee composed of members of the two houses, so that identical bills can be resubmitted for passage in each house and then transmitted to the President for signature.

The above procedure only authorizes the program in a general way. Actual authority to spend funds typically involves a further step—the passage of an appropriations bill again by both houses of Congress, which is then signed by the President. An appropriations bill follows the same general procedure as any substantive legislation, that is, it must pass a subcommittee, then a full committee, and then the full chamber. But for an appropriations bill, no matter what Government agency or subject area is involved, the bill starts its trip in the Appropriations Committee of the House of Representatives before it can be voted upon by the full House and similarly must be passed by the Senate Appropriations Committee before it can be voted on by the full Senate.

[1] Some bills, of course, are passed over a Presidential veto, and a few bills have become law without Presidential signature under the Constitutional provision that, if the President does not sign or veto a bill, it becomes law after ten days provided that Congress is in session.

In effect, then, legislation requiring the spending of money typically goes through two complete rounds of legislative approval—first, the act authorizing the program (with a bill considered first by the subject area committees) and, secondly, the act providing the funds for the program (with a bill originating in the appropriations committees). And it is important to note that the amount of the appropriations bill need not be the whole amount authorized in the legislation setting up the new program (first round). Since control over the scope of any program is ultimately determined by the amount of money made available, it is obvious that the appropriations committees in the two houses occupy a strategic position. Appropriations bills, however, are not the only avenue by which a Government agency can obtain the right to spend, although it is the most important one.

A Government agency acquires the authority to spend money from legislation providing new obligational authority (NOA). The NOA may be given in any of three forms—appropriations, contract authorizations, and authorizations to expend from debt receipts. Only the first two are directly under the control of the appropriations committees of the two houses.

1. Appropriations. These permit an agency to order goods and services and draw funds from the Treasury to pay for these goods and services up to some stated amount. Most spending takes this form. Although appropriations are usually limited to one year, some may cover several years or be "no year" (i.e., available until expended) because of the long-term nature of the project. The Defense Department holds the bulk of these multi-year appropriations. There are also "permanent appropriations," such as for interest on the debt, which do not require new action by Congress when additional funds are needed.

2. Contract authorizations. These allow an agency to contract for goods and services, but payments cannot be made until Congress passes an appropriation to provide funds for the obligations incurred.

3. Authorizations to expend from debt receipts. These allow agencies to borrow money, generally through the United States Treasury, to contract for the purchase of goods and services, and to pay for them with the borrowed funds. This procedure has been called "back door" financing and has been subjected to criticism by some members of Congress because the appropriations committees have no say in establishing the actual amount of spending by the agency under this system. Instead, the authority to borrow from the Treasury—and the amount—are given in the legislation authorizing the program. Under this arrangement, an agency may carry on its activities indefinitely without recourse to any annual appropriations, unless otherwise specified in the law. Many of the Government loan programs have been set up in this fashion since it is usually expected that such programs will sooner or later be self-supporting.

NOA is generally considered the avenue whereby Congress can control the size of the budget. An increase in NOA for a fiscal year above the amount for previous years suggests that Government spending will grow. The failure of NOA to rise, however, may not be significant since Congress may merely have legislated NOA at levels below the amounts needed to pay for commitments under already existing programs. For example, some veterans' programs specify benefit payments to veterans eligible under specified conditions. If NOA for a program of this type is cut without changing the eligibility requirements and claims under the program are greater than projected, then supplementary appropriations must be voted before the end of the fiscal year in order to prevent default on a commitment made by the Government.

NOA, including carry-overs from prior years, represents the potential level of spending for a particular program. By contrast, obligations are commitments already made which will require spending of funds —funds available to the agency from obligational authority previously granted.[2] Expenditures are the end of the line which runs from NOA to obligations to expenditures.

Spending in any single fiscal year is always made up of a combination of spending from appropriations carried over from previous years as well as from appropriations newly legislated. In fiscal 1966, for example, the Administration's recent budget document estimates that $27.6 billion will be spent from the pool of previously authorized NOA—to pay for those parts of long-range programs now under way which will be completed during fiscal 1966. An additional $72 billion will be spent in fiscal 1966 from part of the NOA that the President is asking for in his budget message. Thus, total spending (in the administrative budget) is expected to amount to $99.7 billion—part out of existing multiyear appropriations and part out of new appropriations to be voted this year, which will include some new multiyear appropriations to be spent over several successive fiscal years, roughly at the pace that the programs are carried out.

Expenditures usually take the form of the issuance of a check which, when cashed, will reduce the Treasury's balance at a Federal Reserve Bank. But there are exceptions. Sometimes an expenditure takes the form of the issuance of a security, as in the case of payments of subscriptions to the International Monetary Fund (IMF), which raises

[2] Obligations, particularly of the Department of Defense, have sometimes been interpreted as a good approximation of a "new orders" series. Such an interpretation is incorrect, because obligations also include commitments for expected disbursements for the wages and salaries of Government employees.

the debt but does not reduce the Treasury's cash or bank balances. Since payment by issuance of a security does not affect the Treasury's cash balance, it is therefore not a cash budget expenditure; it is, however, listed as an expenditure in the administrative budget and raises the debt. It does not become a cash expenditure until the security is redeemed (by the IMF in the illustration cited). At that time, the cash balance will be reduced and the transaction will also reduce the amount of outstanding Government debt. Ordinarily, retirement of Federal debt is not counted as an expenditure but as a debt transaction, which is similar to private accounting practice in distinguishing between "current" transactions and balance-sheet transactions.[3]

Administrative Budget

When reference is made to "*the* budget" in the press or in the halls of Congress, it almost invariably means the administrative budget. The President is required by the Budget and Accounting Act of 1921 to submit this budget to Congress every January in order to initiate a new round in the legislative process authorizing funds to support the activities of the regular Government agencies. These agencies are "controlled" by Congress through the power of the purse, i.e., Congress determines how much each agency shall have to spend by specifically approving dollar amounts for various purposes in an appropriations bill (which may lump together a number of agencies).

The administrative budget covers only those agencies for which Congress makes regular appropriations. Prior to the 1930's, this budget was a good measure of total Government activities. However, with the establishment and growth of self-financing agencies—whose operations are not included in the administrative budget—this budget has become an increasingly less adequate measure of the Federal Government sector. Government activities excluded from the administrative budget are the trust funds (of which the best known are the various social insurance funds) and quasi-public agencies, such as the Federal Home Loan Banks. These additional activities in recent years have added some $25 billion to $30 billion a year to Federal receipts and expenditures, as recorded in the cash budget.[4]

In addition to the direct exclusion of certain activities from the ad-

[3] Of course, debt operations—selling or retirement of securities—will change the level of the Treasury's cash balance but will not be recorded as a receipt or expenditure. In other words, transactions in United States Government debt instruments are usually classified as debt operations and are not included in budget transactions.

[4] Many of these activities (trust accounts) are financed by special earmarked taxes, while others (lending agencies) are financed, at least in part, by borrowing from the Treasury or in the market.

ministrative budget, there are some accounting conventions in this budget which must be recast in constructing the cash budget and the Federal budget in the national income and product accounts. An example of these conventions can be seen in the treatment of interest payments. Interest payments for fiscal 1964 totaled $10.8 billion in the administrative budget, while actual cash outlays for interest payments totaled only $8.0 billion.[5] The bulk of some $3 billion of *noncash* interest is accounted for by "bookkeeping" payments by the Government to itself (intragovernmental transactions) for securities held by the trust funds and by the accrual of interest on outstanding Government securities, most notably savings bonds and Treasury discount bills, which becomes a cash expenditure when the savings bonds or Treasury bills are turned in for payment. Other intragovernmental transactions are included in the administrative budget figures, both for receipts and expenditures, in order to allocate these expenses and receipts more properly to the individual agencies concerned. This procedure raises the total of Government receipts and expenditures above the amount shown for the same agencies in the cash budget (described in the next section), because the cash budget eliminates intragovernmental transactions. However, the difference between the cash and the administrative figures for a particular agency in any given year is likely to be relatively small, compared with the total, except for interest payments as just discussed and for those agencies whose operations include trust fund functions, most notably the Department of Health, Education, and Welfare.

Despite its incomplete coverage of the Federal sector, the administrative budget is a source of valuable data to persons interested in knowing how much is spent by a "regular" Government agency and its major divisions. Data for this budget are published in the Treasury's *Monthly Statement of Receipts and Expenditures of the United States Government*, approximately three weeks after the end of each month (see Table I). To the extent that a Government agency, or activity, can be closely identified with a specific activity or segment of the economy (for example, the National Park Service, Rivers and Harbors and Flood Control, or Military Construction), these outlays as summarized in the *Monthly Statement* indicate the current scope of Government activities in the area concerned. Perhaps the most widely used data in the *Monthly Statement*, other than the summary budget totals, are those which give the breakdown of Defense Department spending by functional category—such as research and development, military construction, etc. (More detailed spending and order data are released directly by the Defense Department

[5] Net interest paid in the Federal sector of the national income accounts for fiscal 1964 was $8.1 billion, compared with $8.0 billion in the cash budget. Usually the difference in levels for interest payments in the cash and national income budgets has been larger than that shown for fiscal 1964. The reason for differences in levels is discussed in the section devoted to the national income version of the budget.

Table I. Federal Budgets and Their Data Sources Fiscal 1964 and Fiscal 1965 (In billions of dollars)

Item	Administrative Budget		Cash Budget (DTS basis)*		Consolidated Cash Budget—Receipts from and Payments to Public		National Income Account Budget	
	Fiscal 1964 (actual)	Fiscal 1965 (estimate)	Fiscal 1964 (actual)	Fiscal 1965 (estimate)	Fiscal 1964 (actual)	Fiscal 1965 (estimate)	Fiscal 1964 (actual)	Fiscal 1965 (estimate)
Receipts	89.5	91.2	121.6	‡	115.5	117.4	114.7	116.0
Expenditures	97.7	97.5	125.6†	‡	120.3	121.4	118.5	121.0
Surplus (+) or deficit (−)	−8.2	−6.3	−4.0	‡	−4.8	−4.0	−3.9	−5.0
Type of data	Monthly seasonally unadjusted, available with a three-week lag. The Budget projects annual data for the current fiscal year and the next fiscal year based on the Administration's economic assumptions and proposed programs.		Daily and monthly seasonally unadjusted, available with a three- to four-day lag.		Monthly and quarterly unadjusted, quarterly seasonally adjusted available with a one-month lag. The Budget projects annual data for the current fiscal year and the next fiscal year based on the Administration's economic assumptions and proposed programs.		Quarterly seasonally adjusted, available with a two-month lag (complete expenditure data and all receipts data except corporate profits tax accruals available with a one-month lag). Quarterly unadjusted, available in February and July. The Budget projects annual data for the current fiscal year and the next fiscal year	

				based on the Administration's economic assumptions and proposed programs.
Sources of data	Treasury Department: Monthly Statement, Treasury Bulletin Survey of Current Business Federal Reserve Bulletin Economic Indicators The Budget	Treasury Department: Daily Statement*, Treasury Bulletin	Treasury Department: Monthly Statement, Treasury Bulletin Survey of Current Business Federal Reserve Bulletin Economic Indicators The Budget	Survey of Current Business Economic Indicators The Budget

Note: Because of rounding, figures do not necessarily add to totals.
* Daily Statement of the United States Treasury (DTS).
† Includes clearing account.
‡ Full reconciliation to DTS basis for estimates is not available.
Source: *The Budget of the United States Government*, Fiscal 1966.

but are typically available with a much longer time lag than the administrative budget data). Annual data classified by broad functional categories are given in the budget itself and usually in the budget review, generally issued after each Congressional session; current data appear in the *Monthly Statement* and in the *Treasury Bulletin.*

Cash Budget

The cash budget is the most comprehensive budget statement issued by the United States Government and is designed to show the cash flows between the Federal Government and other sectors of the economy. Unlike the administrative budget, it covers not only the activities of the regular Government agencies found in the administrative budget but also the cash flows associated with the activities of the trust funds (such as social security) and Government-sponsored enterprises (such as the Federal Home Loan Bank Board). Like the administrative budget, it also covers the purchase and sale of assets (both "real," such as buildings, and "financial," such as mortgages and other loans). However, as noted earlier, certain items, e.g., interest payments, treated as accrual items in the administrative budget are placed on a cash basis. For many years a substantial number of economists have regarded the cash budget as the best measure of the total impact of the Federal Government on the economy.

Total expenditures and receipts in the cash budget are larger than in the administrative budget, since the cash budget includes a wide range of Government activities omitted from the administrative budget. Nevertheless, because the cash budget eliminates many transactions of Government agencies with each other (intragovernmental transactions), it records certain Government activities at lower levels than the administrative budget (for example, interest payments, as noted previously). The total of cash budget expenditures, however, does understate the full magnitude of the cash flows between the Government and the private sector, as some agencies are listed only on a *net* basis on the expenditures side. The Post Office, for example, is recorded as having spent $600 million in fiscal 1964—but this amount represents only "net expenditures" obtained after deducting postal receipts (sale of stamps, etc.) from total postal expenditures. Government corporations are also typically recorded only on a net basis. The device of netting, incidentally, is not restricted to the cash budget; it also affects some of the data reported in the administrative budget, as mentioned earlier, and in the national income budget.

The cash budget in the form of "receipts from and payments to the public" is also called "the consolidated cash budget." Annual data giving a functional breakdown for receipts and expenditures are published in the budget and in the budget review (with some exceptions). Monthly data are also available (with functional breaks) for this cash budget in the *Federal Reserve Bulletin*, but seasonally adjusted data are available only on a quarterly basis and only for total cash income, total cash outgo, and the resultant cash surplus or deficit.

Detail for some ten categories of receipts and expenditures are available in a variant of the cash budget known as the *Daily Statement of the United States Treasury* (DTS) which excludes a few Government corporations whose accounts are not commingled with the Treasury's. These DTS data, *not* seasonally adjusted, are published for each working day, with a lag of about three or four days and are cumulated to a monthly total and for the fiscal year to date. The DTS data are used by analysts who are particularly interested in the level of, or changes in, the Treasury's cash balances and by those who need current data (daily and monthly totals) for major categories of Government receipts and expenditures and for debt operations.

A comparison of the consolidated cash budget and administrative budget expenditures on a functional basis is shown in Table II. Differences for the same function, if large, are likely to reflect differences in coverage and in the treatment of intragovernmental transactions in the two budgets. In addition, relatively small differences in amount arise for functions called by the same general name in the two budget accounts because of differences in the accounting techniques used in recording these expenditures for the different budget accounts.

The surplus or deficit of the cash budget (not the administrative budget) will determine how the balances held by the Treasury will change. When a surplus is generated, the balances rise and Government debt held by the public may be retired. On the other hand, cash deficits, depending upon the level of the cash balance, may require that the Government borrow from the public in order to pay its bills. Consequently, the net flows as recorded in the cash budget are one of the major determinants of Government debt operations.

But there is no one-to-one correspondence between cash deficits and Government debt operations. A deficit can be financed simply by running down the cash balance. Alternatively, the Government may borrow to build up its cash balance rather than to meet a deficit in the cash budget. Moreover, the average level of balances maintained by the Government varies from time to time by substantial amounts which are determined by operating and policy considerations not directly related to the cash surplus or deficit.

Table II. Federal Expenditures and Receipts, Administrative Budget and Consolidated Cash Budget Fiscal 1964-66 (In billions of dollars)

Type of Transaction	Administrative Budget			Consolidated Cash Budget		
	Actual	Estimate		Actual	Estimate	
	1964	1965	1966	1964	1965	1966
Receipts						
Individual income taxes	48.7	47.0	48.2	48.7	47.0	48.2
Corporation income taxes	23.5	25.6	27.6	23.5	25.6	27.6
Excise taxes (net)	10.2	10.7	9.8	13.7	14.4	13.7
Employment taxes	–	–	–	16.8	16.7	18.7
Estate and gift taxes	2.4	2.8	3.2	2.4	2.8	3.2
Customs	1.3	1.4	1.5	1.3	1.4	1.5
Deposits by states, unemployment insurance	–	–	–	3.0	3.0	2.9
Veterans' life insurance premiums	–	–	–	0.5	0.5	0.5
Other budget and trust receipts	–	–	–	5.6	6.1	7.1
Miscellaneous budget receipts	4.1	4.5	4.7	–	–	–
Interfund transactions	–0.7	–0.8	–0.6	–	–	–
Total	89.5	91.2	94.4	115.5	117.4	123.5
Expenditures by function						
National defense	54.2	52.2	51.6	54.5	52.8	52.5
International affairs and finance	3.7	4.0	4.0	3.5	3.6	4.2
Space research and technology	4.2	4.9	5.1	4.2	4.9	5.1
Agriculture and agricultural resources	5.6	4.5	3.9	5.8	4.6	4.1
Natural resources	2.5	2.7	2.7	2.6	2.8	2.9
Commerce and transportation	3.0	3.4	2.8	6.5	7.4	6.5
Housing and community development	–0.1	–0.3	*	1.7	–0.2	0.7
Health, labor, and welfare	5.5	6.2	8.3	27.3	28.9	34.1
Education	1.3	1.5	2.7	1.3	1.5	2.6
Veterans' benefits and services	5.5	5.4	4.6	6.1	6.0	5.1
Interest	10.8	11.3	11.6	8.0	8.5	8.8
General government	2.3	2.4	2.5	2.2	2.4	2.4
Unallocated and interfund transactions	–0.7	–0.7	–0.1	–	–	–
Deposit funds (net)	–	–	–	–0.6	*	*
Undistributed adjustments	–	–	–	–2.9	–1.8	–1.6
Total	97.7	97.5	99.7	120.3	121.4	127.4

Note: Because of rounding, figures do not necessarily add to totals.
* Less than $50 million.
Source: *The Budget of the United States Government*, Fiscal 1966.

The National Income Account Budget

The Federal budget in the national income and product accounts (NIP) records the receipts and expenditures of the Government sector as an integrated part of the recorded activities of other sectors of the economy. The national income accounts, sometimes called "the GNP accounts," are a measure of current output (both goods and services) in the economy. The Federal sector data have gained wide currency in the last three years, since the President's Council of Economic Advisers has often used this version of the budget for its analyses of Federal fiscal impact.

Like the cash budget, the Federal sector account is a more comprehensive statement than the administrative budget. It differs from the cash budget, however, by restricting itself to receipts and expenditures which reflect the direct impact of Government spending and tax programs on the flow of current income and output, as measured by the national income accounts. A broader measure of the economic impact of the Government would include not only the direct impact but also influences on asset holdings and liquidity—which may indirectly affect income and output. Thus, such a measure would allow for the effect of all transactions involving existing assets, as well as any assets of a purely financial character (bonds, mortgages, loans, etc.).

On the expenditures side, the cash budget records spending at the time of payment, but in the NIP concept spending is typically recorded when delivery is made to the Government sector—which often does not coincide with the time of payment.[6] On the receipts side, the national income budget differs from the cash budget most importantly in recording corporate profits taxes when the tax liability is incurred rather than when the tax payment is made.

Expenditures in the Federal sector account are classified into five categories (see Table III) which identify the basic economic import of the expenditures. The largest single category, accounting for more than half of the total, is "purchases of goods and services." Such purchases are one of the major components of total GNP as viewed from the product side—the others being personal consumption expenditures, domestic investment, net exports, and state and local government purchases. The next largest category of Federal expenditures, approximately one fourth of the total, is "transfer payments," defined as payments for

[6] It should be noted that the "delivery" concept for recording purchases (or spending) is the standard national income accounts treatment for purchases made by all sectors of the economy (and not only the Government sector). Goods produced, but not yet delivered, show up in the inventory component of gross national product (GNP).

which no goods or services have been rendered in exchange. These are mainly made to individuals and include such items as old-age pensions and unemployment benefits. Although transfer payments are not directly included in GNP, they do affect GNP indirectly because they add to

Table III. Federal Receipts and Expenditures in the National Income Accounts Fiscal 1964-66 (In billions of dollars)

Type of Transaction	Actual 1964	Estimate 1965	Estimate 1966
Receipts			
Personal tax and nontax	51.4	50.3	52.2
Corporate profits tax accruals	23.5	23.9	24.7
Indirect business tax and nontax accruals	16.0	16.8	16.1
Contributions for social insurance	23.8	25.0	28.0
Total	114.7	116.0	121.0
Expenditures			
Purchases of goods and services	66.1	65.9	66.7
Transfer payments	30.4	31.8	35.2
Grants-in-aid to state and local governments	9.8	10.7	13.0
Net interest paid	8.1	8.5	8.6
Subsidies less current surplus of Government enterprises	4.1	4.1	3.5
Total	118.5	121.0	127.0
Surplus (+) or deficit (−)	−3.9	−5.0	−6.0

Note: Because of rounding, figures do not necessarily add to totals.
Source: *The Budget of the United States Government*, Fiscal 1966.

disposable personal income which in turn strongly affects personal expenditures on goods and services. The other three items, accounting for less than one fifth of total Federal expenditures, are (1) "grants-in-aid to state and local governments," which increase the receipts of these governmental units and, in turn, are spent by these units for goods and services or for transfer payments; (2) "net interest paid," which adds to personal income but is not counted as part of GNP,[7] and (3) "subsidies less current surplus of Government enterprises," a category which records the net of subsidy payments to private business offset by any profits made by Government agencies.

This five-part classification is very useful in differentiating broadly, and in a way not available from any other source, between analytically distinct types of Government spending. Moreover, additional details for some of these categories, available on an annual basis only, further enrich our understanding of the composition of Government spending. However, the delivery basis for recording Government expenditures on goods and services sometimes fails to identify properly the time period when the Government is significantly influencing the level of private employment and output. This is particularly troublesome when the level of Government orders is subject to wide variation, as was the case at the beginning and end of the Korean war. This timing problem is one illustration of the need for different budget concepts: it is not possible to construct a single series which is appropriate for all uses.

The Federal sector data are available quarterly on a seasonally adjusted annual rate basis. The figures are released about one month after the quarter is over, except for corporate profit tax accruals which lag by about two months.[8] The data are revised as more information is obtained for the period. While individual adjustments of components are generally small, in combination they sometimes shift the budget from an originally estimated deficit to a surplus.

A comprehensive view of how the administrative, cash, and national income budgets are related is shown in Table IV. In summary, moving from the administrative budget to the cash budget primarily entails adding to the administrative budget a total for the trust funds plus Government-sponsored enterprises while eliminating from the administrative budget a total for intragovernmental transactions. The transition from cash to NIP transactions (with some qualifications) primarily involves: (1) timing adjustments (mainly to an accrual basis on the receipts side and to a delivery basis on the expenditures side), (2) the elimination of assets transactions included in the cash figures, and (3) the elimination of lending activities included in the cash figures.

[7] Interest paid by the Federal Government is considered as part of personal income, though, unlike private interest payments, it is not included in total GNP because Federal Government interest payments are not viewed as income arising out of current production. Government interest in NIP excludes intragovernmental payments (similar to the cash budget) but treats certain items, such as interest on Treasury bills and savings bonds, on an accrual basis (similar to the administrative budget). Therefore, the interest total in NIP is likely to be different from that in the cash and the administrative budgets.

[8] Seasonally unadjusted figures are also available in the February and July issues of the *Survey of Current Business,* United States Department of Commerce.

Table IV. Reconciliation of Administrative Budget and Cash Budget to Federal Receipts and Expenditures in the National Income Accounts Fiscal 1964 (In billions of dollars)

Type of Transaction	Administrative Budget Total	Adjustments from Administrative to Cash Budget	Consolidated Cash Budget Total	Adjustments from Cash to National Income Account Budget	National Income Account Budget Total
Receipts					
Administrative budget receipts	89.5	—	—	—	—
Less: Intragovernmental transactions	—	4.2	—	—	—
Receipts from exercise of monetary authority	—	0.1	—	—	—
Plus: Trust fund receipts	—	30.3	—	—	—
Equals: Federal cash receipts from the public	—	—	115.5	—	—
Adjustments for agency coverage:					
Less: District of Columbia revenues	—	—	—	0.3	—
Adjustments for netting and consolidation:					
Less: Interest and other earnings	—	—	—	1.4	—
Plus: Contributions to Federal employees' retirement funds, etc.	—	—	—	2.0	—
Adjustments for timing:					
Plus: Excess of corporate tax accruals over collections, personal taxes, etc.	—	—	—	–0.7	—
Adjustments for capital transactions:					
Less: Realization upon loans and investments, sale of Government property, etc.	—	—	—	0.6	—
Equals: Receipts—national income budget	—	—	—	—	114.7

Expenditures					
Administrative budget expenditures	97.7	—	—	—	—
Less: Intragovernmental transactions	—	4.2	—	—	—
Accrued interest and other noncash expenditures	—	2.0	—	—	—
Plus: Trust fund expenditures (including Government-sponsored enterprise expenditures net)	—	28.9	—	—	—
Equals: Federal cash payments to the public	—	—	120.3	—	—
Adjustments for agency coverage:					
Less: District of Columbia expenditures	—	—	—	0.3	—
Adjustments for netting and consolidation:					
Less: Interest received and proceeds of Government sales	—	—	—	1.4	—
Plus: Contributions to Federal employees' retirement funds, etc.	—	—	—	2.0	—
Adjustments for timing:					
Plus: Excess interest accruals over payments on savings bonds and Treasury bills	—	—	—	0.9	—
Excess of deliveries over expenditures and other items	—	—	—	1.5	—
Less: Commodity Credit Corporation foreign currency exchange	—	—	—	0.6	—
Adjustments for capital transactions:					
Less: Loans—Federal National Mortgage Association secondary market mortgage purchases, redemption of International Monetary Fund notes, foreign assistance, etc.	—	—	—	3.4	—
Purchases of land and existing assets	—	—	—	0.5	—
Equals: Expenditures—national income budget	—	—	—	—	118.5
Surplus (+) or deficit (−)	−8.2	—	−4.8	—	−3.9

Note: Because of rounding, figures do not necessarily add to totals.
Sources: *Economic Report of the President*, January 1965; *The Budget of the United States Government*, Fiscal 1966.

The Budget and the Economy

Because of its sheer size, the Federal Government inevitably exerts a potent influence on the functioning of the economy. Budget data provide the raw material for analyzing this influence, but each form of the budget statement is not equally useful for this purpose. Typically, the administrative budget is not used for assessing the Government's impact on the economy, because it does not cover the full range of Government activities. Instead, the Government sector in the economy is usually analyzed with the data from the cash budget or the NIP budget.

A lively controversy has been going on for a number of years over the relative merits of the cash versus the NIP budget as the best measure of the Government's impact on the economy. When annual data are used, disagreement over the relative merits of the two comprehensive budget statements is not great. Although there are some differences in the levels of receipts and expenditures and the size of the surplus or deficit, the general trends observed in using either of the two comprehensive budget measures by and large will be similar. When quarterly data are used, however, the problems of choosing between the two measures become more troublesome because there are often conflicts both as to the direction and the magnitude of changes. Much depends on the particular problem under investigation, and often data from both budgets are needed to obtain a rounded picture.

The popular view of budget impact is that a surplus is a contractionary influence, that a deficit is an expansionary influence, and that a balanced budget is neutral. This popular view is, at best, only a partial view of the role of fiscal policy in the economy; a fuller perspective of the role of Government impact is somewhat more complex. In its simplest form, this popular view may be called the "cross section" approach. Taking the economy for a fixed period, a balance sheet of each of the sectors is compiled to show how each is affecting the economy. In this view, a Government deficit of $2 billion for the period is expansionary, because the Government is adding to the demand side of the economy more than it is taking out in taxes. But this is far from the full story. Another dimension is added by the "time series" approach which looks at the change in budget position between two periods. For example, a deficit of $2 billion may be considered restrictive in the second period if it follows a deficit of $7 billion in the first period, whereas it may be held to be expansionary compared with a previous surplus. In other words, if the $2 billion deficit followed a period when the budget deficit has been larger, say $7 billion, then the budget is exerting an effect in a contractionary direction. Given the change in strength of the forces at work in other sectors of the economy, this reduction in the amount of stimulus from the Government sector may be just the right amount of restraint for the economy, if high levels of activity are to be maintained and if potential excesses are to be curbed before they develop.

Both of the approaches described above, however, by measuring the fiscal impact of the Government in terms of the over-all budget surplus and deficit ignore the fact that for any given budget structure (the combination of spending programs and tax programs), the budget outcome depends not only on the specific character of these programs but also on the level of operation of the economy itself. Thus, for any particular year, an economy operating at full employment may give a budget surplus, while the same economy operating at 6 per cent unemployment, with the same expenditure and tax programs, will probably show a sizable budget deficit. As a correction for the distortion introduced by the impact of the economy itself on the realized net budget position, the concept of the full-employment budget surplus has been developed.

The full-employment budget surplus is an estimate of the budget outcome for any given budget structure, assuming that the economy is at full employment. (In theory, there could of course be a full-employment deficit.) By estimating the net surplus or deficit of different budget structures for the assumed full-employment level of activity for any year, it is possible to measure the relative restrictiveness of these different structures, i.e., the budget structure with higher full-employment surplus is taken to be more restrictive than budgets with smaller surpluses (or deficits). While the full-employment surplus concept is a highly suggestive addition to the other techniques of analysis, estimates of the precise magnitude of "full employment" and of the budget outcome at that level of activity are not particularly easy, and there are also some problems in the analytical interpretation of the estimates. The development of this concept, however, is indicative of the imaginative way new analytical tools are being forged to advance the art of fiscal analysis.

The full Government impact, of course, depends not only on the absolute levels of its receipts and expenditures, or how they change, but also on the further changes in spending by the private sector induced by the impulses emanating from the Government. Furthermore, different kinds of Government spending may affect the economy differently—for example, increases in Government purchases of goods may not have the same impact on the economy as an equal dollar increase in old-age payments. Similarly, an increase or decrease in income taxes will affect the economy differently from an equal dollar change in excise taxes. But what is not yet known with much certainty is the quantitative extent of these differences and how they may themselves vary under different economic conditions. Thus, a less aggregative approach also will have to be developed eventually to provide greater information.

The Government sector influences the economy in many different ways—by its spending programs, by its tax programs, by its credit programs, by its debt management actions, by its monetary policy, and by other actions which do not fit neatly into any of the foregoing classifications. Only part, although a very important part, of all this economically significant behavior is encompassed by the data typically found in the various budget documents. Much, however, is still to be learned. In part, improved insights will come from advances in the analytical tools applied to the public finance field. In part, advances also will depend on improvements of the kinds and quality of data available, for data provide the raw materials for the application of the analytical tools.

23

The new look in tax and fiscal policy

Paul A. Samuelson

I

There is much talk about taxes. When I flick on the dial of my radio in the morning, I hear a Congressman quoted on how our high level of taxes is ruining the Nation or a Senator's tape-recorded alarm over the unfair burden the poor man has to carry because the administration has been favoring big business. My morning paper at breakfast brings me the view of its editor that the United States has been pursuing unsound fiscal policy for the last 25 years. Scratch the barber who cuts my hair and you find a philosopher ready to prescribe for the Nation's monetary ills.

This is as it should be. We expect sweeping statements in a democracy. We hope that out of the conflict of extreme views there will somehow emerge a desirable compromise. Yet such sweeping statements have almost no validity from a scientific, or even from a leisurely commonsense point of view: spend as little as a year going over the factual experience of American history and of other economies, devote as little as a month to calm analysis of probable cause and effect, or even spend a weekend in a good economics library—and what will you find? Will you find that there breathes anywhere in the world an expert so wise that he can tell you which of a dozen major directions of policy is unquestionably the best? You will not. Campaign oratory aside, the more assuredly a man asserts the direction along which salvation is alone to be found, the more patently he advertises himself as an incompetent or a charlatan.

* Reprinted from *Federal Tax Policy for Economic Growth and Stability,* Joint Committee on the Economic Report, 84th Cong., 1st Session (Washington, D.C., U.S. Government Printing Office, 1955), pp. 229-234. Paul A. Samuelson, Massachusetts Institute of Technology.

The plain truth is this, and it is known to anyone who has looked into the matter: The science of economics does not provide simple answers to complex social problems. It does not validate the view of the man who thinks the world is going to hell, nor the view of his fellow idiot that ours is the best of all possible tax systems.

I do not wish to be misunderstood. When I assert that economic science cannot give unequivocal answers to the big questions of policy, I do not for a moment imply that economists are useless citizens. Quite the contrary. They would indeed be useless if any sensible man could quickly infer for himself simple answers to the big policy questions of fiscal policy. No need then to feed economists while they make learned studies of the obvious. It is precisely because public policy in the tax and expenditure area is so complex that we find it absolutely indispensable to invest thousands of man-years of scholarly time in scholarly economic research in these areas.

Make no mistake about it. The arguments that we all hear every day of our lives on the burning partisan issues have in every case been shaped by economists—by economists in universities, in business, in Government, and by that rarest of all birds, the shrewd self-made economist. What economists do not know about fiscal policy turns out, on simple examination, not to be known by anyone.

II

With this necessary preamble out of the way, let me record the general views that studies have led me to, about the current state of our fiscal system. This will clear the way for a more detailed analysis of taxes and growth, taxes and stable full employment, taxes and equity, taxes and the level of public expenditure programs.

Here then are the major facts about our system as I see them.

1. The postwar American economy is in good shape. There is nothing artificial or unsound about its underpinnings. For more than a decade we have had generally high employment opportunities. Our production efficiency has been growing at a steady rate that compares well with anything in our history or in the history of countries abroad. For all this we must, in our present-day mixed economy, be grateful to both public and private institutions.

2. The existing structure of Federal, State, and local taxes is in its broad features highly satisfactory. Repeatedly at the polls and through all the legitimate processes of government, the citizens of this Republic have indicated that they want our present type of fiscal structure—its substantial dependence at the Federal level on personal and corporate income taxes, its eclectic dependence on selective excises, on payroll

levies for social security, on property and sales taxes at the local levels. If the consensus of citizens in our democracy were to be other than it is—toward less or more equalitarianism, toward less or more local autonomy—there is no reason that the careful analytic economist can see why our fiscal system is not capable of being altered in the desired direction. In other words, there is nothing in the mechanics of a modern economy which makes it impossible or difficult for the citizenry to get the kind of a tax system that they want; our tax system has plenty to give, plenty of room for adaptation and change.

All the above does not imply that we are living in a new era of perfection. The American economy now faces, and will continue to face, many tough problems, many hard decisions. And, to be sure, there are numerous imperfections, inconsistencies, and loopholes in the present tax structure; these do need improving.

What the optimistic diagnosis of the modern-day economist does contradict is the following:

1. The view that America has long since departed from an orthodox fiscal policy and that it is only a matter of time until a grim Mother Nature exacts retribution from us for our folly in departing from the narrow line of fiscal rectitude. (This is a philosophical position that any dissenter from current trends is free to assume; but it is not a factually verifiable view about reality that dispassionate study of statistics and acts can substantiate.)

2. The view, shared in by the extremes of both left and right wings, that our economy generally is moving in unsound directions so that we must ultimately end up in some unnamed disaster or convulsion. (In terms of business-cycle stability and efficient growth, the United States has in the last dozen years dramatically refuted the sour expectations both of those who look back on a fictitious past golden age and of collectivists who look forward to a golden age that only a revolution can usher in.)

III

Turning now to the goals of any tax system, we can ask: What tax structure will give us the most rapid rate of growth? What tax system will give us the highest current standard of living? What tax structure will make our system most immune to the ups and downs in employment and prices that make American families insecure? What tax structure will realize most closely the community's sense of fairness and equity? What tax structure will have the least distorting effects on our use of economic resources, instead of maximizing the efficiency with which we produce what our citizens most want?

Upon careful thought it will be obvious that there cannot exist a tax system which will simultaneously maximize these five quite different goals of social life.

It is easy to see that high current living standards and rapid growth of our ability to produce are conflicting ends: you have only to look at a collectivized society like the Soviet Union, which decides to sacrifice consumption levels of the current generation in favor of a crash program of industrialization; you have only to reflect that historically in the slums of Manchester working families might have lived longer in the 19th century if England and the other nations had during the industrial revolution slowed down their rates of material progress; you have only to consider the problem of conserving scarce exhaustible natural resources to realize that every society must all the time be giving up higher future resource potentials in favor of keeping current generation consumption as high as it is.

You can imagine a society that decides to devote its income in excess of the bare physiological existence level 100 per cent to capital formation. You can imagine it—but there never has been such a society. Nor would any of us want to live in such a one. It should be obvious, therefore, that no sane person would ever seek a tax program which literally maximized our rate of economic growth. (Yet how many times over the chicken a la king have we all heard speakers reiterate this nonsensical goal.) It is just as obvious that no sane person would want to maximize present living levels if this meant eating up all our capital on a consumption bender that would leave us an impoverished Nation.

There is no need to go through all the other pairs of the five listed goals to show their partial imcompatibility. If we are willing to frame a tax system that strongly favors thrifty men of wealth, we may thereby be able to add to our rate of current growth; if we encourage a gentle rate of inflation, we may be able to increase the profits in the hands of the quick-reacting businessman, perhaps thereby stepping up our rate of growth. So it goes, and one could easily work through the other permutations and combinations.

But not all of our five goals are necessarily competing. Some when you realize them, help you to realize the others. If we succeed in doing away with the great depressions that have dogged the economic record, we may thereby add to our rate of growth. If we shape a graduated-tax system that enables lower income groups to maintain minimum standards of life, we may ease the task of stabilizing business activity. If we replace distorting taxes by less distorting alternatives, the fruits of the resulting more efficient production can add to our current consumption and to our rate of progress in capital formation.

I shall not prolong the discussion of the degree to which the diverse goals of tax policy are competing or complementary. For it will turn out that we can formulate proper policies without having to measure these important, but complicated, relationships.

IV

Upon being told by the economist that it is absurd for Congress to aim at the most rapid rate of growth possible and that it is equally absurd for Congress to aim at the highest possible current level of consumption, the policymaker may be tempted to say: "I understand that. Won't you therefore as an economist advise us as to just what is the best possible compromise between these extremes?"

A good question but, unfortunately, not one that the expert economist can pretend to give a unique answer to. If he is honest, he must reply: "The American people must look into their own hearts and decide on what they consider to be the best compromise rate of growth."

Just because I have advanced degrees in economics and have written numerous esoteric works in the field, I am not thereby empowered to let my personal feelings, as to how much the present generation ought to sacrifice in favor of generations to come, become a prescription for society. It would be as presumptuous for me to offer such specific advice as to let my family's notions about dental care determine how much the typical American family ought to spend on toothpaste. But it is legitimate for me as an economist to say this: "Whatever rate of capital formation the American people want to have, the American system can, by proper choice of fiscal and monetary programs, contrive to do." This can be shown by an example.

Suppose the vast majority of the American people look into the future or across the Iron Curtain at the rate of progress of others. Suppose they decide that we ought to have a more rapid rate of capital formation and technological development than we have been having recently. Then the economist knows this can be brought into being (a) by means of an expansionary monetary policy that makes investment funds cheaper and easier to get. Admittedly, such an expanded investment program will tend, if it impinges on an employment situation that is already full and on a price level that is already stationary, to create inflationary price pressures and overfull employment—unless something is done about it. What would have to be done about this inflationary pressure? Clearly (b) a tight fiscal policy would be needed to offset the expansionary monetary policy: By raising taxes relative to expenditure, we would reduce the share of consumption out of our full employment income, releasing in this way the real resources needed for investment. (It should be unnecessary to go through the reverse programs which would be called for if the national decision were to slow down the rate of capital formation as compared to that of recent years.[1])

From these remarks it will be clear that economic science is not only neutral as to the question of the desired rate of capital accumulation—it is also neutral as to the ability of the economy to realize any decided-on rate of capital formation.

I repeat: With proper fiscal and monetary policies, our economy can have full employment and whatever rate of capital formation and growth it wants.[2]

V

The optimistic doctrine that our economy can have stability and the rate of growth it wants may seem rather novel. Perhaps even a little shocking. But there are worse surprises yet to come.

[1] The fact that variations in the overall defict or surplus of the Government can, if properly reinforced by monetary policy, determine the rate of society's capital formation puts a sobering responsibility on democratic governments. Ordinarily, we assume that each individual is to be the best judge of whether he will spend the income society leaves him after taxes on more butter or on more margarine. We do not ordinarily assume that I, as an individual, am free to determine the amount of smoke my chimney can eject into the public air; I am willing to enter into a compact with my neighbors whereby we all decide democratically how our liberty or license is to be curbed in order to further the good of each one of us. A nation's saving seems to be treated by most 20th century nations as something in between these 2 polar cases; to some degree we all act as if we consider ourselves trustees for future generations, and we desist from using up all the irreplaceable resources of nature. In both the advanced and the unerdeveloped parts of the globe, citizens act at the polls as if they do not completely approve of the saving-investment decisions that they would make in private life; they reinforce and alter these decisions by voting public fiscal and monetary policies which increase (or decrease) the capital formation which private thrift would by itself dictate. Why do they do this? Often they do so implicitly. But often explicitly because, technically speaking, they attach qualified weight to their own changeable ex ante indifference curves between present and future. If full ethical primacy were to be given to these indifference curves and if short-run irregularities were ignored, the proper goal of social policy might be a constantly balanced budget accompanied by an active monetary policy that maintains full employment.

[2] Space does not permit me to give the needed qualifications to this simplified exposition. I have elsewhere explained at some length what might be called the important neoclassical synthesis, which combines the essentials of traditional economics pricing theory with the essentials of the modern theory of income determination and which underlies the asserted proposition. See my chapter entitled "Full Employment Versus Progress and Other Economic Goals," appearing in (Max F. Millikan, editor) *Income Stabilization for a Developing Democracy*, Yale University Press, 1953, pp. 547-580. Also see my related discussion entitled "Principles and Rules in Modern Fiscal Policy: A Neo-Classical Reformulation," in *Essays in Honor of John Williams*, the Macmillian Co., New York, 1951, pp. 157-176. The third edition of my *Economics*, McGraw-Hill, 1955. ch. 29 (Interest and Capital), contains an elementary exposition to show how fiscal and monetary policy interact in the determination of alternative mixes of consumption and investment at full employment.

The reader may think that my argument rests on something like the following reasoning:

Suppose that political party R is more concerned with progress than political party D, which shows a greater concern for the little man, with security, and with current consumption. Then if the Nation gives its approval to the general policy goals of R, the Government will have to change its emphasis away from reducing taxes on individuals—particularly rapid-spending lower-income people; and it will have to change its emphasis toward reducing taxes on business, in an attempt to bolster the incentives toward investment. In short, it is by changing the qualitative pattern of taxation, by sacrificing equity to incentive, that the community succeeds in getting higher levels of capital formation when it desires such higher levels.

I predict that much of the testimony before this subcommittee will proceed along these lines. Certainly much of the political discussion of the last 3 years, when it has had the courage to be frank, has been along these lines.

But this is not at all the train of thought that I wish to emphasize in my testimony. I want to cap the daring doctrine that an economy can have the rate of capital information it wants with a doctrine that may seem even more shocking. Naturally, I cannot here develop all of the underlying reasoning, nor give all the needed qualifications. But I do in advance want to stress the earnestness with which I put it forward, and to underline that it does spring from careful use of the best modern analyses of economics that scholars here and abroad have over the years been able to attain. The doctrine goes as follows:

A community can have full employment, can at the same time have the rate of capital formation it wants, and can accomplish all this compatibly with the degree of income-redistributing taxation it ethically desires.

This is not the place to give a detailed proof of the correctness of this general proposition. It will suffice to illustrate it with two extreme examples.[3]

[3] I do not recall ever seeing mathematical economics in congressional committee hearings. This drought can be ended by the following brief proof of the reasoning underlying my basic proposition. To the initiated the symbols will be almost self-explanatory; to the uninitiated no harm is meant.

Let Y=real national product, y=disposable income in real terms=Y—taxes. Let I and C stand for investment and consumption, G for Government expenditure on goods and services. Let i stand for the cost (and the availability) of borrowing for investment purposes. Let m be a parameter indicating the degree to which the tax structure is income distributing toward the poor and, possibly harmful to investment incentives: the tax structure can be summarized by $T=T(Y,m)$. Our whole system can be defined by the conditions:

$$Y=C(y,m,....)+I(Y,i,m,....)+G, \text{ where } y=Y-T(Y,m)$$

For prescribed levels of G and m, there will always be a level of i and a level of the tax function T that simultaneously leads to full employment and to any desired

In the first, suppose that we desire a much higher rate of capital formation but stipulate that it is to be achieved by a tax structure that favors low-income families rather than high-income. How can this be accomplished? It requires us to have an active expansionary policy (open-market operations, lowering of reserve requirements, lowered rediscount rates, governmental credit agencies of the FHA and RFC type if desired) which will stimulate investment spending. However, with our taxes bearing relatively lightly on the ready-spending poor, consumption will tend to be high at the same time that investment is high. To obviate the resulting inflationary pressure, an increase in the overall tax take with an overly balanced budget would be needed.

Alternatively, suppose the community wants a higher level of current consumption and has no wish to make significant redistributions away from the relatively well-to-do and toward the lower income groups. Then a tighter money policy that holds down investment would have to be combined with a fiscal policy of light taxation relative to expenditure. But note that in this case, as in the one just above, any qualitative mix of the tax structure can be offset in its effects by appropriate changes in the overall budget level and in the accompanying monetary policy.

VI

My discussion has covered a great deal of ground and has necessarily been brief. But I shall be glad to enlarge on the subject if that should be desired.

---

ratio I/Y. (The dots in the functions will permit one to add stocks of wealth or money as further variables and also to make various wage and price level assumptions.)

24

The balanced budget

Arthur Smithies

I

For over a quarter of a century, economists, or the majority of them, have been protesting against the dogma that the annually balanced budget is the path of financial virtue. I regret to have to report that we have made remarkably little headway at the high political levels. Despite the economists, or perhaps because of them, every President has clung tenaciously to the dogma. President Roosevelt's papers clearly reveal that he regarded budget deficits as an evil that had to be tolerated in order to achieve a greater good. The published views of President Truman make his views on the subject abundantly clear. The spoken utterances of President Eisenhower leave no doubt about where he stands on the matter. But despite our failure to demolish this pillar of the financial temple, there seems to be general acceptance of the view that deficits, though evil, are inevitable during depressions. The 12 billion dollar deficit in 1958 does not seem to have been grist to anyone's political mill—perhaps because everyone participated in creating it. But 1959 and 1960 have seen desperate if not ruthless efforts not only to achieve a balanced budget but to achieve balance at the pre-existing level of taxation. I can easily visualize the tortured sessions in the Budget Bureau and the Treasury that produced a surplus of 100 million dollars in the President's budget for fiscal 1960. I imagine similar sessions are going on this minute with respect to fiscal 1961.

Adherence to the dogma is so strong that we are prepared to delay vital defense programs in order to pay lip service to it. The government is also willing to impair the budgetary process itself in order to preserve the semblance of balance. As one spectacular example, I can find no

Reprinted with permission from *The American Economic Review,* Vol. L, No. 2 (May, 1960), pp. 301-309. Arthur Smithies, Harvard University.

indication of the capital cost of post office construction in the President's 1959 budget. Even though the government has decided to buy its post offices on time, surely the public is entitled to know how many post offices are being bought. If rationality rather than dogma dictated attitudes towards the budget, there might be less incentive for deception.

The survival of the balanced budget rule, however, is not entirely a matter of dogma. Individuals and groups with no dogmatic convictions have a strong interest in keeping the dogma alive. The classical objection to government debt was a natural reaction to the consequences of government extravagance during the seventeenth and eighteenth centuries.[1]

The requirement of a balanced budget was and still is the simplest and clearest rule to impose "fiscal discipline" and to hold government functions and expenditure to a minimum. Those who still entertain this desire as an overriding objective may be well advised not to retreat from the general rule until they are reasonably sure that the retreat will not become a rout.

The advocates of unbalanced budgets have not been reassuring from the conservative point of view. The unbalanced budget usually means fiscal freedom, borrowing, and deficits, and not deficits or surpluses as the occasion demands. The New Deal deficits were associated not simply with recovery but with recovery and reform; and when the New Deal was in full flower, the President took pains to insist that recovery was inseparable from reform.

Even an avowedly countercyclical fiscal policy is believed to give rise to an upward trend in expenditures that might not otherwise occur. The expenditures undertaken to counteract a depression are unlikely to be discontinued in the succeeding boom. If the boom is countered at all, the measures taken will be credit restriction or increased taxation; and then further expenditure programs will be taken to offset the next depression. The increased expenditures hastily undertaken to meet the 1958 recession indicate that this possibility is by no means academic.

The discipline of the balanced budget is not necessarily the right degree of discipline. It is generally agreed that in time of war the unwilling taxpayer should not be allowed to hamper the defense of the country. The taxpayer is supposed to come into his own in times of peace. But the present situation is neither peace nor war. Despite amiable conversations among heads of state, a permanent state of military readiness for the indefinite future will be imperative. Organized groups of taxpayers have not shown a clear appreciation of the situation. The President and the leaders of Congress must have some freedom to act

[1] For an admirable survey of the history of the balanced budget doctrine, see Jesse V. Burkhead, "The Balanced Budget," reprinted in *Readings in Fiscal Policy* (A.E.A., 1955).

even though they cannot pay the bills from current revenue. But if all notions of fiscal discipline and budget balance were removed and no alternative was provided, there can be no doubt that expenditures would increase to a level that was economically undesirable and politically demoralizing. However rich we become, public and private wants are likely to increase more rapidly than the means of satisfying them; and in our complex political system some rules of financial conduct that are simple enough to survive in a political context seem to me to be desirable.

II

Nevertheless, the rule that the budget should be balanced annually is inadequate to secure the proper allocation of resources between the public and the private sectors. The objections to it have been stated time and again. I shall therefore confine myself to a brief summary of those I consider the most important.

First, to attempt to balance the budget on an annual basis is inconsistent with the long-range character of many government programs. Research, development, and procurement for defense purposes inevitably involve activities extending over a number of years. If the programs are well conceived in the first place, waste and inefficiency will result from disrupting them in order to achieve particular budget results. I remember that on one occasion during the Korean war the government deferred payments to contractors for the sake of the appearance of the budget and naturally had to pay a high rate of interest to them as compensation for waiting for their money. Again, it is wasteful to suspend work on a battleship for the sake of avoiding disbursements at a particular time. Perhaps the battleship should not have been started; but to leave it half-finished for a time simply adds to its cost.

Another case where the requirement of annual balance is disruptive is the foreign aid program. This program is the favorite target for indignant charges of waste and inefficiency. But there is no surer way to waste and ineffectiveness than to expose our own program to such vicissitudes and uncertainties that the receiving countries are unable to mesh their own activities with it. Everyone who has examined the problem with understanding and sympathy has stressed the need for continuity.

With respect to the question of "annuality," the economists and the accountants are in league against effective government operations. The accountants like to clean up their books every year and hence stress the need for annual control. The economists take the same point of view because they want a flexible fiscal system whose impact on the economy can be varied from year to year as a contribution to general economic stability. Some compromise between the programming and the annual points of view is clearly needed. Neither can be ignored. But a satisfactory compromise, in my opinion, requires less strict adherence than we now attempt to the annual point of view.

A second objection to the balanced budget rule is that stress on the balanced budget as a criterion tends to give the misleading impression that the government is well managed if the budget is balanced. The examples I have just given illustrate this point, but, more generally, there is no indication that some over-all rule will secure efficiency down the line. When budget requests are cut to conform to the rule, the programs most likely to suffer are the new ones designed to meet new situations; and those most likely to survive are those that have acquired the support of powerful vested interests inside or outside the government. Not all new activities are necessarily more meritorious than the old, but some of them are. The way we now seem to be placidly accepting the Russian lead in space exploration—presumably for budgetary reasons—is a vivid illustration of my point.

Government efficiency cannot be achieved by budget ceilings imposed at the behest of hardheaded budget directors and appropriations committees. While some discipline of this kind is probably inevitable, the solution must lie in application of the economics of choice, subject to budget constraints, at every level of government. Public administrators traditionally do not learn economics, and vested interests have a strong interest in avoiding the application of economic principles.

The third objection relates to the effect of the balanced budget on economic fluctuations. Surely it is now agreed by economists that attempts, especially successful ones, to balance the budget every year worsen economic fluctuations. If governments curtail their expenditures when they are short of revenues and expand them when yields rise as a result of economic prosperity, their activities will be cyclical rather than countercyclical. It may be argued that I am stressing income effects and ignoring the monetary consequences of the balanced budget. The pre-Keynesian view was that depression cuts of expenditures released funds for the private economy. But the decisive objection to this point of view is that a central bank can do the same thing, so that the country can have the benefit of both income and monetary effects.

My final objection is that the balanced budget will not necessarily be the policy needed for achieving desired rates of economic growth. One of the unhappy ironies of the present time is that although the country is richer than it has ever been, further growth is becoming an explicit objective of policy—at a time when we should be enjoying the euphoria of John Stuart Mill's stationary state. We are not prepared to get the additional resources needed for national security and social welfare by cutting back on consumption. That would mean higher taxes. We must therefore grow in order to obtain more resources. Some eminent authorities maintain that the American economy must grow at 5 per cent a year instead of its traditional 3 per cent. If accelerated growth is required, it seems to me very likely that the total rate of national saving must be increased, and the only practicable way to increase total saving is through the generation of budget surpluses. Budgetary doctrine in this country has hardly begun to contemplate this possibility.

III

A more general objection to the balanced budget or any other budgetary rule is that it places unnecessary restrictions on ability to achieve a variety of economic policy objectives. In terms of Tinbergen's now famous proposition, the requirement of balance may leave the government with fewer instruments than it has targets; and consequently may mean that objectives more important than balance must be ignored or that new instruments must be discovered.

The point can be illustrated very simply. Let us ignore for the moment pressing issues such as inflation and the balance of payments and assume that the government has only three policy objectives: first, it must spend enough to give effect to foreign and domestic policy objectives; second, it must maintain full employment, and, third, it must ensure that private investment will, in each year, be carried out at the rate required to maintain a given rate of economic growth for the economy as a whole. Thus every year it has three fixed targets: national income or output, private investment, and government expenditures. With present institutional arrangements, convictions, and predilections, it is virtually restricted to three instruments; namely, appropriations, taxation, and general credit expansion or contraction.

If the government has freedom to use these three instruments, it can attain the three objectives. If in addition it must balance the budget or maintain any prescribed relationship between expenditures and revenues, it has set for itself a fourth objective and is consequently one instrument short.

So long as it possesses only the three instruments, some other must give way: growth, full employment, or the government's own programs. As a matter of fact, during the last few years the government has placed even more severe restrictions on itself. It has attempted not only to balance the budget but to balance it at existing levels of taxation. This means that it has denied itself the use of one instrument. The expenditure objective necessarily gives way to this requirement (insofar as the requirement is met), and the government is left with general credit policy to achieve both full employment and a satisfactory rate of growth—a task that it is logically and practically impossible for the harassed monetary authorities to perform. Their difficulties are compounded when in addition they are expected to help correct the balance of payments and to prevent inflation.

If the government is short of instruments, it must acquire new instruments if it is to attain its objectives. Such new instruments could be selective credit controls, selective tax measures, and various kinds of direct controls. It would take me too far afield to discuss these possibilities in detail. Suffice it to say that many of them are pure anathema to those who must vehemently support the balanced budget doctrine. They are likely to be required to pay a high price for the dogma.

IV

We are unlikely to achieve full coherence in the formation of fiscal and budget policy. Some incoherence is likely to remain so long as there is separation of powers between the President and Congress and between the powerful committees of the Congress. Nevertheless, considerable improvement is possible and feasible. To be optimistic about that, one only has to reflect on the extent to which economic thinking has penetrated the government since World War II, largely as an outcome of the Employment Act of 1946 and the institutions set up under it. I therefore consider it worth while to offer some suggestions concerning the directions that improvement might take.

First of all, the President should transmit his budget to Congress as part of a comprehensive economic program. This is not done at the present time. The present Budget Message is notable for its lack of economic analysis. The President's Economic Report, on the other hand, is equally notable for its lack of an analysis of the economic impact of the budget. However much they may consort in private, the Budget Bureau and the Council of Economic Advisers do not embrace in public. The President's program would analyze economic policy as a whole in terms of the variety of objectives to be attained and the instruments to be employed.

With respect to the budget itself, the President would recommend a surplus, balance, or a deficit, depending on economic conditions. If a deficit were proposed, this should be proposed as a positive recommendation, not as a confession of failure to balance the budget combined with a wistful hope that balance will be achieved next year.

This approach could have the same disciplinary value as the balanced budget. If the President were prepared to give the weight of his authority to the need for a surplus or a deficit of a certain amount, that should have the same disciplinary value as balance from the point of view of the Congress and the executive departments.

For this suggestion to be as effective as possible, the Congress would have to co-operate. In particular, the Joint Economic Committee should join with the Appropriations Committee and the Ways and Means Committee in considering the President's program and in formulating Congressional economic and budgetary policy. But such a change in Congressional procedure is unlikely to come about unless the President takes the lead.

Second, the President's economic program should distinguish between long-run economic policy and the policy needed to counteract particular episodes of boom and recession. The long-run policy should contemplate continuity in government operations and continued growth of the economy. Budgetary policy in particular should be designed to conform with the requirements of long-run growth.

Of course long-run policy would be revised from year to year, to take account of changing circumstances and to correct errors in diagnosis. But in the absence of violent changes, say in defense requirements, it seems unlikely that abrupt changes in the relation of government expenditures to revenues would be required. Consequently, some simple budgetary rule that should apply in normal circumstances may be feasible. In times of full employment without inflation, it could be said that the budget should have a surplus or a deficit of some known order of magnitude.

It follows that the basic revenue and expenditure estimates should be made and published with reference to a full employment situation rather than to the situation actually predicted. This is the stabilizing budget approach that has long been advocated by the Committee for Economic Development, but which has made very little headway in official circles.

I suggest, also, that if the government's policy is to keep a stable price level, the expenditure and revenue estimates should be made in stable prices. This procedure provides an automatic check on inflation. It would tend to prevent inflationary increase in revenues from being regarded as a substitute for taxation. It would also put some pressure on the spending agencies in the event of inflation. They should make some contribution by attempting to curtail their activities. But if they consider that impossible, they should demonstrate the fact in requests for supplemental appropriations.

Thirdly, the question of countercyclical policy should be dealt with in a separate chapter of the President's program. This would include a discussion of the effects of recessions or booms on the budget and recommendations concerning the budgetary measures needed as correctives.

In view of what I said above, long-range government procurement programs should be interfered with as little as possible for cyclical policies. Nor should new programs that will last for a number of years be hastily adopted merely for the sake of relieving a single recession. This, however, does not mean that all public works should be continued at the same rate regardless of booms and depressions. Highway construction and many items authorized by the Rivers and Harbors Bill can be adjusted to short-term economic needs.

However, if the main emphasis were placed on changes in taxation and transfer payments for purposes of short-run stability, the inefficiencies connected with abrupt alterations in expenditure programs could be avoided. The 1958 recession furnishes a good example. The government refused to use tax reduction as its fiscal weapon, and consequently got large and ill-considered increases in expenditures which will continue long after the antidepression need for them has passed. Neverthelesss, I believed at the time and still believe that the tax route would have been wrong. Taxes once reduced are notoriously hard to restore. In fact it is hard to think of any tax increases during the last thirty years that were not undertaken in response to emergency situations. Even the tax increases of the New Deal come under that category. If the existing tax rates are likely to be needed for long-run purposes, it is of questionable wisdom to reduce them for short-run reasons, unless the reduction can be of an explicitly temporary character.

This leads me, and has led many others, to the conclusion that short-run stability should be achieved as far as possible through "built-in flexibility" of the budget and through monetary measures that can be readily reversed.

Built-in flexibility has increased appreciably as a result of social security, unemployment compensation, and agricultural support on the expenditure side and through the automatic operation of the tax system. But such measures—even in conjunction with vigorous credit measures—are unlikely to meet the requirements of a severe recession. There is need for further automatic measures. Consequently, I venture to repeat a proposal in which I participated some years ago.[2] Under certain specified signs of recession, there should be an automatic reduction in the first bracket of the income tax. The reduction should be restored automatically when recovery has reached a prescribed point. To guard against inappropriate use of the remedy, its application should be subject to veto by the President. Devices such as this could give reasonable assurance that anything but the deepest depression would be corrected and would help materially to avoid the psychological conditions that might produce depressions of the catastrophic kind.

Our proposal was considered in the chancelleries of the world and was unanimously rejected by respectable opinion. Had it been in effect it would have been very serviceable in 1958. Automatic reversible de-

[2] See the United Nations Report, *National and International Measures for Full Employment* (1950).

vices are the most effective way to avoid the radical political consequences of a flexible fiscal policy, and thus to allay the fears of those who cling to the balanced budget rule on rational rather than superstitious grounds.

To return finally to the long-run question, I have suggested that surpluses rather than deficits may be needed in the future—if the government pursues an economic policy that is consonant with national and international needs. But surpluses are hard to achieve. Senator Taft once remarked that in his long experience, surpluses and debt retirement occurred only through inadvertence. If that is true, perhaps the balanced budget doctrine has some long-run merit after all.

25

Budgetary reform: Notes on principles and strategy

F. M. Bator

The point I wish to make is not concerned with the budget as an aid in program formulation, execution, and control, or as a device for financial planning. It has to do rather with the role of the budget in informing and guiding congressional and public consideration of the large questions of fiscal policy: the balance between total demand and potential output, and the division of output between public and private uses and between consumption and investment.[1]

It is evident that as regards these large issues, upside-down economics still has the better of it in our public discussions. To be sure, the quality of conversation has improved during the last several years, as shown by the currency, among the more sophisticated people who read and write the nontechnical commentary on economic affairs, of a number of therapeutic notions: for instance, that at times deficits are not a bad thing; that if one cares to assess the fiscal impact of the budget one should look to the "cash" or the "national income and product" (NIP) version and not the "administrative budget"; and—a quite subtle notion—that the "full-employment surplus (deficit)," on NIP or cash account, is a better measure of the weight of the fisc than the

Reprinted by permission of the publishers from Francis M. Bator THE REVIEW OF ECONOMICS AND STATISTICS Cambridge, Mass.: Harvard University Press, Copyright, 1963, by the President and Fellows of Harvard College, Vol. XLV (May, 1963), pp. 115-120. F. M. Bator is now Deputy Special Assistant to the President for National Security Affairs.

[1] I would not, of course, deny the importance of the former aspects of budgeting. The budget is a multi-purpose instrument; we must have an integrated system of sub-budgets. If this note appears to ignore that fact, it is only because the suggestion made below is entirely consistent with multi-purpose budget design. There is no conflict.

actual realized surplus or deficit. However, a little sophistication can be a mixed blessing. None of the above notions shifts attention from the "deficit"—some deficit, perhaps a benign deficit—as the crucial magnitude. Yet, as we know, the beginning of wisdom in these matters is that no deficit, however measured, is an indicator of the expansionary or contractionary effect of the budget on total demand. Even the "full-employment deficit," while it is less misleading than the uncorrected, observed deficit, can easily lead down the garden path.[2]

This is not merely an intellectual quibble. As long as the debate is about which deficit is the true deficit and about when it is and when it is not all right to run a deficit, any weakening of attachment to the strict balance-the-budget rule is likely to be accompanied by the hardening of new doctrine, not quite as rigid but equally arbitrary and possibly, because more readily applied, almost as hobbling. One likely candidate —in some quarters it is already full-fledge dogma—is the proposition that deficits are appropriate in "recession" but not during expansion. Another, less likely to gain immediate acceptance but a good bet if the more knowledgeable members of Congress and the newspaper people begin to take up the notion of the full-employment surplus, is that fiscal policy should assure a full-employment surplus of no less than zero. In a situation where private spending propensities are weak, and public resistance makes it impossible to expand government spending fast enough to take up the slack, such a rule would shift to *monetary* policy much of the burden for keeping total demand from falling behind potential output. If investment and spending on consumer durables should happen to be relatively unresponsive to monetary ease—a likely contingency in a sluggish climate—or if the money managers are frustrated by a very elastic demand for money (liquidity trap) or feel themselves hobbled

[2] Two budgets with the same full-employment surplus or deficit, and the same revenue structure and expenditure composition, will have markedly different effects on total demand according to whether the level of expenditure (revenue) is $x billion or $2x billion or $1.03x billion. (The "full-employment surplus" notion has the great merit of distinguishing between shifts in schedules and income change-induced movements along schedules. With a specified level of full-employment income, and if government purchases are constant, changes in the "full-employment surplus" provide a fair, if approximate [one parameter], measure of changes in the "weight" of the net-receipts schedule. For quantitative analysis, however, it is probably better to work directly with the "first-round" effects on the income flow, or if it is government purchases that are changed, on the flow of spending. If there is a *parallel* shift in the total spending schedule, it makes no difference. If, however, the schedule "twists"—and if as a tentative working assumption one posits exact or near-linearity *nearby* and approximates the result by multiplying the autonomous change by some multiplier—then it is a better strategy, because of lags, to work "forward" and not "backward": later effects come later and the magnitude of subsequent re-spending effects is likely to be more uncertain. Moreover, there is less chance of cumulating error. (If the spending schedule is assumed to exhibit appreciable curvature, then one cannot use a constant multiplier anyway but must re-solve the entire system using the new schedule.)

by hot (footloose) money, the burden would be too heavy. There would result a chronic shortfall in total demand relative to potential output. Moreover, even if there were no troubles with sticky interest rates and inelastic investment spending or with the balance of payments, and hence if monetary policy could be made to work, there would be no cause for satisfaction. As a community we would be making our choices between investment and consumption blindly, as it were, uninfluenced by the relative desirability and the terms of trade between consumption now *versus* growth for consumption later. By imposing an arbitrary rule on our fiscal and monetary managers we will have lost an important degree of freedom with regard to the allocation of resources and/or, the purist might point out, the distribution of income (perhaps). (Needless to say, the bias of the zero full-employment surplus rule is not necessarily deflationary. In a situation of brisk private demand and fast expanding government spending, it would stack the cards in favor of inflation and/or tight-money-and-low-investment.)

The dilemma is plain. In our current situation a doctrine calling for a zero full-employment surplus would provide support for a badly needed reduction in the net fiscal load carried by the economy: for an increase in expenditures or a cut in tax rates or both. The "deficits are all right in recession" rule, in turn, could turn out to be useful during the next recession. Perhaps, like Churchill, we should welcome help irrespective of pedigree. However, one had better keep reminding oneself of the perils of such alliances. Employing specious cryptorules in defense of sensible measures today may make it much more difficult to do the things that need to be done tomorrow.

It would be tempting, but wrong, to blame the problem entirely on the national neurosis about deficits. Wrong, because the truth that good fiscal doctrine cannot be based on the difference between expenditures and revenue would pose a thorny problem for fiscal strategy even in the absence of deficit fixation. There would still be a need for budgetary principles which are simple enough to be persuasive at the level of serious newspaper discussion. The President, the Budget Bureau, the Treasury, and the Council can and do work with relatively complicated, unobvious rules; the President, if he will, has the ultimate say. But the Executive has to have a compelling rationale to justify the budget in relation to output, employment, and growth, in the Congress and the country at large. That need would not vanish even if Poor Richard and the fallacy of composition lost their hold.

Could the budget message be used to provide such a rationale? It is fair to say that the attempt has not been made.[3] Even in the 1963 *Budget*, a much more informative and sophisticated document than its predecessors, the assertion that the projected deficit or surplus is appro-

[3] This is written before the publication of the 1964 *Budget*.

priate in terms of the goals of full employment and price stability has to be accepted on its face. In its discussion of the connection between revenues and total output the emphasis of the message is entirely on the revenue-yield of the forecast level of GNP ($570 billion). The much more important inverse relation—what level of taxes will help achieve the desired volume of total demand and output with the given expenditure plans—receives very little attention (see pp. 7-9, 19, 24-25). The budget makes it easy not to discuss that, the critical question, and to talk rather about the rights and wrongs of the surplus or deficit as such.[4]

It would be foolish to think that sensible and explicit treatment of fiscal policy in the budget would be sufficient to assure congressional and public enlightenment. But would it not be a step in the right direction if the President were to present, as the centerpiece of his budget message, a quantitative exposition of the fiscal policy rationale of the budget? This would involve the presentation of the results of a trial-run "nation's economic budget" (NEB) exercise, which would show the expected pattern of net receipts and purchases of households, businesses, the rest of the world, state and local governments, and the federal government, in relation to the target level of "desired" output. The exposition could be relatively brief and refer to a full-length version in the *Economic Report*. It would have to emphasize the tentative, trial-run character of the exercise, its hybrid proposal-program-forecast nature, the large margins of error, the desirability of working with ranges of values (it might give ranges), and the need for revision, month by month, of the estimates, and more often than not, of the proposed policies. But if presented in the right way, such an exercise would make it possible to substitute reasoned discussion for *ex cathedra* pronouncement in the message itself and would help to provide the basis for more intelligent congressional and public consideration of the fiscal policy aspects of the budget.[5]

[4] In fairness, it should be said that last year the government did publish a sophisticated discussion of the 1962 prospects in the *Economic Report* (pp. 62-68), and that the budget message contains a reference to the *Report* (p. 19). However, it is the budget message which sets the tone and provides the grist for the newspaper people and politicians, and hence we cannot afford to relegate sensible discussion of fiscal policy to the *Report* and leave the budget message to treat fiscal policy as though it had to do primarily with raising money to pay for expenditures. The *Economic Report* should be considered as back-up for the interested few who are willing to read. (Ten pages stuck in the middle of Chapter 1 of the *Economic Report* are not likely to be given much publicity. In contrast, the entire budget message is published in the *New York Times*. To be sure, so is the President's report proper, and last year, for once, that did address the issues right-side up (pp. 11 and 12). Nonetheless, I would think that the point stands.)

[5] It would be superfluous, here, to describe such an exercise, especially since the objections to its use, as proposed, do not generally concern matters of technical detail. (For comments on some technical issues, see the appended *Notes*.) The basis for the exercise would lie in a host of executive decisions about programs and targets, and in a quarter-by-quarter forecast of total demand and its major components. The quantita-

What are the objections to such a procedure? The objection that an NEB exercise involves illegitimate "planning" is, of course, untenable. If the government is to do an adequate job in terms of almost any widely held notion of the national interest, in terms of the requirements imposed on it by law, or in terms of its political future, it must take into account the fiscal impact of the budget and hence plan in precisely the sense of such an exercise. Arthur Burns could not avoid, any more than can Walter Heller, making estimates about the major components of total demand. He could not even avoid contamination with some notion of desirable, or tolerable, or "potential" output.

The serious objections have to do not at all with the internal use by the executive of such an exercise, but with the wisdom of a strategy of exposure, of publishing the quantitative results and using these to justify the government's fiscal program. A published NEB exercise will certainly draw fire. It will be attacked as a milestone down the road to serfdom. The target figures for total output and the major components will be attacked as too high, too low, inconsistent, and as reflecting dictatorial tendencies. As forecast, the exercise will be alleged to reflect both technical incompetence and foolish arrogance about the government's ability to predict spending behavior (even its own). Moreover, the betting odds are overwhelming that many of the allegations of inconsistency and bad forecasting will turn out—six or twelve or eighteen months later—to have been justified.

It could be argued, with some truth, that the government cannot hide that it has in fact planned, and cannot avoid revealing many of the specific estimates on which the budget rests, whether or not it publishes an NEB exercise. Even a pure money-raising, Byrd approach to the budget will get involved in the politics of target fixing and target achievement; the revenue estimate implies a target level of GNP. If the $570 billion had not been published, it would almost certainly have "leaked" and provided the basis for jeers by mid-summer and for pressure, desirable I think, for a large cut in taxes. Nonetheless, it is a fair guess that the use of an NEB exercise to justify the budget will stimulate rather than quiet controversy.

But would that be bad? The trouble with the present procedure is not that it gives rise to debate but that it fails to pose the true issues

---

tive results could be summarized in a "G. C. table" (for G. C. read Gerhard Colm and see the first page of any issue of *Economic Indicators*) and presented in the fiscal policy section of the budget message (which should be moved forward to precede the sections on the composition of expenditures and revenues). The text would consist in a commentary on the choice of the target level of output, on the role of the budget in achieving that level of total demand, on the changes in policy that would be required if the projections were to turn out to be too low or too high, etc. Appropriate reference would, of course, have to be made to the fuller discussion in the *Economic Report*.

and hence makes it difficult to engage in sensible debate. We should welcome controversy about the appropriate balance between personal consumption and various kinds of public consumption, or between consumption and investment for growth, as well as about whether the government's target for total demand is too high or too low in the light of the expected consequences for capacity utilization, the labor market, the price level, and the balance of payments. Vigorous public discussion of these issues, informed by quantitative presentation by the government of its own position in the budget message and the *Economic Report,* would perhaps begin to make inroads on the fixation on deficits and spending and taxes as evil in themselves. Not that sensible presentation of the issues in the budget message will result in miracles. But the current mode of presentation discourages and hinders nonmiraculous, slow improvement.[6]

There remains what is perhaps the most serious objection at the level of strategy and tactics: that if the government publishes an NEB exercise, or even if it merely commits itself to a target rate of output, it will thereby impair its freedom of maneuver. This, the flexibility issue, is not without its peculiar side. It not only involves flexibility to adjust policy when off (or on) target, but also flexibility *not* to adjust policy when the economy is off target. One can certainly sympathize with the discomfiture of a Secretary of the Treasury whose own targets and projections of January are being used to pressure him into action in August, action which he may judge undesirable and/or politically unfeasible.

However, the coin has another side. If the government decides in January that, say, $570 billion is a desirable and reasonable target in the light of its estimate of potential output, inflationary pressures, the balance of payments, and a host of political judgments, then it is conceivable that it might even welcome pressure on itself to take corrective action six or nine months later, should performance be substantially off the target. Certainly from the point of view of the national interest, if not the comfort of ministers, it is not evident that the pressure for action which a shortfall or overshoot would generate would be necessarily counter-productive. Moreover, if under such circumstances the

[6] Nor would it be a bad thing if, as a consequence of greater public exposure, the technical quality of the forecasts and projections underlying the budget were subject to more systematic criticism from the outside. Apart from stimulating improvement in the state of the art, exposure might help to strengthen resistance, in the face of political pressure, against excessive "distortion."

Concerning the danger that an NEB strategy will inflame ideological controversy and divert attention from the true issues, I am prejudiced enough to think that the case for the defense is so strong—and that it can be made so simple, compelling, and even interesting—that one might almost welcome attack and a "great debate" as an opportunity for powerful rebuttal and useful public education about the role of taxes and government spending in relation to price-market institutions.

President should decide in favor of corrective fiscal action, he would surely find it much easier to make his case with a target and an underlying NEB exercise on the books. Current practice, since it does not exhibit the quantitative links between the budget and the original half-avowed GNP target, does not provide the ingredients for a convincing brief. An intelligently designed exercise published in January, one which takes account of the inevitable margins of error, would be especially helpful at times when the need is for fiscal action which would make larger deficits (or surpluses) considerably larger.

None of this is to deny the importance of flexibility. Unforeseen price pressures, changes in the balance of payments or in defense requirements, and the like, might well require changes in targets and appropriate adjustments (or non-adjustments) in policy. However, in most such situations it is not only possible but desirable for the Executive to articulate the need for the change. A case can be made that the government should not employ a tactic of comfortable silence even in situations in which it decides to obviate any need for changing policy by passively adopting a change in targets.[7]

All the above is relevant to what may be the real political sticking point. The more explicit is the government's commitment to an output target, and the more explicit are the calculations which underlie such a target, the more clubs one gives away to one's political opponents. Moreover, and quite apart from political warfare, there is always the danger that poor performance—and every now and then performance is bound to be very poor—will be used by the ideologists to discredit systematic quantitative fiscal planning.

Unfortunately, it is not clear that there exists a strategy for the improvement of our fiscal politics which will avoid these dangers. Moreover, sensible and sophisticated presentation of an NEB exercise can do much to blunt irresponsible attack. By emphasizing the tentative quality of the projections and the need to keep re-examining and re-adjusting both the projections and, quite likely, the proposed politics; by avoiding *point* estimates; by scheduling and publishing a re-estimation in June and then quarterly during the entire fiscal year; by indicating the stand-by policies that might have to be invoked, should the gap

[7] It could be argued that the requirement to make a public commitment to a target will distort the choice of the target, that is, that the government will want to play safe in January and pick a target which is relatively easy to attain. On its face, it would not appear obvious which way, if at all, the countervailing temptations of ambition and caution are likely to bias the choice. In general, if the anticipated pressure to live up to target makes the President and his advisers even more careful and sober about what target they pick than they would be otherwise, that is probably all to the good. Sober seriousness is not, of course, equivalent to a preference for a lower rather than a higher output target. To pick a "low" target is to acquiesce in advance—and, given an NEB exercise, in public—in a loss of output, wages, and profits; in unemployment and sluggish growth.

between the performance and target turn out to be large—by all such means the government can both build a strong defense against the charge of naive crystal-ball gazing and, more important, provide for itself a position from which to recommend and undertake changes in policy as circumstances warrant (and pin blame on the Congress when congressional intransigence prevents needed adjustments in policy).[8]

Last, it is perhaps not idle to hope that a strategy of planning and operating fiscal policy rather more in the open will help to create public support for giving the Executive some limited discretion to vary tax rates. As people get used to the notion that taxes and spending are like the brakes and the throttle of a car—that to lock them in place is dangerous, and that small marginal adjustments, made in time, may avoid the need for drastic adjustment—resistance to giving the Executive some freedom of fiscal maneuver is likely to decline.*

[8] Sensible presentation of an NEB exercise, and especially the use of ranges of values, can help to get around the problem that the executive budget is only a proposal to the legislature and hence contains items on the enactment of which the betting odds are poor.

* *Notes on Technical Issues of Budget Design*

1. Offhand, I can think of no strong reason why the summary NEB exercise—which would appear in the fiscal policy section of the budget message and the purpose of which would be to guide congressional and public discussion—should not follow National Income Division conventions and procedures. Moreover, there is good reason why it should do so, e.g., the need for coherence between the federal sector and the other sectors as regards timing, transactions coverage and exclusions, etc. (My preference for NIP timing [receipts by and large on accrual and purchases on delivery] is not motivated by strategic calculation of how to minimize the damaging effects of deficit-fixation. It is not so motivated because I am not clear on strategic grounds where I would come out. I wish I were more convinced by the Schultze argument—or its opposite.)

2. When it comes to the use of the budget figures for quantitative analysis of the effects of the budget on total demand, the situation is more complicated:

a. As regards agency coverage, the administrative budget is inferior to the Cash and NIP Budgets. We need a consolidated statement for the federal government as a whole.

b. As regards the inclusion or exclusion of "asset-transactions," there is no clear answer and neither the cash nor the NIP version alone will do. Loans and purchases of old assets will have important portfolio-effects on total demand and hence the budget statement should provide information on the government's plans with regard to both categories. The NIP Budget is, on this account, incomplete. On the other hand, it would be a mistake not to try to maintain the distinction between "fiscal" and "monetary" effects and simply to lump, as does the Cash Budget, bona fide loans with expenditures—the more so since much the larger part of the government's capital operations (debt management and Federal Reserve open-market) are anyway not included. Why not add two more non-NIP columns to the five now provided in the NIP version (purchases, transfers, interest, grants-in-aid, subsidies less surplus)? The two new columns could be ignored in the unavoidable calculation of the NIP deficit. (Loan guarantee and other indirect subsidy programs had best be covered in connection with the particular functions and activities they are designed to support.)

c. On the question of *timing,* it is best to be eclectic and not make a categorical choice between "cash" and "accrual" on the income/receipts side, or for that matter, between "obligations (orders)", "deliveries" and "cash" on the product/expenditure side. All five time-profiles will be significant for the analysis of the effect of the budget on total demand (no doubt differentially so according to cyclical circumstances and, on the expenditure side, according to what is being bought). To the extent feasible, the budget should provide estimates on all five flow-rates. (I suppose if I were made to choose, for purposes of back-of-the-envelope fiscal analysis I should follow NID and use accruals in timing the impact effect on the income stream, and rather more doubtfully, deliveries on the expenditure side. Needless to say, the choice depends on one's notions about the determinants of various components of private expenditure, production lags, and the like.)

3. It would be useful if the fiscal policy section of the budget message were to point out what the NEB exercise shows to be the estimated effect of the budget on the public share, federal and state-local, in GNP, and on the non-defense public share, and if it were to make some comparisons with prior years. (Appropriate reference would have to be made to subsidies as reflecting some degree of federal absorption of output and also to federal finance of state and local purchases).

4. The fiscal policy section of the budget message should also point out the implications of the "fiscal policy plan" for the investment-consumption mix in the economy as a whole.

5. The lead table on budget expenditures by function should be supplemented by a larger table in which the expenditure figure for each function is broken down into purchases of goods and services, outright subsidies, transfer payments to individuals, grants-in-aid, interest, loans, and purchases of old assets.

6. The above seven-way split should be carried through in the detailed discussion of the "Federal Program by Function."

7. I would think it a bad idea—on grounds both of concept and of strategy—for the federal government to adopt a two-budget system involving a full-fledged capital account. Such a system would almost certainly result in the enthronement of the shibboleth that it is all right to debt-finance capital expenditures but the current budget should be balanced or in surplus. Except under very strong "classical" assumptions, such a rule would not assure "neutrality" as regards the saving-consumption choice, and would lose us an important degree of freedom, making it much more difficult to achieve through fiscal and monetary measures whatever total demand, income distribution, consumption-investment mix, and public-private balance we might desire. (A related secondary danger is that "investment" would come to be defined as bricks and mortar or the purchase of self-liquidating assets and, in particular, that investment in education, public health, etc., would not qualify.)

8. The above does not imply that we should continue to lump together public consumption and public investment. It would be most useful to distinguish, on the expenditure side of the budget, between consumption-type expenditures, investment in tangible assets, and what Musgrave calls "expenditures for future benefit not resulting in acquisition of assets." The burden of (7) above is only that we should not associate particular receipts with particular types of expenditure.

26

The changing views on debt management

William E. Laird

In the period before the Great Depression debt management was not a topic of active controversy. There was virtual unanimity regarding the elements of "sound" debt management policy and very little debt to manage. An interesting similarity of opinion was observed among academic economists and between the economists and the Treasury.

The events of the 1930's and 1940's created problems for the Treasury; the prolonged depression and the demands of war finance resulted in a greatly expanded federal debt which led the Treasury to retreat from its older and simpler precepts of debt management. Some observers were concerned about the debt and the possibility that it might permanently lay to rest independent and flexible monetary policy. Economists set about reconsidering their views on the relationships of debt management, monetary policy, and economic stabilization. The revision of Treasury policy did not coincide with the advance of the newer doctrines. Academic economists disagreed among themselves, and usually they disagreed with the Treasury. Conflicting policy positions developed; proponents described them as sound, and they were judged by various standards to contribute to stability.

Currently there is some dispute about every major aspect of debt management. This paper delineates the changing views on debt management, contrasts the older views with the various newer concepts of policy, and points out the current, and very significant, divergence of thought on debt policy.

Older Views on Debt Management

In the pre-Keynesian, pre-fiscal policy era the benchmark of sound finance was to be found in the concept, and practice, of funding the

Reprinted with permission from *The Quarterly Review of Economics and Business,* Vol. III, (Autumn, 1963), pp. 7-17. William E. Laird, Florida State University.

debt. Short-term or floating debt was not looked upon with favor by the Treasury, and sound policy avoided excessive reliance on shorter-term securities. The objectives of debt policy were all related to these fundamental principles of debt management. Important advantages were seen to accrue from funding the debt and minimizing reliance on floating debt. A longer debt was less likely to expose the Treasury to the mercy of the market, since the Treasury would face fewer holders of maturing debt at any one time. The Treasury would not be tied so closely to the market, and shifts in the market would not be so serious from the Treasury's point of view. Refinancings could be smaller with longer debt. Also, a funded debt could more easily be adapted to plans for debt retirement, and debt retirement was considered a worthy endeavor.

Treasury policy was concerned with the interest burden of the debt, but the Treasury contemplated neither a program of inflation to cope with the debt nor extensive reliance on floating debt. Interest costs were to be kept down by retirement and by refinancing at lower rates.

Pre-Keynesian Treasury policy did not relate debt management to the cycle; neither did pre-Keynesian writers on public finance and debt management. One may observe the interesting similarity of academic view and Treasury policy. Shirras writes:

> It is therefore necessary to reduce the floating debt within manageable proportions, so that it can never be a source of great danger. It is better to borrow for a long time and to pay a higher rate of interest than to be perpetually at the mercy of holders of Treasury bills for repayment.[1]

C. F. Bastable's earlier statement is similar.

> As a general principle of finance it is unquestionable that the floating debt should be kept within the narrowest limits possible. . . . A growth of floating charges is at best a mark of weakness in the treatment of the state liabilities. . . . The great evil of a floating debt is its uncertainty.[2]

Others, including Henry C. Adams, had expressed similar views.[3]

Academic opinion favored debt retirement. It was judged to strengthen the national credit, to facilitate further borrowing should the need arise, and to increase the capital available for industrial growth.

[1] G. Findlay Shirras, *Science of Public Finance,* Vol. 2 (3rd ed.; London: Macmillan, 1936), p. 799.

[2] C. F. Bastable, *Public Finance* (3rd ed.; London: Macmillan, 1903), pp. 694-95.

[3] Henry C. Adams, *Public Debts, An Essay in the Science of Finance* (New York: D. Appleton, 1893), p. 148.

Debt reduction increased confidence and gave a favorable tone to government finance.

The Great Depression, World War II, and After

The Great Depression intensified interest in economic stabilization. Full employment and stability became virtually synonymous, and as a policy goal full employment had no peer. While monetary policy, interpreted as interest rate policy, was relegated to a position of insignificance, fiscal policy emerged as the new and respected tool of analysis and policy. As the Depression lengthened, more attention was paid to this newly discovered weapon. Fiscal deficits became more than respectable. In certain groups deficits were considered the principal defense against secular stagnation.

As pump-priming shaded into compensatory finance, with deficits becoming more a rule than an exception, the growth of the debt appeared certain. Active compensatory finance had as its adjunct a growing federal debt, which would almost inevitably become a problem in its own right, as debt must be managed in some fashion. Since it was widely assumed that money was impotent, monetary policy would increasingly be directed toward "managing the debt." It was assumed that in any monetary policy-debt management conflict, debt would win.

It was World War II rather than the Depression which brought the monetary policy-debt management conflict to the surface, and for a while it appeared that the war debt had completely submerged the final traces of monetary policy. At the end of the war there seemed to be relatively little enthusiasm for ending the Federal Reserve's bond support program. Opinion within the Reserve System gradually solidified against the policy, and the Treasury reluctantly consented to ending the program.

About the time of the Accord, analysis had begun turning to a more sophisticated reinterpretation of classical and neoclassical economics and policy measures tended to reflect this development. Economic stability remained an important objective while discussion and analysis turned to the pressing question of means to attain that end.

It was difficult to ignore the role of liquidity in the postwar inflation and attention was turned to that vast conglomeration of liquid assets, the federal debt. In an era preoccupied with stabilization it is not surprising that any controllable sector in the economy that showed promise as a tool of policy came under consideration. Post-Accord discussion pointed to the possibility of the debt contributing to the goal.

The Debt Management Controversy

The countercyclical approach

Post-Accord discussion relating debt management to economic stabilization, although a break with tradition, was not entirely without precedent. One version of stabilization via debt had been forcefully presented some years earlier by Henry Simons, at a time (1944) when few were prepared for such a view of the debt.[4]

Stabilization via Composition of Debt including Money

Simons recognized the debt as exerting an influence on economic stability, and he believed the real danger of the debt to be inflation. Understanding the climate of opinion of the postwar years, he foresaw the inflation to come and stated quite specifically in 1946 that "we probably shall have, in the near future, no substantial protection against inflation save that which debt policy affords."[5]

In 1944, he had clearly linked debt policy to economic stabilization. Conceiving of debt to be either paper money or consols (having neither call nor maturity features), he prescribed a simple rule of action. "The rule for policy as to consols and currency, that is, the *composition* of the debt including money, is simply stabilization of the value of money."[6] The correct combination depends upon the particular circumstances of the time. "Converting money into consols is an anti-inflation measure; converting consols into money is a reflationary or anti-deflation measure . . ."[7]

Should inflation be the problem, the appropriate action is to sell consol bonds—convert money into consols—which reduces liquidity in the economy, retards spending, and stabilizes the price level. With the economy under deflationary pressure the contrary action is appropriate. Simons assumed that actions taken to stabilize the price level work automatically toward stabilizing the economy. Thus, price level stability provided a clear and serviceable criterion of performance.

Within his framework it is proper for the Treasury to pursue that policy indicated by the general condition of the economy regardless of

[4] Henry C. Simons, "On Debt Policy," *Journal of Political Economy*, Vol. 52, No. 4 (December, 1944).

[5] Henry C. Simons, "Debt Policy and Banking Policy," *Review of Economic Statistics*, Vol. 28, No. 2 (May, 1946), pp. 85-89. Reprinted in his *Economic Policy for a Free Society* (Chicago: University of Chicago Press, 1948), p. 235.

[6] Henry C. Simons, "On Debt Policy," *loc. cit.*, p. 223.

[7] *Ibid.*

interest cost. In fact, *within that framework* it is legitimate to say that the Treasury should pay as much interest as possible. Since transactions would take place in a competitive market and be subject to a price level stabilization rule, maximizing interest payments is equivalent to saying "pay only enough to achieve the goal."

Several comments about Simons' debt policy are in order. First of all, there is some logic to the position that Simons actually had no debt policy in the conventional sense, but rather merely translated his monetary policy into the language of the federal debt. His debt policy is contained in the opening sentence of "On Debt Policy." "I have never seen any sense in an elaborate structure of federal debt."[8] His debt policy per se consists of transforming all federal debt into pure consol bonds. His program is directed toward assuring that the proper quantity of money will be in circulation. He proposes countercyclical monetary policy couched in terms of the federal debt. It may be no more than simple semantic exercise to discuss his concept of countercyclical debt management, as he is actually taking the back door to a flexible monetary policy and merely emphasizing the point that the size of the debt is of secondary importance. A more conventional or "front door" approach would speak explicitly in terms of the money supply and its variations. At the time Simons wrote, fear of the large federal debt was effectively frustrating any real control of the money supply. By approaching the stabilization problem in this unusual fashion, he apparently was attempting to free monetary policy by abating fears of the federal debt. Essentially, he was expressing confidence that the federal debt would not crush the economy while warning that continued fear of the debt would lead to an inflation of the price level. He struck at the debt phobia then existing by posing the alternatives of inflation or interest payments.

Stabilization via Conventional Debt Composition (Excluding Money)

In 1954 the Committee for Economic Development published a study entitled *Managing the Federal Debt,* which clearly related debt management to economic stabilization, as the following statement shows:

> Debt management is important primarily because it affects the economic stability of the country—whether we have high employment and economic growth and price stability, or inflation or depression. The main test of debt management is whether it contributes as much as it can to stability of employment and production at a high and rising level without inflation.[9]

[8] *Ibid.*, p. 220.

[9] Committee for Economic Development, *Managing the Federal Debt* (New York: 1954), pp. 13-14.

According to the CED approach, the composition of the debt is to be varied in a generally countercyclical manner, with the Treasury operating in long-, intermediate-, and short-term debt. A great variety of debt is to be utilized, excluding money. The CED does not conceive of debt management influencing the *size* of the debt, nor the supply of money; those items are classified as "budget policy" and "monetary policy." Consistent with the current institutional arrangements, debt policy determines only the composition of the debt.

The Treasury is to sell long-term securities in boom periods in order to reduce liquidity in the economy, thus contributing to stability. Short-term highly liquid securities are to be sold during deflationary periods in order to increase the over-all liquidity of the economy. There is a minor qualification to the general policy. The CED retains some of the classical flavor in that it favors longer debt; lengthening the debt (at least in 1954), or selling long-term, gives a more "desirable debt structure."[10]

Interest costs would not be minimized, as higher-yielding long-term securities would be sold in larger quantities just at the times when these rates would tend to be highest. It is conceded that interest costs are important, "but reducing the interest cost of the debt is only a secondary objective, to be pursued insofar as it is consistent with a debt policy that conforms to the needs of economic stability."[11] The policy obviously breaks with the older views.

Debt policy is viewed as a potentially valuable supplement to conventional monetary and fiscal policies, not as a substitute for them. Ideally monetary, fiscal, and debt management policies would be coordinated. It is argued that debt policy can increase the flexibility and range of influence of fiscal and monetary policy.

It is worth stressing that the CED position is the antithesis of Simons' policy position on the structure of the debt. Unlike Simons the CED pictures the *structure* of the debt changing in a countercyclical fashion but the *size* of the debt not changing so far as debt policy itself is concerned. Simons pictures the *structure* of the debt remaining unchanged during all phases of the cycle, but the absolute *size* of the debt varying in a countercyclical manner, being simply the reflection of flexible monetary policy. Simons carefully avoids the uncertainties associated with Treasury near-moneys, whereas the CED policy proposal is based on a flexible use of near-moneys in order to influence total spending. What Simons eliminates from consideration, the CED converts into a policy instrument. Whereas Simon views short- and intermediate-

[10] The CED thinks that debt can become "too short" and states that "every opportunity to lengthen the debt without seriously affecting economic stability should be taken." This longer debt "would contribute to stability." *Ibid.*, pp. 23-24.

[11] *Ibid.*, p. 14.

term government debt as creating intolerable monetary uncertainty and economic instability, the CED views such debt as an instrument of stabilization policy.

Stabilization via Debt Size and Composition

In 1957 Earl Rolph developed another system of countercyclical debt management.[12] His system in a sense combines the Simons and CED approaches in that it involves the manipulation of the size of the debt *as well as* the composition of the debt. He states:

> Our first main proposition is that an increase in the size of the net debt of a national government, given the debt composition, has the effect of *decreasing*, and a decrease in the net debt has the effect of *increasing* GNP expenditures. It is elementary that the sale of government securities by a central bank is a deflationary policy. We simply generalize this observation to sales of government debt by any official agency.
>
> The defense of this proposition is identical with the defense of monetary policy.
>
> . . .
>
> A shift in the composition of an outstanding public debt of a given size that reduces its average maturity increases private expenditures, and vice versa for increases in its average maturity. Like any empirical generalization, this proposition does not hold for all circumstances.[13]

Debt policy is to operate in a countercyclical manner; during recession periods the size of the debt as well as its terms may be reduced. Both of these debt operations would tend to increase the over-all liquidity of the economy. With inflationary pressures the size of the debt as well as its term might be increased.[14] Debt management may be viewed as the purchase of illiquidity.

Obviously such a policy is consistent with the debt growing either larger or smaller, and either longer or shorter, depending upon secular tendencies in the economy. Unlike the CED, Rolph does not appear to

[12] Earl Rolph, "Principles of Debt Management," *American Economic Review*, Vol. 47, No. 3 (June, 1957), pp. 302-20.

[13] *Ibid.*, pp. 305-8. A very similar statement of policy is made by Richard A. Musgrave. "A given degree of restriction [stabilization] may be obtained through various combinations of public debt differing in composition and total amount. The problem, then, is to find that combination which secures the desired degree of restriction at least cost." *The Theory of Public Finance* (New York: McGraw-Hill, 1959), p. 601.

[14] As originally presented by Rolph, the scheme involved a unique solution to debt management policy and interest minimization. However, this involved a minor slip in logic and proved untrue. Multiple possibilities are present rather than one single solution. For purposes at hand this is of little significance and will not be pursued.

have any particular preference for longer-term debt. There is no single correct composition of the debt. What is correct depends on the circumstances of the economy; thus what is correct now may be mischievous at a later date.

In Rolph's system interest is minimized only in the sense that an efficient stabilization policy is pursued. For a given amount of stabilization the lowest cost combination of debt size and structure is chosen. Interest is minimized relative to the stabilization goal.

While the Rolph system of policy is a hybrid of sorts, it is worth while to point out briefly the contrast with the CED and Simons. Rolph would vary the *size* of the debt in the same manner as Simons, but he violates Simons' debt structure rule by using alleged near-moneys.

Thus three variations on the countercyclical debt management theme are found. These positions in part conflict with, and in part reinforce, one another. Yet they are only part of the debt management controversy. Two important positions remain to be discussed.

The pro-cyclical approach to stabilization

Most prominent among those taking the pro-cyclical approach to stabilization are United States Treasury spokesmen. The Treasury position is of interest for two significant reasons. First, the Treasury is actually in charge of debt management operations. Second, Treasury policy is in sharp contrast with the countercyclical approaches just discussed.

Treasury experts believe that debt management can most effectively contribute to economic stabilization by following policies that will allow those directly charged with stabilization to pursue vigorous and appropriate countercyclical programs. The debt should be managed so as to minimize interference with responsible monetary policy. Treasury spokesmen assert that the difficulties involved in debt management make the debt best considered a problem in itself, capable of generating substantial difficulties for the government and the market if not properly handled. Poor debt management operations are capable of creating insability in the economy. They do not view countercyclical debt management favorably, believing that it would accentuate the stabilization problem through a build-up of short-term debts over time and would greatly increase costs.[15]

Treasury authorities do not contemplate debt management as influencing the money supply, but regard that as the proper power of the Federal Reserve System. This view reflects the contemporary division of

[15] A reasonably good statement of Treasury policy in recent years can be found in U.S. Congress, Joint Economic Committee, *Employment, Growth, and Price Levels*, Part 6C, *The Government's Management of Its Monetary, Fiscal, and Debt Operations*, 86th Cong., 1st Sess., 1959.

existing powers and the Treasury's acceptance of this institutional arrangement.

Treasury policy directly contradicts countercyclical debt management. The Treasury "tailors the debt to the market," which in practice has meant that longer-term issues have been offered during periods of recession and shorter-term issues during prosperous times. Obviously this results in liquidity restriction during periods of recession and liquidity ease during more prosperous times. This considered by itself is obviously pro-cyclical and antagonistic to the previously discussed positions. However, spokesmen argue that this policy is the most practical policy to follow in stabilizing the economy, because this allows the Treasury to maintain a longer debt and thus lessen the danger of debt management interfering with monetary policy. Absence of the Treasury from the market as much as possible gives the Federal Reserve more latitude in executing its policies. Treasury policy contributes "to the amount of free time which the Federal Reserve has to take effective monetary action without always having to be concerned with a new issue of securities which is still in the process of being lodged with the eventual holders of the securities."[16] The Treasury's continued presence in the market might bias the Federal Reserve toward an easier monetary policy than it would otherwise follow.

The policy of tailoring new securities to the particular needs of the market enables the Treasury to secure necessary funds at lower cost than would otherwise be possible. The Treasury has a practical political interest in lowering the interest burden of the debt, and officials believe that countercyclical debt management would greatly increase interest costs. "Economical borrowing is an important goal of Treasury debt management."[17] This is not doctrinaire interest minimization. "The goal of holding down interest cost on the public debt, although important, does not take precedence over other major goals of debt management."[18] On the other hand, Treasury authorities assert that such policies as the CED recommends would significantly increase interest costs.

Thus, lengthening the debt and minimizing interest costs are both important goals of Treasury policy. These goals are reconciled in the Treasury's tailoring procedures. This tailoring is, in principle, destabilizing in its impact on the economy. Yet officials insist that this policy aids in stabilizing the economy because it lessens the danger of debt management disrupting monetary policy. They conclude that Treasury policy

[16] Remarks by Secretary of the Treasury Anderson, April 7, 1958, at the opening of the "Share in America" savings bonds campaign, New York City, New York. Reprinted in the *Annual Report of the Secretary of the Treasury for the Fiscal Year Ended June 30, 1958* (Washington: U.S. Government Printing Office, 1959), p. 263.

[17] U.S. Congress, Joint Economic Committee, *Employment, Growth, and Price Levels,* Part 6C, p. 1723.

[18] *Ibid.,* pp. 1723-24.

is in practice a program for economic stability and that the opposite policies in practice would be destabilizing, costly, and impractical.

It is apparent that on the policy level the Treasury spokesmen are in direct conflict with the proponents of countercyclical debt management, although both relate debt management to the cycle and to problems of economic stabilization. The policies border on being black and white contrasts, yet their proponents declare them both practical and stabilizing.

The Treasury has retreated somewhat from the simpler precepts of classical debt management. Policies have changed, but much of the old lies beneath the surface of the new. Older precepts did not relate debt operations to the cycle, and they laid greater stress on funding than does present Treasury policy. On the other hand, the current policy definitely retains a bias in favor of longer-term debt. Treasury policy has not departed entirely from the older views, but there has been a change in emphasis and mood. Some believe expediency has come to play a larger role.

Warren Smith develops a debt policy along unmistakably pro-cyclical lines,[19] thus, in effect, defending the essential aspects of Treasury policy: Long-term securities would be sold during recession periods to take maximum advantage of low interest rates. As interest rates rose during periods of expansion, the Treasury would gradually shift to the short-term market. The advantage is to minimize the interference of the Treasury's debt management operations with freedom of action by the Federal Reserve during periods of inflation and/or tight credit. Debt lengthening during recessions would reduce the frequency of the Treasury's presence in the market. Smith states that this approach would require a concomitant flexible monetary policy. If debt managers overshoot the mark in raising long-term funds during a recession, with the result that recovery is impeded, the Federal Reserve should be prepared for offsetting action.

Smith attacks countercyclical debt policy as "mystical." He argues that shifts in debt length do not significantly affect the liquidity of the debt and that "neither the interest rate nor the liquidity effects of marginal changes in the debt structure appear to be very important."[20] Treasury spokesmen often imply that such shifts are important. Smith opposes using liquidity shifts for stabilization and states that "to the

[19] Warren Smith, *Debt Management in the United States,* Study Paper No. 19 for U.S. Congress, Joint Economic Committee, Materials prepared in connection with the Study of Employment, Growth, and Price Levels, 86th Cong., 2nd Sess., 1960. Herbert Stein also defends pro-cyclical policy. In doing so he clearly distinguishes between debt length and temporal structure and calls for careful control of the amount of debt coming due in any year. "Managing the Federal Debt," *Journal of Law and Economics,* Vol. 1 (October, 1958).

[20] Smith, *op. cit.,* p. 8.

extent that such changes do have a net effect on the public's aggregate spending, it would appear that similar effects could be produced by the use of monetary policy. For this reason, it is difficult to see what can be accomplished by contracyclical debt management policy that cannot be accomplished more efficiently by Federal Reserve monetary policy."[21] Further, such a policy would tend to maximize interest costs.

Smith explicitly recognizes the existing debt as an automatic stabilizer, whereas this is largely implicit in the Treasury position. He also clearly states that the benefits of maintaining the debt (rather than inflating it away) must be weighed against the cost of the debt. Treasury officials have said little about this.

Thus the Treasury is not alone in defending pro-cyclical debt management. However, the Treasury's theorists and academic supporters appear subject to the criticism that there is no questioning of the Federal Reserve's ability (or willingness) to adapt monetary policy quickly and accurately to unmeasurable liquidity shifts within the debt. This is of particular interest in light of the apparently divergent assumptions regarding the importance of such shifts.

The advocates of pro-cyclical management in the Treasury and their academic supporters constitute another major division of the debt management controversy. They disagree with the proponents of countercyclical debt management, and the final faction in the controversy disagrees with both groups.

The third and most recent major group taking part in the debt management controversy sets forth a debt policy which is neither countercyclical nor pro-cyclical, but rather aims at "neutrality" and tends to stress the simplification of debt operations. This position divorces debt management policy from the cycle.

The neutrality doctrine

Milton Friedman, representative of the neutrality position, in 1959 published *A Program for Monetary Stability*. He set forth the view that our main need regarding debt management is "to simplify and streamline, in such a manner as to keep debt operations from themselves being a source of instability, and to ease the task of coordinating Treasury debt operations and [Federal] Reserve open market operations."[22] Debt operations should be "regular in timing, reasonably stable in amount, and predictable in form."[23] Friedman would reduce the variety of debt instruments, retaining the tap issues (savings bonds) plus two standard debt forms, a short-

[21] *Ibid.*

[22] Milton Friedman, *A Program for Monetary Stability* (New York: Fordham University Press, 1959), p. 60.

[23] *Ibid.*, p. 65.

term (possibly 90-day) bill and a moderately long-term security (8 to 10 years). Both of these securities would be sold at regular and frequent intervals, and amounts would be kept reasonably stable as a policy goal.

Friedman is critical of the Treasury's tailoring, which he states has resulted in

> a bewildering maze of securities of different maturities and terms, and lumpiness and discontinuity in debt operations, with refunding of major magnitude occurring on a few dates in the year. Instead of proceeding at a regular pace and in a standard way to which the market could adjust, debt management operations have been jerky, full of expedients and surprises, and unpredictable in their impact and outcome. As a result they have been a continuing source of monetary uncertainty and instability.[24]

Tailoring is also criticized on the grounds that in reality it does not lower costs and that it implies that the government is more efficient than the market in the conduct of a particular class of financial operations. Friedman rejects this notion.

He also points out that interest minimization is more complex than the Treasury assumes because "it is necessary to take into account not only interest-bearing debt but also non-interest-bearing debt—Treasury currency and Federal Reserve notes and Federal Reserve deposits."[25] If the debt is made longer and thereby less liquid, it can be reduced in size without inflationary pressures appearing.

> It is difficult if not impossible in the present state of knowledge to predict whether one or another pattern of securities will involve or did involve lower costs, correctly interpreted; hence there is no real basis for judging or improving performance.[26]

Friedman takes the position that minor changes in the length of the debt have only slight influence on the demand for money; hence, changes in the length of the debt are not viewed as a promising tool of stabilization policy. He points out that "shifts in maturity add nothing to open market operations." Furthermore, open market operations seem "likely to be more consistent and predictable in . . . impact."[27] Thus debt management operations, in the strict sense, are not technically good instruments for economic stabilization activity. Countercyclical debt management is rejected.

Tilford C. Gaines develops a position similar in some respects to Friedman's, stating that neutrality should be the object of policy and that Treasury operations should be put on a more orderly basis. Debt operations should be on a routine basis, and the amount of securities

[24] *Ibid.*, p. 60.
[25] *Ibid.*, p. 62.
[26] *Ibid.*
[27] *Ibid.*, p. 61.

offered at any time carefully controlled relative to the market's absorptive capacity. This policy would tend to stabilize the liquidity of the debt and introduce greater certainty into the market. Gaines argues that lengthening the debt "whenever possible" (tailoring) creates a great deal of uncertainty about Treasury operations and tends to disturb the market. Tailoring is a "massive source of instability in the capital market,"[28] and has probably involved added costs because of higher interest rates associated with uncertainty. Further, it tends to make debt operations work counter to monetary policy. Countercyclical debt policy is also rejected because of the inadequate present state of technical knowledge. It is not deemed feasible.

This analysis divorces debt management from the cycle and advocates leaving stabilization to monetary policy, which is judged technically better and more appropriate for that purpose. It argues that debt policy should be based upon simplicity, regularity, and predictability. Both pro-cyclical and countercyclical policies are rejected as inappropriate. The neutrality approach conflicts with the other positions and differs from the classical precepts which stressed funding and interest minimization. It is reminiscent of the classical attitude, however, as debt management is not related to the cycle, and it has a tone of simplicity and certainty. Still, neutrality remains a distinct policy differing from the older and conflicting with the newer views.

Concluding Remarks

Before the Great Depression and World War II there was virtual unanimity regarding the elements of "sound" debt policy, and there could be observed a striking similarity of opinion among academic economists and between the economists and the Treasury. Events since that time have shattered this picture, and there is now much disagreement over what constitutes sound policy.

At the present time three general positions are recognizable, and there is some divergence of thought within each of the basic categories. These can be termed *countercyclical* (Simons, the CED, and Rolph), *pro-cyclical* (the Treasury and Smith), and *neutral* (Friedman and Gaines). All of these positions diverge in some degree from, or conflict with, the older views.

The newer positions are clearly in basic conflict with one another and tend to rest on contradictory assumptions regarding the nature of the stabilization problem and the technical impact of debt operations. They reflect basic disagreement about (1) the importance of interest

[28] T. C. Gaines, *Techniques of Treasury Debt Management* (New York: The Free Press of Glencoe, 1962), p. 266. See especially Chapter 8.

minimization, (2) the best way to secure lowest cost on debt operations, (3) the relevant sense in which interest is to be minimized, (4) what constitutes a desirable or sound temporal structure for the debt, (5) how frequently debt operations should be carried on, (6) how long the debt should be, and (7) the technical impact of changes in debt length.

There is even more disagreement than indicated here, because the topic of institutional reorganization has not been considered. There is disagreement regarding the necessity and/or desirability of changing our institutional framework, and of course a number of plans are suggested. Hence, there is even disagreement as to *who* should manage the debt.

It is obvious that the debt management controversy has many sides, and there seems to be some dispute regarding every major aspect of debt management at the present time. This controversy is one facet of the more general stabilization debate that arose in the 1930's and the 1940's. That debate continues today in a somewhat abated form, although the area of disagreement seems to have narrowed, and emphasis has again drifted toward classical and neoclassical interpretations. The debt controversy may yet follow the same pattern and work toward a more sophisticated reinterpretation of older doctrines. If that is the case, it is likely to mean the eventual triumph of the neutrality school of debt management. For that pattern of thought stressing simplicity and predictability and divorcing debt management from the cycle has much in common with earlier views, though it is more clearly and explicitly formulated and differs somewhat in emphasis. Its eventual impact may well be to bring the old views up to date on such topics as debt length, interest minimization, and debt structure.

27

Debt management: Its relationship to monetary policy, 1951-1962

Thomas R. Beard

Since the famous Treasury-Federal Reserve Accord, the relationship between debt management and monetary policy has undergone certain basic changes.[1] In general, three major periods may be distin-

Reprinted with permission from *The National Banking Review*, Vol. II (September, 1964), pp. 61-76.

Thomas R. Beard is an Assistant Professor of Economics at Louisiana State University and is currently on leave with the Board of Governors of the Federal Reserve System.

The author wishes to acknowledge the earlier financial assistance of the Ford Foundation and the more recent financial assistance of the L. S. U. Foundation during the preparation of this paper. Miss Janice Krummack and Mr. Normand Bernard offered several helpful suggestions. Of course, the author alone is responsible for the views expressed.

[1] For the purposes of this paper, debt management is distinguished as clearly as possible from monetary and fiscal policy. All actions of the Federal government which directly affect the composition and terms of the publicly held Federal debt (debt held outside the Federal Reserve and U. S. Government trust funds and agency accounts), whether initiated by the Treasury or by the Federal Reserve, are included within this definition. However, debt management does not involve the size of the debt (which is a matter of fiscal policy) or the net amount of debt bought and sold by the central bank in the conduct of its open market operations (which is a matter of monetary policy). Our definition of debt management does not include so-called non-interest bearing "debt," either in the form of currency held outside the Federal government or non-Federal government deposits at the Federal Reserve banks. Rather, these latter items are related to monetary policy.

A similar definition of debt management has frequently been utilized. See, for example, Warren L. Smith, *Debt Management in the United States*, Study Paper No. 19 for the Joint Economic Committee, Washington, D. C.: Government Printing Office, 1960. By way of contrast, a broader definition which includes non-interest bearing "debt" has recently been suggested by James Tobin, "An Essay on Principles of Debt Management," *Fiscal and Debt Management Policies*, Englewood Cliffs, N. J.: Prentice-Hall, 1963, pp. 143-218.

guished: (1) the 1951-1952 period of the "transition to free markets"; (2) the 1953-1960 era of "free markets"; and (3) the most recent period of "twisting" or "nudging." In view of the continuing controversy over the proper role of debt management,[2] it is instructive to review the debt management policies followed and their relationship to monetary policies during each of these three periods. Some tentative conclusions can be drawn concerning the probable impact of debt management in each period.

I. The Background

Prior to March 1951, considerations of debt management effectively tied the hands of the monetary authorities. At the end of World War II, it was widely feared that the large war debt might not be held willingly by investors and that debt markets were highly susceptible to destabilizing speculative fluctuations. In some quarters it was feared that instability of security prices would destroy public confidence in the government's credit and undermine investment incentives. These fears at first led to the view that even moderately small increases in interest rates could not be permitted. With the passage of time, the views of Federal Reserve spokesmen changed. While they felt unpegging could take place only gradually, they wished to move steadily in the direction of greater reliance on flexible monetary policy. Treasury officials, however, fought a delaying action.[3]

Continuously pegged long-run interest rates persisted well into the post-war years—long after any legitimate need for rigid pegs had seemingly disappeared. While central bank officials may have been guilty of overemphasizing the extent of debt monetization that took place,[4] effective central bank control over the money supply was greatly hampered.[5]

[2] For a brief summary of alternative views on debt management, see William E. Laird, "The Changing Views on Debt Management," *Quarterly Review of Economics and Business*, III, Autumn, 1963, pp. 7-17.

[3] An excellent account of the evolving positions of Treasury and Federal Reserve officials during the war and early post-war years is found in Henry C. Murphy, *The National Debt in War and Transition*, New York: McGraw-Hill, 1950.

[4] For comments along this line, see Alvin Hansen, "Monetary Policy," *Review of Economics and Statistics*, XXXVII, May 1955, especially pp. 116-117. Kareken has argued that the large scale investor selling of certain maturities which at times did take place—particularly from June 1950, to March 1951—cannot be disassociated from the Treasury-Federal Reserve struggle and the effect of this conflict on market expectations. He has suggested that the actions of the monetary authorities resulted in a more rapid debt monetization than would otherwise have taken place. John H. Kareken, "Monetary Policy and the Public Debt: An Appraisal of Post-War Developments in the U. S. A.," *Kyklos*, X, Fasc. 4, 1957, especially pp. 418-423.

[5] Numerous critical views on pre-Accord monetary-debt management policies can be found in United States Congress, Subcommittee on Monetary, Credit, and Fiscal Policies of the Joint Committee on the Economic Report, *Monetary, Credit, and*

In effect, monetary policy was largely relegated to the task of insuring the liquidity of the public debt. The stabilization of government security prices rendered all maturities virtually as liquid as the shortest-term instruments. At the same time, the initiative in changing bank reserves was transferred to holders of government securities.

Finally, in the Accord of March 1951, the advocates of flexible monetary policy were triumphant. This victory, however, was only one step in the gradual process of regaining central bank "independence."

II. The Transition to Free Markets, 1951-1952

The first major period following the Accord lasted for roughly two years. During most of this period, the predominant atmosphere was one of uncertainty. The Federal Reserve as well as the Treasury wished to avoid any actions that might precipitate disorderly conditions in the securities market. At the same time, it was widely felt that inflation, rather than depression, was the important danger.

A reconciliation of these conflicting claims on central bank policy was facilitated by a new theory of credit control. In short, the theory had been developed within the Federal Reserve that central bank policy could be an effective stabilizer even though fluctuations in interest rates were moderate.[6] The impact of Federal Reserve policy was said to be transmitted primarily through direct effects on lenders and the "availability of credit," rather than through variations in interest rates or in the money supply. Large-scale central bank operations were unnecessary in an environment characterized by a large and widely held public debt spread over a wide range of maturities. Such a debt facilitated the rapid transmission of credit policies between the various segments of the market. Other institutional developments, such as the growth of nonbank financial institutions characterized by conservative portfolio management, were also seen as giving open market operations a significant impact with relatively small changes in interest rates.[7]

Fiscal Policies, Hearings, Statements, and Report, 81st Congress, 1st Session, 1949-1950, and United States Congress, Subcommittee on General Credit Control and Debt Management of the Joint Committee on the Economic Report, *Monetary Policy and the Management of the Public Debt, Their Role in Achieving Price Stability and High Level Employment, Replies, Hearings, and Report*, 82nd Congress, 2nd Session, 1952. These volumes are referred to, respectively, as the Douglas and Patman Documents.

[6] See, in particular, R. V. Roosa, "Interest Rates and the Central Bank," *Money, Trade, and Economic Growth*, New York: The Macmillan Co., 1951, pp. 270-295, and portions of the testimony of Chairman Martin in *Patman Replies*, pp. 207-632.

[7] For an analytical discussion of this theory, including the so-called "locking-in" effect, "value-of-portfolio" effect, credit rationing" effect, etc., see Assar Lindbeck, *The "New" Theory of Credit Control in the United States*, Stockholm: Almqvist and Wiksell, 1959. See, also, Warren L. Smith, "On the Effectiveness of Monetary Policy," *American Economic Review*, XLVI, September 1956, pp. 588-606, and John H. Kare-

As part of the transition to free markets, the Federal Reserve continued until the last refunding of 1952, its previous policy of lending direct support to Treasury refunding operations. This support took the form of central bank purchases of "rights" to new issues, new issues on a "when issued" basis, and outstanding securities in the market comparable to the new issues. Despite the availability of central bank underwriting, there was little enthusiasm for debt lengthening by the Treasury. Treasury spokesmen contended that "no significant volume" of long-term funds was available for the purchase of government bonds. In addition, they continued to emphasize the importance of maintaining "confidence in the Government's credit" and a "sound" securities market—both objectives being correlated with relatively stable security prices.[8] This emphasis worked to discourage all but a few minor attempts at financing outside the short-term (under 1-year) area. Consequently, the maturity structure of the debt continued its short downward drift, as had been the case continuously since the end of the Second World War.

Even after monetary policy tightened in the spring and summer of 1952, attrition (cash-ins of maturing securities) never became especially serious, owing primarily to the rather sizable support purchases by the central bank. Although the debt managers' problems were clearly eased by central bank underwriting, it appears that the monetary authorities were anxious to abandon such support as quickly as feasible. As it was later argued in a subcommittee report on the functioning of the government securities market, a firm commitment to support refundings "seriously hampers freedom of action in effectuating general credit policy." [9] The most persuasive arguments seemed to be that the ability of the Federal Reserve to contain credit expansion is limited by the neces-

---

ken, "Lenders' Preferences, Credit Rationing, and the Effectiveness of Monetary Policy," *Review of Economics and Statistics,* XXXIX, August 1957, pp. 292-302.

It is evident that central bank reliance on the "availability doctrine" in its original form has diminished with the passage of time. For a documentation of this change in emphasis as reflected in one significant publication of the Board of Governors, see Lawrence S. Ritter, "Official Central Banking Theory in the United States, 1939-61: Four Editions of *The Federal Reserve System: Purposes and Functions,*" *Journal of Political Economy,* LXX, February 1962, pp. 14-29. Ritter notes a number of changes with respect to the theory of monetary control from the 1954 to the 1961 editions. Among these changes were the following: the large and widely distributed Federal debt and highly developed financial system were no longer viewed as "unmixed blessings" since they facilitate mobilization of idle balances and credit creation; the lending behavior of non-bank financial institutions was no longer viewed as "highly susceptible" to System control; the "locking-in" effect had been downgraded, etc.

[8] See portions of the report submitted by Secretary Snyder in *Patman Replies,* pp. 1-206, and the *Annual Report of the Secretary of the Treasury, 1952,* pp. 1-13.

[9] Report of the three member subcommittee appointed to investigate the impact of open market techniques on the functioning of the government securities market in: United States Congress, Subcommittee on Economic Stabilization of the Joint Committee on the Economic Report, *United States Monetary Policy: Recent Thinking and Experience, Hearings,* 83rd Congress, 2nd Session, 1954, p. 303. This document is called the *Flanders Hearings.*

sity of making frequent and often substantial support purchases and that support purchases at least temporarily establish a "pegged" market.[10]

The importance of these arguments has been questioned, however.[11] While free reserves often registered sizable increases (or net borrowed reserves registered decreases) for a period of several weeks following large support operations, the seriousness of this situation depended both upon the use made of the additional reserves and the speed with which these reserves were utilized. In general, in any period of refunding support, if banks rapidly utilize their new reserves to make loans rather than to reduce their indebtedness to the central bank, it is probably more difficult for the monetary authorities to offset the initial effect of the support purchases. When banks make new loans before the System has had time to reabsorb the reserves, conditions in the money market, if measured by free reserves,[12] may appear as tight as the conditions existing before the support operation was undertaken. But loan expansion has occurred. On the other hand, commercial banks may consider reserves created in support operations as extremely temporary —especially if the System could make it known that reserves created under such circumstances do not reflect a permanent easing of monetary policy. The "traditional" view of discounting, i.e., that banks are reluctant to remain in debt to the central bank, suggests that the bulk of the newly created reserves would be used to reduce bank indebtedness. In

[10] The full text of the subcommittee report is included in the *Flanders Hearings*, pp. 257-307. The subcommittee also argued that support purchases encourage unrealistic (low) rates on new Treasury issues and result in a "frozen" central bank portfolio heavily weighed with the securities acquired in the support operations.

Neither argument appears persuasive. It is possible that a general commitment to support all refunding operations regardless of the terms set by the Treasury could force massive purchases on the central bank. However, there is no reason why the Federal Reserve should agree to support any operation in which the Treasury insisted on terms that were unrealistically low in the System's view. In such cases—which would undoubtedly become infrequent—the Treasury would have to assume full responsibility. Similarly, the "frozen" portfolio argument is less than completely convincing. The System is not forced to maintain any given security in its portfolio over time. Securities purchased in support operations can later be sold, though perhaps gradually, with the potential effect on bank reserves being offset by purchases of Treasury bills if necessary. This argument reduces to the more basic desire to minimize market intervention, whatever the purpose. For a similar emphasis, see Deane Carson, "Federal Reserve Support of Treasury Refunding Operations," *Journal of Finance*, XII, March 1957, pp. 51-63.

[11] See, in particular, Carson, *op. cit.* Hansen argued that the subsequent decision to abandon support operations amounted to "a partial abdication of the Reserve System with respect to one of its functions. Private flotations have the benefit of underwriting support, and one should imagine that a central bank would have as one of its primary functions the obligation to act as the underwriter for Treasury flotations." Hansen, *op. cit.*, p. 111.

[12] It is, of course, difficult to estimate the importance of the free reserves concept in the formulation and interpretation of central bank policy. It is evident, however, from various pronouncements that the concept does carry substantial weight. See, for example, "The Significance and Limitations of Free Reserves," *Federal Reserve Bank of New York Monthly Review*, XL, November 1958, pp. 162-167.

this case, figures for free reserves reflect the easing conditions brought about by support operations, and therefore the easing conditions are more readily offset.

Furthermore, one should distinguish sharply between so-called "temporary pegs" during a period of refunding support, and a policy of rigid, continuous support of security prices.[13] With temporary pegs, interest rates are free to move between major Treasury operations. It is essential, however, that the Federal Reserve does not purchase securities over a period much longer than the subscription period—as might be done if the purpose were to guarantee purchasers that they could not better their return for an appreciable period of time. If the period of market stabilization is short and support purchases are limited, the bulk of the funds released could be reabsorbed quickly. During the transition period, the Open Market Committee was able to accomplish a certain volume of offsetting sales of other securities concurrently with support purchases, and to mop up substantial reserves between refundings.[14] Apparently, however, the monetary authorities were not satisfied with the net result. It is problematical to what degree their dissatisfaction was attributable to an aversion to interference with the "free market."[15]

III. The Era of Free Markets, 1953-1960

The second major period of debt management corresponds roughly to the eight years of the Eisenhower administration. With the inauguration of "bills only" in March 1953, the Federal Reserve completed its move toward "central bank independence" and "free markets"—a policy which, as one critic put it, came to represent the "fetish of laissez-faire."[16] In addition to the elimination of Treasury refunding support, swapping operations were specifically disallowed. Open market operations were to be confined, except in the correction of disorderly markets, to short-term securities, preferably bills.[17] "Bills only" was felt to be consistent with the objectives of "neutrality" and "minimum intervention." By con-

[13] For an excellent discussion of various approaches to "pegs," see R. V. Roosa, "Integrating Debt Management and Open Market Operations," *American Economic Review, Papers and Proceedings*, XLII, May 1952, especially pp. 218-223.

[14] A tabulation of Federal Reserve transactions in Treasury securities both during periods of refundings and between refunding periods is found in the *Flanders Hearings*, p. 265.

[15] Carson argued that while more market intervention was required and the timing of credit policy was occasionally disrupted because of the support policy, the central bank was able to maintain a "reasonably effective" degree of restraint. By "traditional standards," monetary policy was successful. Carson, *op. cit.*

[16] Ervin Miller, "Monetary Policies in the United States Since 1950: Some Implications of the Retreat to Orthodoxy," *Canadian Journal of Economics and Political Science*, XXVII, May 1961, p. 218.

[17] See the record of operating policy statements in *Annual Report of the Board of Governors of the Federal Reserve System, 1953*, pp. 86-105. Until its discontinuance in 1961, the "bills only" policy was reaffirmed annually by the Open Market Committee.

fining its operations under normal conditions to Treasury bills, the central bank would keep the impact of its activities as broad and impersonal as possible—neither influencing prices of particular securities directly, nor attempting to maintain any particular rate structure.[18]

At roughly the same time, a distinct change in the philosophy and objectives of Treasury debt management was evidenced as the new Administration assumed office. The necessity of lengthening the debt structure was emphasized repeatedly.[19] This objective was primarily (although not entirely) placed in the context of anti-inflationary policy. Excessive short-term maturities were considered disadvantageous primarily for three reasons: (1) frequent trips to the market by the Treasury would restrict the freedom of the central bank to pursue credit restraint;[20] (2)

[18] In addition to its rather specific objections to support of Treasury refunding operations, the subcommittee's major concern was that all central bank operations outside the shortest end of the market (except those necessary in the correction of disorderly conditions) interfered unduly with the operational excellence of the market, i.e., its "depth, breadth, and resiliency." Perhaps the two best defenses of "bills only" have been given by Winfield W. Riefler, "Open Market Operations in Long-Term Securities," *Federal Reserve Bulletin,* XLIV, November 1958, pp. 1260-1274, and Ralph A. Young and Charles A. Yager, "The Economics of 'Bills Preferably,'" *Quarterly Journal of Economics,* LXXIV, August 1960, pp. 341-373. Young and Yager carefully developed the argument that such a policy contributes to a well-performing securities market. Riefler emphasized the argument that the initial effects of monetary policy on short-term interest rates are rapidly transmitted to the long-term market by a process of arbitrage. It makes little difference for the structure of interest rates whether operations are conducted in Treasury bills or directly in long-term securities.

Criticism of the "bills only" policy and of its theoretical and philosophical underpinnings has apparently constituted one of the favorite pursuits of academic economists. For a brief sample of a vast critical literature, see Joseph Aschheim, *Techniques of Monetary Control,* Baltimore: Johns Hopkins Press, 1961, chapter 4; Deane Carson, "Recent Open Market Committee Policy and Technique," *Quarterly Journal of Economics,* LXIX, August 1955, pp. 321-342; Dudley Luckett, "'Bills Only': A Critical Appraisal," *Review of Economics and Statistics,* XLII, August 1960, pp. 301-306; Warren L. Smith, *Debt Management in the United States,* especially pp. 118-134; and Sidney Weintraub, "The Theory of Open Market Operations: A Comment," *Review of Economics and Statistics,* XLI, August 1959, pp. 308-312. In his survey of monetary theory and policy, Harry Johnson suggested, "The bills only policy . . . appeared to most academic economists as an undesirable renunciation by the central bank of an important technique of monetary control, and the reason given for it . . . as a shallow excuse masking the unwillingness of the Federal Reserve to risk unpopularity with the financial community by overtly subjecting it to capital losses." Harry G. Johnson, "Monetary Theory and Policy," *American Economic Review,* LII, June 1962, p. 374.

[19] For example, see the addresses and statements made during 1953 by Secretary Humphrey and Deputy to the Secretary Burgess that are published in the *Annual Report of the Secretary of the Treasury, 1953,* especially pp. 239-242, 255, and 260-266. The debt lengthening objective was frequently reiterated throughout the entire 1953-1960 period, even during those times when few long-term bonds were actually issued.

[20] Even if the central bank does not underwrite the Treasury, it must inevitably adjust somewhat to the Treasury's financing requirements. The Federal Reserve follows a policy of attempting to maintain an "even keel" in the market during periods of Treasury finance. This policy precludes any actions that might alter basic market relationships or market expectations, and thus is felt to prevent the central bank from making independent policy decisions during periods of large Treasury operations.

shorter-term issues were more likely to be purchased by commercial banks, thus resulting in an inflationary expansion in the money supply;[21] and (3) short-term securities, even when purchased by non-banks, would provide close money substitutes, thus adding to inflationary pressures.

In this new atmosphere, debt management assumed a secondary role. Debt management operations were to be conducted so as to contribute to the effectiveness of monetary policy—thus, in effect, largely reversing their pre-Accord roles. At first, it appeared that the public debt might be managed counter-cyclically as a supplement to counter-cyclical monetary policy. A counter-cyclical debt management policy would involve lengthening the debt structure during periods of monetary restraint and placing only minimum reliance on bond issues during periods of recession. Therefore, debt management would alternately reduce liquidity when monetary policy was tight and expand liquidity when monetary policy was easy. Such a policy was not forthcoming, however.

Initial attempts at debt lengthening in the first half of 1953 encountered serious difficulties.[22] After monetary policy shifted in mid-year to combat the ensuing recession, the Treasury judiciously refrained from selling long-term (over 10-year) bonds so as to avoid competition with private demands for long-term funds. However, the policy of monetary ease created a highly receptive market for government securities outside the short-term area. The Treasury's newly adopted program of optional exchange offerings[23] allowed the public, especially commercial banks, to

[21] The argument that sales of securities to commercial banks are inflationary because of the resulting money creation is clearly invalid. Banks which have no excess reserves can purchase government securities only by disposing of other investments or by reducing outstanding loans. If banks have excess reserves, they are either not in equilibrium or they desire to hold excess reserves for liquidity purposes. Banks in disequilibrium would purchase other assets even if they did not buy the government securities, while banks that hold excess reserves for liquidity purposes would buy the government securities only if they could sell other assets. Thus, in no case are sales to commercial banks inflationary in the sense of money creation. Furthermore, when the proceeds of security sales to banks are added to Treasury balances at the Federal Reserve, bank reserves are decreased.

The Treasury's fear of debt monetization leads to a confusion about the respective roles of the Treasury and Federal Reserve. Changes in total commercial bank assets, which include government securities, are more closely regulated by central bank policy than by Treasury debt management. On the question of debt monetization, see Milton Friedman, *A Program for Monetary Stability*, New York: Fordham University Press, 1960, especially pp. 52-55, and Tilford C. Gaines, *Techniques of Treasury Debt Management*, New York: The Free Press of Glencoe, 1962, pp. 249-251.

[22] In April 1953, the Treasury announced a cash issue of 3¼ percent bonds of 1978-83. Approximately $1.1 billion of these long-term securities were sold to the public. However, the government securities market became somewhat disorganized after this issue. The Treasury was criticized in certain quarters for its decision to press for debt lengthening at that time and for the "excessive" rate of interest paid.

[23] In optional exchange offerings, holders of maturing issues are given an option between at least two new issues of varying maturities. While this technique reduces cash-ins of maturing issues, it is subject to the disadvantage of partially transferring control over the maturity structure of the debt to the public.

purchase substantial quantities of intermediate-term (5-10 year) bonds. On balance, debt management was at least mildly pro-cyclical. From the end of June 1953, to December 1954, publicly held marketable Treasury debt in the 5-10 year category rose from $16.9 billion to $32.2 billion, while that under 1 year fell by $5.6 billion.[24]

With the passage of time, Treasury debt management became progressively more, rather than less, pro-cyclical. In mid-1955, the Federal Reserve moved toward a more restrictive policy and, in general, maintained restraint on commercial bank reserves until the last quarter of 1957. However, from March 1955, to September 1957, the Treasury made only a single offering of bonds. Consequently, public holdings of Treasury issues with less than 1-year to maturity rose by $16.0 billion; 1-5 year issues rose by $11.7 billion; and 5-10 year issues declined by $18.5 billion.[25] In view of its long-standing objective to lengthen the debt structure, the Treasury's failure to sell more long-term bonds during these years was undoubtedly an important factor leading to the excessive reliance on bond financing after the recession began in late 1957. In an 8-month period—from November 1957, through June 1958—there were 6 new issues of bonds. Some $3.2 billion of over 15-year bonds and $12.2 billion of intermediate-term bonds were sold to the public in cash and refunding operations. This vigorous debt lengthening program eventually resulted in the speculative market collapse following the issue of 2⅝ percent bonds in June.[26] Both the Treasury and Federal Reserve intervened on a limited scale to stabilize the market. However, by the end of the summer, at a very early stage of the recovery, long-term bond yields were back at approximately the same levels that had prevailed at the previous cyclical peak. The extent to which these events contributed to the un-

[24] These and subsequent figures on the maturity structure of the debt are taken from various issues of the *Federal Reserve Bulletin*. Admittedly, these figures are only broadly indicative of the liquidity impact of debt management and must be used with discretion. The use of maturity categories for liquidity analysis is subject to notorious shortcomings; i.e., all securities within a given category are not equivalent with respect to liquidity and there is likely to be a greater substitutability between certain issues in different categories than between certain issues in the same category.

[25] Treasury spokesmen stressed the difficulty of selling intermediate- and long-term bonds in the face of restrictive monetary policy and prevailing expectations of rising interest rates. Expectational forces are said to be so strong during such times that buyers cannot be found. Attempts to "force" longer-term securities on unwilling investors could result in chaotic markets. In addition to the practical difficulties of issuing longer-term securities, the higher interest costs involved may have also acted as a deterrent to debt lengthening. For an emphasis on the practical difficulties of selling government bonds, see the remarks by Under Secretary of the Treasury Baird published in *Annual Report of the Secretary of the Treasury, 1958*, pp. 275-279; the testimony of Secretary Anderson in United States Congress, Joint Economic Committee, *Employment, Growth, and Price Levels, Hearings*, Part 6C, 86th Congress, 1st Session, 1959, especialy pp. 1721-1729; and Secretary Anderson, "Financial Policies for Sustainable Growth," *Journal of Finance*, XV, May 1960, especially pp. 133-139.

[26] For a detailed description of this episode, see the *Treasury-Federal Reserve Study of the Government Securities Market*, Parts I-III, Washington, D.C.: Government Printing Office, 1959-1960.

satisfactory character of the 1958-1960 economic expansion is problematical. After the upturn in economic activity, monetary policy changed —too quickly according to many critics—[27] to "fostering . . . balanced economic recovery" (August 1958) and finally to "restraining inflationary credit expansion . . ." (May 1959).[28] During this economic expansion, the Treasury, for reasons somewhat beyond its control, again concentrated on short-term financing.[29]

During a large part of the 1953-1960 period, for whatever reasons, debt management operations exerted a pro-cyclical liquidity impact. Thus, whatever contribution the Treasury made with respect to the effectiveness of monetary policy must have taken the form of minimizing interference with central bank operations. Partly for this reason, Treasury officials continued to emphasize the importance of "tailoring" issues to the needs of the market and, beginning in 1958 and 1959, extended the auction technique to 6-month and 1-year bills and instituted non-par pricing of new issues. Following the withdrawal of direct underwriting support by the central bank, however, average attrition rates rose during successive phases of rising interest rates.[30] Based on this evidence, it has been suggested that in appraising the effects of attrition, the Treasury widened its limits of tolerance in order to avoid direct assistance from the central bank.[31]

In retrospect, the Federal Reserve's "bills only" policy contributed to pro-cyclical debt management in two respects; yet, ironically, it is likely that the effectiveness of monetary policy was hampered at times by the pro-cyclical liquidity impact of debt management operations.[32]

[27] See, for example, Warren L. Smith, "Monetary Policy, 1957-1960: An Appraisal," *Review of Economics and Statistics,* XLII, August 1960, pp. 269-272, and Sidney Weintraub, "Monetary Policy, 1957-59: Too Tight, Too Often," *Review of Economics and Statistics,* XLII, August 1960, pp. 279-282.

[28] *Annual Report of the Board of Governors of the Federal Reserve System, 1958,* p. 59; *Annual Report, 1959,* p. 44. In December 1958, the wording of the Open Market Committee's directive had been changed to indicate a greater degree of restraint than that indicated in August 1958.

[29] In particular, the Treasury's hands were tied for a significant part of this period by the legal interest rate ceiling of 4¼ percent on new issues of bonds. In addition, cash requirements were exceptionally large because of the large cash deficits and the net redemption of non-marketable Treasury issues.

[30] As one study found, average attrition rates were higher from December 1954, to August 1957, than from October 1952, to June 1953, and highest of all from August 1958, to August 1959 (the cut-off date). Burton C. Hallowell and Kossuth M. Williamson, "Debt Management's Contribution to Monetary Policy," *Review of Economics and Statistics,* XLIII, February 1961, pp. 81-84.

Carson has pointed out that debt management was more "successful" in minimizing attrition from June 1951, to September 1952, when Federal Reserve support was utilized, than in the "comparable period" of monetary restraint, November 1955, to July 1957, when Treasury support through purchases for the trust funds was utilized. Deane Carson, "Treasury Open Market Operations," *Review of Economics and Statistics,* XLI, November 1959, pp. 438-442.

[31] Hallowell and Williamson, *op. cit.*

[32] For an emphasis on the desirability of counter-cyclical debt management, see

First, except in rare cases, the central bank restricted its own operations in public debt to the short end of the market. By this decision, the Federal Reserve voluntarily passed up opportunities to buy intermediate and long-term issues during recessions and to sell similar maturities during inflationary periods. Second, and more important because of the much larger scale of the Treasury's operations, the lack of dependable underwriting support from the central bank may have led the Treasury into undue conservatism in its choice of securities. During periods of rising interest rates, especially in 1955-1957, the Treasury was reluctant to experiment with new bond issues in the face of prevailing market expectations. By attempting to "force" bond issues on the market, the Treasury felt that it would risk a very substantial attrition. In turn, this reluctance led to exploitation of market expectations in the following recession. While it is quite unrealistic to assume that direct underwriting support by the Federal Reserve would have solved all of the Treasury's problems, it is also quite likely that the performance of debt management would have been somewhat better with such support.

IV. The Twisting or Nudging of the Interest Rate Structure

The most recent of the three post-Accord periods of debt management policy should include, perhaps, the final portion of the 1960-1961 recession as well as the low-level recovery and expansion of 1961-1962.[33] Although the 1960-1961 decline was relatively mild, it was especially disturbing because the economy had never completely recovered from the 1957-1958 recession. While the rise in unemployment was more moderate than in the two preceding recessions, this increase was superimposed on the already high level of unemployment that had persisted at the peak of the previous expansion. Even after the trough of February 1961, unemployment remained disappointingly high—a situation apparently requiring expansionary policies of greater strength and duration than those utilized following the previous upturns in economic activity. On the other hand, the recession developed against the background of

J. M. Culbertson, "A Positive Debt Management Program," *Review of Economics and Statistics*, XLI, May 1959, pp. 89-98; R. L. Bunting, "A Debt Management Proposal," *Southern Economic Journal*, XXV, January 1959, pp. 338-342; and Thomas R. Beard, "Counter-Cyclical Debt Management—A Suggested Interpretation," *Southern Economic Journal*, XXX, January 1964, pp. 244-252.

Both Gaines and Friedman, for somewhat different reasons, have proposed a "neutral" debt management policy. "Neutrality" involves maintaining a constant maturity structure of the debt regardless of economic conditions. Gaines, *op. cit.*, especially chapter 8, and Friedman, *op. cit.*, especially pp. 52-65.

[33] While operations in longer-term issues were not sanctioned until February 1961, open market operations were extended to include issues in the 9-15 month range in the fall of 1960 in order to avoid downward pressure on bill rates. Also, in the last half of 1960 there was some expansion in short-term issues by the Treasury.

a persistent deficit in the balance of payments, a condition considerably aggravated in both 1960 and 1961 by large outflows of short-term capital. The authorities felt that the international differentials in short-term interest rates that developed in 1960 contributed to this state of affairs. Therefore, international considerations began to influence monetary and debt management policies to an extraordinary extent.[34] The main focus of these policies concerned finding methods of combatting the persistent balance of payments deficits, including the outflows of short-term capital, while, at the same time, maintaining conditions conducive to more rapid recovery and expansion.[35]

A major result of this dual concern was the inauguration of "operation nudge," which involved official attempts at "twisting" the interest rate structure. Upward pressure on short-term rates (but not the establishment of a particular floor) was considered desirable for external purposes. Downward pressure on long-term rates was sanctioned for domestic purposes. Debt management was thus elevated to a new level of importance as the major instrument of economic policy most capable of twisting the rate structure in the desired manner. This new approach to debt management presumably involved the cooperation of both the Treasury and the Federal Reserve. The decision by the Federal Reserve to conduct open market operations outside the short-term area constituted a partial abandonment of the "free market" doctrine—but only partial in that direct support of Treasury refunding was not resumed. This decision also represented some apparent change in official thinking with respect to the Federal Reserve's ability to alter the rate structure.[36]

Partly because of the unusual economic conditions, it is not surprising to find considerable debate concerning the degree of ease or tightness of monetary policy in both recession and recovery. According to public statements and testimony, the intentions of the monetary authorities were on the side of relative ease.[37] Supporters of the "easy money"

[34] The extraordinary degree of importance attached to international considerations by historical standards is emphasized in the *Annual Report of the Federal Reserve Bank of New York, 1961*, pp. 6-7.

It is stated, for example,

"The hard facts of recent balance-of-payments developments, in the context of the international role of the dollar, have revised the basic framework for monetary policy in the United States. As an objective of monetary policy, the defense of the international value of the dollar has come to occupy a position alongside the goal of stable domestic growth" (p. 7).

[35] See, for example, R. V. Roosa, "Reconciling Internal and External Financial Policies," *Journal of Finance*, XVII, March 1962, pp. 1-16.

[36] In contrasting the new position with previous statements as to the System's inability to alter the rate structure, one critic was prompted to remark, "How can the Fed influence the interest rate structure for international purposes, if it cannot do so for domestic purposes?" James R. Schlesinger, "The Sequel to Bills Only," *Review of Economics and Statistics*, XLIV, May 1962, p. 185, fn. 3.

[37] See, for example, the testimony of Chairman Martin in United States Congress, Joint Economic Committee, *State of the Economy and Policies for Full Employ-*

interpretation have pointed to a number of indicators: the continuation of positive free reserves throughout 1961 and 1962; the unusual stability of intermediate- and long-term interest rates during the recovery and expansion; the rapid rate of growth in time and savings deposits in commercial banks; etc. Those who support the "tightness" view have emphasized an allegedly inadequate rate of growth in the money supply (demand deposits plus currency);[38] the high level of long-term interest rates, especially during the recession; etc.

Similarly, the impact of debt management on liquidity and the structure of interest rates is debatable. Analyzing the impact of debt management is especially difficult for at least two reasons. First, during this period, the public's holdings of marketable debt in *both* the shortest-term (under 1-year) and the longest-term (over 20-years) maturity categories registered increases. Second, the debt management operations of both the Treasury and the Federal Reserve must be considered, and this creates uncertainties. Treasury and Federal Reserve operations which have identical impacts on the maturity structure of the publicly held debt could still differ in their respective impacts on the interest rate structure. This differential impact can exist if the central bank's operations have a more pronounced effect on market expectations, and if expectations are important in determining the term structure of rates.

With respect to short-term issues, the public's holdings of under 1-year securities rose by roughly $8.6 billion (including tax anticipation issues) in the last half of 1960, by a significant $8.4 billion in 1961, and by another $2.4 billion in 1962. While increases in 1962 were moderate, the authorities emphasized the fact that public holdings of Treasury bills rose by $5.2 billion, being only partially offset by declines in coupon-bearing issues with less than one year to maturity. The bulk of the expansion in short-term issues resulted from Treasury operations. However, the central bank also made a significant contribution, especially from the point of view of the short-term interest rate objective. If the Federal Reserve had followed earlier procedures of supplying reserves only through purchasing bills, it would have been a net buyer of $3.5 billion of short-term issues in 1961-1962, rather than a net seller of about $950 million.[39]

With respect to the capital market, a considerable amount of debt

ment, Hearings, 87th Congress, 2nd Session, 1962, pp. 601-648, and in United States Congress, Joint Economic Committee, *January 1963 Economic Report of the President, Hearings,* 88th Congress, 1st Session, 1963, pp. 337-379.

[38] See, for example, the testimony of J. M. Culbertson in *State of the Economy and Policies for Full Employment, Hearings,* pp. 417-430, and the testimony of Allan H. Meltzer in *January 1963 Economic Report of the President, Hearings,* especially pp. 595-605.

[39] "Changes in Structure of the Federal Debt," *Federal Reserve Bulletin,* XLIX, March 1963, p. 305.

extension was undertaken.[40] Due largely to the Treasury's "senior" advance refunding operations—a debt management technique first utilized in October 1960—outstanding maturities in excess of 20 years rose in the range of $1.5 billion to $2.5 billion per year from 1960-1962. For largely the same reason, the arithmetic average maturity of the total marketable debt rose slightly in both 1960 and 1962, while remaining roughly unchanged in 1961. In these "senior" advance refundings, holders of outstanding 2½ percent bonds with from roughly 6½-11 years to maturity were given the option of exchanging them for new issues in the range of 19-38 years to maturity.[41]

What was the impact of "senior" advance refunding on liquidity and long-term interest rates? While little analysis of advance refunding has yet appeared, it is quite possible that the impact of debt lengthening could be either overestimated or underestimated.

The sale of long-term debt is likely to have a more *immediate* impact on liquidity and long-term interest rates in a cash or regular refunding operation than in a "senior" advance refunding. When the Treasury sells long-term government bonds for cash, the total amount of bonds is increased and additional long-term funds must be found for their purchase—funds which would otherwise be available for the purchase of alternative types of long-term private issues. Likewise, the immediate market impact of regular refundings is similar even though the total volume of government debt is not increased. The maturing securities are held in large measure by short-term investors who are not interested in purchasing long-term bonds. Therefore, investors who wish to exchange for the long-term bonds will ordinarily purchase "rights" to the new issues or "when-issued" securities from dealers (who in turn acquire "rights"), and this transfer of ownership involves a considerable amount of market "churning." The "senior" advance refunding operations, however, minimize changes in ownership and the amount of new cash funds required to purchase the long-term bonds. Presumably, a large portion of the long-term bonds taken in "senior" advance refundings are substituted for the eligible issues of shorter maturity which are held by the typically long-term investors—thus minimizing the *immediate* market impact of the operation.[42]

[40] One of the most persistent critics of debt lengthening operations during this period has been J. M. Culbertson. In addition to his testimony cited earlier, see his paper, "The Recent Policy Mix and Its Implications," *Proceedings of the Business and Economic Statistics Section of the American Statistical Association, 1963,* pp. 362-366.

[41] The Treasury has also conducted "junior" advance refundings and pre-refunding operations. "Junior" advance refundings involve outstanding issues with 1-5 years to maturity and pre-refundings involve issues with less than 1-year to maturity. Normally, a significantly smaller extension in average maturity is accomplished in these operations than is accomplished in advance refundings of the "senior" variety.

[42] Secretary Dillon has defended advance refunding as an important technique for improving the debt structure with a minimum impact on long-term rates and the flow of funds into long-term private investment. See the testimony of Secretary Dillon

On the other hand, the above argument tends to understate the impact of "senior" advance refundings by concentrating narrowly on the immediate market reaction. Even if the purchase of long-term governments does not immediately reduce (significantly) the flow of funds into corporates, municipals and mortgages, this undesirable effect may only be delayed or spread out over time. As the passage of time moves outstanding government bonds toward the intermediate-term range, "long-term" investors normally dispose of many of these securities in order to lengthen their portfolios through the purchase of corporates, municipals, and mortgages. By inducing investors to lengthen their portfolios through the purchase of long-term government bonds, it is likely that over a period of time the flow of funds from the intermediate sector into private long-term securities is reduced.

A related issue concerns the impact of the Federal Reserve's operations in intermediate- and long-term securities. Despite the celebrated abandonment of "bills only," it is clear that the volume of central bank operations outside the short-term area was not large. In both 1961 and 1962 combined, the Federal Reserve made net purchases outside the short-term area of only $4.4 billion, of which less than $1.2 billion were in over 5-year maturities.[43] By way of contrast, the Treasury engaged in substantial bond sales to the public in both regular and advance refundings. It is unlikely that the Federal Reserve can exert much downward pressure on interest rates—even if that were its intention—in those market sectors in which the Treasury is greatly expanding its debt operations. Furthermore, there is good evidence that the relatively small amount of bond purchases by the Federal Reserve was not normally undertaken with the specific purpose of exerting downward pressure on rates, but rather was conducted "in a manner intended to exert minimum direct influence on prevailing prices and yields."[44] Typically, the Trading Desk did not solicit offerings from dealers, but only purchased some of the securities offered at the dealers' initiative. Purchases of longer-term issues were geared largely to the objective of cushioning downward pressures on short-term rates,[45] rather than creating direct downward pressures on intermediate- and long-term rates.

---

in United States Senate, Committee on Finance, *Advance Refunding and Debt Management, Hearings,* 87th Congress, 2nd Session, 1962, pp. 1-86, and in *State of the Economy and Policies for Full Employment,* especially pp. 666-668, 677-681, and 685-686. See, also, "Debt Management and Advance Refunding," U.S. Treasury Department, September 1960, and "Advance Refunding: A Technique of Debt Management," *Federal Reserve Bank of New York Monthly Review,* XLIV, December 1962, pp. 169-175.

[43] "Changes in Structure of the Federal Debt," *Federal Reserve Bulletin,* March 1963, p. 307.

[44] "Federal Reserve Open Market Operations in 1962," *Federal Reserve Bulletin,* XLIX, April 1963, p. 432.

[45] Apparently, the System bought intermediate- and long-term issues for two major reasons: (1) to supply reserves without exerting direct downward pressure on

For these various reasons, the extent of debt management's contribution to the remarkable stability of intermediate- and long-term rates over the 1961-1962 expansion is somewhat questionable. Clearly, during the latter year, the impact of the weak recovery on expectations and the desire of commercial banks for longer-term assets[46] were both significant developments in exerting downward pressure on long-term rates. Likewise, the precise impact of debt management on liquidity is difficult to assess. Nevertheless, even with these uncertainties, one conclusion seems readily apparent—that debt management made a greater contribution to the objectives of monetary policy than was the case in the earlier era of "free markets" when debt management operations were decidedly pro-cyclical. Even if our present international difficulties should largely (and perhaps miraculously) disappear, it would be unwise for debt management to retreat to the policies of the "free market" era.

V. Concluding Observations

Admittedly, we do not have complete answers to all the important questions bearing on the proper relationship of debt management and monetary policy. In recent years, debt management has often been heralded as offering a unique tool for combatting external balance of payments problems without sacrificing the domestic objective of more rapid growth. Clearly, the actions of the authorities with respect to short-term rates were taken largely for external purposes. With respect to international considerations, the important questions are two-fold: (1) the extent to which debt management can push up short-term rates (especially bill rates) to higher levels than would otherwise exist; and (2) the significance of international interest rate differentials with respect to short-term capital flows.

Perhaps even more important, we need to know more about whether a *genuine* policy of "twisting" by both the Treasury and Federal Reserve is desirable for *purely domestic* purposes. Consider an expansionary debt management policy. Increasing the public's holdings of highly liquid short-term issues (to exert upward pressure on short-term rates) while reducing the public's holdings of less liquid longer-term issues (to exert downward pressure on longer-term rates) would have a desirable liquidity effect and reduce the demand for cash balances. But the strength of this effect depends upon the relative "degree of moneyness" of gov-

short-term rates; and (2) to offset the reserve effect of bill sales made to cushion downward pressure on short-term rates.

[46] Higher interest rates on time and savings deposits channeled a large volume of short-term funds into the banking system. Banks tended to invest these funds in medium- and long-term assets, thereby contributing to the actual declines in intermediate- and long-term rates during 1962.

ernment securities of varying maturities—about which little is known of a precise nature.

Similarly, the extent to which debt management can actually "twist" the interest rate structure is important.[47] A policy of attempting to twist the rate structure for domestic purposes will be more desirable to the extent that: (1) long-term rates can be reduced even in the presence of higher short-term rates; (2) short-term rates are much less important than long-term rates with respect to private investment decisions; and (3) higher short-term rates discourage the holding of cash balances while lower long-term rates encourage credit expansion by financial institutions because of the greater capital values of their asset holdings.

One final observation seems relevant. The advance refunding technique has enhanced the flexibility of Treasury debt management and increased the Treasury's ability to sell long-term bonds. It is far from certain, however, what course the Treasury would pursue in the event of a future resurgence of inflationary pressures coupled with generally rising interest rates. Past experience suggests that it may become necessary for the central bank to extend limited underwriting support to the Treasury in an attempt to avoid the consequences of a serious shortening of the debt structure.[48] In such circumstances, the Federal Reserve would

[47] The degree to which changes in the debt structure affect the structure of interest rates partly depends on the strength of other forces that influence the pattern of rates. Some economists consider expectations about future rates to be the predominant influence. If the expectational factor is dominant, a change in the maturity structure of the debt would by itself have little effect on the structure of interest rates. Recent empirical work by Meiselman supports the expectations approach. David Meiselman, *The Term Structure of Interest Rates*, Englewood Cliffs, N.J.: Prentice-Hall, 1962. Meiselman's approach, however, does not completely rule out any effect of debt management on the interest rate structure. Changes in the amounts of various maturities outstanding can cause expectations to change, and thus affect the prevailing interest rate structure. See John H. Wood, "Expectations, Errors, and the Term Structure of Interest Rates," *Journal of Political Economy*, LXXI, April 1963, pp. 160-171.

Some recent empirical work by Okun for the Commission on Money and Credit indicates that the effect of changes in relative supplies of various securities on the rate pattern was small for the period 1946-1959. Arthur M. Okun, "Monetary Policy, Debt Management and Interest Rates: A Quantitative Appraisal," *Stabilization Policies*, Englewood Cliffs, N.J.: Prentice-Hall, 1963, pp. 331-380. However, for an approach which gives considerable weight to the relative supplies of various securities in the market, see J. M. Culbertson, "The Term Structure of Interest Rates," *Quarterly Journal of Economics*, LXXI, November 1957, pp. 485-517. For a discussion of the great variety of factors which can shape the interest rate structure, see Stephen H. Axilrod and Ralph A. Young, "Interest Rates and Monetary Policy," *Federal Reserve Bulletin*, XLVIII, September 1962, pp. 1110-1137.

[48] Central bank underwriting, of course, cannot solve the problems created by the legal interest rate ceiling on government bonds in periods when market rates reach the critical level. The solution to this problem lies in either one of two directions. Either Congress must remove this artificial barrier to flexible debt management policy, or the Treasury must be willing to risk unpopularity and circumvent the ceiling by issuing bonds at a sufficient discount to be in line with existing market rates.

face a dilemma that could be framed in the following terms. To what extent would the central bank's ability to contain monetary and credit expansion be negated by a policy of supporting Treasury debt management operations? But on the other hand, in the absence of support operations, would a decidedly pro-cyclical Treasury debt management policy seriously complicate the task of monetary restraint through the piling up of highly liquid money substitutes?

28

Debt management in the United States[1]

Warren L. Smith

Summary

The Treasury's problems in managing the public debt have been the subject of much attention and concern recently. "Debt management," as we shall define the term, is different from both fiscal policy and monetary policy, although it is closely related to both. To a considerable extent, fiscal policy sets the framework within which debt management is conducted, while the kind of monetary policy being followed affects the Treasury's problems of debt management. At the same time, the debt management policies of the Treasury may interfere with the freedom of the Federal Reserve in conducting monetary policy, and the structure of the debt may significantly influence the way in which monetary controls function. Moreover, under our definition, the Federal Reserve has some powers and responsibilities which come under the heading of debt management.

I. The Federal Debt in Perspective

The size of the debt

There are several concepts of the public debt which are employed in discussions of debt management. On June 30, 1959, the "total gross debt" or "total Federal securities outstanding" amounted to $284.8 billion.

Reprinted from *Study of Employment, Growth, and Price Levels,* Joint Economic Committee, 86th Cong., 2d Session (Washington, D.C., U.S. Government Printing Office, 1960), pp. 1-15. Study Paper No. 19. Warren L. Smith is Professor of Economics, University of Michigan.

[1] I am indebted to Messrs. James E. Sutton and Kyung Mo Huh for research help in the preparation of this paper. I have also benefited from discussions with Prof. David I. Fand.

The gross debt reached a level of $279 billion in February 1946 at the zenith of borrowing connected with the financing of World War II, then declined to $252.4 billion in June 1948 as a result of the immediate postwar budget surpluses and debt retirement. From mid-1948 to mid-1949 the gross debt grew by $32.4 billion.

However, the gross public debt does not represent the true debt of the Federal Government. At the end of the fiscal year 1959, $54.6 billion of Treasury securities was held by Government agencies and trust funds, i.e., within the Federal Government itself. A further $26 billion was held by the Federal Reserve System. Since purchases and sales of Government securities by the Federal Reserve are made for the purpose of controlling bank reserves and the money supply in the interest of maintaining financial and economic stability and since approximately 90 percent of the interest payments made to the Federal Reserve are returned to the Treasury at the end of each year, this portion of the debt is also essentially intragovernmental. The debt that is significant for most aspects of economic analysis is the publicly held debt—that is, debt held by households, business firms, commercial banks, and other financial institutions of the country. Changes in the amount and composition of the publicly held debt affect interest rates and the liquidity of spending units, and these effects may influence the level and composition of private expenditures. The publicly held debt amounted to $204.2 billion on June 30, 1959, and increased by only $9 billion between mid-1948 and mid-1959, compared with the $32.4 billion increase in the gross public debt.

Definition of debt management

We shall define debt management to include all actions of the Government, including both the Treasury and the Federal Reserve, which affect the composition of the publicly held debt. When defined in this way, debt management includes: (1) decisions by the Treasury concerning the types of debt to be issued to raise new money, (2) decisions by the Treasury concerning the types of debt to be issued in connection with the refunding of maturing securities, (3) decisions by the Federal Reserve concerning the types of debt to be purchased and sold in the conduct of open market operations.

It should be noted that under this definition of debt management, the amount of new securities to be sold by the Treasury to cover budget deficits or to be retired with the proceeds of budget surpluses is not a matter of debt management but of fiscal policy. Moreover, decisions by the Federal Reserve which change the publicly held money supply, including changes in reserve requirements and the amount (but not the composition) of open market purchases and sales, fall under the heading of monetary policy.

The debt in relation to other economic magnitudes

Not only has the publicly held debt not grown greatly in absolute amount in the last decade, but it has actually declined substantially in relation to other relevant economic magnitudes. The publicly held debt was equal to 86.9 percent of GNP at the end of 1947; by 1958, due to the rise in GNP, the percentage had fallen to 44.2. Of course a substantial part of this decline in the percentage is the result of inflation, but the fact remains that the debt is much smaller relative to our productive capacity than formerly, and to the extent that this is a measure of our ability to carry the debt, it should sit much more lightly on our shoulders than it did a decade ago. Furthermore, between 1947 and 1958 total net public and private debt outstanding rose from $394.8 billion to $743.9 billion. As a result of this large increase, the publicly held Federal debt declined from 50.8 percent of the total in 1947 to 27.7 percent in 1958. Thus, the relative importance of the Federal debt in our debt structure has declined very substantially.

Interest on the debt

Net interest paid by the Federal Government as shown in the national income accounts is the best measure of the interest cost to the Treasury, since it excludes the intragovernmental transfers involved in payments of interest to the trust funds. Net interest paid increased from $4.2 billion in 1947 to $5.5 billion in 1958 as a result of a steady upward trend in interest rates; however, due to rising incomes, net interest payments fell from 2.1 percent of national income in 1947 to 1.5 percent in 1958. Increases in interest payments have weak effects on the level of income, because the marginal propensity to spend out of interest receipts is relatively low and because such payments are subject to rather high marginal rates of taxation.

Nevertheless, interest costs on the public debt do represent a sizable sum and are a matter for concern. And since the administrative budget is frequently used as a tool of fiscal policy, for some purposes the interest included in this budget is the important thing. For fiscal 1960, interest payments in this budget are estimated at $9 billion, more than 11 percent of total budget expenditures, nearly three times the estimated expenditures of the Department of Health, Education, and Welfare, and nearly 40 percent larger than those of the Department of Agriculture.

With the present emphasis on balancing the budget without raising taxes, a rise in the interest burden tends to cut into other badly needed types of Federal expenditures. Thus, there is good reason for trying to avoid unnecessarily heavy interest costs on the public debt. That is, unless the increased interest payments serve some useful economic function, we should try to reduce them.

Volume of debt operations

In addition to exaggerating the size of the debt itself and of the interest payments on it, the statistics commonly used overstate the magnitude of current debt operations. For example, in the calendar year 1958, the total amount of certificates, notes, and bonds issued by the Treasury both for cash and in exchange for maturing securities amounted to $61.2 billion. However, out of this total, $22 billion represented securities issued to the Treasury trust accounts and Federal Reserve banks—almost entirely automatic (and fictitious) transactions involving no problems of debt management—so that the amount of securities issued to the public amounted to only $39.2 billion. Similar large differences are present in other years. Proper evaluation of the current problems of managing the debt requires that the transactions within the Government be eliminated from the calculations.

Ownership of the debt

Holdings of Treasury securities by various investor groups have undergone substantial changes in recent years. With respect to debt ownership, investors may be divided into three broad categories.

1. Investors whose holdings have declined steadily. This category includes insurance companies and mutual savings banks. Holdings of mutual savings banks declined by $4.7 billion from 1948 to 1959, while holdings of insurance companies declined by $10.8 billion during the same period. As a result of the prosperous conditions and heavy savings of the war period, these institutions grew rapidly during the war, and due to the limited private demand for funds, as well as pressures to assist the Treasury in war financing, most of the inflow of funds was invested in Government securities. In response to the heavy demands for funds which have characterized the postwar period, both of these types of institutions have steadily liquidated Government securities in order to shift their funds into more lucrative private investments—chiefly mortgages in the case of mutual savings banks and corporate bonds and mortgages in the case of life insurance companies. Liquidation of governments by these institutions has not shown any particularly strong tendency to speed up during periods of tight credit. The rate of liquidation appears to have slowed down somewhat as total portfolios have become smaller.

2. Investors whose holdings have increased steadily. Several classes of investors have steadily added to their holdings of Government securities in recent years. These include State and local governments, savings and loan associations, and foreign accounts and international agencies.

3. *Investors whose holdings have fluctuated substantially.* Investments in Government securities by commercial banks and by non-financial corporations have exhibited substantial fluctuations from year to year with no discernible trend during the last decade. Fluctuations in the holdings of these two groups have shown a systematic pattern related to changes in monetary policy and credit conditions, which has made the task of conducting monetary policy more difficult for the Federal Reserve.

Composition of the debt

In June 1959, out of a total publicly held debt of $204.2 billion, $5 billion represented convertible bonds and $54.2 billion represented all other nonmarketable and miscellaneous debt, chiefly savings bonds. The remaining $145 billion was marketable securities including Treasury bills, certificates of indebtedness, notes, and bonds. While there are problems connected with the savings bond program, our main concern is the management of the marketable portion of the debt. The percentage of the debt maturing in 1 year had risen from 24.6 in 1946 to 35.4 in 1959, while at the other end of the scale the percentage maturing beyond 10 years had fallen from 33.9 to 17.7. The maturity composition tends to shorten if nothing is done merely due to the passage of time, while debt operations in the form of cash borrowing, refunding operations, and debt retirement, introduce elements of irregularity into the behavior of the composition. Each time the Treasury refunds a maturing security by offering a new issue in exchange, the average maturity of the debt increases at least a little because securities having a maturity of zero are removed from the debt and replaced by other securities. Cash borrowings may increase or decrease the average maturity of the debt, depending on whether the securities being issued have a maturity longer or shorter than the existing average. Cash retirement of maturing securities lengthens the average maturity, because the securities removed from the debt have a maturity of zero. Consequently, the irregular pattern of debt operations makes the maturity structure and the average maturity behave in somewhat unpredictable fashion. Nevertheless, it is quite clear that the maturity of the debt, however measured, has declined substantially in recent years.

II. Present Debt Management Techniques

Bill financing

The Treasury bill, which may have a maturity up to 1 year, has proved to be a very effective and useful debt instrument. Bills are sold at auction on a discount basis, and the bill auctions seem to interfere very little with the Federal Reserve's freedom of action. Until recently regular bill offerings were made only with maturity of 3 months. The Treasury has within the last year extended bill maturities first to 6 months and then to 1 year. At the present time, the Treasury has outstanding 13 issues of 3-month bills, 13 issues of 6-month bills, these 2 sets forming a pattern in which 1 issue matures and is replaced by a new bill offering each week. In addition, there are now four issues of 1-year bills maturing once each quarter in January, April, July, and October. The total amount of these regular bills was $31 billion at the end of July 1959, and this portion of the debt has been placed on a periodic rollover basis, which is efficient and economical and minimizes interference with Federal Reserve monetary policies. In addition to regular bill issues, the Treasury has recently been relying mainly on bills in its tax anticipation financing to meet seasonal gaps between receipts and expenditures.

Fixed price issues

The Treasury also borrows by issuing certificates of indebtedness, notes, and bonds, both to raise new cash to cover budgetary deficits and to refund maturing securities. Although refunding could be handled by selling new securities for cash and using the cash to retire the maturing securities, in practice refunding is almost always handled by means of exchange offerings. Certificates, notes, and bonds are sold on a fixed price basis.

Several decisions must be made before a fixed-price issue can be offered to the public. These include the choice of a maturity, other provisions such as call or redemption options, and the selection of the coupon rate to be placed on the securities. In deciding upon the maturity and terms of a particular offering, the Treasury is guided by the advice from market experts—particularly the advisory committees of the American Bankers Association and the Investment Bankers Association—by potential investors, and by its own independent study of market conditions. The choice of the coupon rate is made by examining the yield curve at the time of the offering. However, it is necessary to set the interest rate on the new security somewhat above the yield on outstanding debt at the same maturity in order to induce the market to absorb a substantial offering.

Underwriting of short-term cash offerings

The Treasury does not make use of formal underwriting in marketing its issues, such as is provided by investment banking syndicates in the case of corporate offerings. However, it is customary in the case of short-term cash offerings, such as certificates and shorter term notes, to permit commercial bank subscribers to pay for the issue by means of credits to Treasury tax and loan accounts, which means that banks are, in effect, able to obtain the securities by paying only a portion of the price equal to their reserve requirements until such time (commonly 2 to 3 weeks later) as the Treasury transfers the funds to its accounts at the Federal Reserve banks. The use of Treasury tax and loan account credits provides a kind of indirect underwriting.

The banks serve essentially as underwriters, reselling or distributing securities to other investors. The Treasury limits or discourages bank subscriptions on longer term issues apparently on the ground that such subscriptions are unsuitable investments for banks. In restricting bank subscriptions to longer term issues, the Treasury is probably denying itself important support that could be of great help at times. The main underwriting device used by the Treasury to market long-term debt is to offer a rate sufficiently attractive to achieve the required sales.

Refunding operations

Maturing securities are short-term liquid instruments and are likely to be in the possession of investors who are holding them for liquidity reasons. The securities being issued in exchange, on the other hand, if they are of intermediate or long maturity, are more likely to appeal to investors who want either permanent investments or prospective short-term speculative gains. The success of a refunding operation often depends, therefore, on the extent to which maturing securities have been shifted from their normal owners to investors desiring to obtain the new securities. This may require that the terms of the new security be sufficiently attractive to create a premium on the "rights" (i.e., the maturing issue) in order to induce the transactions in these "rights" that are needed to put them in the hands of investors who want the new issue.

Government security dealers buy and sell "rights," thus facilitating their distribution, and as soon as the subscription books open, the securities begin to trade on a "when issued" basis. During the subscription period dealers buy "rights" and sell "when issued" securities. These dealer operations, which contribute to the success of the exchange and the proper placement of the new offering, are the closest thing there is to systematic underwriting in connection with refunding operations.

III. Recent Debt Management Problems

Shortening of debt maturities

The shortening of debt maturities has been a matter of concern to Treasury officials, and debt management policy in the last few years has concentrated on trying to lengthen maturities. The orthodox theory of debt management calls for the issuance of long-term securities during periods of inflation in order to preempt funds from the capital market and reduce liquidity, and the issuance of short-term securities in recession periods in order to increase liquidity and leave the maximum amount of funds available for long-term investment. However, the Treasury has had little success in following the precepts of orthodox debt management theory and has been forced—or induced—to sell long-term securities in recessions. Thus, such debt lengthening as has occurred in the last few years has taken place largely in the recession or early recovery periods of 1953-55 and 1957-58.

The competitive position of Government securities

In recent years, the Treasury has had considerable difficulty in selling long-term bonds. During the period of nearly 7 years since the present administration came into office with the intention of extending debt maturities, only $9.4 billion of bonds with a maturity of more than 10 years has been sold to the public altogether, both for cash and in exchange operations. Thus, the average is less than $1.5 billion per year. Nearly all the investor groups—including savings banks, life insurance companies, pension funds, etc.—who have traditionally shown an interest in Treasury bonds, have been reducing their holdings steadily or at most increasing them only very slowly. Certainly one important aspect of our debt management difficulties appears to be the declining popularity of Government securities, particularly of the longer term variety.

There are several possible explanations of the apparent deterioration of the competitive position of long-term Treasury securities. One is the greatly increased variability in the prices of Government securities as monetary policy has been employed more vigorously. This increased variability has lowered the liquidity, particularly of longer term Treasury securities, and reduced their attractiveness to many investors. Another reason is the increased attractiveness of corporate securities as investors' assessments of the risks associated with these securities have been reduced as a result of continued relatively prosperous business conditions. The increased importance of FHA-insured and VA-guaranteed mortgages has also cut into the market for longer term Government securities, since these mortgages are about as safe investments as Governments and yield the investor higher net returns. The fact that yields on Government securities have risen relative to those on private debt during the last decade, at the same time that the size of the publicly held Federal debt has been declining relative to the amount of private debt outstanding, appears to corroborate the view that Government securities have become less attractive to investors.

Other problems

Interest rates have shown an increasing tendency to undergo rapid changes at turning points in business activity, as investors have become more aware of the implications of flexible monetary policy. This is rather troublesome to the Treasury, particularly at times when an improvement in business activity begins while the Treasury is still operating at a large deficit requiring heavy cash borrowing, as in mid-1958. In addition, speculation in Government securities has had a disorganizing effect on the Government securities market, especially in the case of the 2⅝-percent bonds of February 1965, which were issued in exchange for maturing securities in June 1958, at approximately the time that the outlook for business activity and monetary policy was changing from recession to recovery.

IV. Principles of Debt Management

Economic effects of debt operations

As indicated earlier, we define debt management to include all operations which affect the composition of publicly held debt. On this definition, all debt management operations are reduced, in effect, to the sale of one type of security and the use of the proceeds to retire another type. Intelligent debt management requires that operations of this kind be conducted with a view to their effects on the economy. Suppose, for example, that the Treasury sells long-term bonds and uses the proceeds to retire short-term debt. According to the orthodox and widely accepted theory of debt management, such an operation would have restrictive or anti-inflationary effects, which may be classified under two headings: interest-rate effects and liquidity effects.

1. Interest-rate effects. A sale of long-term bonds and use of the proceeds to retire short-term debt would force up long-term interest rates and lower short-term interest rates. There would be secondary readjustments in the rate structure which would depend upon the nature of investors' expectations, but the net result would very likely be somewhat higher long-term rates and somewhat lower short-term rates. Whatever restrictive effect such an operation might have by way of interest rates would depend upon the restrictive effects of rising long-term interest rates exceeding the stimulative effects of falling short-term interest rates. Such evidence as is available suggests that interest sensitivity of expenditures is rather low. Since it is doubtful whether within moderate limits an increase in the general level of interest rates has strong effects, it becomes even more dubious whether the net effect of raising one rate and lowering another would amount to much. In fact, the presence of a net restrictive effect assumes that the interest elasticity of expenditures with respect to long-term interest rates is greater than the elasticity with respect to short-term interest rates, and, while one might suspect that this is true, there is really little evidence to support it. As far as the interest-rate effects are concerned, debt management involves second-order adjustments of variables whose first-order importance is open to question.

2. *Liquidity effects.* Since the liquidity of an asset depends upon the variability of its price and since prices of short-term securities ordinarily fluctuate less than prices of long-term securities, an operation of the type we are considering would decrease liquidity somewhat by reducing the liquidity of the buyers of long-term securities and increasing the liquidity of sellers of short-term securities by a lesser amount. Apart from the interest-rate effects referred to above, however, it is not entirely clear why such changes in liquidity would produce changes in income-generating expenditures. Moreover, the importance of such restrictive effects as may be present is doubtful, since the total volume of liquid assets in the economy, as ordinarily defined, is not changed. It is merely the degree of the liquidity of assets that is affected. Empirical studies that have been made of the determinants of expenditures, including consumption and investment, have not produced clear evidence that the stock of liquid assets is an important variable affecting spending under normal conditions. This being the case, it is even more doubtful whether changes in the degree of liquidity of a given stock of assets are likely to have important effects. Again, as in the case of interest rates, the liquidity effects produced by debt management are of the second order of importance.

Thus, neither the interest rate nor the liquidity effects of marginal changes in the debt structure appear to be very important. Moreover, to the extent that such changes do have a net effect on the public's aggregate spending, it would appear that similar effects could be produced by the use of monetary policy. For this reason, it is difficult to see what can be accomplished by contracyclical debt management policy that cannot be accomplished more efficiently by Federal Reserve monetary policy. Debt management is a cumbersome instrument of stabilization policy, because it is difficult to time in a flexible way, and because the Treasury is almost unavoidably concerned about its success in raising money.

Debt management as a form of selective control

It seems better to think of marginal changes in the debt structure as a species of selective controls, since a change in the structure of interest rates (and also the structure of liquidity) probably has some effects on the pattern of expenditures; that is, the expenditures stimulated by a fall in short-term interest rates are likely to be different from those discouraged by a rise in long-term interest rates. Unfortunately, however, our knowledge concerning the effects on the expenditure pattern of changes in the interest rate structure is quite unsatisfactory. Moreover, to the extent that we may desire to use changes in the rate structure as a selective control, it is much more sensible to leave such operations to the Federal Reserve. It may be noted that under our definition, to the extent that the Federal Reserve departs from the prevailing bills-only policy and engages in operations which change the maturity structure of its portfolio in order to produce selective effects, it is engaging in debt management.

The existing debt as an automatic stabilizer

It is useful to distinguish between the debt structure at any particular time and marginal changes in the debt structure. The above discussion deals entirely with marginal changes and suggests that their importance may not be particularly great. However, the debt structure at any particular time conditions the way in which the economy and particularly the financial system react to external disturbances. For example, if the public debt consists entirely of short-term securities, monetary controls may not take effect very strongly, because it is easy for financial institutions and other economic units to mobilize funds for spending through transactions in short-term securities. Since these securities are close substitutes for money, it is likely to be possible to find buyers for them among holders of idle cash balances without producing sufficient changes in interest rates to have a strong restrictive effect on expenditures.

If investors hold predominantly long-term securities, their ability to shift their holdings to someone else when they need funds to lend or spend is likely to be somewhat less. The fact that long-term securities fluctuate in price more than short-term securities as interest rates change means that there is somewhat more friction set up to slow down this process of shifting. Thus, a debt consisting predominantly of long-term securities may act as a kind of automatic stabilizer, contributing to the stability of the economy and the effectiveness of monetary policy.

This suggests that those responsible for debt management should concentrate more on trying to maintain a debt structure which contributes to economic stability without worrying so much about the timing of the marginal adjustments necessary to achieve and maintain this structure. It is impossible, however, in our present state of knowledge, to specify a principle which would help us determine the optimum debt structure. But at the present time it is quite clear that we have too much short-term debt and that the debt should be lengthened.

While the structure of the debt is a matter of some importance, its influence should not be exaggerated. As suggested above, long-term debt may make the economy more resistant to external disturbances because it is more resistant to shifting, but we must remember that long-term debt can be shifted also. If expenditures are insensitive to interest rate changes, capital losses on sales of long-term debt by financial institutions can be compensated for by charging a higher interest rate on private debt. There is probably something to the "locked-in" effect, and the likelihood that investors will be locked in will be somewhat greater if they hold long-term debt. But it is a matter of degree.

Interest cost of the debt as a policy consideration

As indicated earlier, the interest cost on the public debt is a matter of some importance, and unless the economic effects produced by the debt are worth the cost, there is no reason why interest has to be paid since it is always possible to turn debt into money. Consequently, debt management policy must in some sense measure the interest cost on the debt against its economic effects.

The problem of selecting the techniques for debt management which would reduce the Treasury's interest cost to a minimum is a difficult one, because it is necessary for the Treasury in this connection to look ahead and try to foresee future changes in interest rates. For example, from the point of view of minimizing interest costs, it would not necessarily be wise for the Treasury to engage in long-term borrowing at a particular time merely because long-term interest rates were below short-term interest rates. If the interest rate level was expected to fall shortly, it would be better to postpone long-term borrowing until the level had fallen, because long-term borrowing fixes interest cost for many years into the future.

The interest rate structure in periods of recession and easy money tends to be one in which the short-term interest rate is substantially below the long-term. As business conditions improve, credit tightens and interest rates rise, short-term interests rates normally rise substantially more than long-term interest rates so that in times of prosperity and tight credit it often happens, as has been the case recently, that short-term interest rates are higher than long-term. Although the movements of the interest rate structure are quite complex and difficult to predict in detail, when interest rates move in this way one thing is reasonably clear, namely, that from the standpoint of minimizing interest costs the Treasury should attempt to sell long-term securities and lengthen the debt in periods of recession when interest rates are low. Strictly speaking, of course, it should probably be raising some funds in many maturity sectors at most times since cost minimization requires the equalization of marginal costs of raising funds in all sectors, but minimization of interest costs would appear to require an increased emphasis on long-term borrowing when interest rates are low.

Combining economic stabilization and cost minimization

While our knowledge of the economic effects of the debt and of the costs associated with various time patterns of debt operations is not sufficient to permit the promulgation of highly specific rules governing debt management, the above considerations suggest that it would be desirable to lengthen debt maturities in order to achieve and maintain a more satisfactory structure and that the timing of marginal changes in the debt structure is not of major importance. There is something to be said for emphasizing debt lengthening operations in periods of recession when interest rates are low in order to keep down the Treasury's interest costs. Moreover, there would be advantages in reducing the emphasis on long-term borrowing in periods of prosperity and inflation in order to minimize interference with Federal Reserve policy, since short-term borrowing is less likely to require support from the Federal Reserve or relaxation of a restrictive monetary policy.

Federal Reserve open-market policy

If debt management conducted along the above lines should interfere with the achievement of economic growth and stability, the Federal Reserve should be prepared to intervene by means of open-market purchases or sales in whatever maturity sector seemed most appropriate. Since 1953, the Federal Reserve has adhered to the so-called "bills-only" policy under which it has confined its open-market operations to short-term securities. This policy was adopted partly to minimize the extent of interference with market forces and partly to encourage the development of a stronger Government securities market. The philosophy of minimum intervention is inappropriate under present conditions, and the "bills-only" policy has not succeeded in strengthening the Government securities market. Accordingly, it would appear desirable for the Federal Reserve to abandon the "bills-only" policy and be prepared to intervene in any maturity sector of the market when such intervention would help to minimize undesirable speculative activity, prevent meaningless short-run fluctuations in Government security prices, or help to achieve economic growth and stability. Most of the Federal Reserve's open-market operations which are merely designed to keep the money market on an even keel would, of course, continue to involve purchases and sales of short-term securities.

V. Suggestions for Improving Debt Management

The analysis presented herein suggests (1) that the Treasury seek to extend the maturity of the public debt, (2) that efforts in this regard be conducted with a view to keeping down interest costs, which means emphasizing long-term borrowing in periods of recession and low interest rates, (3) that the Federal Reserve be prepared to intervene in a flexible and effective manner if the effects of Treasury debt operations along these lines should prove undesirable, and (4) that the Federal Reserve be assigned full responsibility for managing the interest rate structure for the purposes of achieving economic stability and a smoothly functioning capital market. This does not mean that the Treasury should make no effort to borrow at long term during prosperous periods. It is merely suggested that if other policies are properly coordinated with debt management, long-term borrowing in recessions is likely to do little harm and will save the Treasury interest money.

Relation of debt management to other policies

The magnitude of the Treasury's debt management problems depends to a considerable extent upon the kinds of monetary and fiscal policies that are being followed. Three aspects of this relationship are worthy of consideration.

1. The mix of monetary and fiscal policies. Within limits, monetary policy and fiscal policy are substitutes for purposes of economic stabilization. However, the allocation of resources between consumption, private investment, and the production of Government services will be affected by the combination of monetary and fiscal policies chosen. For example, if we want to achieve a more rapid rate of growth, reduction in interest rates to encourage investment, compensated for by increases in taxes which reduce consumption, will contribute to our objectives. Consequently, under present conditions, there is much to be said for an easier monetary policy and a tighter fiscal policy. In addition to encouraging growth, such a change in our policies would reduce the magnitude of our debt problems in two ways: (1) larger surpluses and/or smaller deficits in the cash budget would result in a lower rate of growth (or perhaps even a decline) in the size of the publicly held debt, and (2) lower interest rates resulting from easier monetary policies would save interest costs to the Treasury and would make effective debt management easier to achieve.

2. General versus selective monetary controls. Under present conditions, there is much to be said for greater reliance on selective credit controls directed at some of the sectors of the economy which have exhibited excessive instability. For example, serious consideration should be given to selective controls in the area of consumer credit including housing, and perhaps more effective control over bank lending to stabilize inventory fluctuations. In addition to helping us to maintain stability of growth and employment, these controls might mitigate the inflation problem by reducing the magnitude of shifts in demand. In addition, more reliance on selective controls should simplify our debt management, since most types of selective controls exert their impact by reducing the demand for credit directly, rather than through interest rates. That is to say, a monetary policy relying more on selective controls would presumably require smaller increases in interest rates to achieve a given restrictive effect than would a policy relying entirely on general controls. If this should prove to be the case, there would be some saving in interest costs to the Treasury.

It is very important to emphasize, however, that we should not adopt policies of the kinds just discussed merely because they save the Treasury interest money, reduce the public debt, and simplify the problems of debt management. We should select the proper combination of fiscal and monetary policies, on the one hand, and the proper mix of selective and general monetary controls, on the other, on the basis of the impact these policies have upon the economy. Debt management, while a matter of some importance, is distinctly subsidiary to the selection of proper monetary and fiscal policies. It just happens that, under present conditions, the adjustments in the policy mix that seem to be called for would incidentally reduce interest costs and simplify our debt management problems somewhat.

3. Open-market operations versus reserve requirement changes. In the conduct of its general monetary policy directed at control of the money supply, the Federal Reserve has a choice between the use of open-market operations and changes in reserve requirements. In recent years, the System has relied on open-market operations for short-run stabilization of the economy, but appears to be engaged in a program of secular reduction of member bank reserve requirements. Reserve requirements have been lowered several times during the recessions of 1953-54 and 1957-58, while they have not been increased since 1951. Thus, reserve requirements have been adjusted downward particularly during recession periods, apparently for the purpose of supplying the reserves needed to support economic growth.

The use of open-market purchases of Government securities to supply reserves to the banking system has an advantage, from the standpoint of the Treasury, over reductions in reserve requirements, since open-market purchases absorb securities into the Federal Reserve System's portfolio and since most of the interest on that portfolio is returned to the Treasury at the end of the year. There are, of course, other differences between open-market purchases and lowering of reserve requirements. Lower reserve requirements tend to result in larger profits for the commercial banking system. In addition, lower reserve requirements increase the amount of money and credit that can be created per dollar of additional reserves and thereby increase the leverage of Federal Reserve policy somewhat. Aside from these factors, it is difficult to see that there are any significant observable differences in the impact of these two credit control weapons. On the other hand, it appears that there would be significant savings in interest to the Treasury if the Federal Reserve would desist from further lowering of reserve requirements and supply the reserves needed to support growth by open-market operations. Unless it can be demonstrated that the other effects on bank earnings or on the leverage of monetary policy would be unduly harmful, there is much to be said for relying on open-market operations to supply reserves in future years, leaving reserve requirements at their present levels.

Possible improvements in debt management technique

The techniques used in debt management should be, insofar as possible, the ones which permit the Treasury to sell the desired securities at minimum cost under any given circumstances. Several changes in technique which might be worthy of consideration can be suggested.

1. Auctioning of longer term securities. The auction technique has proved to be highly successful in connection with the sale of Treasury bills, and there might be some advantages in extending it to long-term securities. One possible advantage would be that each block of securities would presumably be sold at the highest price its buyer would be willing to pay, and as a result the Treasury's interest cost might be reduced. There are other advantages and some disadvantages, one of which might be the greater risk imposed on the investor, which might mean that the auctions would be dominated by skilled market professionals so that the market would be narrowed and collusive bidding might develop. Despite difficulties, the device seems promising enough to be worth extending to securities of longer maturity than bills, at least on an experimental basis.

2. Frequent small offerings. It is possible that small offerings of longer term securities made at frequent intervals would help the Treasury to secure a larger share of the current flow of saving. There are difficulties related to the fact that, in order to keep the number of issues from multiplying inordinately, it would be necessary to reopen existing issues, but there should be ways of solving this problem. At first glance, it might appear that such an approach to debt management would mean that the Treasury would be interfering with the Federal Reserve's freedom of action most of the time. However, the opposite might well be true—that the smallness of the offerings would result in a minimum of interference.

3. More effective underwriting. One of the Treasury's difficulties has been that it has had inadequate underwriting support, much less than is used in private financing. One possibility deserving of consideration would be to have the Federal Reserve banks perform the underwriting function for the Treasury, buying up part of a Treasury issue of long-term securities and reselling it to the public over a period of time. A procedure along these lines has been used successfully in England, where the amounts of long-term issues not subscribed by the public are subscribed by the Issue Department of the Bank of England and then gradually resold to the public. In the event it is felt that the Federal Reserve should confine itself primarily to economic stabilization, perhaps some other institutional arrangement could be made.

4. Advance refunding. Advance refunding means offering to holders of an existing security the option of turning it in for a newly issued security before maturity. As longer term securities approach maturity, they frequently fall into the hands of investors who are interested in them as liquidity instruments, and when they mature it is difficult to interest such investors in exchanging them for long-term securities. Judicious advance refunding would catch these securities before they left the hands of those who were holding them as investments and offer a new longer-term security in exchange at that point. Legislation passed in the last session of Congress eliminated technical obstacles and paved the way for the introduction of advance refunding. The Treasury has expressed strong interest in advance refunding as a means of dealing with a large volume of intermediate-term debt scheduled to mature in a few years. Used carefully and in moderation, advance refunding could be a useful way of attaining a more viable debt structure.

5. Call features. The presence of a call feature in a Treasury security gives the Treasury greater possibilities of being able to take advantage of favorable movements of interest rates in future years. The Treasury has issued no callable securities in the last few years. In view of the fact that call features are commonly included in corporate securities, it seems quite possible that inclusion of such a feature might frequently be well worth the extra immediate cost involved.

6. Better selling organization. A more vigorous program for promoting the sale of Government securities by the Treasury might pay big dividends in broadening the market and reducing the Treasury's interest cost. The recent spectacular success of the 5 percent note maturing in August 1964, which attracted over 100,000 subscriptions of less than $25,000 each, aggregating nearly $1 billion, suggests that there is a market that has not been adequately tapped. The development of a more extensive selling organization by the Treasury to attract the interest of small investors as well as of smaller financial institutions far removed from the main centers of finance, would probably be well worth the cost. The facilities of the Federal Reserve banks might also be used.

7. Purchasing-power bonds. It has been suggested recently that the Government might issue bonds whose redemption value (and interest payments, if any) are tied to the Consumer Price Index. The best candidate for this would be savings bonds, which have been a source of substantial cash drains to the Treasury in the last few years. The issue here is not one of saving interest cost, since there is no assurance that savings would result. Rather, it is a question of balancing equity considerations against the dangers of setting up expectations which might intensify the problem of inflation. On balance, there is much to be said for the view that it is a proper function of the Government to provide the small, unsophisticated investor with a form of investment which contains protection against the erosion of his wealth through inflation.

VI. The Interest-Rate Ceiling

In recent months there has been much discussion of the desirability of raising or eliminating entirely the interest-rate ceiling of 4¼ percent applicable to marketable Treasury securities having a maturity of more than 5 years. The controversy concerning the interest-rate ceiling is the culmination of a period of nearly 10 years during which interest rates have drifted steadily upward with only brief interruptions. The tight-money periods of early 1953, 1955-57, and 1958-59 have greatly overbalanced the effects of easier money in the intervening periods. Even during 1958, while the economy was still in the midst of a recession, speculative activity in the Government securities market at midyear was permitted to drive interest rates up sharply, and restrictive monetary policy has driven them even higher in recent months. Clearly, tight money has not been effective in achieving its objective of stopping inflation. At the same time, even in such prosperous years as 1956 and 1957, growth has been slow or nonexistent.

There has been a tendency recently to view high and rising interest rates as a result of the working of inexorable economic laws. There is insufficient recognition of the fact that there are other methods besides general monetary policy which can be used to control inflation; for example, we could place more reliance on fiscal policy and selective credit controls, a combination which would achieve a given restrictive effect with lower interest rates. Nor is enough attention given to the contention that general monetary controls have an uneven impact and that under present conditions such controls may slow down economic growth without stopping inflation.

The interest-rate ceiling is an arbitrary limitation with no analytical justification, and it should accordingly be repealed in order to give the Treasury more freedom to manage the debt effectively. At the same time, however, it is very important that our present stabilization policies be thoroughly reexamined.

29

The significance of the national debt

Roger L. Bowlby

The United States was born with a debt arising from the Revolution and the inability of the Continental Congresses to cover the expenses of that war by taxation. One of the important issues facing Washington's first administration was whether or not the new federal government should assume the debts of the states that made up the infant republic. During this controversy there were opposing views about the significance of national debt. Broadly speaking, Hamilton favored and Jefferson opposed the debt. Hamilton's views prevailed, the federal government assumed the debt of the states, the note issues of the Continental Congresses were honored, and the national debt reached $83 million during Washington's second administration. By 1815 it had risen to $127 million. In following years the debt was virtually eliminated, and fell to a low of only $38 thousand in 1834. By 1837, it again began to assume significant proportions, and reached $61 million in 1851.

During the Civil War there was a phenomenal rise; the debt exceeded a billion dollars for the first time, and by 1866 was about $2.8 billion. For about thirty years after the war, the trend was downward, and the debt dipped briefly below a billion in 1892. It was stable at somewhat more than a billion dollars until World War I. That war saw another spectacular increase in the debt, which reached $25.5 billion in 1919. From that peak, the debt declined in every year through 1930, by which time it had fallen to $16.2 billion. Increases followed in every year through 1946. At the close of 1941, the debt was about $49 billion, and by 1946 it reached a peak of $269 billion. Since then there have been ups and downs, with rather more ups than downs. The debt was

Reprinted from *Business Topics,* Vol. VII, No. 4 (Autumn, 1960), pp. 63-73. By permission of the Bureau of Business and Economic Research, Graduate School of Business Administration, Michigan State University, publishers of *Business Topics.* Roger L. Bowlby is now at the Department of Economics, University of Tennessee.

$289 billion as of August 31, 1960; this is only a little below the all-time high of $291 billion, which came in October, 1959.

The Debt Controversy

This short history of the federal debt could be matched by a history of the controversy that has accompanied it. The controversy is as old as the debt and the nation; it began with Hamilton and Jefferson and continues to the present. In current popular literature, it is possible to find the debt described as a sign of impending national bankruptcy, a generator of inflation, Lenin's favorite strategy for the destruction of capitalism, a tool of the bankers for enriching themselves at the expense of the poor, a burden on helpless children as yet unborn, and a harmless set of bookkeeping entries indicating only the distribution of money among various pockets. Clearly somebody is wrong, and it is probably true that notwithstanding our long experience with the debt, and the years of discussion and argumentation that have accompanied it, the national debt remains one of the most widely misunderstood economic magnitudes.

The purpose of this article is to separate the true and false views of the national debt so far as possible. It is not to advocate debt or to oppose it, but to attempt an evaluation of its significance. It is difficult to make the evaluation without becoming deeply involved in closely related issues, such as government spending and taxation or the need for national military strength. If these issues seem to be slighted in this discussion, it is because attention is focused on the debt itself, rather than the fiscal policies that gave rise to it or might have prevented it.

Some Sources of Confusion

Much of the confusion that accompanies current thinking about the national debt, and leads to erroneous conclusions about its significance, can be traced to two failings. The first of these is failure to distinguish between individual and social debt (or, what comes to the same thing, between internal and external debt). This is not to say that either the debts of individuals or the debts of governments are insignificant, for they are both important. Their significance, however, is different, and the attempt to attribute the same characteristics to government and private debt leads to serious errors.

The second failing arises when the budget and the debt are confused. Again it is obvious that both the budget and the debt are significant, but they are significant in different ways, so that it is important to distinguish between the two. Either the budget or the debt may give us troubles (at present, for example, the budget is balanced and the debt is high) but they are different sorts of troubles.

Individual debt and social debt

Most of us have had some experience with debt in our own households, and appreciate its inherent hazards. We know that debt permits us to attain higher levels of living, but that its repayment involves a lowering of living levels. Debt also creates the possibility of default if it is overextended or if unforeseen circumstances impair the ability to repay. In bygone days such default was commonly punished by prison; today it can lead to bankruptcy proceedings. Since personal debt has these hazards, we are justified in looking at its level with certain misgivings.

There is a strong intuitive appeal for applying the same standards to public debt, but this is simply a case where our intuition is wrong. Some economists have suggested that individual debt and social debt are so different that there ought to be different words for them. Whether or not this is so, it is true that the two crucial elements of individual debt, the possibility of shifting real burdens to a future time and the possibility of bankruptcy, are absent in the case of public debt. Analogies with private household management may consequently have some significance for state and local governments or the business enterprise, but they are not appropriate for considering the significance of the internal debt of a sovereign nation.[1]

The Shifting of Burdens

In discussing the burden of debt, it is useful to distinguish between *real* burdens and *financial* burdens. The difference is analogous to the frequently made comparison between real wages and money wages. A real burden occurs when consumption is foregone or effort expended without adequate compensation. A financial burden occurs when money payments are made and received.

It is widely believed that the national debt makes it possible to shift burdens over time, so that people now living are bearing some of the burden of the Spanish-American War and children not yet born will bear some of the burden of World War II. So far as real burdens

[1] A sovereign nation, for purposes of the present discussion, may be defined as a governmental unit possessed of the power to coin money. (This power is given the federal government by our constitution.) The fifty American states have some aspects of sovereignty, but not the financial one. An internal debt is simply the debt of a sovereign nation held (as is the U.S. debt) by residents of that nation, who accept and make payments in its currency. This is true of the American debt.

are concerned, this is incorrect: an internal social debt, unlike an individual debt, does not involve any shifting of real burdens over time for the society as a whole.

Real burdens: An example

A simple example should make this clear. Let us consider a real burden that many of us now have. Many trees in my neighborhood are shedding their leaves. To rake and burn them is a real burden. I cannot avoid this burden, but private debt enables me to shift the burden to a future time. Suppose that I float a twenty-year mortgage, and use the proceeds to hire someone to do the raking. I avoid the inconvenience of leaf raking by assuming the inconvenience of a mortgage; its repayment will make it necessary for me to rake a lot of leaves (or assume some other real burden) in 1980. If I can manage a fifty-year mortgage (or perhaps a bond issue, by forming an immortal corporation) I may even be able to avoid the burden entirely by passing it on to my son (or the shareholders of the future) who will be living in 2010, although I will not.

A moment's reflection should make it clear that society cannot transfer real burdens over time by this technique. After all, if my leaves are to be raked, they must be raked by someone living in 1960. From the viewpoint of society, the burden is not avoided, it is merely transferred from one individual to another. Just as the borrowing does not really avoid a burden, the repayment does not create a real burden, for if I am inconvenienced by being forced to rake leaves in 1980, another member (or members) of society in 1980 will enjoy the luxury of having his leaves raked without effort on his part. From the social viewpoint the two transactions cancel out; there is no change in the aggregate of real burdens assumed by all people living in 1980.

In summary, individual debt makes it possible to avoid one real burden by assuming another at a different time, while social debt only rearranges burdens among individuals at the same point in time. It is easy to lose sight of these facts when we look at financial burdens, and not real burdens.

Financial burdens: An example

Now let us move to financial burdens, and take a more significant example. As an individual (and a taxpayer) I will pay some of the cost of the Spanish-American War in 1960, since part of my current income must be allocated to taxation, some of which is used to pay pensions to veterans and widows of veterans. The real costs of the war were paid in 1898, by soldiers who took time out from their normal pursuits to storm San Juan Hill and civilians who produced war material rather than civilian goods that they would have enjoyed more. But part of the financial cost is still with us. So far as society is concerned, however, the losses I and my fellow taxpayers are suffering in 1960 are offset by the gains of veterans and widows:[2] they do not impoverish the 1960 society when the entire society is considered collectively. This means that the financial burdens of 1960, while they undoubtedly exist, are quite unlike the real burdens of 1898, for they are burdens only to certain people, and not to the entire society. They involve a redistribution of income, but not a reduction in its total.

The Problem of Bankruptcy

The second element of private debt is the possibility of bankruptcy. If we define bankruptcy as the inability to meet a specific financial commitment when it comes due, it is simply impossible for an internal debt to cause bankruptcy for a sovereign nation. This follows from our definitions of bankruptcy, internal debt, and sovereignty. If one has the legal authority to create the means of payment, he can fail to pay only if he mismanages things so horribly that he does not take the trouble to calculate his accounts payable, his cash on hand, and his orders for new cash. In other words, the government can pay any dollar debt by printing dollars, if it is inconvenient to collect taxes or sell more bonds. If it does not pay a debt, it must be either because it has decided not to pay (this is repudiation, not bankruptcy) or because it has mismanaged its finances.

If bankruptcy is defined in a broader sense, it becomes possible even for a sovereign state with full rights of coinage. Let us agree to call a nation bankrupt if it does not wish to repudiate a debt, but is forced to resort to printing presses to repay, and in so doing depreciates the value of its money so far as to bring about virtual repudiation. The case of Germany after World War I is perhaps the classic instance of such bankruptcy, and in more recent times some underdeveloped countries bent on rapid industrialization have experienced a related phenomenon.

Printing press bankruptcy

Is this type of bankruptcy possible in the case of the American debt? While it is dangerous to call anything impossible, it seems most

[2] Insofar as there are administrative costs associated with the payment of pensions, government bureaucrats who keep records, locomotive engineers who drive the trains that carry checks, bank clerks who clear checks, and others will share some of these gains with the veterans and widows. Since these people, too, are realizing their gains in 1960, and not in 1898, there is still no shift of burdens over time.

unlikely. In the first place, analogies with Germany during the twenties or with contemporary underdeveloped countries are not appropriate. The debt that started the printing presses in Germany was an external debt, owed to non-Germans. As such it possessed the two crucial characteristics of household debt: its repayment did in fact make Germany poorer, and default was possible. The debts of the underdeveloped countries in the postwar era are also partly external debts held by foreign nationals and governments. They are more comparable to individual debt than to internal debt such as the American debt of 1960.

In considering the likelihood that we will be unable to pay off the debt as it comes due by collecting taxes and selling bonds, the national income is probably the most reasonable indicator of our capacity to tax and borrow. The United States, during 1945, proved that it could maintain an internal debt one and one-half times the annual national income without serious currency depreciation. The present national debt is less than three-quarters of national income; if our past experience is a reliable guide, it could be doubled without resort to the printing presses. At the close of the Napoleonic Wars, Great Britain's internal debt was well over two years' worth of national income, and yet it did not impair the stability of the government or the value of the monetary unit. This would indicate that if we have a Secretary of the Treasury who is as skilled in debt management as was the British Chancellor of the Exchequer in 1818, the United States could currently maintain a national debt of a trillion dollars without inflation.[3]

Such a debt would cause difficulties, but the experience of the past indicates that it should be possible to carry the debt by taxation and refunding without bankruptcy or resort to the expedient of inflation. Such a debt would probably increase the possibility of financial difficulties arising from mismanagement, but on the other hand, mismanagement could bring about financial collapse even if the national debt did not exist.

Two historical cases

In this connection it is interesting to examine the two cases from American economic history that came closest to the classic model of printing press inflation. The first of these was the note issues of the Continental Congresses during the American Revolution. These became so depreciated in value that they added the phrase "not worth a Continental" to the American language as a synonym for worthless. In this case, currency depreciation did not result from a high debt, for before the government resorted to the printing presses, it had no debt. The

[3] In 1819, as a matter of fact, British currency was effectively *deflated* by the resumption of specie payments which had been suspended during the war.

inflation did not result because of a large debt, but because of governmental weakness, and perhaps a lack of wisdom. The Continental Congresses could not collect taxes because the thirteen colonies retained financial sovereignty; they decided to issue non-interest bearing notes, rather than bonds, so that no one had any incentive to hold them, and they all passed immediately into circulation, which drove up prices.

Our second experience with printing press inflation came in the Confederacy during the Civil War. Again it is true that the Confederacy had no established debt, so we can hardly blame currency depreciation on the existence of a large debt of long standing. The Confederate states mistrusted strong central government, so they created a weak government without power to tax the states. The inflation can be attributed partly to the weakness of the government and partly to unwise decisions made by government officials responsible for finance. While the Union experienced a certain amount of inflation, it was never a really serious problem, although the national debt increased to more than forty times its prewar level. In large measure, the Union avoided inflation because it increased the national debt, and the Confederacy experienced inflation because the government was partly unwilling and partly unable to incur debt.

These two American case histories indicate that printing press inflation of a destructive type is more likely to be caused by fear of debt, by avoidance of debt, or by unwillingness to assume debt rather than by the existence of debt.

The Budget and the Debt

As a new accounting period approaches, the fiscal authority makes two separate sets of decisions. One is on his budget: he must estimate expenditures and receipts. The other is on the debt: he must decide whether it is to be increased, reduced, or maintained. In the United States, these decisions are separated at the federal level. Congress passes one set of laws appropriating money, another set requiring taxes, and still another limiting the size of the debt. In the public mind, these separate decisions are often confused. A decision to incur a budgetary deficit is thought of as an automatic decision to increase the debt, and a projected budget surplus is equated with debt reduction. These relationships do not always hold true, for a budget surplus need not be devoted to debt retirement (it can be held as Treasury cash) and a reduction of Treasury cash or the issuance of new Treasury currency can be used to offset deficit spending without any increase in the national debt.

These are not far-fetched theoretical arguments. In eight years of the past 86 (most recently in 1916) the national debt increased while the budget held a surplus; in six years of the past 56 (most recently in 1914) the debt fell although the budget showed a deficit. In more recent years, while the direction of changes has been as expected, the magnitudes have varied substantially. During fiscal 1959, for example, a budget deficit of $12.4 billion was accompanied by a debt increase of only $8.4 billion; in 1958 a budget deficit of $2.8 billion occurred along with a debt increase of $5.8 billion. If we look at the debt, rather than the budget, we will misinterpret the role of the government in the economy during fiscal year 1959 by almost 50 percent, and during 1958 by over a hundred percent.

Nor is there a direct short-run relationship between the government's extravagance (or parsimony) and the size of the national debt. Our present government may be too extravagant, or too frugal, but we cannot pass judgment on this point by looking at the level of the national debt or the changes in its level since the present administration took office. We must look at the budget, at the wisdom of specific expenditures, and at our tax laws. We must consider the external threats to our military security, the role of the underdeveloped countries in the world's economic and political future, and a host of other variables. Looking at the national debt may be an easy way to make judgments, but it cannot be an accurate one.

Excessive spending and debt

The size of the national debt does not even give us an accurate picture of the collective extravagance or irresponsibility of all past governments. Every wasteful expenditure that has ever been made might have been financed by non-debt means (most notably money creation) and much of the existing debt (for example, the bonds held as reserves by the Social Security Administration) has not yet been utilized to finance any government spending at all, wasteful or otherwise. It is important that government spending be subjected to scrutiny in order that avoidable waste may be prevented, but an examination of the national debt makes no contribution to this scrutiny.

A further confusion between the budget and the debt sometimes arises when fiscal policy is discussed. Under some conditions, deficit spending may serve a useful social purpose in helping to cure a depression, and a budget surplus may be desirable as an inflation stopper. In neither case does the national debt influence the level of economic activity—it is the government spending that helps cure the depression and the taxation that helps check inflation. As a matter of fact, the budget deficit would produce the biggest increase in aggregate demand if it were covered by new money creation with no debt increase, and the budget surplus would be most effective in checking inflation if it were held in idle cash and not devoted to debt reduction. We can neither promote full employment by running up the national debt (although we can do so with government spending which may incidentally increase the debt) nor check inflation by reducing the debt (although we can do so with taxation, which may incidentally lessen the debt).

Inflation and the Debt

Inflation is a complicated process, and it cannot be identified with a single variable. The national debt can contribute to inflation only in a round-about way. When resources are fully employed, a budget deficit will tend to increase the national debt, and will also tend to generate inflation. This is commonly the case during wartime. In this process, inflation and debt are the joint results of another factor; it is the deficit, and not the debt that is the causal agent. There seem to be other agents at work, too, for we have experienced inflation while the budget held a surplus (this happened in 1947 and 1948) and deflation while a substantial deficit existed (these were striking characteristics of the period from 1930 to 1933).

At any rate, the level of the national debt cannot have much to do with inflation. It is the changes in the national debt that are significant. The deficits from 1860 to 1865 probably caused just as much inflation as the deficits from 1942 to 1945, even though the national debt was only a hundredth as high. There is no reason to believe that a high national debt will cause inflation as long as the budget is balanced.

Debts and Credits

The discussion thus far has largely been a negative one, aimed at showing what the debt cannot do, rather than what it can do. It is not a burden to future generations (when the individuals making up the generation are considered collectively), nor is it a likely cause of government bankruptcy, an indicator of irresponsible extravagance on the part of government, or an unqualified cause of inflation. In making a positive answer, and showing the significance of the national debt, it is necessary to consider that every debt involves a credit.

This means that the people, corporations (including banks and insurance companies) and units of government in the United States now hold about $288 billion of government bonds and certificates of indebtedness, which they regard as valuable assets. If no national debt existed, all agencies now holding government bonds would have to operate quite differently. Banks and private insurance companies, for example, would have to find alternatives to government bonds promising equal security, liquidity, and rates of return or hold less desirable assets. If they could not find equally desirable assets in the same volume (and the magnitudes involved are quite large) they would be forced to suffer lower profits, or even losses, or else increase service charges and premium rates to their customers.

Reliance on the debt

Many government agencies would also be required to make sweeping changes in their operations if no debt existed. The Social Security Administration, for example, would lose income that it currently derives from interest paid on the government bonds in its trust fund. If there were no federal debt, it would be forced to invest the fund in private securities (which is now forbidden by law, and would lead to very considerable government ownership of the means of production), lower the scale of benefits, or seek an increase in payroll tax rates or an overt government subsidy.

Fraternal and educational institutions would likewise be forced to make drastic changes in their finances, as would many units of state and local government. If the debt were reduced rapidly, the results could be chaotic, particularly so far as banks and insurance companies are concerned, but even if the change came about gradually some fundamental adjustments would be required.

In these facts we have a partial affirmative answer to the question of the national debt's significance: it has become an integral part of our whole financial structure, and permits our financial institutions to operate the way they do. Of course banks, insurance companies and the Social Security Administration could operate in the absence of a national debt, but they could not operate in the manner to which we have become accustomed, and the complete retirement of the national debt would produce many changes, some of them truly revolutionary and all of them hard to foresee.

The Transfer Problem

In the example of the Spanish-American War given earlier, it was emphasized that the financial burden of the debt involved a transfer of income, if not a reduction of total income. This transfer of income may now be considered as a problem. It constitutes another positive significance of the debt.

Interest on the federal debt is currently something like $9 billion per year, or more than the total of the debt in any year before World War I. During 1960, because we have a national debt, this sum will be taken away from some people and given to others. It is most difficult to establish the identities of these two groups. The first group, those who lose, can probably be identified with the taxpayers (especially if the budget is in balance, so that the interest payments are met by taxation, and not by further borrowing). But who pays the taxes is a difficult question in an economy where taxes are shifted and hidden. The second group, those who gain, consists of bondholders. But there are problems even here. Banks, for example, are bondholders and so must gain. But the gains are shared in uncertain proportions by bank stockholders (higher dividends), bank managers (higher salaries), and bank customers (lower service charges, higher interests on deposits, and lower interest rates on loans).

Transfers and income distribution

Most economists argue that the transfers of income caused by the interest payments on the national debt are regressive, which is to say that the people who lose are relatively poor and the people who gain are relatively rich, so that the transfer distributes income less equally than it would be if there were no transfers. This conclusion cannot be advanced with certainty, but there is a high probability that the national debt has this effect, and many people would find this undesirable.

Most economists would also argue that the transfer of $9 billion per year from taxpayers to bondholders tends to reduce the national product. This is because those from whom the money is taken lose a certain incentive to produce, and those to whom it is given receive no heightened incentive. As such, the interest on the national debt produces the same undesirable results as any sterile transfer of income not matched by productive services.

Should the national debt be reduced by future taxation, this reduction would also bring about transfers of income with the same general characteristics as the transfers involved in the annual interest payments, the taxpayers would lose and the bondholders would gain. If the transfers are bad, they can be eliminated only by bringing on other transfers which may be as bad as the existing ones.

Debt Management and Monetary Policy

Over the years, dealing in government securities has become the major instrument of monetary policy utilized by the Federal Reserve system. The Federal Reserve banks effect a tight money policy partly by selling government bonds, thus absorbing bank reserves and restricting the money supply, and they encourage easy money partly by buying bonds, thus giving banks higher reserves and permitting credit expansion. The banks are consequently involved in the government bond market on a continuing basis. The Treasury too is actively involved in the bond market, for it is constantly engaged in refunding operations, which amount to turning the debt over by replacing old bonds with new ones.

These two agencies have different responsibilities and objectives. The Treasury is under the executive control of the President. It seeks a stable value for government bonds in order to maintain its credit and the lowest interest rate possible to minimize the burden of the debt. The Federal Reserve is a creation of Congress with a considerable degree of legal autonomy and a long tradition of independent action. One of its goals is overall price stability.

A source of conflict

It is easy to imagine situations in which the operations of one agency make the task of the other more difficult, and such conflicts have occurred. The probability of conflict is greatest when the Federal Reserve enforces monetary stringency; insofar as it does so by selling government bonds, it drives their prices down, increasing interest rates and the interest burden on the taxpayer and the Treasury, and making it more difficult to sell new issues. A conflict of this sort, between the Treasury and the Federal Reserve, reached serious proportions in 1951. It was finally resolved by a gentlemen's agreement which leaves the door open for future conflict.

A different, but related, conflict, this time involving the President and Congress as well as the Federal Reserve Board, arose in recent months. After a long period of credit restriction drove interest rates to a point approaching the ceiling for long-term government bonds, the President asked Congress to raise the ceiling. Congress refused to do so. While recent money market changes have deprived this issue of its vitality, there is certainly a possibility that this difficulty may arise again.

These conflicts, past and future, point up the fact that a large national debt, requiring extensive refunding operations, creates problems that would not otherwise exist. This is not to say that they cannot be solved, but the fact that they have not yet been solved means that a large national debt, even if it is unchanging and accompanied by a balanced budget, cannot be dismissed as a matter of no significance.

Conclusions

The most common errors made in assessing the national debt consist of assigning too much importance to it, and worrying too much about its catastrophic results that are impossible or improbable. While these positions should be rejected, it is an error to dismiss the national debt as a matter of no significance because we owe it all to ourselves. The national debt has, during its 171-year history, become an integral part of our financial structure. For better or worse, our institutions have adjusted to it, and any drastic changes in the debt would lead to changes in our financial institutions that we cannot fully anticipate. The debt makes necessary a continual flow of income transfers. It is quite possible that these are regressive. It is also possible that an attempt to eliminate them would do more harm than good. Finally, the national debt impedes monetary and fiscal policy, probably makes it less flexible, and engenders intragovernmental conflict.

VI

Inflation, Stability, and Growth

The objectives set for monetary policy, as well as fiscal policy and debt management, have been presented in earlier sections as "economic goals." The relationship of policies and tools to objectives such as growth and stability, as well as to each other, have been discussed in many of the readings. It is now necessary to place these policies and tools in the broader context of general macro-economic theory, and to consider the goals themselves in the same context. For instance, to what degree is growth in real terms affected by a fiscal policy that attempts to balance the budget? What other economic forces may be relevant to this problem? William J. Baumol and Maurice H. Peston examine these questions in the first article. Similarly, what are the causes of inflation, and what is the relationship of inflation to employment and national growth? Will a policy that stabilizes the price level also cause unemployment and inhibit economic growth? Paul A. Samuelson and Robert M. Solow suggest some answers to these questions.

An understanding of the workings of the economy in the aggregate is needed if we are to assign monetary and fiscal policies their appropriate roles. Warren L. Smith presents a versatile model of the economy in the third article that permits us to study the manifold effects of changes in policy and the ways in which they may contribute to the attainment of economic goals. In the final article Evsey D. Domar discusses the rate of growth in national income that is necessary to maintain full employment. His analysis and its conclusions are of great importance if monetary and fiscal policies are to be used to attain growth and full employment.

30

More on the multiplier effects of a balanced budget

*William J. Baumol and Maurice H. Peston**

It seems to have become widely accepted among economic theorists[1] that an increase in government expenditures on goods and services matched by an equal rise in taxes, will, subject to qualifications to be indicated presently, tend to result in a rise in national income equal to the tax-expenditure change.[2] In this paper we shall maintain that this argument is likely to be misleading in two respects. First, it suggests that there is something unique about the theory of the balanced budget multiplier which differentiates it sharply from other multiplier theory, in particular, because, as we shall see, the public's marginal propensity to save is alleged to be irrelevant for the process.[3] Second, the argument is misleading in that it appears by a feat of magic to be able to determine an empirical magnitude (the value of the multiplier) without the

Reprinted with permission from *The American Economic Review*, Vol. XLV, No. 1 (March, 1955), pp. 140-148.

* The authors are respectively professor of economics and research assistant in economics at Princeton University. [*Editors' note:* Maurice H. Peston is now at Queen Mary College (London, England).]

1 Most of the substance of this paper we owe to Professor Viner and Mr. Turvey. The former spent much time and effort in convincing one of us of the error of his views on this matter and a considerable part of our analysis can be traced to a rather obscure passage in the cited article by Turvey (pp. 284-86 [see end of this paper for all references]) where he had come to the same conclusions. While they both made highly useful suggestions on an earlier draft of this note they can, of course, not be held responsible for what happened to the baby after we adopted it, any more than can Professor Chandler who was our severest and most helpful critic. We must also express our gratitude for very useful comments to those who attended the faculty-graduate meetings of the economics departments of the University of California at Berkeley and Stanford University where this paper was given.

2 For an extensive list of references see Samuelson's article, p. 140, footnote 5.

3 But see Gurley's paper for a unified treatment which fits the balanced budget case directly into the general analysis.

use of any empirical material. We believe that in this respect, though it involves correct deduction from its peculiar premises, the unit multiplier argument is really likely to be an irrelevant tautology because, as we shall argue, in practice there is very little assurance that unity is even a rough approximation to the multiplier associated with any balanced-budget expenditure program which a government may be expected to undertake.

Indeed, we have no reliable evidence on which to preclude even negative balanced budget multipliers or multipliers considerably in excess of unity. To a large extent this is the result of "leakages" in governmental expenditures, some of which have already been noted in the literature implicitly or explicitly. But we believe the full extent of their ability to affect the magnitude of the multiplier has not been recognized.

Our argument will be seen to apply equally well, *mutatis mutandis,* to all multiplier analysis, but it seems to us that the misunderstanding has been greatest in the balanced budget case.

We can perhaps best review the argument in question by quoting the following passage from Wallich (pp. 79-80), one of the original expositors.

> Suppose that national income is at a level of 130 billion dollars, and that this leaves without employment resources capable of producing ten billion dollars worth of goods and services. Now the personal income tax is raised to yield an extra ten billion dollars, and this money is employed, say, to build more roads, provide more free education, etc., previously idle resources thus finding employment. Incremental taxes and expenditures are so adjusted that the consumption and saving habits of taxpayers as a group are the same as those of the newly employed who are the recipients of incremental government expenditures. It is true that the additional tax reduces by ten billion dollars the purchasing power of those who are employed to begin with; this purchasing power, however, is not destroyed, but merely transferred to the hands of the newly employed. Since in their joint effect the additional taxes and expenditures are, by assumption, non-progressive, the consumption demand exercised by the newly employed will be the same as that which would have been exercised by the initially employed, had they not had to pay the tax. Aggregate private demand and private output therefore remain unchanged, but meanwhile there is an increase in government output equal to the ten billion dollars produced by the previously unemployed, which is paid for out of taxes and therefore is not dependent upon the level of aggregate demand. Thus national income has been raised to a total of 140 billion dollars.[4]

[4] Note that this argument attributes no peculiar powers to government. If it is valid, the same effect could, for example, be produced by a firm which simultaneously increased its saving and its investment expenditure by the same amount (*cf.* Samuelson, p. 151). It might be thought (as we did) that an equal change in imports and exports will also have a unit multiplier effect. But this is not so because, while a governmental balanced budget purchase will (at least initially) reduce the excess supply of goods available to the private sector, imports will increase supplies by just the amount they are decreased by the change in exports.

There are several well-known provisos attached to the argument:

1. The expenditure in question must involve a governmental purchase of goods and services rather than a transfer payment.[5] For a transfer payment merely redistributes income, and adds no government effective demand for output to that of the public.

2. The propensity to consume of the recipients of the expenditures must, on balance, be the same as those of the payers of the tax.[6] Otherwise, even a transfer payment can affect national income if it takes money out of the hands of nonspenders and gives it to spenders. And governmental balanced-budget expenditure on goods and services will affect national income by changing private demands as well as by increasing the government's demands, and so increase national income by more or less than the amount the government spends.

3. The government's program must not affect the level of private investment. In particular it must not affect the businessmen's views on the profitability of investment.[7] If, for example, they are led to fear increased governmental competition or if the taxes affect their incentives adversely, their reduced investment could offset some or even all of the expansionary effects of a governmental balanced-budget expenditure program. Similarly, businessmen's confidence in the program's ability to increase effective demand might further increase its effectiveness.

4. A closed economy is sometimes assumed. The reasons for this are not explained, though they are perhaps regarded as obvious.[8]

All of these qualifications already suggest that in practice a balanced budget might cause national income to go up by considerably more or less than the amount spent by the government. Later considerations will indicate that income might sometimes even be reduced.[9]

Though the effects of taxes have been mentioned in the literature and have already been referred to, there is somewhat more to be said

[5] See, *e.g.*, Wallich, p. 82; Samuelson, pp. 142-43.

[6] See, *e.g.*, Haavelmo [1] p. 311; Wallich, p. 80; Samuelson, pp. 140, 141 footnote 6. It is also assumed by several of these writers that household savings is the only leakage in the system. Peston (pp. 129-30) has shown, however, (challenging Turvey on this point) that the balanced budget theorem is consistent with the existence of other leakages, and with indirect taxes in particular.

[7] See, *e.g.*, Wallich, p. 38; Samuelson, pp. 140, 141, footnote 6.

[8] See, *e.g.*, Haavelmo [1] p. 314; but see White, p. 357 for an explicit discussion. Another qualification which has received considerable attention in the literature involves the possibility of lags. Thus if there is a time lag between the receipt of the taxes and their disbursement, only the long-run multiplier will approach unity. See the papers by Haberler, Goodwin, Hagen and Haavelmo [2]. Other types of lag can be handled in the same way—for example there may be a lag between the time a businessman receives the money spent by the government on goods out of his inventory (when part of the money just adds to his liquid capital) and the time he hires labor, etc., to replace this inventory at which time that portion of the money first constitutes income.

[9] Most of the points raised in the next five paragraphs are made in considerably greater detail in White's valuable article where they are raised as part of a general discussion of the problems of empirical multiplier computation. But no mention is made of the consequences for the balanced budget multiplier theorem.

in this connection. The unit multiplier argument takes the collection of T dollars in taxes to be equivalent to that much of a reduction in the public's income. For only then can it be argued simply that if taxpayers and expenditure recipients have the same marginal propensities to consume, the government's tax expenditure program will not affect private consumption demands. But taxes are ordinarily not the poll taxes of pure income redistribution theory. Besides their income effect they usually also have some substitution effects and result in changes in plans other than what might be expected to follow from a simple change in income.

Perhaps most obvious is the case of a general sales tax which, in so far as it is a tax on consumption, may encourage saving at the expense of consumption. Thus a given increase in tax receipts, if obtained in this way, may reduce consumption expenditure more than would the same reduction in disposable income. We may suspect, therefore, that the government's expenditure of the same amount will not raise the public's consumption by as much as its taxes have reduced it. In this case, the multiplier will be less than unity.[10]

For the same reasons we may expect that a tax on savings or a (foreseen) capital levy which is matched by an equal governmental expenditure on goods and services may well have a multiplier effect greater than one as people rush out to increase their consumption in an effort to avoid the taxes. Death duties and capital-gains taxes too may conceivably cause less of a contraction in the public's expenditures than does a simple reduction of disposable income by the same amount. A tax on imports (tariff) may actually even increase consumer demand for domestically produced articles, particularly if the elasticity of demand for imports is very high. Even if the money is collected out of ordinary income taxes, the balanced budget multiplier cannot be presumed to be unity, for people's incentives to earn income may be affected.

Having examined the assumptions on the nature of the tax structure inherent in the unit balanced budget multiplier argument we shall now, in a similar way, question its premises on the nature of governmental expenditures. The passage from Wallich quoted above suggests that the sort of governmental expenditure program to which the unit balanced budget multiplier argument is meant to apply is typified by road building, free education and the like. But such programs usually involve, to at least a small extent, the purchase of items other than domestically currently produced goods and services. While they may involve no transfer payments, they will usually include some purchases on capital

[10] In Turvey's model such a situation yields a multiplier of zero because he assumes a "money illusion" which leads people to keep their money expenditures unaffected by sales taxes. In other words, consumption is made a function of disposable income, which is defined as receipts from factor sales plus transfers minus income taxes; and indirect taxes are made a function of consumption.

account rather than on income account, and they will usually have some import content. For example, road or school-building construction often involves the purchase of land which affects the seller by increasing his liquidity rather than his income. This may induce him to spend more, but presumably not as much as the amount by which the taxpayers have reduced their expenditure.[11] The result will be similar if some of the government's purchases involve imported ingredients or items obtained out of unwanted inventory, old buildings, etc.

The existence of these "leakages" in a governmental tax-expenditure program—the nonredistributional effects on private consumption, the purchase of items on capital account and of goods from abroad—is, of course, well known, though no attempt seems to have been made to take account of them in the literature. Possibly some of the writers had them in mind but considered them unimportant, since presumably none of them considered the unit multiplier figure as more than an approximation to the empirical magnitude in view of the recognized qualifications. The magnitude of the leakages may indeed be rather small, and it is very tempting to conclude that if 10 per cent of the government's balanced budget expenditure is "leaked," the multiplier will be reduced from unity to say about 0.9. However, we argue now that fairly small leakages can even produce a negative balanced budget multiplier.[12]

Suppose, for example, that T dollars end up being collected in taxes all of which are spent in various ways, many of which involve some sort of leakage. Let us say, then, that of this amount only E dollars end up being added to expenditure on domestic goods and services coming out of current production.[13] National income will then end up being increased by the E dollars of the increased income-creating expenditure, but it will be reduced by private expenditures foregone as a result of the $E - T$ dollar reduction in private disposable income. Taking c as the marginal propensity to consume, this reduction in private expenditure, as is well known, is represented by

$$c(E-T)+c^2(E-T)+\ldots+c^N(E-T)+\ldots=\frac{E-T}{1-c}-(E-T)$$
$$=\frac{E-T}{s}-(E-T)$$

[11] Of course, this does involve a change in private investment but not necessarily a change in the inducement to invest.

[12] For a result which is structurally the same as the following see Higgins, pp. 394-95.

[13] Note that T and E represent final taxes and income-creating expenditures. Actually, for example, T dollars may be collected as a result of a different amount, T^*, of new taxes being levied. For, by affecting the private sector's expenditures, the tax measure may affect the amount of taxes collected with any given tax structure. Similarly the legislated increase in expenditures may differ in magnitude from the increase actually attained, *e.g.*, because of the effect on employment compensation payments. For an analysis in terms of the taxes and expenditures initially undertaken, see Turvey.

where s is the private sector's marginal propensity to "save," that is to save, import and purchase items not currently produced or replaced.[14] Thus the total change in income, ΔY, will be given by the sum of this expression and the initial E dollar contribution to national income. This gives

$$\Delta Y = E + \frac{E-T}{s} - (E-T) = \frac{E-T}{s} + T.$$

The multiplier will then be

$$\frac{\Delta Y}{T} = \frac{E-T}{sT} + 1$$

which has a unit value in the special case where E = T.

Let k be the proportion of the tax not being added to income-creating expenditures so that $(1 - k)\ T = E$. Then our multiplier becomes

$$\frac{\Delta Y}{T} = \frac{(1-k)T - T}{sT} + 1 = \frac{1-k-1}{s} + 1 = 1 - \frac{k}{s}.$$

Thus the multiplier will be positive, zero, or negative as the marginal propensity to save and import of the private sector is greater than, equal to, or less than k, the government's "marginal propensity to leak."

We can see now how misleading is Haavelmo when he argues that "... [Kaldor's] statement would seem to convey the idea that taxes equal to public expenditure can create employment only to the extent that they cut down on people's savings. This is not correct. We shall show below that public expenditures covered by taxes have an employment-generating effect which is *independent* of the numerical value of the propensity to consume."[15]

In line with what might be expected intuitively, our analysis indicates that a balanced-budget expenditure has an expansionary effect only if the government's "marginal propensity to leak" is less than the public's, that the expansionary effect will vary directly with the size

[14] It will be noticed that c is not the straightforward marginal propensity to consume of elementary macro-theory. Rather it is the marginal propensity to consume currently domestically produced goods and services which are renewed. The significance of many of the leakages referred to in the text is that sometimes a consumption change involves a change in investment in the form of imports or inventory variation. This, however, is not necessary to ensure a balanced budget multiplier other than unity.

[15] Haavelmo [1] p. 312. The italics are Haavelmo's. See also the approving reference to this in Samuelson, p. 141, who objects to the conclusion only in the unstable case where s = 0.

of the public's marginal propensity to save, import, etc.,[16] and that the unit multiplier analysis is a direct consequence of the assumption that the government's marginal propensity to consume out of current domestic production of goods and services is unity. In this way we have removed the apparent peculiarity from the theoretical structure of the balanced budget multiplier which can now be seen to fit in directly with other multiplier analysis.

Furthermore, the analysis suggests that negative multipliers (or multipliers considerably greater than unity) are by no means out of the question.[17] We do not have adequate empirical information to assert that government's marginal propensities to import are typically smaller than those of the private sectors. Moreover, purchases on capital account may usually constitute a small proportion of public works expenditures but we really know very little about the magnitude of the private sector's marginal propensity to save or purchase on capital account against which this should be balanced. If we follow the usual multiplier analysis in assuming away any effects which occur outside the consumption sector, then business leakages are irrelevant. Of the remaining leakages certainly the average private propensity to save in this country has usually been rather small (about 1/20 to 1/10) and if the corresponding marginal propensity is anywhere in this neighborhood it might not take much in the way of governmental leakages for balanced-budget expenditure programs to exercise a contractionary effect on national income.

If the business sector were included in the discussion, we would add both to the induced expenditures and to the induced leakages. The average propensity to save would then no longer be negligible, but in this situation saving would only be a leakage if it were not undertaken to increase investment, *i.e.*, if it represented hoarding. This suggests that, if the public's marginal propensities to import and buy on capital account were the same as those of the government, the balanced budget multiplier might typically be rather close to zero. For the derivation of the usual proportion of the increased governmental revenue from excise taxation might, by discouraging consumption, tend to offset the small expansionary effect which works through the public's propensity to hoard. Of course, this sort of reasoning is always treacherous, and really insufficient to establish a presumption that the multiplier will in fact

[16] It is perhaps rather paradoxical to conclude that the magnitude of the balanced budget multiplier will be larger the greater is the marginal propensity to "save" of the private sector. The explanation is quite simple. Some of the money the government taxes away will not be returned to the public as income. The damaging effect will then be minimized if the public would have "saved" a large proportion of the income of which it has been deprived.

[17] On this point the exposition in Baumol and Chandler is rather unsatisfactory. See pp. 349-51. But *cf.* p. 371, question 1.

be low. Our aim is merely to indicate that on the information now available this possibility (among others) cannot be ruled out, and to suggest that *we really do not know much about its magnitude.*

In fact, there is no reason to believe that the magnitude of the balanced budget multiplier will remain unchanged with changing time and circumstances. The following remark is presented primarily as an illustration of one way in which such variation may come about.

If it is true that the public's marginal propensities to save and import fall in a depression and rise at higher income levels, the expansionary effect of a governmental tax-balanced expenditure will tend to be reduced during periods of low national income and increased in more prosperous times. For, as we have seen, the higher the private sector's marginal propensity to "save," the lower will be the balanced budget multiplier. Thus an increase in balanced-budget expenditures may have untoward effects at all times, aggravating the cycle at both extremes.[18]

This result may perhaps be supported by an observation which involves the asset position of consumers. During a depression it is not unlikely that some people's consumption expenditure will be financed out of borrowing (which frequently takes the form of unavoidable arrears on rent payments, grocery bills, etc.). If it is assumed that these loans do not involve a decrease in consumption on the part of lenders, then, *ceteris paribus*, their repayment—in so far as it involves directly only an asset variation—will constitute a leakage in any expansion program. In other words, consumption out of a given income will be less than normal while consumers restore their net asset positions. It must be noted that often they do not have much choice in this matter. In addition, this consideration seems to lead to the conclusion that the sooner the expansion program gets under way, the larger will be its multiplier effect although other factors may be working in the opposite direction.

Of course, the government is not entirely powerless in these matters. By deliberately altering the composition of its purchases it can manipu-

[18] A. G. Hart has suggested to one of the authors that the effect of investment on the balanced budget multiplier may work in the opposite way over the course of the cycle. The matched taxes and expenditures can reduce the liquid assets of business firms if they are designed to transfer purchasing power from nonconsumers to consumers. Many firms seem to be willing to invest out of borrowing only when it is supplemented by the flow of their own savings. In such a case their marginal propensity to invest may exceed unity. Thus any associated reduction in the earnings of business firms may, through its effect on investment, substantially reduce the magnitude of the balanced budget multiplier in prosperous, or even in moderately depressed times. In deep depression, argues Hart, investment will be at a minimum. Only absolutely unavoidable expenditures of this sort may be undertaken. Since the demand for investment will then be relatively inelastic the balanced budget multiplier will not be much affected by any reduction in disposable corporate income which results from an equal increase in governmental expenditures and taxes.

late its "marginal propensity to leak." For example, a successful "Buy American" program could raise the balanced budget multiplier.

REFERENCES

W. J. Baumol and L. V. Chandler, *Economic Processes and Policies* (New York, 1954).

R. M. Goodwin, "Multiplier Effects of a Balanced Budget: The Implication of a Lag for Mr. Haavelmo's Analysis," *Econometrica*, Apr. 1946, XIV, 150-51.

J. G. Gurley, "Fiscal Policies for Full Employment," *Jour. Pol. Econ.*, Dec. 1952, LX, 525-33.

T. Haavelmo [1], "Multiplier Effects of a Balanced Budget," *Econometrica*, Oct. 1945, XIII, 311-18.

——— [2], "Multiplier Effects of a Balanced Budget: Reply," *Econometrica*, Apr. 1946, XIV, 156-58.

G. Haberler, "Multiplier Effects of a Balanced Budget: Some Monetary Implications of Mr. Haavelmo's Paper," *Econometrica*, Apr. 1946, XIV, 148-49.

E. E. Hagen, "Multiplier Effects of a Balanced Budget: Further Analysis," *Econometrica*, Apr. 1946, XIV, 152-55.

B. Higgins, "A Note on Taxation and Inflation," *Can. Jour. Econ. Pol. Sci.*, Aug. 1953, XIX, 392-402.

M. H. Peston, "A Note on the Balanced Budget Multiplier," *Am. Econ. Rev.*, Mar. 1954, XLIV, 129-30.

P. A. Samuelson, "The Simple Mathematics of Income Determination," in *Income, Employment, and Public Policy, Essays in Honor of Alvin H. Hansen* (New York, 1948).

R. Turvey, "Some Notes on Multiplier Theory," *Am. Econ. Rev.*, June 1953, XLIII, 275-95.

H. C. Wallich, "Income-Generating Effects of a Balanced Budget," *Quart. Jour. Econ.*, Nov. 1944, LIX, 78-91.

W. H. White, "Measuring the Inflationary Significance of a Government Budget," *Internat. Monetary Fund Staff Papers*, Apr. 1951, I, 355-78.

31

Analytical aspects of anti-inflation policy

Paul A. Samuelson and Robert M. Solow

I

Just as generals are said to be always fighting the wrong war, economists have been accused of fighting the wrong inflation. Thus, at the time of the 1946-48 rise in American prices, much attention was focused on the successive rounds of wage increases resulting from collective bargaining. Yet probably most economists are now agreed that this first postwar rise in prices was primarily attributable to the pull of demand that resulted from wartime accumulations of liquid assets and deferred needs.

This emphasis on demand-pull was somewhat reinforced by the Korean war run-up of prices after mid-1950. But just by the time that cost-push was becoming discredited as a theory of inflation, we ran into the rather puzzling phenomenon of the 1955-58 upward creep of prices, which seemed to take place in the last part of the period despite growing overcapacity, slack labor markets, slow real growth, and no apparent great buoyancy in over-all demand.

It is no wonder then that economists have been debating the possible causations involved in inflation: demand-pull versus cost-push; wage-push versus more general Lerner "seller's inflation"; and the new Charles Schultze theory of "demand-shift" inflation. We propose to give a brief survey of the issues. Rather than pronounce on the terribly difficult question as to exactly which is the best model to use in explaining the recent past and predicting the likely future, we shall try to emphasize the types of evidence which can help decide between the conflicting theories. And we shall be concerned with some policy implications that arise from the different analytical hypotheses.

Reprinted with permission from *The American Economic Review,* Vol. L, No. 2 (May, 1960), pp. 177-194. Paul A. Samuelson and Robert M. Solow, Massachusetts Institute of Technology.

History of the debate: The quantity theory and demand-pull

The preclassical economists grew up in an environment of secularly rising prices. And even prior to Adam Smith there had grown up the belief in at least a simplified quantity theory. But it was in the neoclassical thought of Walras, Marshall, Fisher, and others that this special version of demand determination of the absolute level of money prices and costs reached its most developed form.

We can oversimplify the doctrine as follows. The real outputs, inputs, and relative prices of goods and factors can be thought of as determined by a set of competitive equations which are independent of the absolute level of prices. As in a barter system, the absolute level of all prices is indeterminate and inessential because of the "relative homogeneity" properties of these market relations. To fix the absolute scale factor, we can if we like bring in a neutral money. Such money, unlike coffee or soap, being valued only for what it will buy and not for its intrinsic utility, will be exactly doubled in demand if there is an exact doubling of all prices. Because of this important "scale homogeneity," fixing the total of such money will, when applied to our already determined real system of outputs, factors, and relative prices, fix the absolute level of all prices; and changes in the total of such money must necessarily correspond to new equilibria of absolute prices that have moved in exact proportion, with relative prices and all real magnitudes being quite unaffected.[1]

As Patinkin and others have shown, the above doctrines are rather oversimplified, for they do not fully analyze the intricacies involved in the demand for money; instead they ignore important (and predictable) changes in such proportionality coefficients as velocity of circulation. But by World War I, this particular, narrow version of demand-pull inflation had more or less triumphed. The wartime rise in prices was usually analyzed in terms of rises in the over-all money supply. And the postwar German inflation was understood by non-German economists in similar terms.

But not all economists ever agree on anything. Just as Tooke had eclectically explained the Napoleonic rise in prices partially in terms of the war-induced increase in tax, shipping, and other costs, so did Harold G. Moulton and others choose to attribute the World War I price rises to prior rises in cost of production. And it is not without

[1] But as Hume had early recognized, the periods of rising prices seemed to give rise to at least transient stimulus to the economy as active profit seekers gained an advantage at the expense of the more inert fixed-income, creditor, and wage sectors. The other side of this Hume thesis is perhaps exemplified by the fact that the post-Civil War decades of deflation were also periods of strong social unrest and of relatively weak booms and long periods of heavier-than-average depressions—as earlier National Bureau studies have suggested.

significance that the great neoclassical Wicksell expressed in the last years of his life some misgivings over the usual version of wartime price movements, placing great emphasis on movements in money's velocity induced by wartime shortages of goods.

Of course, the neoclassical writers would not have denied the necessary equality of competitive costs and prices. But they would have regarded it as superficial to take the level of money costs as a predetermined variable. Instead, they would argue, prices and factor costs are simultaneously determinable in interdependent competitive markets; and if the level of over-all money supply were kept sufficiently in check, then the price level could be stabilized, with any increases in real costs or any decreases in output being offset by enough backward pressure on factor prices so as to leave final money costs and prices on the average unchanged.

Many writers have gone erroneously beyond the above argument to untenable conclusions such as the following: A rise in defense expenditure matched by, say, excise taxes cannot raise the price level if the quantity of money is held constant; instead it must result in enough decrease in wage and other factor costs to offset exactly the rise in tax costs. Actually, however, such a fiscal policy change could be interpreted as a reduction in the combined public and private thriftiness; with M constant, it would tend to swell the volume of total spending, putting upward pressure on interest rates and inducing a rise in money velocity, and presumably resulting in a higher equilibrium level of prices. To roll back prices to their previous level would take, even within the framework of a strictly competitive neoclassical model, a determined reduction in previous money supply. (This illustrates the danger of going from the innocent hypothesis, that a balanced change in all prices might in the long run be consistent with no substantive changes in real relations, to an overly simply interpretation of a complicated change that is actually taking place in historical reality.)

While the above example of a tax-induced price rise that takes place within a strict neoclassical model might be termed a case of cost-push rather than demand-pull, it does not really represent quite the same phenomena that we shall meet in our later discussion of cost-push. This can perhaps be most easily seen from the remark that, if one insisted on holding prices steady, conventional demand reduction methods would work very well, within the neoclassical model, to offset such cost-push.

Demand-pull *à la* Keynes

Aside from the neoclassical quantity theory, there is a second version of demand-pull associated with the theories of Keynes. Before and during the Great Depression, economists had become impressed with the institutional frictions and rigidities that made for downward inflexibilities in wages and prices and which made any such deflationary movements socially painful. Keynes's *General Theory* can, if we are willing to oversimplify, be thought of as a systematic model which uses downward inflexibility of wages and prices to convert any reduction in money spending into a real reduction in output and employment rather than a balanced reduction in all prices and factor costs. (This is overly simple for at least the following reasons: in the pessimistic, depression version of some Keynesians, a hyperdeflation of wages and prices would not have had substantive effects in restoring employment and output, because of infinite elasticity of liquidity preference and/or zero elasticity of investment demand; in the general form of the *General Theory*, and particularly after Pigou effects of the real value of money had been built in, if you could engineer a massive reduction in wages and costs, there would have been some stimulating effects on consumption, investment, and on real output; finally, a careful neoclassical theory, which took proper accounts of rigidities and which analyzed induced shifts of velocity in a sophisticated way, might also have emerged with similar valid conclusions.)

While the Keynesian theories can be said to differ from the neoclassical theories with respect to analysis of deflation, Keynes himself was willing to assume that attainment of full employment would make prices and wages flexible upward. In *How to Pay for the War* (1939), he developed a theory of inflation which was quite like the neoclassical theory in its emphasis upon the demand-pull aggregate spending even though it differed from that theory in its emphasis on total spending flow rather than on the stock of money. His theory of "demanders' inflation" stemmed primarily from the fact that government plus investors plus consumers want, in real terms among them, more than 100 per cent of the wartime or boomtime available produceable output. So prices have to rise to cheat the slow-to-spend of their desired shares. But the price rise closes the inflationary gap only temporarily, as the higher price level breeds higher incomes all around and the real gap reopens itself continually. And so the inflation goes on, at a rate determined by the degree of shifts to profit, the rapidity and extent of wage adjustments to the rising cost of living, and ultimately by the extent to which progressive tax receipts rise enough to close the gap. And, we may add, that firmness by the central bank in limiting the money supply might ultimately so increase credit tightness and so lower real balances as to bring consumption and investment spending into equilibrium with available civilian resources at some higher plateau of prices.

Cost-push and demand-shift theories of inflation

In its most rigid form, the neoclassical model would require that wages fall whenever there is unemployment of labor and that prices fall whenever excess capacity exists in the sense that marginal cost of the output that firms sell is less than the prices they receive. A more eclectic model of imperfect competition in the factor and commodity markets is needed to explain the fact of price and wage rises before full employment and full capacity have been reached.

Similarly, the Keynes model, which assumes stickiness of wages even in the fact of underemployment equilibrium, rests on various assumptions of imperfect competition. And when we recognize that, considerably before full employment of labor and plants has been reached, modern prices and wages seem to show a tendency to drift upward irreversibly, we see that the simple Keynesian system must be modified even further in the direction of an imperfect competition model.

Now the fact that an economic model in some degree involves imperfect competition does not necessarily imply that the concepts of competitive markets give little insight into the behavior of relative prices, resources allocations, and profitabilities. To some degree of approximation, the competitive model may cast light on these important real magnitudes, and for this purpose we might be content to use the competitive model. But to explain possible cost-push inflation, it would seem more economical from the very beginning to recognize that imperfect competition is the essence of the problem and to drop the perfect competition assumptions.

Once this is done, we recognize the qualitative possibility of cost-push inflation. Just as wages and prices may be sticky in the face of unemployment and overcapacity, so may they be pushing upward beyond what can be explained in terms of levels and shifts in demand. But to what degree these elements are important in explaining price behavior of any period becomes an important quantitative question. It is by no means always to be expected that by observing an economy's behavior over a given period will we be able to make a very good separation of its price rise into demand and cost elements. We simply cannot perform the controlled experiments necessary to make such a separation; and Mother Nature may not have economically given us the scatter and variation needed as a substitute for controlled experiments if we are to make approximate identification of the causal forces at work.

Many economists have argued that cost-push was important in the prosperous 1951-53 period, but that its effects on average prices were masked by the drop in flexible raw material prices. But again in 1955-58, it showed itself despite the fact that in a good deal of this period there seemed little evidence of over-all high employment and excess demand. Some holders of this view attribute the push to wage boosts engineered unilaterally by strong unions. But others give as much or more weight to the co-operative action of all sellers—organized and unorganized labor, semimonopsonistic managements, oligopolistic sellers in imperfect commodity markets—who raise prices and costs in an attempt by each to maintain or raise his share of national income, and who, among themselves, by trying to get more than 100 per cent of the available output, create "seller's inflation."

A variant of cost-push is provided by Charles Schultze's "demand-shift" theory of inflation. Strength of demand in certain sectors of the economy—e.g., capital goods industries in 1955-57—raises prices and wages there. But elsewhere, even though demand is not particularly strong, downward inflexibility keeps prices from falling, and market power may even engineer a price-wage movement imitative in a degree of the sectors with strong demand. The result is an upward drift in average prices—with the suggestion that monetary and fiscal policies restrictive enough to prevent an average price rise would have to be so very restrictive as to produce a considerable level of unemployment and a significant drop in production.

II

Truths and consequences: The problem of identification

The competing (although imperfectly competing) theories of inflation appear to be genuinely different hypotheses about observable facts. In that case one ought to be able to distinguish empirically between cost and demand inflation. What are the earmarks? If I believe in cost-push, what should I expect to find in the facts that I would not expect to find were I a believer in demand pull? The last clause is important. It will not do to point to circumstances which will accompany any inflation, however caused. A test must have what statisticians call power against the main alternative hypotheses.

Trite as these remarks may seem, they need to be made. The clichés of popular discussion fall into the trap again and again. Although they have been trampled often enough by experts, the errors revive. We will take the time to point the finger once more. We do this because we want to go one step further and argue that this problem of identification is exceedingly difficult. What appear at first to be subtle and reliable ways of distinguishing cost-induced from demand-induced inflation turn out to be far from airtight. In fact we are driven to the belief that aggregate data, recording the *ex post* details of completed transactions, may in most circumstances be quite insufficient. It may be necessary first to disaggregate.

Common fallacies

The simplest mistake—to be found in almost any newspaper discussion of the subject—is the belief that if money wages rise faster than productivity, we have a sure sign of cost-inflation. Of course the truth is that in the purest of excess-demand inflation wages will rise faster than productivity; the only alternative is for the full increase in the value of a fixed output to be siphoned off into profits, without this spilling over into the labor market to drive wages up still further. This error is sometimes mixed with the belief that it is possible over long periods for industries with rapid productivity increase to pay higher and increasingly higher wages than those where output per man-hours grows slowly. Such a persistent and growing differential is likely eventually to alter the skill- or quality-mix of the labor force in the different industries, which casts doubt on the original productivity comparison.

One sometimes sees statements to the effect that increases in expenditure more rapid than increases in real output necessarily spell demand inflation. It is simple arithmetic that expenditure outrunning output by itself spells only price increases and provides no evidence at all about the source or cause of the inflation. Much of the talk about "too much money chasing too few goods" is of this kind.

A more solemn version of the fallacy goes: An increase in expenditure can come about only through an increase in the stock of money or an increase in the velocity of circulation. Therefore the only possible causes of inflation are M and V and we need look no further.

Further difficulties

It is more disconcerting to realize that even some of the empirical tests suggested in the professional literature may have little or no cutting power in distinguishing cost from demand inflation.

One thinks automatically of looking at the timing relationships. Do wage increases seem to precede price increases? Then the general rise in prices is caused by the wage-push. Do price increases seem to precede wage increases? Then more likely the inflation is of the excess-demand variety, and wages are being pulled up by a brisk demand for labor or they are responding to prior increases in the cost of living. There are at least three difficulties with this argument. The first is suggested by replacing "wage increase" by "chicken" and "price increase" by "egg." The trouble is that we have no normal initial standard from which to measure, no price level which has always existed and to which everyone has adjusted; so that a wage increase, if one occurs, must be autonomous and not a response to some prior change in the demand for labor. As an illustration of the difficulty of inference, consider average hourly earnings in the basic steel industry. They rose, relative to all manufacturing from 1950 on, including some periods when labor markets were not tight. Did this represent an autonomous wage-push? Or was it rather a delayed adjustment to the decline in steel wages relative to all manufacturing, which took place during the war, presumably as a consequence of the differential efficiency of wage control? And why should we take 1939 or 1941 as a standard for relative wages? And so on.

A related problem is that in a closely interdependent economy, effects can precede causes. Prices may begin to ease up because wage rates are expected to. And more important, as wage and price increases ripple through the economy, aggregation may easily distort the apparent timing relations.

But even if we could find the appearance of a controlled experiment, if after a period of stability in both we were to notice a wage increase to a new plateau followed by a price increase, what could we safely conclude? It would be immensely tempting to make the obvious diagnosis of wage-push. But consider the following hypothetical chain of events: Prices in imperfect commodity markets respond only to changes in costs. Labor markets are perfectly competitive in effect, and the money wage moves rapidly in response to shifts in the demand for labor. So any burst of excess demand, government expenditure, say, would cause an increased demand for labor; wages would be pulled up; and only then would prices of commodities rise in response to the cost increase. So the obvious diagnosis might be wrong. In between, if we were clever, we might notice a temporary narrowing of margins, and with this information we might piece together the story.

Consider another sophisticated inference. In a single market, price may rise either because the demand curve shifts to the right or because the supply curve shifts to the left in consequence of cost increases. But in the first case, output should increase; in the second case, decline. Could we not reason, then, that if prices rise, sector by sector, with outputs, demand-pull must be at work? Very likely we can, but not with certainty. In the first place, as Schultze has argued, it is possible that certain sectors face excess demand, without there being aggregate pressure; those sectors will indeed show strong price increases and increases in output (or pressure on capacity). But in a real sense, the source of inflation is the failure of other sectors, in which excess capacity develops, to decrease their prices sufficiently. And this may be a consequence of "administered pricing," rigid markups, rigid wages and all the paraphernalia of the "new" inflation.

To go deeper, the reasoning we are scrutinizing may fail because it is illegitimate, even in this industry-by-industry way, to use partial equilibrium reasoning. Suppose wages rise. We are led to expect a decrease in output. But in the modern world, all or most wages are increasing. Nor is this the first time they have done so. And in the past, general wage and price increases have not resulted in any decrease in aggregate real demand—perhaps the contrary. So that even in a single industry supply and demand curves may not be independent. The shift in costs is accompanied by, indeed may bring about, a compensating shift in the subjectively-viewed demand curve facing the industry. And so prices may rise with no decline and possibly an increase in output. If there is anything in this line of thought, it may be that one of the important causes of inflation is—inflation.

The need for detail

In these last few paragraphs we have been arguing against the attempt to diagnose the source of inflation from aggregates. We have also suggested that sometimes the tell-tale symptoms can be discovered if we look not at the totals but at the parts. This suggestion gains force when we recognize, as we must, that the same general price increase can easily be the consequence of different causes in different sectors. A monolithic theory may have its simplicity and style riddled by exceptions. Is there any reason, other than a desire for symmetry, for us to believe that the same reasoning must account for the above-average increase in the price of services and the above-average increase in the price of machinery since 1951 or since 1949? Public utility prices undoubtedly were held down during the war, by the regulatory process; and services ride along on income-elastic demand accompanied by a slower-than-average recorded productivity increase. A faster-than-average price increase amounts to the corrective relative-price change one would expect. The main factor in the machinery case, according to a recent Joint Economic Committee study, appears to have been a burst of excess demand occasioned by the investment boom of the mid-fifties. And to give still a third variant, Eckstein and Fromm in another Joint Economic Committee study suggest that the above-average rise in the wages of steelworkers and the prices of steel products took place in the face of a somewhat less tight labor and product market than in machinery. They attribute it to a joint exercise of market power by the union and the industry. Right or wrong, it is mistaken theoretical tactics to deny this possibility on the grounds that it cannot account for the price history in other sectors.

Some things it would be good to know

There are at least two classical questions which are relevant to our problem and on which surprisingly little work has been done: One is the behavior of real demand under inflationary conditions and the other is the behavior of money wages with respect to the level of employment. We comment briefly on these two questions because there seems to us to be some doubt that ordinary reversible behavior equations can be found, and this very difficulty points up an important question we have mentioned earlier: that a period of high demand and rising prices molds attitudes, expectations, even institutions in such a way as to bias the future in favor of further inflation. Unlike some other economists, we do not draw the firm conclusion that unless a firm stop is put, the rate of price increase must accelerate. We leave it as an open question: It may be that creeping inflation leads only to creeping inflation.

The standard way for an inflationary gap to burn itself out short of hyperinflation is for the very process of inflation to reduce real demands. The mechanisms, some dubious, some not, are well known: the shift to profit, real-balance effects, tax progression, squeeze on fixed incomes. If price and wage increases have this effect, then a cost-push inflation in the absence of excess demand inflicts unemployment and excess capacity on the system. The willingness to bear the reduced real demand is a measure of the imperfectness of markets permitting the cost-push. But suppose real demands do not behave in this way? Suppose a wage-price rise has no effect on real demand, or a negligible one, or even a slight positive one? Then not only will the infliction not materialize, but the whole distinction between cost-push and demand-pull begins to evaporate. But is this possible? The older quantity theorists would certainly have denied it; but the increase in velocity between 1955 and 1957 would have surprised an older quantity theorist.

We do not know whether real demand behaves this way or not. But we think it important to realize that the more the recent past is dominated by inflation, by high employment, and by the belief that both will continue, the more likely it is that the process of inflation will preserve or even increase real demand, or the more heavily the monetary and fiscal authorities may have to bear down on demand in the interests of price stabilization. Real-income consciousness is a powerful force. The pressure on real balances from high prices will be partly believed by the expectation of rising prices, as long as interest rates in an imperfect capital market fail to keep pace. The same expectations will induce schoolteachers, pensioners, and others to try to devise institutions to protect their real incomes from erosion by higher prices. To the extent that they succeed, their real demands will be unimpaired. As the fear of prolonged unemployment disappears and the experience of past full employment builds up accumulated savings, wage earners may also maintain their real expenditures; and the same forces may substantially increase the marginal propensity to spend out of profits, including retained earnings. If there is anything to this line of thought, the empirical problem of verification may be very difficult, because much of the experience of the past is irrelevant to the hypothesis. But it would be good to know.

The fundamental Phillips schedule relating unemployment and wage changes

Consider also the question of the relation between money wage changes and the degree of unemployment. We have A. W. Phillips' interesting paper on the U.K. history since the Civil War (our Civil War, that is!). His findings are remarkable, even if one disagrees with his interpretations.

In the first place, the period 1861-1913, during which the trade-union movement was rather weak, shows a fairly close relationship between the per cent change in wage rates and the fraction of the labor force unemployed. Due allowance must be made for sharp import-price-induced changes in the cost of living, and for the normal expectation that wages will be rising faster when an unemployment rate of 5 per cent is reached on the upswing than when it is reached on the downswing. In the second place, with minor exceptions, the same relationship fits for 1861-1913 also seems to fit about as well for 1913-48 and 1948-57. And finally Phillips concludes that the money wage level would stabilize with 5 per cent unemployment; and the rate of increase of money wages would be held down to the 2-3 per cent rate of productivity increase with about 2½ per cent of the labor force unemployed.

Strangely enough, no comparably careful study has been made for the U.S. Garbarino's 1950 note is hardly a full-scale analysis, and Schultze's treatment in his first-class Joint Committee monograph is much too casual. There is some evidence that the U.S. differs from the U.K. on at least two counts. If there is any such relationship characterizing the American labor market, it may have shifted somewhat in the last fifty to sixty years. Secondly, there is a suggestion that in this country it might take 8 to 10 per cent unemployment to stabilize money wages.

But would it take 8 to 10 per cent unemployment forever to stabilize the money wage? Is not this kind of relationship also one which depends heavily on remembered experience? We suspect that this is another way in which a past characterized by rising prices, high employment, and mild, short recessions is likely to breed an inflationary bias—by making the money wage more rigid downward, maybe even perversely inclined to rise during recessions on the grounds that things will soon be different.

There may be no such relation for this country. If there is, why does it not seem to have the same degree of long-run invariance as

Phillips' curve for the U.K.? What geographical, economic, sociological facts account for the difference between the two countries? Is there a difference in labor mobility in the two countries? Do the different tolerances for unemployment reflect differences in income level, union organization, or what? What policy decisions might conceivably lead to a decrease in the critical unemployment rate at which wages begin to rise or to rise too fast? Clearly a careful study of this problem might pay handsome dividends.

III

A closer look at the American data

In spite of all its deficiencies, we think the accompanying scatter diagram in Figure 1 is useful. Where it does not provide answers, it at least asks interesting questions. We have plotted the yearly percentage changes of average hourly earnings in manufacturing, including supplements (Ree's data) against the annual average percentage of the labor force unemployed.

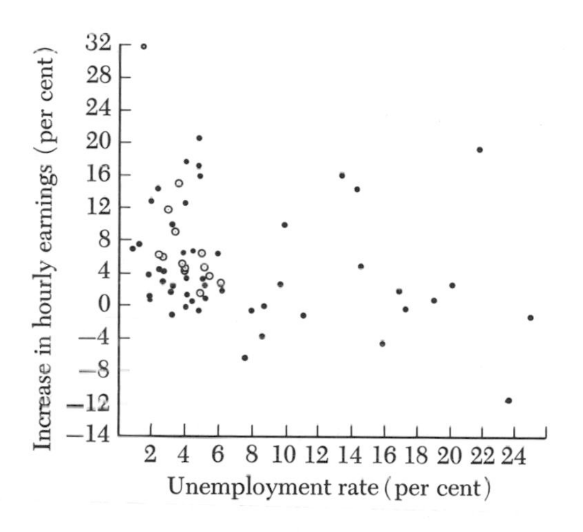

Figure 1

Phillips Scatter Diagram for U.S.
The circled points are for recent years.

The first defect to note is the different coverages represented in the two axes. Duesenberry has argued that postwar wage increases in manufacturing on the one hand and in trade, services, etc., on the other, may have quite different explanations: union power in manufacturing and simple excess demand in the other sectors. It is probably true that if we had an unemployment rate for manufacturing alone, it would be somewhat higher during the postwar years than the aggregate figure shown. Even if a qualitative statement like this held true over the whole period, the increasing weight of services in the total might still create a bias. Another defect is our use of annual increments and averages, when a full-scale study would have to look carefully into the nuances of timing.

A first look at the scatter is discouraging; there are points all over the place. But perhaps one can notice some systematic effects. In the first place, the years from 1933 to 1941 appear to be *sui generis*: money wages rose or failed to fall in the face of massive unemployment. One may attribute this to the workings of the New Deal (the 20 per cent wage increase of 1934 must represent the NRA codes); or alternatively one could argue that by 1933 much of the unemployment had become structural, insulated from the functioning labor market, so that in effect the vertical axis ought to be moved over to the right. This would leave something more like the normal pattern.

The early years of the first World War also behave atypically although not so much so as 1933-39. This may reflect cost-of-living increases, the rapidity of the increase in demand, a special tightness in manufacturing, or all three.

But the bulk of the observations—the period between the turn of the century and the first war, the decade between the end of that war and the Great Depression, and the most recent ten or twelve years—all show a rather consistent pattern. Wage rates do tend to rise when the labor market is tight, and the tighter the faster. What is most interesting is the strong suggestion that the relation, such as it is, has shifted upward slightly but noticeably in the forties and fifties. On the same hand, the first decade of the century and the twenties seem to fit the same pattern. Manufacturing wages seem to stabilize absolutely when 4 or 5 per cent of the labor force is unemployed; and wage increases equal to the productivity increase of 2 to 3 per cent per year is the normal pattern at about 3 per cent unemployment. This is not so terribly different from Phillips' results for the U.K., although the relation holds there with a greater consistency. We comment on this below.

On the other hand, from 1946 to the present, the pattern is fairly consistent and consistently different from the earlier period. The annual unemployment rate ranged only narrowly, from 2.5 per cent in 1953 to 6.2 per cent in 1958. Within that range, as might be expected, wages rose faster the lower the unemployment rate. But one would judge now that it would take more like 8 per cent unemployment to keep money wages from rising. And they would rise at 2 to 3 per cent per year with 5 or 6 per cent of the labor force unemployed.

It would be overhasty to conclude that the relation we have been discussing represents a reversible supply curve for labor along which an aggregate demand curve slides. If that were so, then movements along the curve might be dubbed standard demand-pull, and shifts of the curve might represent the institutional changes on which cost-push theories rest. The apparent shift in our Phillips' curve might be attributed by some economists to the new market power of trade-unions. Others might be more inclined to believe that the expectation of continued full employment, or at least high employment, is enough to explain both the shift in the supply curve, if it is that, and the willingness of employers (conscious that what they get from a work force is partly dependent on its morale and its turnover) to pay wage increases in periods of temporarily slack demand.

This latter consideration, however, casts real doubt on the facile identification of the relationship as merely a supply-of-labor phenomenon. There are two parties to a wage bargain.

U.S. and U.K. compared

A comparison of the American position with Phillips' findings for the U.K. is interesting for itself and also as a possible guide to policy. Anything which will shift the relationship downward decreases the price in unemployment that must be paid when a policy is followed of holding down the rate of wage and price increase by pressure on aggregate demand.

One possibility is that the trade-union leadership is more "responsible" in the U.K.; indeed the postwar policy of wage restraint seems visible in Phillips' data. But there are other interpretations. It is clear that the more fractionated and imperfect a labor market is, the higher the over-all excess supply of labor may have to be before the average wage rate becomes stable and the less tight the relation will be in any case. Even a touch of downward inflexibility (and trade-unionism and administered wages surely means at least this) will make this immobility effect more pronounced. It would seem plausible that the sheer geographical compactness of the English economy makes its labor market more perfect than ours in this sense. Moreover, the British have pursued a more deliberate policy of relocation of industry to mop up pockets of structural unemployment.

This suggests that any governmental policy which increases the mobility of labor (geographical and industrial) or improves the flow of information in the labor market will have anti-inflationary effects as well as being desirable for other reasons. A quicker but in the long run probably less efficient approach might be for the government to direct the regional distribution of its expenditures more deliberately in terms of the existence of local unemployment and excess capacity.

The English data show a quite clear nonlinear (hyperbolic) relation between wage changes and unemployment, reflecting the much discussed downward inflexibility. Our American figures do not contradict this, although they do not tell as plain a story as the English. To the extent that this nonlinearity exists, as Duesenberry has remarked, a given average level of unemployment over the cycle will be compatible with a slower rate of wage increase (and presumably price increase) the less wide the cyclical swings from top to bottom.

A less obvious implication of this point of view is that a deliberate low-pressure policy to stabilize the price level may have a certain self-defeating aspect. It is clear from experience that interregional and interindustrial mobility of labor depends heavily on the pull of job opportunities elsewhere, more so than on the push of local unemployment. In effect the imperfection of the labor market is increased, with the consequences we have sketched.

IV

We have concluded that it is not possible on the basis of a priori reasoning to reject either the demand-pull or cost-push hypothesis, or the variants of the latter such as demand-shift. We have also argued that the empirical identifications needed to distinguish between these hypotheses may be quite impossible from the experience of macrodata that is available to us; and that, while use of microdata might throw additional light on the problem, even here identification is fraught with difficulties and ambiguities.

Nevertheless, there is one area where policy interest and the desire for scientific understanding for its own sake come together. If by deliberate policy one engineered a sizable reduction of demand or refused to permit the increase in demand that would be needed to preserve high employment, one would have an experiment that could hope to distinguish between the validity of the demand-pull and the cost-push theory as we would operationally reformulate those theories. If a small relaxation of demand were followed by great moderations in the march of wages and other costs so that the social cost of a stable price index turned out to be very small in terms of sacrificed high-level employment and output, then the demand-pull hypothesis would have received its most important confirmation. On the other hand, if mild demand repression checked cost and price increases not at all or only mildly, so that considerable unemployment would have to be engineered before the price level updrift could be presented, then the cost-push hypothesis would have received its most important confirmation. If the outcome of this experience turned out to be in between these extreme cases—as we ourselves would rather expect—then an element of validity would have to be conceded to both views; and dull as it is to have to embrace eclectic theories, scholars who wished to be realistic would have to steel themselves to doing so.

Of course, we have been talking glibly of a vast experiment. Actually such an operation would be fraught with implications for social welfare. Naturally, since they are confident that it would be a success, the believers in demand-pull ought to welcome such an experiment. But, equally naturally, the believers in cost-push would be dead set against such an engineered low-pressure economy, since they are equally convinced that it will be a dismal failure involving much needless social pain. (A third school, who believes in cost-push but think it can be cured or minimized by orthodox depressing of demand, think that our failure to make this experiment would be fraught with social evil by virtue of the fact that they expect a creep in prices to snowball into a trot and then a gallop.)

Our own view will by now have become evident. When we translate the Phillips' diagram showing the American pattern of wage increase against degree of unemployment into a related diagram showing the different levels of unemployment that would be "needed" for each degree of price level change, we come out with guesses like the following:

1. In order to have wages increase at no more than the 2½ per cent per annum characteristic of our productivity growth, the American economy would seem on the basis of twentieth-century and postwar experience to have to undergo something like 5 to 6 per cent of the civilian labor force's being unemployed. That much unemployment would appear to be the cost of price stability in the years immediately ahead.

2. In order to achieve the nonperfectionist's goal of high enough output to give us no more than 3 per cent unemployment, the price index might have to rise by as much as 4 to 5 per cent per year. That much price rise would seem to be the necessary cost of high employment and production in the years immediately ahead.

All this is shown in our price-level modification of the Phillips curve, Figure 2. The point *A*, corresponding to price stability, is seen to involve about 5½ per cent unemployment; whereas the point *B*, corresponding to 3 per cent unemployment, is seen to involve a price rise of about 4½ per cent per annum. We rather expect that the tug of war of politics will end us up in the next few years somewhere in between these selected points. We shall probably have some price rise and some excess unemployment.

Aside from the usual warning that these are simply our best guesses we must give another caution. All of our discussion has been phrased in short-run terms, dealing with what might happen in the next few years. It would be wrong, though, to think that our Figure 2 menu that relates obtainable price and unemployment behavior will maintain its same shape in the longer run. What we do in a policy way during the next few years might cause it to shift in a definite way.

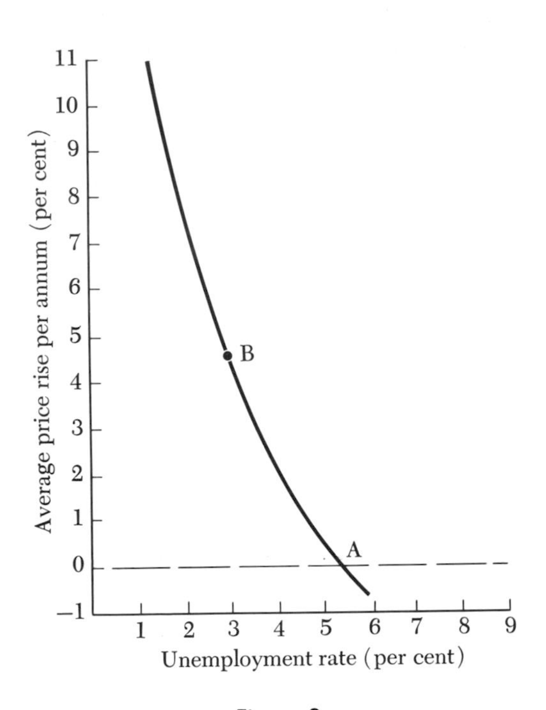

Figure 2

Modified Phillips Curve for U.S.

This shows the menu of choice between different degrees of unemployment and price stability, as roughly estimated from last twenty-five years of American data.

Thus, it is conceivable that after they had produced a low-pressure economy, the believers in demand-pull might be disappointed in the short run; i.e., prices might continue to rise even though unemployment was considerable. Nevertheless, it might be that the low-pressure demand would so act upon wage and other expectations as to shift the curve downward in the longer run—so that over a decade, the economy might enjoy higher employment with price stability than our present-day estimate would indicate.

But also the opposite is conceivable. A low-pressure economy might build up within itself over the years larger and larger amounts of structural unemployment (the reverse of what happened from 1941 to 1953 as a result of strong war and postwar demands). The result would be an upward shift of our menu of choice, with more and more unemployment being needed just to keep prices stable.

Since we have no conclusive or suggestive evidence on these conflicting issues, we shall not attempt to give judgment on them. Instead we venture the reminder that, in the years just ahead, the level of attained growth will be highly correlated with the degree of full employment and high-capacity output.

But what about the longer run? If the per annum rate of technical progress were about the same in a low- and high-pressure economy, then the initial loss in output in going to the low-pressure state would never be made up; however, in relative terms, the initial gap would not grow but would remain constant as time goes by. If a low-pressure economy could succeed in improving the efficiency of our productive factors, some of the loss of growth might be gradually made up and could in long enough time even be more than wiped out. On the other hand, if such an economy produced class warfare and social conflict and depressed the level of research and technical progress, the loss in growth would be compounded in the long run.

A final disclaimer is in order. We have not here entered upon the important questions of what feasible institutional reforms might be introduced to lessen the degree of disharmony between full employment and price stability. These could of course involve such wide-ranging issues as direct price and wage controls, antiunion and antitrust legislation, and a host of other measures hopefully designed to move the American Phillips' curves downward and to the left.

32

A graphical exposition of the complete Keynesian system *

Warren L. Smith

The purpose of this paper is chiefly expository. A simple graphical technique is employed to exhibit the working of several variants of the Keynesian model. Many of the issues discussed have been dealt with elsewhere,[1] but it is hoped that the analysis presented here will clarify some of the issues and be useful for pedagogical purposes.

I. The Keynesian System with Flexible Wages

This system can be represented symbolically by the following five equations:

1. $$y = c(y, r) + i(y, r)$$

Reprinted with permission from *The Southern Economic Journal,* Vol. XXIII, No. 2 (October, 1956), pp. 115-125. Warren L. Smith was at Ohio State University and is now Professor of Economics, University of Michigan.

* The development of the technique employed in this paper is a result of discussions with many persons, particularly Professor Daniel B. Suits of the University of Michigan, to whom the writer wishes to express his thanks.

[1] See particularly L. R. Klein, "Theories of Effective Demand and Employment," *Journal of Political Economy,* April 1947, LV, pp. 108-131, reprinted in R. V. Clemence (ed.), *Readings in Economic Analysis,* Vol. 1 (Cambridge, Mass.: Addison-Wesley Press, 1950), pp. 260-283, and *The Keynesian Revolution* (New York: Macmillan Co., 1950), esp. Technical Appendix; F. Modigliani, "Liquidity Preference and the Theory of Interest and Money," *Econometrica,* Jan. 1944, XII, pp. 45-88, reprinted in F. A. Lutz and L. W. Mints (eds.), *Readings in Monetary Theory* (Philadelphia: Blakiston, 1951), pp. 186-239; also V. Lutz, "Real and Monetary Factors in the Determination of Employment Levels," *Quarterly Journal of Economics,* May 1952, LXVI, pp. 251-272; L. Hough, "The Price Level in Macroeconomic Models," *American Economic Review,* June 1954, LXIV, pp. 269-286.

2. $$\frac{M}{p} = L(y, r)$$

3. $$y = f(N)$$

4. $$\frac{w}{p} = f'(N)$$

5. $$N = \varphi\left(\frac{w}{p}\right).$$

Here y = real GNP (at constant prices), r = an index of interest rates, M = money supply (in current dollars), p = index of the price level applicable to GNP, N = the volume of employment (in equivalent full-time workers), w = the money wage. The model represents a theory of short-run income determination with capital stock fixed and labor the only variable factor of production.

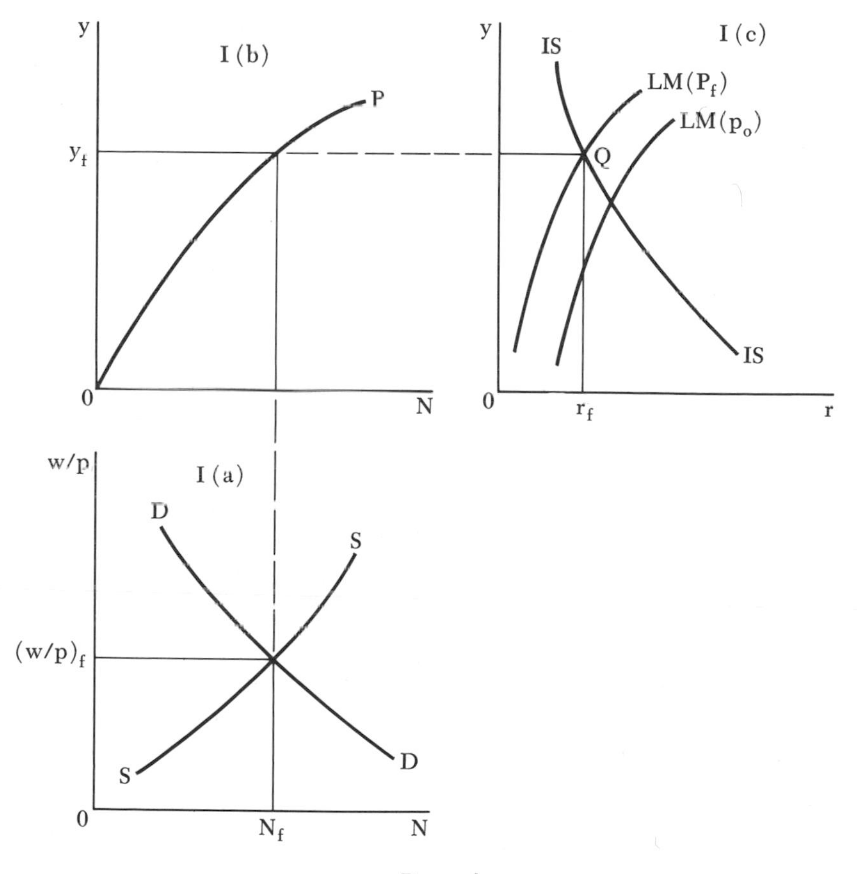

Figure I

The working of this model is illustrated in Figure I. Figure I should be studied in clockwise fashion, beginning with Chart I (a) in the lower lefthand corner. In I (a), *DD* represents the demand for labor [equation (4)], and *SS* represents the supply of labor [equation (5)]. The level of employment and the real wage are determined at the full employment levels, N_f and $\left(\frac{w}{p}\right)_f$. Proceeding to I (b), the curve *OP* represents the aggregate production function [equation (3)], its shape reflecting diminishing returns.[2] With employment of N_f, y would be at the level y_f, indicated in I (b).

Chart I (c) is the type of diagram developed by Hicks and utilized by others to depict the condition of monetary equilibrium in the Keynesian system.[3] The *IS* curve in I (c) depicts equation (1) and indicates for each possible level of the interest rate (r) the equilibrium level of income (y) which would prevail after the multiplier had worked itself out fully.[4] We treat the stock of money as an exogenous variable determined by the monetary authority. Given M, the *LM* curves in I (c), of which there would be one for each possible price level (p) which might prevail, represent equation (2) in our model. For example, if the price level were held constant at p_0, the curve $LM(p_0)$ depicts the different interest rates that would be required to preserve equilibrium in the money market at different income levels. The fact that rising income levels are associated with higher interest rates reflects the presumption that as income rises, transactions cash requirements are larger, leaving less of the fixed (in real terms) quantity of money to satisfy demands for idle balances, thus pushing up the interest rate.

If prices and wages are flexible and the situation is as depicted in Figure I, full employment will automatically be maintained, since the price level will adjust to the level p_f, establishing the *LM* curve in the

2 According to the mathematical formulation of our model in equations (1)—(5), the curve DD in I (a) is the derivative of curve OP in I (b), the relation reflecting the operation of the marginal productivity law under competitive conditions. This precise condition is not important, however, and we shall make no attempt to draw the curves in such a way as to fulfill it. For one thing, the presence of monopoly in the economy or failure of entrepreneurs to seek maximum profits would destroy the precision of the equations, but relations of the type depicted in Figure I would in all probability continue to hold.

3 For a detailed discussion of this diagram, see J. R. Hicks, "Mr. Keynes and the 'Classics': A Suggested Interpretation," *Econometrica,* April 1937, V, pp. 147-159; also A. H. Hansen, *Monetary Theory and Fiscal Policy* (New York: McGraw-Hill, 1949), Chap. 5. The reader's attention is directed to the fact that we have reversed the axes of the Hicks diagram; we measure the interest rate on the horizontal axis and income on the vertical axis.

4 It should be noted that the formal analysis in this paper falls entirely in the category of comparative statics, that is, it refers to conditions of equilibrium and changes in the equilibrium values of the variables brought about by changes in data or exogenous variables and does not pretend to describe the *paths* followed by the variables as they move from one equilibrium position to another.

position $LM(p_f)$ where it will intersect the *IS* curve at point *Q* which corresponds to the full employment level of income (y_f). If, for example, the real wage is initially above $\left(\frac{w}{p}\right)_f$, money wages will fall due to the excess supply of labor. This will reduce costs, resulting in increased output and employment and lower prices. Falling prices shift the *LM* curve upward by increasing the real value of cash balances $\left(\frac{M}{p}\right)$, thus lowering the interest rate and expanding aggregate demand to the point where the market will absorb the output corresponding to full employment.[5]

Two important and related propositions can be set down concerning interest and money in the above model:

1. The rate of interest is determined solely by saving and investment and is independent of the quantity of money and liquidity preference.

2. The quantity theory of money holds for this model—that is, a change in the quantity of money will bring about an equal proportional change in the price level and will have no effect on real income or employment.

In other words the quantity of money and liquidity preference serve not to determine the interest rate, as alleged by Keynes, but the price level. As can readily be seen from Figure I, income is established at the full employment level [I (a) and I (b)], the interest rate adjusts to equate saving and investment [on the *IS* curve in I (c)] at this income level, and the price level adjusts so as to satisfy liquidity requirements at this interest rate [establishing the *LM* curve at the appropriate position in I (c)].

It is a comparatively simple matter to modify the analysis of Figure I to take account of the possible effect of changes in the real value of liquid assets on consumption (the Pigou effect).[6] The real value of the stock of liquid assets would be included in equation (1), and falling prices would then shift the *IS* curve to the right, thus strengthening the tendency toward full employment equilibrium. This suggests the question: Does the introduction of the Pigou effect give the quantity of money the power to change the rate of interest when prices and

[5] We abstract from the possibility of dynamic instability which may arise due to falling prices if the public has elastic expectations. See D. Patinkin, "Price Flexibility and Full Employment," *American Economic Review*, Sept. 1948, XXXVII, pp. 543-564, reprinted with slight modification in Mints and Lutz, *op. cit.*, pp. 252-283.

[6] On the Pigou effect, see A. C. Pigou, "Economic Progress in a Stable Environment," *Economica*, New Series, August, 1947, XIV, pp. 180-188, reprinted in Lutz and Mints, *op. cit.*, pp. 241-251; Patinkin, *op. cit.;* G. Ackley, "The Wealth-Saving Relationship," *Journal of Political Economy*, April 1951, LIX, pp. 154-161; M. Cohen, "Liquid Assets and the Consumption Function," *Review of Economics and Statistics*, May 1954, XXXVI, pp. 202-211; and bibliography in the latter two articles.

wages are flexible? The answer to this question cannot be deduced from the curves of Figure I, but it is not difficult to find the answer with aid of the following simple model:

$$y = c(\bar{y}, r, a) + i(\bar{y}, r)$$

$$\frac{M}{p} = L(\bar{y}, r)$$

$$a = \frac{A}{p}.$$

Here $a =$ the real value of liquid assets which is included in the consumption function and $A =$ their money value. The last three equations of our original model are assumed to determine the real wage, employment, and real income. These equations are dropped and y is treated as a constant (having value $\bar{y}$) determined by those equations. We can now treat M and A as parameters and r, a, and p as variables, differentiate these three equations with respect to M, and solve for $\frac{dr}{dM}$. This gives the following expression:

$$6. \qquad \frac{dr}{dM} = \frac{\frac{c_a}{i_r}\frac{A}{M}(1 - \eta_{AM})}{p\left(1 + \frac{c_r}{i_r} + \frac{A}{M}\frac{L_r c_a}{i_r}\right)}.$$

In this expression, the subscripts refer to partial derivatives, e.g., $c_a = \frac{\partial c}{\partial a}$. Normally, the following conditions would be satisfied: $c_a > 0$, $i_r < 0$, $L_r < 0$. We cannot be sure about the sign of c_r, but it is likely to be small in any case. The coefficient η_{AM} has the following meaning:

$$\eta_{AM} = \frac{M}{A}\frac{dA}{dM} = \frac{\frac{dA}{A}}{\frac{dM}{M}}.$$

For example, if a change in M is brought about in such a way as to produce an exactly proportionate change in A, η_{AM} will be unity. Or if the change in M is not accompanied by any change in A, η_{AM} will be zero. It is apparent from the above expression that a change in the quantity of money will not affect the rate of interest if $\eta_{AM} = 1$, while an increase (decrease) in the quantity of money will lower (raise) the rate of interest if $\eta_{AM} < 1$.[7] Thus, the way in which changes in the quantity of

[7] We assume that $c_r \leq 0$, or if $c_r > 0$, $1 + \frac{A}{M}\frac{L_r c_a}{i_r} > \left|\frac{c_r}{i_r}\right|$, so that the denominator of 6 is positive.

money affect the rate of interest depends upon what asset concept is included in the consumption function (i.e., what is included in A) and how the volume of these assets is affected by monetary change. If M itself is the appropriate asset concept to include in the consumption function (i.e., if $A = M$), changes in M will not affect the interest rate, since in this case η_{AM} is equal to unity. However, the consensus of opinion seems to be that some other aggregate, such as currency, deposits, and government securities held by the non-bank public minus the public's indebtedness to the banks, is more appropriate.[8] If this concept is employed, most of the usual methods of increasing the money supply will ordinarily either leave A unchanged ($\eta_{AM} = 0$) or cause it to increase less than in proportion to the increase in M ($0 < \eta_{AM} < 1$).[9] We may conclude that the Pigou effect gives monetary changes power to influence the rate of interest, even if wages and prices are flexible. An increase (decrease) in the quantity of money will ordinarily lower (raise) the rate of interest and also increase (decrease) investment and decrease (increase) consumption, but will not change income and employment which are determined by real forces (the last three equations of our complete model).[10, 11]

II. Possibilities of Underemployment Disequilibrium

There are several possible circumstances arising from the shapes of the various schedules which might produce a situation in which, even

[8] The question of what asset concept is appropriate is discussed in Patinkin, *op. cit.*, Cohen, *op. cit.*, and J. Tobin, "Asset Holdings and Spending Decisions," *American Economic Review Papers and Proceedings*, May 1952, XLII, pp. 109-123.

[9] Open market purchases of government securities by the central bank from the non-bank public will leave A unchanged, since the initial purchase transaction will result in a decline in the public's security holdings and an equal increase in M, while any induced expansion of loans and investments by the banks will result in an increase in M offset by an equal increase in the public's indebtedness to the banks. On the other hand if the Treasury prints currency and gives it to the public, A will be increased by the same absolute amount as M but the increase in A will be proportionately smaller than the increase in M (provided the public's holdings of government securities exceed its indebtedness to the banks so that $A > M$).

[10] The fact that the existence of a wealth effect on savings may confer upon the quantity of money the power to affect the rate of interest even with flexible wages is demonstrated in L. A. Metzler, "Wealth, Saving, and the Rate of Interest," *Journal of Political Economy*, April 1951, LIX, pp. 93-116. Metzler's conclusions, which differ from those given here, can be attributed to assumptions that he makes, particularly the assumption that the only assets are money and common stock.

[11] If the supply of labor is affected by the real value of wealth held by workers, changes in the quantity of money may affect output and employment by shifting the SS curve in Figure I (a). Also, even though monetary change does not affect the *current* level of income and employment, if, due to the operation of the Pigou effect, it changes the interest rate and thereby investment, it may affect the *future* level of employment, since the change in capital stock will ordinarily shift the demand for labor [DD curve in Figure I (a)] at a future date. Both these points are mentioned in V. Lutz, *op. cit.*

though the relations in the above model held true, it might be impossible, at least temporarily, for equilibrium (full employment or otherwise) to be reached. The most widely discussed of these possibilities is depicted in Figure II.

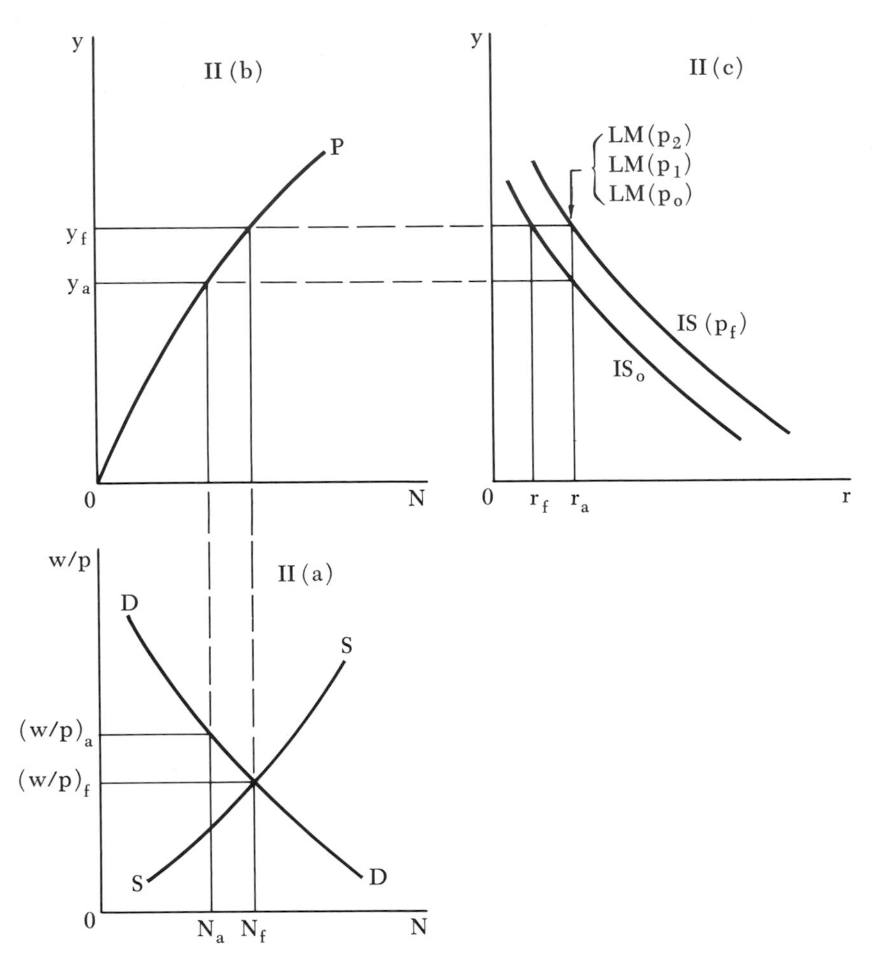

Figure II

II (a) and II (b) are similar to I (a) and I (b). However, the *LM* curves in I (c) are drawn to reflect the much-discussed possibility mentioned by Keynes[12] that the liquidity preference schedule might become infinitely elastic at some low level of interest rates [r_a in II (c)], due either to the unanimous expectations of investors that interest rates

[12] J. M. Keynes, *General Theory of Employment, Interest, and Money* (New York: Harcourt, Brace and Co., 1936), pp. 201-204.

would rise when they reached this extremely low level relative to future expectations or to the cost of investments. In the case depicted, full employment (N_f) would involve a level of income of y_f. If the *IS* curve were at the level IS_0, the interest rate required to make investment equal to saving at income y_f would be r_f. But the infinite elasticity of the *LM* schedule prevents the interest rate from falling below r_a. The result would be that employment and income would be prevented from rising above the level N_a and y_a by inadequate effective demand. The real wage would hold at the level $\left(\frac{w}{p}\right)_a$ which is above the full employment level $\left(\frac{w}{p}\right)_f$. Competition for employment would reduce money wages, costs, and prices. But the falling price level, although it would increase the quantity of money in real terms, would not affect the interest rate, hence would not increase investment. As prices fell, the *LM* curve would take successive positions, such as $LM(p_0)$, $LM(p_1)$, $LM(p_2)$, etc., leaving the interest rate unaffected.[13]

A special case of the situation depicted in Figure II may arise if a negative interest rate is required to equate investment to full employment savings. In this case, the *IS* curve would cut the *y*-axis and lie to the left of it at an income corresponding to full employment. Then, even if there were nothing to prevent the rate of interest from approaching zero, it could not go below zero,[14] and the *LM* curve would have a floor at a zero rate, thus preventing full employment from being attained.

It is interesting to note that if the Pigou effect is operative, a full employment equilibrium may be attainable even in the case illustrated in Figure II. As prices fall, the real value of liquid assets increases. If this increases consumption expenditures, the *IS* curve will shift to the right until it attains the position $IS(p_f)$, where a full employment equilibrium is reached.

Certain other conceivable situations which might lead to an underemployment disequilibrium are worthy of brief mention. One possibility is that the supply of labor might exceed the demand at all levels of

[13] Equations (1)–(5) above apply to the situations covered in both Figure I and Figure II. In the latter case, however, the equations are mathematically inconsistent and do not possess a solution. Mathematics does not tell us what will happen in this case (although the additional conditions necessary to describe the results could be expressed mathematically). The statements made above concerning the results (i.e., that income will be y_a, prices and wages will fall together, etc.) are propositions in economics.

[14] Since the money rate of interest cannot be negative, as long as it costs nothing to hold money. In fact, a zero rate of interest would be impossible, since in this case property values would be infinite; however, the rate might *approach* zero. The *real* rate of interest, *ex post*, may be negative due to inflation, but this is not relevant to our problem. On this, see I. Fisher, *The Theory of Interest* (New York: Macmillan Co., 1930), Chaps. II, XIX, and pp. 282-286.

real wages. Such a situation seems very improbable, however, since there is reason to believe that the short-run aggregate labor is quite inelastic over a considerable range of wage rates and declines when wage rates become very low.[15]

Disequilibrium situations could also arise if (a) the demand curve for labor had a steeper slope than the supply curve at their point of intersection, or (b) the *IS* curve cut the *LM* curve in such a way that *IS* lay to the right of *LM* above their intersection and to the left of *LM* below their intersection in Figure I (c) or II (c). Actually, these are situations of unstable equilibrium rather than of disequilibrium. However, in these cases, a slight departure from equilibrium would produce a cumulative movement away from it, and the effect would be similar to a situation of disequilibrium.

III. Underemployment Equilibrium Due to Wage Rigidity

Next we may consider the case in which the supply of and demand for labor are essentially the same as in Figures I and II, but for institutional or other reasons the money wage does not fall when there is an excess supply of labor.[16] This rigidity of money wages may be due to various factors, including (a) powerful trade unions which are able to prevent money wages from falling, at least temporarily, (b) statutory provisions, such as minimum wage laws, (c) failure of employers to reduce wages due to a desire to retain loyal and experienced employees and to maintain morale,[17] or (d) unwillingness of unemployed workers to accept reduced money wages even though they would be willing to work at lower real wages brought about by a rise in prices.[18]

A situation of this kind is depicted in Figure III. The fixed money wage is designated by $\overline{w}$. In order for full employment (N_f) to be attained, the price level must be at p_f (such as to make $\frac{\overline{w}}{p_f}$ equal to the real wage corresponding to full employment), income will be y_f, and the interest rate must reach r_f. However, in the case shown in Figure III, the quantity of money, M, is such that when p is at the level p_f, the *LM* curve [$LM\ (p_f)$] intersects the *IS* curve at an income (y_0) below the full employment level and an interest rate (r_0) above the full employment

[15] On the probable shape of the short-run aggregative supply of labor, see G. F. Bloom and H. R. Northrup, *Economics of Labor Relations* (Homewood, Ill.: Richard D. Irwin, 1954), pp. 250-253.

[16] We will assume that this rigidity does not prevail in an upward direction—i.e., money wages will rise when there is an excess demand for labor.

[17] See A. Rees, "Wage Determination and Involuntary Unemployment," *Journal of Political Economy*, April 1951, LIV, pp. 143-153.

[18] Keynes, *op. cit.*, Chap. 2; J. Tobin, "Money Wage Rates and Employment," in S. E. Harris (ed.), *The New Economics* (New York: Knopf, 1947), pp. 572-587.

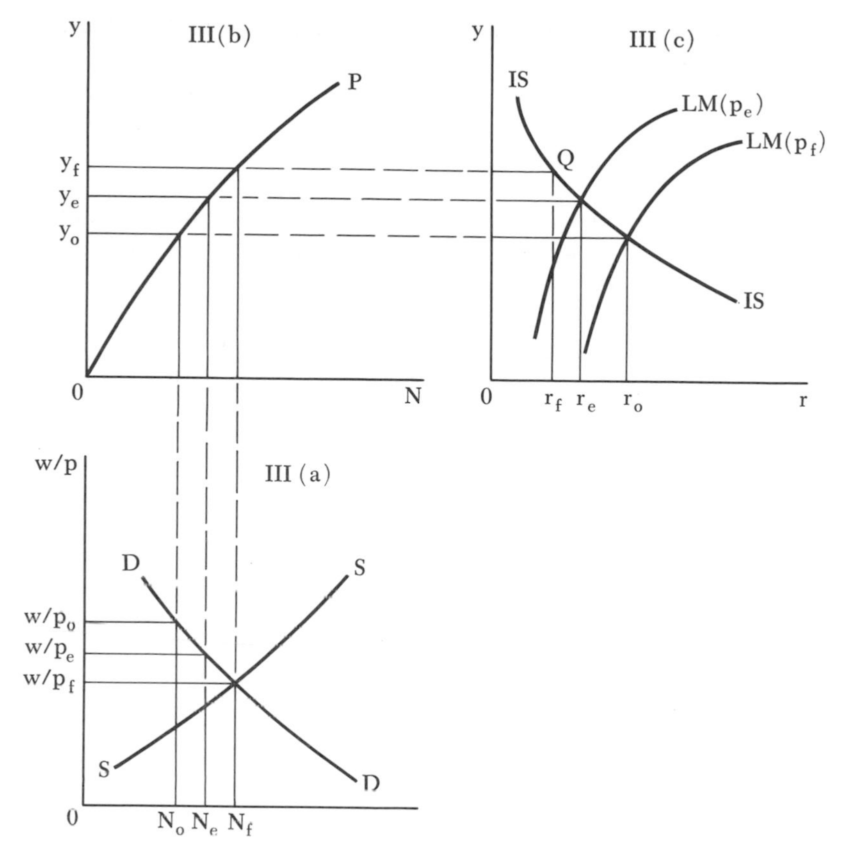

Figure III

level. Hence full employment cannot be sustained due to inadequate effective demand. On the other hand, if production and employment are at y_0 and N_0, with a price level such (at p_0) as to establish a real wage appropriate to this volume of employment, the LM curve will be at a level above $LM(p_f)$. This is because p_0 must be less than p_f in order to make $\overline{w}/p_0$ higher than $\overline{w}/p_f$. In this case production and employment will tend to rise because aggregate demand exceeds output. Therefore, income must be between y_f and y_0, employment between N_f and N_0, the interest rate between r_f and r_0, the price level between p_f and p_0. An equilibrium will be reached somewhere between these limits, say at N_e, y_e, p_e, and r_e.[19]

[19] In the case depicted in Figure III, an additional equation $w = \overline{w}$ is added to equations (1) — (5) above. This gives six equations and only five unknowns (y, N, p, w, and r). Such a system of equations is *overdetermined* and does not in general, possess a solution. If the quantity of money is treated as a variable which is adjusted so as to maintain full employment, we have six equations and six unknowns and there will be a solution (unless the equations are inconsistent).

This is a case of underemployment equilibrium. It should be noted that full employment can be attained by an increase in the quantity of money (M) sufficient to shift the $LM(pf)$ curve to the position where it will intersect the IS curve at point Q. Two propositions can be set down here to be contrasted with the two stated in connection with Figure I.[20]

1. Changes in the quantity of money cause changes in both the price level and the level of output and employment, and the quantity theory of money does not hold true.[21]

2. An increase (decrease) in the quantity of money causes a decrease (increase) in the rate of interest. In this case, the interest rate is determined by the interaction of all the relations in the model. Saving, investment, liquidity preference, and the quantity of money all have a hand in its determination.

Introduction of the Pigou effect into Figure III would not prevent the occurrence of an underemployment equilibrium, although it would somewhat complicate the process of adjustment since changes in p or M would cause changes in the IS curve as well as the LM curve.

To summarize, our analysis of Figures I and III indicates that rigidity of money wages is, in general, a necessary condition for (a) the occurrence of an underemployment equilibrium, (b) the quantity of money to have an effect on the level of real income and employment. The rate of interest will not be affected by the quantity of money and liquidity preference unless (a) there is rigidity of money wages or (b) the Pigou effect is operative with $n_{AM} \neq 1$. Monetary theories of the rate of interest, whether of the loanable funds or liquidity preference variety, ordinarily assume rigidity (or at least stickiness) in the structure of money wages.[22]

IV. Concluding Comments

In conclusion, we would like to call the reader's attention to further uses to which our graphical technique can be put. With appropriate modifications to suit the occasion, it can be used to analyze other vari-

[20] See p. 4, *supra*. [*Editor's note:* Text following note 5.]

[21] In the limiting case in which the DD curve has a horizontal stage which includes the current level of employment, the entire effect of an increase in M is on y, with no change in p. A considerable part of Keynes' *General Theory* (prior to the discussion of wages and prices in Book V) has reference primarily to this situation.

[22] The relative merits of loanable funds and liquidity preference types of monetary interest theories we do not consider, except to say that when appropriately formulated, the two are equivalent.

ations of the Keynesian model.[23] Additional factors affecting the income, employment, and price levels, such as those suggested by Hough[24] and Lutz[25] can be quite easily introduced into the analysis through appropriate shifts in the schedules shown in our system of graphs. Fiscal policy and its relation to monetary policy can be dealt with, since fiscal policy influences the level and shape of the *IS* curve. Finally, it provides a useful starting point for the study of economic growth. Factors affecting the rate of growth, such as capital accumulation, population growth, technological change, etc., can be brought in by allowing for their effects on the various schedules.

[23] For example, the models with which Modigliani begins his analysis (*op. cit.*, pp. 46-48 in original, pp. 187-190 in *Readings in Monetary Theory*). Analysis of these models requires some alteration in the graphical technique, since he assumes that consumption, investment, and the demand for money, all in current dollars, depend upon money income and the rate of interest, thus introducing "money illusions" into his scheme at several points.

[24] *Op. cit.*

[25] *Op. cit.*

33

Expansion and employment[1]

*Evsey D. Domar**

"A slow sort of a country," said the Queen. "Now, *here,* you see, it takes all the running *you* can do, to keep in the same place. If you want to get somewhere else, you must run at least twice as fast as that."

Lewis Carroll: *Through the Looking Glass*

In these days of labor shortages and inflation, a paper dealing with conditions needed for full employment and with the threat of deflation may well appear out of place. Its publication at this time is due partly to a two-year lag between the first draft and the final copy; also to the widely held belief that the present inflation is a temporary phenomenon, and that once it is over, the old problem of deflation and unemployment may possibly appear before us again.

. . . .

Our comfortable belief in the efficacy of Say's Law has been badly shaken in the last fifteen years. Both events and discussions have shown that supply does not automatically create its own demand. A part of income generated by the productive process may not be returned to it; this part may be saved and hoarded. As Keynes put it, "Unemployment develops . . . because people want the moon; men cannot be

Reprinted with permission from *The American Economic Review,* Vol. XXXVII, No. 1 (March, 1947), pp. 34-55.

* The author is assistant professor of economics at the Carnegie Institute of Technology. [*Editors' note:* He is now Professor of Economics, Massachusetts Institute of Technology.]

[1] This paper forms a sequence to my article on "The 'Burden' of the Debt and the National Income," published in this *Review*, Vol. XXXIV, No. 5 (Dec., 1944), pp. 798-827. Though their titles seem different, the two papers are based on the same logical foundation and treat a common subject: the economic role of growth.

employed when the object of desire (*i.e.*, money) is something which cannot be produced. . . ."[2] The core of the problem then is the public's desire to hoard. If no hoarding takes place, employment can presumably be maintained.

This sounds perfectly straight and simple; and yet it leaves something unexplained. Granted that absence of hoarding is a *necessary* condition for the maintenance of full employment, is it also a *sufficient* condition? Is the absence of hoarding *all* that is necessary for the avoidance of unemployment? This is the impression *The General Theory* gives. And yet, on a different plane, we have some notions about an increasing productive capacity which must somehow be utilized if unemployment is to be avoided. Will a mere absence of hoarding assure such a utilization? Will not a continuous increase in expenditures (and possibly in the money supply) be necessary in order to achieve this goal?

The present paper deals with this problem. It attempts to find the conditions needed for the maintenance of full employment over a period of time, or more exactly, *the rate of growth of national income* which the maintenance of full employment requires. This rate of growth is analyzed in Section I. Section II is essentially a digression on some conceptual questions and alternative approaches. It may be omitted by the busy reader. Section III is concerned with the *dual* character of the investment process; that is, with the fact that investment not only generates income but also increases productive capacity. Therefore the effects of investment on employment are less certain and more complex than is usually supposed. In Section IV a few examples from existing literature on the subject are given, and Section V contains some concluding remarks. The most essential parts of the paper are presented in Sections I and III.

As in many papers of this kind, a number of simplifying assumptions are made. Most of them will become apparent during the discussion. Two may be noted at the outset. First, events take place simultaneously, without any lags. Second, income, investment and saving are defined in the *net* sense, *i.e.*, over and above depreciation. The latter is understood to refer to the cost of replacement of the depreciated asset by another one of *equal* productive capacity. These assumptions are not entirely essential to the argument. The discussion could be carried out with lags, and, if desired, in gross terms or with a different concept of depreciation. Some suggestions along these lines are made in Section II. But it is better to begin with as simple a statement of the problem as possible, bearing in mind of course the nature of assumptions made.

[2] John M. Keynes, *The General Theory of Employment Interest and Money* (New York, 1936), p. 235.

I. The Rate of Growth

It is perfectly clear that the requirement that income paid out should be returned to the productive process, or that savings be equal to investment, or other expressions of the same idea, are simply formulas for the retention of the income *status quo*. If underemployment was present yesterday, it would still remain here today. If yesterday's income was at a full employment level, that *income level* would be retained today. It may no longer, however, correspond to full employment.

Let yesterday's full employment income equal an annual rate of 150 billion dollars, and let the average propensity to save equal, say, 10 per cent. If now 15 billions are annually invested, one might expect full employment to be maintained. But during this process, capital equipment of the economy will have increased by an annual rate of 15 billions —for after all, investment *is* the formation of capital.[3] Therefore, the productive capacity of the economy has also increased.

The effects of this increase on employment will depend on whether or not *real income* has also increased. Since money income has remained, as assumed, at the 150 billion annual level, an increase in real income can be brought about only by a corresponding fall in the general price level. This indeed has been the traditional approach to problems of this kind, an approach which we shall have to reject here for the following reasons:

1. The presence of considerable monopolistic elements (in industry and labor) in our economy makes unrealistic the assumption that a falling *general* price level could be achieved without interfering with full employment. This of course does not exclude *relative* changes among prices. As a matter of fact, if industries subject to a faster-than-average technological progress do not reduce their prices to some extent, a constant general price level cannot be maintained.

2. For an economy saddled with a large public debt and potentially faced (in peacetime) with serious employment problems, a falling price level is in itself undesirable.

3. A falling price level can bring about a larger real income only in

[3] The identification of investment with capital formation is reasonably safe in a private economy where only a small part of resources is disposed of by the government. When this part becomes substantial, complications arise. This question will be taken up again in Section II. Meanwhile, we shall disregard it and divide total national income, irrespective of source, into investment (*i.e.*, capital formation) and consumption.

The term "national income" is understood here in a broad sense, as total output minus depreciation, and does not touch on current controversies regarding the inclusion or exclusion of certain items. Perhaps "net national product" would be more appropriate for our purposes.

the special case when prices of consumers' goods fall more rapidly than those of investment goods. For otherwise (with a constant propensity to save) money income will be falling as fast or faster than the price level, and real income will be falling as well. To prevent money income from falling so rapidly, the volume of real investment would have to keep rising—a conclusion which will be presently reached in the more general case.

4. Finally, the assumption of a falling general price level would obscure—and I believe quite unnecessarily—the main subject we are concerned with here.

For these reasons, a *constant general price level* is assumed throughout this paper. But, from a theoretical point of view, this is a convenience rather than a necessity. The discussion could be carried on with a falling or a rising price level as well.

To come back to the increase in capacity. If both money and real national income thus remain fixed at the 150 billion annual level, the creation of the new capital equipment will have one or more of the following effects: (1) The new capital remains unused; (2) The new capital is used at the expense of previously constructed capital, whose labor and/or markets the new capital has taken away; (3) The new capital is substituted for labor (and possibly for other factors).

The first case represents a waste of resources. That capital need not have been constructed in the first place. The second case—the substitution of new capital for existing capital (before the latter is worn out, since investment is defined here in the net sense)—takes place all the time and, in reasonable magnitudes, is both unavoidable and desirable in a free dynamic society. It is when this substitution proceeds on a rather large scale that it can become socially wasteful; also, losses sustained or expected by capital owners will make them oppose new investment—a serious danger for an economy with considerable monopolistic elements.

Finally, capital may be substituted for labor. If this substitution results in a *voluntary* reduction in the labor force or in the length of the work week, no objections can be raised. Such a process has of course been going on for many years. But in our economy it is very likely that at least a part of this substitution—if carried on at an extensive scale—will be involuntary, so that the result will be unemployment.

The tools used in this paper do not allow us to distinguish between these three effects of capital formation, though, as will appear later, our concepts are so defined that a voluntary reduction in the number of man-hours worked is excluded. In general, it is not unreasonable to assume that in most cases all three effects will be present (though not in constant proportions), and that capital formation not accompanied by an increase in income will result in unemployed capital and labor.

The above problems do not arise in the standard Keynesian system because of its explicit assumption that employment is a function of national income, an assumption which admittedly can be justified only over short periods of time. Clearly, a full employment income of 1941 would cause considerable unemployment today. While Keynes' approach—the treatment of employment as a function of income—is a reasonable first approximation, we shall go a step further and assume instead that *the percentage of labor force employed is a function of the ratio between national income and productive capacity.* This should be an improvement, but we must admit the difficulties of determining productive capacity, both conceptually and statistically. These are obvious and need not be elaborated. We shall mean by productive capacity the total output of the economy at what is usually called full employment (with due allowance for frictional and seasonal unemployment), such factors as consumers' preferences, price and wage structures, intensity of competition, and so on being given.

The answer to the problem of unemployment lies of course in a growing income. If after capital equipment has increased by (an annual rate of) 15 billions an income of 150 billions leaves some capacity unused, then a higher magnitude of income can be found—say 155 or 160 billions—which will do the job. There is nothing novel or startling about this conclusion. The idea that a capitalist economy needs growth goes back, in one form or another, at least to Marx. The trouble really is that the idea of growth is so widely accepted that people rarely bother about it. It is always treated as an afterthought, to be added to one's speech or article if requested, but very seldom incorporated in its body. Even then it is regarded as a function of some abstract technological progress which somehow results in increasing productivity per man-hour, and which takes place quite independently of capital formation. And yet, our help in the industrialization of undeveloped countries will take the form not only of supplying technical advice and textbooks, but also of actual machinery and goods. Certainly the 80 odd billion dollars of net capital formation created in the United States in the period 1919-29 had a considerable effect on our productive capacity.[4]

A change in productive capacity of a country is a function of changes in its natural resources (discovery of new ones or depletion of others), in its labor force (more correctly, man-hours available), capital and the state of technique.[5] Since changes in natural resources and technique are very difficult concepts, we can express changes in total capacity via changes in the quantity and productivity of labor or of

[4] This figure, in 1929 prices, is taken from Simon Kuznets, *National Income and Its Composition,* Vol. I (New York, 1941), p. 268. The actual figure was 79.1 billion dollars.

[5] Taking other conditions listed on p. 37 as given.

capital. The traditional approach builds around labor. The several studies of the magnitude of total output corresponding to full employment, made in the last few years, consisted in multiplying the expected labor force (subdivided into several classes) by its expected average productivity.[6] This procedure did not imply that the other three factors (natural resources, technology and capital) remained constant; rather that their variations were all reflected in the changes in productivity of labor.

It is also possible to put capital in the center of the stage and to estimate variations in total capacity by measuring the changes in the quantity of capital and in its productivity, the latter reflecting changes currently taking place in natural resources, technology and the labor force. From a practical point of view, the labor approach has obvious advantages, at least in some problems, because labor is a more homogeneous and easily measurable factor. But from a theoretical point of view, the capital approach is more promising and for this reason: the appearance of an extra workman or his decision to work longer hours *only* increases productive capacity without, however, generating any income to make use of this increase. But the construction of a new factory has a *dual* effect: *it increases productive capacity and it generates income*.

The emphasis on this dual character of the investment process is the essence of this paper's approach to the problem of employment. If investment increases productive capacity and also creates income, what should be the magnitude of investment, or at what rate should it grow, in order to make the increase in income equal to that of productive capacity?[7] Couldn't an equation be set up one side of which would represent the increase (or the rate of increase) of productive capacity, and the other—that of income, and the solution of which would yield the required *rate of growth*?

We shall attempt to set up such an equation. It will be first expressed in symbolic form, and later (on p. 465) illustrated by a numerical example.

Let investment proceed at an annual rate of I, and let annual productive capacity (net value added) per dollar of newly created capital be equal to the average to s. Thus if it requires, say, 3 dollars of capital to produce (in terms of annual net value added) one dollar of output, s will equal one-third or 33.3 per cent per year. It is not meant that s is the same in all firms or industries. It depends of course on the nature of capital constructed and on many other factors. Its treatment here as a

[6] See for instance E. E. Hagen and N. B. Kirkpatrick, "The National Output at Full Employment in 1950," *Amer. Econ. Rev.*, Vol. XXXIV, No. 4 (Sept., 1944), pp. 472-500.

[7] This statement of the problem presupposes that full employment has already been reached and must only be maintained. With a small extra effort we could begin with a situation where some unemployment originally existed.

given magnitude is a simplification which can be readily dispensed with.

The productive capacity of I dollars invested will thus be Is dollars per year. But it is possible that the operation of new capital will take place, at least to some extent, at the expense of previously constructed plans, with which the new capital will compete both for markets and for factors of production (mainly labor). If as a result, the output of existing plants must be curtailed, it would be useless to assert that the productive capacity of the *whole economy* has increased by Is dollars per year.[8] It has actually increased by a smaller amount which will be indicated by $I\sigma$.[9] σ may be called the *potential social average productivity of investment.* Such a long name calls for an explanation.

1. As stated above, σ is concerned with the increase in productive capacity of the whole society and not with the productive capacity per dollar invested in the new plants taken by themselves, that is with s. A difference between s and σ indicates a certain misdirection of investment, or—more important—that investment proceeds at too rapid a rate as compared with the growth of labor and technological progress. This question will be taken up again in Section II.

2. σ should not be confused with other related concepts, such as the traditional marginal productivity of capital. These concepts are usually based on a *caeteris paribus* assumption regarding the quantity of other factors and the state of technique. It should be emphasized that the use of σ does not imply in the least that labor, natural resources and technology remain fixed. It would be more correct therefore to say that σ indicates the increase in productive capacity which *accompanies* rather than which is caused by each dollar invested.

3. For our purposes, the most important property of σ is its *potential character.* It deals not with an increase in national income but with that of the *productive potential* of the economy. A high σ indicates that the economy *is capable* of increasing its output relatively fast. But whether this increased capacity will actually result in greater output or greater unemployment, depends on the behavior of money income.

The expression $I\sigma$ is the supply side of our system; it is the increase in output which the economy *can* produce. On the demand side we have the multiplier theory, too familiar to need any elaboration, except for the emphasis on the obvious but often forgotten fact that, with any given marginal propensity to save, to be indicated by α, an increase in national income is not a function of investment, but of the *increment* in investment. If investment today, however large, is equal to that of yesterday, national income of today will be just equal and not any larger

[8] These comparisons must of course be made at a full employment level of national income. See also pp. 468-470.

[9] We are disregarding here external economies obtained by existing plants from the newly constructed ones.

than that of yesterday. All this is obvious, and is stressed here to underline the lack of symmetry between the effects of investment on productive capacity and on national income.

Let investment increase at an absolute annual rate of ΔI (e.g., by two billion per year), and let the corresponding absolute annual increase in income be indicated by ΔY. We have then

$$1. \qquad \Delta Y = \Delta I \frac{1}{\alpha},$$

where $\frac{1}{\alpha}$ is of course the multiplier.

Let us now assume that the economy is in a position of a full employment equilibrium, so that its national income equals its productive capacity.[10] To retain this position, income and capacity should increase at the same rate. The annual increase in potential capacity equals $I\sigma$. The annual increase in actual income is expressed by $\Delta I(1/\alpha)$. Our objective is to make them equal. This gives us the fundamental equation

$$2. \qquad \Delta I \frac{1}{\alpha} = I\sigma.$$

To solve this equation, we multiply both sides by α and divide by I, obtaining

$$3. \qquad \frac{\Delta I}{I} = \alpha\sigma.$$

The left side of expression (3) is the absolute annual increase (or the absolute rate of growth) in investment—ΔI—divided by the volume of investment itself; or in other words, it is the relative increase in investment, or the annual percentage rate of growth of investment. Thus the maintenance of full employment requires that investment grow at the annual percentage rate $\alpha\sigma$.

So much for investment. Since the marginal propensity to save—α—is assumed to be constant, an increase in income is a constant multiple of an increase in investment (see expression [1]). But in order to remain such a constant multiple of investment, income must also grow at the same annual percentage rate, that is at $\alpha\sigma$.

To summarize, the maintenance of a continuous state of full employment requires that *investment and income grow at a constant annual percentage* (*or compound interest*) *rate* equal to the product of the marginal propensity to save and the average (to put it briefly) productivity of investment.[11]

[10] See note 7.

[11] The careful reader may be disturbed by the lack of clear distinction between increments and rates of growth here and elsewhere in the text. If some confusion exists, it is due to my attempt to express these concepts in non-mathematical form. Actually they all should be stated in terms of rates of growth (derivatives in respect

This result can be made clearer by a numerical example. Let $\sigma =$ 25 per cent per year, $\alpha = 12$ per cent, and $Y = 150$ billions per year. If full employment is to be maintained, an amount equal to $150 \times \frac{12}{100}$ should be invested. This will raise productive capacity by the amount invested times σ, *i.e.*, by $150 \times \frac{12}{100} \times \frac{25}{100}$, and national income will have to rise by the same annual amount. But the relative rise in income will equal the absolute increase divided by the income itself, *i.e.*,

$$4. \qquad \frac{150 \times \frac{12}{100} \times \frac{25}{100}}{150} = \frac{12}{100} \times \frac{25}{100} = \alpha\sigma = 3 \text{ per cent.}$$

These results were obtained on the assumption that α, the marginal propensity to save, and σ, the average productivity of investment, remain constant. The reader can see that this assumption is not necessary for the argument, and that the whole problem can be easily reworked with variable α and σ. Some remarks about a changing α are made on pp. 472-474.

The expression (3) indicates (in a very simplified manner) conditions needed for the maintenance of full employment over a period of time. It shows that it is not sufficient, in Keynesian terms, that savings of yesterday be invested today, or, as it is often expressed, that investment offset saving. Investment of today must always exceed savings of yesterday. A mere absence of hoarding will not do. An injection of new money (or dishoarding) must take place every day. Moreover, this injection must proceed, in absolute terms, at an accelerated rate. The economy must continuously expand.[11a]

II. The Argument Re-examined

The busy reader is urged to skip this section and proceed directly to Section III. The present section is really a long footnote which re-examines the concepts and suggests some alternative approaches. Its purpose is, on the one hand, to indicate the essential limitations of the preceding discussion, and on the other, to offer a few suggestions which may be of interest to others working in this field.

---

to time). For a more serious treatment of this point, as well as for a more complete statement of the logic of the paper, see my article "Capital Expansion, Rate of Growth, and Employment," *Econometrica*, Vol. XIV (Apr., 1946), pp. 137-47.

[11a] After this paper was sent to the printer, I happened to stumble on an article by R. F. Harrod, published in 1939, which contained a number of ideas similar to those presented here. See "An Essay in Dynamic Theory," *Econ. Jour.*, Vol. XLIX (Apr., 1939), pp. 14-33.

It was established in Section I that the maintenance of full employment requires income and investment to grow at an annual compound interest rate equal to $\alpha\sigma$. The meaning of this result will naturally depend on those α and σ. Unfortunately neither of them is devoid of ambiguity.

The marginal propensity to save—α—is a relatively simple concept in a private economy where only a small part of resources is handled by the government. National income can be divided, without too much trouble, into investment and consumption, even though it is true that the basis for this distinction is often purely formal.[12] But on the whole it sounds quite reasonable to say that if marginal propensity to save is α, then an α fraction of an increase in income is saved by the public and invested in income-producing assets.

When a substantial part of the economy's resources is disposed of by the government, two interpretations of the marginal propensity to save, or of savings and investment in general, appear possible. The first is to continue dividing the total output, whether produced by government or by private business, into consumption and investment. This method was implicitly followed in this paper. But a question arises regarding the meaning and stability of α. It makes sense to say that a person or the public save, in accordance with the size of their incomes, their habits, expectations, etc., a certain, though not necessarily constant, fraction of an increment in their *disposable* (*i.e.*, after income and social security taxes) income, but can a similar statement be made regarding total national income, a good part of which is not placed at the disposal of the public? Also it is not easy to divide government expenditures into consumption and investment.

The other method would limit α to indisposable income only, and then provide for government expenditures separately. It would be necessary then to find out the effects of these expenditures on productive capacity.

Depreciation raises another problem. Since all terms are defined here in the net sense, the meaning and magnitude of α will also depend on those of depreciation, irrespective of the choice between the above two methods. Depreciation has been defined here (see page 459) as the cost of replacement of a worn out asset by another one with an equal productive capacity. While this approach is about as bad or as good as any other, the difficulty still remains that businesses ordinarily do not use this definition, and therefore arrive at a different estimate of their net incomes, which in turn determine their propensity to save.

I do not have ready answers to these questions, though I do not

[12] Thanks are due to George Jaszi for his persistent efforts to enlighten me on this subject. The division of national income into investment and consumption is really a more difficult task than my text might imply.

consider them insurmountable. I am mentioning them here partly in order to indicate the limitations of the present argument, and also as obstacles which will have to be overcome if a more exact analysis is undertaken.

σ is even more apt to give rise to ambiguities. s, from which it springs, has been used, in one form or another, in economic literature before, particularly in connection with the acceleration principle.[13] Here it indicates the annual amount of income (net value added) which can be produced by a dollar of newly created capital. It varies of course among firms and industries, and also in space and time, though a study recently made seems to indicate that it has been quite stable, at least in the United States and Great Britain, over the last 70 years or so.[14] Whether s has or has not been relatively stable is not essential for our discussion. The real question is whether such a concept has meaning, whether it makes sense to say that a given economy or a plant has a certain capacity. Traditional economic thinking would, I fear, be against such an approach. Unfortunately, it is impossible to discuss this question here. I believe that our actual experience during the last depression and this war, as well as a number of empirical studies, show that productive capacity, both of a plant and of the whole economy is a meaningful concept, though this capacity, as well as the magnitude of s, should be treated as a *range* rather than as a single number.

In some problems s may be interpreted as the minimum annual output per dollar invested which will make the investment worth undertaking. If this output falls below s, the investor suffers a loss or at least a disappointment, and may be unwilling to replace the asset after it has depreciated.

All these doubts apply to σ even more than to s. As explained on pages 463-464, σ differs from s by indicating the annual increment in capacity of the *whole economy* per dollar invested, rather than that of the newly created capital taken by itself. The possible difference between s and σ is due to the following reasons:

1. The new plants are not operated to capacity because they are unable to find a market for their products.

2. Old plants reduce their output because their markets are captured by new plants.

As productive capacity has no meaning except in relation to consumers' preferences, in both of the above cases productive capacity of the country is increased by a smaller amount than could be produced

[13] See for instance Paul A. Samuelson, "Interactions between the Multiplier Analysis and the Principle of Acceleration," *Rev. Econ. Stat.*, Vol. XXI (May, 1939), pp. 75-79; also R. F. Harrod, *The Trade Cycle* (Oxford, 1936). These authors, however, used not the ratio of income to capital, but of consumption to capital, or rather the reciprocal of this ratio.

[14] See Ernest H. Stern, "Capital Requirements in Progressive Economies," *Economica,* n.s. Vol. XII (Aug., 1945), pp. 163-71.

by the new plants; in the limiting case it is not increased at all, and $\sigma=0$, however high s may be. But it must be made clear that the test of whether or not σ is below s can be made only under conditions (actual or assumed) of full employment. If markets are not large enough because of insufficiency of effective demand due to unemployment, it cannot yet be concluded that σ is below s.

3. The first two cases can take place irrespective of the volume of current investment. A more important case arises when investment proceeds at such a rapid rate that a shortage of other factors relative to capital develops. New plants may be unable to get enough labor, or more likely, labor (and other factors) is transferred to new plants from previously constructed ones, whose capacity therefore declines. In its actual manifestation, case 3 can hardly be separated from cases 1 and 2, because to the individual firm affected the difference between s and σ always takes the form of a cost-price disparity. The reason why we are trying to separate the first two cases from the third lies in the bearing of this distinction on practical policy. The first two cases arise from an error of judgment on the part of investors (past or present) which is, at least to some extent, unavoidable and not undesirable. The struggle for markets and the replacement of weaker (or older) firms and industries by stronger (or newer) ones is the essence of progress in a capitalist society. The third case, on the other hand, may result from poor fiscal policy. It constitutes an attempt to invest too much, to build more capital than the economy can utilize even at full employment. Such a situation can develop if an economy with a high propensity to save tries to maintain full employment by investing all its savings into capital goods. But it should be made clear that the expressions "too much capital" or "high propensity to save" are used in a relative sense—in comparison with the growth of other factors, that is natural resources, labor and technology.

The use of σ certainly does not imply that these factors remain fixed. As a matter of fact, it would be very interesting to explore the use of a more complex function as the right side of expression (2) instead of $I\sigma$, a function in which the growth of labor, natural resources, and technology would be presented explicitly, rather than through their effects on σ.[15] I did not attempt it because I wished to express the idea of growth in the simplest possible manner. One must also remember that in the application of mathematics to economic problems, diminishing returns appear rapidly, and that the construction of complex models requires so many specific assumptions as to narrow down their applicability.

And yet it may be interesting to depart in another direction,

[15] Some work along these lines has been done by J. Tinbergen. See his "Zur Theorie der langfristigen Wirtschaftsentwicklung" in the *Weltwirtschaftliches Archiv*, Vol. LV (May, 1942), pp. 511-49.

namely to introduce lags. In this paper both the multiplier effect and the increase in capacity are supposed to take place simultaneously and without any lag. Actually, the multiplier may take some time to work itself out, and certainly the construction of a capital asset takes time. In a secular problem these lags are not likely to be of great importance, but they may play an essential rôle over the cycle. We shall return to this question on pages 474-475.

Finally, it is possible to approach the problem of growth from a different point of view. It was established here that the rate of growth required for a full employment equilibrium to be indicated by r is equal to

$$r = \alpha\sigma$$

so that if α and σ are given, the rate of growth is determined. But the equation (5) can also be solved for α in terms of r and σ, and for σ in terms of r and α. Thus if it is believed that r should be treated as given (for instance by technological progress), and if it is also decided to keep σ at a certain level, perhaps not too far from s, then it is possible to determine $\alpha = r/\sigma$, as being that marginal propensity to save which can be maintained without causing either inflation or unemployment. This approach was actually used by Ernest Stern in his statistical study of capital requirements of the United Kingdom, the United States and the Union of South Africa.[16] I also understand from Tibor de Scitovszky that he used the same approach in a study not yet published.

It is also possible to treat r and α as given and then determine what $\sigma = r/\alpha$ would have to be. Each approach has its own advantages and the choice depends of course on the nature of the problem in hand. The essential point to be noticed is the relationship between these three variables r, α, and σ, and the fact that if any two of them are given, the value of the third needed for the maintenance of full employment is determined; and if its actual value differs from the required one, inflation in some cases and unused capacity and unemployment in others will develop.

III. The Dual Nature of the Investment Process

We shall continue the discussion of growth by returning to expression (2) on page 465.

$$\Delta I \frac{1}{\alpha} = I\sigma,$$

which is fundamental to our whole analysis. As a matter of fact, the statement of the problem in this form (2) appears to me at least as im-

[16] Stern, *Economica*, n.s. Vol. XII, pp. 163-71.

portant as its actual solution expressed in (3). To repeat, the left part of the equation shows the annual increment in national income and is the demand side; while the right part represents the annual increase in productive capacity and is the supply side. Alternatively, the left part may be called the "multiplier side," and the right part the "σ side."

What is most important for our purposes is the fact that investment appears on both sides of the equation; that is, it has a *dual effect*: on the left side it generates income via the multiplier effect; and on the right side it increases productive capacity—the σ effect. The explicit recognition of this dual character of investment could undoubtedly save much argument and confusion. Unless some special assumptions are made, the discussion of the effects of investment on profits, income, employment, etc., cannot be legitimately confined to one side only. For the generation of income and the enlargement of productive capacity often have diametrically opposed effects, and the outcome in each particular case depends on the special circumstances involved.[17]

Analyzing expression (2) further, we notice that even though investment is present on both its sides, it does not take the same form: for on the σ side we have the *amount* of investment as such; but on the multiplier side we have not the amount of investment but its annual increment, or its absolute *rate of increase.*

The amount of investment (always in the net sense) may remain constant, or it may go up or down, but so long as it remains positive (and except for the rare case when $\sigma \leqq 0$) productive capacity increases. But if income is to rise as well, it is not enough that just any amount be invested: *an increase in income is not a function of the amount invested; it is the function of the increment of investment.* Thus the whole body of investment, so to speak, increases productive capacity, but only its very top—the increment—increases national income.

In this probably lies the explanation why inflations have been so rare in our economy in peacetime, and why even in relatively prosperous periods a certain degree of underemployment has usually been present. Indeed, it is difficult enough to keep investment at some reasonably high level year after year, but the requirement that it always be rising is not likely to be met for any considerable length of time.

Now, if investment and therefore income do not grow at the required rate, unused capacity develops. Capital and labor become idle. It may not be apparent why investment by increasing productive ca-

[17] The effects of labor saving machinery on employment of labor is a good case in point. Some economists, particularly those connected with the labor movement, insist that such machines displace labor and create unemployment. Their opponents are equally sure that the introduction of labor saving devices reduces costs and generates income, thus increasing employment. Both sides cite ample empirical evidence to prove their contentions, and neither side is wrong. But both of them present an incomplete picture from which no definite conclusion can be derived.

pacity creates unemployment of labor. Indeed, as was argued on page 461, this need not always be the case. Suppose national income remains constant or rises very slowly while new houses are being built. It is possible that new houses will be rented out at the expense of older buildings and that no larger rents will be paid than before; or that the new houses will stand wholly or partly vacant with the same result regarding the rents.[18] But it is also possible, and indeed very probable, that the complete or partial utilization of the new buildings which are usually better than the old ones, will require the payment of larger rents, with the result that less income will be left for the purchase of, say clothing; thus causing unemployment in the clothing trades. So the substitution of capital for labor need not take the obvious form of labor-saving machinery; it may be equally effective in a more circuitous way.

The unemployment of men is considered harmful for obvious reasons. But idle buildings and machinery, though not arousing our humanitarian instincts, can be harmful because their presence inhibits new investment. Why build a new factory when existing ones are working at half capacity? It is certainly not necessary to be dogmatic and assert that no plant or house should ever be allowed to stand idle, and that as soon as unused capacity develops the economy plunges into a depression. There is no need, nor is it possible or desirable, to guarantee that every piece of capital ever constructed will be fully utilized until it is worn out. When population moves from Oklahoma to California, some buildings in Oklahoma will stand idle; or when plastics replace leather in women's handbags, the leather industry may suffer. Such changes form the very life of a free dynamic society, and should not be interfered with. The point is that there be no vacant houses while prospective tenants are present but cannot afford to live in them because they are unemployed. And they are unemployed because income and investment do not grow sufficiently fast.

The extent to which unused capacity, present or expected, inhibits new investment greatly depends on the structure of industry and the character of the economy in general. The more atomistic it is, the stronger is competition, the more susceptible it is to territorial, technological and other changes, the smaller is the effect of unused capacity on new investment. One firm may have an idle plant, while another in the same industry builds a new one; steel may be depressed while plastics are expanding. It is when an industry is more or less monopolized, or when several industries are financially connected, that unused capacity presents a particularly serious threat to new investment.

Strictly speaking, our discussion so far, including equation (2), was

[18] It is worth noticing that in both cases the construction of the new houses represents a misdirection of resources, at least to some extent. But a complete avoidance of such misdirection is perfectly impossible and even undesirable.

based on the assumption that α remained constant. If α varies within the time period concerned, the relation between investment and income becomes more involved. What the left side of the equation (2) requires is that *income* increase; and investment must grow only in so far as its growth is necessary for the growth of income. So if α declines sufficiently fast, a growing income can be achieved with a constant or even falling investment. But years of declining α have evidently been offset by others of rising α, because whatever information is available would indicate that over the last seventy years or so prior to this war the percentage of income saved was reasonably constant, possibly with a slight downward trend.[19] Therefore, in the absence of direct government interference, it would seem better not to count too much on a falling α, at least for the time being.

In general, a high α presents a serious danger to the maintenance of full employment, because investment may fail to grow at the required high rate, or will be physically unable to do so without creating a substantial difference between s and σ. This difference indicates that large numbers of capital assets become unprofitable and their owners suffer losses or at least disappointments (see pages 468-469). Space does not permit me to develop this idea at greater length here.[20] But it must be emphasized that what matters is not the magnitude of α taken by itself, but its relation to the growth of labor, natural resources, and technology. Thus a country with new resources, a rapidly growing population, and developing technology is able to digest, so to speak, a relatively large α, while absence or at least a very slow growth of these factors makes a high α a most serious obstacle to full employment.[21] But the problem can be attacked not only by lowering α, but also by speeding up the rate of technological progress, the latter solution being much more to my taste. It must be remembered, however, that technological progress makes it *possible* for the economy to grow, without guaranteeing that this growth will be realized.

In a private capitalist society where α cannot be readily changed, a higher level of income and employment at any given time can be achieved only through increased investment. But investment, as an employment creating instrument, is a mixed blessing because of its σ effect. The economy finds itself in a serious dilemma: if sufficient investment is not forthcoming today, unemployment will be here today. But if enough is invested today, still more will be needed tomorrow.

It is a remarkable characteristic of a capitalist economy that while,

[19] See Simon Kuznets, *National Product since 1869,* National Bureau of Economic Research (mimeo., 1945), p. II-89. I do not mean that we must always assume a constant α; rather that we lack sufficient proof to rely on a falling one.

[20] See my paper, *Econometrica,* Vol. XIV, particularly pp. 142-45.

[21] *Cf.* Alvin H. Hansen, *Fiscal Policy and the Business Cycle* (New York, 1941), particularly Part IV.

on the whole, unemployment is a function of the difference between its actual income and its productive capacity, most of the measures (*i.e.*, investment) directed towards raising national income also enlarge productive capacity. It is very likely that the increase in national income will be greater than that of capacity, but the whole problem is that the increase in income is temporary and presently peters out (the usual multiplier effect), while capacity has been increased for good. So that as far as unemployment is concerned, investment is at the same time a cure for the disease and the cause of even greater ills in the future.[22]

IV. An Economic Excursion

It may be worth while to browse through the works of several economists of different schools of thought to see their treatment of the σ and of the multiplier effects of investment. It is not suggested to make an exhaustive study, but just to present a few examples.

Thus in Marshall's *Principles* capital and investment are looked upon as productive instruments (the σ effect), with little being said about monetary (that is, income or price) effects of investment.[23] The same attitude prevails in Fisher's *Nature of Capital and Income*,[24] and I presume in the great majority of writings not devoted to the business cycle. It is not that these writers were unaware of monetary effects of investment (even though they did not have the multiplier concept as such), but such questions belonged to a different field, and the problem of aggregate demand was supposed to be taken care of by some variation of Say's Law.

In the business cycle literature we often find exactly an opposite situation. The whole Wicksellian tradition treated economic fluctuations as a result of monetary effects of excessive investment. It is curious that all this investment did not lead to increased output which would counteract its inflationary tendencies. Indeed, as one reads Hayek's *Prices and Production*, one gets an impression that these investment projects never bear fruit and are, moreover, abandoned after the crisis. The σ

[22] That income generating effects of investment are temporary and that new and larger amounts must be spent to maintain full employment, has been mentioned in economic and popular literature a number of times. Particular use has been made of this fact by opponents of the so-called deficit financing, who treat government expenditures as a "shot in the arm" which must be administered at an ever increasing dose. What they fail to realize is that exactly the same holds true for private investment.

[23] Marshall was very careful, however, to distinguish between the substitution of a particular piece of machinery for particular labor, and the replacement of labor by capital in general. The latter he regarded impossible, because the construction of capital creates demand for labor, essentially a sort of a multiplier effect. See *Principles of Economics*, 8th ed. (London, 1936), p. 523.

[24] Irving Fisher, *The Nature of Capital and Income* (New York, 1919).

effect is entirely absent, or at least appears with such a long lag as to make it inoperative. Prosperity comes to an end because the banking system refuses to support inflation any longer.[25]

σ fares better in the hands of Aftalion.[26] His theory of the cycle is based upon, what I would call, a time lag between the multiplier and the σ effects. Prosperity is started by income generated by investment in capital goods (the multiplier effect), while no increase in productive capacity has taken place as yet. As investment projects are completed, the resulting increase in productive capacity (the σ effect) pours goods on the market and brings prosperity to an end.

A similar approach is used by Michal Kalecki. The essence of his model of the business cycle consists in making profit expectations, and therefore investment, a function (with appropriate lags) of the relation between national income and the stock of capital. During the recovery, investment and income rise, while the accumulation of capital lags behind. Presently, however, due to the structure of the model, the rise of income stops while capital continues to accumulate. This precipitates the downswing.[27]

Space does not allow us to analyze the works of a number of other writers on the subject, among whom Foster and Catchings should be given due recognition for what is so clumsy and yet so keen an insight.[28] I am also omitting the whole Marxist literature, in which capital accumulation plays such an important rôle, because that would require a separate study. The few remaining pages of this section will be devoted to Hobson and Keynes.

Hobson's writings contain so many interesting ideas that it is a great pity he is not read more often.[29] Anti-Keynesians probably like him not much more than they do Keynes, while Keynesians are apt

[25] Friedrich A. Hayek, *Prices and Production* (London, 1931). I don't mean to say that Professor Hayek is not aware that capital is productive; rather that he did not make use of this fact in his theory of the business cycle. See, however, his "The 'Paradox' of Saving," *Economica*, Vol. XI (May, 1931), pp. 125-69.

[26] Albert Aftalion, "The Theory of Economic Cycles Based on the Capitalistic Technique of Production," *Rev. Econ. Stat.*, Vol. IX (Oct., 1927), pp. 165-70. This short article contains a summary of his theory.

[27] Michal Kalecki, *Essays in the Theory of Economic Fluctuations* (New York, 1939). See particularly the last essay "A Theory of the Business Cycle," pp. 116-49. What Mr. Kalecki's model shows in a general sense is that accumulation of capital cannot proceed for any length of time in a trendless economy (*i.e.*, an economy with a secularly constant income). His other results depend upon the specific assumptions he makes.

[28] William T. Foster and Waddill Catchings, *Profits* (Boston and New York, 1925). This book is the most important of their several published works. It is interesting to note that they did come to the conclusion that ". . . as long as capital facilities are created at a sufficient rate, there need be no deficiency of consumer income. To serve that purpose, however, facilities must be increased at a constantly accelerating rate" (p. 413). This they regarded quite impossible.

[29] I am particularly referring to his *Economics of Unemployment* (London, 1922) and *Rationalization and Unemployment* (New York, 1930).

to regard the *General Theory* as the quintessence of all that was worth while in economics before 1936, and may not bother to read earlier writings. I may say that Keynes's own treatment of Hobson, in spite of his generous recognition of the latter's works, may have substantiated this impression.[30]

Even though both Keynes and Hobson were students of unemployment, they actually addressed themselves to two different problems. Keynes analyzed what happens when savings (of the preceding period) are not invested. The answer was—unemployment, but the statement of the problem in this form might easily give the erroneous impression that if savings were invested, full employment would be assured. Hobson, on the other hand, went a step further and stated the problem in this form: suppose savings are invested. Will the new plants be able to dispose of their products? Such a statement of the problem was not at all, as Keynes thought, a mistake.[31] It was a statement of a different, and possibly also a deeper problem.

Hobson was fully armed with the σ effect of investment, and he saw that it could be answered only by growth. His weakness lay in a poor perception of the multiplier effect and his analysis lacked rigor in general. He gave a demonstration rather than a proof. But the problem to which he addressed himself is just as alive today as it was fifty and twenty years ago.[32]

This discussion, as I suspect almost any other, would be obviously incomplete without some mention of Keynes's treatment of the σ and of the multiplier effects. Keynes's approach is very curious: as a matter of fact, he has two: the familiar short-run analysis, and another one which may be called a long-run one.[33]

Keynes's short-run system (later expressed so admiringly by Oscar Lange[34]) is based on ". . . given the existing skill and quantity of available labor, the existing quality and quantity of available equipment, the existing technique, the degree of competition, the tastes and habits of the consumer . . ."[35] Productive capacity thus being given, employment becomes a function of national income, expressed, to be sure, not in money terms but in "wage units." A wage unit, the remuneration for

[30] See *The General Theory*, pp. 364-71.

[31] *Ibid.*, pp. 367-68.

[32] Contrary to popular impression, Hobson does not advocate a maximum reduction in the propensity to save. What he wants is to reduce it to a magnitude commensurable with requirements for capital arising from technological progress—an interesting and reasonable idea.

[33] This whole discussion is based on *The General Theory* and not on Keynes's earlier writings.

[34] Oscar Lange, "The Role of Interest and the Optimum Propensity to Consume," *Economica*, n.s. Vol. V (Feb., 1938), pp. 12-32. This otherwise excellent paper has a basic defect in the assumption that investment is a function of consumption rather than of the rate of change of consumption.

[35] *The General Theory*, p. 245. See also pp. 24 and 28.

"an hour's employment of ordinary labor" (page 41), is of course a perfect fiction, but some such device must be used to translate real values into monetary and *vice versa,* and one is about as good or as bad as another. The important point for our purposes is the assumption that the amount of equipment (*i.e.,* capital) in existence is given.

Now, the heart of Keynesian economics is the argument that employment depends on income, which in turn is determined by the current volume of investment (and the propensity to save). But investment (in the net sense) is nothing else but the rate of change of capital. Is it legitimate than first to assume the quantity of capital as given, and then base the argument on its rate of change? If the quantity of capital changes, so does (in a typical case) productive capacity, and if the latter changes it can be hardly said that employment is solely determined by the size of national income, expressed in wage units or otherwise. Or putting it in the language of this paper, is it safe and proper to analyze the relation between investment and employment without taking into account the σ effect?

The answer depends on the nature of the problem in hand. In this particular case, Keynes could present two reasons for his disregard of the σ effect. He could assume that the latter operates with at least a one period lag, the period being understood here as the whole time span covered by the discussion.[36] Or he could argue that over a typical year the net addition (*i.e.,* net investment) to the stock of capital of a society, such as England or the United States, will hardly exceed some 3 to 5 per cent; since this increment is small when compared with changes in income, it can be disregarded.[37]

Both explanations are entirely reasonable provided of course that the period under consideration is not too long. A five-year lag for the σ effect would be difficult to defend, and an increase in the capital stock of some 15 or 20 per cent can hardly be disregarded. I am not aware that Keynes did present either of these explanations; but there is just so much one can do in four hundred pages at any one time.

It would be perfectly absurd to say that Keynes was not aware of the productive qualities of capital. In the *long run* he laid great stress on it, possibly too great. All through the *General Theory* we find grave concern for the diminishing marginal efficiency of capital due, in the long run, to its increasing quantity.[38] There is so much of this kind of argument as to leave the reader puzzled in the end. We are told that marginal efficiency of capital depends on its scarcity. Well and good. But scarcity relative to what? It could become less scarce

[36] This again is not quite safe unless some provision for investment projects started in preceding periods and finished during the present period is made.

[37] The second assumption is specifically made by Professor Pigou in his *Employment and Equilibrium* (London, 1941), pp. 33-34.

[38] See for instance pp. 31, 105-106, 217, 219, 220-21, 324, and 375.

relative to other factors, such as labor, so that the marginal productivity of capital in the real sense (*i.e.*, essentially our σ) declined. But then on page 213 we read: "If capital becomes less scarce, the excess yield will diminish, without its having become less productive—at least in the physical sense."

Why then does the marginal efficiency of capital fall? Evidently because capital becomes less scarce relative to income.[39] But why cannot income grow more rapidly if labor is not the limiting factor? Could it be only a matter of poor fiscal policy which failed to achieve a faster growing income? After all we have in investment an income generating instrument; if investment grows more rapidly, so does income. This is *the* multiplier effect of investment on which so much of the *General Theory* is built.

I don't have the answer. Is it possible that, while Keynes disregarded the σ effect in the short-run analysis, he somehow omitted the multiplier effect from the long-run?

V. Concluding Remarks

A traveller who sat in the economic councils of the United States and of the Soviet Union would be much impressed with the emphasis placed on investment and technological progress in both countries. He would happily conclude that the differences between the economic problems of a relatively undeveloped socialist economy and a highly developed capitalist economy are really not as great as they are often made to appear. Both countries want investment and technological progress. But if he continued to listen to the debates, he would presently begin to wonder. For in the Soviet Union investment and technology are wanted in order to enlarge the country's productive capacity. They are wanted essentially as labor-saving devices which would allow a given task to be performed with less labor, thus releasing men for other tasks. In short, they are wanted for their σ effects.

In the United States, on the other hand, little is said about enlarging productive capacity. Technological progress is wanted as the creator of investment opportunities, and investment is wanted because it generates income and creates employment. It is wanted for its multiplier effect.

Both views are correct and both are incomplete. The multiplier is not just another capitalist invention. It can live in a socialist state just as well and it has been responsible for the inflationary pressure which has plagued the Soviet economy all these years, since the first five-year

[39] There is a third possibility namely that income is redistributed against the capitalists, but Keynes makes no use of it.

plan. And similarly, σ is just as much at home in one country as in another, and its effect—the enlarged productive capacity brought about by accumulation of capital—has undoubtedly had much to do with our peacetime unemployment.

But what is the solution? Shall we reduce σ to zero and also abolish technological progress thus escaping from unemployment into the "nirvana" of a stationary state? This would indeed by a defeatist solution. It is largely due to technology and savings that humanity has made the remarkable advance of the last two hundred years, and now when our technological future seems so bright, there is less reason to abandon it than ever before.

It is possible that α has been or will be too high as compared with the growth of our labor force, the utilization of new resources, and the development of technology. Unfortunately, we have hardly any empirical data to prove or disprove this supposition. The fact that private investment did not absorb available savings in the past does not prove that they could not be utilized in other ways (*e.g.*, by government), or even that had private business invested them these investments would have been unprofitable; the investing process itself might have created sufficient income to justify the investments. What is needed is a study of the magnitudes of s, of the difference between s and σ which can develop without much harm and then of the value of α which the economy can digest at its full employment rate of growth.

Even if the resulting magnitude of α is found to be considerably below the existing one, a reduction of α is only one of the two solutions, the speeding up of technological progress being the other. But it must be remembered that neither technology, nor of course saving, guarantee a rise in income. What they do is to place in our hands the *power* and the ability of achieving a growing income. And just as, depending upon the use made of it, any power can become a blessing or a curse, so can saving and technological progress, depending on our economic policies, result in frustration and unemployment or in an ever expanding economy.